Megaflooding on Earth and Mars

Edited by

DEVON M. BURR

University of Tennessee Knoxville, and Carl Sagan Center, SETI Institute, USA

PAUL A. CARLING

University of Southampton, UK

VICTOR R. BAKER

University of Arizona, USA

CAMBRIDGE
UNIVERSITY PRESS

CAMBRIDGE
UNIVERSITY PRESS

University Printing House, Cambridge CB2 8BS, United Kingdom

One Liberty Plaza, 20th Floor, New York, NY 10006, USA

477 Williamstown Road, Port Melbourne, VIC 3207, Australia

4843/24, 2nd Floor, Ansari Road, Daryaganj, Delhi - 110002, India

79 Anson Road, #06-04/06, Singapore 079906

Cambridge University Press is part of the University of Cambridge.

It furthers the University's mission by disseminating knowledge in the pursuit of education, learning and research at the highest international levels of excellence.

www.cambridge.org
Information on this title: www.cambridge.org/9781108447072

First published 2009
First paperback edition 2017

A catalogue record for this publication is available from the British Library

Library of Congress Cataloging in Publication data
Megaflooding on Earth and Mars / editors, Devon Burr, Victor R. Baker, Paul A. Carling.
p. cm.
Includes bibliographical references and index.
ISBN 978-0-521-86852-5 (hardback)
1. Floods. 2. Sedimentation and deposition.
3. Paleoclimatology – Quaternary. 4. Earth. 5. Mars (Planet) – Water.
I. Burr, Devon, 1966– II. Baker, Victor R. III. Carling, Paul.
IV. Title.
GB1399.M44 2009
551.48′9 – dc22 2009019732

ISBN 978-0-521-86852-5 Hardback
ISBN 978-1-108-44707-2 Paperback

Megaflooding on Earth and Mars

Megaflooding is the sudden discharge of exceptional volumes of water that significantly alter the regional landscape. Megafloods have occurred repeatedly on both Earth and Mars where they may have acted as triggers for climate change. Research into megaflooding has progressed greatly over the past several years: on Earth, real-time measurements of contemporary floods in Iceland complement multifaceted research into older and even larger terrestrial floods; while on Mars, terabytes of data from several spacecraft orbiting that planet are dramatically revising our view of flooding there.

Beginning with an historical overview of flood science, this volume presents sections on morphology and mechanisms, flood sedimentology, and modelling, each illustrated with examples from Earth and Mars. By presenting terrestrial and Martian research together, this volume creates a unique synthesis to further our understanding of these enormous palaeoflood events. It is an invaluable reference for researchers and students of hydrology, geomorphology, sedimentology and planetary science, as well as environmental and hydraulic engineers.

DEVON M. BURR received her education at the United States Naval Academy, St John's College, and the University of Arizona, where she was a Fulbright Fellow. After her Ph.D. in 2003, she worked as the Eugene M. Shoemaker Fellow at the United States Geological Survey Astrogeology Branch before moving to the Carl Sagan Center for the Study of Life in the Universe. She is currently Assistant Professor in the Earth and Planetary Sciences Department at the University of Tennessee Knoxville, where she specialises in planetary geomorphology and fluid sediment transport.

PAUL A. CARLING obtained his first degree in Geography from the University of Leicester and a Ph.D. from the University of Wales (Swansea) for research into intertidal sedimentation. He worked for what is now known as the UK Centre for Hydrology and Ecology from 1977 to 1994, before taking up a professorship in geography at the University of Lancaster. He took up the post of Professor of Physical Geography at the University of Southampton in 2000, and is currently chief editor of the journal *Sedimentology*.

VICTOR R. BAKER holds degrees from Rensselaer Polytechnic Institute and the University of Colorado. He was a faculty member at the University of Texas before joining the University of Arizona, where he is presently Regents' Professor of Hydrology and Water Resources, Geosciences, and Planetary Sciences. He is author/editor of 14 other books on flooding and geomorphology, and currently serves on editorial boards for the journals *Geomorphology* and *Zeitschrift für Geomorphologie*.

Contents

	List of contributors		*page* viii
1	Overview of megaflooding: Earth and Mars *Victor R. Baker*		1
		Summary	1
	1.1	Introduction	1
	1.2	Historical and philosophical overview	1
	1.3	The Great Scablands Debate	2
	1.4	Terrestrial megafloods	3
	1.5	Martian megafloods	6
	1.6	Global consequences	6
	1.7	Modern controversies	7
	1.8	Discussion	9
		Acknowledgements	9
		References	9
2	Channel-scale erosional bedforms in bedrock and in loose granular material: character, processes and implications *Paul A. Carling, Jürgen Herget, Julia K. Lanz, Keith Richardson and Andrea Pacifici*		13
		Summary	13
	2.1	Introduction	13
	2.2	Historical perspective	14
	2.3	Processes of bedrock erosion	14
	2.4	A typology of channel-scale erosional bedforms in bedrock	15
	2.5	Example of a very large bedrock channel complex: Kasei Valles, Mars	23
	2.6	Conclusions	27
		References	27
3	A review of open-channel megaflood depositional landforms on Earth and Mars *Paul A. Carling, Devon M. Burr, Timothy F. Johnsen and Tracy A. Brennand*		33
		Summary	33
	3.1	Introduction	33
	3.2	Geomorphological considerations	34
	3.3	Theoretical background to sediment transport and deposition	36
	3.4	Megaflood depositional landforms	38
	3.5	Discussion and conclusions	45
		References	45

4 Jökulhlaups in Iceland: sources, release and drainage 50
Helgi Björnsson
Summary 50
4.1 Introduction 50
4.2 Subglacial lakes 50
4.3 Marginal lakes 58
4.4 Subglacial eruptions as a source of jökulhlaups 59
4.5 Conclusions 61
Notation 61
References 62

5 Channeled Scabland morphology 65
Victor R. Baker
Summary 65
5.1 Introduction 65
5.2 Regional patterns 66
5.3 Macroscale erosional surface forms 68
5.4 Macroscale depositional surface forms 70
5.5 Mesoscale erosional surface forms 71
5.6 Mesoscale depositional surface forms 73
5.7 Palaeohydraulic implications 74
5.8 Discussion 75
Acknowledgements 75
References 75

6 The morphology and sedimentology of landforms created by subglacial megafloods 78
Mandy J. Munro-Stasiuk, John Shaw, Darren B. Sjogren, Tracy A. Brennand, Timothy G. Fisher, David R. Sharpe, Philip S. G. Kor, Claire L. Beaney and Bruce B. Rains
Summary 78
6.1 Introduction 78
6.2 Microforms 80
6.3 Mesoforms 85
6.4 Macroforms 95
6.5 Megaforms 96
6.6 Subglacial hydrology 97
6.7 Concluding remarks 99
Acknowledgements 100
References 100

7 Proglacial megaflooding along the margins of the Laurentide Ice Sheet 104
Alan E. Kehew, Mark L. Lord, Andrew L. Kozlowski and Timothy G. Fisher
Summary 104
7.1 Introduction 104
7.2 Dams and triggers 105
7.3 Erosional processes and landforms 106
7.4 Depositional processes and landforms 108
7.5 Palaeohydrologic considerations 109
7.6 Anatomy of a glacial lake megaflood: the glacial Lake Regina outburst 110
7.7 Megafloods west of Lake Agassiz 111
7.8 Lake Agassiz and megaflood outflows 111
7.9 Southern megaflood drainage from the Great Lakes basins 115
7.10 Eastern megaflood drainage from the Great Lakes basins 118
7.11 Conclusions 123
Acknowledgements 123
References 123

8 Floods from natural rock-material dams 128
Jim E. O'Connor and Robin A. Beebee
Summary 128
8.1 Introduction 128
8.2 Floods from breached valley blockages 129
8.3 Floods from breached basins 132
8.4 Flood magnitude and behaviour 137
8.5 Erosional and depositional features from natural dam failures 145
8.6 Concluding remarks 147
Acknowledgements 147
References 163

9 Surface morphology and origin of outflow channels in the Valles Marineris region 172
Neil M. Coleman and Victor R. Baker
Summary 172
9.1 Introduction 172
9.2 Outflow channels that emerged from the chasmata or other large basins 174
9.3 Outflow channels that emerged from discrete chaotic terrain or fault zones 184
9.4 Conclusions 191
Acknowledgements 191
References 191

10 Floods from fossae: a review of Amazonian-aged extensional–tectonic megaflood channels on Mars 194
Devon M. Burr, Lionel Wilson and Alistair S. Bargery
Summary 194
10.1 Introduction 194
10.2 Ages 194
10.3 Morphology 196
10.4 Mechanisms triggering water release 201
10.5 Thermal and mechanical aspects of water release 203
10.6 Summary and implications 204
Acknowledgements 205
References 205

11 Large basin overflow floods on Mars 209
Rossman P. Irwin III and John A. Grant
Summary 209
11.1 Introduction 209
11.2 Peak discharge from damburst floods on Mars 210
11.3 Basin influence on Martian outflow channels 211
11.4 Intermediate-scale basin overflows on Mars 214
11.5 Conclusions 220
References 221

12 Criteria for identifying jökulhlaup deposits in the sedimentary record 225
Philip M. Marren and Matthias Schuh
Summary 225
12.1 Introduction 225
12.2 Defining critical criteria 226
12.3 The nature of jökulhlaup flooding 226
12.4 Jökulhlaup sedimentation 227
12.5 Skeiðarársandur jökulhlaup bar case study 232
12.6 Discussion: criteria for identifying jökulhlaups in the sedimentary record 237
12.7 Conclusions 237
Acknowledgements 238
References 238

13 Megaflood sedimentary valley fill: Altai Mountains, Siberia 243
Paul A. Carling, I. Peter Martini, Jürgen Herget, Pavel Borodavko and Sergei Parnachov
Summary 243
13.1 Introduction 243
13.2 Methods 244
13.3 Perspective 244
13.4 Sediment characteristics of the giant bars 248
13.5 Discussion 258
13.6 Conclusions 262
Acknowledgements 262
References 263

14 Modelling of subaerial jökulhlaups in Iceland 265
Snorri Páll Kjaran, Sigurður Lárus Hólm, Eric M. Myer, Tómas Jóhannesson and Peter Sampl
Summary 265
14.1 Introduction 265
14.2 Jökulhlaups from Katla and Eyjafjallajökull 266
14.3 Dynamics of subaerial jökulhlaups 266
14.4 Numerical formulation 268
14.5 Model simulations: two examples 269
14.6 Discussion 270
Acknowledgements 271
References 271

15 Jökulhlaups from Kverkfjöll volcano, Iceland: modelling transient hydraulic phenomena 273
Jonathan L. Carrivick
Summary 273
15.1 Introduction, background and rationale 273
15.2 Aim 274
15.3 Study site 274
15.4 Methodology 274
15.5 Results and interpretation 276
15.6 Discussion 283
15.7 Conclusions 286
Acknowledgements 287
References 287

16 Dynamics of fluid flow in Martian outflow channels 290
Lionel Wilson, Alistair S. Bargery and Devon M. Burr
Summary 290
16.1 Introduction 290
16.2 Water sources 290
16.3 Water ascent to the surface 291
16.4 Water release at the surface 293
16.5 Dynamics of water flow 295
16.6 Water flow thermodynamics and rheology 302
16.7 Closing summary 305
Notation 306
Acknowledgements 307
References 307

Index 312
The colour plates are situated between pages 182 and 183

Contributors

Victor R. Baker
Departments of Hydrology and Water Resources, Geosciences, and Planetary Sciences
University of Arizona
Tucson
AZ 85721
USA

Alistair S. Bargery
Environment Centre
Environmental Science Division
Lancaster University
Lancaster
LA1 4YQ
UK

Claire L. Beaney
Department of Geography
University College of the Fraser Valley
45635 Yale Road
Chilliwack
British Columbia
V2P 6T4
Canada

Robin A. Beebee
HDR Alaska, Inc.
Anchorage
AL 99503
USA

Helgi Björnsson
Institute of Earth Sciences
University of Iceland
Building of Natural Sciences
Askja room 329
Sturlugata 7
IS-101 Reykjavík
Iceland

Pavel Borodavko
Laboratory of Glacioclimatology
Tomsk State University
Lenina 36
634050 Tomsk
Russia

Tracy A. Brennand
Department of Geography
Simon Fraser University
8888 University Drive
Burnaby
British Columbia
V5A 1S6
Canada

Devon M. Burr
Earth and Planetary Sciences Department and Planetary Geosciences Institute
University of Tennessee Knoxville
TN 37996
USA

Paul A. Carling
School of Geography
University of Southampton
Highfield
Southampton
SO17 1BJ
UK

Jonathan L. Carrivick
School of Geography
University of Leeds
Leeds
LS2 9JT
UK

Neil M. Coleman
Department of Geology and Planetary Science
University of Pittsburgh at Johnstown
450 Schoolhouse Road
Johnstown
PA 15904
USA

Timothy G. Fisher
Department of Earth Ecological and Environmental Sciences
University of Toledo
2801 W. Bancroft
Toledo
OH 43606
USA

John A. Grant
Center for Earth and Planetary Studies
National Air and Space Museum
Smithsonian Institution
6th St. at Independence Ave. SW
Washington DC 20013
USA

Jürgen Herget
Geographisches Institut der Universität Bonn
Meckenheimer Allee 166
D-53115 Bonn
Germany

Sigurður Lárus Hólm
Vatnaskil Consulting Engineers
Suðurlandsbraut 50
IS-108 Reykjavík
Iceland

Rossman P. Irwin III
Center for Earth and Planetary Studies
National Air and Space Museum
Smithsonian Institution
6th St. at Independence Ave. SW
Washington DC 20013
USA

Tómas Jóhannesson
Icelandic Meteorological Office
Bústaðavegur 9
IS-150 Reykjavík
Iceland

Timothy F. Johnsen
Department of Physical Geography & Quaternary Geology
Stockholm University
SE-10691 Stockholm
Sweden

Alan E. Kehew
Department of Geosciences
Western Michigan University
Kalamazoo
MI 49008
USA

Snorri Páll Kjaran
Vatnaskil Consulting Engineers
Suðurlandsbraut 50
IS-108 Reykjavík
Iceland

Philip S. G. Kor
Ontario Parks
300 Water Street
Peterborough
Ontario
K9J 8M5
Canada

Andrew L. Kozlowski
New York State Museum/Geological Survey
3140 Cultural Education Center
Albany
NY 12230
USA

Julia K. Lanz
Institut für Planetologie
Universität Stuttgart
Herdweg 51
70174 Stuttgart
Germany

Mark L. Lord
Department of Geosciences
Western Carolina University
Cullowhee
NC 28723
USA

Philip M. Marren
Department of Resource Management and Geography
University of Melbourne
Parkville
Victoria 3010
Australia

I. Peter Martini
Land Resource Science
University of Guelph
Guelph
Ontario
N1G 2W1
Canada

Eric M. Myer
Vatnaskil Consulting Engineers
Suðurlandsbraut 50
IS-108 Reykjavík
Iceland

Mandy J. Munro-Stasiuk
Department of Geography
Kent State University
Kent
OH 44242
USA

Jim E. O'Connor
U.S. Geological Survey
2130 SW 5th Ave
Portland
OR 97201
USA

Andrea Pacifici
International Research School of Planetary Sciences
Departimento di Scienze
Universita degli Studi G. de'Annunzio
Via dei Vestini 31
Chieti viale Pindaro 42
Pescara
Italy

Sergei Parnachov
Laboratory of Glacioclimatology
Tomsk State University
Lenina 36
634050 Tomsk
Russia

Bruce B. Rains
Department of Earth and Atmospheric Sciences
1-26 Earth Sciences Building
University of Alberta
Edmonton
Alberta
T6G 2E3
Canada

Keith Richardson
School of Geography
University of Southampton
Highfield
Southampton
SO17 1BJ
UK

Peter Sampl
AVL List GmbH
Hans-List-Platz 1
8080 Graz
Austria

Matthias Schuh
Christophstr. 12
d-72072
Tübingen
Germany

David R. Sharpe
Geological Survey of Canada
601 Booth Street
Ottawa
Ontario
K1A 0E8
Canada

John Shaw
Department of Civil and Environmental Engineering
E6-050 Engineering Teaching & Learning Complex
University of Alberta
Edmonton
Alberta
T6G 2V4
Canada

Darren B. Sjogren
Earth Science Program
Department of Geography
University of Calgary
2500 University Dr. N.W.
Calgary
AB T2N 1N4
Canada

Lionel Wilson
Environment Centre
Environmental Science Division
Lancaster University
Lancaster
LA1 4YQ
UK

1 Overview of megaflooding: Earth and Mars

VICTOR R. BAKER

Summary

After centuries of geological controversy it is now well established that the last major deglaciation of planet Earth involved huge fluxes of water from the wasting continental ice sheets, and that much of this water was delivered as floods of immense magnitude and relatively short duration. These late Quaternary megafloods, and the megafloods of earlier glaciations, had short-term peak flows, comparable in discharge to the more prolonged fluxes of ocean currents. (The discharges for both ocean currents and megafloods generally exceed one million cubic metres per second, hence the prefix 'mega'.) Some outburst floods likely induced very rapid, short-term effects on Quaternary climates. The late Quaternary megafloods also greatly altered drainage evolution and the planetary patterns of water and sediment movement to the oceans. The recent discoveries of Mars missions have now documented the importance of megafloods to the geological evolution of that planet. As on Earth, the Martian megafloods seem to have influenced climate change.

1.1 Introduction

Up until the middle nineteenth century considerable progress was being made in scientific studies of the role of catastrophic flooding in the geological evolution of river valleys. While some of these studies invoked a kind of biblical catastrophism, much of the work merely employed hypotheses of immense floods because these inferences seemed to provide the best explanations for such features as scoured bedrock and accumulations of huge, water-transported boulders. This whole line of scientific inquiry was seriously curtailed because of the popularity of certain of Charles Lyell's (1830–3) logically flawed notions of uniformitarianism. Lyell's methodology for geology resulted in an epistemological stigma for hypotheses invoking types and magnitudes of flood processes not directly observed today. As discussed below, many geologists either ignored or failed to understand the flaws in Lyell's overly prescriptive forms of uniformitarianism, which came to be imbedded into the practice of geology, thereby greatly hindering progress in the understanding of cataclysmic flooding as a geological process (see also, Baker, 1998).

Two developments of the twentieth century produced a renaissance in cataclysmic flood studies. The first was the prolonged scientific controversy over the origin of the Channeled Scabland in the northwestern United States (Baker, 1978, 1981). Extending from the 1920s to the 1970s, the great 'scablands debate' eventually led to a general acceptance of the cataclysmic flood origin for the region that had been championed by J Harlen Bretz, a professor at The University of Chicago. The second important development was the discovery in the early 1970s of ancient cataclysmic flood channels on the planet Mars (e.g. Baker and Milton, 1974; Baker, 1982). The Martian outflow channels were produced by the largest known flood discharges, and their effects have been preserved for billions of years (Baker, 2001).

Relatively recently it has come to be realised that spectacular cataclysmic flooding was a global phenomenon that accompanied the termination of the last major glaciation on Earth, involving huge fluxes of water from land to sea (Baker, 1994). Baker (1997) employed the term 'megaflooding' without definition to describe the association of cataclysmic outburst flooding with the great late Pleistocene ice sheets of North America and Eurasia. The appropriateness of this term was reinforced by the fact that floods of interest generally had peak discharges in excess of one million cubic metres per second (hence the prefix 'mega'). Moreover, this same unit of water flux, designated one 'sverdrup' (named for the pioneering Norwegian oceanographer Harald Sverdrup), is the same measure that is applied to ocean currents, which are the other major short-term water movements on the surface of the Earth (Baker, 2001). When such huge discharges are combined with appropriate energy gradients to produce very high flow velocities, bed shear stresses and unit power expenditures, the result is the phenomenon of high-energy megafloods (Baker, 2002a), which might also deserve the name 'superfloods' (Baker, 2002b). It is such flows that produce distinctive suites of consequent phenomena, many of which are described in other papers in this volume.

1.2 Historical and philosophical overview

According to the *Oxford English Dictionary*, the first reported use of the word 'geology' as pertaining to a branch of science was in the title of the 1690 book *Geologia: or, a Discourse Concerning the Earth Before the Deluge*, written by Erasmus Warren. Warren's *Geologia* concerned the literal truth of the Book of Genesis. However, like biblical literalists before and since, Warren had to resolve a

Megaflooding on Earth and Mars, ed. Devon M. Burr, Paul A. Carling and Victor R. Baker.
Published by Cambridge University Press.

paradox: the mixing in Genesis of two very different accounts of the Noachian Debacle. In one account, The Flood derives from 'foundations of the great deep' (Genesis 7:11). In the other, The Flood derives from 'the windows of heaven', such that it rained continuously for 40 days and 40 nights (Genesis 7:12). Warren privileged the first account, postulating that water burst from great caverns. The irrelevance of the whole question actually had been recognised during the early history of the Christian church by Saint Augustine, who noted that God would not make the mistake of having his revealed creation in conflict with scriptural interpretation. To claim superiority for the latter over what is actually revealed in nature would make the Christian God appear inferior in comparison with the deities of other religions. It is therefore quite ironic that this long Christian tradition of valuing what is revealed in God's creation (nature) is currently being denigrated by the historically recent (Numbers, 1993) development of a so-called 'creationism' based in a biblical literalism that makes the highly dubious claim of being fundamental to its theological underpinnings.

Although the first geology was arguably 'flood geology', it soon became apparent that geologists would not be worthy of the appellation 'scientists' (coined in 1840 by the famous Cambridge polymath William Whewell) if their activity were to consist solely of bearing witness to authoritative pronouncements, including those presumed to come from a deity. It is unfortunate that today 'flood geology' commonly refers to a branch of 'creation science', sharing with that enterprise a misleading use of the word 'science'. Science is no more and no less than an unrestricted inquiry into nature. To have its answers ordained in advance is a restriction on free inquiry and the result is sham reasoning (Haack, 1996), not science. Besides being flawed as science, 'creation science' is also based on dubious theology because of its rejection of the reading by human inquiry of divine creation (nature) in favour of the authority of an imperfect 2000-year-old written account thereof.

By the early nineteenth century, geology had evolved to a science concerned with observations of nature on a path to whatever could be discovered about causal patterns in regard to those observations. The inquiries of geologists had to be free to lead anywhere the observations and their implications required, unconstrained by prior notions of what was true, or even of what might be thought to be proper in the pursuit of that truth. Unfortunately, there also emerged confusion over the last point, and vestiges of that confusion linger even to the present day. This confusion involves the notion of 'uniformitarianism' (another word coined by William Whewell). When it is used as a kind of stipulative prohibition against the valid inference of cataclysmic processes, uniformitarianism is invalid as a concept in science, i.e., it blocks the path of inquiry (Baker, 1998). Indeed, there is nothing wrong, scientifically speaking, with invoking cataclysmic flooding as a natural explanation, if the facts, rather than some preconceived belief, lead the inquiry in that direction.

Prior to Charles Lyell's somewhat misguided advocacy of uniformitarianism in the middle 1800s (Baker, 1998), it was common for genetic hypothesising in natural philosophy to invoke cataclysmic flooding as a mechanism to explain such features as erratic boulders, widespread mantles of boulder clay (so-called 'diluvium'), and wind and water gaps through the ridges of fold mountain ranges (Huggett, 1989). For example, Edward Hitchcock (1835) explained the sediments and landforms of the Connecticut River valley as products of the Noachian debacle. Cataclysmic flooding was generally not invoked for reasons of scriptural literalism. Rather, it was genuinely perceived to provide a reasonable explanation for the phenomena of interest. An example is James D. Dana's (1882) study of the high glacial terraces of the Connecticut River. In addition to Hitchcock's debacle hypothesis, these had been interpreted alternatively to be products of marine submergence or the result of deposition by proglacial outwash streams forming valley fills that were later incised by postglacial river with less sediment load (Upham, 1877). Dana argued that the terraces resulted from a large-scale glacial flood, and he used their heights to infer its high-water surface, averaging about 50 m above the present river profile. From his data one can calculate a maximum paleodischarge for this 'flood' of 2.4×10^5 m^3 s^{-1} (Patton, 1987).

1.3 The Great Scablands Debate

The 1920s/1930s debates over the origin of the Channeled Scabland landscape in eastern Washington, northwestern United States, were critical to the recognition of the geological importance of megaflooding. It was in this region that University of Chicago Professor J Harlen Bretz formulated the hypothesis of cataclysmic flooding (Baker, 1978, 1981). Bretz (personal communication, 1977; Figure 1.1) recalled that he first conceived of the cataclysmic flood hypothesis when in 1909 he saw a topographic map depicting the immense Potholes Cataract, which is now known to have conveyed cataclysmic flood water as part of the Channeled Scabland. However, Bretz forgot about this problem until the summer of 1922 when he led a field party of advanced University of Chicago students into the region. The travels through the Channeled Scabland reminded Bretz of his earlier hypothesis and he decided to devote that field season to an examination of that landscape.

In a paper published in the *Journal of Geology*, Bretz (1923) formally described his hypothesis that an immense late Pleistocene flood had derived from the margins of the

Figure 1.1. Professor J Harlen Bretz (left) at his home with Victor R. Baker (right). Photographed in 1977 by Rhoda Riley.

nearby Cordilleran Ice Sheet. Named the 'Spokane Flood', this cataclysm neatly accounted for numerous interrelated aspects of the Channeled Scabland landscape and nearby regions. In his various papers Bretz noted features that marked the high levels reached by the floodwaters. These included scarps cut into the loess-mantled uplands adjacent to the scabland channels, high-level gravel-bar deposits of the floods, and divide crossings where one scabland channel spilled over into another (Bretz, 1923, 1928).

During the 1920s and 1930s the geological community strongly resisted Bretz's cataclysmic flood hypothesis. The most dramatic confrontation occurred at the 'The Great Scablands Debate' of the 1927 meeting of the Geological Society of Washington. At this meeting Bretz provided an overview of his hypothesis and a detailed listing of the numerous, otherwise anomalous phenomena that were explained by it. His talk was then followed by well-prepared criticisms by selected members of the audience. Almost continuously until 1940 there was nearly unanimous antipathy toward the cataclysmic flooding hypothesis, despite Bretz's eloquent defence thereof. Sceptical attitudes only began to change because of a technical session at the 1940 meeting in Seattle of the American Association for the Advancement of Science (AAAS).

Although the early papers in the AAAS session merely reiterated the various previously published alternatives for the origin of the Channeled Scabland, the eighth paper surprised an audience that expected no one to directly support Bretz's Spokane Flood hypothesis. The speaker was Joseph Thomas Pardee, whose paper was entitled 'Ripple Marks (?) in Glacial Lake Missoula'. Pardee described Camas Prairie, an inter-montane basin in northwestern Montana, on the floor of which were giant 'ripple' marks (probably fluvial dunes) composed of coarse gravel, piled up to 15 m high and spaced up to 150 m apart (Figure 1.2). The ripples had formed in a great ice-dammed lake, glacial Lake Missoula, which Pardee (1910) had earlier documented to have covered an immense area in western Montana. This late Pleistocene lake had held about 2000 cubic kilometres of water, and it was impounded to a maximum depth of about 600 m behind a lobe of the Cordilleran Ice Sheet occupying what is now the modern basin of Lake Pend Oreille in northern Idaho. In his presentation, and in a subsequent extended paper (Pardee, 1942), Pardee presented his new evidence that the ice dam for glacial Lake Missoula had failed suddenly, with a resulting rapid drainage of the lake. Evidence for the latter included the ripple marks, plus severely eroded constrictions in lake basins and giant gravel bars of current-transported debris. Some of the latter accumulated in high eddy deposits, marginal to the lake basins, showing that the immensely deep lake waters were rapidly draining westward toward the heads of the various scabland channels that had been so well described by Bretz.

Though Pardee did not directly claim the connection of glacial Lake Missoula to the Channeled Scabland, his evidence provided needed coherence to Bretz's hypothesis of cataclysmic flooding. Despite this, resistance to Bretz's flood hypothesis remained strong, not subsiding until the accumulated field evidence became overwhelming (e.g. Bretz *et al.*, 1956). When it was eventually demonstrated that the relevant physical processes are completely consistent with that evidence (Baker, 1973), there was no way to continue to invoke a muddled view of uniformitarianism against Bretz's hypothesis.

1.4 Terrestrial megafloods

Continental ice sheets that form during epochs of glaciation exert immense influences on water drainage across the land. Their huge loads depress the underlying land surface, and lakes form in the moats that surround the ice sheets. They can block the lower courses of major rivers, impounding flow, and even diverting it into adjacent drainage basins. Meltwater from glacial margins may introduce huge discharges into land-surface depressions that hold much smaller lakes during non-glacial periods. The lakes may climatically alter water balances, promoting further glaciation by a kind of positive feedback. There are modern examples for all these situations but the relatively small size and different thermal regimes of modern versus ice-age glaciers pose problems of extrapolation to the past conditions of major continental glaciation that favoured the development of glacial megalakes.

1.4.1 Cordilleran Ice Sheet: ice-dammed lakes

Very deep lakes can form when a glacier advances down a mountain valley to block a river. The famous example is glacial Lake Missoula, noted above, which formed

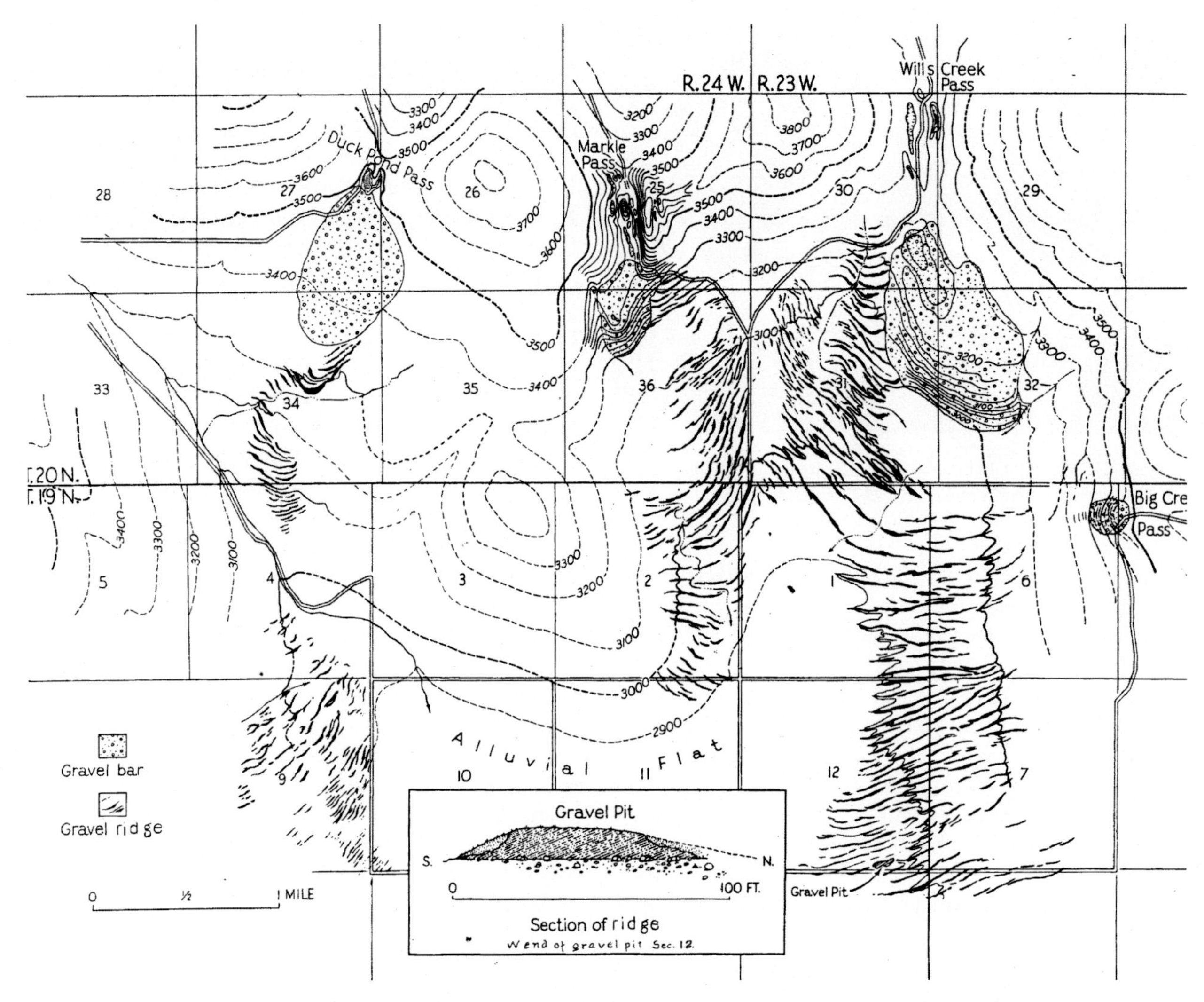

Figure 1.2. Map of giant current 'ripple marks' in the northern part of Camus Prairie Basin of glacial Lake Missoula (Pardee, 1942).

south of the Cordilleran Ice Sheet, in the northwestern United States. The Purcell Lobe of the ice sheet extended south from British Columbia to the basin of modern Pend Oreille Lake in northern Idaho. It thereby impounded the Clark Fork River drainage to the east, forming glacial Lake Missoula in western Montana. At maximum extent this ice-dammed lake covered 7500 km^2, holding a water volume of about 2500 km^3 with a depth of 600 m at the dam. Repeated failures through subglacial tunnels are inferred to have occurred between about 17 500 and 14 500 calendar years ago (Waitt, 1985; Atwater, 1986). Cataclysmic failure of the ice dam impounding this lake resulted in discharges into the Columbia River system of up to about 20 sverdrups (O'Connor and Baker, 1992). As recognised by Baker and Bunker (1985), the multiple outburst events were of greatly differing magnitudes. About 15 exceeded 3 sverdrups, and at least one of these exceeded about 10 sverdrups in discharge (Benito and O'Connor, 2003). The largest failure or failures probably involved a different source mechanism than the subglacial tunnelling envisioned by Waitt (1985) because that mechanism yields discharges of only 1 to 2 sverdrups (Clarke *et al.*, 1984). Upon reaching the Pacific Ocean the Missoula floodwaters continued flowing down the continental slope as hyperpycnal flows and associated turbidity currents (Normark and Reid, 2003). The sediment-charged floodwaters followed the Cascadia submarine channel into and through the Blanco Fracture Zone and out onto the abyssal plain of the Pacific. As much as 5000 km^3 of sediment may have been carried and distributed as turbidites over a distance of 2000 km west of the Columbia River mouth.

1.4.2 Laurentide Ice Sheet: ice-marginal lakes

The largest well-documented glacial megalake formed in north-central North America in association with the largest glacier of the last ice age, the Laurentide Ice

Sheet. This ice sheet is known to have been highly unstable throughout much of its history. Not only did freshwater discharges from the glacier result in ice-marginal lakes; outbursts of meltwater into the Atlantic Ocean may have generated climate changes by influencing the thermohaline circulation of the North Atlantic Ocean (Teller *et al.*, 2002). As the Laurentide Ice Sheet of central and eastern Canada retreated from its late Quaternary maximum extent, it was bounded to the south and west by immense meltwater lakes, which developed in the troughs that surrounded the ice. As the lake levels rose, water was released as great megafloods, which carved numerous spillways into the drainages of the Mississippi, St Lawrence and Mackenzie Rivers (Kehew and Teller, 1994). The greatest megafloods developed from the last of the ice-marginal lakes, a union between Lake Agassiz in south-central Canada and Lake Ojibway in northern Ontario. The resulting megalake held about 160 000 km^3 of water, which was released subglacially about 8200 years ago under the ice sheet and into Labrador Sea via the Hudson Strait (Clarke *et al.*, 2004).

1.4.3 Eurasia ice sheets: river diversions

In Europe, as in North America, there was a long period of controversy before late-glacial megaflooding was broadly recognised as an important geological process. One reason was that much of the megaflooding from the western part of the Fennoscandian Ice Sheet seems to have been conveyed through the English Channel, morphological evidence of which has only recently become compelling (Gupta *et al.*, 2007). In the 1970s the Russian palaeoglaciologist Mikhail G. Grosswald recognised that the Quaternary ice-sheet margins of northern Eurasia, like those of northern North America, held huge proglacial lakes, and great spillways developed for the diversion of drainage. Not only was meltwater diverted to the south-flowing Dnieper and Volga Rivers but also the great north-flowing Siberian rivers, the Irtysh, Ob and Yenisei, were impounded by ice sheets that covered the modern Barents and Kara Seas. Grosswald's original work was highly criticised, but more recent work confirms these impoundments and ice sheets, though there remains considerable controversy over their extent, timing and genesis (Mangerud *et al.*, 2004). Grosswald (1980) interpreted this blockage to be Late Weichselian in age (about 15 000–20 000 years ago). However, more recent work considers the events to have been Early Weichselian (Arkhipov *et al.*, 1995), about 90 000 years ago, when ice-sheet growth was enhanced by the climatic influence of the ice-dammed lakes (Krinner *et al.*, 2004).

The largest lake, formed on the west Siberian plain, was estimated by Mangerud *et al.* (2004) to cover 600 000 km^2 at a surface elevation of 60 m. However, both Arkhipov *et al.* (1995) and Grosswald (1980) postulate a much larger palaeolake, about 1 200 000 km^2 in area, with a volume of about 75 000 km^3 at a surface elevation of 128 m. This west Siberian megalake was the Asian equivalent of Agassiz. It drained southward, through the Turgai divide of north-central Kazakhstan, to the basin of the modern Aral Sea. This megalake had an elevation of 70 or 80 m, higher than the 1960 elevation of 53 m, and an area of 100 000 km^2 compared with 60 000 km^2 in 1960. This palaeolake then drained southwestward through the Uzboi channel into the basin of the modern Caspian Sea. Also fed by glacial meltwater from northern Europe via the Volga, the Caspian expanded to a Late Quaternary size over twice its modern extent. Known as the Khvalyn palaeolake, it inundated an area of 950 000 km^2, holding a volume of 135 000 km^3 at an elevation of 50 m (the modern Caspian level is −28 m). The Khvalyn palaeolake spilled westward through the Manych spillway into the Don valley, then to the Sea of Azov, and through the Kerch Strait to the Euxine Abyssal Plain, which is the floor of the modern Black Sea.

In full-glacial time the Black Sea basin was disconnected from the Aegean and partially filled by freshwater (the New Euxine phase). The glacial meltwater filled the basin to an elevation of −60 m. This basin functioned in two modes during the Quaternary. Its cold-climate mode was a freshwater lake (that may have filled and drained through northwestern Turkey to the Aegean). Its warm-climate phase involved rising global sea level, inducing salt-water invasions through the Turkish straits to form the saline Black Sea, the last of which occurred about 8000 years ago, an inundation claimed by Ryan *et al.* (2003) to have inspired the biblical story of Noah (see further discussion below).

1.4.4 Central Asian mountain areas: ice-dammed lakes

In the mountain areas of central northern Asia, there were several great ice-dammed expansions of modern lakes, such as Issyk-Kul, Kyrgyzstan, the second largest modern mountain lake in the world (6000 km^2 area; 4000 km^3 volume). It was an even larger ice-dammed lake during the last glaciation, rising to 300 m above the present lake level (Grosswald *et al.*, 1994). Spectacular erosion at the full-glacial lake outlet, Boam Canyon, indicates that this palaeolake failed by cataclysmic outburst flooding. In the Altai Mountains, great megafloods emanated from the late Pleistocene ice-dammed Chuja-Kuray palaeolake, covering 12 000 km^2 and holding between 600 and 1000 km^3 of water, perhaps 600 m deep at the ice dam (Baker *et al.*, 1993; Carling *et al.*, 2002, 2008). Downstream of this ice-dammed lake, the Chuja and Katun River valleys are characterised by immense gravel bars, emplaced by the flooding into various valley-side embayments. The bars are up to 5 km long and 120 m in height. The bar surfaces and associated run-up layers of flood-transported gravel indicate maximum flow depths of about 320 m. Flow modelling of the associated palaeoflood discharges retrodicts a peak flow of about 11 sverdrups and mean flow velocities of about 30 m/s (Herget, 2005).

Another region with extensive evidence of Late Quaternary megafloods is Tuva. The Tuvan palaeofloods derived from an ice-dammed lake in the Darkhat depression of north-central Mongolia (Grosswald and Rudoy, 1996). The palaeolake held about 250 km^3 of water with a depth of 200 m at its ice dam. The palaeoflooding entered the upper Yenisei River and it was remarkable for the emplacement of spectacular trains of giant current ripples (gravel dunes) near the Tuva capital city of Kyzyl. Following the Yenisei River valley, the Tuvan floods emerged from canyons of the West Sayan Mountains and entered the Abakan Basin, where they deposited a great fan complex.

1.5 Martian megafloods

Earth and Mars are the only two planets known to have geological histories involving vigorous surface cycling of water between reservoirs of ice, liquid and vapour. Although the current state of the Martian surface is exceedingly cold and dry, there are extensive reservoirs of polar ice and ground ice (Boynton *et al.*, 2002), plus probable immense quantities of groundwater beneath an ice-rich permafrost zone (Clifford, 1993; Clifford and Parker, 2001). As recognised early in the era of spacecraft exploration, channels and valleys extensively dissect the surface of Mars. Channels are elongated troughs that display clear evidence for large-scale fluid flow across their floors and on parts of their walls or banks. Immense channels, with widths of tens of kilometres and lengths of up to a few thousand kilometres, display a suite of morphological attributes that are most consistent with genesis by cataclysmic flows of water and sediment (Baker and Milton, 1974; Baker, 1982). On Earth such flows produced the distinctive landforms of the Channeled Scabland (Baker, 1978). In contrast to terrestrial megafloods, however, the Martian outflow channels involved much larger cataclysmic flood flows (Baker, 2001), and many floods emanated from subsurface sources, mostly during periods of Martian history during the billion years or so after termination of the heavy bombardment phase at about 3.9 Ga. The huge peak discharges implied by the size and morphology of the outflow channels (Baker *et al.*, 1992; Baker, 2001) are explained by several models. A currently popular view holds that a warm, wet climate phase during the heavy bombardment was followed by a progressively thickening ice-rich permafrost zone during subsequent, cold and dry Martian history. The outflow channels result from releases of subsurface water that was confined by this cryosphere (Carr 1979, 2000). However, the geological record shows that volcano–ice–water interactions are commonly associated with outburst flood channels. Thus, episodic heat flow and volcanism (Baker *et al.*, 1991) affords a reasonable alternative to the linear model of progressive pressurisation of confined water by cryosphere thickening.

An important recent discovery is that Martian flood channel activity, involving outbursts of water and associated lava flows, occurred in the Cerberus Plains region on the order of 10 million years ago (Berman and Hartmann, 2002; Burr *et al.*, 2002). The huge discharges associated with these floods and the temporally related volcanism should have introduced considerable water into active hydrological circulation on Mars. The process could have been a trigger for atmospheric water migration leading to ice emplacement (Baker, 2003) in latitudinal zonation for which orbital variations likely acted as the pacemaker (Head *et al.*, 2003).

1.6 Global consequences

1.6.1 Earth

The oceans of the Earth constitute a vast, interconnected body of water that covers about 70% of the surface of the planet. This immense water reservoir is integral to long-term climate change. Ocean currents distribute heat between the equator and the poles. For example, the Gulf Stream flows at discharges of up to 100 sverdrups, involving relatively slow-moving water 1 km in depth and 50 to 75 km wide. As northward-flowing Atlantic Ocean seawater evaporates and becomes more saline, it increases in density, eventually sinking in the northern Atlantic. This process forms a portion of the great global thermohaline circulation pattern, which acts as a conveyor belt for heat in the oceans. However, this global-scale circulation pattern was disrupted during the last glaciation when megafloods introduced relatively low-density, freshwater lids over large areas of ocean surface. The resulting disruption of sea-surface temperatures and density structure drastically altered the meridional transport of heat on a global scale (Clark *et al.*, 2001). The global climate of the Earth was altered on time scales of decades to centuries (Broecker *et al.*, 1989; Barber *et al.*, 1999).

The connections between Pleistocene ice sheets and the oceans are still very poorly understood. An important link is inferred from relatively brief (100- and 500-year) intervals in which thick marine layers of ice-rafted material were widely distributed across the North Atlantic. Called 'Heinrich events', these layers are thought to record episodes of massive iceberg discharge from unstable ice sheets. The youngest Heinrich events date to 17 000 (H1), 24 000 (H2) and 31 000 (H3) years ago. Closely related is the Younger Dryas (YD) event, a global cooling at 12 800 years ago, which is also associated with North Atlantic ice-rafted rock fragments. There is evidence from corals at Barbados that sea level rose spectacularly, about 20 m during H1 and about 15 m just after YD. Such rises would require short-term freshwater fluxes to the oceans respectively of about 14 000 and 9000 km^3/yr. (Fairbanks, 1989). These events are thought to relate to ice-sheet collapse, reorganisation of ocean–atmosphere circulation, and release of subglacial and proglacial meltwater, most likely during episodes of cataclysmic megaflooding.

Terrestrial megafloods influence the global climate system through their interactions with the ocean. A megaflood can enter the ocean either as a buoyant spreading freshwater plume over higher density (salty) seawater (Kourafalou *et al.*, 1996), or as a descending flow of sediment-laden, high-density fluid, known as a hyperpycnal flow (Mulder *et al.*, 2003). The turbidity current deposits of hyperpycnal flows may extend across hundreds, even thousands of kilometres of abyssal plain seafloor, as in the case of the Missoula Floods of the Cordilleran Ice Sheet (Normark and Reid, 2003). Nevertheless, these flows gradually lose momentum as they drop their sediment loads, thereby releasing low-density freshwater from the bottom of deep ocean basins. The buoyant freshwater then moves upward in massive convective plumes (Hesse *et al.*, 2004). These plumes disrupt the thermal structure of the ocean, with consequences for the currents that distribute heat and moderate climates.

1.6.2 Mars 'oceans'

Evidence for persistent standing bodies of water on Mars is abundant (e.g. Scott *et al.*, 1995; Cabrol and Grin, 2001; Irwin *et al.*, 2002). Despite the lack of direct geomorphological evidence that the majority of the surface of Mars was ever covered by standing water, the term 'ocean' has been applied to temporary ancient inundations of the northern plains, which did not persist through the whole history of the planet. Although initially inferred from sedimentary landforms on the northern plains (e.g. Lucchitta *et al.*, 1986), inundation of the northern plains has been controversially tied to identifications of 'shorelines' made by Parker *et al.* (1989, 1993).

Sediment-charged Martian floods from outflow channels (Baker, 1982) would have entered the ponded water body on the northern plains as powerful turbidity currents. This is the reason for the lack of obvious delta-like depositional areas at the mouths of the outflow channels. High-velocity floods, combined with the effect of the reduced Martian gravity (lowering the settling velocities for entrained sediment) promotes unusually coarse-grained washload (Komar, 1980), permitting the turbidity currents to sweep over the entire northern plains. The latter are mantled by a vast deposit, the Vastitas Borealis formation, that covers almost 3×10^7 km^2, or approximately one-sixth of the area of the planet. This sediment is contemporaneous with the post-Noachian outflow channels and it was likely emplaced as the sediment-laden outflow channel discharges became hyperpycnal flows upon entering ponded water on the plains (Ivanov and Head, 2001). In another scenario, Clifford and Parker (2001) envision a Noachian 'ocean', contemporaneous with the highlands valley networks and fed by a great fluvial system extending from the south polar cap, through Argyre and the Chryse Trough, to the northern plains.

The immense floods that initially fed the hypothetical Oceanus Borealis on Mars could have been the triggers for hydroclimatic change through the release of radiatively active gases, including carbon dioxide and water vapour (Gulick *et al.*, 1997). During the short-duration thermal episodes of cataclysmic outflow, a temporary cool-wet climate would prevail. Water that evaporated off Oceanus Borealis would transfer to uplands, including the Tharsis volcanoes and portions of the southern highlands, where precipitation as snow promoted the growth of glaciers and rain contributed to valley development and lakes. However, this cool-wet climate was inherently unstable. Water from the evaporating surface-water bodies was lost to storage in the highland glaciers, and as infiltration into the porous lithologies of the Martian surface.

1.7 Modern controversies

1.7.1 Black Sea

Although Late Quaternary freshwater inundation of the Black Sea is well recognised, it is usually correlated to enhanced proglacial meltwater flow via the many rivers draining southward from the northern Eurasian ice sheets, as noted above. A recent controversy has arisen, however, over the marine influx to the Black Sea that occurred during rising Holocene sea level, which spilled Mediterranean water through the Hellespont and Bosporus, reaching the Black Sea about 9500 years ago. One view (Bryan *et al.*, 2003) holds that the global ocean rose to the level of a relatively shallow sill of the Bosporus outlet and catastrophically spilled into the Black Sea basin, which then was then partly filled with freshwater to a level about 85 m below that of today. The resulting cataclysmic inundation presumably displaced a large human population in a calamity that is equated to the Noachian flood myth. An alternative view is that much of the Bosporus is underlain by freshwater facies of late Pleistocene age, derived from the Black Sea. The minimal erosion of these sediments is not consistent with cataclysmic flooding leading to the overlying Holocene marine facies of Mediterranean origin.

Based on marine science data from the Black Sea, Ryan and colleagues (Ryan *et al.*, 1997) proposed that the Black Sea basin had been catastrophically flooded during the early Holocene, now thought to have been about 8400 years ago (Ryan, 2007). Ryan and Pitman (1998) subsequently elaborated that, prior to this flood, the Black Sea basin held an isolated freshwater lake, which was separated from the world ocean (then at a much reduced sea level) by the mountains of Turkey. Moreover, a large number of people inhabited the shores of this lake.

Rising world sea level eventually resulted in the breaching of the mountain divides that separated the freshwater lake of the Black Sea from the world ocean. As the water burst through the modern Bosporus Strait, the water rose 15 cm per day in the Black Sea, filling its basin in about 2 years. The human population that experienced this cataclysm was forced to disperse, carrying with it a memory of the great flooding, and conveying that story to the many other cultures that were encountered. Given that one of those cultures provided the Mesopotamian influence on the author(s) of Genesis, it was appropriate to label the model for this event, the 'Noah's Flood Hypothesis'.

The current status of this hypothesis, modified from the original by Ryan *et al.* (2003), is defended by Ryan (2007). Hiscott *et al.* (2007) present an alternative model, the 'Outflow Hypothesis', involving a gradual transition in salinity of the late Quaternary Black Sea. These latter authors do not accept the early Holocene evaporative drawdown of the freshwater lake in the Black Sea basin that preceded the 8.4 kyr BP cataclysmic saltwater inundation of the 'Noah's Flood Hypothesis'. However, they do accept one of the modifications made in the original Ryan *et al.* model, specifically that late-glacial, meltwater-induced inflow to the Black Sea basin induced it to spill freshwater through the Bosporus to the Sea of Marmara. This late Pleistocene freshwater flooding was on an immense scale, such that Chepalyga (2007) claims that 'The Flood' was not the 8.4 kyr BP saltwater inundation of the Black Sea basin (derived from the world ocean via the Bosporus, Sea of Marmara, etc.). Instead, there was an earlier, much larger catastrophe in which a cascade of spillings occurred from the Aral to the Caspian basins, and ultimately to the Black Sea via the Manych Spillway. Additional water was supplied by 'superfloods' in the river valleys of European Russia, the Don, Dneiper and Volga.

Was the Black Sea inundation the source of a flood myth, specifically one that inspired western Asiatic peoples to the beliefs that inspired the account of Noah in Genesis? The anthropological implications of the 'Noah's Flood Hypothesis' were greeted with considerable scepticism by many archaeologists. Nevertheless, physical aspects of the Ryan *et al.* model of early Holocene saltwater flooding of the Black Sea basin receive some support from Coleman and Ballard (2007), Algan *et al.* (2007) and Lericolais *et al.* (2007). These studies document spectacular evidence for submerged palaeoshorelines, including drowned beaches, sand dunes and wave-cut terraces. Some radiocarbon dates (Ryan, 2007) support the proposed early Holocene age for these presumed shorelines of the freshwater lake that existed prior to the cataclysmic inflow of marine water. Other studies find no evidence in preserved fauna or sediments that there was a cataclysmic flood (e.g. Yanko-Hombach, 2007).

1.7.2 Subglacial megafloods

The recent documentation of subglacial flood flows as a modern process for Antarctic subglacial lakes (Wingham *et al.*, 2006; Fricker *et al.*, 2007) leads to the tantalising suggestion that the large late-glacial continental ice sheets may also have experienced subglacial flooding. Indeed, the topic of subglacial flooding has been the subject of a major controversy in glacial geomorphology over the last 25 years or so. The controversy has arisen from John Shaw's (1983, 2002) quite original emphasis on flooding as the mechanism for the generation of subglacial landforms. A broad variety of enigmatic landforms, involving water erosion and deposition, developed beneath the Laurentide Ice Sheet. These landforms include drumlins, Rogen moraines, large-scale bedrock erosional flutings and streamlining, and tunnel channels (valleys). Though most commonly explained by subglacial ice and debris-layer deformational processes, the genesis of these features cannot be observed in modern glaciers that are much smaller than their late Quaternary counterparts. Shaw (1996) explains the assemblage of landforms as part of an erosional/depositional sequence beneath continental ice sheets that precedes regional ice stagnation and esker formation with a phase of immense subglacial sheet floods, which, in turn, follows ice-sheet advances that terminate with surging, stagnation and melt-out. Shaw (1996) proposes that peak discharges of tens of sverdrups are implied by the late Quaternary subglacial landscapes of the southern Laurentide Ice Sheet. Release volumes are also huge; Blanchon and Shaw (1995) propose that a 14 m sea-level rise at 15 ka resulted from this megaflooding.

Shoemaker (1995) provides some theoretical support for Shaw's model, though at smaller flow magnitude levels. Arguments against the genetic hypotheses for the landforms are given by Benn and Evans (1998, 2006). Walder (1994) criticises the hydraulics, and Clarke *et al.* (2005) object that the theory requires unreasonably huge volumes of meltwater to be subglacially or supraglacially stored and then suddenly released. Shaw and Munro-Stasiuk (1996) respond to these and other criticisms.

Subglacial cataclysmic flooding has been proposed recently for the Mid-Miocene ice sheet that overrode the Transantarctic Mountains (Denton and Sugden, 2005; Lewis *et al.*, 2006). Spectacular scabland erosion occurs on plateau and mountain areas up to 2100 m in elevation, over an 80 km mountain front. The inferred high-energy flooding could have been supplied from subglacial lakes, such as modern Lake Vostok (Siegert, 2005). Such lakes would have formed beneath the thickest, warm-based portion of the ice sheet, located inland (west) of the mountains (Denton and Sugden, 2005). Thinner ice overlying the mountains was probably cold-based, thereby generating a seal that could be broken when the pressure of water in the lakes, confined by thick overlying ice, reached threshold values. The water from the lakes would move along the ice–rock interface, along the pressure gradient up and across the mountain rim. The high-velocity flooding would be further enhanced to catastrophic proportions as conduits were opened by frictional heat in the subglacial flows (Denton and Sugden, 2005).

Perhaps the most spectacular hypothesised subglacial flooding is envisioned by Grosswald's (1999) proposal that much of central Russia was inundated in the late Quaternary by immense outbursts from the ice-sheet margins to the north. Using satellite imagery to map large-scale streamlined topography and flow-like lineations, Grosswald (1999) infers colossal flows of water from beneath an ice cap that covered the entire Arctic Ocean. His hypothesised floods entered what is now central Siberia from the north and turned westward to follow the Turgai and other spillways noted above, eventually reaching the Caspian and Black Sea basins.

1.8 Discussion

Too much can be made in science of the current philosophical fad of testing (falsifying) hypotheses. As long recognised in geological investigations, hypotheses about past phenomena cannot function as propositions to be experimentally manipulated in a controlled laboratory setting. Because geologists study a past that is inaccessible to experimentation, they follow 'working hypotheses', testing for their consistency and coherence with the whole body of collected evidence. Applying methods described by T. C. Chamberlin, G. K. Gilbert and W. M. Davis (see Baker, 1996), geologists have long used their working hypotheses to advance a path of inquiry toward the truth of the past, while avoiding the blockage of that inquiry by privileging any particular take on that past. It is certainly within this tradition that the various studies in this volume have operated. For both its advocates and detractors, the 'Noah's Flood Hypothesis' of Ryan and colleagues, Shaw's subglacial floods, and Grosswald's ice sheets have provided stimuli to further inquiry, made more productive by having a target to consider for the investigation. That these targets involved considerable inspiration to the popular imagination just makes the inquiry more intense and compelling. For what more could one ask in a scientific controversy?

With much of North America and Eurasia experiencing huge diversions of drainage by glacial meltwater flooding during the period of major ice-sheet decay, it is not surprising that many human cultures developed narrative traditions involving 'worldwide flooding'. Certainly, 'the world' for a local human society of 12 000 years ago involved a much smaller geographical extent than that word would convey to the global human society of today. Thus, there is no mystery that the most impressive event in the lives of many late ice-age cultures would have been 'worldwide flooding', and the collective memory of this experience would be conveyed via oral traditions to later generations, inspiring awe up to the present day.

Acknowledgements

My early studies of megafloods were stimulated and encouraged by William Bradley, Hal Malde, David Snow and, especially, J Harlen Bretz.

References

Algan, O., Ergin, M., Keskin, S. *et al.* (2007). Sea-level changes during the late Pleistocene-Holocene on the southern shelves of the Black Sea. In *The Black Sea Flood Question*, eds. V. Yanko-Hombach, A. S. Gilbert, N. Panin and P. M. Dolukhanov. Dordrecht: Springer, pp. 603–631.

Arkhipov, S. A., Ehlers, J., Johnson, R. G. and Wright, Jr., H. E. (1995). Glacial drainage towards the Mediterranean during the Middle and Late Pleistocene. *Boreas*, **24**, 196–206.

Atwater, B. F. (1986). *Pleistocene Glacial-lake Deposits of the Sanpoil River Valley, Northeastern Washington*. U.S. Geological Survey Bulletin 1661.

Baker, V. R. (1973). *Paleohydrology and Sedimentology of Lake Missoula Flooding in Eastern Washington*. Boulder, Colorado: Geological Society of America Special Paper 144.

Baker, V. R. (1978). The Spokane Flood controversy and the Martian outflow channels. *Science*, **202**, 1249–1256.

Baker, V. R. (Ed.) (1981). *Catastrophic Flooding: The Origin of the Channeled Scabland*. Stroudsburg, PA: Dowden, Hutchinson & Ross, Inc.

Baker, V. R. (1982). *The Channels of Mars*. Austin: University of Texas Press.

Baker, V. R. (1994). Glacial to modern changes in global river fluxes. In *Material Fluxes on the Surface of the Earth*. Washington, DC: National Academy Press, pp. 86–98.

Baker, V. R. (1996). The pragmatic roots of American Quaternary geology and geomorphology. *Geomorphology*, **16**, 197–215.

Baker, V. R. (1997). Megafloods and glaciation. In *Late Glacial and Postglacial Environmental Changes: Quaternary, Carboniferous-Permian and Proterozoic*, ed. I. P. Martini. Oxford: Oxford University Press, pp. 98–108.

Baker, V. R. (1998). Catastrophism and uniformitarianism: logical roots and current relevance. In *Lyell: The Past is the Key to the Present*, eds. D. J. Blundell and A. C. Scott.

London: The Geological Society, Special Publications 143, pp. 171–182.

Baker, V. R. (2001). Water and the Martian landscape. *Nature*, **412**, 228–236.

Baker, V. R. (2002a). High-energy megafloods: planetary settings and sedimentary dynamics. In *Flood and Megaflood Deposits: Recent and Ancient Examples*, eds. I. P. Martini, V. R. Baker and G. Garzon. International Association of Sedimentologists Special Publication 32, pp. 3–15.

Baker, V. R. (2002b). The study of superfloods. *Science*, **295**, 2379–2380.

Baker, V. R. (2003). Planetary science: icy Martian mysteries. *Nature*, **426**, 779–780.

Baker, V. R. and Bunker, R. C. (1985). Cataclysmic late Pleistocene flooding from glacial Lake Missoula: a review. *Quaternary Science Reviews*, **4**, 1–41.

Baker, V. R. and Milton, D. J. (1974). Erosion by catastrophic floods on Mars and Earth. *Icarus*, **23**, 27–41.

Baker, V. R., Strom, R. G., Gulick, V. C. *et al.* (1991). Ancient oceans, ice sheets and hydrological cycle on Mars. *Nature*, **352**, 589–594.

Baker, V. R., Carr, M. H., Gulick, V. C., Williams, C. R. and Marley, M. S. (1992). Channels and valley networks. In *Mars*, eds. H. H. Kieffer, B. Jakosky and C. Snyder. Tucson: University of Arizona Press, pp. 493–522.

Baker, V. R., Benito, G. and Rudoy, A. N. (1993). Paleohydrology of Late Pleistocene superflooding, Altay Mountains, Siberia. *Science*, **259**, 348–350.

Barber, D. C., Dyke, A., Hillaire-Marcel, C. *et al.* (1999). Forcing of the cold event of 8200 years ago by catastrophic drainage of Laurentide lakes. *Nature*, **400**, 344–348.

Benito, G. and O'Connor, J. E. (2003). Number and size of last-glacial Missoula floods in the Columbia River valley between the Pasco Basin, Washington, and Portland, Oregon. *Geological Society of America Bulletin*, **115**, 624–638.

Benn, D. I. and Evans, D. J. A. (1998, 2006). *Glaciers and Glaciation*. London: Arnold.

Berman, D. C. and Hartmann, W. K. (2002). Recent fluvial, volcanic, and tectonic activity on the Cerberus Plains of Mars. *Icarus*, **159**, 1–17.

Blanchon, P. and Shaw, J. (1995). Reef drowning during the last deglaciation: evidence for catastrophic sea-level rise and ice sheet collapse. *Geology*, **23**, 4–8.

Boynton, W. V. and 24 others (2002). Distribution of hydrogen in the near surface of Mars: evidence for subsurface ice deposits. *Science*, **297**, 81–85.

Bretz, J H. (1923). The Channeled Scabland of the Columbia Plateau. *Journal of Geology*, **31**, 617–649.

Bretz, J H. (1928). Channeled Scabland of eastern Washington. *Geographical Review*, **18**, 446–477.

Bretz, J H., Smith, H. T. U. and Neff, G. E. (1956). Channeled Scabland of Washington: new data and interpretations. *Geological Society of America Bulletin*, **67**, 957–1049.

Broecker, W. S., Kennett, J. P., Flower, B. P. *et al.* (1989). The routing of meltwater from the Laurentide ice-sheet during the Younger Dryas cold episode. *Nature*, **341**, 318–321.

Bryan, W. B. F., Major, C. O., Lericolais, G. and Goldstein, S. L. (2003). Catastrophic flooding of the Black Sea. *Annual Reviews of Earth and Planetary Sciences*, **31**, 525–554.

Burr, D. M., Grier, J. A., McEwen, A. S. and Keszthelyi, L. P. (2002). Repeated aqueous flooding from the Cerberus Fossae: evidence for very recently extant, deep groundwater on Mars. *Icarus*, **159**, 3–73.

Cabrol, N. A. and Grin, E. A. (2001). The evolution of lacustrine environments of Mars: is Mars only hydrologically dormant? *Icarus*, **149**, 291–328.

Carling, P. A., Kirkbride, A. D., Parnacho, S., Borodavko, P. S. and Berger, G. W. (2002). Late Quaternary catastrophic flooding in the Altai Mountains of south-central Siberia: a synoptic overview and an introduction to flood deposit sedimentology. In *Flood and Megaflood Processes and Deposits: Recent and Ancient Examples*. International Association of Sedimentologists Special Publication 32, pp. 17–35.

Carling, P. A., Villanueva, V., Herget, J. *et al.* (2008) Unsteady 1-D and 2-D hydraulic models with ice-dam break for Quaternary megaflood, Altai Mountains, southern Siberia. *Global and Planetary Change.*

Carr, M. J. (1979). Formation of Martian flood features by release of water from confined aquifers. *Journal of Geophysical Research*, **84**, 2995–3007.

Carr, M. H. (2000). Martian oceans, valleys and climate. *Astronomy and Geophysics*, **41**, 3.21–3.26.

Chepalyga, A. L. (2007). The late glacial great flood in the Ponto-Caspian basin. In *The Black Sea Flood Question*, eds. V. Yanko-Hombach, A. S. Gilbert, N. Panin and P. M. Dolukhanov. Dordrecht: Springer, pp. 119–148.

Clark, P. U., Marshall, S. J., Clarke, G. K. C. *et al.* (2001). Freshwater forcing of abrupt climate change during the last glaciation. *Science*, **293**, 283–287.

Clarke, J. J., Mathews, W. H. and Pack, R. T. (1984). Outburst floods from glacial Lake Missoula. *Quaternary Research*, **22**, 289–299.

Clarke, G., Leverington, D., Teller, J. and Dyke, A. (2004). Paleohydraulics of the last outburst flood from glacial Lake Agassiz and the 8200 BP cold event. *Quaternary Science Reviews*, **23**, 389–407.

Clarke, G. K. C., Leverington, D. W., Teller, J. T., Dyke, A. S. and Marshall, S. J. (2005). Fresh arguments against the Shaw megaflood hypothesis. A reply to comments by David Sharpe on 'Paleohydraulics of the last outburst flood from glacial Lake Agassiz and the 8200 BP cold event'. *Quaternary Science Reviews*, **23**, 389–407.

Clifford, S. M. (1993). A model for the hydrologic and climatic behavior of water on Mars. *Journal of Geophysical Research*, **98**, 10,973–11,016.

Clifford, S. M. and Parker, T. J. (2001). The evolution of the martian hydrosphere: implications for the fate of a primordial ocean and the current state of the northern plains. *Icarus*, **154**, 40–79.

Coleman, D. F. and Ballard, R. D. (2007). Submerged paleoshorelines in the southern and western Black Sea: implications for inundated prehistoric archaeological sites. In *The Black Sea Flood Question*, eds. V. Yanko-Hombach, A. S. Gilbert, N. Panin and P. M. Dolukhanov. Dordrecht: Springer, pp. 697–710.

Dana, J. D. (1882). The flood of the Connecticut River valley from the melting of the Quaternary glacier. *American Journal of Science*, **123**, 87–97, 179–202.

Denton, G. H. and Sugden, D. E. (2005). Meltwater features that suggest Miocene ice-sheet overriding of the Transantarctic Mountains in Victoria Land, Antarctica. *Geographiska Annaler*, **87A**, 1–19.

Fairbanks, R. G. (1989). A 17,000-year glacio-eustatic sea level record: influence of glacial melting rates on the Younger Dryas event and deep ocean circulation. *Nature*, **342**, 631–642.

Fricker, H. A., Scambos, T., Bindschadler, R. and Padman, L. (2007). An active subglacial water system in West Antarctica mapped from space. *Science*, **315**, 1544–1548.

Grosswald, M. G. (1980). Late Weichselian Ice Sheet of northern Eurasia. *Quaternary Research*, **13**, 1–32.

Grosswald, M. G. (1999). *Cataclysmic Megafloods in Eurasia and the Polar Ice Sheets*. Moscow: Scientific World (in Russian).

Grosswald, M. G. and Rudoy, A. N. (1996). Quarternary glacier-dammed lakes in the mountains of Siberia. *Polar Geography*, **20**, 180–198.

Grosswald, M. G., Kuhle, M. and Fastook, J. L. (1994). Wurm glaciation of Lake Issyk-Kul area, Tian Shan Mts: a case study in glacial history of central Asia. *Geojournal*, **33**, 273–310.

Gulick, V. C., Tyler, D., McKay, C. P. and Haberle, R. M. (1997). Episodic ocean-induced CO_2 pulses on Mars: implications for fluvial valley formation, *Icarus*, **130**, 68–86.

Gupta, S., Collier, J. S., Palmer-Felgate, A. and Potter, G. (2007). Catastrophic flooding origin of shelf valley systems in the English Channel. *Nature*, **448**, 342–345.

Haack, S. (1996). Preposterism and its consequences. In *Scientific Innovation, Philosophy and Public Policy*, eds. E. F. Paul *et al.* Cambridge: Cambridge University Press, pp. 296–315.

Head, J. W., Mustard, J. F., Kreslavsky, M. A., Milliken, R. E. and Marchant, D. R. (2003). Recent ice ages on Mars. *Nature*, **426**, 797–802.

Herget, J. (2005). *Reconstruction of Pleistocene Ice-dammed Lake Outburst Floods in the Altai Mountains, Siberia*. Geological Society of America Special Paper 386.

Hesse, R., Rashid, H. and Khodabkhsk, S. (2004). Fine-grained sediment lofting from meltwater-generated turbidity currents during Heinrich events. *Geology*, **32**, 449–452.

Hiscott, R. N., Aksu, A. E., Mudie, P. J. *et al.* (2007). The Marmara Sea Gateway since ~16 ky BP: non-catastrophic causes of paleoceanographic events in the Black Sea at 8.4 and 7.15 ky BP. In *The Black Sea Flood Question: Changes in Coastline, Climate, and Human Settlement*, eds. V. Yanko-Hombach, A. S. Gilbert, N. Panin and P. M. Dolukhanov. Dordrecht: Springer, pp. 89–117.

Hitchcock, E. (1835). *Report on the Geology, Mineralology, Botany and Zoology of Massachussetts*. Amherst, MA: J. S. and C. Adams.

Huggett, R. (1989). *Cataclysms and Earth History: The Development of Diluvialism* Oxford: Clarendon Press.

Irwin, R. P., III, Maxwell, T. A., Howard, A. D. and Craddock, R. A. (2002). A large paleolake basin at the head of Ma'adim Vallis, Mars. *Science*, **296**, 2209–2212.

Ivanov, M. A. and Head, J. W., III (2001). Chryse Planitia, Mars: topographic configuration, outflow channel continuity and sequence, and tests for hypothesized ancient bodies of water using Mars Orbiter Laser Altimeter (MOLA) data. *Journal of Geophysical Research*, **106**, 3257–3295.

Kehew, A. E. and Teller, J. T. (1994). History of the late glacial runoff along the southwestern margin of the Laurentide Ice Sheet. *Quaternary Science Reviews*, **13**, 859–877.

Komar, P. D. (1980). Modes of sediment transport in channelized flows with ramifications to the erosion of Martian outflow channels. *Icarus*, **42**, 317–329.

Kourafalou, V. H., Oey, L. Y., Wang, J. D. and Lee, T. N. (1996). The fate of river discharge on the Continental shelf. 1. Modeling the river plume and the inner shelf coastal current. *Journal of Geophysical Research*, **101**, 3415–3434.

Krinner, G., Mangerud, J., Jakobsson, M. *et al.* (2004). Enhanced ice sheet growth in Eurasia owing to adjacent ice-dammed lakes. *Nature*, **427**, 429–432.

Lericolais, G., Popescu, I., Guichard, F., Popescu, S.-M. and Manolakakis, L. (2007). Water-level fluctuations in the Black Sea since the Last Glacial Maximum. In *The Black Sea Flood Question*, eds. V. Yanko-Hombach, A. S. Gilbert, N. Panin and P. M. Dolukhanov. Dordrecht: Springer, pp. 437–452.

Lewis, A. R., Marchant, D. R., Kowalewski, D. E., Baldwin, S. L. and Webb, L. E. (2006). The age and origin of the Labyrinth, western Dry Valleys, Antarctica: evidence for extensive middle Miocene subglacial floods and freshwater discharge to the Southern Ocean. *Geology*, **34**, 513–516.

Lucchitta, B. K. Ferguson, H. M. and Summers, C. (1986). Sedimentary deposits in the northern lowland plains, Mars. *Journal of Geophysical Research*, **91**, E166–E174.

Lyell, C. (1830–3). *Principles of Geology*. London: John Murray.

Mangerud, J. and 13 others (2004). Ice-dammed lakes and rerouting of the drainage of northern Eurasia during the last glaciation. *Quaternary Science Reviews*, **23**, 1313–1332.

Mulder, T., Syvitski, J. P. M., Migeon, S., Faugeres, J. C. and Savoye, B. (2003). Marine hyperpycnal flows: initiation behavior and related deposits: a review. *Marine and Petroleum Geology*, **20**, 861–882.

Normark, W. R. and Reid, J. A. (2003). Extensive deposits on the Pacific Plate from late Pleistocene North-American glacial lake bursts. *Journal of Geology*, **111**, 617–637.

Numbers, R. L. (1993). *The Creationists: The Evolution of Scientific Creationism*. Berkeley: University of California Press.

O'Connor, J. E. and Baker, V. R. (1992). Magnitudes and implications of peak discharges from glacial Lake Missoula. *Geological Society of America Bulletin*, **104**, 267–279.

Pardee, J. T. (1910). The glacial Lake Missoula, Montana. *Journal of Geology*, **18**, 376–386.

Pardee, J. T. (1942). Unusual currents in glacial Lake Missoula, *Geological Society of America Bulletin*, **53**, 1569–1600.

Parker, T. J. Saunders, R. S. and Schneeberger, D. M. (1989). Transitional morphology in the west Deuteronilus Mensae region of Mars: implications for modification of the lowland/upland boundary. *Icarus*, **82**, 111–145.

Parker, T. J., Gorsline, D. S., Saunders, R. S., Pieri, D. and Schneeberger, D. M. (1993). Coastal geomorphology of the Martian northern plains. *Journal of Geophysical Research*, **98**, 11,061–11,078.

Patton, P. C. (1987). Measuring the rivers of the past: a history of fluvial paleohydrology. In *History of Hydrology*, eds. E. R. Landa and S. Ince. Washington, DC: American Geophysical Union, History of Geophysics Number 3, pp. 55–67.

Ryan, W. B. F. (2007). Status of the Black Sea flood hypothesis. In *The Black Sea Flood Question: Changes in Coastline, Climate, and Human Settlement*, eds. V. Yanko-Hombach, A. S. Gilbert, N. Panin and P. M. Dolukhanov. Dordrecht: Springer, pp. 63–88.

Ryan, W. B. F. and Pitman, W. C., III (1998). *Noah's Flood: The New Scientific Discoveries about the Event that Changed History*. New York: Simon & Schuster.

Ryan, W. B. F., Pitman, W. C., III, Major, C. O. *et al.* (1997). Abrupt drowning of the Black Sea shelf. *Marine Geology*, **138**, 119–126.

Ryan, W. B. F., Major, C. O., Lericolais, G. and Goldstein, S. L. (2003). Catastrophic flooding of the Black Sea. *Annual Reviews of Earth and Planetary Sciences*, **31**, 525–554.

Scott, D. H., Dohm, J. M. and Rice, J. W. (1995). Map of Mars showing channels and possible paleolakes. U.S. Geological Survey Misc. Inv. Map I-2461.

Shaw, J. (1983). Drumlin formation related to inverted meltwater erosion marks. *Journal of Glaciology*, **29**, 461–479.

Shaw, J. (1996). A meltwater model for Laurentide subglacial landscapes. In *Geomorphology sans Frontieres*, eds. S. B. McCann and D. C. Ford. New York: Wiley, pp. 182–226.

Shaw, J. (2002). The meltwater hypothesis for subglacial landforms. *Quaternary Science Reviews*, **90**, 5–22.

Shaw, J. and Munro-Stasiuk, M. (1996). Reply to Benn and Evans. In *Glacier Science and Environmental Change*, ed. P. G. Knight. Oxford: Blackwell, pp. 46–50.

Shoemaker, E. M. (1995). On the meltwater genesis of drumlins. *Boreas*, **24**, 3–10.

Siegert, M. N. J. (2005). Lakes beneath the ice sheet: the occurrence, analysis and future exploration of Lake Vostok and other Antarctic subglacial lakes. *Annual Reviews of Earth and Planetary Sciences*, **33**, 215–245.

Teller, J. T., Leverington, D. W. and Mann, J. D. (2002). Freshwater outbursts to the oceans from glacial Lake Agassiz and their role in climate change during the last glaciation. *Quaternary Science Reviews*, **21**, 879–887.

Upham, W. (1877). The northern part of the Connecticut Valley in the Champlain and terrace periods. *American Journal of Science*, **114**, 459–470.

Waitt, R. B., Jr. (1985). Case for periodic, colossal jokulhlaups from Pleistocene glacial Lake Missoula. *Geological Society of America Bulletin*, **96**, 1271–1286.

Walder, J. S. (1994). Comments on 'Subglacial floods and the origin of low-relief ice-sheet lobes'. *Journal of Glaciology*, **40**, 199–200.

Wingham, D. J., Siegert, M. J., Shepherd, A. and Muir, A. S. (2006). Rapid discharge connects Antarctic subglacial lakes. *Nature*, **440**, 1033–1036.

Yanko-Hombach, V. (2007). Controversy over Noah's Flood in the Black Sea: geological and foraminiferal evidence from the shelf. In *The Black Sea Flood Question: Changes in Coastline, Climate, and Human Settlement*, eds. V. Yanko-Hombach, A. S. Gilbert, N. Panin and P. M. Dolukhanov. Dordrecht: Springer, pp. 149–203.

2

Channel-scale erosional bedforms in bedrock and in loose granular material: character, processes and implications

PAUL A. CARLING, JÜRGEN HERGET, JULIA K. LANZ, KEITH RICHARDSON and ANDREA PACIFICI

Summary

High-energy fluid flows such as occur in large water floods can produce large-scale erosional landforms on Earth and potentially on Mars. These forms are distinguished from depositional forms in that structural and stratigraphical aspects of the sediments or bedrock may have a significant influence on the morphology of the landforms. Erosional features are remnant, in contrast to the depositional (constructional) landforms that consist of accreted water-borne sediments. A diversity of erosional forms exists in fluvial channels on Earth at a range of scales that includes the millimetre and the kilometre scales. For comparison with Mars and given the present-day resolution of satellite imagery, erosional landforms at the larger scales can be identified. Some examples include: periodic transverse undulating bedforms, longitudinal scour hollows, horseshoe scour holes around obstacles, waterfalls, plunge pools, potholes, residual streamlined hills, and complexes of channels. On Earth, many of these landforms are associated with present day or former (Quaternary) proglacial landscapes that were host to jökulhlaups (e.g. Iceland, Washington State Scablands, Altai Mountains of southern Siberia), while on Mars they are associated with landscapes that were likely host to megafloods produced by enormous eruptions of groundwater. The formative conditions of some erosional landforms are not well understood, yet such information is vital to interpreting the genesis and palaeohydraulic conditions of past megaflood landscapes. Correct identification of some landforms allows estimation of their genesis, including palaeohydraulic conditions. Kasei Valles, Mars, perhaps the largest known bedrock channel landscape, provides spectacular examples of some of these relationships.

2.1 Introduction

Large floods on Earth are known to erode loose granular material and bedrock to form channels. For example, jökulhlaups can erode bedrock (Baker, 1988; Tómasson, 1996; Carrivick *et al.*, 2004). However, many moderate river flows, as well as large floods, are capable of eroding bedrock over prolonged periods of time by abrasion induced by bedload (Sklar and Dietrich, 1998, 2001) and suspended load (Richardson and Carling, 2005). A major challenge will be to determine the relative importance of long-term slow erosion induced by frequent flows and major 'cataclysmic' floods that can induce very noticeable changes in channel forms. By analogy, water floods of a variety of scales should also be capable of eroding granular materials and bedrock on Mars and evidence of such fluvial erosion would strengthen the argument for water-induced geomorphological activity on the Earth's neighbouring planet.

Channel-scale erosional bedforms are worthy of study for at least two reasons. Firstly, correct identification of key features may elucidate the nature of the flow hydraulics and, in some cases, the magnitude of the formative discharge may be determined. Erosional antidunes, for example, (detailed later in this review) can be related to transcritical and supercritical flow regimes. Secondly, in the case of Mars, identification of erosional 'fluvial' bedforms implies the presence of water rather than some other liquid or deformable flowing solid such as ice. However, qualitative and quantitative understanding of the origin of erosional bedforms and their geomorphological significance is immature. Erosional bedforms in loose granular materials have received the most attention especially in relation to flow around non-deformable obstacles (e.g. horseshoe scour around bridge piers) but erosion in bedrock has been little studied. Consequently, this review begins with a brief historical perspective that largely considers the bedrock erosion of the Channeled Scablands of Washington State, and this is followed by a proposed typology of large-scale erosional bedforms that provides a framework for description and discussion. Examples of eroded bedforms on Earth and Mars are considered.

Megaflooding on Earth and Mars, ed. Devon M. Burr, Paul A. Carling and Victor R. Baker.
Published by Cambridge University Press.

2.2 Historical perspective

Many scientists in the nineteenth century were interested in fluvial catastrophism as an agent in transforming the landscape. Whilst most invoked the Noachian biblical flood, a few grasped the importance of large-scale fluvial action caused by natural events rather than a flood induced by a deity. Parker (1838), for example, recognised that large-scale dry valleys, locally termed 'coulees' in Washington state, were fluvially eroded: a theme furthered by others (Russell, 1893; Dawson, 1898; Salisbury, 1901; Calkins, 1905). Baulig (1913), in particular, recognised that the coulees could be water-eroded and identified dry waterfalls, water-scoured rock basins and plunge pools, but attributed these to the normal action of a prior course of the Columbia River. However, he commented specifically on the scale of the features, presumably having some difficulty reconciling the scale with normal fluvial action.

Bretz's (1923a) geomorphological study of the Columbia Plateau in southeastern Washington state may be taken as the benchmark introduction to the study of catastrophic erosional landscapes. A map accompanies the paper of Bretz and it depicts patterns of abandoned anastomosed canyon-like waterways – the coulees. Many of these channels contain seemingly abandoned waterfalls, cataracts, plunge pools, hanging tributary and distributary valleys and large-scale potholes, all eroded into the basalt bedrock. The entire composite became known as the 'Channeled Scabland' (Baker, 1978a) and most of this historical perspective considers the scabland (see also Baker, 2008). Bretz (1924) supported his arguments in favour of vigorous fluvial action by describing large-scale erosional features in the modern Columbia River near The Dalles. Bretz (1923b) also identified streamlined loess hills as flood-eroded remnants. Bretz (1930b) synthesised his ideas inasmuch as he realised that the total suite of erosional features could only be genetically related by invoking a short-duration, large-volume and high-velocity flood. As recorded by Baker (1973b), the scabland erosional features include more than one hundred overfit anastomosing channels, hanging valley junctions, gigantic potholes, large longitudinal grooves up to 5 m deep and 50 m wide, bedrock pool–riffle sequences at 10 to 15 km spacings, streamlined loess hills and flood-sculpted lozenge-shaped residual outcrops of basalt. Malde (1968) described similar features associated with the Pleistocene breakout flood from Lake Bonneville through the Snake River in Idaho.

Although the plucking of bedrock on a large scale by catastrophic floodwaters became understandable with the hydrodynamic study of Baker (1973a) there are antecedents. There is evidence from contemporary correspondence that Pardee had considered as early as 1910 that the Scablands might be related to catastrophic drainage from a large Pleistocene lake. Bretz identified the source of the floodwaters as glacial Lake Missoula in 1928 and although he made his views known, he did not publish at this time. Harding, without consultation or acknowledgement, published aspects of the ideas of Bretz in 1929. However, the intellectual attribution was properly made when the full details of an abrupt ice-dam failure were published (Bretz, 1930a, 1932). Pardee published his ideas linking glacial lake Missoula with the floods in 1942.

However, it was not until later in the twentieth century that the hydraulic formation of large-scale channels in the scablands was considered in detail (Baker, 1973a, 1973b). Better conceptualisation of the progress of floodwaters across the landscape included the recognition that floodwaters might be hydraulically ponded at constrictions and bendways in the flood course, a concept that can be tested using modern hydraulic models of flood progression and field evidence. Such constrictions, causing local elevation of the water surface, are sometimes marked by supposed scour lines in the landscape, such as scarps eroded into loess that are presumed high-water marks. This scenario was described by Waitt (1972, 1977a, b) for late Pleistocene flooding down the Columbia River. In the case of Siberian Altai Quaternary floods, on a scale similar to the Missoula floods, ponding might also have occurred in the Katun River valley at Big Ilgumen (Carling *et al.*, 2002c). The Altai floods have left little evidence of erosional landforms; rather, it is the erosion of the Yakima basalts in the scabland by the Missoula flood that has left the most visually impressive erosional topography. The distinctive rock-jointing habit of the basalt means that colonnade and entablature structures are well developed, often as columns of prismatic jointed blocks. Baker (1973a) demonstrated that large flood-transported blocks and cobbles emanated from these columns. Bretz (1924, 1932; Bretz *et al.*, 1956) recognised that macroturbulence and the distinctive jointing patterns in the basalt when combined could lead to ready bedrock erosion and Embleton and King (1968) suggested that cavitation might have caused the scour of deep Scabland rock basins. Often plucking was concentrated at tension-jointed anticlinal crests, usually well represented at divide crossings that are common in the Scablands. Water-scoured crossings consist of narrow, generally parallel-sided, short, flat-floored and steep-walled bedrock channels (Baker, 1978c) where water-filled valley flows spilled over into adjacent channel systems, often truncating rock slip-off slopes leading to truncated bluffs that Bretz (1928) called 'trenched spur buttes'. Divide crossings also are important because they are often a component in the anastomosed channel pattern.

2.3 Processes of bedrock erosion

There are several accounts of the mechanisms of fluvial erosion in bedrock channels, and it is not necessary to describe the processes in detail here. The interested reader is referred to the useful works of Allen (1971a), Selby (1985), Hancock *et al.* (1998), Wohl (1998), Wende (1999) and Whipple *et al.* (2000). However, brief definitions of the various processes are provided here for reference (see also Richardson and Carling, 2005).

1. Plucking, also known as quarrying and jacking, involves the removal of whole blocks of rock from the boundary by lift and drag forces acting on the block. The blocks are delineated by joints and other structural features of the rock and undergo a period of preparation prior to their entrainment by the flow. Block preparation involves the widening and propagation of cracks and the loosening of blocks by a combination of hydraulic forces, bedload impact, wedging of sediment within cracks and physical and chemical weathering (Whipple *et al.*, 2000). Related to plucking is the process of flaking in foliated rocks such as shales and slates, which occurs on a smaller scale. Flaking involves the removal of thin rock fragments by entrainment in the flow, bedload impact, wind or ice.
2. Abrasion, also known as corrasion, involves the wearing away of a surface by numerous impacts from sediment particles transported in the fluid. These particles may be carried in suspension or as bedload, or both. Each impact, if sufficiently energetic, breaks off a small piece of the boundary material, which may be at the scale of a small part of a grain in the case of suspended load, or as much as a substantial flake or chip of rock in the case of coarse bedload.
3. Fluid stressing, also known as evorsion, refers to the removal of particles by the turbulent stresses exerted directly on the boundary by a clear fluid. At the scale of individual grains and grain aggregates, this process is likely to be important only in clays and poorly consolidated rocks. However, at larger scales, this mechanism is important in flaking and it is the process by which prepared blocks are plucked.
4. Cavitation occurs in very rapid flows when the instantaneous dynamic fluid pressure locally drops below the vapour pressure and bubbles of water vapour appear. The water effectively boils, though energy is supplied by the high-velocity water flow instead of by external heating. The bubbles very rapidly collapse, however, and the pressure shock wave thus generated erodes any nearby boundary. Cavitation is known to occur on engineering structures (Arndt, 1981), but whether it occurs in natural channels is still a matter of debate (Baker, 1973a, b; Sato, *et al.* 1987; Wohl, 1992b; Baker and Kale, 1998; Hancock *et al.*, 1998; Tinkler and Wohl, 1998a; Gupta *et al.*, 1999; Whipple *et al.*, 2000).
5. Corrosion occurs on soluble rocks, where it is also known as dissolution, and on chemically reactive rocks, where it is referred to as chemical weathering. Soluble rocks include limestone, marble and evaporites. Other rocks, such as quartzites, extrusive igneous rocks, granitic rocks and rocks with a high calcareous content such as marl, are slightly soluble or have soluble components and may also show evidence of dissolution. Chemically reactive rocks have unstable components, such as feldspars, olivine and sulphide minerals, which break down into various decay products and are subsequently easily removed. Corrosion may occur subaqueously or subaerially. It acts both directly in removing the rock or components of it, and indirectly in physically weakening the rock and assisting its removal by the other erosion mechanisms. It is possible that the growth of algae, moss and lichen, etc. on the rock enhances corrosion by inhibiting evaporation and maintaining a film of water in contact with the rock when surfaces elsewhere are dry.
6. Physical weathering acts indirectly in fluvial erosion by physically weakening the rock (Whipple *et al.*, 2000) and increasing its surface area as a result of crack propagation, thereby enhancing erosion by the mechanisms listed above. Perhaps the most important physical weathering processes are wetting/drying cycles, which may be particularly important in the processes of flaking and weathering due to the crystallisation of salts (Sparks, 1972), and freeze/thaw cycles, which may play a role in crack propagation, block loosening and rock disaggregation. Blocks of rock can also be removed directly by ice lift, when a layer of ice forms on the rock surface during low flow in winter, and rises to the surface during a subsequent flood, carrying with it small attached blocks of rock (see Carling and Tinkler (1998) for a discussion on the effect of ice).

2.4 A typology of channel-scale erosional bedforms in bedrock

Bedrock channel morphologies have traditionally been considered to be determined by the physical and structural properties of the substrate (Ashley *et al.*, 1988), and to have boundaries that are essentially imposed on the flow and cannot be adjusted by hydraulic processes because of their high resistance (Baker and Pickup, 1987) or are adjusted only rarely by high-magnitude, low-frequency events in which the fluid forces exceed the high resistance threshold. This is in contrast to alluvial rivers, whose boundaries have long been known to be deformable by relatively low-magnitude, high-frequency events, and which are known to adjust their geometries according to hydraulic and sediment conditions (Wolman and Miller, 1960). However, in recent decades there has been an increasing realisation that bedrock channel boundaries should also be considered deformable (presumably over long time scales in response to low-frequency, high-magnitude events) and that channel adjustments are conditioned both by hydraulic controls and by any imposed bedrock structural controls. This is shown by the development of bedforms in bedrock channels apparently analogous to those found in alluvial channels (e.g. Keller and Melhorn, 1978; Baker and Pickup, 1987; Ashley *et al.*, 1988; Wohl, 1992a; Wohl and Grodek, 1994; Koyama and Ikeda, 1998; Wohl and Ikeda, 1998), and by the recognition that definable channel width–area and slope–area relationships exist for bedrock rivers (Montgomery and Gran, 2001; Kobor and Roering, 2004). Furthermore, the conclusion that rivers in bedrock can adjust their channel boundaries is a logical consequence of the observation that many bedrock channels are extensively covered with water-eroded bedforms (Richardson and Carling, 2005).

Erosional bedforms in bedrock fluvial channels on Earth are extremely diverse in morphology, orientation and scale. Herein, the only features considered in detail are channel-scale features that might be identified in plan from aerial photographs and satellite images. Erosional bedforms are identified with respect to both positive and negative bed elevations including extensive fields of undulating forms. Erosional bedforms are therefore residual surfaces developed within a substratum. The morphology of negative features may be simple, representing a single locus of enhanced erosion, or compound, consisting of multiple centres of erosion. Planform geometries range from near-circular, large-scale potholes to extended linear furrows. Longitudinal, transverse, obliquely orientated and unorientated bedforms occur. In terms of spacing, erosional bedforms may exist as isolate individuals, as conjugate individuals with shared boundaries, or as coalesced individuals (Richardson and Carling, 2005).

An important consideration in any discussion of erosional bedforms in bedrock relates to the range of features to be considered and the processes by which they are generated. The dominant process will depend principally on the structure of the substrate; rock that has closely spaced fractures or planes of weakness will be dominated by the plucking of blocks and fragments, while other more ‘homogeneous’ rocks will generally be dominated by abrasion (Bretz, 1924; Whipple *et al.*, 2000). Cavitation is often reported as being of possible importance in extreme events. This was an idea promulgated in early flood studies (e.g. Dahl, 1965; Embleton and King, 1968) and one that has persisted (e.g. Baker and Kale, 1998; Gupta *et al.*, 1999; Whipple *et al.*, 2000). Without gainsaying this supposition, there is little overt evidence for the action of cavitation in fluvial systems (Barnes, 1956) and this matter is not considered further in this chapter. Corrosion is generally the most important agent of erosion in carbonate rocks, and may also be significant in rocks prone to chemical weathering, such as granite. The shapes of plucked fragments and the spaces they leave behind are defined by the geometry of the structural elements of the substrate, while features scoured by abrasion in relatively homogeneous surfaces reflect local hydrodynamic patterns. However, at the channel scale it should be remembered that plucking of blocks may result in channel-scale features that are morphologically similar to channel-scale abraded features. A hypothetical example would be series of step–pool pairs that might be clearly definable in a plucked bedrock system although therein the mesoscale roughness of the bed would reflect the structural control (e.g. a joint pattern). A tangible example of such a feature is the *c.* 40 m diameter pothole eroded through plucking by the catastrophic Pleistocene Missoula floods described by Baker (1973b).

This section illustrates the range of erosional bedforms in bedrock described on Earth. Published descriptions come mainly from observations of active forms in modern rivers rather than relict superflood bedforms. However, it is assumed that the fundamental processes responsible for bedforms operative in modern rivers would also be operative in extreme events and that most sculpted forms essentially exist as a self-similar suite of features over a wide range of scales, and in flows over a wide range of discharges (Kor *et al.*, 1991; Kor and Cowell, 1998). Importantly, some bedforms have dimensions directly related to flow velocity, for example (small-scale) scallops (Curl, 1966; Blumberg and Curl, 1974). Similarly, longitudinal furrows described by Ikeda (1978), amongst others, seem to scale with flow depth. Bedrock pool–riffle sequences (Keller and Melhorn, 1978) appear to scale with the width of the flow-field. Deriving such relationships for channel-scale bedforms would provide potentially powerful tools to reconstruct aspects of palaeofloods (Springer and Wohl, 2002).

2.4.1 Concave features

Potholes Near-circular potholes are a well-known member of a suite of concave sculpted forms (terminology from Richardson and Carling, 2005) that ranges from equant plan-view geometries to extended, highly linear furrows. Potholes may be of similar size to the channel width (Richardson and Carling, 2005). Examples, as reported from the Missoula flood tract, can be upwards of 40 m in diameter but of variable depth. Within the eponymous

Potholes Coulee of the Channeled Scabland, small lakes up to 250 m in diameter, which are often quasi-circular, are probably potholes, being associated with smaller but nevertheless readily identifiable large potholes, which are often an integral part of the scoured bed of dry channels. The Potholes cataract complex (Plate 1) also exhibits large pothole-shaped plunge pools. Many modifications on the simple pothole morphology can be found. Large potholes may also be breached as they enlarge into neighbouring sculpted forms; extremely convoluted surfaces can result from the coalescence of a group of closely spaced potholes, for instance. Consequently, for studies of relict large-scale features on Earth and Mars, the absence of a classic rounded plan view for a suspect pothole need not preclude the identification of the feature as a pothole.

Flutes Flutes are generally shallow scour hollows with a rim that is typically parabolic in plan view, convex in the upstream direction, and open in the downstream direction. They occur immediately downstream of some defect in the surface, which sets up vortex scour in its wake (Allen, 1971b). First described by Maxson and Campbell (1935), they are well known in sedimentary environments (Dzulynski and Walton, 1965) and have been studied experimentally (Allen, 1971a, b). Flutes are also common in bedrock, although they are morphologically less diverse than their sedimentary counterparts and typically lack a median ridge, a common feature of sedimentary flutes. Nevertheless, flutes in bedrock occur with a wide range of width:length and width:depth ratios. Small-scale flutes with overhanging upstream ends are common in bedrock (e.g. Maxson and Campbell, 1935; Hancock *et al.*, 1998).

Furrows Closed linear concave features (with a length more than twice their width) are termed furrows herein (see Richardson and Carling (2005) for a more detailed typology). Generally, these are orientated longitudinally and again a diverse spectrum of forms exists. Those with a relatively high width:length ratio have elliptical outlines in plan view and are named short furrows. Parallel-sided furrows are more elongated forms whose rims on either side are parallel for some distance. The rims may be cuspate or rounded. Straight, curved, sinuous and bifurcating varieties occur. Longitudinal furrows can be initiated at defects and sometimes occur in the shear zone between the wake of a large obstacle and the free stream (Tinkler, 1997) but generally are unrelated to any such feature. They may often be found in the steeply dipping region leading into a knickpoint. Longitudinal furrows can often be found in groups of parallel individuals, sometimes with regular spacing (Blank, 1958; Ikeda, 1978; Wohl, 1993). Compound longitudinal furrows, in which a very long feature is subdivided into several shorter furrows by ridges, are common. Wohl (1993) described long furrows that contained regularly spaced short depressions. These depressions grew in a downstream direction as the furrow deepened, and became small offset lateral potholes on alternate sides of the furrow. With further distance downstream, these potholes coalesced into an inner channel with undulating walls (see also Wohl *et al.*, 1999).

Not all concave features are orientated with the assumed flow direction. Furrows may slant obliquely down a sloping channel margin in the downstream direction. Other furrows are orientated transverse to the flow direction (e.g. Shrock, 1948; Jennings, 1985; Kor *et al.*, 1991). Transverse furrows are often found in the lee of some defect or line of defects that runs in a cross-stream direction, but they also occur in the absence of any obvious irregularity. Transverse furrows can have simple or compound morphologies; in the latter case, they appear to be constituted by smaller, initially isolate bedforms that have coalesced in the transverse direction. A peculiar form of the transverse furrow is that in which a single well-defined, relatively narrow furrow runs almost the entire width of the channel (Tinkler and Wohl, 1998b).

Small-scale to medium-scale linear or sinuous furrows often occur in open bedrock channels (Richardson and Carling, 2005). However, as a cautionary note, large-scale linear furrows have also been associated with subglacial meltwater (e.g. Bradwell, 2005) and with the action of ice-streams (e.g. Canals *et al.*, 2000; Dowdeswell *et al.*, 2004). Various scale-independent furrows known collectively as s-forms are also associated with subglacial meltwater (Kor *et al.*, 1991). Larger-scale lineation believed to be of fluvial origin has been reported from the Channeled Scabland (see above) and which presumably parallels the palaeoflow. Often these lineations are picked out by the presence of linear lakes that fill the depressions: examples include Lena Lake and Table Lake within the Grand Coulee and as parallel and anastomosed networks of channels immediately upstream of the famous Dry Falls (Plate 2). Patton and Baker (1978) describe scabland grooves up to 300 m wide, 25 m deep and up to 2.5 km in length. Lineation might be caused by longitudinal vortices in the flow (e.g. Coleman, 1969) and this is the preferred model for long linear parallel *depositional* ridges that develop on Earth in shallow coastal seas in subtidal locations (Blondeaux, 2001). Within fluvial systems (Ikeda, 1978) and (less exactly) within coastal waters, the spacings of these ridges tend to scale with flow depth, with spacing variation induced possibly by other subordinate parameters. It is not clear, however, why they should develop so persistently across wide swaths of the flow path, as is illustrated below for erosional examples. Erosional grooves of similar and lesser scale to depositional ridges also develop on Earth in intertidal and subtidal muddy environments, but also occasionally in soft bedrock

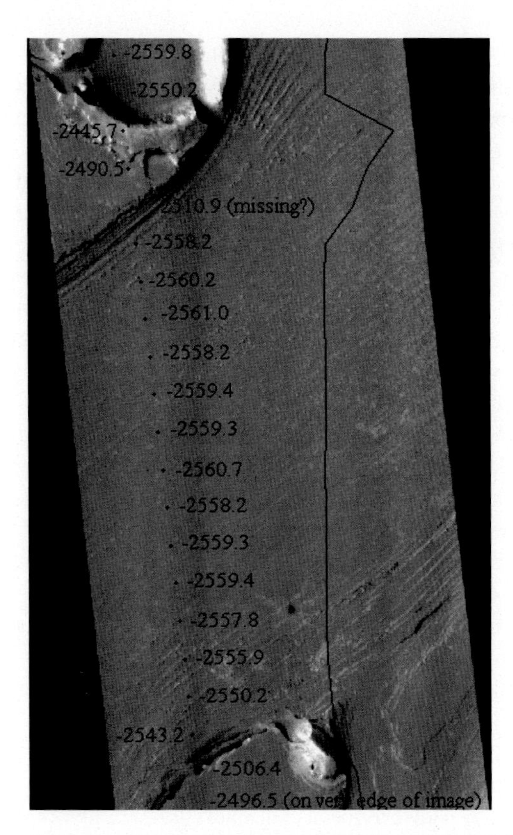

Figure 2.1. A portion of Mars Orbiter Camera (MOC) image M07-00614. The location is in Athabasca Valles, near 9.66 N 204.19 E. The overlying elevational data points are from the Mars Orbiter Laser Altimeter (MOLA); the values are negative because this region is below the mean datum for Mars. The black line on the right graphically represents the elevational data. The image is 3.02 km across. The scene is illuminated from the upper left.

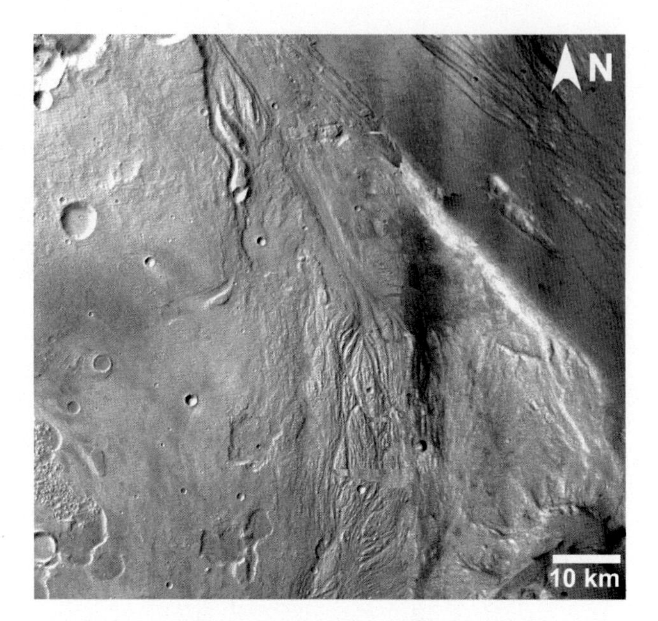

Figure 2.2. Longitudinal grooves and braided channel systems cut into bedrock within Ares Valles, Mars. The view is developed from an HRSC image. (Images courtesy of HRSC. THEMIS Public Data Releases, Mars Space Flight Facility, Arizona State University, 2006, http://themis-data.asu.edu.)

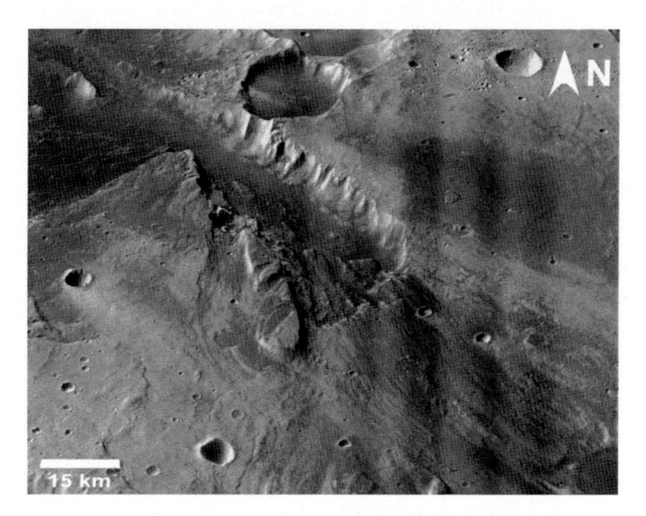

Figure 2.3. 500 m high putative former waterfall in Ares Valles, Mars. Flow bottom right to top left. Note the linear grooves that are well developed close to the lip of the falls and the secondary channels to the left of the main transverse fall. The view is developed from an HRSC image. (Images courtesy of HRSC. THEMIS Public Data Releases, Mars Space Flight Facility, Arizona State University, 2006, http://themis-data.asu.edu.)

(Allen, 1969; 1987). These ridges and grooves may bifurcate, usually with the grooves bifurcating in the direction of the dominant tidal flow, but are persistent over tens of metres to hundreds of metres as distinctive features. More recently, extensive and visually striking parallel ridges and grooves have been identified on the continental rise in the Gulf of Mexico (Plate 3), where they persist for tens of kilometres and parallel measured benthic currents (Bryant, 2000; Bean, 2003), and in the Gulf of Lions, Mediterranean Sea (Canals *et al.*, 2006). These features are very similar in appearance and scale to bedforms recorded on Mars, where strongly parallel linear systems (Figure 2.1) have been identified together with weakly and strongly anastomosed networks of eroded channels (Figures 2.2 and 2.3) that have not been well studied. Often the furrows on Mars are contained within, and are parallel with, distinct canyon walls and so may be formed by flow-induced vorticity, which adjusts to the channel expansions and contractions (compare Plate 3 and Figure 2.1), and diverge 'smoothly' (Figure 2.1) around obstacles similarly to ocean-floor grooves on Earth (Bean, 2003). Elsewhere, parallel-grooved terrain conceivably could form within large lake bodies, such as might have existed on Mars (Parker *et al.*,

1993; Parker, 1994), as wind-driven Langmuir circulation cells can cause large-scale longitudinal vortices to be established within a fluid (Leibovich, 2001), although it is not known if this causes erosion of the bed of the water body.

Large-scale channels, inner channels and knickpoints Distinctive linear bedrock channels, reminiscent of river courses, occur on both Earth and Mars. On Earth, these channels may be dry or they may contain small streams that cannot be responsible for cutting the larger channels. Channels may be relatively straight, sinuous and highly meandering or they may form as a network of interconnecting channels termed anastomosed systems (Figures 2.2, 2.3 and Plate 4). Structural control may be evident locally but otherwise the channels are cut within bedrock with the morphology conditioned by hydraulic action. Lateral benches or terraces may be evident together with longitudinal variation in channel width (beaded) and longitudinal variation in bed elevation, including riffle–pool sequences and steep cascades. Suites of smaller-scale bedforms may be evident within these channels and some of these were described above.

A common channel cross-section morphology is that of the incised inner channel with bedrock benches on either side that may extend for hundreds of metres or indeed many kilometres. Descriptions of inner channels include those of Bretz (1924), Baker (1973b), Nemec *et al.* (1982), Baker and Pickup (1987), Wohl (1992a, b, 1993), Wohl *et al.* (1994), Baker and Kale (1998) and Wohl and Achyuthan (2002). In the Mekong River through Laos, for example, inner channels occur frequently (Gupta, 2004). Very deep and narrow channels are termed 'slot canyons'. A closely related phenomenon to that of the extensive inner channels is the more locally developed, deeply incised knickpoints that occur in vertical waterfalls or inclined cascades or other short and very steep reaches. These knickpoints represent locally enhanced channel incision. In the experience of the authors, inner channels almost always possess a knickpoint at their head, implying that even extensive inner-channel formation occurs through knickpoint retreat (Baker, 1973b; Wohl, 1993; Wohl *et al.*, 1994), although an exception is noted below. However, the formation of inner channels has been observed experimentally through the simple incision of a plane bed (Shepherd and Schumm, 1974; Kodama and Ikeda, 1984; Wohl and Ikeda, 1997; Johnson and Whipple, 2007). A common mechanism of knickpoint retreat in homogeneous rocks is that of pothole growth and coalescence in the reach immediately above the knickpoint (Wohl, 1998; Wohl *et al.*, 1999). This results in inner channels and slot canyons with convoluted wall topography (Elston, 1917, 1918; Jennings, 1985; Wohl, 1999; Kunert and Coniglio, 2002) in which undulations on opposite sides are generally out of phase (Wohl, 1993; Wohl *et al.*, 1999; Wohl and Achyuthan, 2002). However, an inner channel has also been observed with in-phase wall undulations (Richardson and Carling, 2005). A stair-like succession of knickpoints may be termed a cataract (Wohl, 1998). Knickpoint migration is also an important mechanism of erosional escarpment retreat (Weissel and Seidl, 1998).

Deeply incised inner channels are common within the scabland tracts and often emanate from former locations of cataracts. Potholes Coulee and Frenchman Springs are two such channels that seem to have developed by headward retreat of a knickpoint beneath floodwaters. Examples of inner channels in modern rivers are usually associated with knickpoint retreat (Richardson and Carling, 2005). Elsewhere floodwaters within the scablands excavated the preflood stream valleys to leave preflood tributaries joining the main stems as hanging valleys. However, as a cautionary note, it should be recognised that deeply incised inner channels can also result not from floodflow and knickpoint retreat over short periods of time but through tectonic uplift and incision over many thousands of years.

Erosional constrictions and hydraulic jumps Flow through channel constrictions can cause intense erosion of bedrock and modelling studies have indicated that Missoula flood flows through several scabland constrictions (notably Soap Lake, Wallula Gap and Staircase Rapids) would each have been characterised by a hydraulic jump with critical or supercritical flow within the constrictions. Thus, narrow but short lengths of channel that are essentially short canyons (gorges) that lie downstream and upstream of broader sections of channel might be erosional features attributable to critical flow in large floods. The presence of a flood-formed landform in this instance is not readily determined from field study, but through hydraulic modelling. An example is the constriction on the Mae Chaem River in Thailand, where river flow (maximum recorded flow $>1000\,m^3\,s^{-1}$) is constrained to flow through a vertical slot a few metres wide (Kidson *et al.*, 2005). Identification of such sites is important as they can be used to constrain palaeo-discharge estimates.

2.4.2 Convex features, undulating features and composite features

Unlike concave features, there is little agreement on terminology and the identification of channel-scale convex undulating and composite features. There are relatively few published descriptions of large-scale forms in relict flood landscapes. Thus the simple subdivision in this section reflects the limited knowledge base rather than an exhaustive consideration. It is not known how this class of bedform relates to the erosional antidunes discussed in a subsequent section.

Convex features are not as diverse as their concave counterparts, presumably reflecting the relative uniformity of flow patterns that exist around positive features as compared with flow patterns within negative features. The most important feature in terms of abundance is the upstream-facing convex surface. This bedform is all but ubiquitous in bedrock channels with significant roughness relief, and consists simply of the rounding and streamlining of the upstream surfaces of positive features. Other convex features include bladed forms, in which erosion within two or more contiguous scour hollows produces an intervening keeled ridge, faceted obstacles (e.g. Wohl, 1992b; Maxson, 1940; Baker and Pickup, 1987) and streamlined hills, some of which may be analogous to a group of similar features in aeolian environments, often termed yardangs, which are well described on both Earth and Mars (e.g. El Baz *et al.*, 1979; Greeley and Iversen, 1985; Goudie, 1999).

The major group of undulating features is the hummocky forms, of which two types occur. The gently rounded type are found as longitudinal, transverse and non-directional forms with wavelengths for those so far described at the decimetre scale (Richardson and Carling, 2005). Although very common in massive lithologies, they are often overlooked as a bedrock bedform, perhaps because of their subtle appearance. Previous descriptions of hummocky forms include the 'undulating surfaces' of Kor *et al.* (1991) and Tinkler (1997), and the 'water-smoothed undulatory surfaces' of King (1927). Equally common are the second type, the ripple-like and dune-like sharp-crested hummocky forms (SCHFs). These are frequently described, albeit under a variety of names, for example the 'eversion marks' of Ängeby (1951), the 'horns' of Jennings (1985), the 'ripple-like bedforms' of Sevon (1988), Hancock *et al.* (1998) and Whipple *et al.* (2000) and the 'hummocky surfaces' of Wohl (1992b). Crest planforms range from two-dimensional to highly three-dimensional and the troughs may take on complex morphologies.

Cataracts Abandoned near-vertical or inclined waterfalls within dry channels can be impressive evidence of formative flood waters. The most famous of these is Dry Falls shown on the Coulee City Quadrangle map of the Grand Coulee, which is 120 m high and 5.5 km wide and developed in jointed basalt (Plate 2). Below the falls, the former plunge pool is occupied by Dry Falls Lake. Equally impressive are the steep horseshoe-shaped cataracts to the east that plunge into Red Alkali Lake and Castle Lake. Bretz (1932) and Bretz *et al.* (1956) argued that these various falls were developed sub-fluvially rather than by subaerial erosion and knickpoint retreat. This supposition was later supported by evidence of high-water marks (Baker, 1973a). Figure 2.3 shows a 500 m high cataract on Mars. Of note is the well-developed furrow system upstream of the lip of the falls. Many of these furrows become well incised at the lip and continue down the face of the falls as chute and chimney furrows (terminology after Richardson and Carling, 2005).

Step–pool and chute–pool sequences Step–pool sequences consist of a series of cataracts that often are spaced at roughly regular intervals and are found in channels that are steeper than those that support the pool–riffle sequences described below (Wohl, 2000; Wohl and Legleiter, 2003). Steps in such sequences can be very steep, near-vertical drops in the channel floor, but whereas water is in freefall over a step, water remains largely in contact with the rock surface as supercritical flow over an inclined chute. The intervening pools may be deeply scoured into the bedrock. In many cases, such regular variations in gradient need not be structurally controlled (Carling *et al.*, 2005) and where structural control is important (Wohl, 2000) it is often evident in air photographs or satellite images. Kodama and Ikeda (1984), Wohl and Ikeda (1997) and Koyama and Ikeda (1998) have produced repeating steps experimentally in homogeneous cohesive substrates, whilst Wohl and Grodek (1994), Koyama and Ikeda (1998) and Wohl and Ikeda (1998) describe field examples of step–pools in homogeneous substrate. Step–pool sequences may become modified by the process of pool breaching, of which there are two varieties. Breaching may occur when a headcut progresses from the immediate downstream pool through the intervening step (Carling *et al.*, 2005) and such breaches should be visible in aerial photographs and satellite images. Drainage of the upstream pool during low flow is enhanced such that its low-flow volume is reduced, but during high flow ponding occurs such that the pool volume is large. Thus the wetted perimeter of such a series of breached steps and intervening pools, when viewed from the air, appears as a string of 'narrow beads' (pools) and intervening 'threads' (breached steps) during low flow and as a string of 'broad beads' during high flows. More rarely, the base of a pool may be breached, although the step on the downstream side remains intact. This occurs when there is a strong counter-current acting against the headwall of the downstream pool, which in time causes the recession of the base of the headwall until a breakthrough is made into the base of the upstream pool. In this case, the upstream pool is drained completely. The breached step may then constitute a channel scale 'arch', which will only be evident in plan-view images if the lighting is oblique and produces a definitive shadow depicting the nature of the arch. However, in wet systems, the 'disappearance' of water that is clearly visible upstream and downstream of an arch is a good indicator of the possible presence of an arch or other subterranean conduit.

Pool–riffle sequences As noted for step–pools, a common feature of bedrock channels is quasi-periodic variations in gradient. In low-gradient channels this is usually termed a pool–riffle sequence (e.g. Dury, 1970; Keller and Melhorn, 1978; Baker and Pickup, 1987; Wohl, 1992a; Thompson, 2001; Wohl and Legleiter, 2003) whereas in high-gradient channels, step–pool (and chute–pool) sequences result. Riffles may constitute either bedrock highs upstream and downstream of bedrock-floored pools (i.e., riffles are high points in an erosional setting) or isolated accumulations of coarse gravel (depositional setting) separated by rock-floored pools. Based largely upon visual appearance and morphology, most authors reporting these features deem them analogous to pool–riffle sequences in alluvial rivers. However, whereas there has been considerable progress in recent years in understanding the formation and maintenance of alluvial pool–riffle sequences (see Wilkinson *et al.*, 2004), the controlling hydraulic processes in bedrock channels are not well understood. In contrast to alluvial rivers wherein the amplitude of both pools and riffles may adjust at the same time scale, in a bedrock river over relatively short time scales there is a greater opportunity to construct an alluvial riffle but excavation of a pool requires a longer time period. The genetic distinction and any relationship of pool–riffle sequences in bedrock channels to step–pool systems has not been explored. Patton and Baker (1978) reported a crude palaeo-pool–riffle sequence incised into bedrock in the Cheney-Palouse Scabland and this seems to be the only example of a 'fossil' sequence that has been published. Very deep pools (<70 m deep) have been reported from low-gradient bedrock reaches of the River Mekong and bathymetric maps show intervening rocky riffles, but it is not known if these are features of regular spacing. Thus pool–riffle sequences in low-gradient bedrock channels are little reported and controlling mechanisms not elucidated. Keller and Melhorn (1978), however, found that average riffle spacing in bedrock pool–riffle sequences was identical to that in alluvial channels at between five and seven channel widths. Undulating channel beds have been noted within some small run-off Martian channel systems in the southern highlands (Kereszturi, 2005; Lanz *et al.*, 2000) but not in the large outflow channels of the northern highlands.

Erosional antidunes and critical flow Large-scale 'dune-like' transverse ridges have been described from a variety of locations on Earth using satellite, aerial photography and ground and bathymetric surveys (Baker, 1973a; Alt, 2001; Carling *et al.*, 2002c; Fildani *et al.*, 2006), but the majority of these are largely depositional forms. MOC images of the Martian surface show transverse ridges that may be analogous to dune-like bedforms on Earth (Burr *et al.*, 2004). Although sometimes called 'ripples' in the older literature, given the scale of the bedforms and the high turbulence levels of megafloods on Earth it is reasonable to conclude that these features are not ripples, which cannot form in fully turbulent flows (Carling, 1999). Rather they are most likely dunes or antidunes. Both dunes and antidunes develop by erosion of the underlying sediments within the troughs between bedforms and deposition of sediments to form the bedforms. However, these features tend to migrate within the flow such that preserved bedforms consist wholly of deposited sediments (see Carling *et al.* this volume Chapter 3).

Descriptions of erosional wavy surfaces with the morphological characteristics of antidunes are rare (Fildani *et al.*, 2006). Bedforms developed in mobile sediments tend to be limited in size by negative feedback mechanisms associated with the flow and bedform migration and by poorly understood system instabilities that tend to split bedforms more often than merging them. In contrast, the shape and spacing of antidunes eroded into bedrock will be subject to flow controls alone as there is no bedform migration and no bedform splitting and amalgamation, rather only erosion of the crests, troughs and flanks. Over-deepening eventually will be precluded by reduced shear stresses at the bed. However, there are no systematic studies of erosional antidunes that could provide scaling data on bedform geometry and formative flows. Nevertheless, antidunes forming in water flows observed on Earth cover at least two orders of magnitude in scale (0.1 m to 10 m) and are known to increase in dimensions with flow depth (Allen, 1984). Thus it is not improbable to observe larger 'fossil' antidunes (100 m to 1000 m) in the landscape of Earth or indeed Mars that would have scaled with the magnitude of the water floods. Antidunes are especially useful for palaeoflood reconstruction because the wavelength of the spacing of the transverse ridges corresponds to the spacing of the standing water waves that formed them, and this latter spacing can be related to the flow speed (U) of the fluid (Kennedy, 1963; Allen, 1984; Tinkler, 1997). Froude number, Fr, is defined as the flow speed, U, ratioed to the gravity wave speed, $\surd(gh)$, where g is the gravitational acceleration and h is the flow depth. Thus, given that antidunes form in transcritical flows, $Fr > 0.82$ to *c.* 1.2, it is possible to derive a water depth (h). This depth and flow speed, if coupled with information on the limits of the flow width (W), can indicate the discharge (Q) magnitude through the continuity equation: $Q = UhW$.

Possible megaflood erosional antidunes occur near the village of Chagun Uzun in the Chuja Basin, south-central Siberia (Carling *et al.*, 2002c; Herget, 2005). These landforms consist of large-scale (tens of metres high; wavelengths of up to 300 m) sub-parallel transverse waveforms that were formerly locally much more extensive, but have been erased by postglacial river meandering (Plate 5).

These bedforms occur immediately downstream of a hill that would have caused a standing wave to develop in its lee if floodwaters overtopped the hill, or alternatively the location is one where flow would have been channelled into a narrower constriction and possible transcritical flows would have developed locally. Riverside exposures of the internal structure of these bedforms show clearly that the core of the structures is a weakly stratified coarse cobble and boulder glacial diamicton with many angular fragments as well as well-rounded clasts. This diamicton is around 10 to 15 m thick within the highest bedforms. The caps of the bedforms, however, consist of a conformable drape of finer pebble-sized gravel up to 2.5 m thick that is thickest on the crests of the large bedforms and thins and pinches out down both flanks. In the caps many particles are angular, but sub-rounded particles also exist and, although no clear bedding is discernable, the discontinuity between the diamicton and the cap indicates a separate depositional event. On the stoss side of one large bedform, the cap sediments are deformed into a few short-span duneforms up to 2 m high with wavelengths up to 4 m. Taking the dunes as evidence for deposition from water flow, it is concluded that the large-scale bedforms in the diamicton are erosional antidunes developed beneath standing waves within transcritical flows, and that the finer-grained caps were deposited during subcritical waning flow. Transverse boulder ridges with amplitudes less than those of antidunes are usually termed 'transverse ribs'. They may form in the same manner as antidunes but this is disputed (see Carling, 1999). Good examples from Earth have been reported from Modrudalur in Iceland, where the spacings were used to estimate the palaeoflood velocity and water depth (Rice *et al.*, 2002).

Streamlined residual landforms Large-scale bedrock macroforms that might reflect flood sculpting usually are crudely formed. Lozenge-shaped residual outcrops of basalt in the Cheney-Palouse Scablands may represent flood-formed residual massifs and en masse a group of these with their associated intervening channels give an impression of a braided or anastomosed channel (Plate 4). However, although such an association of macroforms might be formed by catastrophic flood flow, similar associations are also noted where structural influence on channel macroforms is important. A particularly good example of the latter is the 'Four Thousand Islands' reach of the Lower Mekong River within Laos, close to the border with Cambodia (Plate 6). Here Quaternary dykes, as young as 5720 yr BP, have been eroded into a myriad of bedrock-cored sandy islands and rocky islets by Late Pleistocene and Holocene discharges that today can exceed 56 000 $m^3\ s^{-1}$ during each annual wet season. The resulting landscape is reminiscent of both furrow complexes and streamlined landforms with an underlying structural control on channel alignment.

Equally striking are the streamlined residual hills cut into scabland loess (Plate 7). Bretz (1923b) recognised that hundreds of isolated loess hills in the eastern scablands were characterised by steep margins with distinctive prows at the up-flow margins. An origin owing to fluvial erosion was posited by Bretz *et al.* (1956), and Baker (1973b) argued that many had been formed sub-fluvially by floodwater emanating from Lake Missoula. Regardless of overall size, many of the streamlined hills are approximately three times as long as they are wide. Measures of maximum length, maximum breadth and total planform area can be used to demonstrate that each hill has a close resemblance to an airfoil in shape, for which form drag is minimised, which supports a hypothesis of streamlining by water (Baker, 1979; Komar, 1983).

Obstacle marks Foremost amongst composite features are obstacle marks, which consist of a concave current-scour crescent curving around the upstream side of some convex obstacle (Peabody, 1947; Karcz, 1968). These scour elements may have some depositional features but are here treated as erosional bedforms. Typical patterns of scour are reproduced at a variety of scales from the very small (e.g. 10^{-2} m: Bunte and Poesen, 1994) to the very large (10^2 m: Fay, 2002). Baker (1978c, his Figure 4.7) provides an example of the very large from the Channeled Scabland where Lenore Canyon and Long Lake occupy scoured depressions that wrap around an obstacle formed by High Hill and Pinto Ridge. A scour hole at a boulder is also described from the Lake Missoula flood (Baker, 1978b; Baker *et al.*, 1987). Other possible examples are considered below.

Earth scientists have used a variety of expressions to describe these features and these include the term 'obstacle mark' (e.g. Karcz, 1968; Reineck and Singh, 1980; Allen, 1984; Russell, 1993) as used herein, as well as 'scour mark' (e.g. Allen, 1965; Richardson, 1968; Baker, 1973b; Elfström, 1987), 'current crescents' (Bridge, 2003), 'crescentic/hairpin erosional marks' (Shaw, 1994), 'obstruction-formed pool' (Hassan and Woodsmith, 2004), 'scour hole' (Baker, 1973a) and 'comet mark' for special large-scale obstacle marks (Werner *et al.*, 1980). The variety of erosional obstacle marks incised into bedrock has been illustrated and considered by Richardson and Carling (2005) and an overview of the variety of current crescents and shadows formed by unidirectional currents is given by Allen (1984). In engineering sciences, scour around isolated obstacles, such as bridge piers, that protrude through the water surface have been studied intensively (e.g. Shen, 1971; Breusers *et al.*, 1977; Breusers and Raudkivi, 1991; Hoffmans and Verheij, 1997; Melville and Coleman, 2000;

Richardson and Davis, 2001). Less well studied are fully submerged obstacles (Carling *et al.*, 2002a, b).

Many factors influence the final shape of the erosional hollow. These include: obstacle shape and alignment to flow, flow speed and flow steadiness, Reynolds number, flow depth, sediment grade or bedrock type and time of development (e.g. see Melville and Coleman, 2000 and Herget, 2005). Unfortunately, the complex interaction of different controlling factors means that it is frequently difficult to recreate the hydraulic conditions associated with obstacle marks within palaeochannels. Some flume-derived relationships for submerged or free-surface obstacles are difficult to apply to obstacle marks in the field as they are valid only for the limited range of flume conditions (e.g. Johnson, 1995) and scale effects are frequently not regarded (Ettema *et al.*, 1998). On the other hand, Johnson (1995) has argued that, by comparing several different approaches, often consistent data sets emerge that tend to point to a narrow range of possible hydraulic conditions. Finally, guidelines and computer software to address the problem are available (e.g. Richardson and Davis, 2001; Landers *et al.*, 1996).

Among the largest obstacle marks on Earth are those related to cataclysmic ice-dammed lake outburst floods from Pleistocene times, such as those from Lake Missoula in the northwestern USA (e.g. Baker and Bunker, 1985; Baker *et al.*, 1993) or in the Siberian Altai Mountains (e.g. Carling *et al.*, 2002c; Herget, 2005). These floods inundated valleys and basins to depths of up to hundreds of metres, while submerged flood-resistant bedrock hills acted as obstacles inducing large-scale local scour around them. One example is the bedrock hill located in the western Chuja Basin of the Altai Mountains, Siberia. During Late Pleistocene times, valley glaciers blocked the course of the River Chuja downstream, generating an ice-dammed lake with a depth of up to 350 m (2100 m a.s.l.) above the bedrock hill (Herget, 2005). During the failure of the ice dam, the currents draining the Chuja Basin formed a connected scour hole with a maximum depth of 8.1 m, a length of 91.5 m and a width of about 400 m in front of the hill with a height of about 50 m. (Plate 8). A palaeohydraulic interpretation of the obstacle scour at Chagun-Uzun is problematic but one has been presented by Herget (2005). Obstacle marks are also found along the pathway of the Lake Missoula Flood in the northwestern USA. A particularly large example is the volcano outcrop ‘Rocky Butte’ in the eastern parts of the city of Portland (Allen *et al.*, 1986) but no hydraulic interpretations have been advanced beyond general descriptions. According to Alt (2001), Locust Hill, located about 100 km northwest of the city of Missoula in Montana, divided the flow reaching it from the east into two branches. The current system in front of the obstacle scoured the bedrock and generated a depression. This scour hole, called Banana Lake due to its characteristic shape in front of the hill, is still filled with water today (see Pardee, 1942).

2.5 Example of a very large bedrock channel complex: Kasei Valles, Mars

The remainder of this review describes an immense landscape on Mars that has probably been subject to water erosion. It is one of many intriguing features on the Martian surface that were discovered in 1971 by the American Mariner 9 probe and termed *outflow channels*, in accordance to their strong resemblance to terrestrial rivers. Outflow channels commonly start abruptly and fully developed, usually from circular to elliptical depressions called *chaotic terrains*, large collapse depressions in which the collapsed material lies tilted and broken in a chaotic assemblage on the depression floor. Some start equally developed at deep fissures, mainly in the vicinity of large volcano-tectonic rises. They have no or few tributaries, a large and relatively constant width to depth ratio and a low sinuosity.

Baker and Milton (1974) were the first to notice the resemblance of the Martian outflow channels to the Channeled Scablands in Washington and Oregon, USA, that have been shaped by Pleistocene catastrophic glacial floods. The sudden outburst of the glacially dammed Lake Missoula stripped away surface materials in a high-velocity turbulent flow, shaping a bizarre and quite unique landscape with an assemblage of erosional features amazingly similar to what are found in the outflow channels on Mars. It is now widely believed (yet not generally accepted) that the outflow channels were formed by similar processes of sudden and catastrophic outbursts of large amounts of water. A common scenario introduced by Carr (1979) is as follows (in summary form):

> Surface water, abundant in the very early history of Mars, oozed away forming a large groundwater system. Globally falling surface temperatures and a developing and growing global permafrost layer trapped the groundwater in a system of large aquifers. The increasing depth of the permafrost basis increased the pore pressure in the aquifers. Sudden groundwater outbursts might have occurred as a result of the pore pressure exceeding the lithospheric pressure which would destabilise the aquifers and open faults and cracks in the overlaying permafrost-rock layers, allowing the highly pressurised water to escape onto the surface. Tectonics or meteorite impacts could also have broken the permafrost seal. Emptying of the aquifers and decreasing discharge rates would refreeze the remaining water, thus closing the disrupted permafrost seal. Carr suggested that the aquifers could have refilled, leading to a repeating cycle of outflow activity and groundwater recharge. As the outflowing water could not have been reintegrated into the groundwater cycle due to the permafrost seal and would have quickly been lost to the thin Martian atmosphere

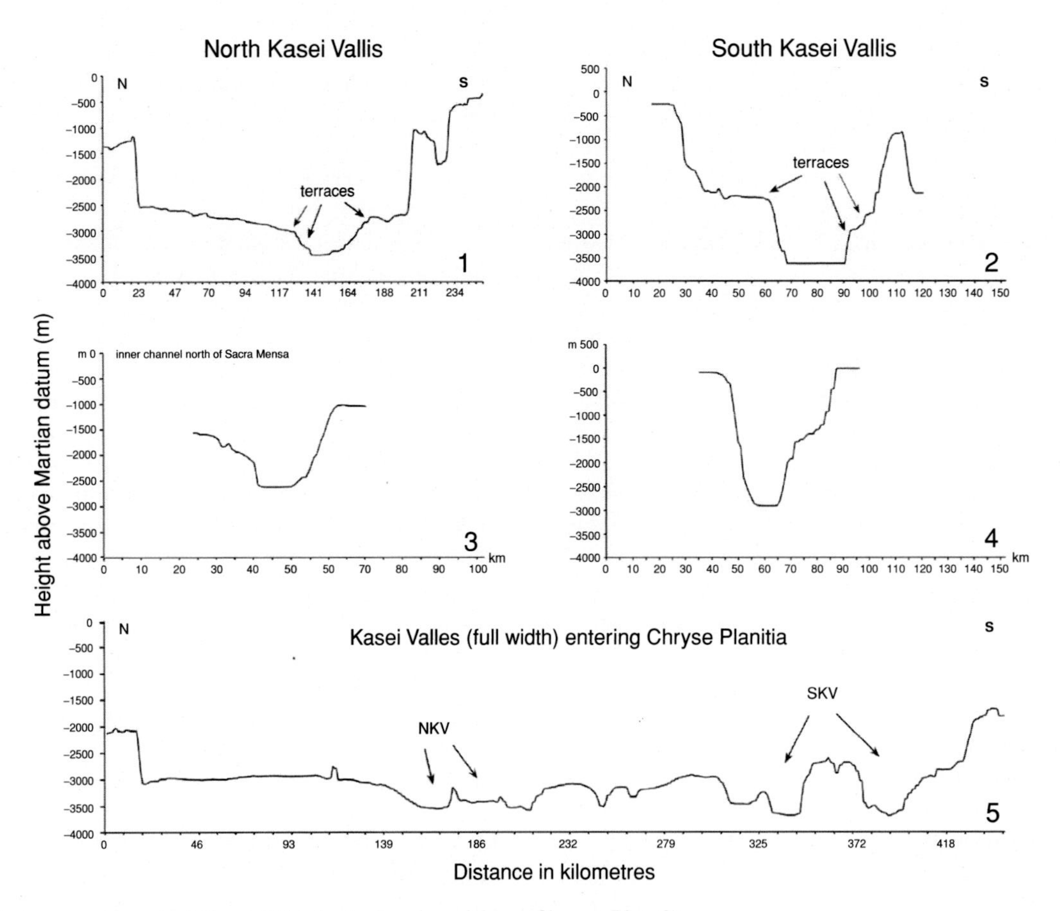

Figure 2.4. Cross-profiles of NKV and SKV (for the location of the profiles see Plate 9).

by evaporation, the loss of water from the system would have been great and the ability of the cycle to repeat itself strongly limited.

Other processes of outflow channel generation have been proposed involving e.g. glacial activity (Lucchitta and Anderson, 1980; Lucchitta *et al.*, 1981; Lucchitta, 1982, 2001), liquefaction (Nummedal, 1978) or gas-supported density flows (Hoffman, 2000) similar to pyroclastic flows on Earth but under cryogenic temperatures and with carbon dioxide being the active agent in the 'floods'. A combination of different processes is imaginable and even likely, especially, regarding the extreme environment on Mars, a combination of fluvial and water-ice processes possibly in the form of ice-covered rivers.

The Kasei Valles lie in the western circum-Chryse region, separating the Luna Planum Highlands from the Tharsis and Tempe Terra volcanotectonic provinces (Plate 9). They are 3000 km long, in parts up to 400 km wide and 3–4 km deep, making them the largest outflow channel system on Mars. They originate in a relatively shallow north–south-oriented depression that adjoins a large tectonic graben closely connected to the Valles Marineris canyon system. Two branches of Kasei Valles are distinctive, North Kasei Vallis (NKV) and South Kasei Vallis (SKV), to the north and south of the Sacra Mensa. The Kasei Valles are one of the most interesting places to study on Mars due to the fact that those parts which are not covered by young lava flows exhibit many typical erosion features mentioned above, though at an immense scale.

2.5.1 Topography

The most prominent topographic features of Kasei Valles are several large terraces (Plate 10). These provide a unique opportunity to study different phases of channel development in more detail. The Mars Orbiter Laser Altimeter (MOLA), an instrument onboard the American Mars Global Surveyor (MGS) mission, measured the Martian topography in great detail and generated a global topography model with a pixel resolution of 3 km. The along-track resolution of single MOLA tracks is even better, with only 300 m between single laser spots and a vertical resolution of 1.5 m (with a relative error in altitude along MOLA profiles given to be 1–10 m). These data allow the generation of cross-profiles and long-profiles of Kasei Valles (Figures 2.4 and 2.5).

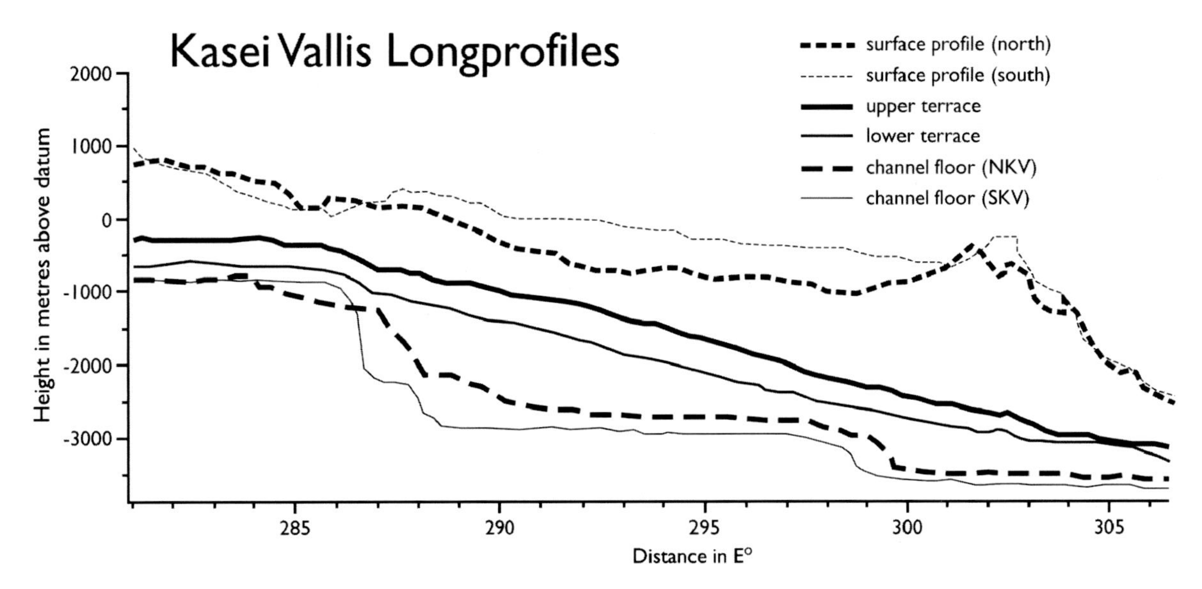

Figure 2.5. Long-profiles of Kasei Valles channel floors and terraces.

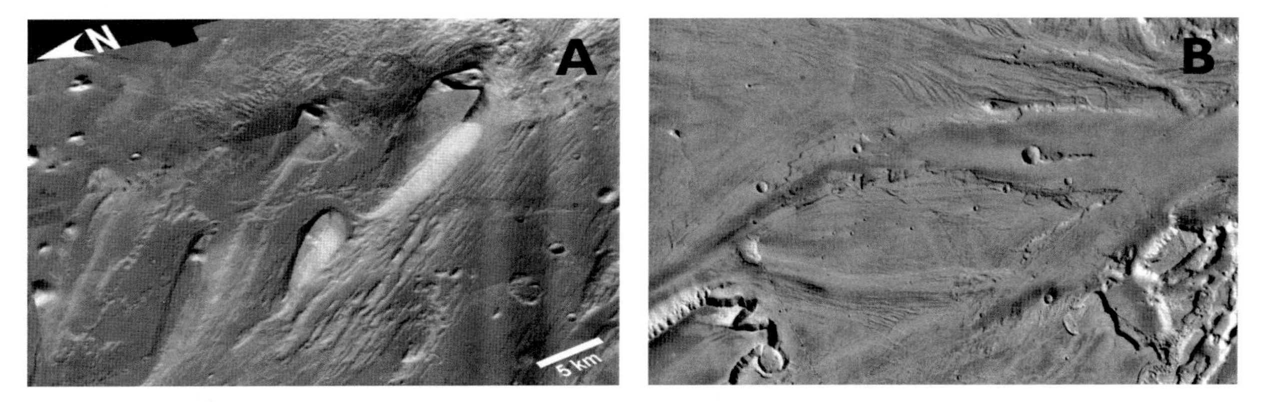

Figure 2.6. (A) Streamlined islands in Ares Valles as depicted within a mosaic of THEMIS Visible images and an HRSC image. The entire mosaic is draped over an HRSC stereo derived DTM. (B) Detail of streamlined island and branching, longitudinal grooves and cataracts in Kasei Valles. (Images courtesy of HRSC. THEMIS Public Data Releases, Mars Space Flight Facility, Arizona State University, 2006, http://themis-data.asu.edu.)

The two branches of Kasei Valles, NKV and SKV (Figure 2.6), show a distinct morphology. NKV is very broad and mostly shallow, showing numerous examples of typical outflow morphology, whereas SKV is rather narrow, very deep (up to 4 km) and morphologically more uniform. In Figure 2.4 several cross-profiles of NKV and SKV are shown, indicating terraces and other channel features, providing a general overview of the different channel shapes of the two Kasei Valles branches. A correlation of terraces based on MOLA data shows that, despite the different appearance of NKV and SKV and a less distinct terracing in SKV, the development of both branches probably was connected and contemporaneous as terrace heights correlate strongly (in the error range given by the instrument). The widths of the deep central parts of both valley branches (Plate 9) are similar and both channels show a noticeable increase in the width to depth ratio towards the channel mouth. The depths of NKV and SKV differ slightly with the channel floor of SKV being generally 200 m deeper than the NKV floor indicating that activity in SKV might have lasted longer.

The long-profiles of Kasei Valles give another insight in channel development. The good MOLA coverage allows generation of long-profiles of the channel floor as well as along prominent valley terraces. Figure 2.5 shows long-profiles of the valley floor of NKV and SKV (lower lines), two main terraces (middle lines) and the adjacent highland surfaces to the north and to the south (upper lines). The long-profiles of the terrace floors are very uniform with steep slopes parallel to the adjacent highland surfaces, whereas along the channel floors the slopes are generally low except along two prominent knickpoints visible in both

NKV and SKV at approximately the same longitude that developed at later stages during the channel evolution. The knickpoints might indicate a change in bedrock resistance or the position of tectonic faults as they lie quasi-parallel to tectonic features that dominate the adjacent highlands. They are not an indication of a change of the erosion level during channel development, as is often the case on Earth.

Based on MOLA data, information can be inferred regarding channel slopes and water depths (bankfull stages) of assumed floods, allowing better calculations of maximum discharge rates of the palaeofloods. Older calculations based on Viking data gave discharge rates between 10^9 m^3 s^{-1} (Baker, 1982) and 2.3×10^9 m^3 s^{-1} (Robinson and Tanaka, 1990) for the Kasei Valles floods (during the initial stages). Calculations based on MOLA data show that the discharge rates were probably more in the range of large floods on Earth or only slightly higher, with values between 5×10^7 m^3 s^{-1} and 10^8 m^3 s^{-1} for the initial stages of channel development and values as low as 5×10^4 m^3 s^{-1} during the end of the outflow activity (Lanz, 2004; Williams *et al.*, 2000).

2.5.2 Morphology

The most prominent morphologic characteristics of Kasei Valles, the streamlined islands and longitudinal grooves (Figure 2.6), have been used as arguments for both glacial and fluvial activity. High-resolution imagery from the Mars Orbiter Camera (MOC) onboard MGS shows details that seem to support the flood hypothesis. Figure 2.7A shows, for example, possible scour marks along the upstream side of a streamlined island in Kasei Valles. Scour marks are typical of erosion around obstacles by flowing water, though they do not necessarily rule out a glacial origin. Another interesting feature is seen in Figure 2.7B. This MOC image shows the downstream side of a streamlined island. It has a 'fretted' appearance with small erosional alcoves on both sides of the island, merging at its tail-end. The fact that these features appear only at the tail-end of the island and no talus deposits can be found rules out younger denudation processes, as they would affect the whole island and should leave clearly visible landslip deposits. It is therefore proposed that they are the result of erosion in turbulent high-velocity flows. Gouges or potholes shaped by glaciers can be similar in appearance but these are usually randomly spaced features rather than aligned along the downstream side of an obstacle. Another fact inconsistent with a purely glacial origin of the outflow channels is that to date no glacial deposits have been identified clearly in any of the outflow channels on Mars and it seems unlikely, especially imagining the size of the glaciers needed to carve these enormous valleys, that they simply vanished leaving no signs of the millions of cubic metres of material they eroded along the way.

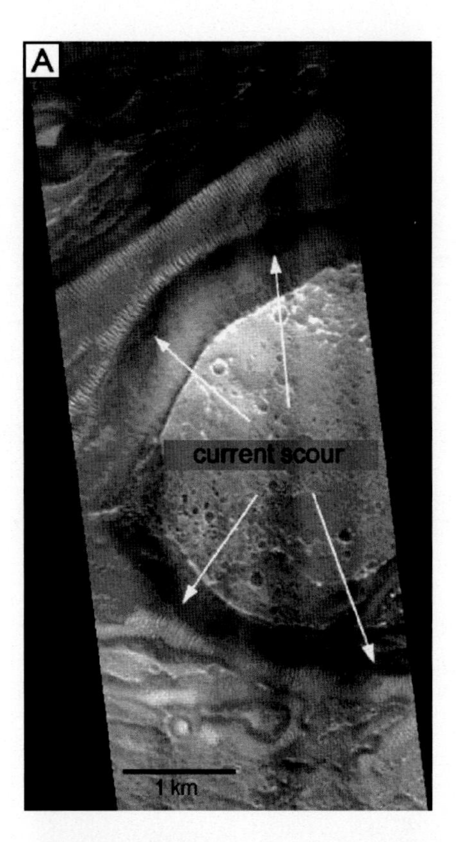

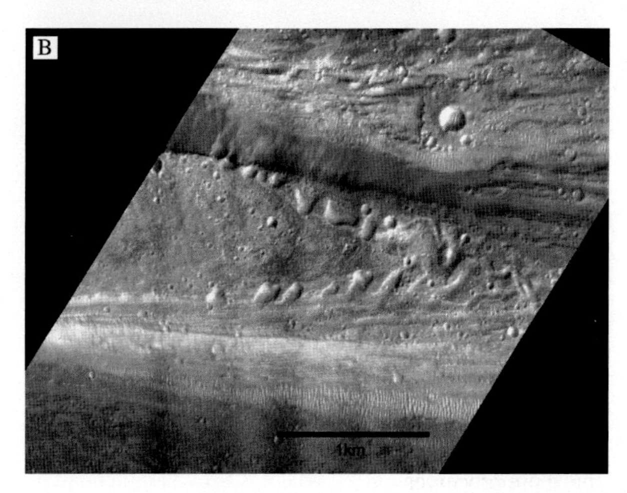

Figure 2.7. (A) Possible current-scour around upstream side of a streamlined island (MOC-Image E1401150). (B) 'Fretted' island in Kasei Valles; the erosion features appear only at the tail end and no talus deposits are visible (MOC SP245505).

2.5.3 The time scale of outflow events in Kasei Valles

Besides the question of what shaped the Martian surface and which agents were involved (e.g. water, ice, carbon dioxide), the timing and sequence of depositional and erosional processes and events needs to be established. Impact crater statistics provide a tool for determining relative and, within model-dependent limitations, absolute time scales (Neukum and Hiller, 1981). The method of defining relative ages of surface units by crater statistics is based on the assumption that all solid planetary surfaces, or rather geologic units inside these surfaces, accumulate more and more craters with time. By measuring and comparing the frequencies of impact craters superimposed on these surfaces it is then possible to determine the relative ages of units. Units exhibiting high crater frequencies are generally older than those with lower crater frequencies as they have been exposed longer to impact cratering processes. If the time dependency of the cratering rate of a planet is known, as is the case for Mars, absolute ages can then be deduced from the crater frequencies (for a detailed discussion of the method see e.g. Neukum and Wise, 1976a, b; Hartmann, 1977; Hartmann and Neukum, 2001; Ivanov, 2001).

The dating of different Kasei Valles units based on Viking, MGS and Mars Express imagery gives results regarding the evolution of the outflow channel. Outflow channel activity started approximately 3.7 billion years (Ga) ago and ended possibly as late as 1.0 Ga ago in the Upper Amazonian period of Mars history (Lanz, 2004; Neukum *et al.*, 2007), giving a time span of up to 2.7 Ga during which the Kasei Valles have been the site of recurring episodic outflow events. The outflow activity seems to centre upon two main phases of increased activity. The first phase between 3.6 and 3.1 Ga ago eroded the broad reaches of the upper terrace floors. The second phase eroded the grooved terrace floors and deep inner channels and started approximately 2.2 Ga ago. Outflow activity during both phases was not continuous but happened in recurring pulses of presumably short time-spans. Nevertheless, these studies show that the erosional activity in the Kasei Valles lasted much longer than previously believed (Neukum and Hiller, 1981; Nelson and Greeley, 1999).

The long (episodic) activity of the Kasei Valles channel system requires a process that periodically triggers the outflow events. Catastrophic outbursts of groundwater as described by Carr (1979) partly meet these criteria. However, these processes are temporally limited due to the loss of water from the cycle by the refreezing permafrost seal and it is unlikely that they could have lasted for almost 3 Ga. In addition, a period of enhanced outflow activity towards the end of the cycle, as indicated by the crater frequency analyses, can also not be explained by these processes. Another possibility is that the fluvial activity was linked to pulses of volcanic activity (e.g. Lanz, 2004; Neukum *et al.*, 2007). Neukum *et al.* extensively re-mapped and dated volcanic and fluvial units in the Kasei Valles and other regions on Mars based on the latest Mars Express data. They found that both volcanic and fluvial activity show common episodic pulses in intensity throughout Mars history. They believe that these pulses are related to the interior evolution of the planet when convection in the asymptotic stationary state changes from the so-called stagnant-lid regime to an episodic behaviour. Therefore, it appears likely that volcanic activity may have triggered the outflow events, e.g. by the melting of ground-ice and/or the mobilisation of subsurface waters.

2.6 Conclusions

Erosional bedrock forms occur over an immense range of scales, from small flutes and furrows of bedrock rivers to the streamlined hills, cataracts, and anastomosing channels of the Channeled Scabland, and their immense counterparts on Mars. All of these forms contain information about generative hydraulic processes, and considerable progress has been made in understanding some of these relationships. However, despite the detailed classification of channel-scale bedforms that can be developed for small modern river systems (Richardson and Carling, 2005), relatively few types of channel-scale erosional bedforms have been described previously and no detailed process-based typology has been developed to define features unambiguously. There is considerable opportunity to advance understanding of channel-scale features using flume-based study of the erosion of materials that mimic bedrock, aided by computer modelling of the development of such features given different hydraulic conditions. Aeolian, glacial and lava-flow processes also can produce erosional landforms that have characteristics very similar to fluvial and megaflood features (e.g. Evans, 2003) and consequently care is required in identification and interpretation. This is especially so when the only information available is drawn from morphological planform data obtained using satellite or aerial photography. Recent studies often allow height data to be derived using techniques such as photoclinometry, or shape-shading (Beyer *et al.*, 2003; Burr *et al.*, 2004). Where self-similar spatially contiguous groups of erosional bedforms occur then suites of morphological geostatistics, such as fractal properties, might help resolve the genetic origins of scale-specific landforms (Evans, 2003). Such a geostatistical approach has yet to be applied to megaflood erosional landforms as usually the population of self-similar features is small (see Carr and Malin (2000) for a perspective).

References

Allen, J. R. L. (1965). Scour marks in snow. *Journal of Sedimentary Petrology*, **35**, 331–338.

Allen, J. R. L. (1969). Erosional current marks of weakly cohesive mud beds. *Journal of Sedimentary Petrology*, **39**, 607–623.

Allen, J. R. L. (1971a). Transverse erosional marks of mud and rock: their physical basis and geological significance. *Sedimentary Geology*, **5**, 167–385.

Allen, J. R. L. (1971b). Bed forms due to mass transfer in turbulent flows: a kaleidoscope of phenomena. *Journal of Fluid Mechanics*, **49**, 49–63.

Allen, J. R. L. (1984). *Sedimentary Structures: Their Character and Physical Basis.* Amsterdam: Elsevier.

Allen, J. R. L. (1987). Streamwise erosional structures in muddy sediments: Severn Estuary, Southwestern UK. *Geografiska Annaler*, **69A**, 37–46.

Allen, J. E., Burns, M. and Sargent, S. C. (1986). *Cataclysms on the Columbia*. Portland: Timber Press.

Alt, D. (2001). *Glacial Lake Missoula and Its Humongous Flood*, Missoula, Montana: Mountain Press Company.

Ängeby, O. (1951). Pothole erosion in recent waterfalls. *Lund Studies in Geography, Series A. Physical Geography*, **2**, 1–34.

Arndt, R. E. A. (1981). Cavitation in fluid machinery and hydraulic structures. *Annual Reviews of Fluid Mechanics*, **13**, 273–328.

Ashley, G. M., Renwick, W. H. and Haag, G. H. (1988). Channel form and process in bedrock and alluvial reaches of the Raritan River, New Jersey. *Geology*, **16**, 436–439.

Baker, V. R. (1973a). *Paleohydrology and Sedimentology of Lake Missoula Flooding in Eastern Washington*. Geological Society of America Special Paper 144.

Baker, V. R. (1973b). Erosional forms and processes for the catastrophic Pleistocene Missoula floods in eastern Washington. In *Fluvial Geomorphology*, ed. M. Morisawa. State University of New York, Binghamton, NY, pp. 123–148.

Baker, V. R. (1978a). The Spokane Flood Debate. In *The Channeled Scabland: A Guide to the Geomorphology of the Columbia Basin, Washington*, eds. V. R. Baker and D. Nummedal. Prepared for the Comparative Planetary Geology Field Conference held in the Columbia Basin June 5–8, 1978, pp. 9–30.

Baker, V. R. (1978b). Large-scale erosional and depositional features of the Channeled Scabland. In *The Channeled Scabland: A Guide to the Geomorphology of the Columbia Basin, Washington*, eds. V. R. Baker and D. Nummedal. Washington: National Aeronautics and Space Administration, pp. 81–115.

Baker, V. R. (1978c). Paleohydraulics and hydrodynamics of scabland floods. In *The Channeled Scabland: A Guide to the Geomorphology of the Columbia Basin, Washington*, eds. V. R. Baker and D. Nummedal. Washington: National Aeronautics and Space Administration, pp. 59–79.

Baker, V. R. (1979). Erosional processes in channelized water flows on Mars. *Journal of Geophysical Research*, **84**, 7985–7993.

Baker, V. R. (1982). *The Channels of Mars*. Austin, TX: University Texas Press.

Baker, V. R. (1988). Flood erosion. In *Flood Geomorphology*, eds. V. R. Baker, R. C. Kochel and P. C. Patton. New York: Wiley, pp. 81–95.

Baker, V. R. (2008). The Spokane Flood debates: historical background and philosophical perspective. In *History of Geomorphology and Quaternary Geology*, eds. R. H. Grapes, D. Oldroyd and A. Grigelis. London: Geological Society, Special Publication 301, pp. 33–50.

Baker, V. R. and Bunker, R. C. (1985). Cataclysmic late Pleistocene flooding from glacial lake Missoula: a review. *Quaternary Science Review*, **4**, 1–41.

Baker, V. R. and Kale, V. S. (1998). The role of extreme floods in shaping bedrock channels. In *Rivers Over Rock: Fluvial Processes in Bedrock Channels*, eds. K. J. Tinkler and E. E. Wohl. Geophysical Monographs 107. Washington D.C.: American Geophysical Union.

Baker, V. R. and Milton, D. J. (1974). Erosion by catastrophic floods on Mars and Earth, *Icarus*, **23**, 27–41.

Baker, V. R. and Pickup, G. (1987). Flood geomorphology of the Katherine Gorge, Northern Territory, Australia. *Geological Society of America Bulletin*, **98**, 635–646.

Baker, V. R., Greeley, R., Komar, P. D., Swanson, D. A. and Waitt, R. B. (1987). Columbia and Snake River plains. In *Geomorphic Systems of North America*, ed. W. L. Graf. Boulder: Geological Society of America, pp. 403–468.

Baker, V. R., Benito, G. and Rudoy, A. N. (1993). Paleohydrology of Late Pleistocene superflooding, Altay Mountains, Siberia. *Science*, **259**, 348–350.

Barnes, H. L. (1956). Cavitation as a geological agent. *American Journal of Science*, **254**, 493–505.

Baulig, H. (1913). Le plateaux de lave du Washington Central et la Grand 'Coulee'. *Annales de Geographie*, **22**, 149–159.

Bean, D. A. (2003). Characteristics of mega-furrows on the continental rise seaward of the Sigbee Escarpment, Gulf of Mexico. Abstract. In *Siltstones, Mudstones and Reservoir Characteristics*, eds. E. D. Scott and A. H. Bouma, p. 111.

Beyer, R. A., McEwen, A. S. and Kirk, R. L. (2003). Meter-scale slopes of candidate MER landing sites from point photoclinometry. *Journal of Geophysical Research*, **108** (E12), 8085. doi:10.1029/2003JE002120.

Blank, H. R. (1958). Pothole grooves in the bed of the James River, Mason County, Texas. *Texas Journal of Science*, **10**, 292–301.

Blondeaux, P. (2001). Mechanics of coastal forms. *Annual Reviews of Fluid Mechanics*, **33**, 339–370.

Blumberg, P. N. and Curl, R. L. (1974). Experimental and theoretical studies of dissolution roughness. *Journal of Fluid Mechanics*, **5**, 735–742.

Bradwell, T. (2005). Bedrock mega grooves in Assynt, NW Scotland. *Geomorphology*, **65**, 195–204.

Bretz, J H. (1923a). Glacial drainage on the Columbia Plateau. *Geological Society of America Bulletin*, **34**, 573–608.

Bretz, J H. (1923b). The Channeled Scabland of the Columbia Plateau. *Journal of Geology*, **31**, 617–649.

Bretz, J H. (1924). The Dalles type of river channel. *Journal of Geology*, **32**, 139–149.

Bretz, J H. (1928). Bars of the Channeled Scabland. *Geological Society of America Bulletin*, **39**, 643–702.

Bretz, J H. (1930a). Lake Missoula and the Spokane flood. *Geological Society of America Bulletin*, **41**, 92–93.

Bretz, J H. (1930b). Valley deposits immediately west of the Channeled Scabland. *Journal of Geology*, **38**, 385–422.

Bretz, J H. (1932). *The Grand Coulee.* American Geographical Society, Special Publication 15.

Bretz, J H., Smith, H. T. U. and Neff, G. E. (1956). Channeled Scabland of Washington; new data and interpretations. *Geological Society of America Bulletin*, **67**, 957–1049.

Breusers, H. N. C. and Raudkivi, A. J. (1991). *Scouring, IAHR Hydraulic Structures Design Manual*, Vol. 2. Rotterdam, the Netherlands: Balkema.

Breusers, H. N. C., Nicollet, G. and Shen, H. W. (1977). Local scour around cylindrical piers. *Journal of Hydraulic Research*, **15**, 211–252.

Bridge, J. S. (2003). *Rivers and Floodplains: Forms, Processes and Sedimentary Record*. Oxford: Blackwell.

Bryant, W. R. (2000). Megafurrows on the continental rise south of the Sigsbee Escarpment, northwest Gulf of Mexico. Abstract. *AAPG Annual Meeting Program*, **9**, A18.

Bunte, K. and Poesen, J. (1994). Effects of rock fragment size and cover on overland flow hydraulics, local turbulence and sediment yield on an erodible soil surface. *Earth Surface Processes and Landforms*, **19**, 115–135.

Burr, D. M., Carling, P. A., Beyer, R. A. and Lancaster, N. (2004). Flood-formed dunes in Athabasca Valles, Mars: morphology, modeling, and implications. *Icarus*, **171**, 68–83.

Calkins, F. C. (1905). *Geology and Water Resources of a Portion of East-central Washington*. U.S. Geological Survey Water-Supply Paper 118.

Canals, M., Urgeles, R. and Calafat, A. M. (2000). Deep sea-floor evidence of past ice streams, off the Antarctic peninsula. *Geology*, **28**, 31–34.

Canals, M., Puig, P., Durrieu de Madron, X. *et al.* (2006). Flushing submarine canyons. *Nature*, **444**, 354–357.

Carling, P. A. (1999). Subaqueous gravel dunes. *Journal of Sedimentary Research*, **69**, 534–545.

Carling, P. A. and Tinkler, K. (1998). Conditions for the entrainment of cuboid boulders in bedrock streams: an historical review of literature with respect to recent investigations. In *Rivers over Rock*, eds. E. Wohl and K. Tinkler. American Geophysical Union, pp. 19–34.

Carling, P. A., Hoffmann, M. and Blatter, A. S. (2002a). Initial motion of boulders in bedrock channels. In *Ancient Floods, Modern Hazards: Principles and Applications of Paleoflood Hydrology*, Water Science and Applications Volume 5, pp. 147–160. American Geophysical Union.

Carling, P. A., Hoffmann, M., Blatter, A. S. and Dittrich, A. (2002b). Drag of emergent and submerged regular obstacles in turbulent flow above bedrock surface. In *Rock Scour due to Falling High-Velocity Jets*, eds. A. J. Schleiss and E. Bollaert. Lausanne: Swets and Zeitlinger/Balkema, pp. 83–94.

Carling, P. A., Kirkbride, A. D., Parnachov, S. V., Borodavko, P. S. and Berger, G. W. (2002c). Late-Quaternary catastrophic flooding in the Altai Mountains of south-central Siberia: a synoptic overview and an introduction to flood deposits sedimentology. In *Flood and Megaflood Processes and Deposits: Recent and Ancient Examples*, eds. P. I. Martini, V. R. Baker and G. Garzon. International Association of Sedimentologists, Special Publication 32, 17–35.

Carling, P. A., Tych, W. and Richardson, K. (2005). The hydraulic scaling of step-pool systems. In *River, Coastal and Estuarine Morphodynamics*, Vol. 1, eds. G. Parker and M. H. Garcia. New York: Balkema, Taylor and Francis, pp. 55–63.

Carr, M. H. (1979). Formation of Martian flood features by release of water from confined aquifers. *Journal of Geophysical Research*, **84**, 2995–3007.

Carr, M. H. and Malin, M. C. (2000). Meter-scale characteristics of Martian channels and valleys. *Icarus*, **146**, 366–386.

Carrivick, J. L., Russell, A. J. and Tweed, F. S. (2004). Geomorphological evidence for jókulhlaups from Kverfjöll volcano, Iceland. *Geomorphology*, **63**, 81–102 doi: 10.1016/j.geomorph.2004.03.006.

Coleman, J. M. (1969). Brahmaputra River: channel processes and sedimentation. *Sedimentary Geology*, **3**, 129–239.

Curl, R. L. (1966). Deducing flow velocity in cave conduits from scallops. *Bulletin of the National Speleological Society*, **36**, 1–5.

Dahl, R. (1965). Plastically sculptured detail forms on rock surfaces in northern Nordland, Norway. *Geografiska Annaler*, **47**, 83–140.

Dawson, W. L. (1898). Glacial phenomena in Okanogan County, Washington. *American Geology*, **22**, 203–217.

Dowdeswell, J. A., O'Cofaigh, C. and Pudsey, C. J. (2004). Thickness and extent of the subglacial till layer beneath an Antarctic ice stream. *Geology*, **32**, 13–16. doi:10.1130/G19864.1.

Dury, G. H. (1970). A re-survey of the Hawkesbury River, New South Wales, after one hundred years. *Australian Geographical Studies*, **8**, 121–132.

Dzulynski, S. and Walton, E. K. (1965). *Sedimentary Features of Flysch and Greywackes*. London: Elsevier.

El Baz, F., Breed, C. S., Grolier, M. J. and McCauley, J. F. (1979). Yardangs on Mars and Earth, *Journal of Geophysical Research*, **84**, 8205–8221.

Elfström, A. (1987). Large boulder deposits and catastrophic floods. *Geografiska Annaler*, **A69**, 101–121.

Elston, E. D. (1917). Potholes: their variety, origin and significance (I). *Scientific Monthly*, **5**, 554–567.

Elston, E. D. (1918). Potholes: their variety, origin and significance (II). *Scientific Monthly*, **6**, 37–51.

Embleton, C. and King, C. A. M. (1968). *Glacial and Periglacial Geomorphology*. New York: St. Martin's Press.

Ettema, R., Melville, B. W. and Barkdoll, B. (1998). Scale effect in pier-scour experiments. *Journal of Hydraulic Engineering*, **124**, 639–642.

Evans, I. S. (2003). Scale-specific landforms and aspects of the land surface. In *Concepts and Modelling in Geomorphology: International Perspectives*, eds. I. S. Evans, R. Dikau, E. Tokunaga, H. Ohmori and M. Hirano. Tokyo: Terrapub, pp. 61–84.

Fay, H. (2002). Formation of ice block obstacle marks during the November 1996 glacier-outburst flood (jökulhlaup), Skeidararsandur, southern Iceland. In *Flood and Megaflood Processes and Deposits: Recent and Ancient Examples*, eds. I. P. Martini, V. R. Baker and G. Garzon. International Association of Sedimentologists, Special Publication 32, 85–97.

Fildani, A., Normark, W. R., Kostic, S. and Parker, G. (2006). Channel formation by flow stripping: large-scale scour features along the Monterey East Channel and their relation to sediment waves. *Sedimentology*, **53**, 1265–1287.

Goudie, A. S. (1999). Wind erosional landforms: yardangs and pans. In *Aeolian Environments, Sediments and Landforms*, eds. A. S. Goudie, I. Livingstone and S. Stokes. Chichester: Wiley, pp. 167–180.

Greeley, R. and Iversen, J. D. (1985). *Wind as a Geological Process: On Earth, Mars, Venus and Titan.* Cambridge: Cambridge University Press.

Gupta, A. (2004). The Mekong River: morphology, evolution and palaeoenvironment. *Journal of Geological Society of India*, **64**, 525–533.

Gupta, A., Kale, V. S. and Rajaguru, S. N. (1999). The Narmanda River, India, through space and time. In *Variety of Fluvial Form*, eds. A. J. Miller and A. Gupta. Chichester: Wiley.

Hancock, G. S., Anderson, R. S. and Whipple, K. X. (1998). Beyond power: bedrock river incision process and form. In *Rivers Over Rock: Fluvial Processes in Bedrock Channels*, eds. K. J. Tinkler and E. E. Wohl. Geophysical Monograph 107. Washington, DC: American Geophysical Union.

Harding, H. T. (1929). Possible supply of water for the channeled scablands. *Science*, **69**, 188–190.

Hartmann, W. K. (1977). Relative crater production rates on planets, *Icarus*, **31**, 260–276.

Hartmann, W. K. and Neukum, G. (2001). Cratering chronology and the evolution of Mars. *Space Science Reviews*, **96**, 165–194.

Hassan, M. A. and Woodsmith, R. D. (2004). Bed load transport in an obstruction-formed pool in a forest gravelbed stream. *Geomorphology*, **58**, 203–221.

Herget, J. (2005). *Reconstruction of Ice-dammed Lake Outburst Floods in the Altai Mountains, Siberia.* Geological Society of America, Special Paper 386. Boulder: Geological Society of America.

Hoffman, N. (2000). White Mars: a new model for Mars' surface and atmosphere based on CO_2, *Icarus*, **146**, 326–342.

Hoffmans, G. J. C. M. and Verheij, H. J. (1997). *Scour Manual.* Rotterdam: Balkema.

Ikeda, H. (1978). Large-scale grooves formed by scour on cohesive mud surfaces. *Bulletin of the Environmental Research Center, University of Tsukuba*, **2**, 91–95 (in Japanese).

Ivanov, B. A. (2001). Mars and Moon cratering ratio estimates. In *Chronology and Evolution of Mars*, eds. R. Kallenbach, J. Geiss, W. K. Hartmann. Space Science Series of ISSI, Space Science Review 96, 87–104.

Jennings, J. N. (1985). *Karst Geomorphology.* Oxford: Blackwell.

Johnson, J. P. and Whipple, K. X. (2007). Feedback between erosion and sediment transport in experimental bedrock channels. *Earth Surface Processes and Landforms*, **32**, 1048–1062.

Johnson, P. A. (1995). Comparison of pier-scour equations using field data. *Journal of Hydraulic Engineering*, **121**, 626–629.

Karcz, I. (1968). Fluviatile obstacle marks from the wadis of the Negev (southern Israel). *Journal of Sedimentary Petrology*, **38**, 1000–1012.

Keller, E. A. and Melhorn, W. N. (1978). Rhythmic spacing and origin of pools and riffles. *Geological Society of America Bulletin*, **89**, 723–730.

Kennedy, J. F. (1963). The mechanics of dunes and antidunes in erodible-bed channels. *Journal of Fluid Mechanics*, **16**, 521–544.

Kereszturi, A. (2005). Cross-sectional and longitudinal profiles of valleys and channels in Xanthe Terra on Mars, *Journal of Geophysical Research*, **110**, E12S17, doi:10.1029/2005JE002454.

Kidson, R., Richards, K. S. and Carling, P. A. (2005). Reconstructing the 1-in-100-year flood in Northern Thailand. *Geomorphology*, **70**, 279–295.

King, P. B. (1927). Corrosion and corrasion on Barton Creek, Austin, Texas. *Journal of Geology*, **35**, 631–638.

Kobor, J. S. and Roering, J. J. (2004). Systematic variation of bedrock channel gradients in the central Oregon Coast Range: implications for rock uplift and shallow landsliding. *Geomorphology*, **62**, 239–256.

Kodama, Y. and Ikeda, H. (1984). An experimental study on the formation of Yakama Potholes. In *Research Report of "Yakama Potholes"*, Special Natural Commemoration. Yakama Research Association, Yanadani Village, Ehime Prefecture, Japan (in Japanese).

Komar, P. D. (1983). Shape of streamlined islands on Earth and Mars: experiments and analyses of the minimum drag form. *Geology*, **11**, 651–654.

Kor, P. S. G. and Cowell, D. W. (1998). Evidence for catastrophic subglacial meltwater sheetflood events on the Bruce Peninsula, Ontario *Canadian Journal of Earth Sciences*, **35**, 1180–1202.

Kor, P. S. G., Shaw, J. A. and Sharpe, D. R. (1991). Erosion of bedrock by subglacial meltwater, Georgia Bay, Ontario: a regional review. *Canadian Journal of Earth Sciences*, **28**, 623–642.

Koyama, T. and Ikeda, H. (1998). Effect of riverbed gradient on bedrock channel configuration: a flume experiment. *Proceedings of the Environmental Research Center, Tsukuba University, Japan*, **23**, 25–34 (in Japanese).

Kunert, M. and Coniglio, M. (2002). Origin of vertical shafts in bedrock along the Eramosa River valley near Guelph, southern Ontario. *Canadian Journal of Earth Sciences*, **39**, 43–52.

Landers, M. N., Mueller, D. S. and Martin, G. R. (1996). *Bridge Scour Data Management System User's Manual.* U.S. Geological Survey Open-File Report 95–754.

Lanz, J. K. (2004). *Geometrische, morphologische und stratigraphische Untersuchungen ausgewählter Outflow Channel der Circum-Chryse-Region, Mars, mit Methoden der Fernerkundung.* Dissertation an der Freien Universität Berlin, Forschungsbericht 2004–02, Deutsches Zentrum für Luft- und Raumfahrt e.V.

Lanz, J. K., Hebenstreit, R. and Jaumann, R. (2000). Martian channels and their geomorphologic development as revealed by MOLA. Abstract. *Second International Conference on Mars Polar Science and Exploration*, p. 102.

Leibovich, S. (2001). Langmuir circulation and instability. Online *Encylopedia Ocean Sciences*, doi:10.1006/rwos.2001.0141.

Lucchitta, B. K. (1982). Ice sculpture in the Martian outflow channels. *Journal of Geophysical Research*, **87**, 9951–9973.

Lucchitta, B. K. (2001). Antarctic ice streams and outflow channels on Mars. *Geophysical Research Letters*, **28**, 403–406.

Lucchitta, B. K. and Anderson, D. M. (1980). Martian outflow channels sculptured by glaciers. *NASA Technical Memorandum*, **81776**, 271–273.

Lucchitta, B. K., Anderson, D. M. and Shojii, H. (1981). Did ice streams carve the martian outflow channels? *Nature*, **290**, 759–763.

Malde, H. E. (1968). *The Catastrophic Late Pleistocene Bonneville Flood in the Snake River Plain, Idaho.* U.S. Geological Survey Professional Paper 596.

Maxson, J. H. (1940). Fluting and faceting of rock fragments. *Journal of Geology*, **48**, 717–751.

Maxson, J. H. and Campbell, I. (1935). Stream fluting and stream erosion. *Journal of Geology*, **43**, 729–744.

Melville, B. W. and Coleman, S. E. (2000). *Bridge Scour.* Highlands Ranch: Water Resources Publications.

Montgomery, D. R. and Gran, K. B. (2001). Downstream variation in the width of bedrock channels. *Water Resources Research*, **37**, 1841–1846.

Nelson, D. M. and Greeley, R. (1999). Geology of Xanthe Terra outflow channels and the Mars Pathfinder landing site. *Journal of Geophysical Research*, **104**, 8653–8669.

Nemec, W., Lorenc, M. W. and Alonso, J. S. (1982). Pothole granite terrace in the Rio Salor valley, western Spain: a study of bedrock erosion by floods. *Tecniterrae*, S-286, October–November 1982.

Neukum, G. and Hiller, K. (1981). Martian ages. *Journal of Geophysical Research*, **86**, 3097–3121.

Neukum, G. and Wise, D. U. (1976a). Mars: a standard crater curve and possible new time scale, *Science*, **194**, 1381–1387.

Neukum, G. and Wise, D. U. (1976b). *Mars: A Standard Crater Size-frequency Distribution Curve and a Possible New Time Scale.* NASA Technical Memorandum, TM X – 74316.

Neukum, G., Basilevsky, A. T. Chapman, M. G. *et al.*, and the HRSC Co-Investigator Team (2007). The geologic evolution of Mars: episodicity of resurfacing events and ages from cratering analysis of image data and correlation with radiometric ages of Martian meteorites. *Lunar and Planetary Science Conference* XXXVIII, abstract 2271.

Nummedal, D. (1978). The role of liquefaction in channel development on Mars. In *Reports of Planetary Geology and Geophysics Program 1977–1978*, NSA TM-79729, 257–259.

Pardee, J. T. (1942). Unusual currents in Glacial Lake Missoula, Montana. *Geological Society of America Bulletin*, **53**, 1569–1600.

Parker, S. (1838). *Journal of an Exploring Tour Beyond the Rocky Mountains*. Published privately, Auburn, NY.

Parker, T. J. (1994). Martian paleolakes and oceans. Ph.D. Dissertation, Department of Geological Sciences, University of Southern California, Los Angles.

Parker, T. J., Gorsline, D. S., Saunders, R. S., Pieri, D. C. and Schneeberger, D. M. (1993). Coastal geomorphology of the Martian Northern Plains. *Journal of Geophysical Research*, **98**, 11061–11078.

Patton, P. C. and Baker, V. R. (1978). New evidence for pre-Wisconsin flooding in the Channeled Scabland of eastern Washington. *Geology*, **6**, 567–571.

Peabody, F. E. (1947). Current crescents in the Triassic Moenkopie Formation. *Journal of Sedimentary Petrology*, **17**, 73–76.

Reineck, H.-E. and Singh, I. B. (1980). *Depositional Sedimentary Environments*, 2nd edn. Berlin: Springer.

Rice, J. W., Parker, T. J., Russell, A. J. and Knudsen, O. (2002). Morphology of fresh outflow channel deposits on Mars. *Lunar and Planetary Science*, XXXIII, 2026.pdf.

Richardson, P. D. (1968). The generation of scour marks near obstacles. *Journal of Sedimentary Petrology*, **38**, 965–970.

Richardson, K. and Carling, P. A. (2005). *A Typology of Sculpted Forms in Open Bedrock Channels*. Geological Society of America Special Paper 392.

Richardson, E. V. and Davis, S. R. (2001). *Evaluating Scour at Bridges*. Hydraulic Engineering Circular HEC 18, 4th edn.

Robinson, M. S. and Tanaka, K. L. (1990). Magnitude of a catastrophic flood event at Kasei Vallis, Mars. *Geology*, **18**, 902–905.

Russell, I. C. (1893). Geological reconnaissance in central Washington. *U.S. Geological Survey Bulletin*, **108**.

Russell, A. J. (1993). Obstacle marks produced by flow around stranded ice blocks during a glacier outburst flood (jökulhlaup) in west Greenland. *Sedimentology*, **40**, 1091–1111.

Salisbury, R. D. (1901). Glacial work in the western mountains in 1901. *Journal of Geology*, **9**, 721–724.

Sato, S., Matsuura, H. and Mayazaki, M. (1987). Potholes in Shikoku, Japan, part 1. Potholes and their hydrodynamics in the Kurokawa River, Ehime. *Memoirs of the Faculty of Education of Ehime University, Natural Science*, **7**, 127–190.

Selby, M. J. (1985). *Earth's Changing Surface*. Oxford: Oxford University Press.

Sevon, W. D. (1988). Exotically sculpted diabase. *Pennsylvania Geology*, **20**, 2–7.

Shaw, J. (1994). Hairpin erosional marks, horseshoe vortices and subglacial erosion. *Sedimentary Geology*, **91**, 269–283.

Shen, H. W. (1971). Scour near piers. In *River Mechanics*, Vol. II, ed. H. W. Shen. Fort Collins, CO.

Shepherd, R. G. and Schumm, S. A. (1974). Experimental study of river incision. *Geological Society of America Bulletin*, **85**, 257–268.

Shrock, R. R. (1948). *Sequence In Layered Rocks*. New York: McGraw-Hill Book Co.

Sklar, L. and Dietrich, W. E. (1998). River longitudinal profiles and bedrock incision models: streampower and the influence of sediment supply. In *Rivers Over Rock: Fluvial Processes in Bedrock Channels*, eds. K. J. Tinkler and E. E. Wohl. Geophysical Monograph 107, Washington, DC: American Geophysical Union, pp. 237–260.

Sklar, L. and Dietrich, W. E. (2001). Sediment and rock strength controls on river incision into bedrock. *Geology*, **29**, 1087–1090.

Sparks, B. W. (1972). *Global Geomorphology*. Harlow, UK: Longman.

Springer, G. S. and Wohl, E. E. (2002). Empirical and theoretical investigations of sculpted forms in Buckeye Creek Cave, West Virginia. *Journal of Geology*, **110**, 469–481.

Thompson, D. M. (2001). Random controls on semi-rhythmic spacing of pools and riffles in constriction-dominated rivers. *Earth Surface Processes and Landforms*, **26**, 1195–1212.

Tinkler, K. J. (1997). Indirect velocity measurements from standing waves in rockbed rivers. *Journal of Hydraulic Engineering*, **123**, 918–921.

Tinkler, K. J. and Wohl, E. E. (1998a). A primer on bedrock channels. In *Rivers Over Rock: Fluvial Processes in Bedrock Channels*, eds. K. J. Tinkler and E. E. Wohl. Geophysical Monograph 107, Washington, DC: American Geophysical Union, pp. 1–18.

Tinkler, K. J. and Wohl, E. E. (1998b). Field studies of bedrock channels. In *Rivers Over Rock: Fluvial Processes in Bedrock Channels*, eds. K. J. Tinkler and E. E. Wohl. Geophysical Monograph 107, Washington, DC: American Geophysical Union, pp. 261–277.

Tómasson, H. (1996). The jókulhlaup from Katla in 1918. *Annals of Glaciology*, **22**, 249–254.

Waitt, R. B. (1972). Revision of Missoula flood history in Columbia Valley between Grand Coulee Dam and Wenatchee, Washington. *Geological Society of America, Abstracts with Programs*, **4**, 255–256.

Waitt, R. B. (1977a). Missoula flood sans Okanogan lobe. *Geological Society of America, Abstracts with Programs*, **9**, 770.

Waitt, R. B. (1977b). *Guidebook to Quaternary Geology of the Columbia, Wenatchee, Peshastin and Upper Yakima Valleys, West Central Washington*. U.S. Geological Survey Open-file Report 77–753.

Weissel, J. K. and Seidl, M. A. (1998). Inland propogation of erosional escarpments and river profile evolution across the southeast Australian passive continental margin. In *Rivers Over Rock: Fluvial Processes in Bedrock Channels*, eds. K. J. Tinkler and E. E. Wohl. Geophysical Monograph 107, Washington, DC: American Geophysical Union, pp. 189–206.

Wende, R. (1999). Boulder bedforms in jointed-bedrock channels. In *Varieties of Fluvial Form*, eds. A. J. Miller and A. Gupta. Chichester, UK: Wiley.

Werner, F., Unsöld, G., Koopmann, B. and Stefanon, A. (1980). Field observations and flume experiments on the nature of comet marks. *Sedimentary Geology*, **26**, 233–262.

Whipple, K. X., Hancock, G. S. and Anderson, R. S. (2000). River incision into bedrock: mechanics and relative efficacy of plucking, abrasion and cavitation. *Geological Society of America Bulletin*, **112**, 490–503.

Wilkinson, S. N., Keller, R. J. and Rutherford, I. D. (2004). Phase-shifts in shear stress as an explanation for the maintenance of pool-riffle sequences. *Earth Surface Processes and Landforms*, **29**, 737–753.

Williams, R. M., Phillips, R. J. and Malin, M. C. (2000). Flow rates and duration within Kasei Valles, Mars: implications for the formation of a martian ocean, *Geophysical Research Letters*, **27**, 1073–1076.

Wohl, E. E. (1992a). Gradient irregularity in the Herbert Gorge of northeastern Australia. *Earth Surface Processes and Landforms*, **17**, 69–84.

Wohl, E. E. (1992b). Bedrock benches and boulder bars: floods in the Burdekin Gorge of Australia. *Geological Society of America Bulletin*, **104**, 770–778.

Wohl, E. E. (1993). Bedrock channel incision along Piccaninny Creek, Australia. *Australian Journal of Geology*, **101**, 749–761.

Wohl, E. E. (1998). Bedrock channel morphology in relation to erosional processes. In *Rivers Over Rock: Fluvial Processes in Bedrock Channels*, eds. K. J. Tinkler and E. E. Wohl. Geophysical Monograph 107, Washington, DC: American Geophysical Union, pp. 133–151.

Wohl, E. E. (1999). Incised bedrock channels. *Incised River Channels*, eds. S. E. Darby and A. Simon. Chichester, UK: Wiley, pp. 133–151.

Wohl, E. E. (2000). Substrate influences on step-pool sequences in the Christopher Creek drainage, Arizona. *Journal of Geology*, **108**, 121–129.

Wohl, E. E. and Achyuthan, A. (2002). Substrate influences on incised-channel morphology. *Journal of Geology*, **110**, 115–120.

Wohl, E. E. and Grodek, T. (1994). Channel bed-steps along Nahal-Yael, Negev Desert, Israel. *Geomorphology*, **9**, 117–126.

Wohl, E. E. and Ikeda, H. (1997). Experimental simulation of channel incision into a cohesive substrate at varying gradients. *Geology*, **25**, 295–298.

Wohl, E. E. and Ikeda, H. (1998). Patterns of bedrock channel erosion on the Boso Peninsula, Japan. *Journal of Geology*, **106**, 331–345.

Wohl, E. and Legleiter, C. J. (2003). Controls on pool characteristics along a resistant-boundary channel. *Journal of Geology*, **111**, 103–114.

Wohl, E. E., Greenbaum, N., Schick, A. P. and Baker, V. R. (1994). Controls on bedrock channel incision along Nahal Paran, Israel. *Earth Surface Processes and Landforms*, **19**, 1–13.

Wohl, E. E., Thomspon, D. M. and Miller, A. J. (1999). Canyons with undulating walls. *Geological Society of America Bulletin*, **111**, 949–959.

Wolman, M. G. and Miller, J. P. (1960). Magnitude and frequency of forces in geomorphic processes. *Journal of Geology*, **68**, 54–74.

3

A review of open-channel megaflood depositional landforms on Earth and Mars

PAUL A. CARLING, DEVON M. BURR,
TIMOTHY F. JOHNSEN
and TRACY A. BRENNAND

Summary

Catastrophic out-bursts of water from lakes impounded by glacial ice or debris such as moraines have caused large freshwater floods on Earth in recent times at least back to the Quaternary. Resultant large-scale depositional sedimentary landforms are found along the courses of these floodwaters. On Mars, similar floods have resulted from catastrophic efflux of water from within the Martian crust. This latter conclusion is based on large-scale and mesoscale landforms that appear similar to those identified in flood tracts on Earth. Both on Earth and on Mars, these landforms include suites of giant bars – 'streamlined forms' – of varying morphology that occur primarily as longitudinal features within the floodways as well as in flooded areas that were sheltered from the main flow. Flow-transverse bedforms, notably giant fluvial dunes and antidunes also lie within the floodways. The flood hydraulics that created these forms may be deduced from their location and morphology. Some other fluvial landforms that have been associated with megafloods on Earth have yet to be identified on Mars.

3.1 Introduction

Exceptionally large freshwater floods on Earth are associated with the catastrophic draining of glacial lakes Missoula and Agassiz amongst others in North America (Teller, 2004). Other glacially related large floods occurred in the mountains of Eurasia, which have only recently received attention (Grosswald, 1999; Montgomery *et al.*, 2004), and geomorphological evidence of other large floods may be discovered in formerly glaciated terrain on other continents. There is a general knowledge about what landforms are associated with flood action but there is relatively little knowledge of what complex and different suites of depositional landforms might result from many floods of different magnitudes. Although several studies have estimated power expenditure by large floods, there has been little critical appreciation of the detail of the distribution of power throughout a flood (Costa and O'Connor, 1995) and the timing of deposition and/or erosion including defining the threshold phenomena (Benito, 1997) required to develop specific landforms in given environments. Similarly, the effect of large sediment loads on the hydraulics has not been explored widely (Carling *et al.*, 2003) and there is little knowledge of the relationship of the hydraulics to the stratigraphic associations, although the spatial variation in the former can be modelled (e.g. Miller, 1995; Clarke *et al.*, 2004) and the latter may be well recorded (e.g. Sridhar, 2007; Duller *et al.*, 2008).

Here, is presented an overview of some key considerations relevant to identifying alluvial landforms on Earth and other planets. These considerations include the mode of sediment transport and the granulometry of the deposits. This synopsis includes depositional landforms and the known conditions of formation and serves as both a summary of past work and a foundation for future research. An analysis of the sedimentology and stratigraphy of landforms that can provide useful insights into formative mechanisms is given elsewhere (see Marren and Schuh, this volume Chapter 12). Firstly, basic terminology is discussed, starting with the issue of scale of flooding.

Having peak discharges of $10^6\,m^3\,s^{-1}$ or larger (Baker, 2002), megafloods might be deemed catastrophic either with respect to origin (e.g. ice-dam, subglacial efflux, rock-dam failures, or volcanic fissure eruptions) or effectiveness in changing the landscape 'irreversibly'. Considering megaflood tracts on Earth and Mars, an exact definition of the term 'large-scale depositional landform' presently cannot be provided owing to a lack of defining criteria. On Earth, alluvial bedforms in river channels are classified as microforms, mesoforms and macroforms (Jackson, 1975) and this typology can be adopted for similar scale forms on Mars. Being unobservable from orbit, microforms fall outside the purview of this chapter. Mesoforms scale with water depth and are typified by subaqueous river dunes. Macroforms scale in a general sense with channel width (Bridge, 2003) although there often is a distinct and important depth limitation to the height of the sediment accumulation. The latter class of bedforms is typified by channel bars that may occur centrally within channels or along the margins of channels. The scale of some mesoforms and macroforms on Earth and Mars and the position

Megaflooding on Earth and Mars, ed. Devon M. Burr, Paul A. Carling and Victor R. Baker.
Published by Cambridge University Press.

of these features relative to the main flood channel mean that the appellation 'bedform' often is not apposite. Rather, for megaflood systems, the term 'landform' is to be preferred as this latter term has neither spatial nor genetic association with a particular portion of a channelway. In a similar sense the adjective 'diluvial' may be a useful precursor to the term 'depositional landform' inasmuch as the term may be used to signify an association with exceptionally large floods. However, some scientists object to the biblical connotations that the word 'diluvial' carries. Such definitions may not apply well to all systems but, with the exception of 'diluvial', are used herein. Further discussion of the unresolved issue of terminology applied to large floods (e.g. catastrophic, cataclysmic etc.) and the effects on sediment transport and landform change is provided by Marren (2005), but see also Papp (2002) and Russell (2005). Closed-conduit flows (such as esker deposition beneath icesheets) and small-scale jökulhlaups, are excluded from consideration.

Figure 3.1A applies to proglacial environments on Earth as the most common terrestrial environment to produce periodic catastrophic floods. The two main observations with respect to Figure 3.1 are: (i) no precise quantitative scale can be added to either axis of the graph (the abscissa is scaled approximately from 10^0 to 10^9 hours), and (ii) that megafloods on Earth are of low frequency and associated with large-scale climate change, affecting the hydrological system at a regional scale (i.e., 10^6 km^2) rather than basin scale (10^2 to 10^4 km^2) (see O'Connor *et al.*, 2002; Hirschboeck, 1988). Yet, two conclusions related to megafloods on Earth result. Firstly, given a frequent climate-change trigger, evidence of megafloods will not be confined to one basin but extend to several basins within a region where conditions allowed water to accumulate. Secondly, geomorphologic changes should occur in the landscape such that the suites of postflood landforms contrast to preflood ones. The large size of many of these landforms resist postflood modification by ordinary processes and these landforms persist in the landscape.

A similar diagram for Mars (Figure 3.1B) lacks information on variability, because satellite images observe the geomorphology integrated over the history of the planet. The time scale is replaced with the frequency of formation, estimated as the ratio of the number of flood channels formed during an epoch to the time length of the epoch. The history of Mars is divided into three time-stratigraphic epochs, which, from oldest to youngest, are the Noachian, Hesperian and Amazonian. Two Noachian-aged Martian flood channels start at large intercrater basins. These basins may have been filled by rainwater and/or groundwater flow (Irwin and Grant, this volume Chapter 11) but the immediate floodwater source was surface-ponded water. The Hesperian saw a peak in flood-channel formation, most of the channels of Mars being formed during this epoch. These flood channels originate either at chasmata, that may have been filled with ponded water, or at chaos terrain interpreted as geothermally triggered groundwater release (summarised by Coleman and Baker, this volume Chapter 9). Thus, the immediate source of floodwater during the Hesperian may have been both surface ponds and groundwater. The four megaflood channels inferred to have flowed during the Amazonian originate at fissures (as discussed by Burr *et al.*, this volume Chapter 10). These fissures may have been induced by dyke injection or by tectonic stresses (see Wilson *et al.*, this volume Chapter 16). In either case, the Amazonian-aged channels originate from groundwater discharges. Like Earth, Mars experiences smaller flows. High-resolution imagery shows small gullies, inferred to have been created recently by flowing water, although the origin of the very youngest examples (formed within the last decade) is less uncertain (Malin *et al.*, 2006; McEwen *et al.*, 2007). High-resolution imagery of slope streaks also permits their formation by flowing liquid, although dry mass wasting is an equal possibility (Chuang *et al.*, 2007; Phillips *et al.*, 2007). The origins of these two types of feature are an area of active research but are included on Figure 3.1B for completeness. Beyond the basic floodwater source, the specific release mechanisms and flow processes for Martian floods are inferred from geomorphology (see Wilson *et al.* this volume Chapter 16). More detailed mapping of depositional bedforms on Mars will provide additional constraints on the surface hydraulics, subsurface groundwater flow and erosion and sedimentary processes. Other issues that require consideration for the correct identification of sedimentary landforms as depositional bedforms are detailed below.

3.2 Geomorphological considerations

Scale Is the size of a sedimentary landform commensurate with formation by megafloods?

Location Firstly, is the landform located where hydraulics and sediment transport suggest a depositional bedform would occur and be preserved? Secondly, is there sufficient accommodation space to allow an alluvial bedform to have been deposited? Accommodation space is especially important where there have been successive flood episodes, as space can be filled already by landforms from earlier events.

Geometry Is the body geometry of the landform recognisable as having been formed by flowing liquid? Full analysis of body geometry considers three dimensions. On Mars, data of the third dimension (i.e., from Mars Orbiter Laser Alimeter (MOLA), photoclinometry or stereo imagery)

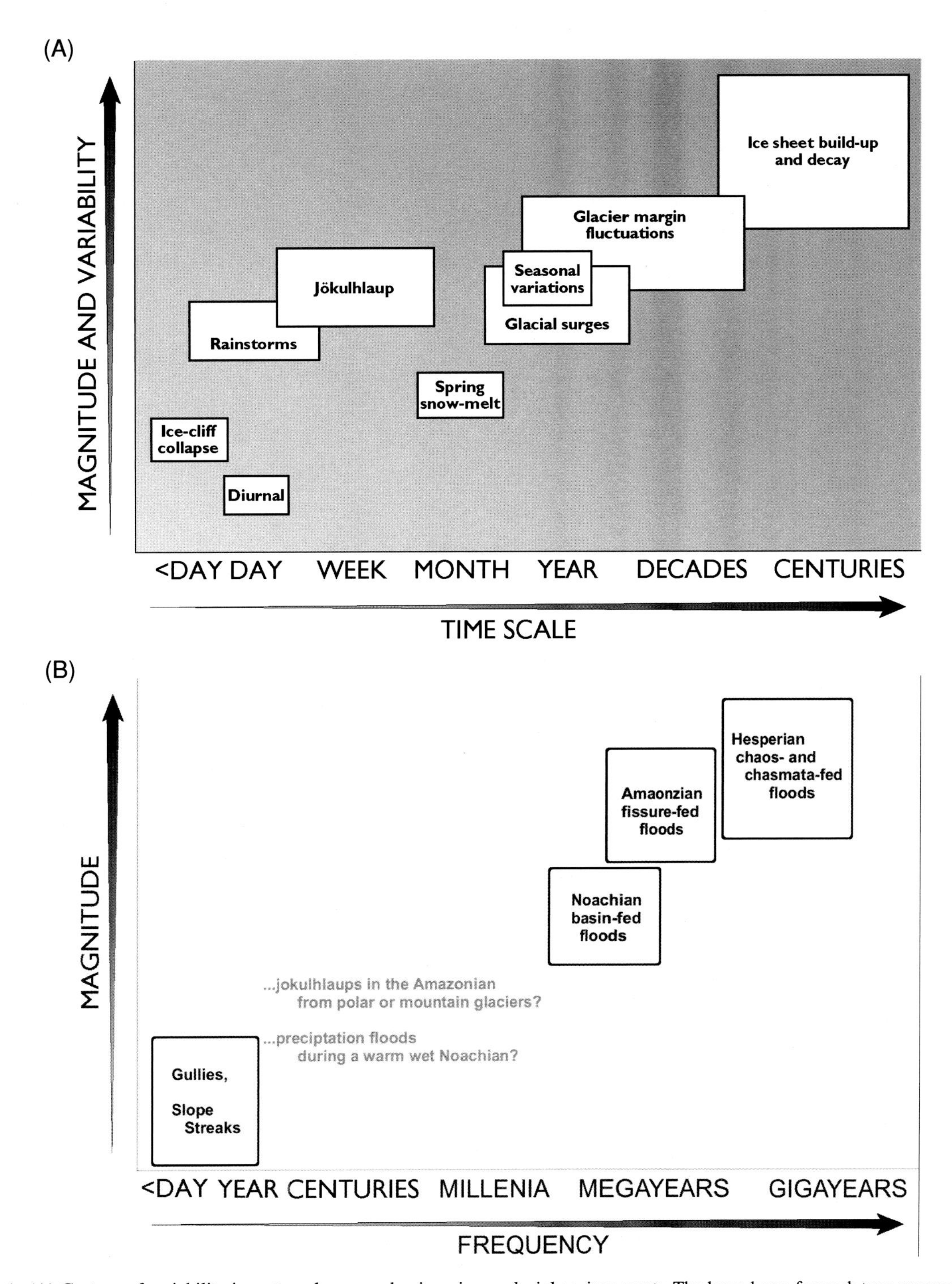

Figure 3.1. (A) Cartoon of variability in water release mechanisms in proglacial environments. The boxed area for each type encompasses the range of time scales and discharges. Increasing magnitudes are associated with increasing variability. (Redrawn from Marren (2005).) (B) Plot showing broad trend with time in Martian flooding. Megaflood channel formation peaks during the Hesperian, when ponded water sourced from chasmata and groundwater sourced from chaos terrain may have combined.

are in some locations sparse or missing, so that often only planform analyses are possible.

Stratigraphy and sedimentology On Earth, the stratigraphy and detailed sedimentology define the formation mechanisms for the landform. For example, grain size, orientation, coarsening/fining trends, regional relationships etc. may indicate the associated fluid and the detailed processes of emplacement (Duller *et al.*, 2008). On Mars, these analyses only occasionally can be explored, but analyses of Martian surface data (e.g. Squyres *et al.*, 2006) suggest the utility of applying terrestrial-style stratigraphic and sedimentological analysis to extraterrestrial landforms.

Association Association refers to whether the sedimentary landform is contiguous to or associated with other landforms, including self-similar forms (e.g. trains of dunes), that would reasonably have resulted during flood formation. The presence or absence of associated landforms has to be compatible and explicable. The current generation of computational fluid dynamics (CFD) models can be used to explore association, although there has been little consideration of detailed and quantified flood hydraulics with respect to landforms to date.

Correct identification of a landform as an alluvial bedform requires that alternative explanations be significantly less likely. This process should entail consideration of the specific processes and fluids responsible, including whether the scales of landforms and fluid flow inferences are compatible. For example, subcritical or supercritical flows may be inferred from flow modelling and the landforms present have to be compatible with such flows.

At the basin scale, alluvial depositional landforms appear rare on Mars in comparison with the frequency of erosional landforms on Mars (Carling *et al.*, this volume Chapter 2) and with the well-described alluvial landforms on Earth (e.g. Baker, 1973a, b; Baker and Nummedal, 1987; O'Connor, 1993; Carling, 1996a; Carling *et al.*, 2002). The rarity of such deposits might suggest that, unlike on Earth, climate change did not drive megafloods on Mars. Additionally, or alternatively, this rarity may be a function of conditions specific to Mars. Under Martian gravity, finer-grained sediment tends to travel as washload in comparison with the same sized sediment under terrestrial gravity (Komar, 1980). The morphology of Martian megaflood channels has been interpreted as indicating that, unless the flow is impeded, this washload sediment remains in suspension for considerable distances, leaving behind fewer depositional landforms (Rice *et al.*, 2003; Burr, 2005; Burr and Parker, 2006). Where alluvial depositional bedforms are identified, their characteristics can provide indications of sediment transport mode or characteristic flow regimes. For example, fluvial dunes and bars are deposited predominantly from bedload with a usually lesser component of suspended load (Bridge, 2003). On Earth the presence of fluvial dunes indicates subcritical flow whereas the presence of antidunes indicates transcritical or supercritical flow (Carling, 1999). In comparison, subaqueous fans, although often constructed from deposition of traction sediment, in some instances may be due to hypopycnal deposition from suspension, with slow net rates of sediment accumulation from 'rain-out' of the very finest material in the water column. These examples illustrate the potential for understanding palaeoflow hydraulics from the location, form, stratigraphy and sedimentology of megaflood landforms on Earth and on Mars.

Komar (1979, 1980) laid the foundation for understanding of flood sediment transport on Mars and for comparing putative water-laid Martian landforms with terrestrial analogues. Komar (1979) drew analogy between fan-like features on Mars and subaqueous fans on Earth. Such interpretations may provide explanation for some of the new SHARAD (radar sounding) data being returned from the Mars Reconnaissance Orbiter (Seu *et al.*, 2004). Chapman *et al.* (2003) argue that some features on Mars are morphologically similar to gravel bars and dunes in Iceland. Burr *et al.* (2004) show that the morphology of channelised Martian dunes is similar to that of terrestrial flood dunes and Burr (2005) argues that clustered streamlined forms in Athabasca Valles are at least in part depositional (see section on 'Large-scale bars').

At a simple level of analysis, depositional landforms result from bedload and/or suspended load transport. A better consideration of the modes of transport that lead to the deposition of given landforms might aid the modelling and interpretation of the hydraulics associated with suites of megaflood landforms. These issues are considered next.

3.3 Theoretical background to sediment transport and deposition

Sediment may be moved by flowing water in three modes. The coarsest sediments move as bedload, with individual grains rolling or sliding along the bed more-or-less in continuous contact with the bed (traction transport). In more energetic flows the bedload may bounce along the bed (saltation transport). Finer sediments move as suspended load, lifted above the bed by turbulence and moved downstream within the water column, making rare contacts with the bed. The finest sediment may remain within the water column for extended downstream distances and this portion of the suspended load has traditionally been termed 'washload'. The term is one of conceptual convenience rather than a mechanistic distinction from suspended load (Woo *et al.*, 1986).

The distinction between bedload, suspension load and washload is useful in understanding flood sedimentation processes. Mode of transport determines the relative speeds of sediment movement. Bedload moves slowly because of its continuing or frequent contact with the bed, whereas suspended sediment travels at practically the rate of the water. An identified transport mode, along with other considerations, may be used to estimate the size of the sediments that are incorporated into depositional bedforms. Most fluvial bedforms consist predominantly of bedload with a variable component introduced from suspension, whilst some bars in slack-water areas may form from suspension fall-out onto weakly mobile beds. Thus identification of a transport mode can help identify the bedforms that are present within a channel and vice versa. In some instances identification can indicate whether the location is proximal or distal relative to the original source of fluid. However, few studies have considered what the presence of specific landforms might indicate in respect of changes in the power distribution through the system and how the landforms might relate to the palaeoflood hydraulics and flood behaviour (e.g. Benito, 1997). This limitation is despite a variety of studies that have calculated the stream power expenditure at a variety of locations within large floodways (e.g. Kale and Hire, 2007).

Paola (reported in Church, 1999) characterised the primary controls in down-channel sedimentation over short time scales as the hydraulics of the transport system, the size distribution of the sediment supply and the distribution of sediment deposition. An example of a system where the Paola controls apply well is the sudden failure of a man-made dam and the consequent deposition of a sediment slug in the valley downstream. Similarly, these controls pertain to individual megafloods, which are singular events with a distinctive source, definitive depositional tract and short duration. Thus down-system sediment sorting and distribution of deposition will be primarily the result of spatial changes in flow hydraulics (Paola *et al.*, 1992; Seal *et al.*, 1997), primarily driven by gradient changes (Gomez *et al.*, 2001) and mediated by temporal changes in the flood hydrograph. In such a simple system, Church (2006) argues that the expected landforms in a channel can be inferred through forward modelling (i.e., from process to landform) or from inverse modelling (from landform to process). In both modelling approaches, the partitioning of the total sediment load between bedload and suspended load may vary down system in accord with the expenditure of the total power (Dade and Friend, 1998; Dade, 2000). A threshold function may be used to delimit when sediment is moving as bedload or in suspension (e.g. Equation (3.1)).

At the simplest, the modes of sediment transport may be distinguished by a ratio of the settling velocity of the particle in still water to the flow frictional shear velocity (a proxy for the turbulence acting to suspend particles):

$$Ro = \frac{w_s}{\rho u_*}, \tag{3.1}$$

where Ro is the Rouse number, ρ is the von Kármán constant, (usually about 0.4, although k values in the range 0.12 to 0.65 have been reported), w_s is the settling velocity, and u_* is the frictional shear velocity.

The frictional shear velocity is given as

$$u_* = \left(\frac{ghS}{\rho}\right)^{\frac{1}{2}} = \left(\frac{\tau}{\rho}\right)^{\frac{1}{2}}, \tag{3.2}$$

where h is the water depth, S is the gradient, τ is the shear stress and ρ is the density of the fluid.

Like u_*, w_s is also a function of the square root of gravity (see e.g. Komar, 1980). Thus, the critical values of ρ for distinguishing modes of sediment transport are the same on any planet. However, because the settling velocity, w_s, varies with gravity, the mode of transport for different grain sizes is not necessarily the same on different planets (Komar, 1980; Dade, 2000; Burr *et al.*, 2006).

Generally (e.g. Valentine, 1987):

- $u_* > 2.5 w_s$, then particles will be entrained; i.e. $\sim 0.4 u_*/w_s > 1$;
- $\tau > \rho 2.5\, w_s^2$, then particles will be entrained.

Thus:

- Bedload: $Ro > 2.5$
- 50% suspended: Ro 1.2 to 2.5
- 100% suspended: Ro 0.8 to 1.2
- Washload: $Ro < 0.8$

Theoretical modelling of flood flows on Mars has indicated that sediments on Mars would tend to move more readily (e.g. with a lower minimum water depth) than for comparable conditions on Earth (Komar, 1980; Burr *et al.*, 2006). The lower gravity on Mars produces a slower water flow, which tends to result in less sediment movement, but also a lower settling velocity, which tends to result in more sediment movement. Modelling shows that, in these two countervailing tendencies, the effect of lower settling velocity predominates (Komar, 1980; Burr *et al.*, 2006). The result is that for a given flow and grain size, the mode of sediment transport is more likely to be suspension or washload on Mars in contrast to bedload on Earth. Conversely, it may be stated that coarser grain sizes are more readily transported on Mars in contrast to Earth. Thus, all other conditions (e.g. channel slope, volumetric discharge, grain size available) being equal, coarser sediment can be expected to constitute Martian depositional bedforms in comparison with terrestrial depositional bedforms and proportionately more material can be expected to have moved in suspension or as a washload. The exact sizes and proportions of material in each transport mode depend on the conditions of a given flow. Although the criteria for the modes of sediment transport have been addressed, curiously, given that sedimentary landforms cannot exist without deposition, little attention has been given to the detail of depositional processes on Earth and Mars. Rather it is assumed that when the threshold for motion defined by the Rouse number or another threshold function (such as the Shields criterion) is not exceeded, deposition will occur.

The detailed controls imposed by the geometry of the flood channel on the deposition process and the relationship of some landforms to deposition from subcritical or supercritical flows are issues that have not been explored widely in the case of superfloods, although Druitt (1998) provides a framework from which approaches might be developed. Within such a context, flood landforms are considered below.

3.4 Megaflood depositional landforms

Meinzer as early as 1918 had noted that the Columbia River had at some time performed exceptional work in transporting large boulders many miles downstream from the source area but it was Bretz in his classic paper of 1923 concerning 'The Channeled Scabland of the Columbia Plateau' who made the first report of megaflood depositional landforms. Bretz made reference to 'great river bars' constructed in 'favourable' situations associated with a monstrous flood, which he termed the Spokane flood. Latterly it was recognised that many floods had occurred and the term Spokane was dropped. Bretz (1925a, b) recognised the process of slack water deposition and also argued that the floods receded rapidly leaving the bars unmodified. During fieldwork in the region of the Channeled Scabland and in the basin of Lake Missoula, Bretz also identified large-scale 'ripple marks' formed in gravel (Bretz *et al.*, 1956), initially identifying puzzling wavy bedforms on the top of bars (Bretz, 1928a). The significance of the forms became clear with their subsequent interpretation as fluvial dunes (Baker, 1973a). Ripples and dunes are distinct bedforms related to specific hydrodynamic conditions (see Carling, 1999). Although the term 'ripples' has been applied loosely to some large-scale megaflood dunes in the older literature, the two bedforms should not be confused as a correct identification can aid in determining the palaeoflow regime. The widespread preservation of the dunes indicated that the flood must have receded rapidly and the presence of dunes indicates that a lower flow regime pertained during the formative flow event (Baker and Nummedal, 1987; see also Carling, 1996a, b).

Pardee (1942) described transverse and arcuate ridges of gravel on Camus Prairie, interpreting them as giant current dunes. He related a progressive change in the height of the dunes to a reduced flow speed from the basin margin towards the basin centre, thus linking landform geometry and location with inferred flood pathways and processes. In similar vein he described flood-deposited expansion bars and indicated mechanisms of deposition. This insight provided important information in identifying megaflood landforms, and influenced acceptance of Bretz' flood origin hypothesis for the Channeled Scabland (see Chapter 2).

As noted by Marren and Schuh (this volume Chapter 12) there is a broad range of sedimentary landforms and bedforms that might be related to large-scale floods but that might also develop for smaller-scale flood events. Thus taken alone, the presence of a particular feature is not necessarily diagnostic of megaflooding. Potentially, there are solutions to this conundrum. Firstly, the scale of the landform, especially the amplitude of the feature, might indicate an exceptional water depth as flood-formed bars frequently develop close to the maximum water depth (Costa, 1984; Carling and Glaister, 1987) and such an assumption has been used to constrain megaflood hydraulic reconstructions (O'Connor, 1993; Burr, 2003; Herget, 2005). Secondly, associations or suites of features may occur, which taken together might be used to make a stronger case for megaflood deposition. At the time of writing there is only a nascent understanding that the consideration of suites of landforms may have diagnostic capacity (Marren and Schuh, this volume Chapter 12).

3.4.1 Large-scale bars

Bars form readily during large-scale floods but those within the main floodway are often severely modified or destroyed on the recession limb of flood hydrographs. In addition, these landforms are subject to further erosion or burial by later small-scale discharges unless they are exceptionally large or composed of unusually large material. For example, the Missoula and Altai Quaternary-flood tracts show only local evidence of such massive landforms completely blocking the main floodway (e.g. the huge Komdodj bar, Altai; Carling *et al.*, this volume Chapter 13). Conversely, examples of coarse component bars that resist further erosion are provided by Fahnestock and Bradley (1973), Russell and Marren (1999), Marren *et al.* (2002) and Marren (2005). Thus it is reasonable to suppose that Holocene reworking, incision and alluviation have effaced most evidence of this kind of landform, leaving only stratigraphic evidence at the base of later valley fill (Carling *et al.*, 2002; Smith, 2006; Carling *et al.*, this volume Chapter 13).

It follows that bars that are useful for diagnostic purposes need to have been deposited within areas of the floodway that are protected from further reworking. Where sediment can be deposited in areas sheltered from the main flow, a variety of large-scale bars may be observed as they persist through time. The literature has suggested a typology of expansion bars, pendant bars (Malde, 1968) and eddy bars (Baker, 1973a; O'Connor, 1993; Maizels, 1997). Expansion bars form in areas where the flood channel widens suddenly downstream of a valley constriction (e.g. Baker, 1973a; Russell and Knudsen, 1999, 2002). Often there is a region of non-deposition between the bar and the valley wall, which was illustrated by Bretz (1928b) and termed a 'fossa' by Bretz *et al.* (1956). Carling (1987, 1989) provides field and experimental flume examples of expansion bar deposition and fossae formation. Expansion bars are often incised or streamlined by subsequent flows. For example, the sedimentology and stratigraphy of bars in the Quincy Basin in the Channeled Scabland indicate formation through deposition of a large deltaic deposit with subsequent incision during the waning stages of the flood or by later floods (Bretz, *et al.* 1956 pp. 969–974; Baker, 1973b pp. 39 *et seq.*). A similar origin was inferred for a cluster of streamlined bars on Mars (Burr, 2005). Pendant bars form in the lee of obstacles, such as bedrock hills in the flood channel (Baker, 1973a; Lord and Kehew, 1987; Rudoy and Baker, 1993; O'Connor, 1993) or impact craters on Mars. Eddy bars form in the re-entrants of sheltered back-flooded tributary valleys but importantly may extend far up the tributary with consequences for the sedimentary signature (Plate 11). All prior descriptions of the morphology of these bars assume that their modern day topography is more or less the same as that which was present after flood recession (except for incision by later floods, e.g. Bretz *et al.*, 1956), and indeed in some cases they may be streamlined to form lemniscate forms that minimise skin and form drag (Baker, 1973b; Baker and Nummedal, 1987; Komar, 1983). However, Carling *et al.* (this volume Chapter 13) suggest that in some cases bars might be remnants of a sediment body that originally completely filled valleys. Later incision then cut out the majority of the valley fill, leaving marginal remnants as apparent 'bars'. A similar idea was also proposed by Baker (1973a) with respect to a fan-complex fill within the Quincy Basin, Washington State (Bretz *et al.*, 1956, Figure 4). The detailed styles of deposition and hence hydraulic environment of deposition might be adduced through detailed study of the bar stratigraphy using geophysical techniques.

Associated with giant bars in the Altai are prominent so-called run-up deposits (Plate 11). These are wedge-like landforms consisting of 'smears' of fine well-rounded fluvial gravel found draping valley side alcoves at elevations up to 100 m above bar tops. Often these deposits have very steep slopes facing the main river channel. They are interpreted to represent the deposition of sediment above the main bar tops and peak water level of floods by initial surges or by the inherently unsteady flow of highly turbulent flood waves circulating around spurs and other headlands in the valley-wall alignments (Herget, 2005). Similar run-up deposits have not been identified in association with the Missoula floods.

Large-scale streamlined bars in the giant, Hesperian-aged circum-Chryse outflow channels were perhaps the most obvious indicator of megaflooding on Mars, but these streamlined forms have been interpreted consistently to be erosional (e.g. Baker, 1979; Baker and Kochel, 1979; Rice *et al.*, 2003). Large-scale bars specifically identified as depositional are more limited in imagery available to date. A medial bar in the Noachian-aged Ma'adim Vallis flood channel (Irwin and Grant, this volume Chapter 11) is interpreted as depositional, based on layering visible in an impact crater on the bar (Irwin *et al.*, 2004). This context is similar to that of giant bars along the Chuja River valley, which were deposited during backflooding of tributaries at their confluence with the main trunk channel (Carling *et al.*, 2002).

A cluster of streamlined bars in the Amazonian-aged Athabasca Valles outflow channel is hypothesised to be largely depositional, formed by sediment deposition during hydraulic damming. This hypothesis is based on (a) a similarity of bar upper elevations with another palaeoflow water-height indicator several kilometres up-channel, which is indicative of ponding; (b) the clustering of bars upslope of a flow obstacle, which is interpreted to reflect hydraulic damming by the obstacle; and (c) the morphology of the bars, which shows a difference between obstacles upslope and finely layered tails downslope (Burr, 2003, 2005). This suggested mechanism is analogous to that which produced the streamlined forms in the Quincy and Pasco Basins during the Channeled Scabland flooding (Baker, 1973a; Baker *et al.*, 1991; Bjornstad *et al.*, 2001). Although the tendency towards suspended sediment transport on Mars might preclude deposition from suspension in locations proximal to flood sources, Martian floods should have carried more coarse sediment than terrestrial floods (Burr and Parker, 2006), as the reduced gravity results in a reduced settling velocity that more than offsets the reduction in flow velocity. Thus, on Mars as on Earth, the coarser bedload component of floods should largely be deposited proximally. Consequently, the apparent lack of giant bars in putative Martian floodways may not be entirely explained by the theoretical arguments presented above. Explanations for the absence of bars might include an absence of a source for coarse bedload in some Martian channels and the effect of local conditions such as the large width-to-depth ratio of channels like the Grjotá Valles (Burr and Parker, 2006)

Figure 3.2. (A) High-angle oblique aerial view of regularly spaced gravel ridges at Modrudalur in Iceland that have been interpreted to be transverse ribs or antidunes. The features have average wavelengths of 27 m, heights of 0.8 m and breadths of 14 m. Shadow is of a light aircraft. (B) View along a ridge shown in A. Largest boulders are 1.8 m long. Spade for scale is *c.* 1 m in length. (Photographs courtesy of Dr. Jim Rice and caption information from Rice *et al.* (2002).)

wherein flood waters would have spread widely and dissipated the transporting power to move coarse sediment.

3.4.2 Transverse ribs, hydraulic jumps and antidunes

Transverse ribs are regularly spaced gravel ridges formed in relatively shallow, high-energy fluvial systems and oriented transversely to the current direction (see Carling (1999) for key references). The crestlines are distinctive, being generally straight and continuous over long distances relative to the breadth of the bedforms, and only locally are crest bifurcations recorded. Koster (1978), amongst others, has interpreted relatively small-scale transverse ribs (i.e. heights <0.3 m and wavelengths *c.* 1 m) to be relict antidune bedforms, which formed within transcritical or supercritical flows, although this interpretation is disputed (Allen, 1983; Whittaker and Jaeggi, 1982). Koster demonstrated that using an antidune analogy (see Equation (3.3)), key palaeohydraulic parameters, such as the mean velocity, mean depth and Froude number, can be calculated from transverse rib data. Rice *et al.* (2002a; see also Chapman *et al.*, 2003, Figure 15b) describe bedforms from Holocene-aged Icelandic jökulhlaups (Waitt, 2002) with average wavelengths of 27 m and average heights of 0.84 m as large-scale transverse ribs and use the antidune analogy to calculate palaeoflood data (Figure 3.2). However, there are no reports of modern transverse ribs of similar scale to use as analogues and sections cut in the Icelandic ridges by one of the present authors (DB) reveal an internal stratification that dips downslope and may be more consistent with dune formation. In addition, these particular examples may be erosional remnants rather than pure accumulative features.

Baker and Nummedal (1987) tentatively ascribed a large fan downstream of the Soap Lake topographic constriction within the Channeled Scabland as a depositional landform established beneath a hydraulic jump but did not provide additional comment. In this respect there is potential to use other landforms developed in association with putative hydraulic jumps and hydraulic drops. For example, push bars are the bars that develop downstream of waterfalls (Levson and Giles, 1990) and some of these may be ‘fossil’, i.e., associated with former flow regimes (Jacob *et al.*, 1999). Such bars may be seen, for example, on topographic maps and satellite images of Dry Falls, near Coulee City, Washington State. Nott and Price (1994) have explored the significance of push bars for deducing palaeoclimate and Carling and Grodek (1994) have used the characteristics of push bars to calculate hydraulic indicators of past floods. However, the utility of push bars for large flood reconstructions remains largely unexplored.

Spectacular standing waves often form in both large and small rivers (Plate 12). Often they develop in steep channels such as at the head of alluvial fans (Zielinski, 1982), in steep or flood-prone rivers and upstream and downstream of major obstacles to flow such as islands. Antidunes are the bedforms that develop beneath standing waves. The formation process involves both erosion and deposition but, within loose granular beds, usually only the morphology owing to accumulation of sediment is evident. The bulk hydraulics of standing waves are well known but the conditions for the preservation of antidunes is less well understood (Carling and Shvidchenko, 2002). On falling-water stages antidunes typically are erased. Antidunes usually occur as trains of self-similar ridges and intervening troughs that are roughly transverse to the main direction of flow. However, for supercritical flows the standing waves can become increasingly three-dimensional with wavy crest-lines or, for higher flow conditions, more isolated steep waves occur known as rooster-tails. Increasing three-dimensionality of the water waves often produces a rhomboid water surface profile, a condition that is most prevalent in those situations where the channel margins are close-by and waves are reflected from the channel margins. The shape of the antidunes beneath these various water waves is of similar form to the waveform, with steep isolated mounds of sediment occurring beneath rooster-tails.

The identification of large-scale antidunes is important in a number of regards. Firstly, antidunes develop in transitional and supercritical flows when the Froude number is greater than about 0.84 and may be greater than unity (Carling and Shvidchenko, 2002). Thus, identification would preclude the occurrence of subcritical flows, which are characterised by fairly even water surface levels during the formative phase of the bedforms; rather surface water instabilities are necessarily present for supercritical flows. Thus, identification of palaeo-antidunes might indicate the presence of an irregular palaeowater surface slope. Secondly, the morphology of the antidunes, i.e, regular-transverse, wavy-transverse, or isolated mounds indicates progressively higher Froude numbers respectively. The spacing of transverse antidunes scales with the Froude number in a manner that permits an estimation of water depth or velocity; this is a powerful tool for palaeoflow reconstruction. Allen (1984) deduced from the work of Kennedy (1963) that the average wavelength ($\overline{L_w}$) of standing waves scales with the depth (h) of the water flow:

$$\overline{L_w} = 2\pi h. \tag{3.3}$$

Tinkler (1997a, b) and Grant (1997) have used this relationship to reconstruct flood hydraulic parameters from the spacing of observed standing waves. Allen (1984) showed using flume data that the average wavelength $\left(\overline{L_a}\right)$ of trains of antidunes was in accord with the average wavelength of the standing waves with which they were associated. Consequently $\overline{L_a}$ can be substituted into Equation (3.3) to estimate flow palaeodepth from 'fossil' antidune wavelengths. Likewise, flow velocity can also be estimated (Kennedy 1963) using:

$$U = \sqrt{\frac{\overline{L_g}}{2\pi}} \tag{3.4}$$

where $\overline{L_g}$ is the average wavelength of either the standing waves (if observed) or the palaeo-antidunes. Given that gravity is accounted for in Equation (3.4), the expression should be applicable to Mars as well as to Earth. Finally, the presence of preserved antidunes, after flood recession, usually indicates rapid recession of floodflow (Alexander and Fielding, 1997). For example, Reddering and Eserhuysen (1987) provide a brief description of a flood on the Mzimvubu River in South Africa, which produced gravel antidunes (Reddering, personal communication, 2006). Although the paper contains no information on the antidunes themselves, the hydrograph was very abrupt, lasting only a few hours and so rapid draw-down might explain their preservation. The morphology of the Mzimvubu antidunes is similar to those reported by Shaw and Kellerhals (1977). Karcz and Hersey (1980) developed formative theories and examples of rhomboid bedforms such as those that develop beneath the rooster-tails noted above, but these ideas have not been applied to 'fossil' bedforms.

A small ($\sim$1 km^2) but significant field of probable antidunes was discovered recently in British Columbia, Canada (Figure 3.3A; Johnsen and Brennand, 2004). These bedforms have wavelengths of 100 to 230 m and heights of 3 to 7 m. They are two-dimensional bedforms, having fairly straight troughs and crest-lines. Their streamwise long profiles are asymmetrical, with steeper upflow (stoss) slopes (Figure 3.3B) than the downflow (lee) slopes. The bedforms were created during the catastrophic drainage of a narrow, valley-filling, ice-dammed glacial lake approximately 12 kyr BP. Ground-penetrating radar profiles across one bedform in the upflow part of the field show foreset truncations and suggest that this bedform may be partially erosional (Johnsen and Brennand 2004, Figure 3). However, the upper 1.5 m is composed of backset bed couplets (dipping $\sim$15–35° upflow) of normally graded, sandy-matrix-supported medium-grained to fine-grained gravel and openwork gravel. In addition, numerous boulders mantle the bedforms and have their long axes strongly orientated transverse to flow (i.e., deposited during traction transport). Deep scours occur in the troughs of some antidunes (Figures 3.3A and 3.4), suggesting late-stage erosion by vertical flow vortices (e.g. Baker and Komar 1987). Bedforms in the upflow part of the field are the product of phases of erosion and deposition associated with rapid and spatially varying water depth and velocity induced by the topography of the bedforms. Downflow the bedforms increase in size (#4, Figure 3.5), and are composed of gravel backset beds (Figure 3.6). The increase in flow depth and hence accommodation space may explain the apparent downflow transition into fully depositional antidunes. Flutes (Figure 3.3B) ornament these downflow antidunes, indicating late-stage erosion by longitudinal flow vortices (e.g. Pollard *et al.*, 1996). Steep stoss slopes (Figure 3.3B) may have been oversteepened through erosion by roller vortices developing at these upflow-facing steps. The rapidly receding floodwaters led to both the creation and preservation of this antidune field. From Equations (3.3) and (3.4) and the range of antidune

Figure 3.3. Antidune bedforms on the distal portion of Deadman delta produced during catastrophic drainage of glacial Lake Deadman, British Columbia, Canada (Johnsen and Brennand 2004). (A) Bedforms in the upflow (eastern) portion of the field (1–3, Figure 3.5). L, inset delta levels; K, kettle or scour holes. (B) The largest antidune (4, Figure 3.5) in the downflow (western) portion of the field. Note the steep stoss slopes and superimposed flutings (flute residuals).

wavelengths in the field, the minimum water depth during antidune formation was ~16 to 36 m (consistent with computer-modelled water depths) and the flow velocity was ~13 to 19 m s^{-1}.

Rice *et al.* (2002b) proposed that linear features, observed on the floor of the Athabasca Valles, on Mars, are transverse ribs based on plan-view morphometric analysis of a MOC image and comparison with the Icelandic features noted above. The transverse rib average wavelength, in this region of Mars, is 53.6 m; rib height at that time could not be determined. Rice *et al.* (2002b) applied Equation (3.3) to the Martian features in order to calculate and constrain the velocity, depth and Froude number of the floods and obtained the following results: mean velocity of 6 m s^{-1}; flow depths ranging from a minimum of 5.5 m to a maximum of 19 m. Froude numbers ranged from 0.7

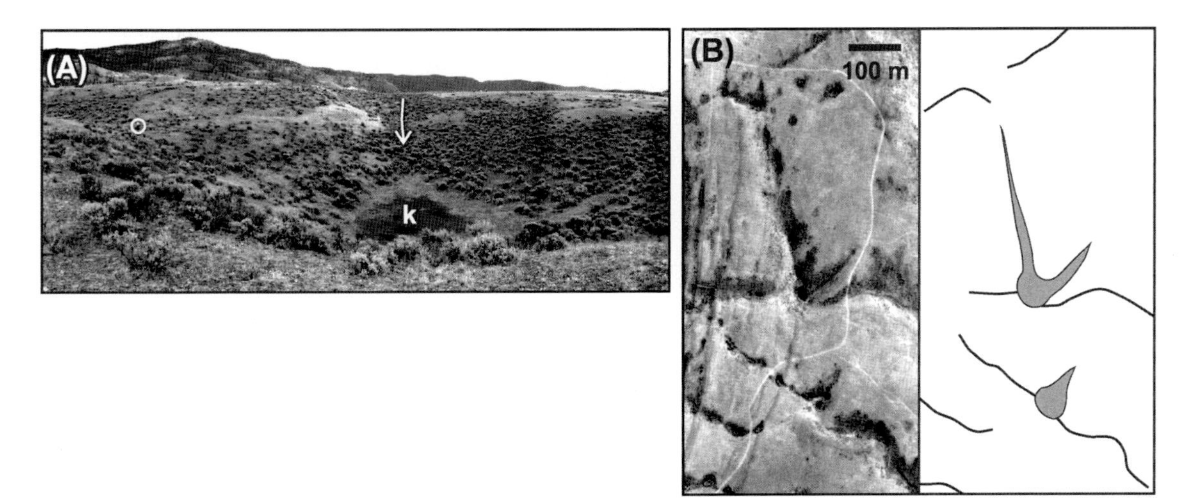

Figure 3.4. (A) Large scour (k) cut into the trough of an antidune in the upflow part of the field (upper scour in B). An associated large groove (arrow) deepens downflow into the scour. A second groove enters obliquely from the right side of the photograph. Person (circled) for scale. (B) Aerial photograph and interpretive sketch of 'scours' (shaded). Solid lines are bedform troughs. Flow from top.

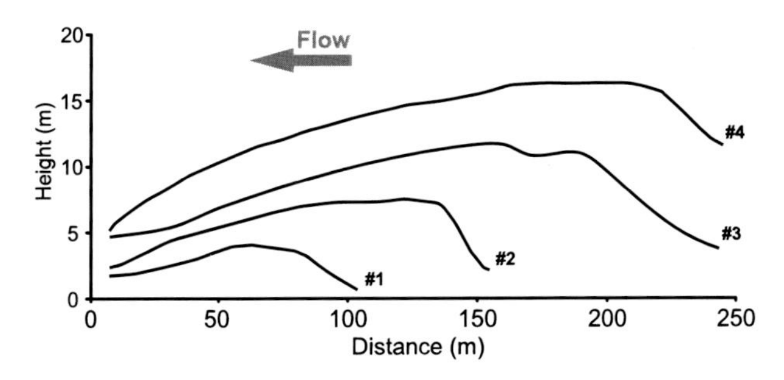

Figure 3.5. Comparison of streamwise long profiles of select bedforms on Deadman delta (5× vertical exaggeration). Average length:height is 30.3, standard deviation is 2.0.

Figure 3.6. Gravely backset beds below the crest of an antidune in the downflow and downslope part of the field. True dips of beds are solid lines and apparent dips are dotted lines. Person is holding a 1 m stick.

to 1.3, which is the range associated with an antidune flow regime. These bedforms probably were deposited during the last and most recent flood through this region. This region of Mars has undoubtedly been subjected to multiple flood episodes (Parker and Rice, 1997). However, using additional MOC images and a photoclinometric technique to measure heights and landform slope angles, Burr *et al.* (2004) have challenged the interpretation of the Athabasca Valles features as antidunes (Rice *et al.*, 2002b), preferring a fluvial dune model. This reinterpretation is based on the clinometric data, which allowed detailed morphological profiles for the Athabasca Valles features to be developed, that were not available to Rice *et al.* (2002b). The analysis of Burr and colleagues shows that the bedforms are strongly asymmetric. Antidunes and transverse ribs are usually symmetrical or show weak upstream or downstream asymmetry. In contrast, fluvial dunes are predominantly asymmetric with respect to the flow, displaying shorter and steeper lee sides in comparison with the longer and less steep stoss slopes (Allen, 1984).

Alt (2001) argues that some of the largest bedforms in the Camus Prairie, Washington State, USA, are antidunes based on their size (height < 10.7 m; wavelength < 91 m) and asymmetry; having their steepest slopes facing upstream. The specific location is also relevant to the interpretation. The bedforms are on steep terrain just downstream of a divide-crossing marked by Markle Pass and Wills Creek Pass between Little Bitterroot Valley and the Camus Prairie. Alt further argues that the stratigraphy illustrated in Pardee (1942) is typical of antidunes. In such a location, the bedforms would readily be preserved as the flood flow would stop abruptly when the water level fell below the level of the passes.

Spectacular putative antidunes occur in the Chuja basin and have been associated with the Altai megafloods

Figure 3.7. Oblique aerial photographs of dunefields produced by catastrophic drainage of Neoglacial Lake Alsek, Yukon, Canada (Clague and Rampton, 1982). (Photographs courtesy of John Clague.)

(Carling *et al.*, 2002; Herget, 2005) but these appear to be erosional features with only a veneer of deposited sediment (see Carling *et al.*, this volume Chapter 2).

3.4.3 Dunes

Well-known examples of large-scale dunes associated with Quaternary megafloods are described by Baker (1973a), Carling (1996a, b), Carling *et al.* (2002) and Clague and Rampton (1982) (Figure 3.7). These examples are developed in cobble-sized gravel and most reports of palaeodunes are from gravel deposits (Carling, 1999). Other published examples of 'fossil' fluvial dunes are less studied than the Missoula and Altai dunefields (see references in Carling, 1999; Carling and Breakspear, 2007) and in some cases the identification as dunes is not verified by detailed study or is disputed (see Munro-Stasiuk and Shaw, 1997; Evans *et al.*, 2006). Other putative palaeodune fields are as follows. Malin (1986) describes enigmatic large 'duneforms' from Antarctica in fine-grained sediments, which he argues could be waterlain deposits. The problem with the latter interpretation is that they occur throughout the TransAntarctic Mountains, occasionally at or near the summits of mountains, without obvious attendant fluvial features. Alternative explanations are that these bedforms are aeolian (Malin, 1986) or formed by subglacial sheetflow (Denton *et al.*, 1984).

On Earth, large-scale dunes related to megafloods are probably more common than the literature would suggest but remain to be identified. For example, Andrea Pacifici of the International Research School of Planetary Science (IRSPS), Italy and one of the authors (PAC) have identified probable long-wavelength 'fossil' fluvial dunefields formed in coarse gravel on the floodplain of the Santa Cruz River at Condor Cliffs, Patagonia, Argentina, downstream of the ice-dammed water body Largo Argentino that is subject to outbreak floods (Pacifici, 2009). These bedforms are visible in satellite images but, being of very low amplitude, are difficult to locate in the field. Schoeneich and Maisch (2003a, b) describe and illustrate 40 m wavelength, 0.5 m high gravel dunes, which they relate to a late glacial outbreak flood near Davos, Switzerland.

The identification of landforms as dunes is significant as dunes can only form in subcritical flow conditions, which prescribes estimates of flow velocities and Froude numbers. Bedform asymmetry may indicate flow direction whilst the width of the dune field may indicate a minimum width of the palaeoflow field. Flow depths clearly cannot have been less than the height of the dunes, and average flood flow depths often scale with dune heights and/or dune wavelengths (Allen, 1984). Thus, a wealth of important palaeohydraulic data may be deduced from the geometry of palaeodunes, which can be of great assistance in estimating flood discharges (e.g. Carling, 1996b).

The most clearly evident depositional bedforms on Mars are sedimentary dunes. The greater majority of these must have formed subaerially as aeolian features. However, dune-like forms seen in the Maja Valles and Athabasca Valles outflow channels have been interpreted as potentially subaqueous in origin (Chapman *et al.*, 2003; Burr *et al.*, 2004). The argument for the aqueous origin of the Maja Valles forms is their qualitative plan-view similarity to flood-formed dunes in the Jökulsá á Fjöllum outflow channel in Iceland (Waitt, 2002). As discussed above, an additional argument for the depositional nature of the Athabasca Valles forms is their dune-like morphology of steeper downslope and shallow upslope slope angles (Burr *et al.*, 2004), a characteristic shared with the gravel flood-dunes from the Altai megafloods (Carling, 1996a, b). However, in other similar channels, where Martian flooding is the suspected cause of erosion, there is a lack of evidence for dunes or other flood features. For example, Gjrotá Valles, which is very similar in age, origin, context and location to Athabasca Valles, does not show any such depositional bedforms (Burr and Parker, 2006).

3.4.4 Longitudinal ridges

A number of young Martian channels show several-kilometre long lineations parallel to the flow direction (Burr *et al.*, 2002; Ghatan *et al.*, 2005; Burr and Parker, 2006). Some of these lineations can be seen to be composed of metre-scale bumps or mounds suggested to be flood-deposited boulders, whose alignment is hypothesised to be due to longitudinal vortices (Burr *et al.*, 2002). Some of these bumps stand on pedestals above the surrounding channel floor and may represent glaciofluvial strata subsequently eroded by flooding (Gaidos and Marion, 2003). In either case, their origin is depositional, although the mechanism of deposition is distinctly different. On Earth, longitudinal ridges have been reported widely in sediments ranging from mud, through sand and gravel, but despite some laboratory studies, which indicate an origin owing to longitudinal vorticity, the exact mechanisms of initiation and maintenance of these bedforms remain elusive (Williams *et al.*, 2008; Carling *et al.*, 2009).

3.5 Discussion and conclusions

Despite the detailed classification of channel-scale bedforms that can be developed for small modern river systems, relatively few types of large-scale depositional landforms have been described on Earth and Mars. The process-based understanding of how these features develop is therefore poor. Subaerial gullying and Rogen moraines can produce features that in plan view resemble giant fluvial dunes and some moraines and kame terraces can resemble giant flood bars. Consequently, care is required in identification and interpretation of landscape features, especially if they can only be considered remotely such as on Mars and the other planets. In some cases, as with large-scale transverse ridges, it is not clear if these are depositional features or erosional remnants of formerly more extensive gravelly-sand deposits. Every putative interpretation of landforms should consider alternative possibilities. This is especially important when the only information available is drawn from morphological planform data obtained using satellite or aerial photography. Recent studies often allow height data to be derived using techniques such as photoclinometry or shape-shading (Beyer *et al.*, 2003; Burr *et al.*, 2004). Where self-similar, spatially contiguous groups of bedforms occur then suites of morphological geostatistics, such as fractal properties, might help resolve the genetic origins of landforms, as some landforms are scale specific (Evans, 2003). However, further scales might also be present, induced for example by the presence of other flood-depositional bedforms (e.g. giant bars in Missoula and in the Altai floodways) as well non-flood features such as moraines, and these may overprint the signatures of bedforms. As far as is known, this geostatistical approach has not been applied to megaflood erosional landforms as usually within any one fluvial system the population of self-similar features is small (see Carr and Malin (2000) for a perspective). The features so far described exist at a range of scales, including the scale of the channel width. Thus, the issues of considering concepts of 'association' and whether the assumed nature of sediment transport and depositional processes are compatible with the suite of landforms present in any floodway requires closer attention.

In conclusion, although it can be difficult to couple models of fluid flow with sediment transport functions, it seems that simple rules can be applied with respect to whether sediment size fractions will be deposited in specific locations within complex channel geometries from bedload or from suspended load. As noted by Church (2006) and as explained in this perspective, such models can be forward or backward predictors such that better insight into megaflood landscapes will be derived from future consideration of the association landforms and the hydraulic controls.

References

Alexander, J. and Fielding, C. (1997). Gravel antidunes in the tropical Burdekin River, Queensland, Australia. *Sedimentology*, **44**, 327–337.

Allen, J. R. L. (1983). A simple cascade model for transverse stone-ribs in gravelly streams. *Proceedings of the Royal Society of London, Series A*, **385**, 253—266.

Allen, J. R. L. (1984). *Sedimentary Structures: Their Character and Physical Basis*. Amsterdam: Elsevier.

Alt, D. (2001). *Missoula and Its Humogous Flood*, Missoula, MT: Montana Press Company.

Baker, V. R. (1973a). *Paleohydrology of Catastrophic Pleistocene Flooding in Eastern Washington*. Geological Society of America Special Paper 144, 1–79.

Baker, V. R. (1973b). Erosional forms and processes for the catastrophic Pleistocene Missoula floods in eastern Washington. In *Fluvial Geomorphology*, ed. M. Morisawa. Publications in Geomorphology, State University of New York, Binghamton, NY, pp. 123–148.

Baker, V. R. (1979). Erosional processes in channelized water flows on Mars. *Journal of Geophysical Research*, **84**, B14, 7985–7993.

Baker, V. R. (2002). High-energy megafloods: Planetary settings and sedimentary dynamics. In *Flood and Megaflood Deposits: Recent and Ancient Examples*, eds. I. P. Martini, V. R. Baker and G. Garzon. International Association of Sedimentologists Special Publication 32, 3–15.

Baker, V. R. and Kochel, R. C. (1979). Martian channel morphology: Maja and Kasei Valles. *Journal of Geophysical Research*, **84**, 7961–7983.

Baker, V. R. and Komar, P. D. (1987). Cataclysmic processes and landforms. In *The Channeled Scabland*, eds. V. R. Baker and D. Nummedal. Washington, DC: National Aeronautics and Space Administration.

Baker, V. R. and Nummedal, D. (Eds.) (1987). *The Channeled Scabland.* Washington, DC: National Aeronautics and Space Administration.

Baker, V. R., Bjornstad, B. N., Busacca, A. J. *et al.* (1991). Quaternary geology of the Columbia Plateau. In *Quaternary Nonglacial Geology, Conterminous U.S. Geology of North America*, ed. R. B. Morrison. Boulder, CO: Geological Society of America, pp. K-2:215– K-2:250.

Benito, G. (1997). Energy expenditure and geomorphic work of the cataclysmic Missoula flooding in the Columbia River Gorge, USA. *Earth Surface Processes and Landforms*, **22**, 457–472.

Beyer, R. A., McEwen, A. S. and Kirk, R. L. (2003). Meter-scale slopes of candidate MER landing sites from point photoclinometry. *Journal of Geophysical Research*, **108** (E12), 8085, doi:10.1029/2003JE002120.

Bjornstad, B. N., Fecht, K. R. and Pluhar, C. J. (2001). Long history of pre-Wisconsin, ice age cataclysmic floods: evidence from southeastern Washington state. *Journal of Geology*, **109**, 695–713.

Bretz, J H. (1923). The Channeled Scabland of the Columbia Plateau. *Journal of Geology*, **31**, 617–649.

Bretz, J H. (1925a). *The Spokane Flood Beyond the Channeled Scablands*. Department of Conservation, Division of Mines and Geology Bulletin No. 45.

Bretz, J H. (1925b). The Spokane flood beyond the Channeled Scablands. *Journal of Geology*, **33**, 97–115, 236–259.

Bretz, J H. (1928a). Bars of the Channeled Scabland. *Geological Society of America Bulletin*, **39**, 643–702.

Bretz, J H. (1928b). The Channeled Scabland of Eastern Washington. *Geographical Review*, **18**, 446–477.

Bretz, J H., Smith, H. T. U. and Neff, G. E. (1956). Channeled Scabland of Washington: new data and interpretations. *Geological Society of America Bulletin*, **67**, 957–1049.

Bridge, J. S. (2003). *Rivers and Floodplains*. Malden, MA: Blackwell Publishing.

Burr, D. M. (2003). Hydraulic modelling of Athabasca Vallis, Mars. *Hydrological Sciences Journal*, **48** (4), 655–664.

Burr, D. M. (2005). Clustered streamlined forms in Athabasca Valles, Mars: Evidence for sediment deposition during floodwater ponding. *Geomorphology*, **69**, 242–252.

Burr, D. M. and Parker, A. H. (2006). Grjotá Valles and implications for flood sediment deposition on Mars. *Geophysical Research Letters*, **33**, L22201, doi:10.1029/2006GL028011.

Burr, D. M., Grier, J. A., McEwen, A. S. and Keszthelyi, L. P. (2002). Repeated aqueous flooding from the Cerberus Fossae: evidence for very recently extant, deep groundwater on Mars. *Icarus*, **159**, 53–73.

Burr, D. M., Carling, P. A., Beyer, R. A. and Lancaster, N. (2004). Flood-formed dunes in Athabasca Valles, Mars: morphology, modeling, and implications. *Icarus*, **171**, 68–83.

Burr, D. M., Emery, J. P., Lorenz, R. D., Collins, G. C. and Carling, P. A. (2006). Sediment transport by liquid overland flow: application to Titan. *Icarus*, **181**, 235–242.

Carling, P. A. (1987). Hydrodynamic interpretation of a boulder-berm and associated debris-torrent deposits. *Geomorphology*, **1**, 53–67.

Carling, P. A. (1989). Hydrodynamic models of boulder-berm deposition. *Geomorphology*, **2**, 319–340.

Carling, P. A. (1996a). Morphology, sedimentology and palaeohydraulic significance of large gravel dunes: Altai Mountains, Siberia. *Sedimentology*, **43**, 647–664.

Carling, P. A. (1996b). A preliminary palaeohydraulic model applied to late Quaternary gravel dunes: Altai Mountains, Siberia. In *Global Continental Changes: The Context of Palaeohydrology*, eds. J. Branson, A. G. Brown and K. J. Gregory. Special Publication Geological Society London, No. 115, 165–179. Bath: Geological Society Publishing House.

Carling, P. A. (1999). Subaqueous gravel dunes. *Journal of Sedimentary Research*, **69**, 534–545.

Carling, P. A. and Breakspear, R. M. D. (2007). Gravel dunes and antidunes in fluvial systems. In *River, Coastal and Estuarine Morphodynamics*, Vol. 2, eds. C. M. Dohmen-Janssen and S. J. M. L. Hulscher, London: Taylor and Francis, pp. 1015–1020.

Carling, P. A. and Glaister, M. S. (1987). Reconstruction of a flood resulting from a moraine-dam failure using geomorphological evidence and dam-break modeling. In *Catastrophic Flooding*, eds. L. Mayer and D. Nash. Boston: Allen and Unwin, pp. 181–200.

Carling, P. A. and Grodek, T. (1994). Indirect estimation of ungauged peak discharges in a bedrock channel with reference to design discharge selection. *Hydrological Processes*, **8**, 497–511.

Carling, P. A. and Shvidchenko, A. B. (2002). The dune:antidune transition in fine gravel with especial consideration of downstream migrating antidunes. *Sedimentology*, **49**, 1269–1282.

Carling, P. A., Kirkbride, A. D., Parnachov, S., Borodavko, P. S. and Berger, G. W. (2002). Late Quaternary catastrophic flooding in the Altai Mountains of south-central Siberia: a synoptic overview and an introduction to flood deposit sedimentology. In *Flood and Megaflood Deposits: Recent and Ancient Examples*, eds. I. P. Martini, V. R. Baker and G. Garzon. International Association of Sedimentologists Special Publication 32, 17–35.

Carling, P. A., Kidson, R., Cao, Z. and Herget, J. (2003). Palaeohydraulics of extreme flood events: Reality and myth. In *Palaeohydrology: Understanding Global Change*, eds. K. J. Gregory and G. Benito. Chichester, UK: J. Wiley & Sons, pp. 325–336.

Carling, P. A., Williams, J. J., Croudace, I. and Amos, C. L. (2009). Formation of mudridge and runnels in the intertidal zone of the Severn Estuary, UK. *Continental Shelf Research*, doi:10-1016/j.csr.2008.12.009.

Carr, M. H. and Malin, M. C. (2000). Meter-scale characteristics of Martian channels and valleys. *Icarus*, **146**, 366–386.

Chapman, M. G., Gudmundsson, M. T., Russell, A. J. and Hare, T. M. (2003). Possible Juventae Chasma subice volcanic eruptions and Maja Valles ice outburst floods on Mars: implications of Mars Global Surveyor crater densities, geomorphology, and topography. *Journal of Geophysical Research*, **108** (E10), 2–1, CiteID 5113, doi:10.1029/2002JE02009.

Christensen, P. R. and 21 co-authors (2003). Morphology and composition of the surface of Mars: Mars Odyssey THEMIS results. *Science* 300(5628), 2056–2061, doi:10.1126/science.1080885.

Chuang, F. C., Beyer, R. A., McEwen, A. S. and Thomson, B. J. (2007). HiRISE observations of slope streaks on Mars. *Geophysical Research Letters*, **34**, L20204, doi:10.1029/2007GL031111.

Church, M. (1999). Sediment sorting in gravel-bed rivers, *Journal of Sedimentary Research*, **69** (1), 20.

Church, M. (2006). Bed material transport and the morphology of alluvial river channels. *Annual Review of Earth and Planetary Science*, **34**, 325–354.

Clague, J. J. and Rampton, V. N. (1982). Neoglacial Lake Alsek. *Canadian Journal of Earth Sciences*, **22**, 1492–1502.

Clarke, G. K. C., Leverington, D. W., Teller, J. T. and Dyke, A. S. (2004). Paleohydraulics of the last outburst flood from glacial Lake Agassiz and the 8200 BP cold event. *Quaternary Science Reviews*, **23**, 389–407.

Costa, J. E. (1984). The physical geomorphology of debris flows. In *Developments and Applications of Geomorphology*, eds. J. E. Costa and P. J. Fleisher. New York: Springer-Verlag, pp. 268–317.

Costa, J. E. and O'Connor, J. E. (1995). Geomorphological effective floods. In *Natural and Anthropogenic Influences in Fluvial Geomorphology*, eds. J. E. Costa, A. J. Miller, K. W. Potter and P. R. Wilcock. Geophysical Monograph 89, pp. 45–56.

Dade, W. B. (2000). Grain size, sediment transport and alluvial channel pattern. *Geomorphology*, **35**, 119–126.

Dade, W. B. and Friend, P. F. (1998). Grain-size, sediment transport regimes, and channel slope in alluvial rivers. *Journal of Geology*, **106**, 661–675.

Denton, G. H., Prentice, M. I., Kellog, D. E. and Kellog, T. B. (1984). Late Tertiary history of the Antarctic ice sheet: evidence from the Dry Valleys. *Geology*, **12**, 263–267.

Druitt, T. H. (1998). Pyroclastic density currents. In *The Physics of Explosive Volcanic Eruptions*, eds. J. S. Gilbert and R. S. J. Sparks. London: Geological Society, Special Publication 145, pp. 145–182.

Duller, R. A., Mountney, N. P., Russell, A. J. and Cassidy, N. C. (2008). Architectural analysis of a volcaniclastic jökulhlaup deposit, southern Iceland: sedimentary evidence for supercritical flow. *Sedimentology*, **55**, 939–964.

Evans, I. S. (2003). Scale-specific landforms and aspects of the land surface. In *Concepts and Modelling in Geomorphology: International Perspectives*, eds. I. S. Evans, R. Dikau, E. Tokunaga, H. Ohmori and M. Hirano. Tokyo: Terrapub, pp. 61–84.

Evans, D. J. A., Rea, B. R., Hiemstra, J. F. and ó Cofaigh, C. (2006). A critical assessment of subglacial mega-floods: a case study of glacial sediments and landforms in south-central Alberta, Canada. *Quaternary Science Reviews*, **25**, 1638–1667.

Fahnestock, R. K. and Bradley, W. C. (1973). Knik and Matanuska rivers, Alaska: a contrast in braiding. In *Fluvial Geomorphology*, ed. M. Morisawa. London: Allen & Unwin, pp. 220–250.

Gaidos, E. and Marion, G. (2003). Geological and geochemical legacy of a cold, early Mars. *Journal of Geophysical Research*, **108** (E6), 5005, doi:10.1029/2002JE002000.

Ghatan, G. J., Head, J. W. and Wilson, L. (2005). Mangala Valles, Mars: assessment of early stages of flooding and downstream flood evolution. *Earth Moon Planets*, **96**, (1–2), 1–57, doi:10.1007/s11038-005-9009-y.

Gomez, B., Rosser, B. J., Peacock, D. H., Hicks, D. M. and Palmer, J. A. (2001). Downstream fining in a rapidly aggrading gravel bed river. *Water Resources Research*, **37** (6), 1813–1823.

Grant, G. E. (1997). Critical flow constrains flow hydraulics in mobile-bed streams: a new hypothesis. *Water Resources Research*, **33**, 349–358.

Grosswald, M. G. (1999). *Cataclysmic Megafloods in Eurasia and the Polar Ice Sheets*. Moscow: Moscow Scientific World (in Russian).

Herget, J. (2005). *Reconstruction of Pleistocene Ice-Dammed Lake Outburst Floods in the Altai Mountains, Siberia*. Geological Society of America Special Paper 386.

Hirschboeck, K. K. (1988). Flood hydroclimatology. In *Flood Geomorphology*, eds. V. R. Baker, R. C. Kochel and P. C. Patton. London: Wiley, pp. 27–49.

Irwin, R. P, Howard, A. D. and Maxwell, T. A. (2004). Geomorphology of Ma'adim Vallis, Mars and associated paleolake basins. *Journal of Geophysical Research*, **109**, E12009, doi:10.1029/2004JE002287.

Jackson, R. G. (1975). Hierarchical attributes and a unifying model of bedforms composed of cohesionless sediment and produced by shearing flow. *Geological Society of America Bulletin*, **86**, 1523–1533.

Jacob, R. J., Bluck, B. J. and Ward, J. D. (1999). Tertiary-age diamondiferous fluvial deposits of the Lower Orange River valley, southwestern Africa. *Economic Geology*, **94**, 749–758.

Johnsen, T. F. and Brennand, T. A. (2004). Late-glacial lakes in the Thompson Basin, British Columbia: paleogeography and evolution. *Canadian Journal of Earth Sciences*, **41**, 1367–1383.

Kale, V. S. and Hire, P. S. (2007). Temporal variations in the specific stream power and total energy expenditure of a monsoonal river: The Tapi River, India. *Geomorphology*, **92**, 134–146.

Karcz, I and Hersey, D. (1980). Experimental study of free-surface flow instability and bedforms in shallow flows. *Sedimentary Geology*, **27**, 263–300.

Kennedy, J. F. (1963). The mechanics of dunes and antidunes in erodible-bed channels. *Journal of Fluid Mechanics*, **16**, 521–544.

Komar, P. D. (1979). Comparisons of the hydraulics of water flows in Martian outflow channels with flows of similar scale on Earth. *Icarus*, **37**, 156–181.

Komar, P. D. (1980). Modes of sediment transport in channelized water flows with ramifications to the erosion of theMartian outflow channels. *Icarus*, **42**, 317–329.

Komar, P. D. (1983). Shapes of streamlined islands on Earth and Mars: experiments and analyses of the minimum-drag form. *Geology*, **11**, 651–654.

Koster, E. H. (1978). *Fluvial Sedimentology*, Canadian Society of Petroleum Geologists Memoir, 5, pp. 161–186.

Levson, V. M. and Giles, T. R. (1990). Stratigraphy and geological settings of gold placers in the Cariboo mining district. *Geological Fieldwork*, Paper 1991–1, 331–352.

Lord, M. L. and Kehew, A. E. (1987). Sedimentology and paleohydrology of glacial-lake outburst deposits in southeastern Saskatchewan and northwestern North Dakota. *Geological Society of America Bulletin*, **99**, 663–673.

Maizels, J. (1997). Jökulhlaup deposits in proglacial areas. *Quaternary Science Reviews*, **16**, 793–819.

Malde, H. E. (1968). The catastrophic late Pleistocene Bonneville Flood in the Snake River Plain, Idaho. *U.S. Geological Survey professional paper* 596, 53pp.

Malin, M. C. (1986). Rates of geomorphic modification in ice-free areas southern Victoria Land, Antarctica. *Antarctic Journal of the United States*, **20** (5), 18–21.

Malin, M. C., Edgett, K. S., Posiolova, L. V., McColley, S. M. and Noe Dobrea, E. Z. (2006). Present-day impact cratering rate and contemporary gully activity on Mars. *Science*, **314**, 1573–1577, doi:10.1126/science.1135156.

Marren, P. M. (2005). Magnitude and frequency in proglacial rivers: a geomorphological and sedimentological perspective. *Earth-Science Reviews*, **70**, 203–251.

Marren, P. M., Russell, A. J. and Knudsen, Ó. (2002). Discharge magnitude and frequency as a control on proglacial fluvial sedimentary systems. In *The Structure, Function and Management Implications of Fluvial Sedimentary Systems*, eds. F. Dyer, M. C. Thoms and J. M. Olley. IAHS Publication Vol. 276, pp. 297–303.

McEwen, A. S., Hansen, C. J., Delamere, W. A. *et al.* (2007). A closer look at water-related geologic activity on Mars. *Science*, **317**, 1706–1709, doi:10.1126/science.1143987.

Meinzer, O. E. (1918). The glacial history of Columbia River in the Big Bend Region. *Journal of the Washington Academy of Science*, **8**, 411–412.

Miller, A. J. (1995). Valley morphology and boundary conditions influencing spatial patterns of flood flows. In *Natural and Anthropogenic Influences in Fluvial Geomorphology*, eds. J. E. Costa, A. J. Miller, K. W. Potter and P. R. Wilcock. Geophysical Monograph, 89, pp. 57–81.

Montgomery, D. R., Hallet, B., Yuping, L. *et al.* (2004). Evidence for Holocene megafloods down the Tsangpo River gorge, southeastern Tibet. *Quaternary Research*, **62**, 201–207.

Munro-Stasiuk, M. J. and Shaw, J. (1997). Erosional origin of hummocky terrain, south-central Alberta, Canada. *Geology*, **25**, 1027–1030.

Nott, J. and Price, D. (1994). Plunge pools and paleoprecipitation. *Geology*, **22**, 1047–1050.

O'Connor, J. E. (1993). *Hydrology, Hydraulics and Geomorphology of the Bonneville Flood.* Special Paper Geological Society of America, 274.

O'Connor, J. E., Grant, G. E. and Costa, J. E. (2002). The geology and geography of floods. *Water Science and Applications*, **5**, 359–385.

Pacifici, A. (2009). The Argentinean Patagonia and the Martian landscape. *Planetary and Space Science*, doi:10.1016/j.pss.2008.11.006.

Paola, C., Heller, P. L. and Angevine, C. L. (1992). The large-scale dynamics of grain-size variation in alluvial basins, 1: theory. *Basin Research*, **4**, 73–90.

Papp, F. (2002). Extremeness of extreme floods. In *The Extreme of the Extremes: Extraordinary Floods*, eds. Á. Snorasson, H. P. Finnsdóttir and M. Moss. IAHS Special Publication, 271, pp. 373–378.

Pardee, J. T. (1942). Unusual currents in glacial lake Missoula, Montana. *Geological Association of America Bulletin*, **53**, 1569–1600.

Parker, T. J. and Rice Jr., J. W. (1997). Sedimentary geomorphology of the Mars Pathfinder landing site. *Journal of Geophysical Research*, **102** (E11), 25,641–25,656.

Phillips, C. B., Burr, D. M. and Beyer, R. A. (2007). Mass movement within a slope streak on Mars. *Geophysical Research Letters*, **34**, L21202, doi: 10.1029/2007GL031577.

Pollard, A., Wakarani, N. and Shaw, J. (1996). Genesis and morphology of erosional shapes associated with turbulent flow over a forward-facing step. In *Coherent Flow Structures in Open Channels*, eds. P. J. Ashworth, S. J. Bennett, J. L. Best and S. J. McLelland. Chichester: Wiley, pp. 249–265.

Reddering, J. and Eserhuysen, K. (1987). The effects of river floods on sediment dispersal in small estuaries: a case study from East London. *Suid-Afrikaanse Tydskrif vir Geologie*, **90** (4), 458–470.

Rice, J. W., Christensen, P. R. , Ruff, S. W. and Harris, J. C. (2003). Martian fluvial landforms: a THEMIS perspective after one year at Mars. *Lunar and Planetary Science*, XXXIV (abstract 2091).

Rice, J. W., Russell, A. J. and Knudsen, Ó. (2002a). Paleohydraulic interpretation of the Modrudalur transverse ribs: comparisons with Martian outflow channels, Abstract. In *The Extreme of the Extremes: Extraordinary Floods*, eds. Á. Snorasson, H. P. Finnsdóttir and M. Moss. International Association of Hydrological Sciences Red Book Publication 271.

Rice, J. W., Parker, T. J., Russell, A. J. and Knudsen, Ó. (2002b). Morphology of fresh outflow channel deposits on Mars. *Lunar and Planetary Science*, XXXIII [CD-ROM], 2026 (abstract).

Rudoy, A. N. and Baker, V. R. (1993). Sedimentary effects of cataclysmic Late Pleistocene glacial outburst flooding, Altay Mountains, Siberia, *Sedimentary Geology*, **85**, 53–62.

Russell, A. J. (2005). Catastrophic floods. In *Encyclopedia of Geology*, eds. R. C. Selley, L. R. M. Cocks and I. R. Primer. Oxford: Elsevier, pp. 628–641.

Russell, A. J. and Knudsen, Ó. (1999). Controls on the sedimentology of the November 1996 jökulhlaup deposits, Skeiðarársandur, Iceland. In *Fluvial Sedimentology*, Vol. VI, eds. N. D. Smith and J. Rogers J. Special Publication Number 28 of the International Association of Sedimentologists. Oxford: Blackwell, pp. 315–329.

Russell, A. J. and Knudsen, Ó. (2002). The effects of glacier-outburst flood flow dynamics and ice-contact deposits, Skeiðarársandur, Iceland. In *Flood and Megaflood Processes and Deposits: Recent and Ancient Examples*, eds. I. P. Martini, V. R. Baker and G. Garsón. Special Publication Number 32 of the International Association of Sedimentologists. Oxford: Blackwell, pp. 67–83.

Russell, A. J., Marren, P. (1999). Proglacial fluvial sedimentary sequences in Greenland and Iceland: a case study from active proglacial environments subject to jökulhlaups. In *QRA Technical Guide Number 7*, eds. A. P. Jones, M. E. Tucker and J. K. Hart. London: Quaternary Research Association, pp. 171–20.

Schoeneich, P. and Maisch, M. (2003a). Témoins géomorphologiques de mégacrues et de débâcles dans les Alpes. *Communication aux journées Nivologie-Glaciologie de la Société Hydrotechnique de France*, 16–17 March 2003, Grenoble.

Schoeneich, P. and Maisch, M. (2003b). A late glacial megaflood in the Swiss Alps: the outburst of the great lake of Davos. *3rd International Paleoflood Workshop*, 1–8 August 2003, Hood River, Oregon.

Seal, R., Paola, C., Parker, G., Southard, J. B. and Wilcock, P. R. (1997). Experiments on downstream fining of gravel. 1: Narrow flume channel runs. *Journal of Hydraulic Engineering*, **123**, 874–884.

Seu, R., Biccari, D., Orosei, R. *et al.* (2004). SHARAD: The MRO 2005 shallow radar. *Planetary and Space Science*, **52** (1–3), 157–166, doi:10.1016/j.pss.2003.08.024.

Shaw, J. and Kellerhals, R. (1977). Paleohydraulic interpretation of antidune bedforms with applications to antidunes in gravel. *Journal of Sedimentary Petrology*, **47**, 257–266.

Smith, L. N. (2006). Stratigraphic evidence for multiple drainings of glacial Lake Missoula along the Clark Fork River, Montana, USA. *Quaternary Research*, **66**, 311–322.

Squyres, S. W. and 53 co-authors (2006). Overview of the Opportunity Mars Exploration Rover Mission to Meridiani Planum: Eagle Crater to Purgatory Ripple. *Journal of Geophysical Research*, **111** (E12), CiteID E12S12. doi:10.1029/2006JE002771.

Sridhar, A. (2007). A mid-late Holocene flood record from the alluvial reach of the Mahi River, Western India. *Catena*, **70**, 330–339

Teller, J. T. (2004). Controls, history, outbursts, and impact of large late-Quaternary proglacial lakes in North America. In *The Quaternary Period in the United States, INQUA Anniversary Volume*, eds. A. Gilespie, S. Porter and B. Atwater. Elsevier, pp. 45–6.

Tinkler, K. J. (1997a). Critical flow in rockbed streams with estimated values of Manning's *n*. *Geomorphology*, **20**, 147–164.

Tinkler, K. J. (1997b). Indirect velocity measurement from standing waves in rockbed rivers. *Journal of Hydraulic Engineering*, **123**, 918–921.

Valentine, G. A. (1987). Stratified flow in pyroclastic surges. *Bulletin of Volcanology*, **50**, 352–355.

Waitt, R. B. (2002). Great Holocene floods along Jökulsá á Fjöllum, north Iceland. *Special Publications International Association of Sedimentologists*, **32**, 37–51.

Whittaker, J. G. and Jaeggi, M. N. R. (1982). Origin of step-pool systems in mountain streams. *American Society of Civil Engineers, Proceedings, Journal of the Hydraulics Division*, **108**, 758–773.

Williams, J. J., Carling, P. A., Amos, C. L. and Thompson, C. (2008). Field investigation of ridge-runnel dynamics on an intertidal mudflat, *Estuarine, Coastal and Shelf Science*, **79**, 213–229.

Woo, H. S., Julien, P. Y. and Richardson, E. V. (1986). Washload and fine sediment load. *Journal of Hydraulic Engineering*, **112**, 541–545.

Zielinski, T. (1982). Contemporary high-energy flows, their deposits and reference to the outwash depositional model. *Geologia*, **6**, 98–108 (in Polish with extended English abstract).

4 Jökulhlaups in Iceland: sources, release and drainage

HELGI BJÖRNSSON

Summary

Jökulhlaups in Iceland may originate from marginal or subglacial sources of water melted by atmospheric processes, permanent geothermal heat or volcanic eruptions. The release of meltwater from glacial lakes can take place as a result of two different conduit initiation mechanisms and the subsequent drainage from the lake occurs by two different modes. Drainage can begin at pressures lower than the ice overburden in conduits that expand slowly over days or weeks. Alternatively, the lake level may rise until the ice dam is lifted and water discharges rise linearly, peaking in a time interval of several hours to 1–2 days. The linearly rising floods are often associated with large discharges and floods following rapid filling of subglacial lakes during subglacial eruptions or dumping of one marginal lake into another. Jökulhlaups during eruptions in steep ice and snow-covered stratovolcanoes are swift and dangerous and may become lahars and debris-laden floods. The jökuhlaups can be seen as modern analogues of past megafloods on the Earth and their exploration may improve understanding of ice–volcano processes on other planets.

4.1 Introduction

The Icelandic word jökulhlaups means glacier-related floods ranging from small bursts to megafloods of enormous landscaping impact. They may originate from marginal or subglacial sources of water melted by atmospheric processes, permanent geothermal heat or volcanic eruptions. They may range from floods consisting almost entirely of water to hyperconcentrated fluid–sediment mixtures and even more destructive gravity-driven mass flow of mostly volcanic materials mixed with water and ice. They have created large Pleistocene river canyons, and have transported and deposited enormous amounts of sediments and icebergs over vast outwash plains, which in Iceland are called 'sandurs'. The jökulhlaups have threatened human populations and farms, damaged cultivated and vegetated areas, disrupted roads on the alluvial plains surrounding the ice cap, and even generated flood waves in coastal waters. Debris transported by jökulhlaups is found in marine sediments.

Studies of modern outburst floods in Iceland, whether they are small or large, contribute to analyses of the conditions and processes that dictate the styles, flow morphologies, release mechanisms, scales and characteristics of glacier outburst floods in general. These jökulhlaups can be seen as modern analogues of past megafloods on the Earth and their exploration may improve understanding of ice–volcano processes on other planets.

The goal of this chapter is to outline and describe the following: (1) the location and geometry of the various jökulhlaup sources; (2) the production and accumulation of water leading to outbursts and the conditions in which they begin; (3) the mechanisms and discharge characteristics of outbursts; and (4) case histories of present day jökulhlaups in Iceland. The chapter does not deal with floods occurring by moraine-dam failure which are discussed in O'Connor and Beebee (this volume Chapter 8).

4.2 Subglacial lakes

Powerful subglacial hydrothermal systems may be generated by interaction of glacial meltwater with magma intrusions at shallow depths in the crust of the Earth. Magma chambers are found at 2 to 3 km depth under the glacier at several of the central volcanoes under the ice caps Vatnajökull and Mýrdalsjökull (Figure 4.1) as witnessed by persistent earthquake activity associated with their alternating deflation and inflation (Einarsson, 1991; Einarsson and Brandsdóttir, 1984; Alfaro *et al.*, 2007). Water percolates in a permeable flood basalt down to magma intrusions at 1200 °C. Upon heating, it rises up to the glacier base where ice is melted and the water sinks again, setting up a circulation of hydrothermal fluid. The circulating fluid, which is probably liquid dominated, attains a base temperature of 300 to 340 °C and obtains its chemical composition by interaction with the ambient rocks (Björnsson and Kristmannsdóttir, 1984). The fluid is probably at the boiling point when it is injected as water and steam through numerous vents scattered over the glacier bed. The rising geothermal fluid starts to boil when the fluid reaches a level where the surrounding hydrostatic pressure is lower than the steam pressure of the fluid. A geothermal fluid at 300 °C has a steam pressure of 86 bars and will thus start to boil when it has ascended sufficiently to achieve a hydrostatic pressure below 86 bars. From there on the fluid will cool itself by boiling so that the steam pressure is always in equilibrium with the hydrostatic pressure. At 40 bars overburden the corresponding boiling temperature is 250 °C (Schmidt, 1979). The resulting heat flow

Megaflooding on Earth and Mars, ed. Devon M. Burr, Paul A. Carling and Victor R. Baker.
Published by Cambridge University Press.

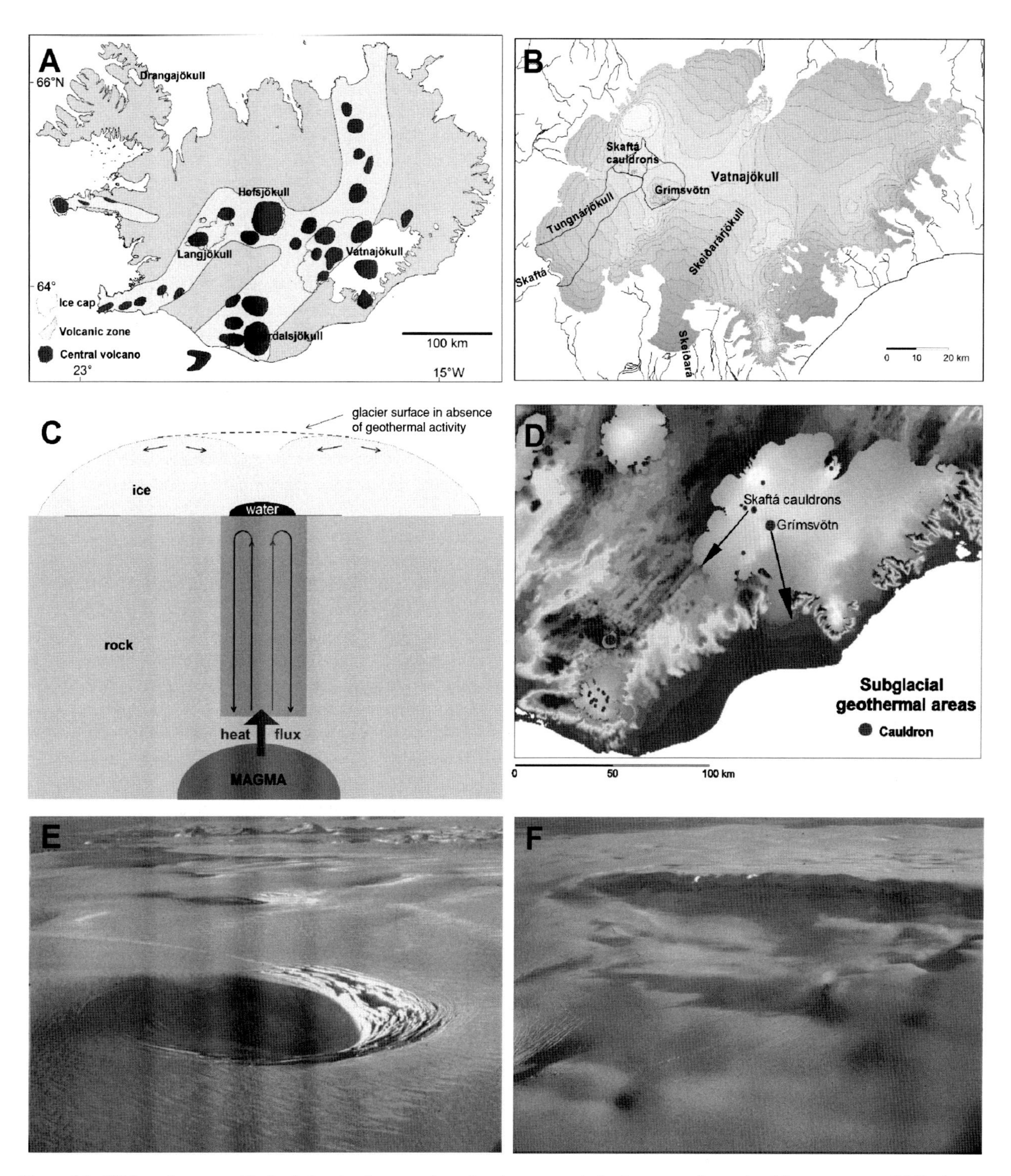

Figure 4.1. (A) Location map of Iceland showing ice caps, the volcanic zone, central volcanoes and associated fissure swarms. (B) Drainage basins of Grímsvötn and Skaftá cauldrons. (C) Schematic cross-section of an ice cap and a subglacial hydrothermal area. (D) Location of subglacial geothermal systems and lakes in Vatnajökull and Mýrdalsjökull. (E) The 150 m deep and 3 km wide eastern Skaftá cauldron, just after a jökulhlaup in January 1982. The crevasse trending from the upper left of the cauldron suggests lifting of the ice dam. (Photo: H. Björnsson.) (F) An oblique photograph of Grímsvötn. View from the N toward the 400 m high southern caldera rim (Mt Grímsfjall). (Photo: H. Björnsson.)

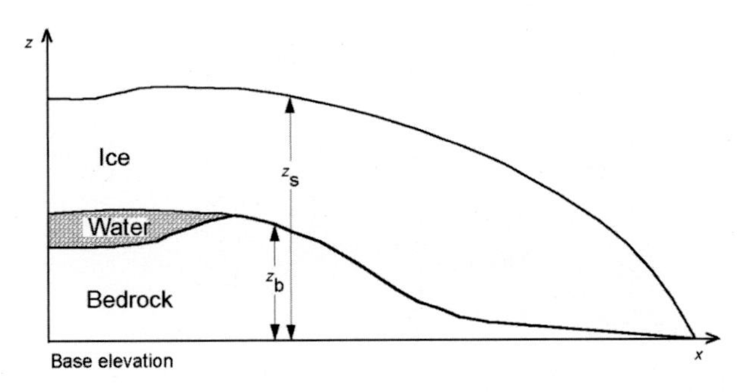

Figure 4.2. Parameters for describing the principles of water flow in glaciers.

continuously melts ice into water at the glacier bed (Figure 4.1C). The meltwater either drains toward the glacier margin or accumulates in subglacial lakes. Thermal plumes ascend into subglacial water bodies and convection brings heat up from the floor to the melting ice. A depression is created in the glacier surface under which the meltwater may accumulate in a lake which is sealed by high ice overburden pressure on a rim around a depression (Figure 4.1D–E). The lakes are situated at positions (Figure 4.2) where water flows toward a minimum in the total fluid potential $\varphi_b = \rho_w g z_b + p_w$. Approximating the basal water pressure (p_w) with the overburden pressure from the ice ($p_i = \rho_i g h_i$) the gradient driving the water can be described (Shreve, 1972) as

$$\nabla\varphi_b = (\rho_w - \rho_i) g \nabla z_b + g \rho_i \nabla z_s. \qquad (4.1)$$

The state in which there is no gradient driving water inside the lake, $\nabla\varphi_b = 0$, can be used to define the location and geometry of a subglacial reservoir. Hence,

$$\nabla z_b = -\rho_i/(\rho_w - \rho_i)\nabla z_s \qquad (4.2)$$

describes the relationship of the slope in the ice/water outline of the lake to the slope at the upper surface of the glacier.

When ice-dammed lakes, regardless of whether they are positioned subglacially or marginally, receive water inflow and gradually expand, basal water pressure will increase and the overlying ice will be raised. Eventually the hydraulic seal of the ice dam will be opened, either suddenly by a rupture or slowly and gradually, and the flood starts, which can be confined to tunnels or dispersed in a sheet at the glacier bed. Seepage starting beneath the ice blockage causes enlargement of the drainage system, initiating a flood under the surrounding ice. The flood could be terminated before the lake is empty as the overlying ice may collapse abruptly into the tunnel and seal the lake again, with water beginning to accumulate again until another jökulhlaup occurs. In general, the frequency of jökulhlaups and the volume of water released depend upon the thickness of the ice barrier.

Jökulhlaups regularly drain six subglacial lakes beneath Vatnajökull, of which Grímsvötn and the Skaftá cauldrons are typical examples (Figure 4.1D, E, F; Thorarinsson, 1953, 1974; Björnsson, 1974, 1977, 1988, 2002). About 15 small cauldrons are at present found on Mýrdalsjökull (Björnsson *et al.*, 2000).

4.2.1 Grímsvötn Lake

Geometry, mass and energy balance At Grímsvötn a 10 km wide and 300 m deep depression has been created in the Vatnajökull ice cap surface (Figures 4.1 and 4.3). The extent of the subglacial Grímsvötn lake is identified by the flat floating ice shelf and the abrupt change in surface slope at the margins. Subglacial topography is known from radio echo-soundings and seismic profiling (Björnsson, 1988; Guðmundsson, 1989, 1992). The lake is situated within the caldera floor of the Grímsvötn central volcano and to the south and west the caldera walls (Mt Grímsfjall) protrude through the glacier surface and confine the lake, but the lake can expand to the north and northeast as the water level rises.

The hydrothermal vents in Grímsvötn are mainly located at ring fractures along the caldera rims that provide open channels for vertical flow of water. On the surface of the glacier the geothermal activity is expressed by small cauldrons over the caldera rims. Downward percolation of water takes place at the cooler parts of the bed between the hydrothermal vents. The chemical composition of the jökulhlaup water (silica content) and assumptions about the likely temperature in the geothermal reservoir signify that about 15% of the total mass in the lake is fluid discharged from the geothermal reservoir that circulates in a closed system beneath the caldera (Sigvaldason, 1965; Björnsson and Kristmannsdóttir, 1984).

The meltwater is cooled down to temperatures close to the melting point before entering the caldera lake. Further, the lake-water density (calculated from chemical measurements), increasing with depth, inhibits thermal convection. Chemical analyses show that the concentration of magnesium in water from Grímsvötn is similar to that in cold groundwater (Björnsson and Kristmannsdóttir, 1984). On the other hand, temporarily the lake temperature may reach well above the melting point due to inflow of meltwater produced during eruptions. This scenario happened during the 1996 Gjálp eruption (see later) and may have occurred in 1938 when a similar subglacial eruption also took place north of the lake.

The ice-shelf thickness has been measured regularly by radio echo-soundings and computed from measured surface elevations, assuming the shelf is floating in hydrostatic equilibrium. The total volume of water drained out of Grímsvötn in jökulhlaups has been derived using known variations in lake level during jökulhlaups, the thickness

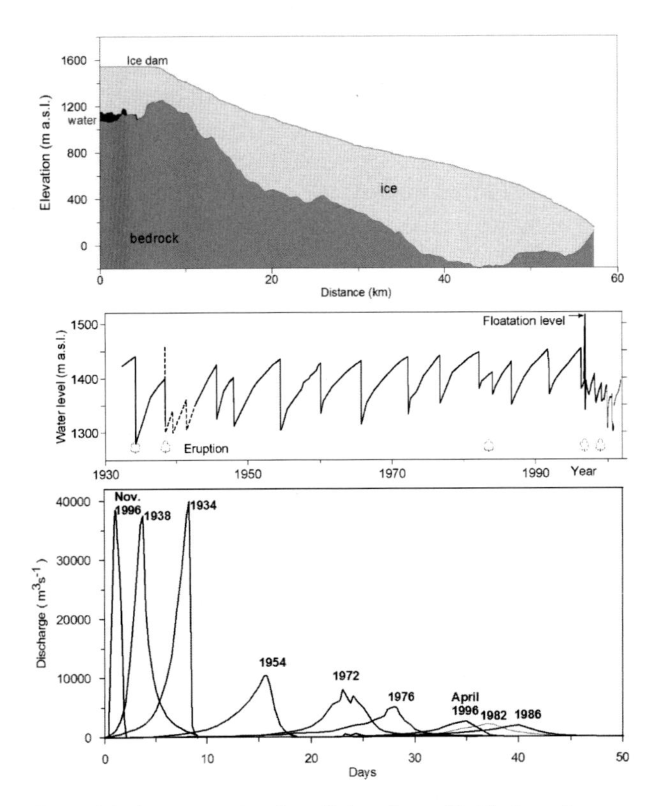

Figure 4.3. A cross-section from Grímsvötn to Skeiðarársandur along the flowpath of jökulhlaups (upper). The lake level of Grímsvötn, 1930–2000 (middle). In 1996 the lake rose to the level required for flotation of the ice dam. Hydrographs of jökulhlaups from Grímsvötn (1934, 1938, 1954, 1976, 1982, 1986 and 1996; lower).

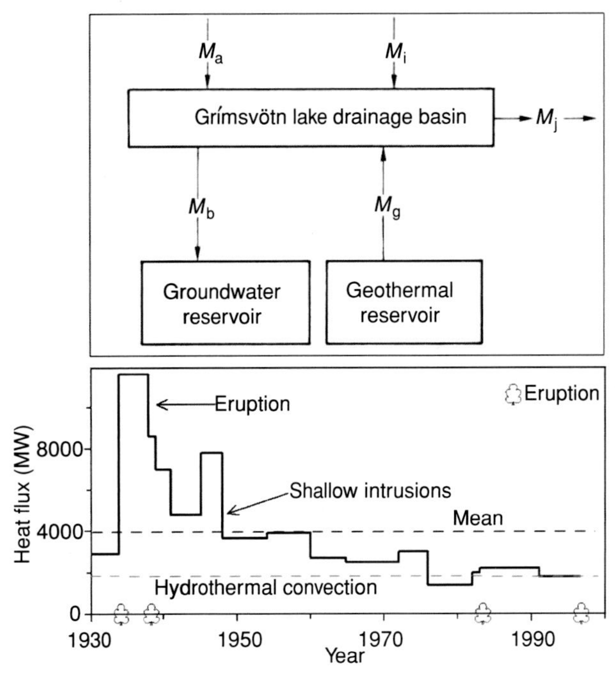

Figure 4.4. Schematic flow chart of the mass balance of the Grímsvötn lake drainage basin (upper). Heat flux of the Grímsvötn caldera 1922–91 (lower) (Björnsson and Guðmundsson, 1993).

of the floating ice cover on the lake and the bedrock topography in the Grímsvötn area (Björnsson, 1988, 1992; Björnsson and Guðmundsson, 1993). The ice shelf gradually thickened from 150 m in the 1950s to 230–260 m in the 1980s due to reduced melting by the hydrothermal system. Hence, the extent and volume of the subglacial lake has gradually been reduced.

Data on the mass balance of the Grímsvötn drainage basin enable an estimation of the heat output of the subglacial geothermal area by using the lake as a natural calorimeter (Figures 4.3 and 4.4). For a long-term average it is assumed that the geothermal reservoir is recharged from the groundwater basin at the same mass rate as geothermal fluid enters the lake, $M_g = M_b$. Hence, the heat release from the subglacial geothermal system, melting the ice (M_i) in excess of the climatically induced melting (M_a), may be derived from *in situ* measurements of the mass of meltwater accumulated in the caldera lake (M_j). Thus, assuming long-term equilibrium, the mass balance of the lake can be expressed as

$$M_a + M_i + M_g = M_b + M_j. \tag{4.3}$$

Heat is brought into the lake principally through the mass flow of geothermal fluid (M_g) although some heat is conducted through the lake floor. The heat flux

$$\psi_G = M_i L_i + (M_j - M_g) c_w (T_l - T_0) \tag{4.4}$$

melts the mass M_i of ice and warms the lake water ($M_j - M_g$) from the melting temperature (T_0) to the average lake temperature (T_l). Except during brief episodes of volcanic eruptions, the observed Grímsvötn bulk lake temperature (T_l) is close to melting point; hence $\psi_G = M_i L_i$.

According to the Grímsvötn Lake calorimeter, the heat released by magma (Figure 4.4) over the period 1922 to 1991 was $(8.1 \pm 1.6) \times 10^{18}$ J, equivalent to the energy released by the solidification and cooling of $2.1 \pm 0.4\,\mathrm{km}^3$ of basaltic magma from $T_{mi} = 1200\,°\mathrm{C}$ to $T_{mf} = 200\,°\mathrm{C}$; i.e., $E_{mt} = \rho_m V_m \{L_m + c_{pm}(T_{mi} - T_{mf})\}$. The overall fluctuations in the heat flux are closely related to volcanic activity and are dominated by a main pulse of 11 600 MW, caused by a major eruption in 1938, gradually declining to 1600 MW in 1976–82. Heat extracted from the roof of a magma chamber, with the aid of hydrothermal convection, may have given a basic contribution to the heat flux of 1500 to 2000 MW. The mass and energy balances require that steam is 20–35% (by mass) of the geothermal fluid that enters the lake (Björnsson and Kristmannsdóttir, 1984). About 45–60% of the total thermal power toward the base of the glacier is transported by steam and the rest by water.

The contribution to the total heat flux was 45% (maximum) from a magma chamber, 35% (minimum) from shallow intrusions, and 20% was released from magma erupted at the base of the glacier. Hence, 80% is melted by heat drawn from magma with the aid of hydrothermal convection. Altogether, magma of the order of 0.4 km^3 (assuming $\rho_m = 2.65 \times 10^3$ kg m^{-3}) was discharged in the eruptions of 1922, 1934, 1938 and 1983 (Björnsson and Guðmundsson, 1993) which is about 20% of the total magma volume calculated. Hence, 80% (1.7 ± 0.4 km^3), or on the average 2.5×10^7 m^3 a^{-1}, have solidified and cooled as intrusive rocks in the crust. This implies that magma at the roof of a deep chamber solidified and cooled at the rate of 1.2–1.6×10^7 m^3 a^{-1} or about 1 km^3 over the period 1922–91. Björnsson *et al.* (1982) have suggested that the observed heat flux can be explained by penetration of water into the hot boundaries of a magma body at shallow depths. Assuming an upper surface area of 10 km^2 for the magma body under Grímsvötn, water penetrating into that body would have to propagate at an average rate of 4 m a^{-1} to yield the observed flux of 4000 MW. The area where signs of elevated heat flux have persisted throughout this period is possibly 50–60 km^2. Hence, the average flux density over the period 1960–91 was of the order of 40 W m^{-2}. The volcanic material that is erupted straight into the atmosphere without contact with ice (estimated 1.5×10^6 m^3 a^{-1}) causes negligible melting and is therefore not detected by the Grímsvötn calorimeter.

Increase in thermal output has frequently been reported for some months after a jökulhlaup (Björnsson, 1988), presumably due to explosive boiling caused by pressure release during the fast lowering of the lake level. The overburden pressure on top of the hydrothermal system drops from 40 bar to typically 25 bar and the boiling water temperature drops from 250 °C to 225 °C (whereas the ambient rock temperature remains unchanged at 250 °C). Explosive boiling will propagate downward in the permeable groundwater reservoir. Hence, the jökulhlaups speed up the cooling of the subglacial magma body. Placing a glacier on top of a shallow magma chamber creates powerful hydrothermal systems. Furthermore, pressure release subsequent to lake subsidence may trigger eruptions by rupturing the roof of shallow magma chambers (Thorarinsson, 1974; Sigmundsson *et al.*, 2004). Seismic tremors suggest that this mechanism may have triggered several small but invisible eruptions under the Skaftá cauldrons after 1985 (Páll Einarsson, personal communication).

Release of lake water In recent years the lake level of Grímsvötn has risen 10–15 m a^{-1} and a jökulhlaup occurs when it rises 80–110 m and reaches a particular threshold (Figure 4.3). The timing of bursts depends on what lake level will provide the subglacial water pressure necessary for breaking the hydraulic seal affected by the overburden pressure at the ice dam. Based on the long-term record of lake levels, the elevation at which a jökulhlaup will begin can be predicted with some precision. In a typical case, though for reasons not yet fully explained, leakage starts before the lake has reached the level that would cause flotation of the ice dam (Björnsson, 1975, 1988; Fowler, 1999). The typical threshold lake level for triggering a jökulhlaup is 60–70 m lower than required for simple flotation of the ice dam. Jökulhlaups from Grímsvötn occur at any time of the year so sudden changes in subglacial drainage due to surface melting do not, in general, trigger jökulhlaups. The onset of lake drainage is marked by ice-quakes and subsidence of the lake level, and the arrival time of lake water to the glacier margin is identified by a sulphurous odour in the glacial river. Throughout the flood the water temperature at the outlet exit has been observed to be at the melting point (Rist, 1955; Björnsson, 1988). Usually, the overlying ice collapses abruptly into the tunnel and seals the lake again.

In one case the lake level has risen until the ice dam floats. This flotation was subsequent to the rapid filling of the lake during the volcanic eruption north of the lake in 1996. In contrast, a few jökulhlaups have occurred from Grímsvötn at lake levels far below the usual threshold. These premature jökulhlaups may have been triggered by the opening of waterways from the lake along the north-eastern slopes of Grímsfjall (higher than the lowest crest at the caldera rim) facilitated by increased localised melting due to hydrothermal or volcanic activity.

Typical discharge of jökulhlaups Jökulhlaups from Grímsvötn have occurred at 1 year to 10 year intervals. Their peak discharge ranges from 600 to $4-5 \times 10^4$ m^3 s^{-1} at the Skeiðarársandur outwash plain, their duration is two days to four weeks, and their total volume at each event is 0.5 to 4.0×10^9 m^3 (Björnsson and Guðmundsson, 1993; Guðmundsson *et al.*, 1995; Björnsson, 1997; Snorrason *et al.*, 1997). The greatest floods peaked in less than one week and subsided in two days, whereas the smaller ones peak in two to three weeks, after which they usually terminate in about a week (Figure 4.3). Jökulhlaups from Grímsvötn flow a distance of some 50 km beneath ice to the glacier terminus. The flow path reaches a depth of 200 m below sea level before it emerges on the outwash plain. The most violent Grímsvötn jökulhlaups flooded the entire Skeiðarársandur, measuring 1000 km^2. All such outburst floods drain through one main ice tunnel, which feeds the river Skeiðará. If discharge exceeds about 3000 m^3 s^{-1}, water starts to drain from other, smaller tunnels at the central part of the terminus. In the most voluminous floods, 10 to 15 high-capacity ice tunnels develop. Occasionally, rapid injection of water creates high water pressures, as is evident from water forced up to the glacier surface through crevasses that formed in 300 m thick ice (Roberts *et al.*, 2000; Björnsson, 2002).

After discharge has begun, pressure from the ice constricts the passageway, and water flow at an early stage in the jökulhlaup correlates primarily with enlargement of the ice tunnel due to heat from friction against the flowing water and to thermal energy stored in the lake. The recession stage of the hydrograph sets in when tunnel deformation begins to exceed enlargement by melting. This development has been explained by classic jökulhlaup theories of flow in a single water-filled tunnel of given roughness located in the glacier base (Röthlisberger, 1972; Nye, 1976; Spring and Hutter, 1981; Clarke, 1982, 2003). The theoretical models are based on the physics of mass conservation, momentum, energy conservation and heat transfer, respectively: frictional energy is generated in the water flow (i.e., the dissipation of potential energy), which is driven by the fluid potential gradient ($\nabla\varphi_b$), and sensible heat is transported from a lake. Thermal energy in the turbulent current is transferred to the tunnel walls melting ice, causing the tunnel to expand in competition with ice closure due to the difference between the water pressure and the ice overburden. The water produced by melting the ice walls is added to the current.

From his general model, Nye (1976) derived an analytical solution, which predicted discharge to rise asymptotically with time as $Q(t) \sim (-1/t)^4$ (where $t = 0$ is chosen as the time when Q becomes infinite), if confinement by the overburden was neglected and expansion of a cylindrical ice tunnel was solely attributed to instantaneous transfer of frictional heat from the flowing water to the enclosing ice. This ignores the present theory of heat transfer within the water flow, which predicts that water draining from the exits of the outlets will have a temperature above the melting point, although the water is observed to be at the melting point. The water flow was assumed to be driven by the fluid potential gradient averaged over the length of the conduit. This assumption gave a good fit to the rising limb of the 1972 hydrograph of Grímsvötn jökulhlaup. Numerical models including the complete set of equations describing the flood along the entire length of the conduit were derived by Spring and Hutter (1981). The best fit to a Grímsvötn hydrograph (for 1972) was obtained by assuming that the temperature of water draining the lake increases from the melting point to 4 °C at the peak discharge and then drops in one day, speeding up the closure of the tunnel. Nonetheless, the lake temperature has repeatedly been measured and found to be close to the melting point and Björnsson (1992) suggested that this simulation may have overestimated advected heat from the lake by underestimating transfer of frictional heat. The rate of heat transfer from the flood water to the surrounding ice is evidently more efficient than suggested by current theory (based on an empirical expression for fluid flow in smooth pipes). Clarke (1982) modified the Nye theory to account for lake temperature and lake geometry and assumed that the evolution of the jökulhlaup was controlled by opening due to melting and creep-closure of a single bottleneck at a given distance from the entrance of the conduit inlet. This model yielded moderately good simulations of some jökulhlaups from Grímsvötn (Björnsson, 1992). In general, the simulated ascending limbs of the hydrograph corresponded to the measured ones (Figure 4.5). The peaks in the computed graphs, however, were not as sharp as the actual climaxes, rendering the simulation of the descending limbs unsatisfactory. Occasionally, jökulhlaup discharge has increased at a rate suggesting that thermal energy stored in the lake contributes to tunnel expansion. The rise of the jökulhlaup in 1934 can be simulated by lake water of 1 °C, and in 1938 by water of 4 °C when a peak of 3×10^5 m^3 s^{-1} was reached in four days (Figure 4.5). Recently Clarke (2003) simulated jökulhlaups from Grímsvötn using a slightly modified form of the Spring–Hutter (1981) equations and pointed out that, as the flood progresses, the location of flow constrictions controlling the flood magnitude may move along the flood path; in the early stages of the flood the bottleneck was located near the conduit outlet and later it jumped to the conduit inlet.

Referring to eleven jökulhlaups from Grímsvötn, Björnsson (1992) described a regression relation $Q_{max} = K_p V_t^b$ between the peak discharge Q_{max} (in terms of m^3 s^{-1}) and the total volume V_t (in 10^6 m^3) of water released from the lake, where $K_p = 4.15 \times 10^{-3}$ s^{-1} m$^{-2.52}$ and $b = 1.84$. The theoretical Nye model has been shown to predict such a power law for the flood culmination, i.e., when the rates of melting by frictional energy and closure balance (Ng and Björnsson, 2003). Clague and Mathews (1973) and Walder and Costa (1996) derived an exponent $b < 1$ when fitting data from many lakes, some of which were drained completely.

Drainage of linearly rising jökulhlaup The jökulhlaup from Grímsvötn in November 1996 was of an extraordinary type. For the first time in observational history the Grímsvötn lake level rose until the ice dam was floated off the bed (Björnsson, 1997, 2002; Björnsson *et al.*, 2001; Jóhannesson, 2002; Flowers *et al.*, 2004). This jökulhlaup was preceded and indirectly triggered by the Gjálp eruption (see later). The meltwater from the eruption collected in the Grímsvötn lake for a month until it drained in a catastrophic jökulhlaup over 4–7 November 1996, in which 3.2 km^3 of water drained from the lake within a period of 40 hours. On 4 November, the lake had risen to the level required for flotation of the ice dam, 1510 m, and ice-quakes marked the onset of lake drainage. About 10.5 hours later water emerged from the margin of Skeiðarárjökull as a flood

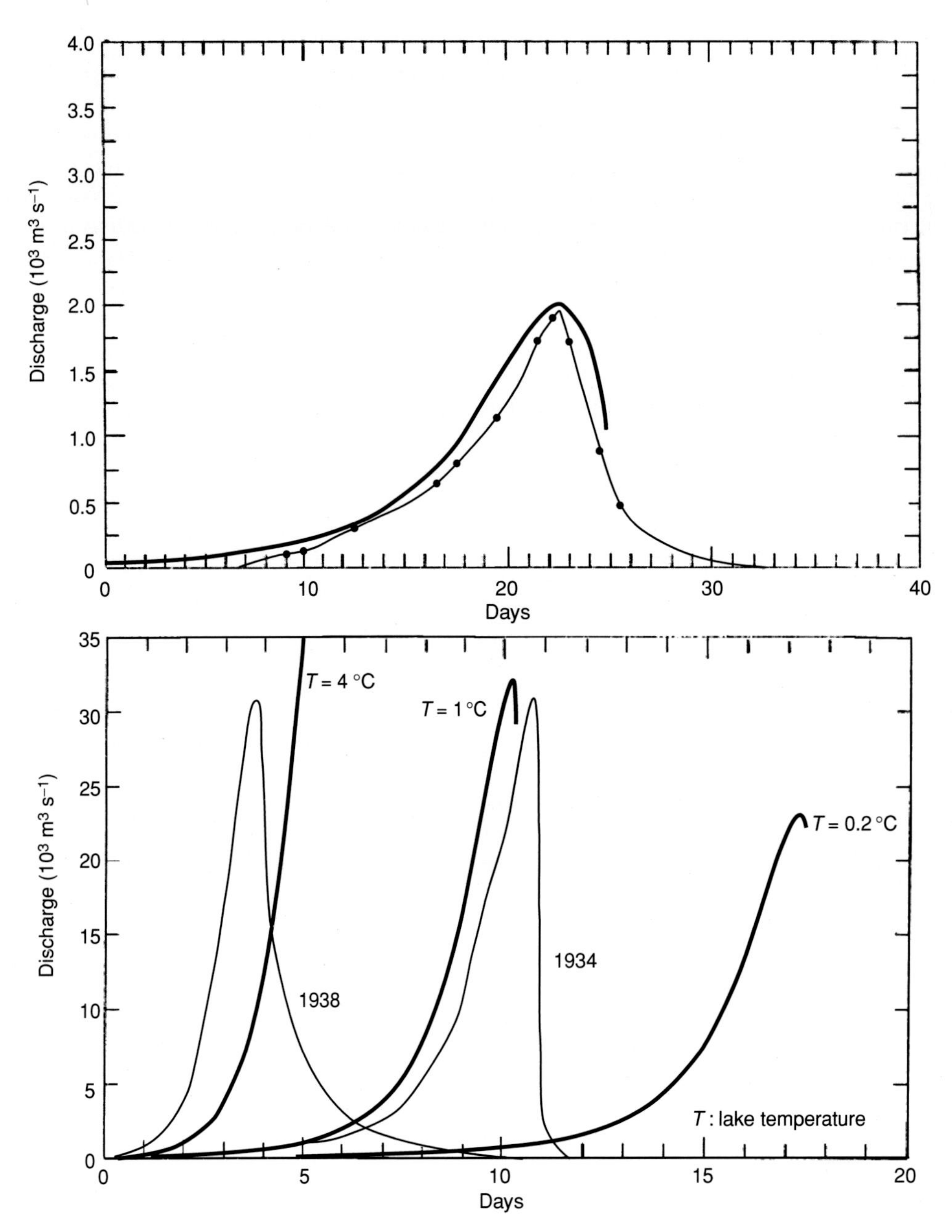

Figure 4.5. Simulation of jökulhlaups from Grímsvötn (Q_L) at the glacier outlet in 1986 (upper), 1934 and 1938 (lower) shown as thin lines. Calculated hydrographs (Q_{out}) for theoretical simulations shown as thick lines. Manning roughness coefficient $n = 0.08\,\mathrm{m}^{-1/3}$ s (Björnsson, 1992).

wave inundating Skeiðarársandur (at 100 m elevation), in the most rapid jökulhlaup ever recorded from Grímsvötn (Figure 4.6). When floodwater started to drain from the glacier margin a volume of 0.6 km^3 of water had accumulated under the glacier. Melting enlargement of the conduit by the frictional heat of the flowing water can only account for a portion (0.01 km^3) of the required conduit volume. The discharge increased faster than can be accommodated by the melting of conduits, and the glacier was lifted along the flow path by water pressure in excess of the overburden as the water forced open space for itself, prior to conduit formation. Longitudinal crevasses were observed above the entire flowpath. Approaching the glacier terminus, basal water burst out on the glacier surface through several hundred metres of ice; the swift flood managed to cause hydrofracturing and force its way englacially from the base of the glacier to its surface (Roberts *et al.*, 2000). Freezing of suspended sediment and formation of frazil

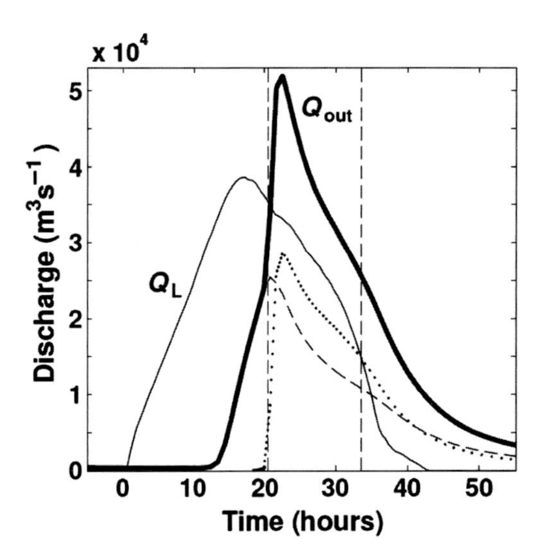

Figure 4.6. The observed discharge hydrograph of the 1996 jökulhlaup from Grímsvötn (Q_L) and the simulated hydrograph on Skeiðarársandur (Q_{out}). Contributions shown from sheet flow (dashed line) and tunnel flow (dotted line) (Flowers *et al.*, 2004).

ice at the glacier terminus indicate that supercooled water exits from depths at the river outlet. Icebergs were broken off the margin and spread over the outwash plain; 15 km of the road across the outwash plain was swept away and the coastal line was advanced seaward by up to 900 m due to sediment transport.

During the lake drainage a 6 km long, 1 km wide and 100 m deep depression was created by collapse of the jökulhlaup flowpath across the ice dam. The volume of the depression was 0.3 km^3. Assuming that all of the thermal energy in the lake was used in the formation of the depression, it can be calculated that the average temperature of the 3.2 km^3 of water released was 8 °C (Björnsson, 2002). This high lake temperature is exceptional and due to inflow of meltwater from the Gjálp eruption. In addition to this we calculate that 0.1 km^3 of water was produced by frictional melting during the descent of the water from Grímsvötn to Skeiðarársandur.

The discharge out of the lake during this jökulhlaup was derived directly from lake-level observations and the known lake hypsometry. Discharge increased linearly with time and reached a peak value of 4×10^4 $m^3 s^{-1}$ in 16 hours and dropped to zero 27 hours later (Figure 4.6). During the drainage the lake surface subsided by 175 m, and the floating ice cover was reduced from 40 km^2 to less than 5 km^2. While the discharge hydrograph for the lake could be derived, the shape of the hydrograph for the outburst from Skeiðarárjökull is unknown and likely to be different since the flood was not drained through a single subglacial conduit. Four points on the hydrograph, two on the rising limb and two during the recession, were estimated but their accuracy is uncertain (Snorrason *et al.*, 1997). Similar jökulhlaups, reaching peak discharge in 3–4 days, also accompanied by a volcanic eruption north of Grímsvötn, may have occurred before, although descriptions do not provide unquestionable evidence for as rapid a rise in discharge (e.g. in 1861, 1867 and 1892; Thorarinsson, 1974; Björnsson, 1988). In 1938 a volcanic eruption took place north of Grímsvötn and meltwater of the order of 2–3 km^3 drained down to the lake (Björnsson, 1988) before a jökulhlaup took place which peaked in three days and receded in a week. The rapid rise suggests that the lake water was above the melting point (Figure 4.5). The slow recession of that jökulhlaup indicates that water may have spread out beneath the glacier before being collected to the river outlets.

The November 1996 jökulhlaup from Grímsvötn cannot be described by the classic theory of outbursts through subglacial tunnels. Rather than initial drainage from the lake being localised in one narrow conduit, the water was suddenly released as a sheet flow, surging downhill and propagating a subglacial pressure wave, which exceeded the ice overburden and lifted the glacier along the water flowpath in order to create space for the water, prior to draining through conduits. To simulate the 1996 jökulhlaup, Flowers *et al.* (2004) described water transport in a coupled sheet-conduit one-dimensional flowline model (Figure 4.6). A laminar/turbulent water sheet and a system of ice-walled conduits of a given spacing coexist in a coupled system and nourish each other, the water exchange depending on the sheet and conduit pressures. The model allows for local hydraulic uplift proportional to the flotation fraction (p_w/p_i), Gaussian around the peak uplift, increasing the volume of the sheet drainage system. The model suggests that the initial flood was characterised by a turbulent sheet flood distributed along the length of the glacier, which acted as a source to the rapid conduit development along the flood path. The sheet flood arriving first broke into conduits. In contrast with the classic jökulhlaups, which are only fed by water at the lake entrance of the tunnel, the rapid conduit growth was facilitated by the distribution of the source water along the length of the flood path.

4.2.2 Lakes under ice cauldrons

Many small ice cauldrons are located over hydrothermal areas at the rims of the Grímsvötn and Katla calderas, but most prominent are the Skaftá cauldrons (Figure 4.1B, F) situated at two central volcanoes on the EW trending Loki ridge (Björnsson and Einarsson, 1991), 10–15 km to the northwest of Grímsvötn. Since 1955, at least 40 jökulhlaups have drained from these cauldrons 30 km under the glacier to the river Skaftá. The period between the drainage events is 2–3 years for each cauldron. The cauldrons are approximately circular, 1–3 km in diameter,

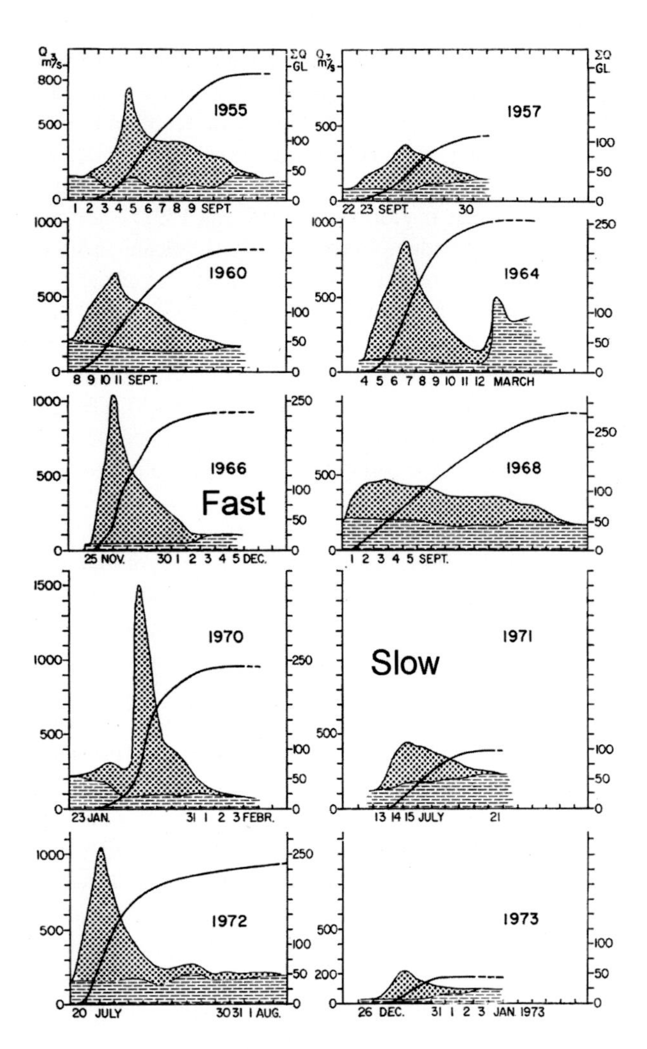

Figure 4.7. Various hydrographs of Skaftárhlaup (Björnsson, 1977).

and their ice cover on top of the lake is 300–400 m thick. The centre of the eastern cauldron subsides by 100–150 m in each jökulhlaup, draining 50–350 × 10^6 m^3 of water; the western cauldron subsides by 50–100 m and 0.05–0.16 km^3 are expelled (Zóphóníasson, 2002). The ice-drainage area of the western and the eastern cauldrons have been estimated at 20 km^2 and 29 km^2, respectively (Björnsson, 1988; Pálsson *et. al.*, 2006). Based on similar mass and energy balance studies as described above for Grímsvötn, the average geothermal heat flow under the cauldrons is estimated at about 1500 MW.

Two types of jökuhlaups have been observed from the Skaftá cauldrons: swift and slow. Typically, the jökulhlaup discharge from the eastern cauldron rises rapidly and recedes slowly (Figure 4.7). The peak discharge from the eastern cauldron was 200–1500 m^3 s^{-1} and was reached in 1–3 days, receding slowly in 1–2 weeks. Throughout the flood, water drains from the glacier at the melting point. The speedy rise may suggest that the water temperature at the reservoir is well above the melting point (Björnsson, 1992), which may not be unexpected because, in contrast to Grímsvötn, the lake is created over a concentrated cluster of hydrothermal vents. However, crevasses observed across the ice dam of the eastern cauldron after jökulhlaups (and sporadic surface elevation data) suggest that they may start with flotation of the glacier making space for a sheet flow, which subsequently feeds the rapidly peaking conduit flow. Complicating this interpretation, the slow recession after the peak suggests that these floods do not drain through a single tunnel but spread out beneath the glacier and later slowly collect into the river outlet. Moreover, occasionally jökulhlaups from the cauldrons rise slowly and maintain relatively high discharge for days, indicating that a distributed drainage system hampers the drainage of the jökulhlaup. This supposition is supported by InSAR analysis (ERS1/ERS2 tandem mission) showing jökulhlaups reducing the coupling of ice with the glacier bed and increasing sliding over a several kilometres wide area in western Vatnajökull (Magnússon *et al.*, 2007).

The current hydrothermal activity in Mýrdalsjökull, creating 12 to 15 small ice cauldrons, is located just inside the Katla caldera rims, where faults allow rapid vertical transport of hydrothermal fluid (Björnsson *et al.*, 2000). Small jökulhlaups drain frequently from some of the cauldrons and persistent hydrogen sulphide odour indicates that meltwater is drained continuously from others. These ice cauldrons are typically 20 to 50 m deep and 500 to 1000 m wide. The total power of the geothermal areas under the cauldrons is about 100 MW.

Two jökulhlaups with extremely rapidly rising discharge have been observed from cauldrons in Mýrdalsjökull. In 1955 a jökulhlaup draining a cauldron at the eastern caldera rim reached a peak of 2500 m^3 s^{-1} in 20 hours (Figure 4.8) and subsided slowly over a period of about three weeks (Thorarinsson, 1957; Rist, 1967). The jökulhlaup was possibly triggered by a small injection of magma to the glacier base (Tryggvason, 1960). The same applies to a jökulhlaup from Sólheimajökull in 1999 (Russell *et al.*, 2000).

4.3 Marginal lakes

At present, jökulhlaups originate from some 15 marginal ice-dammed lakes at all the main ice caps, most of which are located in ice-free tributary valleys (Thorarinsson, 1939; Björnsson, 1976). The thinning of glaciers, which began around the end of the nineteenth century, initiated jökulhlaups from marginal lakes that had drained continuously over a col. During the subsequent glacier recession jökulhlaups have become more frequent (occurring once, even twice a year) and gradually smaller in volume due to the thinning of the ice dams. This trend

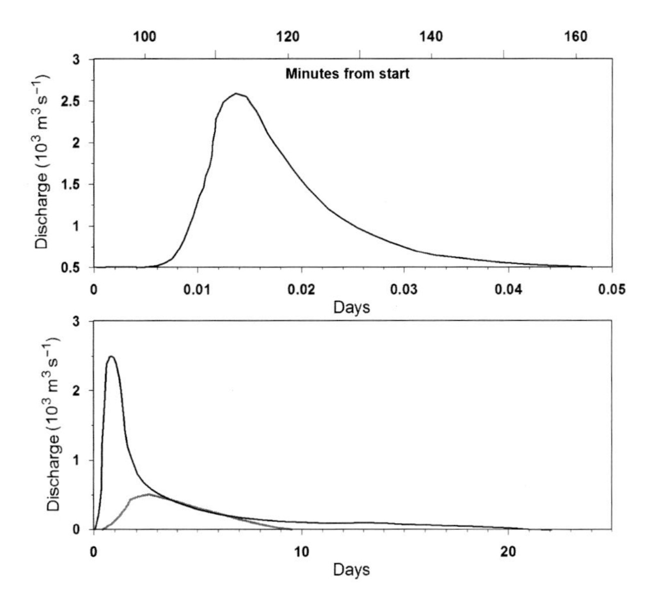

Figure 4.8. Hydrographs of jökulhlaups from Steinsholtsjökull 1967 (upper) and two rivers from Mýrdalsjökull 1955 (lower).

has been manifested in successive lowering of shore lines in marginal lakes. Also, the number of dumping glacier lakes has been greatly reduced. In a few instances, this trend has been interrupted by the thickening of ice dams during surges and temporary formation of new ice-dammed lakes. The short-term advance of some steep and active glacier outlets during a cold spell in the 1970s dammed ravines at the glacier margin, which dumped small jökulhlaups. The best-known marginal lakes are Grænalón (maximum area 18 km^2, volume 2 km^3) and Vatnsdalslón, (2 km^2, 0.1 km^3), both with peak jökulhlaup discharges of 2000–3000 m^3 s^{-1}. The rapid rise of some outbursts indicates that the lake temperature has been some degrees above the melting point (Björnsson, 1992). The jökulhlaups are triggered before flotation of the ice dam can take place (Björnsson, 1988). In one case, an ice dam was lifted because of rapid filling of the lake by inflow of a jökulhlaup from another ice-dammed lake (Björnsson and Eydal, 1999). The most explosive drainage observed of a marginal lake in Iceland occurred when a rockslide suddenly breached an ice dam of Steinsholtsjökull and in ten minutes the flood reached a peak of 2000 m^3 s^{-1} (Figure 4.8; Kjartansson, 1967).

Large jökulhlaups were frequent during the early Holocene from ice-dammed lakes at the margin of the decaying late Pleistocene ice cap, as is evident by the presence of erosion and sedimentary forms (Kjartansson, 1943, 1964; Sæmundsson, 1973; Tómasson, 1973, 1993; Geirsdóttir *et al.*, 2000; Harðardóttir *et al.*, 2001; Waitt, 2002; Carrivick *et al.*, 2004). Their sources changed in size and number with variations in the extent and thickness of the glaciers. The floods were drained subglacially. However, at the end of the last glaciation ice barricades may have been frozen to the bed and breached as lake water spilled over the top of the dam into a supraglacial channel that rapidly melted into a bigger breach conducting swift floods.

4.4 Subglacial eruptions as a source of jökulhlaups

Various information has been gathered about jökulhlaups caused by subglacial volcanic eruptions in the ice caps Mýrdalsjökull and the western part of Vatnajökull, and on the snow- and ice-capped stratovolcanoes, such as Öræfajökull and Eyjafjallajökull (Larsen, 2000, 2002; Larsen *et al.*, 1998; Björnsson and Einarsson, 1991). Jökulhlaups have also been observed associated with Hekla eruptions.

The production of meltwater is controlled by (1) the geological conditions at the eruption site (localised on confined vents, on elongated fissures or inside subglacial lakes), (2) the input rate and chemical composition of magma, and (3) the hydrological conditions at the glacier base (drainage systems, water pressure). All these factors are interrelated and affected by the drainage of the produced meltwater from the eruption site.

The rapid heat transfer observed in subglacial eruptions calls for lava to explode into tiny fragments of pyroclastic glass, rapidly cooled by quenching (Guðmundsson *et al.*, 1997, 2003). The partial pressure of water vapour and carbon dioxide, counteracted by the basal overburden water pressure (~the glacier thickness), predominantly exerts control over gasification of magma and production of explosive volcanic products. If lava is injected into a subglacial lake, explosive boiling of water sustains vigorous heat transfer toward the melting ice surface and rapid cooling of the fragmented volcanic products. There are no observations of a jökulhlaup being triggered during eruption into a subglacial lake. Melting of a floating ice cover does not add mass to the lake; its water level only rises slightly to make space for the volcanic material. In a situation of non-instantaneous heat transfer, the volume of volcanic material injected to the glacier base (with no lake) exceeds the released space due to melting of ice, and water is squeezed out from the eruption site. However, some accumulation of water might take place temporarily at the eruption site because the drainage system cannot expand fast enough to provide transport capacity for all the meltwater.

Injection of magma and tectonic movements at the initiation of a subglacial eruption may create a conduit incised into the basal ice that conducts the meltwater from the vent until it joins pre-existing surrounding watercourses. As it approaches the glacier margin the highly pressurised basal water may escape to the surface through moulins and englacial cracks, opened by hydrofracturing. On steep-sloping ice-clad volcanoes, deep canals, broken up and thermally eroded by water, may extend all the way from the summit down to the glacier margin. Meltwater at the eruption site may overtop the glacier and cut open channels into the surface. The supraglacial water flow may soon find routes down moulins and crevasses to the basal drainage system.

4.4.1 Jökulhlaups from caldera eruptions

Jökulhlaups from Katla Since the time of the settlement of Iceland (AD 870), 20 volcanic eruptions have been traced to the Katla volcanic system in Mýrdalsjökull ice cap. The Katla eruptions take place in a 600–750 m deep ice-filled caldera (containing no lake) that encircles an area of 100 km^2 (Björnsson *et al.*, 2000). Fascinatingly, the eruptions break through an ice cover of 400 m in one or two hours. The swiftest and greatest outbursts of glacial meltwater noted in world history accompany the eruptions along with heavy fallout of tephra that make the volcano the most hazardous in Iceland. The outbursts from Katla have durations of 3–5 days, peak discharges estimated at 10^5–10^6 m^3 s^{-1}, and total volumes of 1–8 × 10^9 m^3 (Thorarinsson, 1957, 1975; Maizels, 1989, 1995; Tómasson, 1996; Larsen, 2000). Calorimetry based on estimated jökulhlaup peak discharge rates would suggest a heat flux of 20–60 TW for the Katla eruptions, but this is an overestimate because the discharge rate is higher than the production rate of meltwater. As they leave the eruption site, the floods break up the ice barrier and flow over it. Immense blocks of ice break off the glacial margins, and a mixture of water, ice, volcanic emissions and sediment surges over the outwash plain at velocities of 5–15 m s^{-1}. During part of the flow event, the water masses may consist of a hyperconcentrated fluid–sediment mixture. The amount of volcanic debris produced per event and carried away by the water has been estimated to range from 0.7 to 1.6 × 10^9 m^3 (Tómasson, 1996; Larsen, 2000). During the last eruption in 1918, the coastline advanced 3 km into the sea. Although many loose deposits have washed away in succeeding years, the affected seashore remains 2.2–2.5 km farther south than its position in 1660. During 18 of the 20 documented historical eruptions, the associated jökulhlaups flowed southeast down to the Mýrdalssandur outwash plain, but in two cases jökulhlaups flowed southwest to the Sólheimasandur. On the northern side of Mýrdalsjökull, the outbursts have eroded deep canyons in the bedrock (in Markarfljót River). The most recent jökulhlaup taking that route took place 1600 years ago (Haraldsson, 1981).

Prehistoric eruptions in the voluminous, ice-filled calderas of Bárðarbunga and Kverkfjöll in Vatnajökull may have caused the largest, most catastrophic jökulhlaups in Iceland, with peak discharges of the order of 10^6 m^3 s^{-1} (Sæmundsson, 1973; Tómasson, 1973; Björnsson, 1988; Björnsson and Einarsson, 1991; Knudsen and Russell, 2002; Waitt, 2002; Carrivick *et al.*, 2004). The floods sweeping down Jökulsá á Fjöllum carved a conspicuous scabland and a deep canyon (Jökulsárgljúfur); erosion by cavitation is considered to have been a significant factor in their creation. Some of these floods may be mudflows or debris flows. The jökulhlaups have been dated at about 7100 BP, 4600 BP, 3000 BP and before 2000 BP, the earliest of which may have originated from a marginal ice-dammed lake, but the others were presumably caused by subglacial volcanic activity.

Jökulhlaups from Öræfajökull In historical times two eruptions in the stratovolcano Öræfajökull in Vatnajökull caused catastrophic jökulhlaups, washing away several farmsteads (Thorarinsson, 1958). Öræfajökull, rising 2000 m above the lowland, has a 500 m deep ice-filled caldera. The 1362 flood was over in less than one day and the peak might have been at least 10^5 m^3 s^{-1}. A contemporary record described

> . . . several floods of water that gushed out, the last of which was the greatest. When these floods were over, the glacier itself slid forwards over the plane ground, just like melted metal poured out of a crucible. The water now rushed down on the earth side without intermission, and destroyed what little of the pasture grounds remained . . . The glacier itself burst and many icebergs were run down quite to the sea, but the thickest remained on the plain at a short distance from the foot of the mountain . . . we could only proceed with the utmost danger, as there was no other way except between the ice-mountain and the glacier that had slid forwards over the plain, where the water was so hot that the horses almost got unmanageable. Thorarinsson, 1958

A less devastating eruption took place in AD 1727.

4.4.2 Jökulhlaup from fissure swarm eruptions

Two fissure eruptions under 500–700 m thick ice north of Grímsvötn in 1938 and 1996 produced 2.7 km^3 of meltwater (0.3–0.5 km^3 of hylaoclastite) and 4.0 km^3 of meltwater, respectively (Björnsson, 1988; Guðmundsson and Björnsson, 1991; Einarsson *et al.*, 1997; Guðmundsson *et al.*, 1997). In both cases meltwater was drained in jökulhlaups to Grímsvötn, where the water accumulated and later drained in Grímsvötn jökulhlaups. Eruptions outside Grímsvötn shortened the period between jökulhlaups from the lake, whereas eruptions inside the lake (1934, 1983, 1998 and 2004) do not. They cut through a 150–200 m thick ice shelf in 10–20 minutes, forming an opening 0.5 km in diameter and only melting small quantities of ice (about 0.1 km^3; Guðmundsson and Björnsson, 1991). The volume of volcanic material injected into the lake is negligible and inflow of ice from outside, compensating for the melted ice, is slow.

During the two-week long 1996 Gjálp eruption, apparently no water was accumulated at the 6 km long eruption fissure. The meltwater that drained from the Gjálp eruption site down to Grímsvötn had a temperature of 15–20 °C, as calculated from the volume of a depression created above the entire tunnel and the discharge rates. The volume of ice melted at the eruption site was derived from the volume of the surface depressions (ΔV_{gl}) above the eruptive fissure, and from the volume of meltwater accumulated in Grímsvötn (Guðmundsson *et al.*, 1997; Einarsson *et al.*, 1997):

$$\Delta V_i = (\rho_w/(\rho_w - \rho_i))(\Delta V_{gl} - \Delta V_m). \tag{4.5}$$

During the eruption ΔV_{gl} was monitored by radar altimetry, and after the eruption the volume of the volcanic material piled up at the eruption site (ΔV_m) was estimated by radio echo-soundings. The volume of material transported into Grímsvötn by meltwater was approximated by seismic soundings of the changes in the bed topography (Guðmundsson *et al.*, 2003). Another estimate of the water volume drained from the eruption site was obtained from lake-level measurements of Grímsvötn. During the first four days, meltwater was produced at a rate of 5×10^3 m^3 s^{-1} and the heat output at the peak of the eruption was 10^{12} W with heat flux density of 5×10^5 W m^{-2} (M. T. Guðmundsson *et al.*, 1997, 2003; S. Guðmundsson *et al.*, 2002). The total volume of ice melted during the first six weeks after the beginning of the eruption was 4.0 km^3, equivalent to 1.1×10^{12} kg of magma cooling from 1000 °C to 0 °C if all the heat were used for melting. After one year (January 1998) the melted volume was 4.7 km^3.

In Vatnajökull no rapid basal sliding on a regional scale has been observed in association with a subglacial eruption. The jökulhlaups are not known to have led to surges of the glacier. During rapidly rising jökulhlaups, water may drain through braided watercourses at high pressures but these passageways quickly develop into high-capacity ice tunnels.

4.5 Conclusions

The release of meltwater from glacial lakes can take place by two different conduit initiation mechanisms and the subsequent drainage from the lake occurs by two different modes. Drainage can begin at pressures lower than the ice overburden (in Grímsvötn, typically 6–7 bar lower) in conduits that expand slowly over days or weeks due to melting of the ice walls by frictional heat in the flowing water and sensible stored lake heat. Otherwise, the lake level may rise until the ice dam is lifted and water pressure in excess of the ice overburden opens the waterways. The discharge rises faster than can be accommodated by melting of the conduits and the glacier is lifted along the flowpath to make space for the water. During rapidly (linearly) rising jökulhlaups (peaking in a time interval of several hours to two days), the initial flood is characterised by a turbulent sheet flood distributed along the length of the glacier, which acts as a source for the rapid conduit development along the flood path. These passageways quickly develop into high-capacity ice tunnels. Icebergs may be broken off the glacier margins and transported by water over the river courses. The linearly rising floods are often associated with large discharges and floods following rapid filling of subglacial lakes during subglacial eruptions or dumping of one marginal lake into another. However, linearly rising jökulhlaups may take place from ice cauldrons that are gradually filled. Jökulhlaups during eruptions on steep ice-covered and snow-covered stratovolcanoes are swift and dangerous and may become lahars and debris-laden floods. Normally the glacier dynamic effects of meltwater production during subglacial eruptions are localised and surges triggered by jökulhlaups have not been witnessed. However, during eruptions in ice-capped stratovolcanoes, distributed drainage of meltwater has led to sliding of the glacier down to the lowland.

Further research should explore the processes initiating jökulhlaups and those controlling their discharge rate, i.e., the different circumstances producing floods at lake-water pressures equal to, versus lower than, those required for flotation of the ice dam. Improved theories are required of the thermodynamics of jökulhlaups describing the rate of heat transfer from the floodwater to the surrounding ice.

Notation

Symbol	Definition	Units
b	coefficient in empirical Clague–Matthew law	(1.84)
c_{pm}	the specific heat capacity of magma	(1.05 kJ kg^{-1} K^{-1})
c_w	specific heat capacity of water	(4.2 kJ kg^{-1} K^{-1})
g	acceleration due to gravity	(9.81 m s^{-2})
E_{mt}	total thermal energy in magma	(J)
h_i	thickness of glacier	(m)
K_p	coefficient in an empirical power law	(4.15×10^{-3} s^{-1}m$^{-2.52}$)
L_i	latent heat of fusion for ice	(333.5 kJ kg^{-1})
L_m	latent heat of fusion for magma	(4.2×10^3 kJ kg^{-1})

M_a	mass of ice melted per unit time on glacier surface	($kg\ s^{-1}$)
M_i	ice melted by geothermal heat	($kg\ s^{-1}$)
M_b	meltwater percolating into groundwater reservoir under lake	($kg\ s^{-1}$)
M_j	mass flow rate from subglacial lake	($kg\ s^{-1}$)
M_g	mass flow rate of geothermal fluid	($kg\ s^{-1}$)
p_i	ice overburden pressure	(Pa), $p_i = \rho_i g h_i$
p_w	water pressure	(Pa)
Q_{max}	peak discharge of jökulhlaup	($m^3\ s^{-1}$)
t	time	(s)
T_i	ice temperature	(°C)
T_w	water temperature	(°C)
T_{mi}	initial temperature of magma	(°C)
T_{mf}	temperature of lava after solidification	(°C)
T_0	ambient temperature of lake water	(°C)
T_l	average lake temperature	(°C)
V	lake volume	(m^3)
V_t	total volume (in $10^6\ m^3$) of ice-dammed lake in Clague–Matthew power law	
ΔV_m	subglacial accumulation of volcanic material	(m^3)
V_m	volume of magma	(m^3)
ΔV_i	volume of ice melted during eruption	(m^3)
ΔV_{gl}	glacier surface subsidence	(m^3)
z_b	elevation of bed surface above datum	(m)
z_s	elevation of ice surface above datum	(m)
φ	fluid potential	(Pa)
φ_b	fluid potential at bed	$\varphi_b = \rho_w g z_b + p_w$ (Pa)
ρ_i	density of ice	($916\ kg\ m^{-3}$)
ρ_m	density of magma	($2.65 \times 10^3\ kg\ m^{-3}$)
ρ_w	density of water	($10^3\ kg\ m^{-3}$)
ψ_G	energy balance for lake water	(W)

References

Alfaro, R., Brandsdóttir, B., Rowlands, D. P., White, R. S. and Guðmundsson, M. T. (2007). Structure of the Grímsvötn central volcano under the Vatnajökull icecap, Iceland. *Geophysical Journal International*, **168**, 863–876.

Björnsson, H. (1974). Explanation of jökulhlaups from Grímsvötn, Vatnajökull, Iceland. *Jökull*, **24**, 1–26.

Björnsson, H. (1975). Subglacial water reservoirs, jökulhlaups and volcanic eruptions. *Jökull*, **25**, 1–14.

Björnsson, H. (1976). Marginal and supraglacial lakes in Iceland. *Jökull*, **26**, 40–51.

Björnsson, H. (1977). The cause of jökulhlaups in the Skaftá river, Vatnajökull. *Jökull*, **27**, 71–78.

Björnsson, H. (1988). Hydrology of ice caps in volcanic regions. *Societas Scientarium Islandica*, **45**, Reykjavík.

Björnsson, H. (1992). Jökulhlaups in Iceland: prediction, characteristics and simulation. *Annals of Glaciology*, **16**, 95–106.

Björnsson, H. (1997). Grímsvatnahlaup fyrr og nú. In *Vatnajökull. Gos og hlaup 1996*, ed. H. Haraldsson. Reykjavík, Iceland: Icelandic Public Roads Administration, pp. 61–77.

Björnsson, H. (2002). Subglacial lakes and jökulhlaups in Iceland. *Global and Planetary Change*, **35**, 255–271.

Björnsson, H. and Einarsson, P. (1991). Volcanoes beneath Vatnajökull, Iceland. Evidence from radio echo-sounding, earthquakes and jökulhlaups. *Jökull*, **40**, 147–168.

Björnsson, H. and Eydal, G. P. (1999). Jökulhlaup í Kverká og Kreppu frá jaðarlónum við Brúarjökul. *Raunvísindastofnun Háskólans*, RH-15-99.

Björnsson, H. and Guðmundsson, M. T. (1993). Variations in the thermal output of the subglacial Grímsvötn Caldera, Iceland. *Geophysical Research Letters*, **20**, 2127–2130.

Björnsson, H. and Kristmannsdóttir, H. (1984). The Grímsvötn geothermal area, Vatnajökull, Iceland. *Jökull*, **34**, 25–50.

Björnsson, H., Björnsson, S. and Sigurgeirsson, Th. (1982). Penetration of water into hot rock boundaries of magma at Grímsvötn. *Nature*, **295**, 580–581.

Björnsson, H., Pálsson, F. and Guðmunsdsson, M. T. (2000). Surface and bedrock topography of the Mýrdalsjökull ice cap, Iceland: the Katla caldera, eruption sites and routes of jökulhlaups. *Jökull*, **49**, 29–46.

Björnsson, H., Rott, H., Guðmundsson, S. *et al.* (2001). Glacier-volcano interactions deduced by SAR interferometry. *Journal of Glaciology*, **47** (156), 58–70.

Carrivick, J. L., Russell, A. J. and Tweed, F. S. (2004). Geomorphological evidence for jökulhlaups from Kverkfjöll volcano, Iceland. *Geomorphology*, **63**, 81–102.

Clague, J. J. and Mathews, W. H. (1973). The magnitude of jökulhlaups. *Journal of Glaciology*, **12** (66), 501–504.

Clarke, G. K. C. (1982). Glacier outburst flood from "Hazard Lake", Yukon Territory, and the problem of flood magnitude prediction. *Journal of Glaciology*, **28** (98), 3–21.

Clarke, G. K. C. (2003.) Hydraulics of subglacial outburst floods: new insights from the Spring-Hutter formulation. *Journal of Glaciology*, **49** (165), 299–313.

Einarsson, P. (1991). Earthquakes and present-day tectonism in Iceland. *Tectonophysics*, **189**, 261–279.

Einarsson, P. and Brandsdóttir, B. (1984). Seismic activity preceding and during the 1983 volcanic eruption in Grímsvötn, Iceland. *Jökull*, **34**, 13–23.

Einarsson, P., Brandsdóttir, B., Guðmundsson, M. T. *et al.* (1997). Center of the Iceland hotspot experiences volcanic unrest. *Eos*, **78** (35), 374–375.

Flowers, G. E., Björnsson, H., Pálsson, F. and Clarke, G. K. C. (2004). A coupled sheet-conduit model of jökulhlaup propagation. *Geophysical Research Letters*, **31**, L05401, doi:10.1029/2003GL019088.

Fowler, A. (1999). Breaking the seal of Grímsvötn. *Journal of Glaciology*, **45** (151), 506–516.

Geirsdóttir, Á., Harðardóttir, J. and Sveinbjörnsdóttir, Á. E. (2000). Glacial extent and catastrophic meltwater events during the deglaciation of Southern Iceland. *Quaternary Science Reviews*, **19**, 1749–1761.

Guðmundsson, M. T. (1989). The Grímsvötn Caldera, Iceland, subglacial topography and structure of caldera infill. *Jökull*, **39**, 1–19.

Guðmundsson, M. T. (1992). The crustal structure of the subglacial Grímsvötn volcano, Vatnajökull, Iceland, from multiparameter geophysical surveys. Ph.D. thesis, University of London.

Guðmundsson, M. T. and Björnsson, H. (1991). Eruptions in Grímsvötn 1934–1991. *Jökull*, **41**, 21–46.

Guðmundsson, M. T., Björnsson, H. and Pálsson, F. (1995). Changes in jökulhlaup sizes in Grímsvötn, Vatnajökull, Iceland, 1934–1991, deduced from in situ measurements of subglacial lake volume. *Journal of Glaciology*, **41** (138), 263–272.

Guðmundsson, M. T., Sigmundsson, F. and Björnsson, H. (1997). Ice-volcano interaction of the 1996 Gjálp subglacial eruption, Vatnajökull, Iceland. *Nature*, **389**, 954–957.

Guðmundsson, M. T., Sigmundsson, F., Björnsson, H. and Högnadóttir, Th. (2003). The 1996 eruption at Gjálp, Vatnajökull ice cap, Iceland: efficiency of heat transfer, ice deformation and subglacial water pressure. *Bulletin of Volcanology*, **66**, 46–65.

Guðmundsson, S., Guðmundsson, M. T., Björnsson, H. *et al.* (2002). Three-dimensional glacier surface motion maps at the Gjálp eruption site, Iceland, inferred from combining InSAR and other ice displacement data. *Annals of Glaciology*, **34**, 315–322.

Haraldsson, H. (1981). The Markarfljót sandur area, southern Iceland. Sedimentological, petrographical and stratigraphical studies. *Striae*, **15**, 1–65.

Harðardóttir, J., Geirsdóttir, Á. and Sveinbjörnsdóttir, Á. E. (2001). Seismostratigraphy and sediment studies of Lake Hestvatn, southern Iceland: implications for the deglacial history of the region. *Journal of Quaternary Science*, **16** (2), 167–179.

Jóhannesson, T. (2002). Propagation of a subglacial flood wave during the initiation of jökulhlaup. *Hydrological Sciences Journal*, **47**, 417–434.

Kjartansson, G. (1943). Jarðsaga. In *Árnesingasaga I.* Reykjavík.

Kjartansson, G. (1964). Ísaldarlok og eldfjöll á Kili. *Náttúrufræðingurinn*, **34**, 9–38.

Kjartansson, G. (1967). The Steinsholtshlaup, central-south Iceland on January 15th, 1967. *Jökull*, **17**, 249–262.

Knudsen, Ó. and Russell, A. R. (2002). Jökulhlaup deposits at Ásbyrgi, northern Iceland: sedimentology and implications of flow type. In *The Extremes of the Extremes. Extraordinary Floods*. Proceedings of a symposium held in Reykjavík, Iceland, July 2000. IAHS Publication 271, 107–112.

Larsen, G. (2000). Holocene eruptions within the Katla volcanic system, Iceland: notes on characteristics and environmental impact. *Jökull*, **50**, 1–28.

Larsen, G. (2002). A brief overview of eruptions from ice-covered and ice-capped volcanic systems in Iceland during the past 11 centuries: frequency, periodicity and implications. In *Ice–Volcano Interaction on Earth and Mars*, eds. J. L. Smellie and M. Chapman. Geological Society of London Special Publication 202, pp. 81–90.

Larsen, G., Guðmundsson, M. T. and Björnsson, H. (1998). Eight centuries of periodic volcanism at the center of the Iceland hot spot revealed by glacier tephrastratigraphy. *Geology*, **26** (10), 943–946.

Magnússon, E., Rott, H., Björnsson, H. and Pálsson, F. (2007). The impact of jökulhlaups on basal sliding observed by SAR interferometry on Vatnajökull, Iceland. *Journal of Glaciology*, **53** (181), 232–240.

Maizels, J. (1989). Sedimentology and paleohydrology of Holocene flood deposits in front of a jökulhlaup glacier, south Iceland. In *Floods: Hydrological, Sedimentological and Geomorphological Implications*, eds. K. Beven and P. Carling. New York: John Wiley and Sons, pp. 239–252.

Maizels, J. (1995). Sediments and landforms of modern proglacial terrestrial environments. In *Modern Glacial Environments*, ed. J. Menzies. Oxford: Butterworth-Heinemann, pp. 365–416.

Ng, F. and Björnsson, H. (2003). On the Clague-Mathews relation for jökulhlaups. *Journal of Glaciology*, **49** (165), 161–172.

Nye, J. F. (1976). Water flow in glaciers: jökulhlaups, tunnels and veins. *Journal of Glaciology*, **17** (76), 181–207.

Pálsson, F., Björnsson, H., Haraldsson, H. H. and Magnússon, E. (2006). Vatnajökull. Mass balance, meltwater drainage and surface velocity of the glacial years 2004–2005. RH-06–2006. Science Institute, University of Iceland.

Rist, S. (1955). Skeiðarárhlaup 1954. *Jökull*, **5**, 30–36.

Rist, S. (1967). Jökulhlaups from the ice cover of Mýrdalsjökull on June 25, 1955 and January 20, 1956. *Jökull*, **17**, 243–248.

Roberts, M. J., Russell, A. J., Tweed, F. S. and Knudsen, Ó. (2000). Ice fracturing during jökulhlaups: implications for englacial floodwater routing and outlet development. *Earth Surface Processes and Landforms*, **25**, 1–18.

Röthlisberger, H. (1972). Water pressure in intra- and subglacial channels. *Journal of Glaciology*, **62**, 177–203.

Russell, A. J., Tweed, F. S. and Knudsen, Ó. (2000). Flash flood at Sólheimajökull heralds the reawakening of an Icelandic subglacial volcano. *Geology Today*, **16**, 102–106.

Sæmundsson, K. (1973). Straumrákaðar klappir í kringum Ásbyrgi. [Grooving on lava surfaces at Ásbyrgi NE-Iceland.] *Náttúrufræðingurinn*, **42**, 81–99.

Schmidt, E. (1979). *Properties of Water and Steam in SI-Units*, ed. U. Grigull. Springer Verlag.

Shreve, R. (1972). Movement of water in glaciers. *Journal of Glaciology*, **62**, 205–214.

Sigmundsson, F., Sturkell, E., Pinel, V. *et al.* (2004). Deformation and eruption forecasting at volcanoes under retreating ice caps: discriminating signs of magma inflow and ice unloading at Grímsvötn and Katla volcanoes, Iceland. *Eos Trans.* AGU 85(47), Fall Meeting Supplement, 608.

Sigvaldason, G. (1965). The Grímsvötn thermal area. Chemical analysis of jökulhlaup water. *Jökull*, **15**, 125–128.

Snorrason, A., Jónsson, P., Pálsson, S. *et al.* (1997). Hlaupið á Skeiðarársandi haustið 1996. Útbreiðsla, rennsli og aurburður. In *Vatnajökull: Gos og hlaup 1996*, ed. H. Haraldsson. Icelandic Public Roads Administration, Reykjavík, pp. 79–137.

Spring, U. and Hutter, K. (1981). Numerical studies of jökulhlaups. *Cold Regions Science and Technology*, **4** (3), 221–244.

Thorarinsson, S. (1939). The ice dammed lakes of Iceland with particular reference to their values as indicators of glacier oscillations. *Geografiska Annaler*, **21**, 216–242.

Thorarinsson, S. (1953). Some aspects of the Grímsvötn problem. *Journal of Glaciology*, **14**, 267–275.

Thorarinsson, S. (1957). The jökulhlaup from the Katla area in 1955 compared with other jökulhlaups in Iceland. *Jökull*, **7**, 21–25.

Thorarinsson, S. (1958). The Öræfajökull eruption of 1362. *Acta Naturalia Islandica*, **2** (2), 100pp.

Thorarinsson, S. (1974). *Vötnin Stríð. Saga Skeiðarárhlaupa og Grímsvatnagosa.* [The swift flowing rivers. The history of Grímsvötn jökulhlaups and eruptions.] Reykjavík: Menningarsjódur.

Thorarinsson, S. (1975). Katla og annáll Kötlugosa. [Katla and annals of Katla eruptions.] *Árbók Ferðafélags Íslands*, Reykjavík, pp. 125–149.

Tómasson, H. (1973). Hamfarahlaup í Jökulsá á Fjöllum. *Náttúrufræðingurinn*, **43**, 12–34.

Tómasson, H. (1993). Jökulstífluð vötbn á Kili og hamfarahlaup í Hvítá í Árnessýslu. *Náttúrufræðingurinn*, **62** (1–2), 77–98.

Tómasson, H. (1996). The jökulhlaup from Katla in 1918. *Annals of Glaciology*, **22**, 249–254.

Tryggvason, E. (1960). Earthquakes, jökulhlaups and subglacial eruptions. *Jökull*, **10**, 18–22.

Waitt Jr., R. B. (2002). Great Holocene floods along Jökulsá á Fjöllum, North Iceland. In *Flood and Megaflood Processes: Recent and Ancient Examples*, eds. I. P. Martini, V. R. Baker and G. Garzón. International Association of Sedimentologists Special Publication 31, 37–51.

Walder, J. S. and Costa, J. E. (1996). Outburst floods from glacier-dammed lakes: the effect of mode of lake drainage on flood magnitude. *Earth Surface Processes and Landforms*, **21** (8), 701–723.

Zópópníasson, S. (2002). *Rennsli í Skaftárhlaupum 1955–2002.* SZ-2002/01, Orkustofnun, Vatnamælingar.

5

Channeled Scabland morphology

VICTOR R. BAKER

Summary

The Channeled Scabland comprises a regional anastomosing complex of overfit stream channels that were eroded by Pleistocene megaflooding into the basalt bedrock and overlying sediments of the Columbia Plateau and Columbia Basin regions of eastern Washington State, USA. Immense fan complexes were emplaced where sediment-charged water entered structural basins. The cataclysmic flooding produced macroforms eroded into the rock (coulees and trenched spur buttes) and sediment (streamlined hills and islands). Several types of depositional bars also are scaled to the channel widths. The erosional mesoforms (scaled to flow depth) include longitudinal grooves, butte-and-basin scabland, potholes, inner channels and cataracts. These make up an erosional sequence that is scaled to levels of velocity, power per unit area and depths achieved by the cataclysmic flooding. Giant current 'ripples' (dunes) developed in the coarse gravel bedload, and large-scale scour marks were formed around various flow obstacles, including rock buttes and very large boulders.

5.1 Introduction

The Channeled Scabland region (Figure 5.1) is that portion of the basaltic Columbia Plateau and Columbia Basin that was subjected to periodic cataclysmic flooding during the late Pleistocene, resulting in a distinctive suite of flood-related landforms. Bretz (1923a, pp. 577–578) defined 'scablands' as 'lowlands diversified by a multiplicity of irregular and commonly anastomosing channels and rock basins eroded into basalt . . .' The term was in local use in reference to chaotically eroded tracts of bare basalt which occur in relatively large channels that the floods cut through the loess cover on the plateau. In a series of papers during the 1920s and 1930s, Bretz described the then amazing assemblage of landforms that included rock basins, anastomosing channel ways, cataracts, gravel bars and coulees. Field relations among many of these features, most notably the multiple levels of divide crossings, the cataracts, gravel bars and rock basins, led him to propose that an immense cataclysmic flood had swept across the Columbia Plateau in late Pleistocene time (Bretz, 1923b, 1928a; Bretz *et al.*, 1956). The source for this flood was eventually established to be glacial Lake Missoula, which had been impounded over a large region of western Montana because of a lobe of the Cordilleran Ice Sheet that extended into northern Idaho (Pardee, 1942).

The central figure in the controversy is J Harlen Bretz, the University of Chicago professor (1882–1981; Figures 1.1 and 5.2) and who formulated the hypothesis of cataclysmic flooding as the origin for the unusual landforms in this region (Baker, 1978a, 1981). Bretz (personal communication, 1977) recalled that he first conceived of the cataclysmic flood hypothesis when he saw a topographic map depicting the immense Potholes Cataract. His hypothesis remained highly controversial for decades (Baker, 1978a) but it gradually gained acceptance in the light of continued discoveries, including the giant current 'ripples' (dunes) on gravel bar surfaces (Pardee, 1942; Bretz *et al.*, 1956), credible mechanical explanations for the landforms (Baker, 1973) and additional examples of megaflood landscapes on Earth (Malde, 1968) and Mars (Baker, 1982). Thus, the current rapidly accelerating research on magaflooding derives much of its historical legacy from the pioneering studies of the Channeled Scabland by Bretz. Indeed, scabland landforms and their genesis are commonly cited in regard to current controversies over processes associated with phenomena as diverse as high-energy tsunami erosion of rocky coastlines (Bryant and Young, 1996), subglacial erosion (Shaw, 2002; Denton and Sugden, 2005; Lewis *et al.*, 2006) and landforms produced by large-scale flooding on Mars (Baker, 1982, 2001; Burr *et al.*, 2002).

Controversy also remains in regard to the number, timing, relative magnitudes and sources of the late Pleistocene floods that impacted the Channeled Scabland region (Baker and Bunker, 1985; Waitt, 1985; Smith, 1993). Recent work from the Columbia Gorge, downstream of the Channeled Scabland, documents at least 25 megafloods (flows with peak discharges of at least $1 \times 10^6\,m^3s^{-1}$) that occurred after about 20 000 years ago (Benito and O'Connor, 2003). However, only six or seven of these exceeded $6.5 \times 10^6\,m^3\,s^{-1}$, while at least one reached $10 \times 10^6\,m^3\,s^{-1}$). One of the smaller megafloods was from Lake Bonneville (O'Connor, 1993). Shaw *et al.* (1999) proposed that flooding emanated from beneath the Cordilleran Ice Sheet, a view that is disputed by Atwater *et al.* (2000).

This chapter will briefly review some of the now-classical morphological elements of the Channeled

Megaflooding on Earth and Mars, ed. Devon M. Burr, Paul A. Carling and Victor R. Baker.
Published by Cambridge University Press.

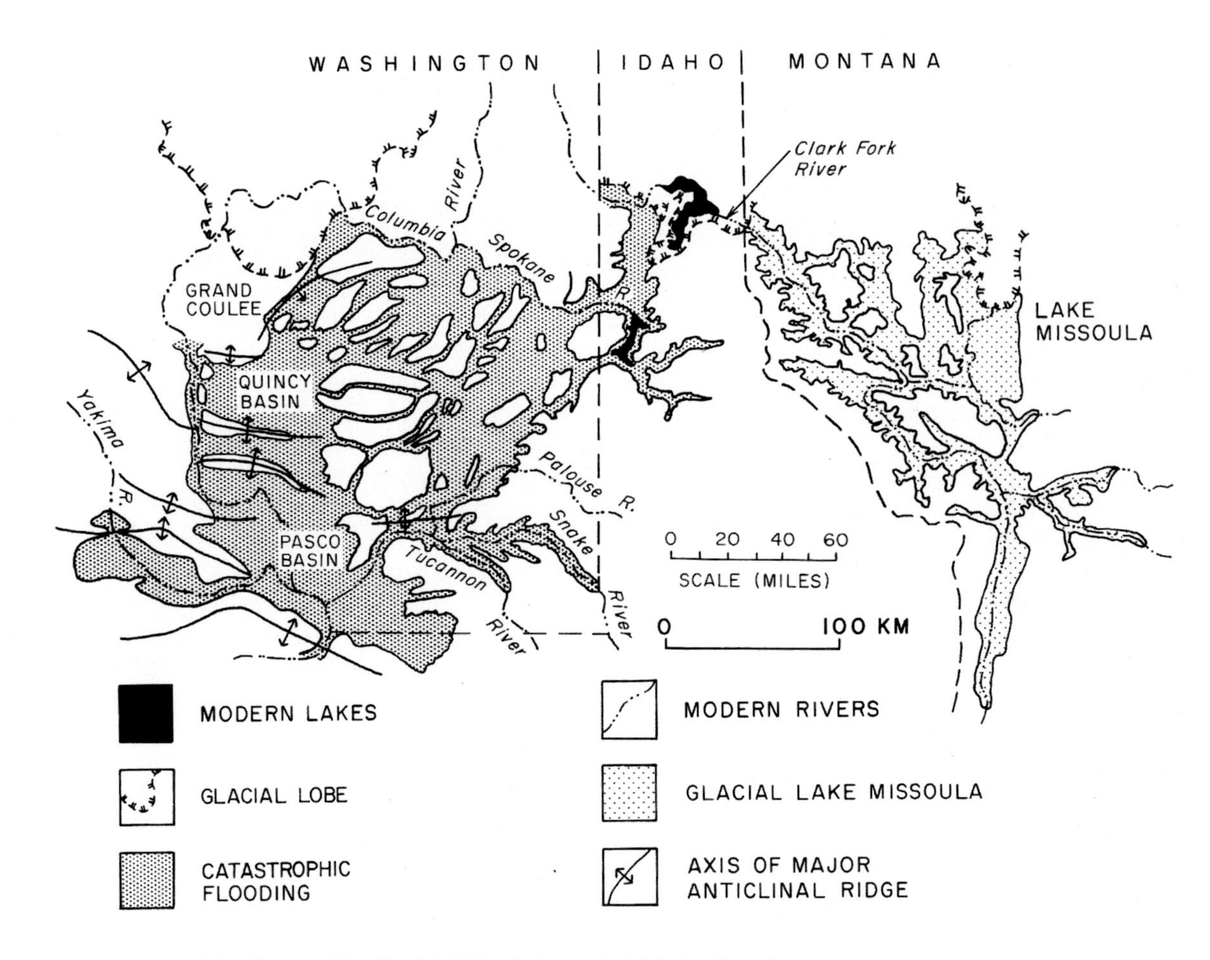

Figure 5.1. Location of the Channeled Scabland (left) in relation to glacial Lake Missoula.

Scabland, which divides into four major scabland tracts (Figure 5.3). From east to west these are (1) the Cheney–Palouse Tract, which heads to the southwest of Spokane and enters the Snake River Canyon with the Palouse River; (2) the Telford–Crab Creek Tract, which heads about 60 kilometres to the west of (1); (3) the Grand Coulee, which makes a single great gash through the divide about 60 kilometres west of (2); and (4) Moses Coulee, another 30 kilometres west of (3), which heads to the south of the divide in the marginal area of the Pleistocene Okanogan Lobe of the Cordilleran Ice Sheet.

5.2 Regional patterns

5.2.1 Anastomosis

The term 'anastomosis' was used by Bretz (1923a) to describe the pattern of scabland channels. Bretz (1923b) used both 'anastomosing' and 'braiding' to describe the pattern of scabland tracts, which he viewed as the individual elements in what he named 'channeled scablands' (Bretz, 1923b, p. 618). However, Bretz (1924, p. 148) seems to prefer the term 'anastomosing channels' to designate the chaotically eroded scabland channels that divide around steep-walled rock islands. Anastomosis occurs in the Channeled Scabland because preflood valleys did not have the capacity to convey the immense flood discharges that were imposed upon them. The floodwater filled the valleys to such an extent that water spilled across divides between the valleys. In this way the preflood valleys were transformed to a complex of dividing and rejoining channelways. This spilling across preflood divides produces a pattern of large-scale dividing and rejoining of channel ways that is cut into the basalt bedrock of the Columbia Plateau (Figure 5.3). Unlike the braided patterns that develop in alluvial rivers, scabland anastomosis does not involve deposition as a primary component of the overall pattern.

Although Bretz (1923a, 1924) clearly applied the term to bedrock channel morphologies in the Channeled Scabland, anastomosing river patterns were commonly confused with braided patterns in alluvial streams. For multichannel alluvial rivers a distinction between braiding and anastomosing is now common (Bridge, 2003), such that the former involves channels splitting around bars or islands, while the latter involves channels that diverge around

Figure 5.2. J Harlen Bretz photographed about the same time as his work in the Channeled Scabland.

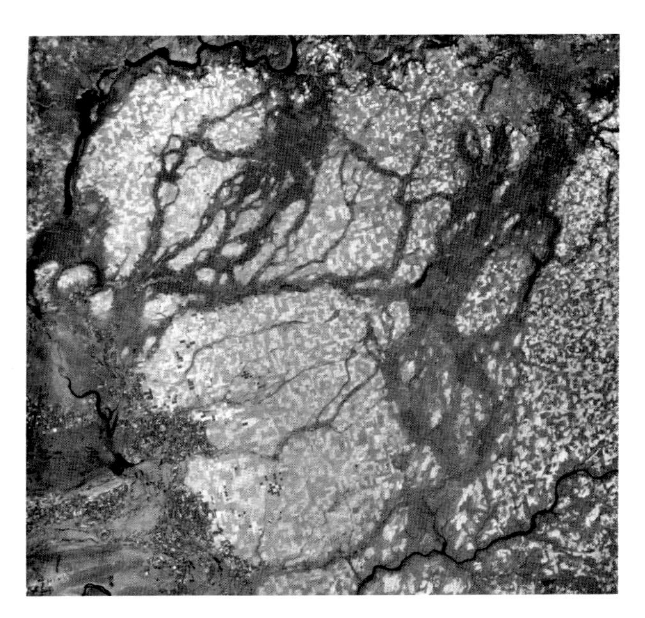

Figure 5.3. Anastomosing pattern of channels on the Columbia Plateau, showing three of the four main scabland tracts. From right to left, these are (1) the Cheney–Palouse tract, (2) the Telford–Crab Creek tract (top, centre), and (3) the Grand Coulee (upper left). This Landsat image (E-1381–18142) was acquired on 8 August 1973. It shows an area 180 × 200 kilometres. The Ephrata Fan (Figure 5.4) is located at the right-centre of the image.

floodplain areas. Moreover, the term 'anabranching', originally applied to multi-channel alluvial streams in Australia (Jackson, 1834, p. 79), is also applied to channels splitting around areas of floodplain (Nanson and Knighton, 1996). Bridge (2003) notes that these terms are not mutually exclusive, such that very large braided rivers, like the Brahmaputra, are both braided and anastomosing.

5.2.2 Overfitness

Misfit streams are either too small or too large for the valleys that they presently occupy (Dury, 1964). Although underfit streams are most often cited in the geomorphological literature, the overfit variety was recognised (Dury, 1964) as the result of sudden increases in discharge that produced such great channel enlargement that the channel became larger than the original valley that previously contained it. The Crab Creek valley near the town of Wilson Creek affords excellent examples of the overfitness generated by the cataclysmic flooding of the Channeled Scabland (Bretz, 1928b; Bretz *et al.*, 1956).

5.2.3 Coulees and hanging valleys

In the northwestern USA, the term 'coulee' is applied to very large steep-walled, trench-like troughs that often contain no stream along the valley floor. These are commonly the spillways and flood channels of the overall scabland plexus and many were parts of preflood fluvial valleys that formerly were more shallowly incised into the basalt plateau. Hanging valleys occur where the tributaries to these valleys are no longer graded to the main valley floor because of its deepening and widening by the cataclysmic flood scour. Examples occur in Moses Coulee and Lenore Canyon, where the preflood tributaries enter cliff faces on the coulee margins at elevations 50 to 100 or more metres above the coulee floor.

5.2.4 Fan complexes

Large fan complexes occur where constricted cataclysmic flood channel ways debouche into large structural basins. A well-developed example occurs where the floodwaters from the lower Grand Coulee expanded into the wide Quincy Basin in the west-central part of the Channeled Scabland (Figure 5.3). This is the Ephrata Fan (Figure 5.4). The deposit is alternatively interpreted as (1) an immense subfluvial expansion bar deposited at maximum flood stage that was modified by subsequent erosive flows (Baker, 1973), or (2) an outwash plain (sandur) formed by the coalescence of multiple bars emplaced by multiple jökulhlaup floods (Rice and Edgett, 1997). Another large fan complex occurs where the cataclysmic flows through the Columbia Gorge debouched into the Willamette lowland. This 'Portland Delta' (Bretz, 1925; Trimble, 1963) is characterised by an immense horseshoe-shaped scour hole developed around

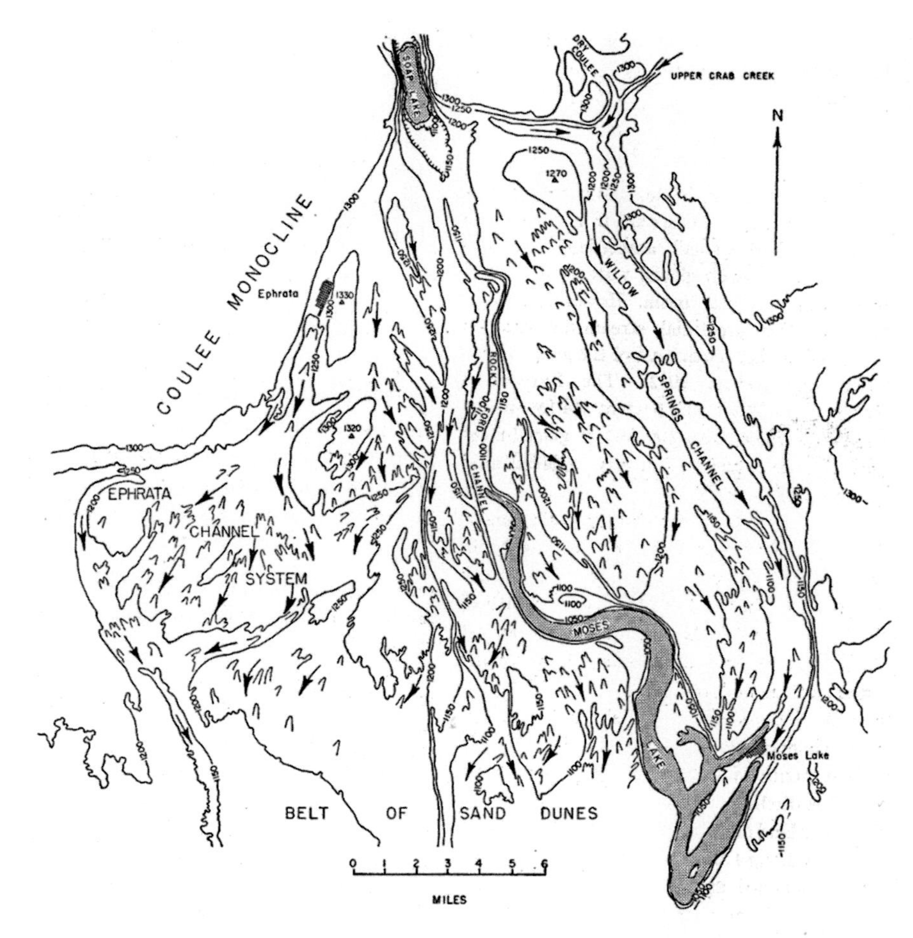

Figure 5.4. Topographic map of the Ephrata Fan complex at the northern end of the Quincy Basin. The arrows show the inferred surface flow directions. (The figure is from Bretz (1959), p. 33.)

the upstream end of large bedrock knob, Rocky Butte, in the northern part of Portland, Oregon.

5.3 Macroscale erosional surface forms

Southard (2003) applies the term 'surface form' to geometrical features that develop on a sedimentary surface by the action of a fluid over that surface. He prefers this term to that of 'bedform', which has come to have ambiguous meaning. Erosional surface forms on rock are termed 'sculpted forms' by Richardson and Carling (2005). The megaflood erosional and depositional surface and sculpted forms of the Channeled Scabland are classified in a hierarchical system, introduced by Baker (1978b), following a system that was applied to alluvial rivers by Jackson (1975). The scabland system (Table 5.1) has both depositional and erosional elements. The macroscale elements are scaled to channel widths. The mesoscale forms are scaled to channel depths.

The large erosional forms of scabland channels consist of eroded channel ways and the residual uplands between them. The channels fit into a regional anastomosing pattern, described above (Figure 5.4); they may locally contain inner channels, cataracts, butte-and-basin scabland and longitudinal grooves, which are their mesoscale erosional forms, to be described below.

5.3.1 Trenched-spur buttes

Prior to the megaflooding, the topography just west of Wilson Creek (Figure 5.5) was that of a gently meandering valley, incised through the loess-mantled uplands into the underlying basalt. Here the valley meandered with a wavelength of about 600 metres. The megaflooding completely filled this pre-flood valley such that water spilled over divides into adjacent valleys. Bretz (1928b) proposed that the flooding could not tolerate the leisurely meandering curves of the deeply incised pre-flood valley. Slip-off slopes of the meander bends were attacked vigorously to produce what he called 'trenched-spur buttes' (Figure 5.5). Huge streamlined bars were deposited downstream of the former valley bends and many of these have giant current 'ripples' on their surfaces (Figure 5.5).

Table 5.1. Erosional and depositional landforms of the Channeled Scabland

Scale	Eroded in rock	Eroded in sediment	Deposition
Regional patterns	Anastomosis		Fan complexes
Macroforms (scaled to flow width)	Coulees trenched spur buttes	Streamlined hills and islands	Longitudinal bars Eddy bars
Mesoforms (scaled to flow depth)	Longitudinal grooves Butte-and-basin scabland Inner channels cataracts Potholes	Scour holes	Giant current ripples (dunes)

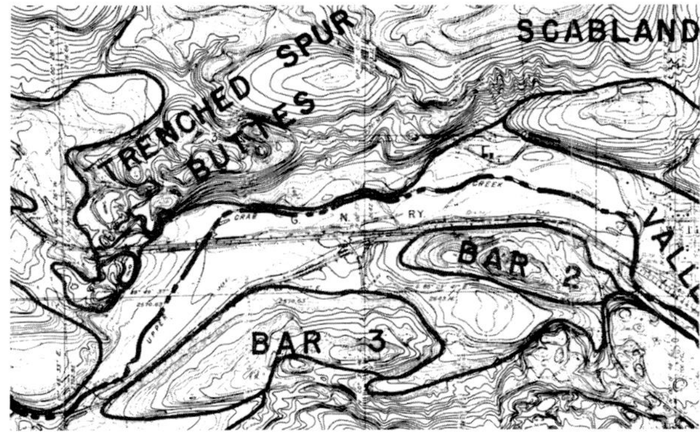

Figure 5.5. Trenched-spur buttes (left centre) and depositional bars immediately west of Wilson Creek, Washington. The photograph (top) was taken 30 June 1961, and it shows an area 2 × 3.5 km. The contour map (bottom) (U.S. Bureau of Reclamation Map G5883) shows the same area (from Bretz (1959), Plate 4). The contour interval is 0.6 metres (2 feet).

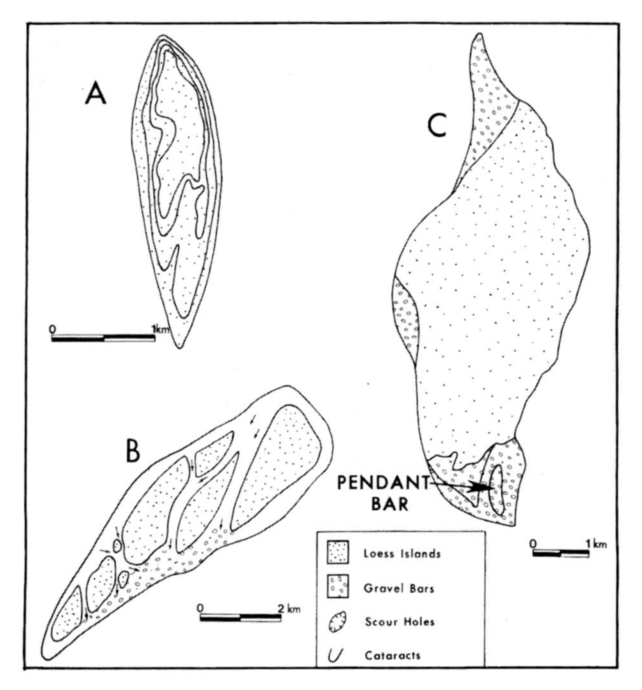

Figure 5.6. Morphologies of various loess islands and flood-modified loess hills in the Cheney–Palouse scabland tract. (A) Loess hill showing highly streamlined morphology that resulted from full submergence during the flooding. (B) Partially submerged loess island showing major divide crossings (arrows) and gravel deposition to form a leeward accumulating bar. (C) Fully emergent loess island with margins shaped by floodwater flow and gravel bar accumulations upstream, downstream and marginal to the island. (The figure is from Patton and Baker (1978), p. 123.)

5.3.2 Streamlined hills and islands

The eastern portions of the Channeled Scabland, particularly the Cheney–Palouse tract (Patton and Baker, 1978), contain spectacular examples of kilometre-scale hills that show distinctive flow streamlining of the edges ('islands') and, in some cases, tops (Figure 5.6). These generally occur in local clusters, organised in a braid-like pattern within the overall scabland complex. The hills are composed predominantly of loess that was not stripped from the underlying basalt. They may also partially contain gravel and cementing petrocalcic horizons that derive from more ancient phases of cataclysmic flooding than those responsible for the streamlining.

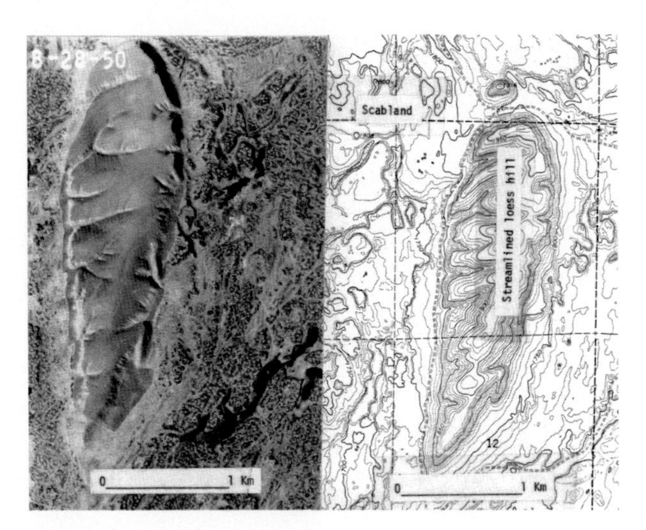

Figure 5.7. Detailed morphology of a streamlined loess 'island' in the Cheney–Palouse scabland tract. Groove and butte-and-basin scabland can be seen on the channel floor marginal to the loess hill. Water velocities were on the order of 12 m/s and flow depths were about 30–40 m during the flooding (Baker, 1973). The contour interval for the map is about 3 m (10 feet). (The figure is from Baker (1978b), p. 90.)

The streamlined hills generally rise up to 50 m above the surrounding scabland areas, and they are commonly 1–4 km long and 0.5 km wide. Bretz (1923b) first recognised their remarkable shape, with steep, ungullied bounding hillslopes that converge upstream to prow-like terminations and downstream to tapering tails (Figure 5.7). Baker (1974) compared the planimetric chapes to the lemniscate form, used by Chorley (1959) to characterise the streamlining of glacial drumlins. In a more detailed analysis of the lemniscate shape, Komar (1984) found that it provides a close representation of symmetrical airfoils (Joukowski sections), and thus serves to characterise the streamlining process.

Streamlining serves to reduce the drag or resistance to a flowing fluid, resulting in distinctive relationships among lengths, widths and areas for the landforms. These have been demonstrated for both streamlined hills in the Channeled Scabland and for analogous forms on Mars (Baker and Kochel, 1978; Baker, 1979). Direct measurements on airfoils in both air and water demonstrate that the minimisation of drag occurs for length-to-width (L/W) ratios in the range of 3 to 4 (Komar, 1983). Baker and Kochel (1978) and Baker (1979) showed that scabland L/W values average 3.25, confirming their minimisation of drag.

5.4 Macroscale depositional surface forms

The term 'bar' is used for a great variety of large-scale depositional landforms in streams and rivers. Here the term is applied to all the depositional macroforms in scabland channels. Because there is no universally accepted classification for fluvial bars, a scheme has evolved for local use in the Channeled Scabland and adjacent areas. The scheme is based on the relationship of the bars to large-scale flow patterns in a local scabland channel reach, as recognised by Bretz (1928b), Bretz *et al.* (1956) and Baker (1973). As with alluvial rivers, scabland bars are macroforms, in the sense of Jackson (1975) and Church and Jones (1982), such that their dimensions scale to channel widths.

Many coarse gravel-transporting alluvial rivers assume a braided pattern in which linguoid and transverse bars form with relatively low profiles. Such rivers have relatively high width-to-depth ratios for their channel cross-sections. Scabland channel ways, in contrast, are relatively narrow and deep, as characteristic of resistant-boundary streams, such as those that are incised into bedrock (Baker, 1984). As consequence of this, scabland bars are commonly tens of metres in height, with an internal structure of foreset bedding (Baker, 1973).

5.4.1 Longitudinal bars

These bars are elongated parallel to the predominant flow direction. They commonly alternate at the bends of the palaeomeanders in scabland valleys that were transformed to flood channel ways (Figure 5.5). Unlike the broad, low-lying bars of wide, shallow braided rivers, the scabland longitudinal bars are high, mounded and streamlined hills of gravel, unusually tens of metres thick. Internal stratification is dominated by foreset bedding that developed on avalanche faces by accretion to the downstream margins of the bar. Many bars developed from multiple episodes of such accretion, either during the course of a single prolonged megaflood, or as the result of multiple floods.

Many longitudinal bars develop immediately downstream of bedrock projections on the scabland channel floors (Figure 5.8). Malde (1968) introduced the term 'pendant bar' for this landform, based on studies of the Pleistocene Bonneville megaflood in Idaho, and Baker (1973) found that similar features were common in the Channeled Scabland.

5.4.2 Eddy bars

These bars occur at the mouths of alcoves or valleys that were tributary to the valleys that were invaded by the megaflooding. Along the eastern margins of the Cheney–Palouse scabland tract nearly every tributary valley is blocked at its mouth by this kind of bar (Bretz, 1929). Subsequent erosion by the tributary stream may breach the blockade but smaller tributaries and alcoves may retain the original morphology of the eddy bar. Excellent examples of these occur in the basin of glacial Lake Missoula

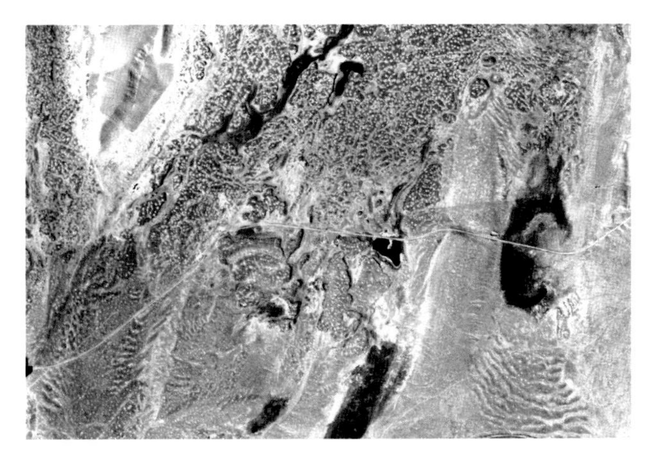

Figure 5.8. Pendant bar (right half of image) and scabland topography from near Macall, Washington. Note the giant current 'ripples' (GCRs) at the bottom left and bottom right. The cataclysmic flows were from top to bottom in this scene. The photograph shows an area 3 × 4.5 kilometres. It was taken 28 August 1950.

Figure 5.9. Eddy bar in the former basin of glacial Lake Missoula.

(Figure 5.9), where Pardee (1942) invoked them as evidence for the cataclysmic drainage of that lake.

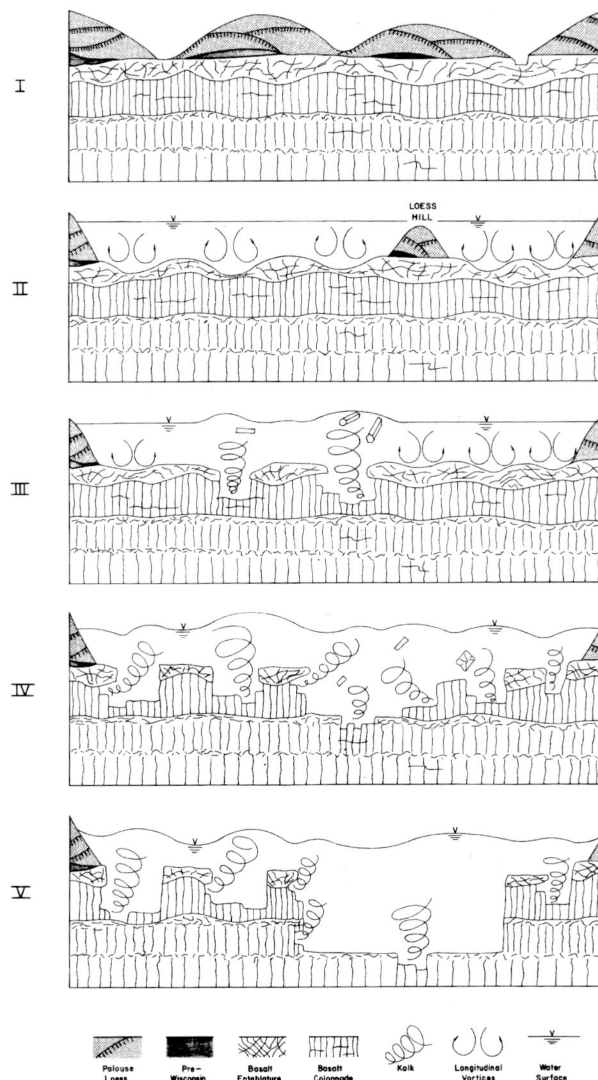

Figure 5.10. Schematic cross-sections showing the inferred sequence of flood erosion of a scabland complex. The numbered stages of erosion are described in the text. They are related to flow hydraulics in Table 5.2. (The figure is from Baker (1978b), p. 105.)

5.5 Mesoscale erosional surface forms

5.5.1 Sequence of erosional forms

A sequence of erosional forms, recognised in flume experiments by Shepherd and Schumm (1974), seems to characterise scabland channel ways (Baker, 1974, 1978b; Baker and Komar, 1987). This sequence can be illustrated with a series of schematic cross-sections (Figure 5.10). The first floodwater to enter the eastern Columbia Plateau regions would encounter the Palouse Hills topography of loess overlying the jointed basalt bedrock (Phase I of Figure 5.10). The high-velocity floodwater would easily remove much of the loess, leaving local remnants as streamlined loess hill or 'islands' (Phase II). Encountering the surface of the uppermost basalt flow, the water would begin by eroding longitudinal grooves (Figure 5.11), probably associated with longitudinal vortex structure in the macroturbulent flow field (Baker, 1979). Continued incision into the resistant basalt flow would eventually encounter columnar jointing, sometimes at the top of a flaring colonnade along a rolling surface created by the inflationary nature of the primary basalt flow emplacement. Hydraulic plucking of the basalt columns (Phase III) would then lead to the development of large-scale, basin-like potholes with intervening buttes. With enlargement and coalescence of potholes, the surface would develop to the butte-and-basin scabland (Phase IV) that typifies many scabland channel ways (Figure 5.12). Continued development eventually

Figure 5.11. Oblique aerial photograph of the Dry Falls cataract complex showing longitudinal grooves upstream of the cataract. Flow was from top to bottom in this scene. Note the prominent rock basins at the lower right. These basins are relatively narrow and deep.

Figure 5.12. Butte-and-basin scabland in Lenore Canyon, lower Grand Coulee. Note the dirt roadways for scale.

results in the formation of a dominant inner channel (Phase V), which usually heads at a rock step, or cataract (Figure 5.11).

An excellent example of this whole sequence can be seen at the Dry Falls cataract complex (Figure 5.13). The Dry Falls complex is 5.5 km wide and 120 m high. It consists of two western, horseshoe-shaped head cuts with plunge pools at their base and an eastward-extending inner channel. Dry Falls heads an inner channel that was eroded into the fractured basalt of the Coulee Monocline (Bretz, 1932). Prominent longitudinal grooves are developed on the basalt surface located immediately upstream of the cataract head (Figure 5.11).

Richardson and Carling (2005) restrict the term ‘pothole’ to round, deep depressions that are not eroded by plucking. Scabland rock basins, in contrast, are commonly relatively shallow and wide, though much larger in scale than the potholes usually found in bedrock river channels.

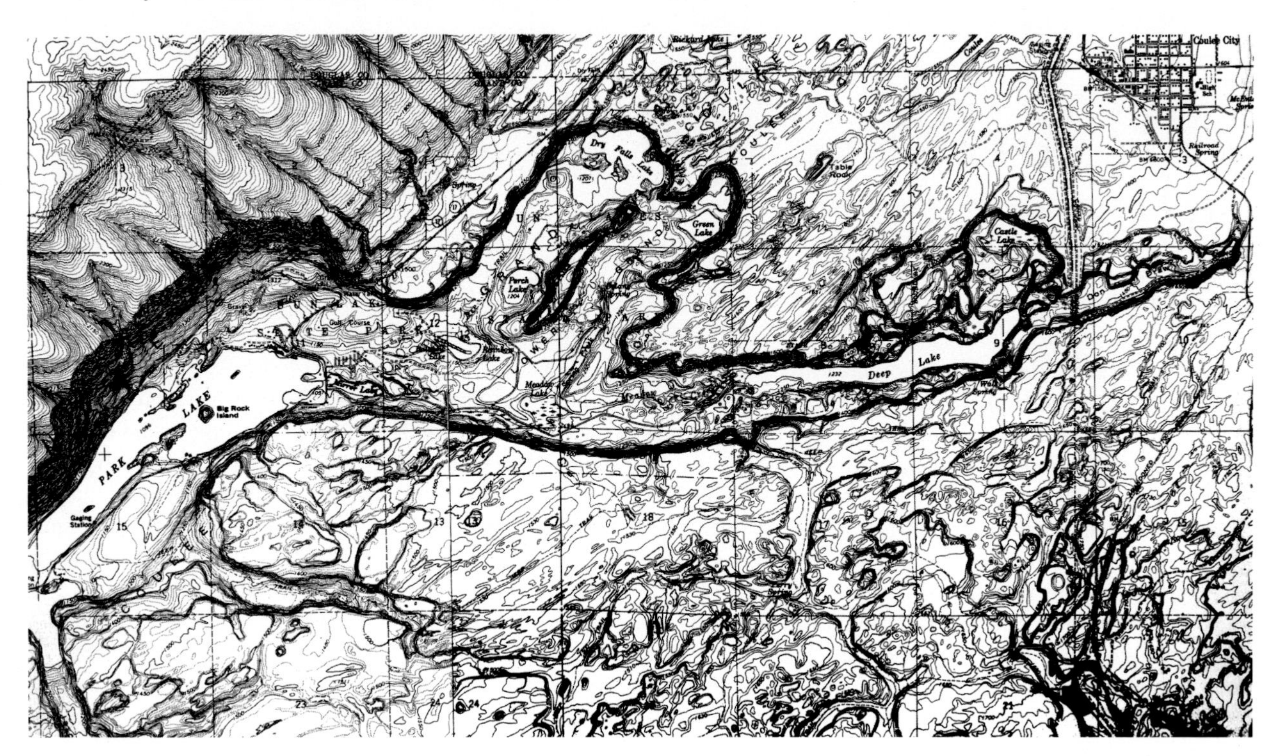

Figure 5.13. Topographic map of the Dry Falls cataract complex (compare to Figure 5.11).

5.5.2 Scour marks

Various obstacles to flow in scabland channels, such as large boulders or bedrock knobs, resulted in deformed flow streamlines for the deep cataclysmic floods. The hydrodynamics of the resulting flow involves large-scale vortex formation at the obstacle front and in its wake. The process is well known in the engineering literature concerned with scour during flooding at bridge piers (Shen, 1971). Baker (1973) described this process in relation to an 18 m long boulder on the Ephrata Fan surface. Herget (2005) describes many similar examples from the Altai cataclysmic flood region of central Asia.

Whether the cataclysmic flows scoured or deposited upstream from obstacles depends on the flow Reynolds number, which is directly proportional to the approach velocity of the flow and the obstacle diameter. At high Reynolds number, erosion occurs for both the 'horsehoe vortex' that forms at the upstream side of an obstacle and the wake vortex system that forms at its downstream end, resulting in the distinctive scour marks. However, at low Reynolds number, deposition may occur. This latter process can result in the pendant bars, described above, and, less commonly, in gravel bars that develop on the upstream margins of scabland flow obstacles and streamlined loess hills (Figure 5.6C).

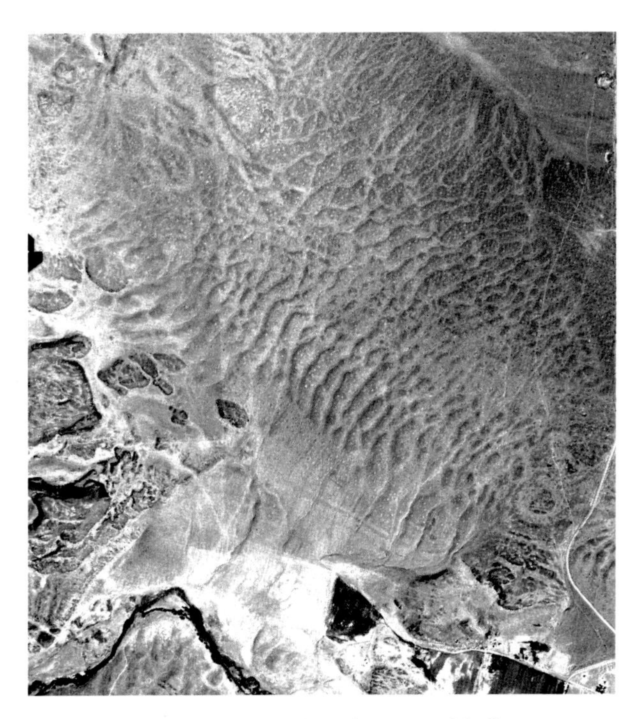

Figure 5.14. Giant current 'ripples' (dunes) near Marlin, Washington. The largest dune forms have heights of about 2–3 metres and a spacing of up to 60 metres (Baker, 1973). The aerial photograph is dated 17 August 1957. It shows a region of about 2.7 by 3 km.

5.6 Mesoscale depositional surface forms

5.6.1 Giant current ripples (dunes)

Bretz *et al.* (1956) applied the name 'giant current ripples' to the mesoscale transverse gravel depositional forms of the Channeled Scabland. Following Pardee's (1942) discovery of similar features in the basin of glacial Lake Missoula, Bretz *et al.* (1956) identified about a dozen examples. The Wilson Creek area was studied in some detail, and several examples were found of 'ripple' trains superimposed on gravel bars (Figure 5.14). Subsequently, Baker (1973) documented 60 of the most prominent sets of 'giant current ripple' (GCR) forms. Unpublished work indicates that there are certainly well over 100 GCR occurrences throughout the regions impacted by Missoula megaflooding.

Current terminological convention (Ashley, 1990) classifies various flow-transverse, mesoscale depositional surface forms as 'dunes'. For gravel compositions, in contrast to sand or silt, dunes form at flow strengths (velocity, shear stress or stream power) just above the threshold for bed movement (Southard, 2003). In the relatively shallow flows that are typical for alluvial rivers, dunes are replaced by antidunes when the flow strengths increase above a few metres per second (Southard, 2003). Like the various bedforms classed as 'large-scale asymmetrical ripples', 'megadunes' and 'sand waves', dunes form in the upper part of the lower flow regime, at Froude numbers less than 1.

The striking appearance of the scabland GCR forms on aerial photographs (Figure 5.14) arises from local post-depositional factors. Deposition of aeolian silt in the swales between the surface-form crests locally results in differences in vegetation cover. In drier regions, the gravelly GCR summits are covered by sagebrush (*Artemisia tridentata*) and the adjacent swales are covered by cheat grass (*Bromus tectorum*). In contrast, a prominent GCR occurrence in the wetter region near Spirit Lake, Idaho, has a second-growth forest cover in which the larger pine trees (*Pinus ponderosa*) occupy the relatively wet swales, but not the drier surface-form summits (Figure 5.15).

It is an interesting historical sidelight that when the Channeled Scabland GCRs were first seen on aerial photographs, H. T. U. Smith, who was accompanying Bretz for the 1952 summer field season, immediately recognised their form as that of dunes. Smith, of course, was presuming from their size and patterns that these were sand dunes. One can imagine the surprise when the field site visits were made, and it was found that the features were composed entirely of gravel, including many boulders (G. E. Neff, 1969, personal communication).

Table 5.2. Relationship of flow hydraulics to stages of scabland channel cross-sectional morphology (Figure 5.10)

Erosional stage	Description	Mean velocity ($m\,s^{-1}$)	Power per unit area (watts m^{-2})	Depth (m)
I–II	Streamlined loess hills	3–5	500–2000	30–100
II–III	Stripped basalt, grooves	3–9	500–3000	35–125
III–IV	Butte-and-basin scabland	7–15	2000–20 000+	100–250
IV–V	Inner channels	15–25	5000–25 000+	100–250+

Hydraulic data are from calculations by Baker (1973, 1978c) and Benito (1997).

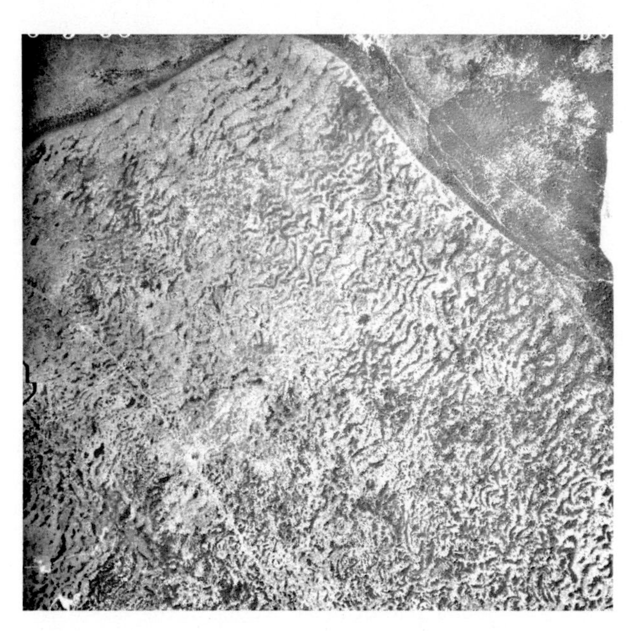

Figure 5.15. Giant current 'ripples' (dunes) near Spirit Lake, northern Idaho. The largest dune forms have heights of up to 7 metres and a spacing of up to 125 meters (Baker, 1973). Flow was from right to left. The aerial photograph is dated 5 August 1958. It shows a scene 4.6 by 4.6 km.

An inventory of these surface forms by Baker (1973) shows that their chords (spacings) generally range from 20 to 200 m and their heights range from 1 to 15 m. The relationship of bedform height H (metres) to chord or spacing λ (metres) is very regular, and similar to that found for relatively straight-crested gravel dunes, approximately as follows (Baker, 1973; Carling, 1999):

$$H = 7 \times 10^{-3}\lambda^{1.5}.$$

In plan view these surface forms have crest lines that look similar to those of sand ripples in rivers. However, the scabland forms are composed of gravel, cobbles and boulders. They have an internal structure of foreset bedding with openwork gravel textures.

5.7 Palaeohydraulic implications

Some evidence for the hydraulic implications of various scabland landforms comes from hydraulic modelling studies. Benito (1997), working in the Columbia River Gorge, found that the morphological sequence described in Figure 5.10 can be related to various measures of flow strength, specifically mean flow velocity, stream power per unit area of bed, and depth (Table 5.2). Note that the levels of flow strength are much larger than what is achieved in most contemporary terrestrial rivers (Baker and Kochel, 1988) but comparable to the values for other ancient megafloods (Baker, 2002).

The palaeohydraulic implications of the GCRs are problematic. Baker (1973) developed correlations for bedform dimensions (chord and height) to maximum flood flow strength variables (mean flow velocity, stream power per unit area of bed, and depth–slope product). Baker's (1973) peak flow velocities at the GCR locations range from 8 to 18 metres per second for flow depths of 20 to 150 metres. Power values for these peak flows fall in the range of 2000 to 20 000 watts per square metre (Baker, 1973). These values are comparable to those determined by Benito (1997) (see Table 5.2) for the butte-and-basin scabland with which the GCRs are commonly associated (e.g. Figures 5.7, 5.8). Of course, it is likely that the GCRs formed after the flood peak, such that they were emplaced at lower values of flow strength. Prominent armouring of GCR surfaces with a lag of coarse particles suggests that they were modified during waning flow stages (Baker, 1973).

In contrast to the results of Baker (1973), studies of the very similar-appearing giant current ripples (GCRs) of the Altai region in Siberia indicate mean flow velocities of 1.5 to 8 metres per second (Carling, 1996), and 5 to 11 metres per second (Herget, 2005), for flow depths of about 20 to 80 metres. This discrepancy probably occurs because the Channeled Scabland GCR morphologies were correlated to peak flow conditions (Baker, 1973), and not to the flows lower than the maxima, which most likely generated the GCRs. The strong correlations found by Baker (1973) probably reflect the properties of the reaches in which the surface forms developed. Both the peak flow hydraulics and the waning flow hydraulics at the time of GCR formation would have correlated to these reach properties.

5.8 Discussion

Bretz once considered the landforms of the Channeled Scabland to be unique in the world. In being so, he reasoned, their cataclysmic origin might be more acceptable to his contemporary geologists who held to an overly rigid form of uniformitarianism. In this conclusion he was wrong. Cataclysmic flood landscapes have now been documented in many parts of the world (Baker, 2002). Spectacular examples of giant current bedforms occur in central Asia (Baker *et al.*, 1993; Carling, 1996; Rudoy, 2005), along with immense gravel bars and scour marks (Herget, 2005). Streamlined hill and bar morphologies occur in the glacial lake spillway channels of central North America (Kehew and Lord, 1986). Moreover, the patterns, forms and processes evident in the Channeled Scabland have helped inform understanding of processes that occur at a smaller scale in modern bedrock channels that are highly influenced by extreme flood processes (Baker, 1977, 1984; Baker and Kale, 1998). This is one controversy in which Bretz would probably have been pleased to concede defeat. What better outcome from his point of view than to have a kind of reverse unformitarianism derive from his famous controversy with the overly strict adherents to the more common form of that logically flawed doctrine (see Baker, 1998)?

Acknowledgements

My understanding of Channeled Scabland morphology owes much to the influences of W. C. Bradley, R. C. Kochel, P. C. Patton and especially J Harlen Bretz. My Channeled Scabland research was initially supported by U.S. National Science Foundation Grant GA-21478, and subsequent work on scabland morphology was supported by NASA.

References

Ashley, G. M. (1990). Classification of large-scale suaqueous bedforms; a new look at an old problem. *Journal of Sedimentary Research*, **60**, 161–172.

Atwater, B. F., Smith, G. A. and Waitt, R. B. (2000). Comment: The Channeled Scabland: Back to Bretz? *Geology*, **28**, 574.

Baker, V. R. (1973). *Paleohydrology and Sedimentology of Lake Missoula Flooding in Eastern Washington*. Boulder, CO: Geological Society of America Special Paper 144.

Baker, V. R. (1974). Erosional forms and processes for the catastrophic Pleistocene Missoula floods in eastern Washington. In *Fluvial Geomorphology*, ed. M. Morisawa. London: Allen and Unwin, pp. 123–148.

Baker, V. R. (1977). Stream channel response to floods with examples from central Texas. *Geological Society of America Bulletin*, **88**, 1057–1070.

Baker, V. R. (1978a). The Spokane Flood controversy and the Martian outflow channels. *Science*, **202**, 1249–1256.

Baker, V. R. (1978b). Large-scale erosional and depositional features of the Channeled Scabland. In *The Channeled Scabland*, eds. V. R. Baker, and D. Nummedal. Washington, DC: National Aeronautics and Space Administration Planetary Geology Program, pp. 81–115.

Baker, V. R. (1978c). Paleohydraulics and hydrodynamics of scabland floods. In *The Channeled Scabland*, eds. V. R. Baker and D. Nummedal. Washington, DC: National Aeronautics and Space Administration Planetary Geology Program, pp. 59–79.

Baker, V. R. (1979). Erosional processes in channelized water flows on Mars. *Journal of Geophysical Research*, **84**, 7985–7993.

Baker, V. R. (Ed.) (1981). *Catastrophic flooding: the origin of the Channeled Scabland*, Stroudsburg, PA: Dowden, Hutchinson & Ross, Inc.

Baker, V. R. (1982). *The Channels of Mars*. Austin: University of Texas Press.

Baker, V. R. (1984). Flood sedimentation in bedrock fluvial systems. In *The Sedimentology of Gravels and Conglomerates*, eds. E. H. Koster and R. J. Steel. Canadian Society of Petroleum Geologists Memoir 10, pp. 87–98.

Baker, V. R. (1998). Catastrophism and uniformitarianism: logical roots and current relevance. In *Lyell: The Past is the Key to the Present*, eds. D. J. Blundell and A. C. Scott. London: The Geological Society, Special Publication 143, pp. 171–182.

Baker, V. R. (2001). Water and the Martian landscape. *Nature*, **412**, 228–236.

Baker, V. R. (2002). High-energy megafloods: planetary settings and sedimentary dynamics, In *Flood and Megaflood Deposits: Recent and Ancient Examples*, eds. I. P. Martini, V. R. Baker and G. Garzon. International Association of Sedimentologists Special Publication 32, pp. 3–15

Baker, V. R. and Bunker, R. C. (1985). Cataclysmic late Pleistocene flooding from glacial Lake Missoula: a review. *Quaternary Science Reviews*, **4**, 1–41.

Baker, V. R. and Kale, V. S. (1998). The role of extreme events in shaping bedrock channels. In *Rivers Over Rock: Fluvial Processes in Bedrock Channels*, eds. K. J. Tinkler and E. E. Wohl. American Geophysical Union Monograph 107, Washington, DC, pp. 153–165.

Baker, V. R. and Kochel, R. C. (1978). Morphometry of streamlined forms in terrestrial and Martian channels. In *Proceedings 9th Lunar and Planetary Science Conference*, Vol. 3. New York: Pergamon Press, pp. 3193–3203.

Baker, V. R. and Kochel, R. C. (1988). Flood sedimentation in bedrock fluvial systems. In *Flood Geomorphology*, eds. V. R. Baker, R. C. Kochel and P. C. Patton. New York: John Wiley and Sons, pp. 123–137.

Baker, V. R. and Komar, P. D. (1987). Cataclysmic flood processes and landforms. In *Geomorphic Systems of North America*, ed. W. L. Graf. Geological Society of America Centennial Special Volume 2, pp. 423–443.

Baker, V. R., Benito, G. and Rudoy, A. N. (1993). Paleohydrology of Late Pleistocene superflooding, Altay Mountains, Siberia. *Science*, **259**, 348–350.

Benito, G. (1997). Energy expenditure and geomorphic work of the cataclysmic Missoula flooding in the Columbia River Gorge, USA. *Earth Surface Processes and Landforms*, **22**, 457–472.

Benito, G. and O'Connor, J. E. (2003). Number and size of last-glacial Missoula floods in the Columbia River valley between the Pasco Basin, Washington, and Portland, Oregon. *Geological Society of America Bulletin*, **115**, 624–638.

Bretz, J H. (1923a). Glacial drainage on the Columbia Plateau. *Geological Society of America Bulletin*, **34**, 573–608.

Bretz, J H. (1923b). The Channeled Scabland of the Columbia Plateau. *Journal of Geology*, **31**, 617–649.

Bretz, J H. (1924). The Dalles type of river channel. *Journal of Geology*, **32**, 139–149.

Bretz, J H. (1925). The Spokane Flood beyond the channeled scablands. *Journal of Geology*, **33**, 97–115, 236–259.

Bretz, J H. (1928a). Channeled Scabland of eastern Washington. *Geographical Review*, **18**, 446–477.

Bretz, J H. (1928b). Bars of the Channeled Scabland. *Geological Society of America Bulletin*, **39**, 643–702.

Bretz, J H. (1929). Valley deposits immediately east of the channeled scabland of Washington. *Journal of Geology*, **37**, 393–427, 505–541.

Bretz, J H. (1932). *The Grand Coulee*. American Geographical Society Special Publication 15.

Bretz, J H. (1959). *Washington's Channeled Scabland*. Washington Division of Mines and Geology Bulletin 45.

Bretz, J H., Smith, H. T. U. and Neff, G. E. (1956). Channeled Scabland of Washington: new data and interpretations. *Geological Society of America Bulletin*, **67**, 957–1049.

Bridge, J. S. (2003). *Rivers and Floodplains: Forms, Processes, and Sedimentary Record*. Oxford: Blackwell.

Bryant, E. and Young, R. (1996). Bedrock-sculpturing by tsunami, South Coast New South Wales, Australia. *Journal of Geology*, **104**, 565–582.

Burr, D. M, Grier, J. A., McEwen, A. S. and Keszthelyi, L. P. (2002). Repeated aqueous flooding from the Cerberus Fossae: evidence for very recently extant, deep groundwater on Mars. *Icarus*, **159**, 3–73.

Carling, P. A. (1996). Morphology, sedimentology and palaeohydraulic significance of large gravel dunes: Altai Mountains, Siberia. *Sedimentology*, **43**, 647–664.

Carling, P. A. (1999). Subaqueous gravel dunes. *Journal of Sedimentary Research*, **69**, 534–545.

Chorley, R. J. (1959). The shape of drumlins. *Journal of Glaciology*, **3**, 339–344.

Church, M. and Jones, D. (1982). Channel bars in gravel-bed rivers. In *Gravel-Bed Rivers*, eds. R. E. Hey, J. C. Bathurst and C. R. Thorne. Chichester: Wiley, pp. 291–324.

Denton, G. H. and Sugden, D. E. (2005). Meltwater features that suggest Miocene ice-sheet overriding of the Transantarctic Mountains in Victoria Land, Antarctica. *Geographiska Annaler*, **87A**, 1–19.

Dury, G. H. (1964). *Principles of Underfit Streams*. U.S. Geological Survey Professional Paper 452-A.

Herget, J. (2005). *Reconstruction of Pleistocene Ice-dammed Lake Outburst Floods in the Altai Mountains, Siberia*. Geological Society of America Special Paper 386.

Jackson, J. R. (1834). Hints on the subject of geographical arrangement and nomenclature. *Royal Geographical Society Journal*, **4**, 72–88.

Jackson, R. G. (1975). Hierarchical attributes and unifying model of bedforms composed of cohesionless material and produced by shearing flow. *Geological Society of America Bulletin*, **86**, 1523–1533.

Kehew, A. E. and Lord, M. L. (1986). Origin of large-scale erosional features of glacial-lake spillways in the northern Great Plains. *Geological Society of America Bulletin*, **97**, 162–177.

Komar, P. D. (1983). Shapes of streamlined islands on Earth and Mars: experiments and analyses of the minimum-drag form. *Geology*, **11**, 651–655.

Komar, P. D. (1984). The lemniscate loop-comparisons with the shapes of streamlined landforms. *Journal of Geology*, **92**, 133–145.

Lewis, A. R., Marchant, D. R., Kowalewski, D. E., Baldwin, S. L. and Webb, L. E. (2006). The age and origin of the Labyrinth, western Dry Valleys, Antarctica: evidence for extensive middle Miocene subglacial floods and freshwater discharge to the Southern Ocean. *Geology*, **34**, 513–516.

Malde, H. E. (1968). *The Catastrophic Late Pleistocene Bonneville Flood in the Snake River Plain, Idaho*. U.S. Geological Survey Professional Paper 596.

Nanson, G. C. and Knighton, A. D. (1996). Anabranching rivers: their cause, character and classification. *Earth Surface Processes and Landforms*, **21**, 217–239.

O'Connor, J. E. (1993). *Hydrology, Hydraulics and Sediment Transport of Pleistocene Lake Bonneville Flooding on the Snake River, Idaho*. Geological Society of America Special Paper 274.

Pardee, J. T. (1942). Unusual currents in glacial Lake Missoula, *Geological Society of America Bulletin*, **53**, 1569–1600.

Patton, P. C. and Baker, V. R. (1978). Origin of the Cheney-Palouse scabland tract. In *The Channeled Scabland*, eds. V. R. Baker and D. Nummedal. Washington, DC: National Aeronautics and Space Administration Planetary Geology Program, pp. 117–130.

Rice, Jr., J. W. and Edgett, K. S. (1997). Catastrophic flood sediments in Chryse Basin, Mars, and Quincy Basin, Washington: application of sandar facies model. *Journal of Geophysical Research*, **102**, 4185–4200.

Richardson, K. and Carling, P. A. (2005). *A Typology of Sculpted Forms in Open Bedrock Channels*. Geological Society of America Special Paper 392.

Rudoy, A. N. (2005). *Giant Current Ripples: History of Research, Their Diagnostics, and Paleogeographical Significance*. Tomsk, Russia (in Russian).

Shaw, J. (2002). The meltwater hypothesis for subglacial landforms. *Quaternary Science Reviews*, **90**, 5–22.

Shaw, J., Munro-Stasiuk, M., Sawyer, B. *et al.* (1999). The Channeled Scabland: Back to Bretz? *Geology*, **27**, 605–608.

Shen, H. W. (1971). Scour near piers. In *River Mechanics*, ed. H. W. Shen. Ft. Colins, CO: Water Resources Publications, pp. 23.1–23.25.

Shepherd, R. G. and Schumm, S. A. (1974). An experimental study of river incision. *Geological Society of America Bulletin*, **85**, 257–268.

Smith, G. A. (1993). Missoula flood dynamics and magnitudes inferred from sedimentology of slack-water deposits on the Columbia Plateau, Washington. *Geological Society of America Bulletin*, **105**, 77–100.

Southard, J. (2003). Surface forms. In *Encyclopedia of Sediment and Sedimentary Rocks*, ed. G. V. Middleton. Dordrecht: Kluwer Academic, pp. 703–712.

Trimble, D. E. (1963). *Geology of Portland, Oregon, and Adjacent Areas*. U.S. Geological Survey Bulletin 1119.

Waitt Jr., R. B. (1985). Case for periodic, colossal jokulhlaups from Pleistocene glacial Lake Missoula. *Geological Society of America Bulletin*, **96**, 1271–1286.

6 The morphology and sedimentology of landforms created by subglacial megafloods

MANDY J. MUNRO-STASIUK, JOHN SHAW, DARREN B. SJOGREN, TRACY A. BRENNAND, TIMOTHY G. FISHER, DAVID R. SHARPE, PHILIP S.G. KOR, CLAIRE L. BEANEY and BRUCE B. RAINS

Summary

Subglacial landforms across various scales preserve the history of movement, deposition and erosion by the last great ice sheets and their meltwater. The origin of many of these landforms is, however, contentious. In this chapter these forms are described both individually and as suites that make up entire landscapes. Their interpretations are discussed with reference to the megaflood hypothesis. A description is provided of individual forms via their size, shape, landform associations, sedimentology and the relationship between landform surfaces and internal sediments. The possible origins of each are then discussed. To simplify the chapter the landforms are categorised by their size (micro, meso, macro and mega), although, importantly, it should be noted that several landforms show similarities across scales. Also discussed is the relevant subglacial hydrology associated with the described forms, especially the volume and discharge rates of megaflood flows, and where water may have been stored prior to the megaflood events.

6.1 Introduction

As early as 1812, Sir James Hall interpreted the famous Castle Rock in Edinburgh, Scotland, a crag and tail, as a landform created by immense, turbulent floods. Likening the hill to features carved in snow by wind, he could only hypothesise that water was responsible; probably giant tidal waves, as, at that time, he knew of no other mechanism that could conceivably create such streamlining. It is now very clear that the streamlined forms first noted by Hall are part of a continuum containing landforms of many shapes and sizes. While many researchers have attributed their formation to subglacial processes involving deformation of the glacier bed, in the last few decades attention has moved back to water as a major landforming agent.

This shift in ideas began when Shaw (1983) noted that drumlins were similar in form to inverted erosional marks produced by turbulently flowing water. He therefore proposed that they may have been formed by meltwater erosion and deposition. Since the publication of this paper an enormous volume of research has been undertaken addressing a meltwater genesis for a broad suite of subglacial landforms including: s-forms, drumlins, fluting, hummocky terrain, Rogen moraine, megaripples, tunnel channels, eskers, streamlined hills, rises and re-entrant valleys. As a continuum, these landforms define broad landscape unconformities which on the largest scale indicate meltwater flowpaths. Adding credence to the meltwater hypothesis is the recognition of other erosional landforms of similar form that are produced by turbulent flow. These include: yardangs (Figure 6.1a), which are widely accepted as the work of wind (Greeley and Iversen, 1985); sastrugi in snow and ice, which also result from wind sculpting (e.g. Herzfeld *et al.*, 2003) (Figure 6.1b); erosional marks produced by tsunami waves (Bryant and Young, 1996) (Figure 6.1c); and the loess hills of the Channeled Scablands (Figure 6.1d), which were formed by enormous outpourings of water from glacial Lake Missoula and the Cordilleran Ice Sheet system (Bretz, 1959; Baker, 1978; Shaw *et al.*, 2000). Common to the forms illustrated in Figure 6.1, and the majority of the subglacial forms discussed in this paper, is that they are considered as the products of turbulent flow that contained vortices. For the forms described here the inferred turbulent flow is subglacial meltwater.

In this chapter, a synopsis is provided of the subglacial landforms that are interpreted to be produced by meltwater, and together they document evidence for megafloods. Their size, shape, landform associations, sedimentology, and the relationship between landform surfaces and internal sediments are all described. It is argued here that all the features described, except for Livingstone drumlins, Rogen moraine and eskers, are exclusively the product of subglacial meltwater erosion. Livingstone drumlins and

Megaflooding on Earth and Mars, ed. Devon M. Burr, Paul A. Carling and Victor R. Baker.
Published by Cambridge University Press.

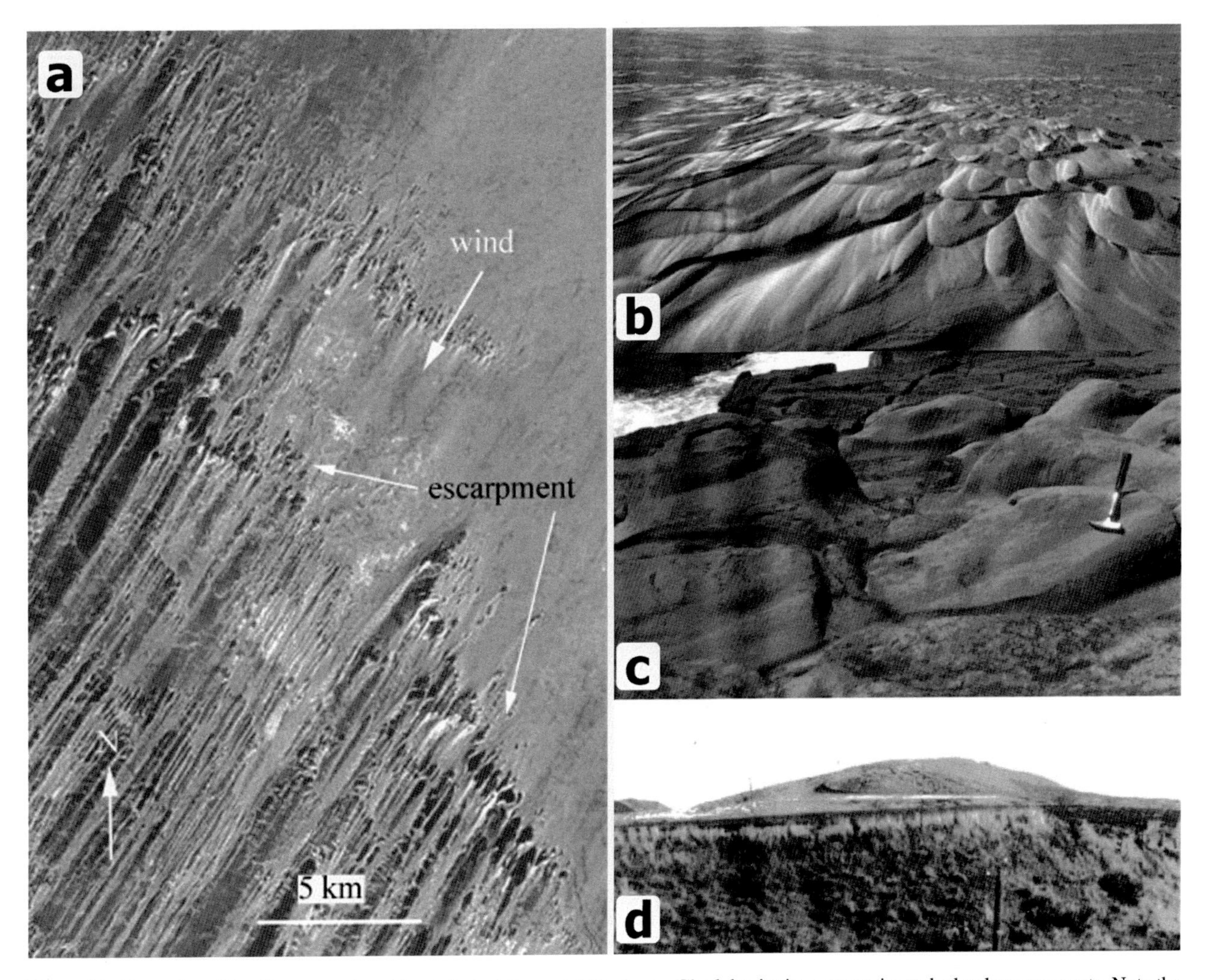

Figure 6.1. Examples of landforms produced by turbulent flows. (a) Yardangs, Chad, beginning at prominent, bedrock escarpments. Note the strong linearity of the bedforms (image from Google Earth); (b) Sastrugi in snow (courtesy the Royal Geographical Society); (c) Erosional marks produced by tsunami waves (courtesy of E. Bryant); (d) Streamlined loess hill in the Channeled Scablands.

Rogen moraine are related to erosion; however, the erosion was up into the base of the ice rather than down into the substratum. The eroded cavities were filled subsequently with sediment during waning flows. Eskers are believed to represent the final phases of floods. Because several of the features discussed cross the boundaries of erosional versus depositional, they are categorised by size (micro, meso, macro and mega) (Table 6.1). Importantly, though, some of these landforms show similarities across several orders of magnitude in scale and this fact is noted where appropriate. All of the described forms are observed as surface or buried features in formerly glaciated areas.

Table 6.1. Classification of subglacial landforms created by meltwater flows (D: depositional, E: erosional)

Scale	Landform
Microforms	s-forms (E)
Mesoforms	Drumlins (E & D)
	Fluting (E)
	Hummocky terrain (E)
	Rogen moraine (E & D)
	Megaripples (E)
	Tunnel channels (E)
	Eskers (D)
Macroforms	Streamlined hills (E)
Megaforms	Landscape unconformities (E)
	Flowpaths (E)

6.1.1 The megaflood controversy

Before discussing megaflood landforms, it is pointed out that the megaflood hypothesis is not accepted universally. There is a significant debate in the literature as to whether many of the features described herein were formed by ice or by water. Arguments in opposition to the meltwater theory has been summarised by Benn and Evans (2006) and Evans *et al.* (2006). In particular, Evans *et al.* (2006) state, 'It is regrettable that (the megaflood hypothesis) has not been subject to systematic scrutiny or testing at local scales of field based enquiry.' This statement is a misrepresentation of the megaflood research as the majority of evidence, in the form of dozens of studies, is from field-based work.

Perhaps the biggest cause for concern among critics of the meltwater hypothesis is the use of form analogy (e.g. Benn and Evans, 2006), one of the original lines of evidence that Shaw (1983) used to develop his hypothesis. For instance, Benn and Evans (1998, p. 446) wrote, 'The use of sole marks beneath turbidites as analogues for drumlins involves a major leap of faith, given the huge difference in the scale of the features.' This assertion is simply wrong. The analogy between drumlins and erosional marks does not always involve huge differences in scale. For instance: Normark *et al.* (1979) described 500 m wide crescentic scour marks on the Navy Submarine Fan; Fildani *et al.* (2006) described scours along the Monterey East Channel that are up to 4.5 km wide, 6 km long and 200 m deep; and Piper *et al.* (2007) described a turbidity current scour on the Laurentian Fan, 2 km long and 100 m deep. More importantly, the analogy is strong because there are a large number of points of similarity between Livingstone drumlins (the ones originally compared to the scour marks) and inverted sole marks. Lorenz (1974) stated that the improbability of coincidental similarity is $1{:}2^{(1-n)}$, where n is the number of points of similarity. In this case, there are conservatively eight points of similarity: (1) orientation with flow; (2) appearance in fields of similar form; (3) steeper at the upstream end; (4) spindle form; (5) parabolic form; (6) transverse asymmetrical form; (7) barchanoid form; and (8) infilling by current-bedded sediment. Thus the improbability of coincidental similarity is 1:128, strongly suggesting that the Livingtone drumlins and scours marks are related.

In addition, Gilbert (1896) noted, 'Hypotheses are always suggested through analogy'. He made it clear that analogy is just a starting point for a hypothesis and explained that additional features to those already observed are necessary before a hypothesis can be considered satisfactory. Much of this chapter concerns additional features that support the original drumlin hypothesis. Also these features expand the scope from the origin of drumlins to the much broader topic of subglacial outburst floods.

6.2 Microforms

Microforms are the smallest of all the landforms that are discussed in this chapter. They typically range from a few millimetres in size to a few tens of metres (e.g. Kor *et al.*, 1991; Munro-Stasiuk *et al.*, 2005). Ljüngner (1930) first described and classified these features, now commonly referred to as s-forms (sculpted forms) after Kor *et al.* (1991) (Figure 6.2). Many microforms, especially muschelbrüche, sichelwannen, spindle flutes, and other scour marks have been reproduced in flumes (e.g. Allen, 1971; Shaw and Sharpe, 1987) and identical forms have been noted in fluvial environments (e.g. Maxson, 1940; Karcz, 1968; Baker and Pickup, 1987; Wohl, 1992; Tinkler, 1993; Baker and Kale, 1998; Hancock *et al.*, 1998; Wohl and Ikeda, 1998; Gupta *et al.*, 1999; Whipple *et al.*, 2000; Richardson and Carling, 2005). It is known, therefore, that these can be eroded by water. Also, the presence of sharp upper rims on many of these forms typically is representative of flow separation (e.g. Allen, 1982; Sharpe and Shaw, 1989) and the ever-present crescentic and hairpin furrows represent the generation of horseshoe vortices around obstacles encountered by the flow (Peabody, 1947; Dzulynski and Sanders, 1962; Karcz, 1968; Baker, 1973; Allen, 1982; Sharpe and Shaw, 1989; Shaw, 1994; Lorenc *et al.*, 1994).

A recent thorough synopsis of sculpted forms by Richardson and Carling (2005) divides strictly fluvial forms into three topological types: (1) concave features that include potholes and furrows; (2) convex and undulating forms such as hummocky and undulating forms; and (3) composite forms such as compound obstacle marks. However, here a simpler classification is used, based on Kor *et al.* (1991), that groups forms into three types (Table 6.2) based on their orientation relative to the flow direction: (1) transverse s-forms are usually wider than they are long, where length is measured parallel to the formative flow direction; (2) longitudinal forms are generally longer than they are wide, with their long axes lying parallel to the flow direction; and (3) non-directional forms have no obvious relationship to flow direction, although their location amongst other s-forms may indicate their origin. Importantly, these forms typically occur en echelon (Figure 6.3a).

6.2.1 Transverse s-forms

Transverse s-forms that originate transverse to the flow direction include muschelbrüche, sichelwannen, comma forms and transverse troughs.

Muschelbrüche (singular muschelbrüch) are shallow depressions that resemble the inverted casts of the shells of mussels, with sharp, convex upflow rims and indistinct, downflow margins merging imperceptibly with the adjacent rock surface (Figure 6.3b). The proximal slope is steeper than the distal slope.

Sichelwannen (singular sichelwanne) are sickle-shaped marks (Figure 6.3c) that resemble the classical transverse erosional marks (flutes) of Allen (1971,

Table 6.2. Classification of s-forms based on Kor *et al.* (1991)

Group	Type	Description
Transverse forms	Muschelbrüche	Mussel-shell-shaped depressions with sharp, convex upflow rims
	Sichelwannen	Sickle-shaped convex upflow marks, with arms extending around medial ridge
	Comma forms	Similar to sichelwannen, but with one arm
	Transverse troughs	Relatively straight troughs perpendicular to flow
Longitudinal forms	Spindle flutes	Narrow, long, shallow spindle-shaped marks with sharp rims on upflow side
	Cavettos	Curvilinear undercut channels
	Stoss-side furrows	Shallow linear depressions on stoss side of bedrock rises
	Furrows	Long linear troughs adorned by many smaller s-forms
Non-directional forms	Undulating surfaces	Non-directional bed undulations
	Potholes	Near-circular depressions with spiralling flow patterns on walls

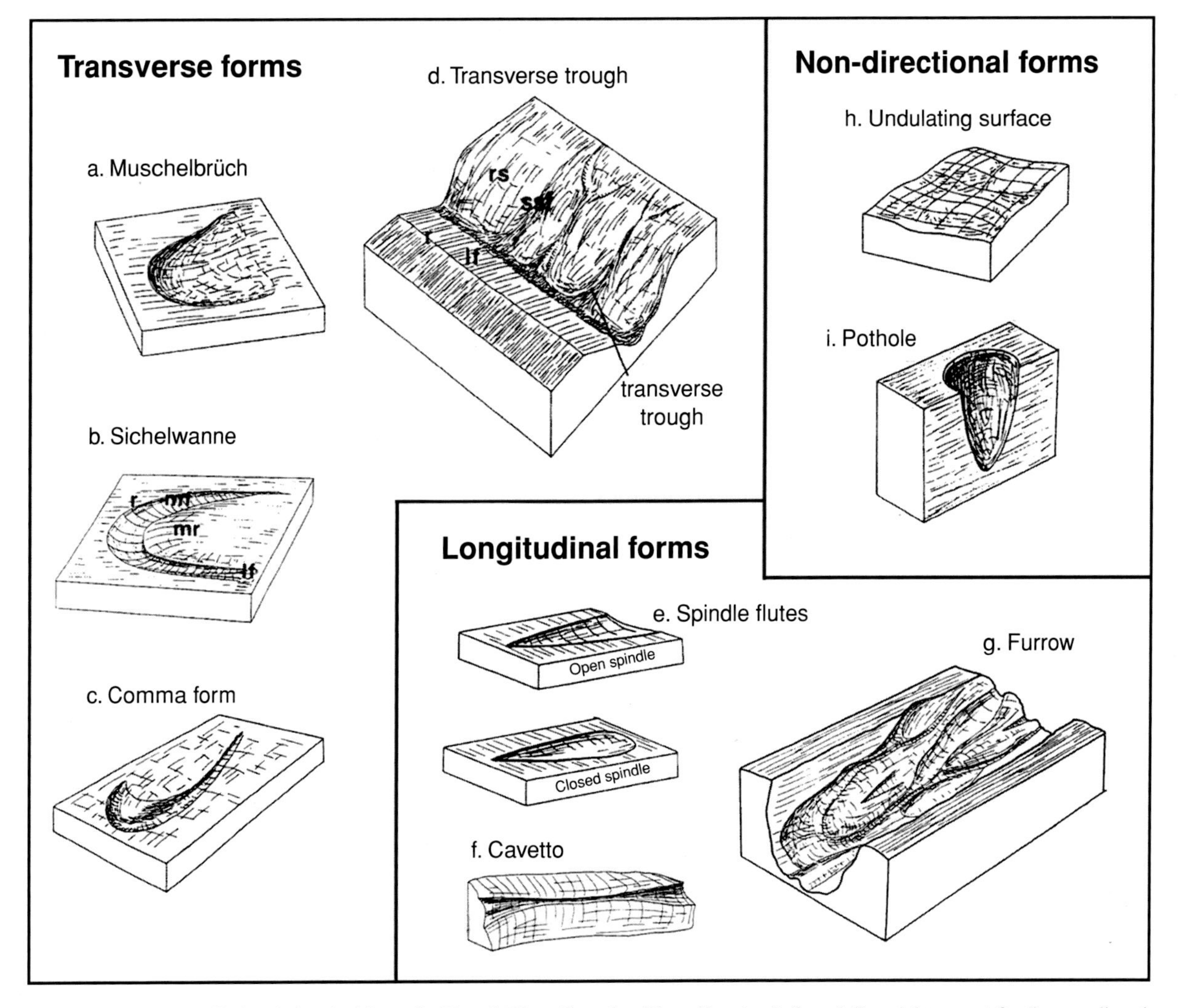

Figure 6.2. S-forms identified and sketched from the French River Complex. Flow direction is from left to right except for the non-directional forms. (Modified from Kor *et al.* (1991).)

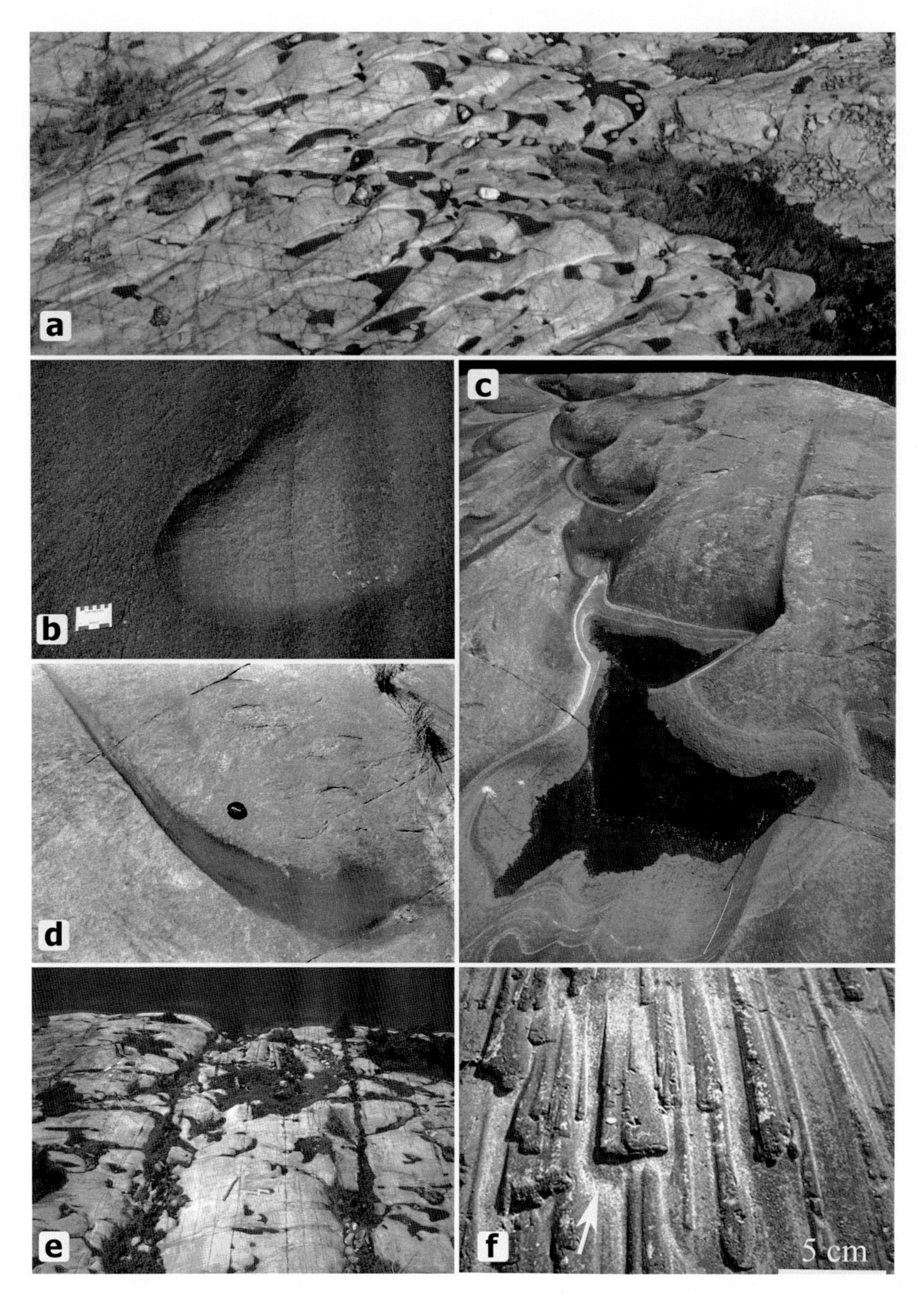

Figure 6.3. Photographs of each s-form type. All images are from the French River Complex, Georgian Bay, Ontario, unless otherwise noted. (a) S-forms typically occur in suites; (b) sichelwanne; (c) muschelbrüche; (d) comma form; (e) transverse troughs; (f) rat-tails at Marysville, Ontario, Canada; (g) closed spindles; (h) cavetto at Kelleys Island, Ohio; (i) stoss-side trough; (j) the large furrow at Kelleys Island – contained within it are many other types of s-forms; (k) undulating surface; (l) large pothole at Key River, Georgian Bay.

Figure 6.3. (*cont.*)

Figure 1). They have sharp, convex upflow rims and a crescentic main furrow, extending downflow into arms wrapped around a median ridge. Lateral furrows may flank the main furrow. In an en-echelon system, the 'arms' merge and bifurcate downflow into other sichelwannen, or may extend downflow into comma forms. Munro-Stasiuk *et al.* (2005) noted that these forms sometimes occur as multiple-trough forms having as many as ten individual semi-parallel troughs.

Sichelwannen sometimes blend into comma forms or vice versa and they are thus part of a continuum. Morphologically they are almost identical to sichelwannen except there is only one well-developed 'arm', or exit furrow; the other arm is either missing or poorly formed (Figure 6.3d) (Shaw and Kvill, 1984).

Transverse troughs are relatively straight troughs arranged perpendicular to flow, with widths much greater than lengths (Figure 6.3e). They commonly have a steep, relatively planar upflow slope or lee face below a relatively straight rim. Potholes often occupy this upflow slope. The downflow or riser slope is gentler and normally eroded by shallow, stoss-side furrows, which produce sinuous slope contours. Large transverse troughs are normally compound forms enclosing numerous, smaller-scale s-forms.

6.2.2 Longitudinal s-forms

Longitudinal s-forms have a distinct orientation in the direction of the forming flow. They include rat-tails, spindle flutes, cavettos, stoss-side furrows and furrows.

Rat-tails are positive residual bedforms defined by hairpin scours (crescentic scours with arms extending downflow) wrapped around their stoss ends (Figure 6.3f). They extend in the downflow direction from resistant obstacles. As the lateral scours widen and shallow downflow, the rat-tail tapers and becomes lower. Rat-tails occur at the millimetre scale up to several kilometres around seamounts and islands in tidal estuaries (Allen, 1982). Gilbert *et al.* (2003) drew the same analogy between small-scale features in Ontario and kilometre-scale rat-tails on the Antarctic shelf.

Spindle flutes are narrow, shallow, spindle-shaped marks much longer than they are wide and with sharp rims bounding the upflow side and, in some cases, the downflow margins (Figure 6.3g) (Allen, 1971). They are pointed in the upflow direction and broaden downflow. Whereas open spindle flutes merge indistinctly downflow with the adjacent rock surface, closed spindles have sharp rims closing at both the upflow and downflow ends. Spindle flutes may be asymmetrical, with one rim more curved than the other.

Cavettos are curvilinear, undercut channels eroded into steep, commonly vertical or near-vertical rock faces (Figure 6.3h) (Johnsson, 1956; Kor *et al.*, 1991; Munro-Stasiuk *et al.*, 2005). The upper lip is usually sharper than the lower one, although both may be sharp. Cavetto forms vary greatly in size. Small-scale, long sinuous channels are seen on the flanks of bedrock promontories already sculpted with other s-forms (Figure 6.3h). Others adorn large vertical rock promontories making them difficult to detect in the field.

Stoss-side furrows are shallow, linear depressions on the stoss side of bedrock rises, giving a regular, gently curving, sinuous contour to the slope (Figure 6.3i). They have rounded rims and are open at both ends. These furrows commonly form on the distal slopes of transverse troughs.

Furrows are linear troughs, much longer than wide, that carry a variety of s-forms and remnant ridges on their beds and walls (Figure 6.3j). Rims are remarkably straight when viewed over the full length of furrows but are usually sinuous in detail, due to sculpting into the trough walls by smaller s-forms. These forms vary greatly in size. Commonly they are recognised easily at outcrop scale but may be so large that they are only recognised from the air or on topographic maps. Remarkable examples of furrows at this scale are the huge re-entrant valleys that mark the northern portion of the Niagara Escarpment (Straw, 1968; Kor and Cowell, 1998). Munro-Stasiuk *et al.* (2005) suggested that some furrows may be longitudinal troughs that lie between larger-scale fluting ridges (see section on fluting).

6.2.3 Non-directional s-forms

Non-directional s-forms record no discernable flow direction and include undulating surfaces and potholes.

Undulating surfaces are smooth, non-directional, low-amplitude undulations found on gentle lee slopes of rock rises (Figure 6.3k). These low-amplitude features are very common to broad areas of low-relief terrain, usually located peripheral to the main sheetflow events, and in association with the other s-forms described above.

Potholes are near-circular, deep depressions that may show spiralling, descending flow elements inscribed on their walls (Figure 6.3l) (Gilbert, 1906; Alexander, 1932). Potholes typically occur in bedrock terrains on the summits, flanks and, notably, on the lee sides of bedrock knolls and ridges. They range in size (diameter and depth) from a few centimetres to many metres and commonly have a circular to slightly oval surface expression. The most spectacular potholes occur in the lee of bedrock obstacles or obstructions and are often missing their distal walls. Potholes may occur singly or in small groups, or may occur in large concentrations, often showing an en echelon pattern. Some potholes contain a lag of boulders or other debris.

When potholes occur on the lee side of obstructions and they are associated with abundant sculpted forms on bedrock surfaces, they may be interpreted as resulting from

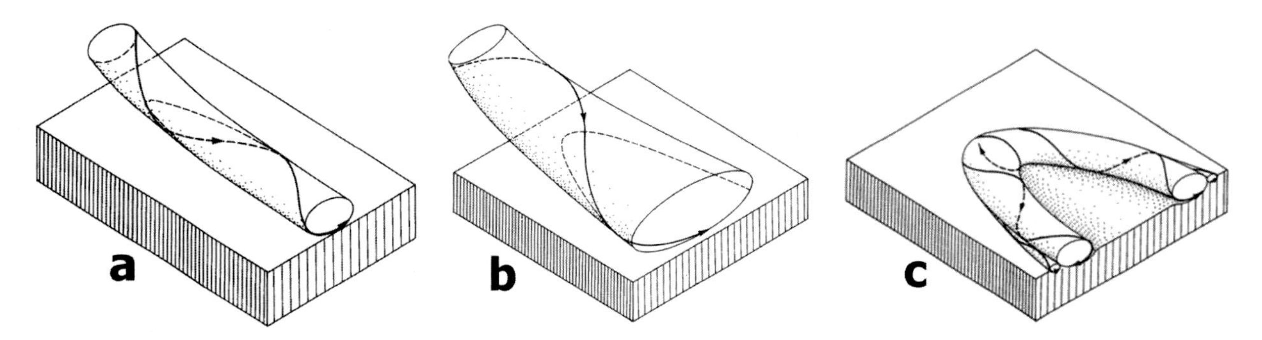

Figure 6.4. Vortices responsible for some s-forms. (a) Low-angle plunging vortex responsible for creating spindle flutes; (b) high-angle plunging vortex responsible for creating muschelbrüche; (c) vortices generated by a bed defect responsible for creating sichelwannen.

catastrophic flood release of meltwater from beneath an active ice sheet (e.g. Embleton and King, 1975; Sugden and John, 1976; Kor *et al.*, 1991). In this association, potholes are considered to be a type of s-form (Kor *et al.*, 1991).

6.2.4 Flow characteristics related to s-forms

S-forms are typically formed by preferential erosion by vortices. Muschelbrüche and spindle forms are considered to result from vortex impingement on the rock bed (Kor *et al.*, 1991). Spindle forms result when the vortex impingement is at a low angle (Figure 6.4a) and muschelbrüche form when impingement is at a high angle (Figure 6.4b). By contrast, sichelwannen result from separation at a convex upstream rim creating counter-rotating vortices, which erode longitudinal troughs that wrap around medial ridges (Figure 6.4c). In some cases the primary vortices generate secondary, lateral vortices that erode lateral furrows.

Transverse troughs also involve flow separation that gives rise to rollers along the upstream slope. It is probable that rollers become counter-rotating longitudinal vortices, which are swept over the stoss side of downstream ridges (Pollard *et al.*, 1996). If this happens with regular spacing a series of stoss-side furrows would form. Furrows would grow as they captured vortex pairs. This process is analogous with the formation of re-entrant valleys at escarpments to be discussed. Longitudinal forms (furrows and cavettos) are formed by longitudinal vortices although the flow is frequently complex, resulting in the formation of transverse features as ornamentation (Munro-Stasiuk *et al.*, 2005). Potholes are formed by vortices spiralling down the walls and stretched vortices emerging from the core of the flow. Undulating surfaces are not related clearly to a particular structure, though a three-dimensional wave structure in the flow may be responsible for this bedform type.

Observations in flumes show that rat-tails are eroded by horseshoe vortices generated by an interaction between the obstacle and the boundary layer. Flow separation gives rise to two counter-rotating vortices wrapping around the upstream sides of obstacles and extending down the flanks

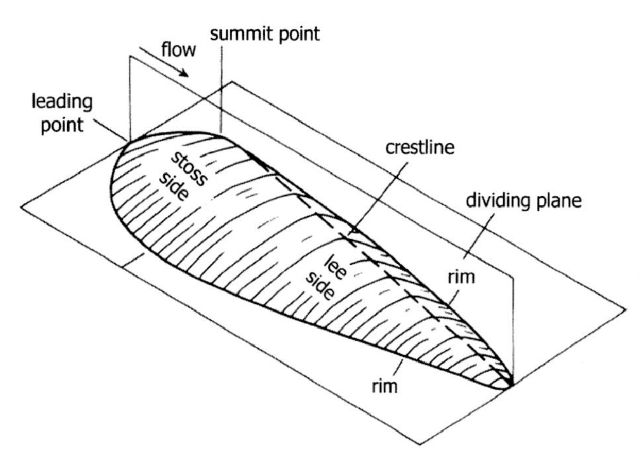

Figure 6.5. The classical streamlined drumlin with blunt stoss side and tapered lee side.

of the rat-tail. As the vortices increase in diameter and weaken, the hairpin scours on either side of the rat-tail become wider and shallower so that it is pointed downstream and merges with the bed. The flow structure for the formation of rat-tails is scale independent.

6.3 Mesoforms

Mesoforms are perhaps the most widely recognised yet least understood of all the landforms discussed in this chapter. They are on the scale of tens of metres to tens of kilometres in length and tens to hundreds of metres in width. Despite a larger scale, their common erosional origins mean that they have much in common with s-forms.

6.3.1 Drumlins

Classical drumlins (Figure 6.5) are streamlined and are arranged in fields, some containing tens of thousands of individual landforms. They are asymmetrical in plan, highest at their proximal blunt ends and taper in a downflow direction on their lee sides. There have been several theories of drumlin formation proposed including but not limited to: subglacial deformation (Boulton, 1987), the dilatancy theory (Smalley and Unwin, 1968) and subglacial meltwater

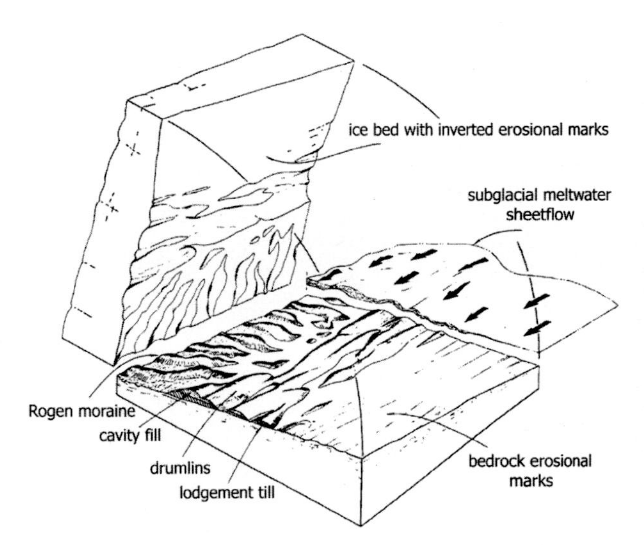

Figure 6.6. Landform formation by broad, subglacial meltwater flow. Note erosional and depositional features.

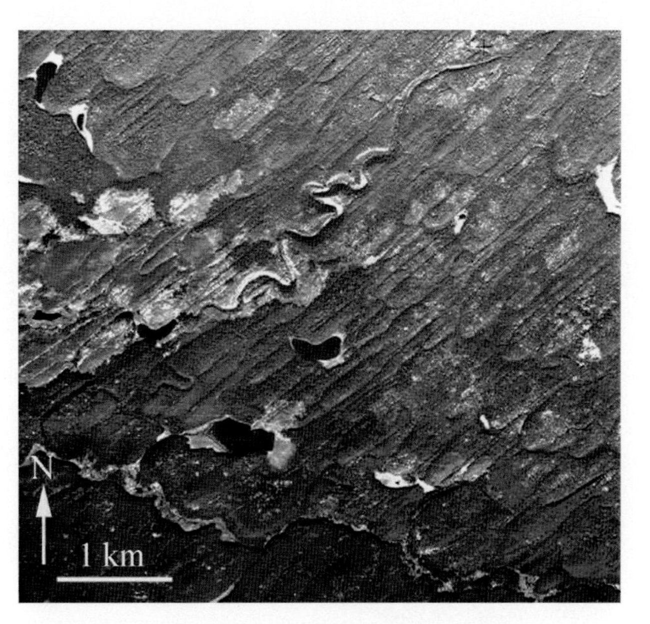

Figure 6.7. Erosional drumlins, Nechako Plateau, British Columbia. Flow from the southwest. Individual drumlins are outlined by narrow, hairpin troughs (dark) wrapped around their stoss ends.

erosion and deposition (Shaw *et al.*, 1989). As shown in Figure 6.6 the meltwater hypothesis for drumlin formation holds that drumlins are created either by direct fluvial erosion of the bed (Beverleys), or by fluvial infilling of cavities formed as erosional marks in glacier beds (Livingstones) (e.g. Shaw, 1996).

It is generally thought that drumlin streamlining represents minimum resistance to flow but this is only true for flows of high Reynolds numbers, i.e., turbulently flowing water (Shapiro, 1961). Hairpin furrows or horseshoe-shaped scours commonly are wrapped around the proximal ends of Beverleys (Figure 6.7) and crag and tails (e.g. Shaw, 1994), in a similar fashion to the s-forms previously described. Analogous forms in turbulent flow are produced by vortex erosion around bridge piers (Dargahi, 1990) and on the upstream side of scour remnant ridges (Allen, 1982). Thus, streamlining and hairpin troughs support the hypothesis of drumlin formation by turbulent meltwater.

The classical drumlin shape is represented by Beverleys (Figure 6.7). Beverleys are erosional remnants composed of pre-existing material. They, therefore, can contain sorted and stratified sediment, diamicton (till or debris flow deposits), soil horizons and even bedrock (e.g. Habbe, 1989).

Livingstones have distinctive forms that are different from Beverleys: spindle, parabolic and transverse asymmetrical forms (Figure 6.8) (Shaw 1983). When inverted, these three types have analogues in erosional marks produced by turbulently flowing water and air (e.g. Shaw, 1983). Livingstones are mainly composed of fluvial sediment, some of which may have been resedimented by mass movement on steep, primary slopes. Exposures in Livingstones in the Livingstone Lake drumlin field in northern Saskatchewan show mainly sorted and stratified sediments with large boulders (Shaw *et al.*, 1989). Boulder and cobble gravels interbedded with graded and bedded sand give way distally to fine, graded sand with pebbly diamicton, although there are striated boulders in proximal and distal positions. Bedding in all excavations is conformable with the land surface. Gravel clasts in the Livingstone Lake drumlins are predominantly from the local Athabasca Sandstone, indicating a short distance of transport (Shaw *et al.*, 1989).

Deposition of Livingstones is explained best as a combination of fluvial sedimentation and mass movement in cavities eroded into the underside of ice sheets with intermittent emplacement of boulders released from cavity roofs.

Drumlins and their relationships with adjacent landforms Because Beverleys occur on interfluves between tunnel channels and both are part of an unconformity (Sharpe *et al.*, 2004), it is suggested that they are products of erosion by the same agent. The channels are regarded widely as meltwater forms (see section on tunnel channels) and it has been argued above that drumlins are formed by a turbulent fluid, probably meltwater. Drumlins and fluting commonly extend from forward-facing steps where turbulent fluids form streamwise zones of high and low velocity related to longitudinal vortices (e.g. Tinkler and Stenson, 1992; Pollard *et al.*, 1996; Wilhelm *et al.*, 2003). Downflow walls of tunnel channels lying oblique or transverse

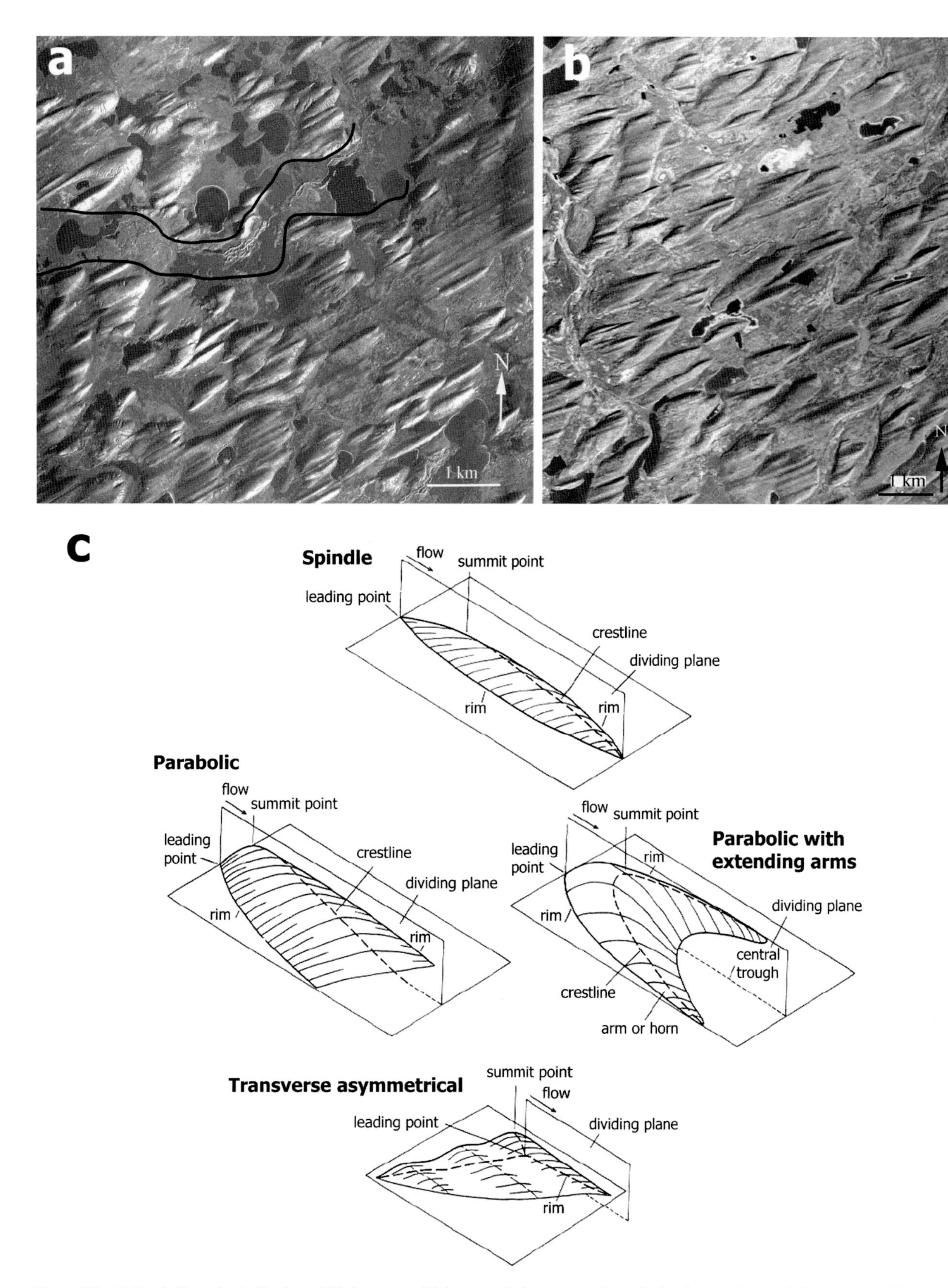

Figure 6.8. (a) Parabolic and spindle-shaped Livingstones, Livingstone Lake area, northern Saskatchewan. A tunnel channel containing eskers is accentuated. Flow from northeast. (b) Transverse asymmetrical drumlins, Livingstone Lake area, northern Saskatchewan. Flow from northeast. (c) Forms of Livingstone drumlins. To view these as erosional marks, turn the page upside down.

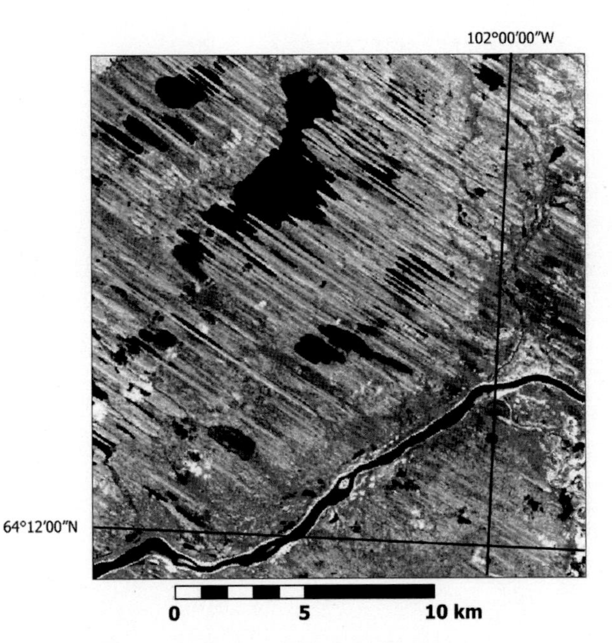

Figure 6.9. Landsat TM image of fluting initiated on a forward-facing slope at Dubawnt Lake, Northwest Territories, Canada. Flow from southeast.

to the regional flow are forward-facing steps along which drumlins and flutings are typically clustered (Shaw *et al.*, 2000; Munro-Stasiuk and Shaw, 2002).

The close association of drumlins and tunnel channels and eskers points to a sequence: formation of drumlins by broad meltwater flows; flow concentration in channels and the formation of tunnel channels; followed by concentration of flow into conduits eroded upwards into the ice bed and the formation of eskers. Together, these make up the broader landscape.

6.3.2 Large-scale fluting

Sometimes called megascale glacial lineations, large-scale fluting can be tens of kilometres long and a few hundred metres wide. Fluted terrain comprises ridges with lateral troughs. Crescentic troughs commonly wrap around the leading edges of ridges (Figure 6.9). Fluting occurs together with drumlins in the same field and they are part of a continuum. High ground downstream from forward-facing slopes or steps is commonly fluted (Pollard *et al.*, 1996; Shaw *et al.*, 2000; Munro-Stasiuk and Shaw, 2002). This characteristic is shared with yardangs (Figure 6.1a) and furrows associated with volcanic blasts, and is explained by the formation of streamwise vortices in turbulent flow in those locations (Pollard *et al.*, 1996; Wilhelm *et al.*, 2003). The internal structure of fluting is typically truncated by the land surface and a boulder lag commonly lies on this surface (Shaw *et al.*, 2000; Munro-Stasiuk and Shaw, 2002), indicating erosional formation by turbulent fluid. For subglacial environments, this fluid would have been meltwater.

6.3.3 Hummocky terrain

Unlike drumlins and fluting, hummocky terrain, known traditionally as hummocky moraine, was long thought to be interpreted easily. Although its origin has been attributed to many glacial processes, it is generally thought to have originated by supraglacial deposition during let-down at, or near, ice margins (e.g. Gravenor and Kupsch, 1959). Therefore this terrain has been used to delineate recessional stages of glaciation (e.g. Clayton and Moran, 1974; Dyke and Prest, 1987). While this interpretation may hold true in some regions, observations in the southwest sector of the Laurentide Ice Sheet demonstrate that hummocky terrain was formed by erosion, specifically, subglacial meltwater erosion (Munro and Shaw, 1997).

The major north–south trending hummocky belts in Alberta and Saskatchewan specifically have been identified previously as hummocky moraines. Munro-Stasiuk and Sjogren (2006) grouped hummocks into six types based on shape (both plan and cross-section): I – mounds with no discernable orientations or patterns; II – mounds with central depressions; III – linked mounds with central depressions; IV – ridged mounds; V – elongate mounds; and VI – moraine plateaux (Figure 6.10). Materials in the hummocks include lodgement and melt-out till, *in-situ* and disturbed lake sediments, and *in-situ* and thrust local bedrock and preglacial sediments (e.g. Kulig, 1985; Tsui *et al.*, 1989; Munro and Shaw, 1997; Sjogren, 1999; Munro-Stasiuk, 2003; Evans *et al.*, 2006).

The presence of *in-situ* bedrock in some of these forms demonstrates conclusively that they are the products of erosion rather than deposition. In addition, exposures clearly show that intact regional lithostratigraphies and local sedimentary beds are truncated by hummock surfaces (Figure 6.11). Hummock surfaces, therefore, form a landscape unconformity (see section on landscape unconformities). Sedimentary observations point to a subglacial origin for the erosion: (1) subglacial eskers overlie the hummocks (Munro and Shaw, 1997); and (2) the youngest recorded unit in the hummocks is a subglacial melt-out till with strongly oriented clast fabrics, which are up to 70° at variance from the alignments noted in hummocks (Munro-Stasiuk, 2000). Thus, erosion of the hummock surfaces was not related directly to the processes that deposited the underlying till; it clearly occurred after till deposition.

While the dynamics of flow for hummocky terrain currently are poorly understood, several lines of evidence suggest erosion by fluvial processes that removed sediment grain by grain, thus scouring underlying sediment but leaving beds undisturbed: (1) surface boulders at many locations are best explained as fluvial lags resulting from lower

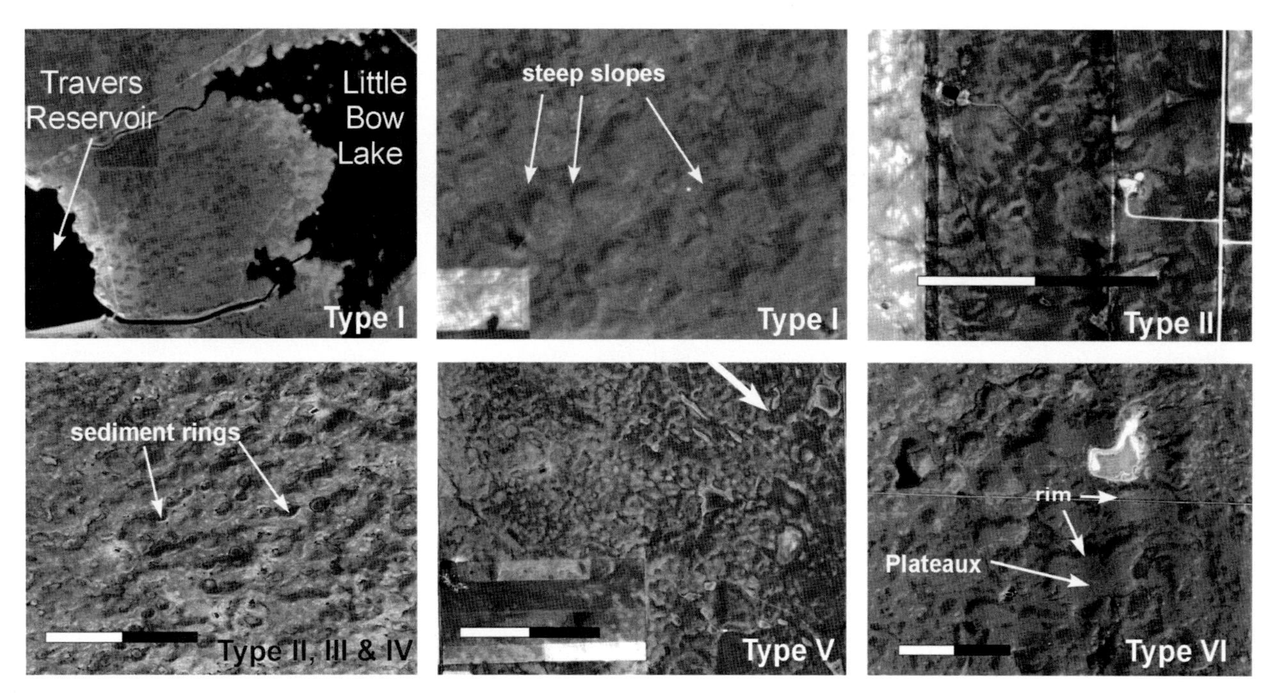

Figure 6.10. Hummock types observed in central and southern Alberta.

flow competence in some areas; (2) sorting of the lags suggests fluvial transport; (3) many boulders are heavily pitted with percussion marks attesting to clast on clast collisions; (4) type IV hummocks (ridged mounds) resemble fluvial bedforms (see Allen, 1982), erosional marks produced on the underside of river ice (Ashton and Kennedy, 1972) and may in fact be a type of Rogen moraine (see section on Rogen moraine); and (5) horseshoe-shaped troughs are wrapped around the upstream sides of some mounds suggesting scouring by horseshoe vortices generated at obstacles in the flow (e.g., Shaw, 1994), as for some drumlins and s-forms.

6.3.4 Rogen moraine

Rogen moraine, also known as ribbed moraine, consists of transverse ridges that formed perpendicular to flow. They form a continuum with drumlins and are accepted by most researchers as having formed subglacially, although there is still debate as to their specific origin. Based on both form analogy and sedimentology, Fisher and Shaw (1992) preferred a meltwater origin for forms that they observed on the Avalon Peninsula, Newfoundland. Rogen moraine ridges were held to have formed in cavities eroded upwards into the glacier bed. Identical-shaped cavities of smaller scale are produced by turbulent water flowing under river ice (Fisher and Shaw, 1992). These transverse cavities (Figures 6.12a, b and d) when inverted display all the geometric aspects of Rogen ridges including broad crescentic shapes, convex upflow, anastomosing patterns of ridges, and individual hummocks between ridges. Ashton and Kennedy (1972), who also examined river ice, noted a fluted morphology with small scallops aligned perpendicular to flow making up asymmetrical ripples that resemble drumlins in form.

Sediments in the Rogen ridges are typical of both glaciofluvial deposition and debris flows (Figures 6.12e and f) (Fisher, 1989; Fisher and Shaw, 1992). Clasts overlying granule beds record cut and fill by flows, and the silt and sand stringers probably were elutriated during type I and II debris flows (see Lawson, 1979). Gravel and sand beds may record meltwater deposition between debris flow events. All sediments appear to record primary deposition with only minor shearing. This observation is consistent with deposition in subglacial cavities. Similar descriptions and interpretations have been provided for Swedish Rogen moraine by Möller (2006).

Fisher and Shaw (1992) presented a model whereby transverse cavities were eroded into the base of a glacier by a broad sheet flow, similar to the transverse cavities noted by Ashton and Kennedy (1972). These cavities then filled with glaciofluvial sediment, some of which was reworked by debris flows. There is similarity in both the environment and processes of deposition for Rogen ridges and Livingstones.

6.3.5 Megaripples

Features identified as megaripples have rarely been described although they appear prominently on DEMs of

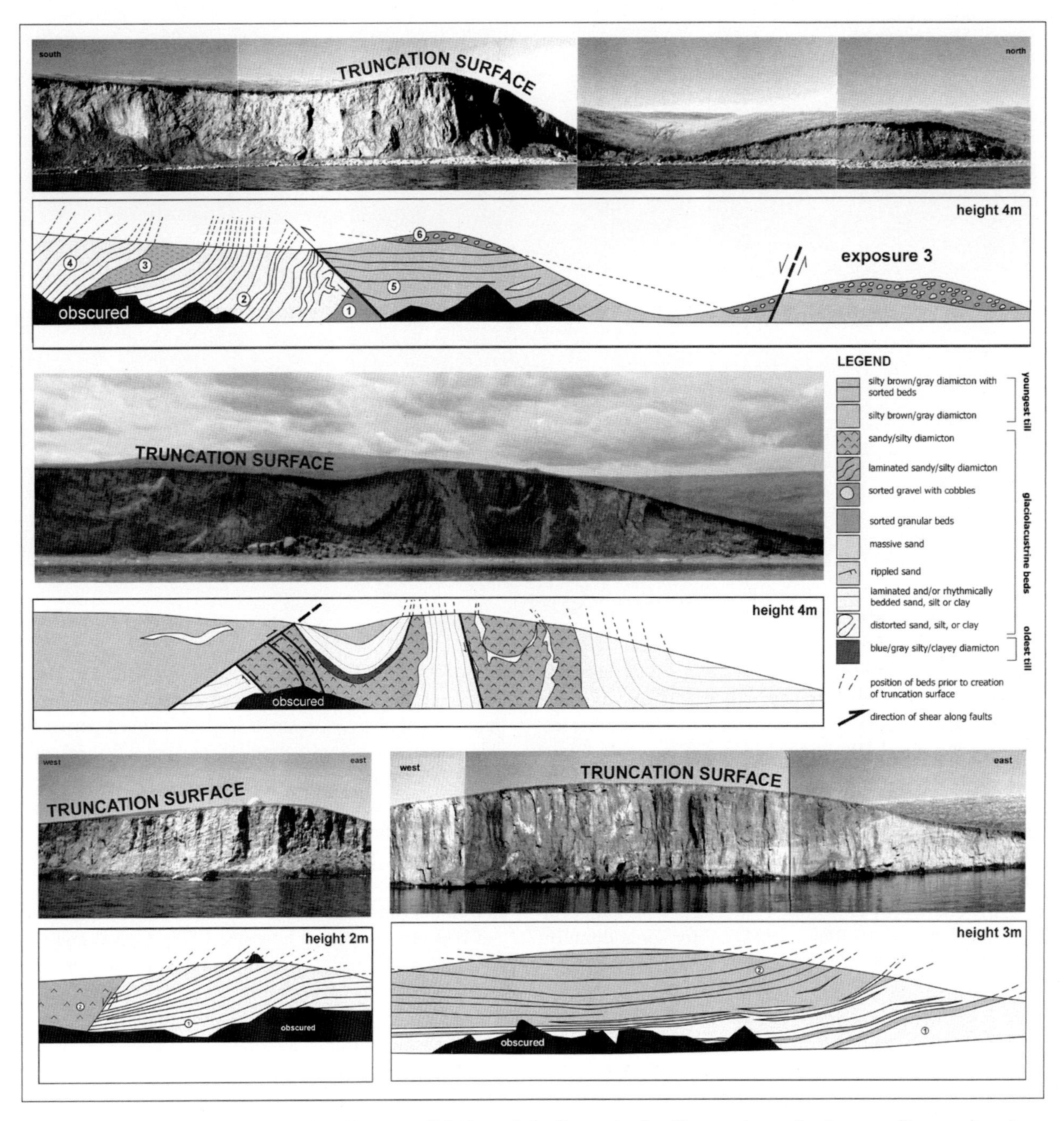

Figure 6.11. Representative exposures along Travers and McGregor Lake Reservoirs that illustrate the erosional nature of hummock surfaces.

previously glaciated regions (Figure 6.13). Megaripples are identical in form to regular sand ripples except they are orders of magnitude larger. Beaney and Shaw (2000) identified erosional megaripples on the preglacial Milk River drainage divide in southeast Alberta. These were previously interpreted as glaciotectonic thrust ridges of Cretaceous Shale (see Shetsen, 1987). Megaripples are part of a suite of landforms, including fluted terrain, scoured bedrock and tunnel channels, all interpreted to be the product of meltwater erosion. Beaney and Shaw noted the similarity between the ridge forms and transverse s-forms (Kor *et al.*, 1991), erosional megaripples on the Laurentian Fan (Hughes Clarke *et al.*, 1990), and antidunes formed on river beds (Shaw and Kellerhals, 1977; Allen, 1982). The ridges have crests up to 10 km long, with wavelengths ranging from 400 m to 1300 m, and heights ranging from 2 m to 15 m (Figure 6.13). The ridges have a well-developed asymmetric profile, with steeper downflow sides. These

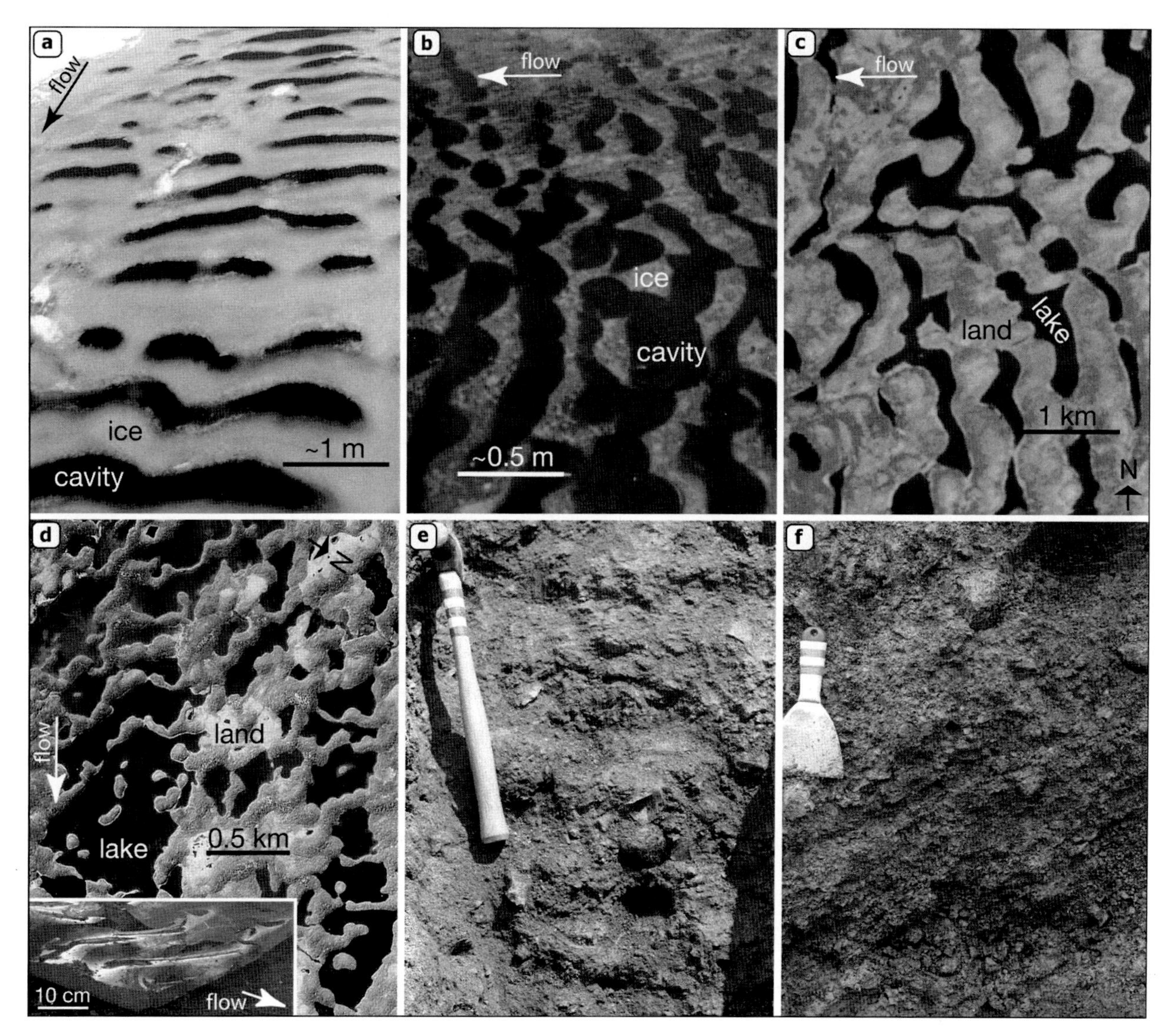

Figure 6.12. (a) Ripple cavities on the underside of river ice on the Sturgeon River, NW Ontario. (b) Ripples eroded into the base of river ice. (Courtesy of G. Ashton) (Modified from Fisher and Shaw, 1992.) (c) Ribbed moraine from Boyd Lake area, Northwest Territories, Canada. Compare the outline of ribbed moraine in (c) with the outline of the cavities in river ice (b). (Modified from Fisher and Shaw (1992).) (d) Ribbed moraine, Avalon Peninsula, Newfoundland, with inset photograph of overturned river ice from the Dogpound Creek, Alberta, revealing sinuous ripple cavities, compared with sinuous ribbed moraine above. (Aerial photograph A-25619 #16 copyrighted National Aerial Photograph Library, Energy Mines and Resources, Canada, reproduced with permission.) (e, f) Representative sediment exposures from Colinet and Brigus Junction respectively showing interbedded diamictons and poorly sorted, gravelly, muddy sand with high proportions of granules.

landforms occur in three forms: (1) 3D waveforms similar to hummocky terrain; (2) 2D nested waveforms that bifurcate in places; and (3) rhomboidal interfering forms. The crests of many of the ridges are superimposed by small fluting, an observation also made by Munro-Stasiuk and Bradac (2002) in NW Pennsylvania (Figure 6.13).

Ground-penetrating radar of the subsurface architecture (Beaney and Shaw, 2000) also showes that while folded strata are common within the Alberta ridges, anticlines in the folds do not coincide with ridge form and beds are invariably truncated by the landscape surface, demonstrating erosional modification after folding. The only surface sediment on or between the ridges is a scattering of boulders resting on the erosional surface. Also, tunnel channels cross-cut these ridges and dissect the preglacial divide.

Beaney and Shaw (2000) offered two hypotheses for the ridge formation. Firstly, the ridges may be polygenetic

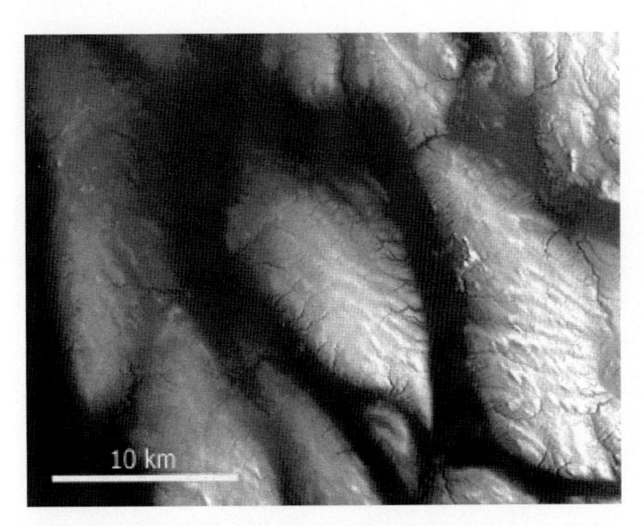

Figure 6.13. Transverse ridges on the Ohio–Pennsylvania border with superimposed fluting.

with structure inherited from glaciotectonic processes and then modified by subglacial meltwater. Glaciotectonic thrusting likely resulted from increased basal shear stresses, where ice grounded on the preglacial divide as a result of subglacial meltwater ponding upflow. Once the ponded water was released it would have flowed across the drainage divide eroding the thrust terrain and forming megaripples. Alternatively, the ridges may represent erosional features associated with internal stationary waves related to abrupt density gradients in hyperconcentrated subglacial meltwater flows.

6.3.6 Tunnel valleys/channels

There is little debate that valleys or scoured corridors that truncate subglacial bedforms such as drumlins and megaripples (see above), that contain eskers and/or underfit streams and have upslope paths, are the geomorphological expression of large subglacial meltwater flows that efficiently evacuated meltwater and sediment from beneath past ice sheets (e.g. Wright, 1973; Rampton, 2000). The tunnel valleys/channels are generally long (tens of kilometres), wide (kilometres), flat-bottomed, over-deepened (tens to hundreds of metres), radial or anabranched valley systems. They can be incised into bedrock or sediment, terminate in ice marginal fans and be empty, partially filled or buried by sediments. In recent years, there has been considerable debate as to the precise mechanism(s) by which such large valleys formed (e.g. O'Cofaigh, 1996) and their implications for ice sheet dynamics and hydrology. Two dominant hypotheses for their formation have emerged. (1) The piping (bed deformation) hypothesis (e.g. Boulton and Hindmarsh, 1987) states that tunnel valleys formed at below bankfull conditions in a headward progression as a saturated substrate dewatered and formed pipes at the ice margin. As sediment was flushed from these pipes and with ice margin retreat, tunnel valleys gradually evolved, growing deeper and longer. (2) The channelised underburst hypothesis (e.g. Brennand and Shaw, 1994) states that tunnel channels were incised under bankfull conditions by channelised underbursts (jökulhlaups or megafloods) draining a subglacial or supraglacial meltwater reservoir. Thermal conditions at the glacier sole may have facilitated reservoir growth and drainage (Cutler *et al.*, 2002). While the piping hypothesis may hold true for some sediment-walled tunnel valleys formed at the ice margin, many observations from tunnel channels/valleys associated with the Laurentide and Cordilleran ice sheets support the channelised underburst hypothesis (e.g. Brennand and Shaw, 1994; Sjogren and Rains, 1995; Beaney and Shaw, 2000; Cutler *et al.*, 2002; Fisher *et al.*, 2005). This assertion is illustrated by exploring the character of the central southern Ontario (Canada) tunnel channel network.

Central southern Ontario exhibits a regional Late Wisconsinan unconformity (see section on landscape unconformities) that truncates Palaeozoic bedrock and a thick Quaternary sediment cover (e.g. Sharpe *et al.*, 2004). This unconformity is composed of drumlins, s-forms and valleys. The valleys form a dense, anabranched NE–SW-oriented network (Figure 6.14). The valleys are assigned to five classes based on their geomorphology, and their absolute and stratigraphic depth of incision. Bedrock valleys are structurally controlled, steep-sided, ornamented by s-forms (Shaw 1988) and occur headward (north and east) of sediment-walled valleys. Sediment-walled valleys (classes 1–4) (Figure 6.14) extend from bedrock valleys and dissect drumlinised terrain. Classes 1–4 identify progressively smaller and/or shallower sediment-walled valleys (Brennand *et al.*, 2006). Valley fills are up to ~150 m thick and generally fine upward and sometimes they include eskers (e.g. Russell *et al.*, 2003).

This integrated, anabranched valley network is inferred to record a tunnel channel system that was hydromechanically eroded and/or re-utilised by a channelised turbulent, meltwater underburst in the Late Wisconsinan. This is because valleys: (1) are incised into and contain Late Wisconsinan sediment; (2) have undulating floors and upslope paths; (3) locally contain eskers and are filled by sediments indicative of rapid sedimentation (e.g. sandy hyperconcentrated flow deposits); (4) exhibit no evidence of convergent sediment deformation along their margins; (5) are cut to elevations below Lake Ontario base level and fail to terminate in deltas or fans at proglacial or modern shorelines; and (6) contain modern underfit streams up to an order of magnitude narrower than valleys (e.g. Brennand and Shaw, 1994; Russell *et al.*, 2003).

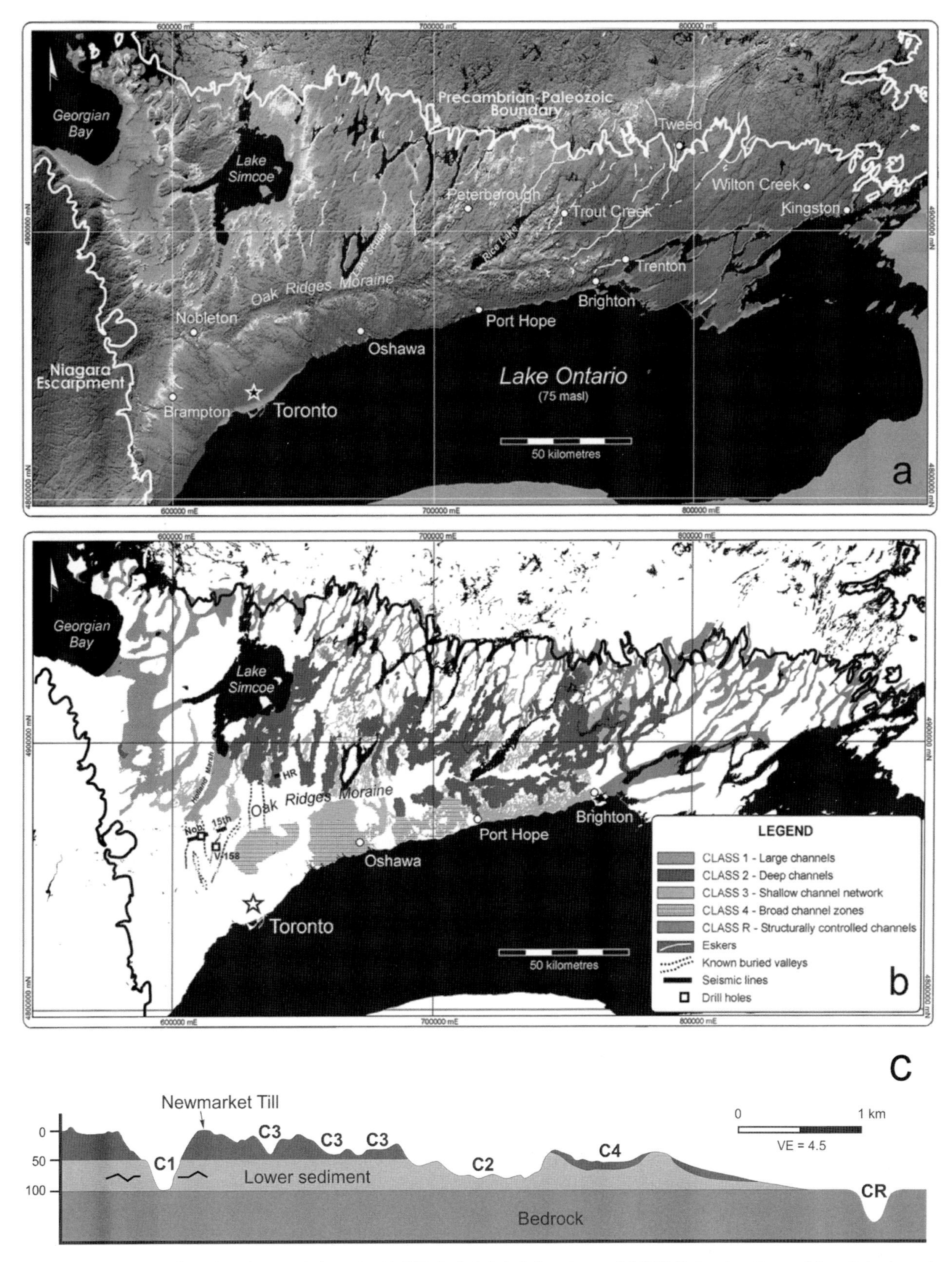

Figure 6.14. Central southern Ontario tunnel channels. (a) Hill-shaded digital elevation model (DEM) showing dissected drumlinised terrain, the Oak Ridges Moraine, escarpments and eskers. (b) Distribution of five tunnel channel classes, buried valleys, seismic lines and drill holes between the Precambrian–Palaeozoic boundary and the Niagara escarpment. (c) Simplified stratigraphy of central southern Ontario showing relative depths of incision of the five tunnel channel classes (C1–4, CR). Newmarket till is a regional Late Wisconsinan till sheet. Wiggly lines denote piping. (Modified from Brennand *et al.* (2006).)

The spatial variation in valley character suggests progressive channelisation of the underburst from a regional shallow channel network (class 3) to progressively fewer, larger channels (class 2 then 1) as flow concentrated and waned; the bedrock channels may have been antecedent and re-utilised. Bedrock structure, the gap width between the basal ice and the bed (Brennand and Shaw, 1994), enhanced scour at thread confluences and hydraulic jumps (Russell *et al.*, 2003), and reservoir location (e.g. Evatt *et al.*, 2006) likely controlled tunnel channel location. Groundwater flow and piping at depth through sandy beds of the lower sediment (Figure 6.14) (Russell *et al.*, 2003) may explain the depth of class 1 valleys. Channel fills record rapid and voluminous sedimentation as flow waned (e.g. Brennand and Shaw, 1994; Russell *et al.*, 2003).

The regional extent, density and character of tunnel channels in central southern Ontario and their intimate association with drumlins ascribed to sheet-flow underburst erosion (Shaw and Sharpe, 1987) strongly suggest that these valleys record the channelised, waning flow phase of the same underburst (the Algonquin event; Shaw and Gilbert, 1990). This inference is supported by recent observations at Skeiðarárjökull, where the channelisation of a sheet-flow underburst feeding the November 1996 jökulhlaup resulted in tunnel channels (Russell *et al.*, 2007).

6.3.7 Eskers

Eskers are sinuous ridges of stratified sand and gravel: the casts of past ice-walled streams (Banerjee and McDonald, 1975). They occur in a range of sizes (up to tens of metres high, hundreds of metres wide, and hundreds of kilometres long). They can be located in valleys or follow upslope paths. They may occur in isolation or in groups forming subparallel, de-ranged (not aligned with regional ice flow) or dendritic patterns. The presence and distribution of eskers is determined by a combination of factors: (1) meltwater supply (location of crevasses, moulins, reservoirs); (2) sediment supply; (3) the nature of the basal substrate as it controls the style of the subglacial plumbing system (Clark and Walder, 1994); and (4) the presence and/or drainage of proglacial water bodies (Brennand, 2000). It is also possible that antecedent underbursts may have reorganised ice surface slopes and hydrology sufficiently to dictate R-channel, and hence esker, location (Brennand and Shaw, 1996).

Eskers associated with past ice sheets mainly record subglacial streams or ice-walled streams close to the glacier sole and/or margin, because these environments contain a ready sediment supply and allow for ridge preservation during ice melt (Brennand, 2000). The fact that many eskers exhibit upslope paths and occur within tunnel channels further suggests that such eskers record closed-conduit (water filled) conditions (Brennand, 1994).

Figure 6.15. Rhythmic sand and gravel couplets, Norwood esker, Ontario, Canada. Metre rod for scale. (From Brennand (1994).)

Some eskers may record 'normal' melt-related flows, others flood events of various scales; some may record both. Where eskers terminate in subaqueous fans that contain varves they likely record an integrated plumbing system (with supraglacial to subglacial connection) in the melt zone with seasonal flows through a persistent conduit. The presence of proglacial water bodies may have facilitated conduit maintenance over multiple seasons (Brennand, 2000). Such conduits may have experienced both normal melt-related flows and floods associated with precipitation events and/or drainage of meltwater reservoirs. Yet the eskers resulting from such flows may record only the last few flow events through the conduit due to the flushing effects of floods; the fans record a more continuous record of flow variation (Brennand, 1994).

Pleistocene esker sediments provide significant evidence of flood flows through conduits (e.g. Brennand, 1994; Brennand and Shaw, 1996). Eskers are composed of sand and gravel lithofacies (Figure 6.15), which frequently contain cobbles and boulders and exhibit structures indicative of deposition from both fluidal and hyperconcentrated flows. High flow velocities (floods) are inferred from boulder size (up to 1 m *b*-axis) and shape (rounding and sphericity), and the presence of gravelly hyperconcentrated flow deposits. Esker lithofacies are arranged into identifiable architectural elements or macroforms (barforms). The sediments in these macroforms conform to ridge shape: composite macroforms occur at wider higher portions of the ridge, and pseudo-anticlinal macroforms occur at ridge constrictions. The presence of macroforms suggests that conduit filling was a dynamic process driven by powerful flows down non-uniform conduits. Within macroforms, sand and gravel lithofacies often form rhythmic couplets (up to four to five couplets within any exposure) (Figure 6.15). These couplets may record autocyclic or allocyclic processes. Autocyclicity may result from episodic storage and transport of sediment in a non-uniform conduit. Allocyclicity may result from seasonal or episodic changes in water supply and hence flow velocity. Flood hydrographs related to precipitation events, or the drainage of subglacial or supraglacial reservoirs, can readily explain the rhythmic couplets. Gravel macroforms and rhythmic couplets have been observed within a recently exposed esker in Iceland, the product of deposition in an ice-walled channel during the waning flow of the November 1996 jökulhlaup at Skeiðarárjökull (Russell *et al.*, 2001).

Although eskers may form under active ice (Shreve, 1972), the preservation of extensive and long dendritic networks of Laurentide eskers suggests their formation in association with stagnant ice (Brennand, 2000). Meltwater underbursts may have resulted in a thin, stagnant ice sheet more likely to disappear by backwasting and downwasting in the absence of regional ice flow ice, thus favouring esker preservation.

6.4 Macroforms

Macroforms tend to be larger, typically on the order of at least tens of kilometres long, by tens of kilometres wide. Here the streamlined hills and bedrock rises are discussed. These are often present in the centre of flowpaths (see section on megaforms) and re-entrant valleys, which commonly cut into large escarpment slopes.

6.4.1 Streamlined hills

Large streamlined hills lie within, or at the junction between, swales (elongated depressions) within fluted terrain in many regions. A good example is the Hand Hills in Alberta. Young *et al.* (2003) refer to the Hand Hills as a megadrumlin because the hills, as a complex, resemble a drumlin in shape and profile (Figure 6.16). In this case the megadrumlin rises 70 m above the prairie land surface and has a length of approximately 3 km (smaller than most streamlined hills but significantly larger than drumlins). Like many drumlins of usual dimensions this megadrumlin even has a large horseshoe-shaped scour at the base of its upflow-facing side. While many streamlined hills have somewhat flattened tops, indicating that water likely never overtopped them, the Hand Hills likely were submerged fully in the flow. This supposition does not indicate that water was 70 m deep, merely that water was following a steep hydraulic pressure gradient causing it to be driven over the hills between the ice and its bed. A hanging tunnel valley on top of the Hand Hills contains an esker complex. This channel was formed subglacially, probably during the waning stages of a flood that overtopped the Hand Hills and sculpted them into a streamlined megadrumlin.

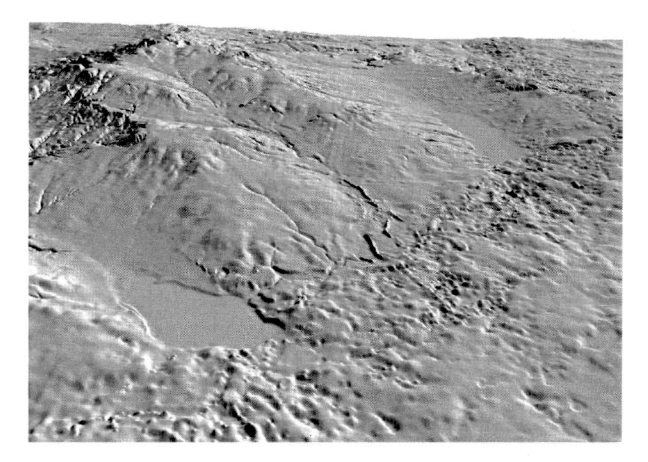

Figure 6.16. Digital elevation model of the streamlined Hand Hills, central Alberta.

6.4.2 Bedrock rises

Rises are erosional forms, much larger than fluting. They are narrow and relatively clearly defined at their upstream ends and splay downstream where they merge gradually with the surrounding land surface. They may be as much as 100 km long and 10–20 km wide. They rise to heights of about 20 m which give them such low relief that they are barely discernable from the ground. However, they show up very clearly on DEMs (Figure 6.17).

6.4.3 Re-entrant valleys

Where escarpments faced into the inferred flows they are commonly cut by re-entrant valleys that often contain lakes (Gilbert and Shaw, 1994). The Finger Lakes in northern New York State are good examples (Shaw, 1996). In a detailed survey of re-entrant valleys along the southern margin of the Canadian Shield, Gilbert and Shaw (1994) showed that these overdeepened valleys narrow and deepen towards escarpments and they commonly have a ridge running along their axes.

Figure 6.17. Hill-shade from the Shuttle Radar Topography Mission (SRTM) of Alberta, Canada, showing bedrock rises that sit within the eastern flowpath of the inferred Livingstone Lake Event.

Tributaries are rare and those that do occur hang above the re-entrant valley floors. The valleys usually are oriented in approximately the same direction and are equally spaced. Although some valleys are associated with structural lineations, not all of them are.

The absence of lineations in some valleys rules out valley formation under structural control. Overdeepening, hanging tributaries and the medial ridges make a fluvial genesis for the valleys unlikely. A glacial origin is contradicted by the medial ridges and the equal spacing of the valleys. However, they are well explained in terms of vortex action in turbulent flow over forward-facing steps (escarpments) as described by Pollard *et al.* (1996) and Wilhelm *et al.* (2003). In this explanation, periodic distortion of flow structures generated along scarp slopes produces counter-rotating longitudinal vortices that erode valleys. This periodicity explains the equal spacing. In time, valleys funnel vortices and concentrate erosion. Vortices are expected to be most erosive where they curve into valleys, which explains the maximum depth at this point. Zones of rising flowlines along valley axes, which would experience relatively low rates of erosion, are sites of medial ridges. Rare or hanging tributaries indicate that the re-entrant valleys were cut in the absence of tributaries or when they were inactive. This arrangement, which is not expected for subaerial drainage systems, is in keeping with subglacial meltwater flow. The broad flow structures producing re-entrant valleys and stoss-side furrows were probably very similar and this is a further example of scale independence.

6.5 Megaforms

Megaforms are the largest of all the features discussed in this chapter. They typically cover hundreds of square kilometres and make up entire landscapes (Figure 6.17). Consequently, they can be a single large form or a complex series of smaller forms that, together, make the larger landscape.

6.5.1 Landscape unconformities: regional erosion surfaces

Landforms and associated flow patterns are used to reconstruct regional erosion surfaces or unconformities, particularly on land (Sharpe *et al.*, 2004) and recently in offshore terrain (e.g. Fulthorpe and Austin, 2004). This technique has been used to infer either subglacial deformation resulting from ice-induced shear stress (e.g. Boyce and Eyles, 1991), or meltwater (e.g. Shaw and Gilbert, 1990; Sharpe *et al.*, 2004) as the primary agent of erosion. Mapped landform relationships and analysis of event sequences derived from seismic and drillcore data allow a systematic field test of unconformities and inferred regional meltwater processes. For example, a late-glacial unconformity is identified across the Great Lakes region based on evidence of scoured bedrock tracts, upland drumlin fields, channel networks, boulder lags and coarse-sediment channel fills (Figure 6.18). This evidence constrains inferences on landscape-forming processes: (i) ice surging, (ii) deforming beds (e.g. Boyce and Eyles, 1991), and (iii) large meltwater discharges (e.g. Shaw and Gilbert, 1990; Brennand and Shaw, 1994; Mullins *et al.*, 1996). Seismic profiles linked to cored boreholes define erosion surfaces marked by the surfaces of erosional drumlins, the intervening troughs and adjacent meltwater channels. Sculpted sediment and bedrock, the presence of boulder lags and coarse-grained deposits on erosional surfaces, as well as undulating channel profiles provide decisive support for meltwater erosion of unconformities. Inter-regional extent can be inferred from mapping contiguous elements (e.g. local: s-forms, drumlins; to mid range: scoured bedrock, transverse features, channels; to regional: mega-lineation and flowpaths) (Figure 6.18).

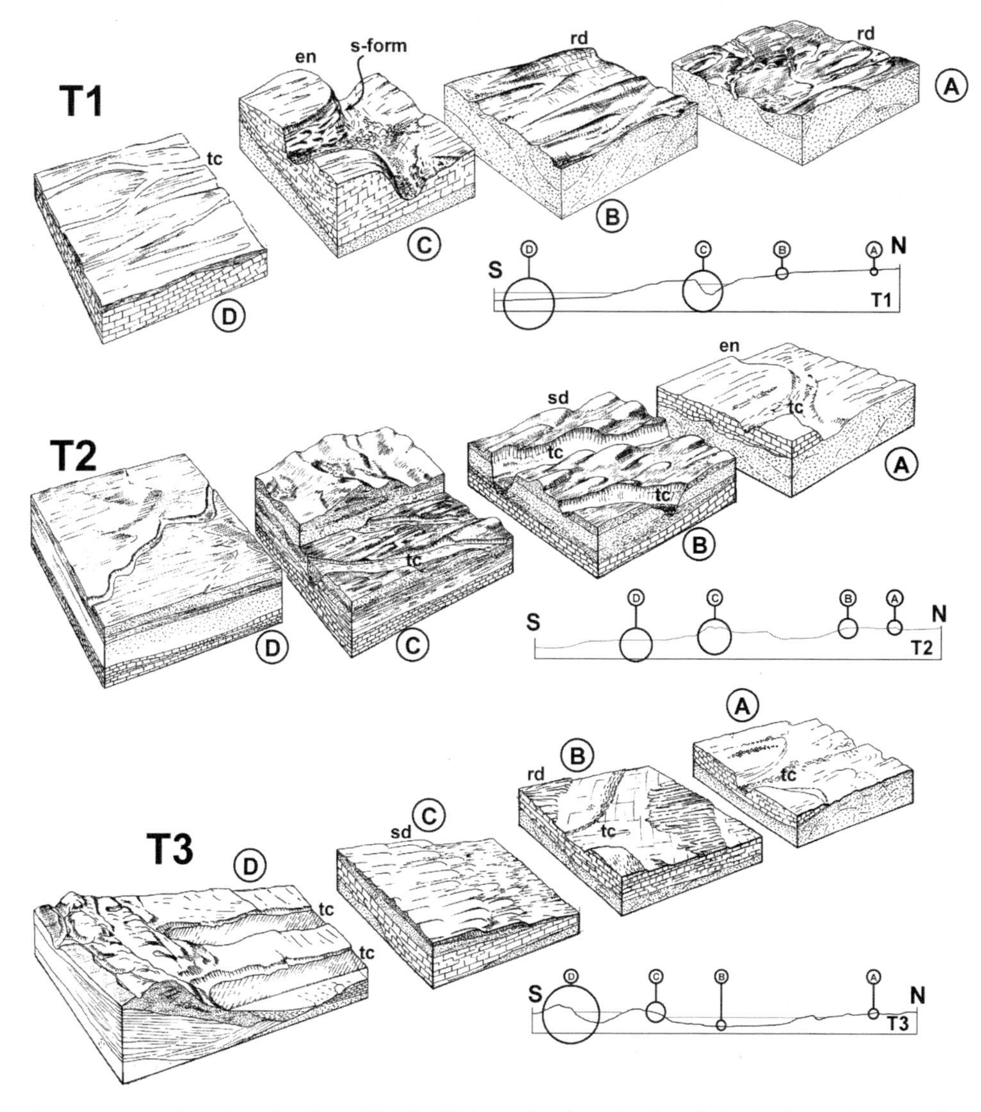

Figure 6.18. Landscape transect along three flowlines (T1, T2, T3) in erosional terrain, Great Lakes basin, e.g. T1: (A) s-forms, stubby rock drumlins (rd); (B) streamlined forms, spindle rock drumlins; (C) s-forms, escarpment noses (en) and crescentic furrows, channels on Palaeozoic scarps; (D) streamlined sediment drumlins (sd) cross-cut by tunnel channels (tc). (From Sharpe *et al.* (2004).)

Similarly, regional erosion surfaces have been identified in the prairies (e.g. Rains *et al.*, 2003; Munro-Stasiuk and Sjogren, 2006). The collective pattern is illustrated by flowpaths dominated by drumlins or fluting cross-cutting hummocky terrain (Figs. 6.17 and 6.19) (e.g. Rains *et al.*, 1993) (see discussion below).

6.5.2 Flowpaths

Distinct flowpaths are associated with the Laurentide Ice Sheet. These paths are hundreds of kilometres long and up to one hundred kilometres wide. They are recognised as tracts of low relief, typically dominated by bedrock, bordered by higher relief, commonly hummocky terrain (Rains *et al.*, 1993; Shaw *et al.*, 1989). Flowpaths are characterised by two types of landform: large-scale fluting and rises (Figure 6.17). Detailed observations on fluting sedimentology show that fluting ridges are erosional remnants, following the removal of material from the intervening troughs. Flutings are found most commonly downstream from upstream-facing slopes within flowpaths (Shaw *et al.*, 2000; Munro-Stasiuk and Shaw, 2002) (Figure 6.9). Longitudinal vortices in this location are predicted from fluid mechanics (Pollard *et al.*, 1996: Wilhelm *et al.*, 2003) and such vortices explain the erosional pattern for fluting formation.

6.6 Subglacial hydrology

Given that the topic of this volume is megafloods, it is important to discuss the relevant subglacial hydrology

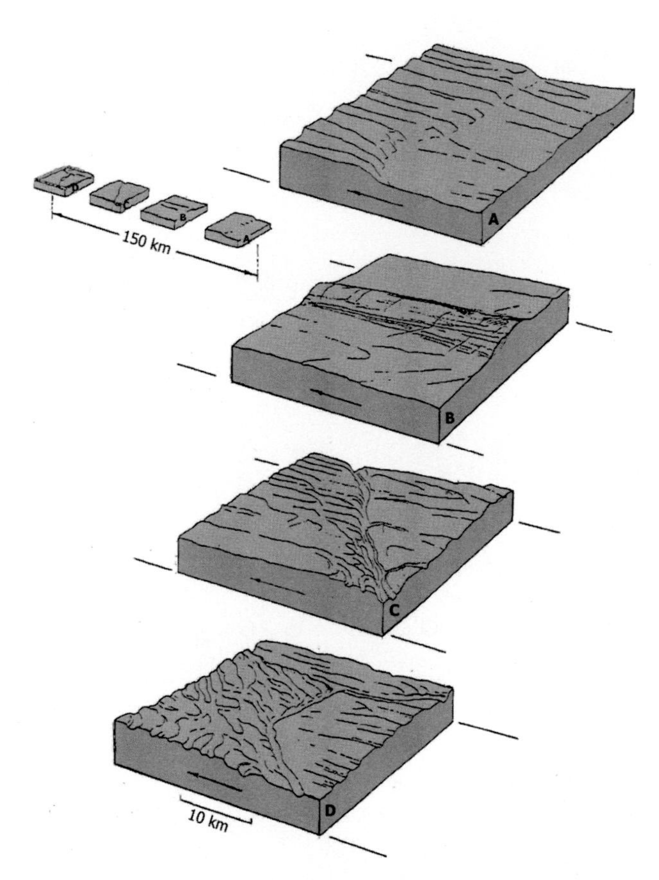

Figure 6.19. Landscape transect (flowpath) across erosional terrain in central Alberta: (A) bedrock ridge with escarpment noses, furrows and elongate rises; (B) subdued escarpment noses, crescentic furrows, elongate rises and drumlins; (C) downflow evolution to elongate streamlined forms with crescentic furrows and incipient hummocks; (D) flow path is transitional from streamlined forms to hummocky forms in places, and truncates hummocky terrain in others.

associated with bedforms described in this chapter. Reconstructed hydrology from these bedforms is controversial and has attracted a lot of attention recently (e.g. Sharpe, 2005; Clarke *et al.*, 2005; Benn and Evans, 2006).

Murray (1988) first concluded that subglacial meltwater bedforms indicate broad meltwater flows extending in width over tens of kilometres. Kor *et al.* (1991) mapped a coherent pattern of bedrock erosional marks (s-forms) over a distance of about 75 km across the flow in the French River area, Ontario. There, s-forms about 20 m in height were submerged in broad flows, and boulders averaging 0.3 m in diameter accumulated in depressions and sheltered positions. The maximum height of the erosional bedforms suggests a depth of flow of about 20 m, and the size of boulders on the bed suggests a flow velocity of about 5 m s^{-1} (Sundborg, 1956). Using the continuity equation:

$$Q = wdv, \tag{6.1}$$

where Q is discharge ($m^3\,s^{-1}$), w is width (m), d is depth (m) and v is mean velocity ($m\,s^{-1}$), and substituting the above values, gives a discharge of $7.5 \times 10^6\ m^3\,s^{-1}$. Drumlin-forming flows were wider and deeper than flows eroding s-forms. It is, however, more difficult to estimate the discharge of drumlin-forming flows.

Clearly, the discharge magnitude presented above could not have been sustained in a steady state by rates of ice melting. The sources of landscape-forming floods must have been large subglacial or supraglacial reservoirs. In the absence of direct evidence on reservoir volumes or of flow duration, Shaw *et al.* (1989) estimated the volume of ice ablated from the ice bed during the formation of inverted erosional marks at three times the volume of cavity-fill drumlins in the Livingstone Lake field. The average volume of individual drumlins is $1.57 \times 10\,m^3$, and the drumlin density $0.6\,km^{-2}$. Nye (1976) presented the discharge equation:

$$q[(1 - \gamma)dp/dx] = Lm, \tag{6.2}$$

where q ($m^2\,s^{-1}$) is discharge per unit width, dp/dx (Pa/m) is the pressure gradient, x is the distance down glacier in metres, m is the mass of ice melted by meltwater in kg, γ is a dimensionless constant, and L ($J\,kg^{-1}$) is the latent heat of melting for ice. Using low estimates of the ice surface slope (Mathews, 1974) to obtain dp/dx, the total flow for formation of the Livingstone lake drumlin field is estimated at $84\,000\,km^3$. This volume is problematical: Shoemaker (1995) argued that it is probably too large by an order of magnitude and Clarke *et al.* (2005) pointed out that it is difficult to produce and store that volume of water. These problems are compounded when it is realised that the Livingstone Lake drumlin field is one of many occupying different flow paths.

Shoemaker (1995) noted that Equation (6.2) does not take into account abrasion of ice by suspended solids and suggested that the above estimate could be reduced by a factor of 10. Shoemaker (1995) and Clarke *et al.* (2005) considered the potential sources of meltwater and showed that subglacial melting was inadequate to feed such a large reservoir. However, Shoemaker (1995) and Clarke *et al.* (2005) pointed out that supraglacial melting would have been more than adequate to feed such reservoirs, and Zwally *et al.* (2002) presented observations on the Greenland Ice Sheet demonstrating that surface meltwater can penetrate to the bed through more than one thousand metres of ice. Clarke *et al.* (2005) showed that supraglacial melt would take 724 years to fill a reservoir, drainage of which would be capable of raising sea level by 2 m and of sustaining drumlin-forming floods.

Clarke *et al.* (2005) pointed out that the findings of Zwally *et al.* (2002) make it highly unlikely that large volumes of water could be stored in supraglacial reservoirs – the reservoirs would drain long before they became large enough to sustain flows of 10^6 m^3/s. Consequently, if these flows were to be sustained, the reservoirs must have been subglacial. Clarke *et al.* (2005) noted several difficulties with this suggestion. The principal one comes from an analysis of potential gradient in an ice sheet, which dictates the following relationship for the existence of a subglacial lake:

$$dZ_B/dx > -8dZ_S/dx, \tag{6.3}$$

where Z_B is the height of the bed above an arbitrary datum and Z_S is the height of the ice surface above the datum. Clarke *et al.* (2005) claimed that this relationship prohibits the formation of large subglacial lakes other than those enclosed in basins because it requires much steeper bed slopes than surface slopes and that steep-sided subglacial lake basins are unlikely. However, low glacier surface slopes would be expected for floating ice, and bed slopes such as those in the Hudson Bay catchment, enhanced by isostatic depression, would be steep enough for the formation of large subglacial lakes if the ice surface slope was on the order of 1/1000 (Alley *et al.*, 2006). Clarke *et al.* (2005) also dismissed large-scale subglacial lakes even though they presented captured lakes as a plausible mechanism for creating them. Erlingsson (1994), Alley *et al.* (2006) and Domack *et al.* (2006) independently provided the case for large lakes originating by 'capture' as an ice shelf advances across a proglacial lake. During advance of the Laurentide Ice Sheet, such lakes would have been far larger than Lake Superior, which is the size of captured lakes proposed by Clarke *et al.* (2005). Erlingsson (1994) and Alley *et al.* (2006) explained how grounding of the ice at the far side of the lake causes a rise on the ice surface with a reverse slope on the up-ice side of the rise. This rise dams the lake. Alley *et al.* (2006) suggested that freeze-on at the bed of grounded ice could act as a further damming mechanism. If the floating ice becomes thicker, the water becomes pressurised and deepens until it is sufficiently deep to lift the ice off the bed, at which point a jökulhlaup occurs. Both Erlingsson (1994) and Alley *et al.* (2006) refer to erosional landforms as evidence of such an outburst. Such drainage of subglacial megalakes is exactly what is required for the formation of the subglacial flood landscapes described in this chapter.

Perhaps the most telling aspect of the debate on subglacial hydrology comes from observed lakes beneath the Antarctic ice sheets. Wingham *et al.* (2006) noted that subglacial lakes drained rapidly from one to the other. Fricker *et al.* (2007) noted similar behaviour of lakes beneath Antarctic ice streams and Bell *et al.* (2007) observed large lakes at the onset of an East Antarctic ice stream. The importance of these lakes is that they support inferences of intermittently draining lakes beneath Pleistocene ice sheets. Munro-Stasiuk (1999, 2003) inferred the presence of subglacial reservoirs in small basins under the western portions of the Laurentide Ice Sheet but also noted they predated any megaflood events. Bell *et al.* (2007) also showed that lakes may grow anywhere along the paths of ice streams. This observation indicates that meltwater reservoirs may fill along flood paths and their release as outburst floods would then produce the suites of landforms described in this chapter. However, larger subglacial lakes may have been captured by glacial ice overriding proglacially dammed lakes. This capture is most likely to have occurred in places like Hudson Bay, an already well-established basin and all of the Great Lakes basins, as these all trapped water against the uphill advancing ice-sheet. If the water was deep enough, the ice may have started floating but continued to advance, therefore trapping the water subglacially.

6.7 Concluding remarks

In summary, there is a wide range of landforms at a range of scales, containing a variety of internal materials. These internal materials include till, glaciofluvial deposits, glaciolacustrine deposits and bedrock. The landforms can be entirely *in situ*, can be slightly modified by glaciotectonism, or can be pervasively deformed. It is argued here that all features described, except for Livingstone drumlins, Rogen moraine and eskers, are exclusively the product of subglacial meltwater erosion. These erosional forms record erosion over broad surfaces by highly turbulent, high-velocity and, for the most part, sediment-laden meltwater flows. Livingstones and Rogen moraine are also the initial product of erosion. However, the erosion was up into the base of the ice rather than down into the substratum. Sorted sediments filled up the basal cavities as the ice lowered back down onto the bed. As flows started to channelise, tunnel channels were formed, and eskers represent the final stages of flood events when the ice had recoupled with all other areas of its bed.

While there is no doubt that many of these landforms are erosional in nature and there is significant evidence pointing to subglacial meltwater erosion, more research on the dynamics and nature of the erosion is required. For instance, the exact mechanisms for hummocky terrain formation are poorly understood. There are many different hummock types, yet it is known only that each is the product of erosion. However, the velocity, flow depth, duration and vorticity of the formative flows are all poorly constrained. Flume experiments may help determine flow characteristics.

It is also clear that steady-state flow conditions cannot account for the magnitude of events associated with the erosional landforms here described; there must have been reservoir storage and release. However, the nature of the water storage is unknown at this time. Direct observations on modern ice sheets is a promising approach to explaining the hydrology of past subglacial lakes (Wingham *et al.*, 2006; Fricker *et al.*, 2007).

Regardless of questions that must still be answered, it is suggested here that the geomorphological and sedimentological evidence presented in this chapter for megafloods is compelling. Thus a radical revision of the contemporary view of Pleistocene ice sheets will be necessary to provide understanding of the glacial hydrology that gave rise to these floods.

Acknowledgements

We thank Keith Richardson and Robert Gilbert for their reviews, which greatly improved this chapter.

References

Alexander, H. S. (1932). Pothole erosion. *Journal of Geology*, **40**, 305–337.

Allen, J. R. L. (1971). Transverse erosional marks of mud and rock: their physical basis and geological significance. *Sedimentary Geology*, **5**, 167–385.

Allen, J. R. L. (1982). *Sedimentary Structures: Developments in Sedimentology, Volume 2*. Amsterdam: Elsevier.

Alley, R. B., Dupont, T. K., Parizek, K. B. R. *et al.* (2006). Outburst flooding and the initation of ice-stream surges in response to climate cooling: an Hypothesis. *Geomorphology*, **75**, 76–89.

Ashton, G. D. and Kennedy, M. J. F. (1972). Ripples on underside of river ice covers. *Journal of the Hydraulics Division, Proceedings of the American Society of Civil Engineers*, **98**, 1603–1624.

Baker, V. R. (1973). Erosional forms and processes for the catastrophic Pleistocene Missoula floods in Eastern Washington. In *Fluvial Geomorphology*, ed. M. Morisawa. Proceedings volume of the 4th annual geomorphology symposia series held at Binghamton, New York.

Baker, V. R. (1978). The Spokane Flood controversy and the Martian outflow channels. *Science*, **202**, 1249–1256.

Baker, V. R. and Kale, V. S. (1998). The role of extreme floods in shaping bedrock channels. In *Rivers Over Rock: Fluvial Processes in Bedrock Channels*, eds. K. J. Tinkler and E. E. Wohl. American Geophysical Union, Geophysical Monograph 107.

Baker, V. R. and Pickup, G. (1987). Flood geomorphology of the Katherine Gorge, Northern Territory, Australia. *Geological Society of America Bulletin*, **98**, 635–646.

Banerjee, I. and McDonald, B. C. (1975). Nature of esker sedimentation. In *Glaciofluvial and Glaciolacustrine Sedimentation*, eds. A. V. Jopling and B. C. McDonald. Society of Economic Paleontologists and Mineralogists Special Publication 23, pp. 304–320.

Beaney, C. L. and Shaw, J. (2000). The subglacial geomorphology of southeast Alberta: evidence for subglacial meltwater erosion. *Canadian Journal of Earth Sciences*, **37**, 51–61.

Bell, R. E., Studinger, M., Shuman, C., Fahnestock, M. A. and Joughin, I. (2007). Large subglacial lakes in East Antarctica at the onset of fast-flowing ice streams. *Nature*, **445**, 904–907.

Benn, D. I. and Evans, D. J. A. (1998). *Glaciers and Glaciation*. London: Arnold.

Benn, D. I. and Evans, D. J. A. (2006). Subglacial megafloods: outrageous hypothesis or just outrageous? In *Glacier Science and Environmental Change*, ed. P. G. Knight. Oxford: Blackwell Publishing Ltd, pp. 42–50.

Boulton, G. S. (1987). A theory of drumlin formation by subglacial sediment deformation. In *Drumlin Symposium*, ed. J. Rose. Leiden: Balkema, pp. 25–80.

Boulton, G. S. and Hindmarsh, R. C. A. (1987). Sediment deformation beneath glaciers: rheology and geological consequences. *Journal of Geophysical Research*, **92**, 9059–9082.

Boyce, J. I. and Eyles, N. (1991). Drumlins carved by deforming till streams below the Laurentide ice sheet. *Geology*, **19**, 787–790.

Brennand, T. A. (1994). Macroforms, large bedforms and rhythmic sedimentary sequences in subglacial eskers, south-central Ontario: implications for esker genesis and meltwater regime. *Sedimentary Geology*, **91**, 9–55.

Brennand, T. A. (2000). Deglacial meltwater drainage and glaciodynamics: inferences from Laurentide eskers, Canada. *Geomorphology*, **32**, 263–293.

Brennand, T. A. and Shaw, J. (1994). Tunnel channels and associated landforms: their implications for ice sheet hydrology. *Canadian Journal of Earth Sciences*, **31**, 502–522.

Brennand, T. A. and Shaw, J. (1996). The Harricana glaciofluvial complex, Abitibi region, Quebec: its genesis and implications for meltwater regime and ice-sheet dynamics. *Sedimentary Geology*, **102**, 221–262.

Brenannd, T. A., Russell, H. A. J. and Sharpe, D. R. (2006). Tunnel channel character and evolution in central southern Ontario. In *Glacier Science and Environmental Change*, ed. P. G. Knight. Oxford: Blackwell Publishing Ltd, pp. 37–39.

Bretz, J H. (1959). *Washington's Channeled Scabland*. Washington Department of Conservation, Division of Mines and Geology Bulletin, 45.

Bryant, E. A. and Young, R. W. (1996). Bedrock-sculpturing by tsunami, South Coast of New South Wales, Australia. *Journal of Geology*, **104**, 565–582.

Clark, P. U. and Walder, J. S. (1994). Subglacial drainage, eskers, and deforming beds beneath the Laurentide and Eurasian ice sheets. *Geological Society of America Bulletin*, **106**, 304–314.

Clarke, G. K. C., Leverington, D. W., Teller, J. T., Dyke, A. S. and Marshall, S. J. (2005). Fresh arguments against the Shaw megaflood hypothesis. A reply to comments by D. Sharpe, Correspondence. *Quaternary Science Reviews*, **24**, 1533–1541.

Clayton, S. L. and Moran, S. R. (1974). A glacial process-form model. In *Glacial Geomorphology*, ed. D. R. Coates. New York: University of New York, pp. 89–119.

Cutler, P. M., Colgan, P. M. and Mickelson, D. M. (2002). Sedimentologic evidence for outburst floods from the Laurentide Ice Sheet in Wisconsin, USA: implications for tunnel-channel formation. *Quaternary International*, **90**, 23–40.

Dargahi, B. (1990). Controlling mechanism of local scouring. *Journal of Hydraulic Engineering*, **116**, 1197–1214.

Domack, E., Amblàs, D., Gilbert, R. *et al.* (2006). Subglacial morphology and glacial evolution of the Palmer deep outlet system, Antarctic Peninsula. *Geomorphology*, **75**, 125–142.

Dyke, A. S. and Prest, V. K. (1987). Late Wisconsinan and Holocene history of the Laurentide ice sheet. *Géographie physique et Quaternaire*, **41**, 237–263.

Dzulynski, S. and Sanders, J. E. (1962). Current marks on firm mud bottoms. *Transactions of the Connecticut Academy of Arts and Sciences*, **42**, 57–96.

Embleton, C. and King, C. A. M. (1975). *Glacial Geomorphology*. London: Edward Arnold.

Erlingsson, U. (1994). The 'captured ice shelf' hypothesis and its applicability to the Weichselian Glaciation. *Geografiska Annaler*, **76A**, 1–12.

Evans, D. J. A., Rea, B. R., Hiemstra, J. F. and O'Cofaigh, C. (2006). A critical assessment of subglacial mega-floods: a case study of glacial sediments and landforms in south-central Alberta, Canada. *Quaternary Science Reviews*, **25**, 1638–1667.

Evatt, G. W., Fowler, A. C., Clark, C. D. and Hulton, N. R. J. (2006). Subglacial floods beneath ice sheets. *Philosophical Transactions of the Royal Society A*, **364**, 1769–1794.

Fildani, A., Normark, W. R., Kostic, S. and Parker, G. (2006). Channel formation of flow-stripping large-scale scour features along the Monterey East Channel and their relation to sediment waves. *Sedimentology*, **53**, 1265–1287.

Fisher, T. G. (1989). Rogen moraine formation examples from three distinct areas within Canada. Unpublished M.Sc. thesis, Queen's University at Kingston, Ontario, Canada.

Fisher, T. G. and Shaw, J. (1992). A depositional model for Rogen moraine, with examples from the Avalon Peninsula, Newfoundland. *Canadian Journal of Earth Sciences*, **29**, 669–686.

Fisher, T. G., Jol, H. M. and Boudreau, A. M. (2005). Saginaw Lobe tunnel channels (Laurentide Ice Sheet) and their significance in south-central Michigan, USA. *Quaternary Science Reviews*, **24**, 2375–2391.

Fricker, H. A., Scambos, T., Bindschadler, R. and Padman, L. (2007). An active subglacial water system in West Antarctica mapped from space. *Science*, **315**, 1544–1548.

Fulthorpe, C. and Austin, Jr. J. A. (2004). Shallowly buried, enigmatic seismic stratigraphy on the New Jersey outer shelf: evidence for latest Pleistocene catastrophic erosion? *Geology*, **32**, 1013–1016.

Gilbert, G. K. (1896). The origin of hypotheses, illustrated by the discussion of a topographic problem. *Science*, **3**, 1–13.

Gilbert, G. K. (1906). Crescentic gouges on glaciated surfaces. *Geological Society of America Bulletin*, **72**, 303–316.

Gilbert, R. and Shaw, J. (1994). Inferred subglacial meltwater origin of lakes on the southern border of the Canadian Shield. *Canadian Journal of Earth Sciences*, **31**, 1630–1637.

Gilbert, R., Chong, A., Domack, E. W. and Dunbar, R. B. (2003). Sediment trap records of glacimarine sedimentation at Mueller Ice Shelf, Lallemand Fjord, Antarctic Peninsula. *Arctic, Antarctic, and Alpine Research*, **35**, 24–33.

Gravenor, C. P. and Kupsch, W. O. (1959). Ice-disintegration features in western Canada. *Journal of Geology*, **67**, 48–64.

Greeley, R. and Iversen, R. E. (1985). *Wind as a Geological Process*. Cambridge: Cambridge University Press.

Gupta, A., Kale, V. S. and Rajaguru, S. N. (1999). The Narmada River, India, through space and time. In *Varieties of Fluvial Form*, eds. A. J. Miller and A. Gupta. Chichester: Wiley and Sons Ltd.

Habbe, K. A. (1989). The origin of drumlins of the south German Alpine Foreland. *Sedimentary Geology*, **62**, 357–369.

Hancock, G. S., Anderson, R. S. and Whipple, K. X. (1998). Beyond power: bedrock river incision process and form. In *Rivers over Rock: Fluvial Processes in Bedrock Channels*, eds. K. J. Tinkler and E. E. Wohl. American Geophysical Union, Geophysical Monograph 107.

Herzfeld, U. C., Caine, N., Erbrecht, T., Losleben, M. and Mayer, H. (2003). Morphogenesis of typical winter and summer snow surface patterns in a continental alpine environment. *Hydrological Processes*, **17**, 619–649.

Hughes Clarke, J. E., Short, A. N., Piper, D. J. W. and Mayer, L. A. (1990). Large-scale current-induced erosion and deposition in the path of the 1929 Grand Banks turbidity current. *Sedimentology*, **37**, 631–646.

Johnsson, G. (1956). *Glacialmorfologiska studier i soedra Sverige, med saerskild haensyn till glaciala riktningselement och periglaciala frostfenomen. [Glacial morphology in southern Sweden, with special reference to glacial orientation elements and periglacial cryoturbation.]* Lund University Geographical Institute Meddelanden.

Karcz, I. (1968). Fluviatile obstacle marks from the wadis of the Negev (southern Israel). *Journal of Sedimentary Petrology*, **38**, 1000–1012.

Kor, P. S. G. and Cowell, W. (1998). Evidence for catastrophic subglacial meltwater sheetflood events on the Bruce Peninsula, Ontario. *Canadian Journal of Earth Sciences*, **35**, 1180–1202.

Kor, P. S. G., Shaw, J. and Sharpe, D. R. (1991). Erosion of bedrock by subglacial meltwater, Georgian Bay, Ontario: a regional view. *Canadian Journal of Earth Sciences*, **28**, 623–642.

Kulig, J. J. (1985). A sedimentation model for the deposition of glacigenic deposits in west-central Alberta: a single (Late Wisconsinan) event. *Canadian Journal of Earth Sciences*, **26**, 266–274.

Lawson, D. E. (1979). *Sedimentological analysis of the western terminus region of the Matanuska Glacier, Alaska.* United States Army, Corps of Engineers, Cold Regions Research and Engineering Laboratory, Report 79–9, Hanover, NH.

Ljüngner, E. (1930). Spaltektonik und morphologie der schwedishen Skagerakk-Kusts, Tiel III, Die erosienformen. *Bulletin of the Geological Society of the University of Uppsala*, **21**, 255–475.

Lorenc, M. W., Barco, P. M. and Saavedra, J. (1994). The evolution of potholes in granite bedrock, western Spain. *Catena*, **22**, 265–274.

Lorenz, K. (1974). Analogy as a source of knowledge. *Science*, **185**, 229–234.

Mathews, W. H. (1974). Surface profiles of the Laurentide ice sheet in its marginal area, *Journal of Glaciology*, **13**, 37–44.

Maxson, J. H. (1940). Fluting and faceting of rock fragments. *Journal of Geology*, **48**, 717–751.

Möller, P. (2006). Rogen moraine; an example of glacial reshaping of pre-existing landforms. *Quaternary Science Reviews*, **25**, 362–389.

Mullins, H. T., Anderson, W. T., Dwyer, T. R. *et al.* (1996). Seismic stratigraphy of the Finger Lakes: a continental record of Heinrich event H-1 and Laurentide ice sheet instability. *Special Paper, Geological Society of America*, **311**, 1–35.

Munro, M. and Shaw, J. (1997). Erosional origin of hummocky terrain, south-central Alberta, Canada. *Geology*, **25**, 1027–1030.

Munro-Stasiuk, M. J. (1999). Evidence for storage and release at the base of the Laurentide Ice Sheet, south-central Alberta. *Annals of Glaciology*, **28**, 175–180.

Munro-Stasiuk, M. J. (2000). Rhythmic till sedimentation: evidence for repeated hydraulic lifting of a stagnant ice mass. *Journal of Sedimentary Research*, **70**, 94–106.

Munro-Stasiuk, M. J. (2003). Subglacial Lake McGregor, south-central Alberta, Canada. *Sedimentary Geology*, **160**, 325–350.

Munro-Stasiuk, M. and Bradac, M. (2002). Digital elevation models and satellite imagery as tools for examining glaciated landscapes: examples from Alberta, Ohio, and Pennsylvania. *Abstracts with Programs, Geological Society of America*, **34**, Boulder, CO.

Munro-Stasiuk, M. J. and Shaw, J. (2002). The Blackspring Ridge Flute Field, south-central Alberta, Canada: evidence for subglacial sheetflow erosion. *Quaternary International*, **90**, 75–86.

Munro-Stasiuk, M. J. and Sjogren, D. B. (2006). The erosional origin of hummocky terrain, Alberta, Canada. In *Glaciers and Earth's Changing Environment*, ed. P. G. Knight. Oxford: Blackwell Publishing.

Munro-Stasiuk, M. J., Fisher, T. G. and Nitzsche, C. R. (2005). The origin of the western Lake Erie grooves, Ohio: implications for reconstructing the subglacial hydrology of the Great Lakes sector of the Laurentide ice sheet. *Quaternary Science Reviews*, **24**, 2392–2409.

Murray, E. A. (1988). Subglacial erosional marks in the Kingston, Ontario, Canada region: their distribution, form and genesis. Unpublished M.Sc. thesis, Queen's University, Canada.

Normark, W. R., Piper, D. J. W. and Hess, G. R. (1979). Distributary channels, sand lobes and mesotopography of Navy Submarine Fan, California Borderland with application to ancient sediments. *Sedimentology*, **26**, 749–774.

Nye, J. (1976). Water flow in glaciers: jökulhlaups, tunnels and veins. *Journal of Glaciology*, **17**, 181–205.

O'Cofaigh, C. (1996). Tunnel valley genesis. *Progress in Physical Geography*, **20**, 1–19.

Peabody, F. E. (1947). Current crescents in the Triassic Moenkopi Formation. *Journal of Sedimentary Petrology*, **17**, 73–76.

Piper, D. J. W., Shaw, J. and Skene, K. I. (2007). Stratigraphic and sedimentological evidence for late Wisconsinan sub-glacial outburst floods to Laurentian Fan. *Palaeogeography, Palaeoclimatology, Palaeoecology*, **246**, 101–119.

Pollard, A., Wakarini, N. and Shaw, J. (1996). Genesis and morphology of erosional shapes associated with turbulent flow over a forward-facing step. In *Coherent Flow Structures in Open Channels*, eds. P. J. Ashworth, S. Bennett, J. L. Best and S. McLelland. New York: Wiley.

Rains, B., Kvill, D., Shaw, J., Sjogren, D. and Skoye, R. (1993). Late Wisconsin subglacial megaflood paths in Alberta. *Geology*, **21**, 323–326.

Rampton, V. N. (2000). Large-scale effects of subglacial water flow in southern Slave Province, Northwest Territories, Canada. *Canadian Journal of Earth Sciences*, **37**, 81–93.

Richardson, K. and Carling, P. A. (2005). *A Typology of Sculpted Forms in Open Bedrock Channels*. Geological Society of America Special Publication 392.

Russell, A. J., Gregory, A. R., Large, A. R. G., Fleisher, P. J. and Harris, T. D. (2007). Tunnel channel formation during the November 1996 jökulhlaup, Skeiðarárjökull, Iceland. *Annals of Glaciology*, **45**, 95–103.

Russell, A. J., Knudsen, Ó., Fay, H. *et al.* (2001). Morphology and sedimentology of a giant supraglacial, ice-walled, jökulhlaup channel, Skeiðarárjökull, Iceland: implications for esker genesis. *Global and Planetary Change*, **28**, 193–216.

Russell, H. A. J., Arnott, R. W. C. and Sharpe, D. R. (2003). Evidence for rapid sedimentation in a tunnel channel, Oak Ridges Moraine, southern Ontario, Canada. *Sedimentary Geology*, **160**, 33–55.

Shapiro, A. H. (1961). *Shape and Flow: The Fluid Dynamics of Drag*. London: Heineman.

Sharpe, D. R. (2005). Comments on "Paleohydrology of the last outburst flood from glacial Lake Agassiz and the 8200 B. P. cold event." *Quaternary Science Reviews*, **28**, 1529–1532.

Sharpe, D. R. and Shaw, J. (1989). Erosion of bedrock by subglacial meltwater, Cantley, Quebec. *Geological Society of America Bulletin*, **101**, 1011–1020.

Sharpe, D., Pugin, A., Pullan, S. and Shaw, J. (2004). Regional unconformities and the sedimentary architecture of the Oak Ridges Moraine area, Southern Ontario. *Canadian Journal of Earth Sciences*, **41**, 183–198.

Shaw, J. (1983). Drumlin formation related to inverted melt-water erosional marks. *Journal of Glaciology*, **29**, 461–479.

Shaw, J. (1988). Subglacial erosional marks, Wilton Creek, Ontario. *Canadian Journal of Earth Sciences*, **25**, 1442–1459.

Shaw, J. (1994). Hairpin erosional marks, horseshoe vortices and subglacial erosion. *Sedimentary Geology*, **91**, 269–283.

Shaw, J. (1996). A meltwater model for Laurentide subglacial landscapes. In *Geomorphology sans frontiers*, eds. S. B. McCann and D. C. Ford. Chichester: Wiley, pp. 181–236.

Shaw, J. and Gilbert, R. (1990). Evidence for large-scale subglacial meltwater flood events in southern Ontario and northern New York State. *Geology*, **18**, 1169–1172.

Shaw, J. and Kellerhals, R. (1977). Paleohydraulic interpretation of antidune bedforms with applications to antidunes in gravel. *Journal of Sedimentary Petrology*, **47**, 257–266.

Shaw, J. and Kvill, D. R. (1984). A glaciofluvial origin for drumlins of the Livingstone Lake area, Saskatchewan. *Canadian Journal of Earth Sciences*, **12**, 1426–1440.

Shaw, J. and Sharpe, D. R. (1987). Drumlin formation by subglacial meltwater erosion. *Canadian Journal of Earth Sciences*, **24**, 2316–2322.

Shaw, J., Kvill, D. and Rains, B. (1989). Drumlins and catastrophic subglacial floods. *Sedimentary Geology*, **62**, 177–202.

Shaw, J., Munro-Stasiuk, M. J., Sawyer, B. *et al.* (1999). The Channeled Scabland: back to Bretz? *Geology*, **27**, 605–608.

Shaw, J., Faragini, D. M., Kvill, D. R. and Rains, R. B. (2000). The Athabasca fluting field: implications for the formation of large-scale fluting (erosional lineations). *Quaternary Science Reviews*, **19**, 959–980.

Shetsen, I. (1987). *Quaternary Geology of Southern Alberta*. Map scale 1:500 000. Edmonton: Alberta Research Council.

Shoemaker, E. M. (1995). On the meltwater genesis of drumlins. *Boreas*, **24**, 3–10.

Shreve, R. L. (1972). Movement of water in glaciers. *Journal of Glaciology*, **11**, 205–214.

Sjogren, D. B. (1999). *Formation of the Viking Moraine, East-central Alberta: Geomorphic and Sedimentary Evidence*. Unpublished Ph.D. thesis, University of Alberta.

Sjogren, D. B. and Rains, R. B. (1995). Glaciofluvial erosion morphology and sediments of the Coronation-Spondin Scabland, east-central Alberta. *Canadian Journal of Earth Sciences*, **32**, 565–578.

Smalley, I. J. and Unwin, D. J. (1968). The formation and shape of drumlins and their distribution and orientation in drumlin fields. *Journal of Glaciology*, **7**, 377–390.

Straw, A. (1968). Late Pleistocene glacial erosion along the Niagara Escarpment of southern Ontario. *Geological Society of America Bulletin*, **79**, 889–910.

Sugden, D. E. and John, B. S. (1976). Glaciers and landscape; a geomorphological approach. London: Edward Arnold.

Sundborg, Å. (1956). The river Klarälven. A study of fluvial processes. *Geografiska Annaler*, **38**, 127–316.

Tinkler, K. J. (1993). Fluvially sculpted rock bedforms in Twenty Mile Creek, Niagara Peninsula, Ontario. *Journal of Geology*, **105**, 263–274.

Tinkler, K. J. and Stenson, R. E. (1992). Sculpted bedrock forms along the Niagara Escarpment, Niagara Peninsula, Ontario. *Geographie Physique et Quaternaire*, **46**, 195–207.

Tsui, P. C., Cruden, D. M. and Thomson, S. (1989). Ice-thrust terrains in glaciotectonic settings in central Alberta. *Canadian Journal of Earth Science*, **6**, 1308–1318.

Whipple, K. X., Hancock, G. S. and Anderson, R. S. (2000). River incision into bedrock: mechanics and relative efficacy of plucking, abrasion and cavitation. *Geological Society of America Bulletin*, **112**, 490–503.

Wilhelm, D., Härtel, C. and Kleiser, L. (2003). Computational analysis of the two-dimensional to three-dimensional transition in forward-facing step flow. *Journal of Fluid Mechanics*, **489**, 1–27.

Wingham, D. J., Marshall, G. J., Muir, A. and Sammonds, P. (2006). Mass balance of the Antarctic ice sheet. *Philosophical Transactions of the Royal Society. Mathematical, Physical and Engineering Sciences*, **364**, 1627–1635.

Wohl, E. E. (1992). Bedrock benches and boulder bars: floods in the Burdekin Gorge of Australia. *Geological Society of America Bulletin*, **104**, 770–778.

Wohl, E. E. and Ikeda, H. (1998). Patterns of bedrock channel erosion on the Boso Peninsula, Japan. *Journal of Geology*, **106**, 331–345.

Wright Jr., H. E. (1973). Tunnel valleys, glacial surges and subglacial hydrology of the Superior Lobe, Minnesota. In *The Wisconsinan Stage*, eds. R. F. Black, R. P. Goldthwait and H. B. Williams. Geological Society of America, Memoir 36.

Young, R. R., Sjogren, D. B., Shaw, J., Rains, B. R. and Monroe-Stasiuk, M. (2003). A hanging tunnel channel/esker/megadrumlin/hummocky terrain complex on the high plains of Alberta. *XVI INQUA Congress*, Reno, Nevada.

Zwally, H. J., Adalati, W., Herring, T. *et al.* (2002). Surface melt-induced acceleration of Greenland ice-sheet flow. *Science*, **297**, 218–222.

7

Proglacial megaflooding along the margins of the Laurentide Ice Sheet

ALAN E. KEHEW, MARK L. LORD,
ANDREW L. KOZLOWSKI
and TIMOTHY G. FISHER

Summary

Megafloods from glacial lakes were common along the margin of the Laurentide Ice Sheet during the last deglaciation. These outbursts resulted in a complex network of spillways with characteristic erosional and depositional forms. Meltwater was impounded in front of and beneath the ice margin as the Late Wisconsin Laurentide Ice Sheet melted back from its maximum extent. Lakes that formed in this dynamic environment drained completely or partially. Various factors aided in the impoundment of meltwater lakes, including isostatic depression of the land surface in the vicinity of the retreating margin, topography that sloped toward the ice margin, glacial erosion of trough-like depressions by ice streaming in major outlet lobes, and moraine ridges formed when the ice was more extensive. Subaerial megafloods were triggered probably by the failure of ice-cored or sediment-cored moraine dams or rapid incision of outlets caused by incoming subaerial and/or subglacial megafloods. Complete drainage of lakes was likely where non-resistant glacial or bedrock materials made up the outlet region.

Proglacial megaflood discharges were typically 0.1–1.0 Sv, short in duration, and, in some places, achieved velocities in excess of 10 m s^{-1}. Outburst flows were highly erosive and carved a suite of small-scale to large-scale erosional forms, including potholes, longitudinal grooves, streamlined hills, transverse bedforms, anastomosing channels and spillways. Not all these forms are present along all megaflood pathways, except for spillways, which are ubiquitous; they are trench-shaped, 1–4 km wide, and tens of metres deep. Spillways occur individually or in networks connecting lake basins along former ice margins; some of them also lead to the oceans. The presence or absence of individual types of erosional forms in megaflood pathways is dependent on, among other things: flow duration and magnitude, number of flood events, underlying lithology, distance from the outlet, presence/absence of ice, and pre- and post-event history. Many of the erosional forms are unique to megafloods, but few, if any, appear to be unique to proglacial or subglacial outbursts. Upslope valley gradients, however, are associated with subglacial outbursts. Because the megafloods were dominantly erosive, the vast majority of sediment was dumped in lake basins and marine environments as anomalously coarse-grained deposits. Within the spillway systems, deposition was limited mainly to isolated, bouldery gravel bars and coarse traction carpets.

Megafloods drained to successively lower impoundments along the ice margin until they reached drainage divides leading to the Arctic Ocean, the Gulf of Mexico or the Atlantic Ocean. In the Great Plains region of the USA and Canada, networks of megaflood channels routed meltwater first to the Gulf of Mexico and then to glacial Lake Agassiz, the largest proglacial lake along the ice margin, and also the source of periodic megafloods. The routing, chronology and linkages with oceanic circulation and climate associated with these megaflood events are currently topics of active research. Ice-dammed lakes in the basins of the modern Great Lakes drained first from southern outlets to the Gulf of Mexico, and also to the North Atlantic through outlets leading to the St Lawrence, Mohawk, Hudson and Susquehanna Valleys. The morphology of these outlets and spillways indicate formation from megaflood events. As the Laurentide Ice Sheet retreated northward, glacial Lakes Agassiz and Ojibway coalesced to form a superlake that drained catastrophically one or more times under the wasting ice sheet to the Tyrrell Sea and the North Atlantic. This megaflood has been linked to the 8200 ka cooling event.

7.1 Introduction

At its maximum extent, the Laurentide Ice Sheet covered approximately 10% of the land area of the Earth, contained an estimated ice volume of 15.9–37 × 10^6 km^3 (Licciardi *et al.*, 1998), and achieved a thickness of more than 3000 m (Dyke *et al.*, 2002). Ice domes surrounding the Hudson Bay region of Canada defined the centre of the ice sheet as well as the area of greatest isostatic crustal depression. As the ice sheet expanded outward from its central ice domes, it merged with the Cordilleran Ice Sheet along its western margin, formed a mostly marine margin along its northwestern and northeastern sectors, merged with the Innuitian Ice Sheet along part of its northern boundary (Dyke *et al.*, 2002), and formed a long terrestrial margin to

Megaflooding on Earth and Mars, ed. Devon M. Burr, Paul A. Carling and Victor R. Baker.
Published by Cambridge University Press.

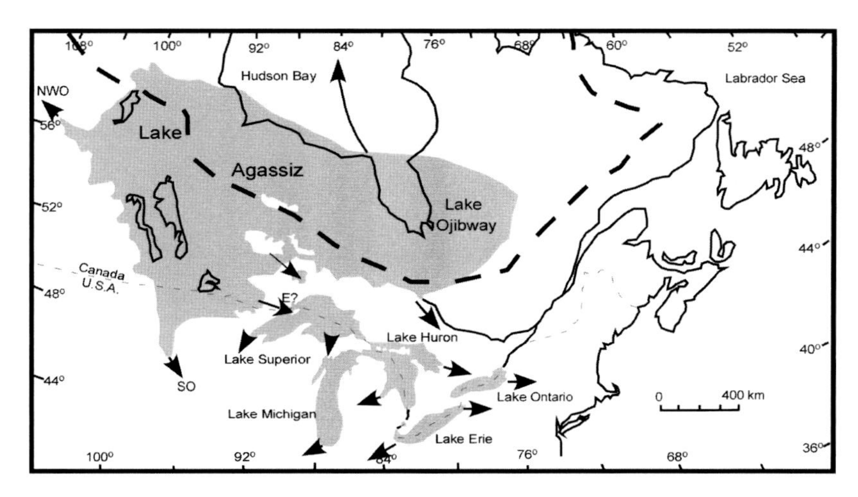

Figure 7.1. Maximum extent of glacial Lakes Agassiz and Ojibway. Major outlets formed by megafloods from glacial Lake Agassiz and the Great Lakes as indicated by arrows. Great Lakes discussed later in text. Heavy dashed line shows approximate position of margin of Laurentide Ice Sheet at 9 ka ^{14}C BP. SO: southern outlet; NWO: northwestern outlet; E?: eastern outlet. Question mark reflects uncertainty of existence of a Younger Dryas outlet at its location. (Modified from Teller *et al.* (2002).)

the south. Following the Late Glacial Maximum (LGM), numerous ice-dammed lakes developed along the ice sheet during its retreat to small isolated remnants in the Canadian Arctic.

During its advance, erosion and deposition produced widespread and dramatic changes to the landscape by diversion of streams, scouring of troughs, basins and channels, and construction of a multitude of glacial landforms. A huge volume of meltwater produced by ablation of the ice sheet, which was greatest along the southern margin, therefore was forced to find new pathways to the oceans. Initially, most of the meltwater from the southern margins flowed southward, down the Mississippi Valley and its tributaries to the Gulf of Mexico. As the ice sheet retreated north of the Mississippi drainage divide and downslope toward the isostatically depressed Hudson Bay lowland, huge proglacial lakes ponded against the receding ice sheet. Depending on the topography, proglacial lakes either expanded or contracted based on the location of outlets. If during a glacial readvance an outlet was closed, the lake would rise to the elevation of a higher existing outlet or would spill over its divide to rapidly incise a new outlet, thereby lowering or draining the lake. Outlets were also opened as ice retreat exposed topographically lower drainage routes. In either case, initial release of meltwater was likely to be catastrophic. Over the past several decades, increasing attention has been focused on the erosional and depositional processes of these megafloods, as well as their role in global climate change.

The purpose of this paper is to review the erosional and depositional mechanisms of these cataclysmic events, illustrated by examples chosen from among the many that have been described or observed. Chronology of these events is an area of current research, especially with respect to Lake Agassiz (Figure 7.1). Although a complete review of the drainage chronology of all known Laurentide proglacial lakes is beyond the scope of this paper, approximate ages of known or hypothesised megafloods are reviewed.

7.2 Dams and triggers

Proglacial lakes form behind dams of ice, glacial debris or bedrock. In mountainous terrain supporting a network of valley glaciers, ice-dammed lakes form in a variety of settings (Tweed and Russell, 1999). Along much of the margin of the Laurentide Ice Sheet, most proglacial lakes at least were impounded partially by dams of glacial drift or non-resistant sedimentary rock. In the Great Lakes region, the major lobes of the Laurentide Ice Sheet exploited weak lithologic zones in the bedrock and eroded trough-like basins. Erosion may have been accentuated by streaming flow, for example by the Lake Michigan and Huron–Erie lobes (Clayton *et al.*, 1985; Shoemaker, 1992; Clark, 1994; Marshall *et al.*, 1996; Kehew *et al.*, 2005). Arcuate moraine ridges were constructed of debris transported by the glacier at and beyond the distal margins of these basins. Buried ice probably was common in these ridges. As the ice retreated, lakes were ponded between it and the moraine dams.

The formation of an outlet for the proglacial lakes would take place at the topographically lowest point around the margin. The size and geomorphological form of these outlets (Plate 13) reflect the very large discharges released from proglacial lakes. It is uncertain whether these outlets and the channels that lead away from them, known as spillways, represent catastrophic outlet development and full or partial lake drainage, or whether they indicate high,

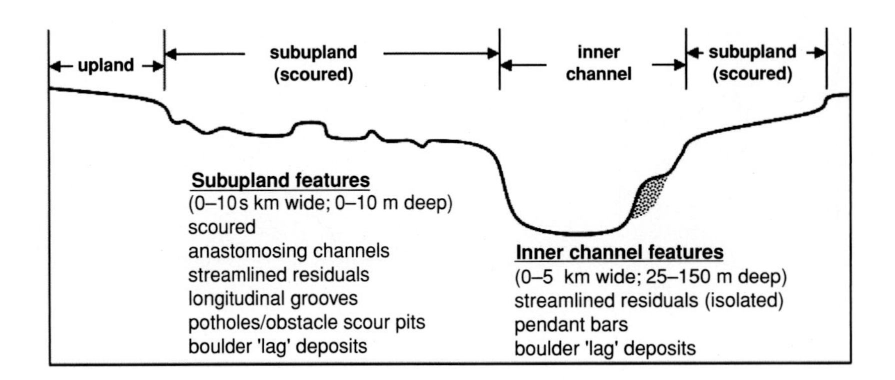

Figure 7.2. Schematic diagram of spillway cross-section with erosional and depositional features. (Modified from Kehew and Lord (1986).)

but sustained discharges of meltwater. The most likely scenario is that outlet development was rapid and that lake levels declined from one stable level to another in a very short amount of time. This conclusion is based upon several factors. Firstly, the debris dams were inherently weak. Broad spillover along the crest of a moraine dam would quickly become channelised, initiating a positive feedback process in which higher velocities as the channel is excavated would lead to enlargement of the channel and still higher discharge and velocity. This process would continue as long as a sufficient head was present behind the dam or until a resistant substrate was encountered by the eroding flow. At this point, greater stability of the channel would prevail until the lake was lowered to the elevation of this sill. Boulder armours, produced by boulder concentration during downcutting through till, stabilised the outlet in some spillways (Bretz, 1951, 1955; Matsch and Wright, 1966; Kehew, 1993). The occurrence of palaeoshorelines converging to the outlets at the stable levels, as well as their absence between stable levels, supports the hypothesis of episodic outlet downcutting. This scenario is essentially the outlet control model of Bretz (1951, 1955) for changes in lake level in the Lake Michigan basin (glacial Lake Chicago). Alternative sill models for Lake Chicago have been suggested by Hansel and Mickelsen (1988).

Rapid outlet incision also probably requires an appropriate triggering mechanism that would cause the lake level to rise and/or discharge in the outlet to increase. Such a mechanism could involve a surge of the glacier into the lake or, more likely, a large meltwater discharge entering the lake at some point. Opportunities for a sudden influx of water likely were abundant and include subglacial discharges into an ice-dammed lake or extra-basinal meltwater input. The movement of outburst floods from lake basin to lake basin (domino effect) is well documented in the literature (Kehew and Clayton, 1983; Kehew and Lord, 1986, 1987; Lord, 1991; Colman *et al.*, 1994; Kehew and Teller, 1994a; Sun and Teller, 1997). A meltwater influx triggering mechanism is not necessary, however, if the moraine dam simply failed due to piping, melting of buried ice, or slope failure. Utilisation of a given outlet persists until ice-margin retreat uncovers a lower outlet. The opening of these outlets also would occur catastrophically, if the thinning or retreat of the ice reached a critical threshold to initiate one of the mechanisms suggested by Tweed and Russell (1999).

7.3 Erosional processes and landforms

Megafloods along the margins of the Laurentide Ice Sheet had a tremendous erosional capacity and created a characteristic set of channels and associated landforms. The erosional effects of the proglacial lake megafloods were controlled by the hydraulic characteristics of the floods, the lithologic properties of the substrate materials, and the pre-event and post-event history of the region. Along the southern margin of the Laurentide Ice Sheet, the substrate materials are dominated by unconsolidated glacial drift and poorly lithified Mesozoic and Tertiary sedimentary rocks. Consequently, short-lived, high-velocity flows encountered little resistance to vertical incision; exceptions occurred where resistant Palaeozoic or Precambrian rocks were encountered (e.g. parts of the Minnesota River spillway). These settings contrast with the Channeled Scabland (Baker and Nummedal, 1978; Baker *et al.*, 1987), where basalt provided a more resistant barrier to vertical incision once megafloods had cut through the surface loess deposits.

The most characteristic landforms formed by megafloods along the Laurentide Ice Sheet are the spillways; they are trench-shape channels with steep sides and relatively uniform widths (1–4 km) and depths (25–150 m) (Plate 13 and Figure 7.2) (Kehew and Lord, 1987). Most spillways begin and end at glacial lake basins, whereas those that do not have an obvious lacustrine source and may have been eroded by the emergence of subglacial outbursts

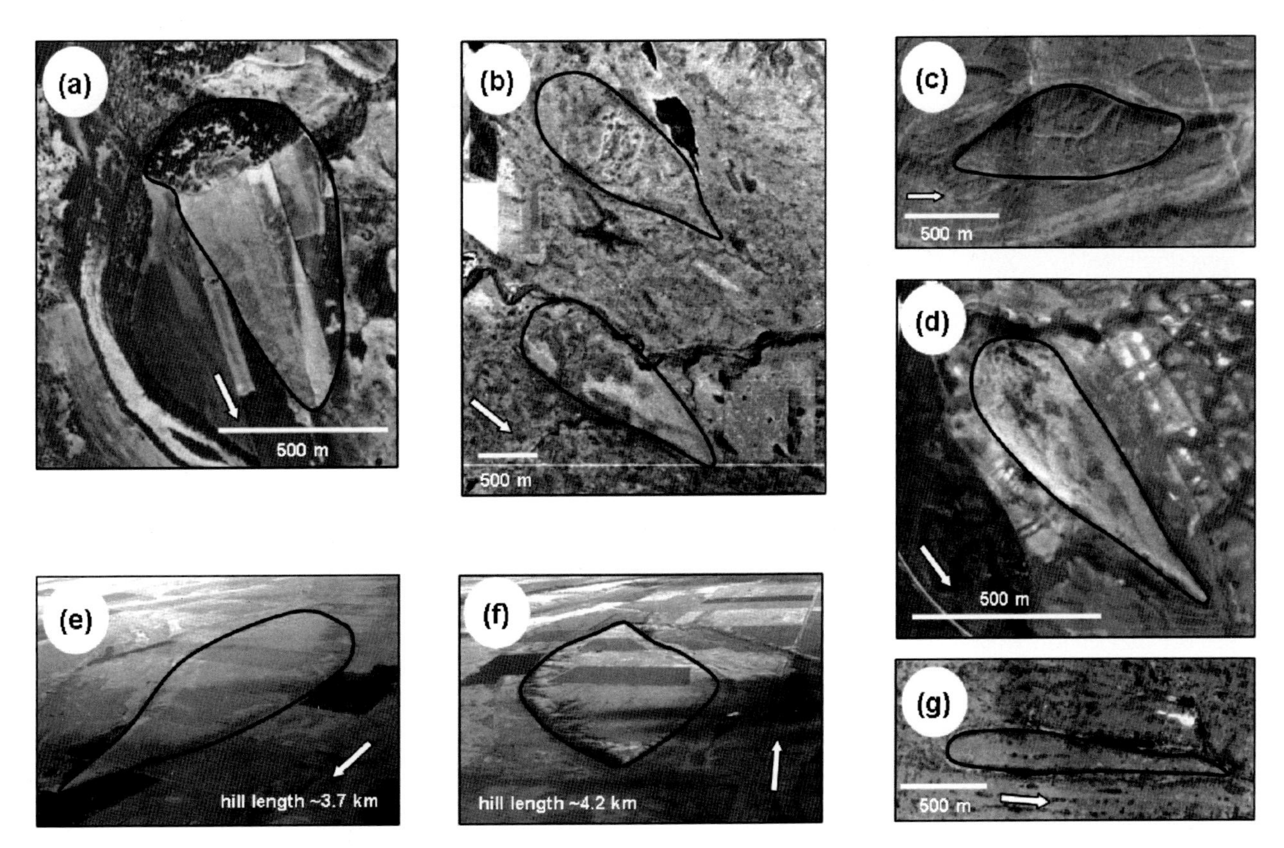

Figure 7.3. A variety of streamlined hill types and shapes from northern Great Plains spillways; arrows are flow directions. (a) Hill of sand and gravel downstream of bedrock knob, inner channel of Minnesota spillway. (b) Erosional residuals in scoured upland of Souris spillway near glacial Lake Regina outlet. Upper streamlined form was covered with dead ice for duration of flood; lower form was shaped by water on top and sides. (c) Erosional remnant between anastomosing channels; Souris spillway, near Estevan, Saskatchewan. (d) Sand and gravel streamlined hill adjacent to the inner channel of Des Lacs spillway near Minot, North Dakota. (e) and (f) interchannel erosional residuals in the Souris-Hind spillway, near Melita, Saskatchewan. (g) Erosional remnant formed by convergence of surrounding longitudinal grooves in scoured upland, Souris spillway, near Midale, Saskatchewan.

(e.g. Kehew and Teller, 1994b; Kozlowski *et al.*, 2005). Spillways are present in all megaflood drainages described in later sections of this chapter. In many places, the inner channels of spillways are flanked by a scoured subupland surface, probably a vestigial channel form from an early megaflood stage carved out before channel enlargement could accommodate the outburst flows. The scoured subupland is incised below the spillway and contrasts with the non-scoured terrain adjacent to the spillway. Additional megaflood erosional forms are anastomosing channels (Plate 14), streamlined erosional forms (Figure 7.3), longitudinal grooves, transverse bedforms, potholes and obstacle scour depressions (Plate 13 and Figure 7.2).

The spillways and related landforms represent different stages of channel development in drainage tracts that were either created entirely by, or overwhelmed by megaflood discharges (Kehew and Lord, 1986, 1987; Lord and Schwartz, 2003). Initial stages of erosion carve a wide, shallow tract, commonly with anastomosing channels. Further erosion leads to channel deepening with more organised flow resulting in longitudinal grooves. A central large, deep inner channel results from coalescence of longitudinal grooves and capture of lateral flow. Where the duration of the megaflood was long enough and the bed material was erodable, the spillways likely enlarged to convey the entire flow. This model of a megaflood-eroded landscape is generally consistent with the very well studied Channeled Scabland (e.g. Baker *et al.*, 1987), spillways in the northern Great Plains (e.g. Kehew and Lord, 1987), and experimental flume studies (Shepherd and Schumm, 1974). Not all spillways developed this way or show these traits. In places where the flood pathways were able to utilise large pre-existing spillways or river valleys, the spillway form may be obscure. Similarly, postflood event deposition by glacial or postglacial activity may alter the form of the spillways. For example, Holocene streams and lakes substantially infilled many mid-continent spillways (Kehew and Boettger, 1986).

Analysis of streamlined hills formed by megafloods, which are present along all margins of the Laurentide Ice Sheet, also support the evolution of spillway systems described above. Kehew and Lord (1986) performed morphometric analysis of 168 streamlined hills from five spillway systems in the mid-continent region. The streamlined hills are polygenetic (Figure 7.3). Most of the streamlined forms originated by erosional modification of interchannel areas of anastomosing networks; less common are streamlined hills that developed in isolation or due to deposition. All forms, regardless of origin or material, have a remarkably consistent relationship among length, width and area, indicating that these ratios are established very early in flow and maintained with subsequent erosion. Streamlined hills in the Channeled Scabland and in Martian outflow channels also share similar relationships (Baker, 1982; Komar, 1983; Kehew and Lord, 1986; Burr, 2005). However, the relative position of the maximum width of streamlined hills does vary in the mid-continent spillway systems. Poorly developed streamlined forms have a position of maximum width at about the midpoint of the hill, whereas more mature forms have the position of maximum width closer to that of the ideal, minimum resistant streamlined form, a lemniscate loop (Komar, 1983; Kehew and Lord, 1986).

7.4 Depositional processes and landforms

Megafloods from glacial lakes along the margins of the Laurentide Ice Sheet were highly erosive, leaving little depositional record within the spillway systems. Whereas some eroded materials were redeposited in spillway systems, especially distal from the source of flood waters, almost all fine-grained sediments were deposited in large glacial lakes and the marine offshore (Kehew and Lord, 1987; Kehew and Teller, 1994a; Teller, 2004). Within the spillways, there are two broad categories of deposits: (1) large, boulder bars within the inner channel of the spillway, and (2) scattered boulders that mantle much of the scoured subupland and, in some places, spillway bottoms incised into bedrock (e.g. Kehew and Clayton, 1983; Kehew and Lord, 1986; Fisher, 2004).

Large-scale, cobble or boulder bars formed at point bar positions or zones of flow expansion. In the northern Great Plains, these bars are commonly 2 km in length, 0.5 km wide, and up to 20 m thick (Figure 7.4a) (Kehew and Lord, 1987). Most boulder bars are massive, unstructured and matrix supported, and are generally interpreted to have been deposited by hyperconcentrated flows (Lord and Kehew, 1987; Smith and Fisher, 1993; Fisher, 2004). Hyperconcentrated flows were probably more common where the underlying materials were dominantly fine grained, such as clay-rich tills, or poorly lithified, clay-rich bedrock. Gravel bars in some locations exhibit a mix of large-scale cross-bedding, massive structure and planar bedding (Wilson and Muller, 1981; Fraser and Bleuer, 1988) and likely were deposited by turbulent fluid flows. All of these facies types have been described from historic and prehistoric jökulhlaup deposits (Maizels, 1997). More extensive valley fills consisting of megascale cross-sets are present in several spillways. Fraser and Bleuer (1988) attributed a fill in the Wabash Valley of Indiana to a jökulhlaup event originating in a stagnating glacier followed by an outburst from Lake Maumee. Johnson *et al.* (1998) described a fill of this type in the Minnesota River valley, deposited by Lake Agassiz drainage in a reach where the spillway widens.

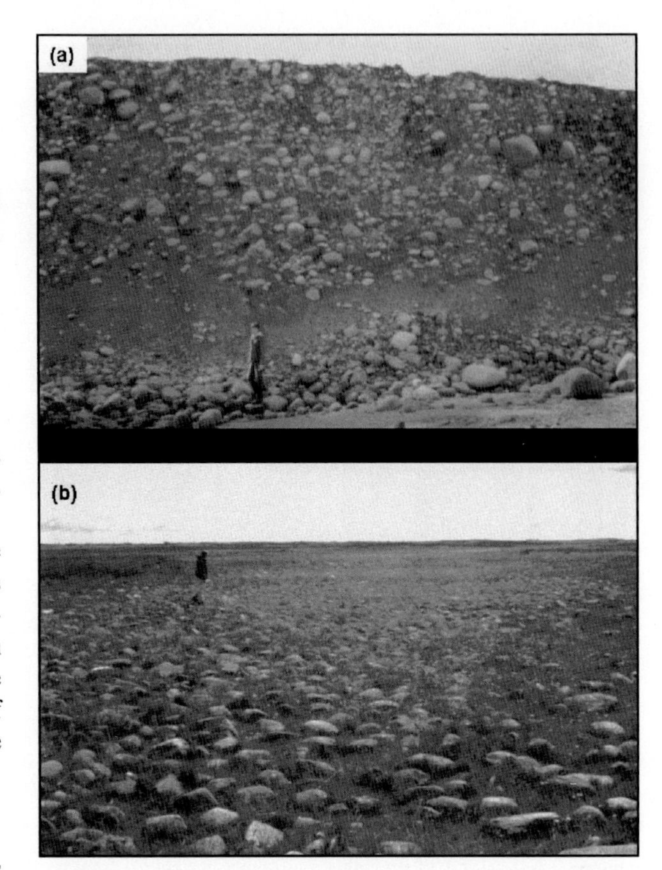

Figure 7.4. (a) Exposure of a glacial Lake Regina outburst deposit; the material is a structureless, poorly sorted, bouldery gravel with some clasts greater than 1 m in diameter. This bar formed in the Souris spillway about 150 km downstream from the outlet of glacial Lake Regina. (b) A single-particle thickness of well-sorted boulders and cobbles in the centre of a longitudinal groove in the scoured upland of the Souris spillway; view is downstream.

Unlike scarce large-scale gravel bars, boulder lag concentrations are ubiquitous where megafloods eroded sediments containing resistant boulders (i.e. tills) or resistant bedrock units. In some places the boulders are up to several metres in diameter, poorly sorted and irregularly spaced. In this setting, the boulders represent a true lag, or erosional residual. In other places, however,

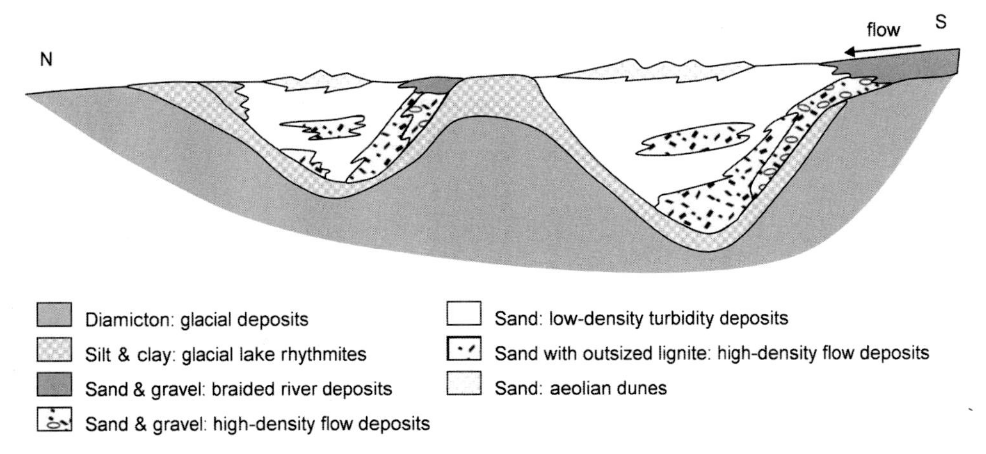

Figure 7.5. Diagrammatic longitudinal cross-section through the glacial Lake Souris basin showing distribution of facies. (Modified from Lord (1991).) Megaflood deposited sediments are coarse grained (mostly sand) and were mainly deposited by high-density currents.

boulders are well sorted, evenly spaced and their patterns are clearly related to large-scale erosional bedforms (Figure 7.4b) (Kehew and Teller, 1994b; Lord and Schwartz, 2003; Fisher, 2004). In these cases, the boulders are not a lag deposit, but rather a bedload deposit.

Lake basins that received megaflood waters, like outburst deposits in spillways, have a sedimentary record that differs from glaciolacustrine basins not inundated with megafloods. Outburst-deposited sediments can be distinguished from normal meltwater deposits by their thick, coarse-grained, homogeneous vertical sections that systematically fine with distance from the inlet (Kehew and Clayton, 1983; Kehew and Lord, 1987; Lord, 1991; Wolfe and Teller, 1993; Sun and Teller, 1997). The flood-deposited sediments, mostly sands, are very well sorted and are interpreted to have been deposited by low-density and high-density turbidity currents (Figure 7.5) (Kehew and Clayton, 1983; Lord, 1991). Aeolian sand dunes now mark the location of flood-deposited fan sediments in several glacial lake basins (e.g. Agassiz, Souris and Hind) because the well-sorted sands were readily reworked by wind. Although silt and clay-size particles made up most of the material eroded by the megafloods, they are largely absent in smaller glacial lake basins because the floodwaters triggered the outlet downcutting and drainage of successive lakes in a domino fashion (Kehew and Clayton, 1983). Fine-grained flood sediments were deposited in larger lakes, such as Lake Agassiz and the Great Lakes, and the marine offshore.

7.5 Palaeohydrologic considerations

Numerous studies of glacial lake outbursts along margins of the Laurentide Ice Sheet have resulted in estimates of maximum discharges and velocities, a complete review of which is beyond the scope of this contribution, but many are reported in subsequent sections of this chapter. Most discharge estimates range from 0.2 to 1.0 Sv, but values up to 5.0 Sv have been estimated for the last outburst flood from glacial Lake Agassiz into Hudson Bay (Teller *et al.*, 2002; Clarke *et al.*, 2004). Typical maximum velocities reported range from 5 to $12\,\mathrm{m\,s^{-1}}$, but velocities up to $30\,\mathrm{m\,s^{-1}}$ have been suggested for the southern spillway of Lake Agassiz (Fisher, 2004). Several approaches have been used to estimate the palaeohydrology of megafloods, including Manning's equation, HEC-2 modelling, maximum clast size competency-based methods, and sophisticated numerical simulations based on ice margins, topography and ice physics (see Fisher (2004) for review of methods used for parts of Lake Agassiz).

All palaeohydrologic methods have uncertainties that can severely limit the precision and accuracy of the values reported. These estimates are constrained by the understanding of the processes that are being evaluated, the quality of the erosional and depositional record preserved, and the limitations and simplifications inherent to any method that attempts to simulate complex natural processes (Baker, 1973; Lord and Kehew, 1987; O'Connor and Baker, 1992; Kehew and Teller, 1994b; Herget, 2005). Part of the complexity relates to the uncertainty of key variables for many approaches, such as energy slope (difficult to evaluate under ideal conditions, but complicated by differential isostatic rebound), channel margins (high-water marks – some megafloods were initially ice-walled or subglacial) specific weight of water (varies with sediment water concentration and some floodwaters were hyperconcentrated) and channel roughness (palaeohydraulic estimates are sensitive to channel roughness and can vary significantly depending on the mobility of the bed). Importantly, the evolution of spillway cross-sectional geometry and its relationship to peak flow conditions remain poorly understood, especially for spillways that were carved in highly erodable sediment or weak bedrock.

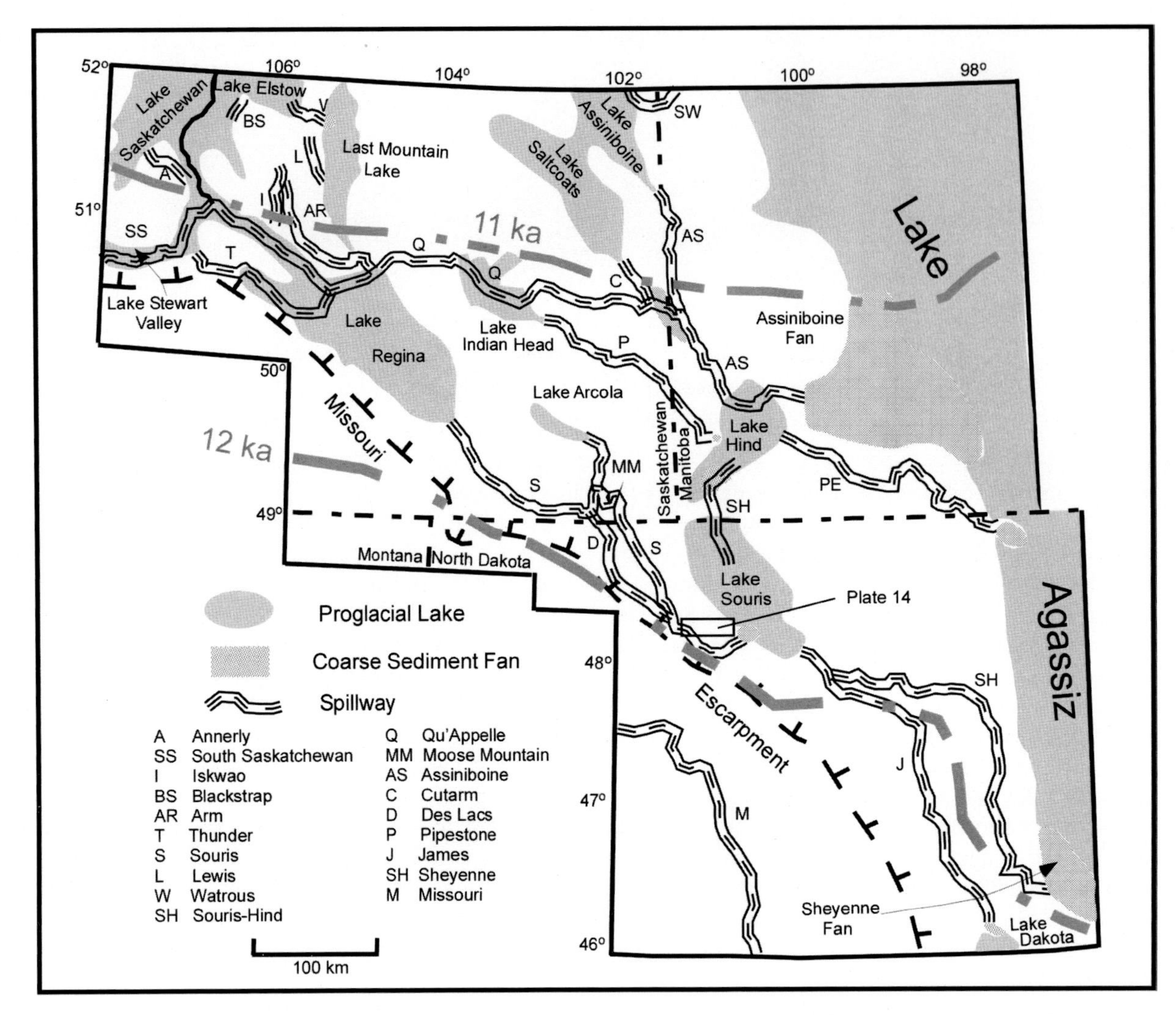

Figure 7.6. Network of proglacial lake basins and spillways formed by megaflood events west of glacial Lake Agassiz. Thick grey lines represent approximate ice marginal positions. Box shows location of Plate 14. (After Kehew and Teller (1994a).)

Spillway depth, for example, is probably not a good indicator of flow depth because of the rapid rate of channel incision. These broad, relatively straight, valleys appear to be non-equilibrium flood channels, the dimensions of which probably represent, at a minimum, the width of the flows. Kehew and Lord (1986, 1987) showed that the Souris spillway in Saskatchewan evolved from a broad flow scouring subupland surfaces to a narrower flow completely contained within the deepening spillway. As this incision progressed, it is likely that water depth was successively lowered along with channel bottom elevation. Spillway depths provide, at best, a maximum estimate of flow depth and, if a spillway was utilised by more than one megaflood event, then bankfull width and depth of the spillway almost certainly overestimate the flow parameters of an individual megaflood.

7.6 Anatomy of a glacial lake megaflood: the glacial Lake Regina outburst

The glacial Lake Regina outburst, first hypothesised by Kehew (1982), produced the Souris spillway, which is characterised by a well-defined inner channel; a wide (~10–15 km) scoured subupland, longitudinal grooves, transverse erosional bedforms, anastomosing channels and streamlined hills (Plate 15, Figure 7.6) (Lord and Schwartz, 2003). In addition, a widespread bedload deposit one boulder thick directly relates to the distribution of erosional bedforms (Figure 7.4b). This outburst resulted in a well-preserved, single-event array of large-scale erosional bedforms. The outburst initiated with the breach of a stagnant or dead ice dam that released about 7.4×10^{10} m^3 of water from glacial Lake Regina generating a discharge of about 0.5 Sv and velocities from 3 to 12 m s^{-1} (Kehew and

Clayton, 1983; Lord and Kehew, 1987; Lord and Schwartz, 2003). Ice was present along some of the flow margins and as isolated blocks within the spillway and, although some persisted for the duration of the flood, the bedforms developed without ice cover.

The outburst was highly erosive, and was competent to transport almost all of the material eroded. Because of the clear-water source of the floodwaters, Lake Regina, it is likely the sediment–water concentration in the Souris spillway was low near the lake outlet but increased rapidly downstream until at some point, in time and/or space, it became hyperconcentrated (Lord and Kehew, 1987).

The bedform trends define spatial and temporal changes in fluid secondary circulation patterns and velocity both down and across the spillway (Lord and Schwartz, 2003). In the central portion of the spillway at the lake outlet, flow was likely smooth-turbulent, and, with distance from the outlet became fully turbulent, dominated by transverse rollers producing transverse erosional bedforms (Plate 15A). Farther downstream, streamwise vortices developed, causing helical flow and the formation of longitudinal grooves. The bedform field was dynamic during the flood as erosion and flow convergence caused upstream migration of the longitudinal flow field (Lord and Schwartz, 2003). Continued deepening and upstream migration of the inner channel eventually captured all outburst flow from the upland surface at distances greater than 10 km downstream from the outlet (Plate 15). Flow ceased abruptly as is shown by the well-preserved bedforms and bedload deposits.

Glacial Lakes Souris, Hind and Agassiz (Figure 7.6), downstream glacial lakes that received these outburst flows, contain a depositional record consistent with inferences made from analysis of the outlet zone. Most notably, the outburst of glacial Lake Regina was a high-magnitude event that began abruptly, eroded and conveyed tremendous amounts of coarse-grained sediment in high-density flows, and ended abruptly (Figure 7.5) (Kehew and Clayton, 1983; Lord, 1991; Sun and Teller, 1997).

7.7 Megafloods west of Lake Agassiz

The southwestern quadrant of the Laurentide Ice Sheet is dissected by a network of channels cut by glacial meltwater, ranging from fresh channels dating from the Late Wisconsin glaciation (Figure 7.6) to valleys from earlier glaciations that are completely buried. Channels that begin at one of the lake basins shown on Figure 7.6 can unambiguously be identified as proglacial lake spillways. Channels not associated with a specific proglacial lake or buried valleys of unknown origin could be either spillways, subglacial tunnel channels, or composite channels, as described by Kozlowski *et al.* (2005). Spillways are eroded during deglaciation and partially filled by interglacial sediment derived from dissection of the valley walls and from fluvial and lacustrine systems active in the valleys. Reoccupation by meltwater during successive glaciations was probable.

The spillway network that developed during the Late Wisconsin deglaciation is characterised by rapid shifts in channel usage as the ice retreated across drainage divides to the north, exposing lower terrain that could be exploited by eastward-draining meltwater (Kehew and Teller, 1994a). Prior to approximately 12 ka ^{14}C BP, the ice extended south of the Missouri escarpment and lakes along the margin spilled over into the Missouri drainage basin. Kehew and Teller (1994a) grouped the proglacial drainage events into four phases between approximately 12 and 10.7 ka ^{14}C BP, as the ice margin retreated north of the Missouri Escarpment (Table 7.1). Each phase involved a major spillway or spillway system leading to Lake Agassiz. The first of these, the James phase, represents the last drainage into the Missouri drainage basin. The next three phases, the Sheyenne, the Souris-Pembina and the Qu'Appelle-Assiniboine, involve at least 13 proglacial lakes and 20 spillways. Each phase represents a period of no more than several hundred years. A major drainage reorganisation occurred at approximately 11.2 ka ^{14}C BP, when the Qu'Appelle spillway (Figure 7.6), one of the largest in the Canadian prairies, captured drainage over a 500 000 km^2 area from the Rocky Mountains east, including spillways in Alberta (Evans, 2000) and Saskatchewan. At the time of this amalgamation, the Qu'Appelle spillway carried meltwater combined with surface runoff from 1000 km of the Cordilleran ice margin and 2000 km of the Laurentide Ice Sheet (Clayton and Moran, 1982). Although the valley must have conveyed sustained high discharges of water for some period of time, its size and morphology indicate that outburst floods were critical to its formation. The effects of these bursts of water upon Lake Agassiz are unknown, but may have triggered outlet downcutting and the release of Lake Agassiz megafloods.

7.8 Lake Agassiz and megaflood outflows

Glacial Lake Agassiz was the largest proglacial lake in North America (Figure 7.1), eventually occupying approximately 1 500 000 km^2 over a 5000-year period (Teller and Leverington, 2004). Fluvial input to the lake was mainly along its western margin from the Sheyenne, Pembina and Assiniboine Rivers, where the largest deltas are found. The northern border of the lake was the Laurentide Ice Sheet. Large kame deposits such as the Sandyland Moraine in southern Manitoba and stratified moraines in northwest Ontario (Sharpe and Cowan, 1990) record large volumes of meltwater entering the lake, which at times may have been catastrophic. Such large inputs of meltwater would be dampened in the large Agassiz basin, but

Table 7.1. Megaflood chronology

Region	Spillway(s)	Figure no.	Age (ka ^{14}C BP)	Q (Sv)	Source waters	Megaflood features	References
Western	James	7.6	11.9–11.7	?	Lake Souris	IC, BB	Kehew and Teller, 1994a
	Sheyenne	7.6	11.8	?	Lake Souris	IC, BB, SSU	Kehew and Teller, 1994a; Lepper *et al.*, 2007
	Des Lacs	7.6	11.7–11.2	?	Lake Regina	IC, BB, SSU	Kehew and Lord, 1986, 1987; Lord and Kehew, 1987
	Souris	7.6	11.4–11.2	~1	Lake Regina	IC, BB, SSU, SER	Kehew and Lord, 1986, 1987; Lord and Kehew, 1987
	Pembina	7.6	11.4–11.1	?	Lake Souris/Hind	IC, BB, SSU	Kehew and Lord, 1987; Kehew and Teller, 1994a
	Moose Mountain	7.6	11.4–11.2	?	Lake Arcola	IC	Kehew and Lord, 1987; Kehew and Teller, 1994a
	Pipestone	7.6	11.3–11.2	?	Lake Indian Head	IC	Kehew and Teller, 1994a
	South Saskatchewan	7.6	11.3–10.7	?	?	IC	Scott, 1971; Kehew and Teller, 1994a
	Thunder	7.6	11.3–11.1	?	Lake Stewart Valley; Lakes Beechy/Birsay (?)	IC, SSU, BB, SER	Kehew and Lord, 1987; Kehew and Teller, 1994a
	Upper Qu'Appelle	7.6	11.2–10.8	?	?	IC, SSU	Kehew and Teller, 1994a
	Lower Qu'Appelle, Lower Assiniboine	7.6	11.2–10.8	?	?, Lake Saltcoats, Lake Assiniboine	IC	Kehew and Teller, 1994a
	Annerly	7.6	11.1–11.0	?	Lake Saskatchewan	IC, SER	Kehew and Teller, 1994a
	Arm	7.6	11.0–10.9	?	?	IC	Kehew and Teller, 1994a
	Blackstrap	7.6	11.0–10.9	?	Lake Saskatchewan	IC	Kehew and Teller, 1994a
	Watrous	7.6	11.0–10.8	?	Lake Elstow	IC, BB	Kehew and Teller, 1994a
	Last Mountain	7.6	11.0–10.8	?	Lake Last Mountain	IC	Kehew and Teller, 1994a
	Cutarm	7.6	11.2–10.8	?	Lake Saltcoats	IC	Kehew and Teller, 1994a
	Upper Assiniboine	7.6	11.2–10.8	?	Lake Saskatchewan, Lake Melfort	IC, SSU	Christiansen, 1979; Wolf and Teller, 1993; Kehew and Teller, 1994a
	Swan	7.6	10.8–10.7	?	?	?	Kehew and Teller, 1994a
Agassiz	Minnesota	7.7	12–10.65, 9.5–9.4	Q_{max} 0.1–0.36	Agassiz	IC, BB, SER, SSU, RIF	Fisher, 2003; 2004; Lepper *et al.*, 2007

(cont.)

Table 7.1. (cont.)

Region	Spillway(s)	Figure no.	Age (ka ^{14}C BP)	Q (Sv)	Source waters	Megaflood features	References
	Eastern	7.1	post 9.4	~0.1	Agassiz	IC, BB, SSU, SER	Teller and Thorleifson, 1983; Fisher, 2003; Teller and Leverington, 2004
	Clearwater–Lower Athabasca	7.8	9.9–9.6	Q_{peak} 2.16	Agassiz and/or other proglacial lake	IC, BB, SSU, SER	Fisher *et al.*, 2002; Waterson *et al.*, 2006
	Hudson Bay	7.1	8.5–8.3 ka cal yr	0.5–5	Agassiz	?	Clarke *et al.*, 2004
Western Great Lakes	Wabash	7.9	~14.0	?	Lake Maumee	IC, BB, SER	Eschman and Karrow, 1985; Fraser and Bleuer, 1988; Lewis *et al.*, 1994
	Illinois Valley	7.10	16.0–14.0	?	Kankakee Valley and others	IC, BB, SSU, SER, RIF	Kehew and Lord, 1987; Hajic, 1990
	Grand Valley, Michigan	Plate 13, 7.9	13–12.5	?	Lake Saginaw	IC, BB, SER	Kehew, 1993
	Sag Channel, Des Plains Channel, Illinois Valley	7.9, Plate 16, 7.10	13–12.5	?	Lake Chicago; Glenwood II Phase	IC, BB, SSU, SER, RIF	Bretz, 1951, 1955; Kehew, 1993; Hajic, 1990
	Sag Channel, Des Plains Channel, Illinois Valley	7.9, Plate 16, 7.10	10.9–10.4(?)	?	Lake Chicago; Calumet Phase	IC, BB, SSU, SER, RIF	Kehew, 1993; Colman *et al.*, 1994
	Portage, Brule, St Croix	7.9, 7.11	11.5–10.0	?	Lake Duluth	IC, BB, SSU, SER, RIF	Farrand and Drexler, 1985; Kehew and Lord, 1987; Johnson, 2000
Eastern Great Lakes	Holtwood Gorge	7.12	Pre-Wisconsin	~0.2	Lake Lesley	IC, BB, RIF, PH	Kochel and Parris, 2000; Potter, 2001; Thompson and Sevon, 2001
	Batavia–Genesee	7.12, Plate 18	~12.6–12.4(?)	~0.3	Lake Warren	IC, BB, SSU, SER	Wilson, 1980; Wilson and Muller, 1981; Muller and Prest, 1985
	Syracuse Channels	7.12, 7.13	?	0.2	Lake Warren (II)?	IC, BB	Hand and Muller, 1972; Thrivikramaji, 1977; Muller and Prest, 1985
	Mohawk Valley	7.12, Plate 19	11.2?	0.6	Lake Iroquois	IC, BB, SER, SSU, PH	Wall, 1995; Wall and LaFleur, 1996
	Covey Hill	Plate 17	10.9	0.85	Lake Iroquois	IC, SSU, BB	Franzi *et al.*, 2002; Rayburn *et al.*, 2005
	Hudson Valley	7.12, Plate 17	?	1.5	Lake Iroquois, Lake Vermont	IC, SSU, PH, RIF	Franzi *et al.*, 2002; Rayburn *et al.*, 2005
	Gulf of St Lawrence	7.12, Plate 17	?	1.5	Lake Iroquois	IC, SSU	Franzi *et al.*, 2002; Rayburn *et al.*, 2005

Note: Megaflood dates are approximations. Abbreviations: IC: spillway form, inner channel; BB: boulder bar; SSU: scoured subupland; SER: streamlined erosional residual; RIF: remnant inner channel fills; PH: pot holes.

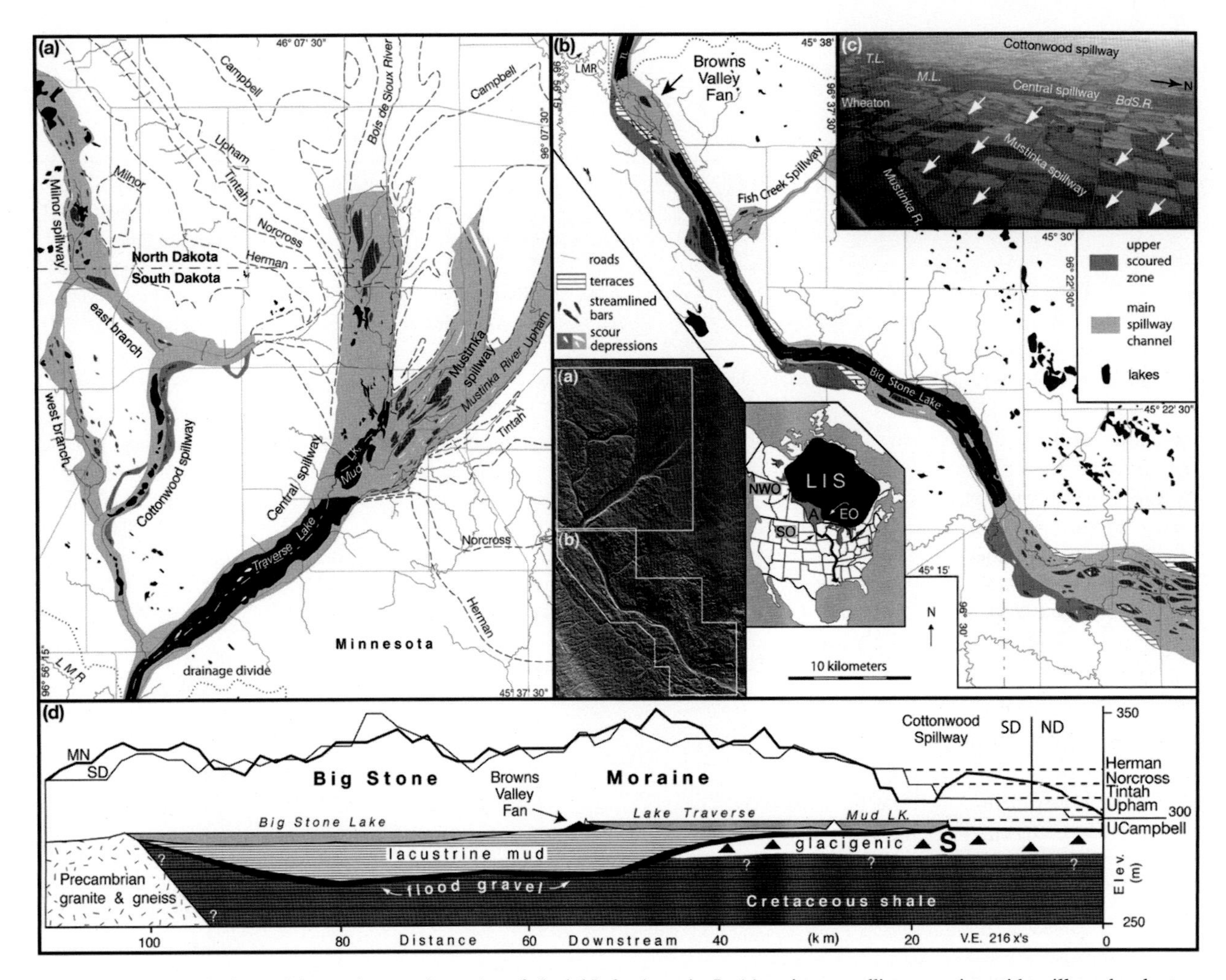

Figure 7.7. Geomorphology of the southern outlet region of glacial Lake Agassiz. In (a) notice strandlines merging with spillway heads at different elevations, from highest (oldest) to lowest (youngest) the Milnor, Cottonwood, Mustinka and Central spillways. Widening of the spillway south of Traverse Lake (b) likely records pre-existing topographic depression across the Big Stone Moraine and former routing of meltwater from the Fish Creek spillway (b). The spillways are cut into a variety of materials (d), all of which host streamlined residual hills. Black arrows on (c) identify streamlined bars and scour depressions. T.L., M.L. BdS. R. refer to Traverse Lake, Mud Lake and Bois de Sioux River, respectively. (Figures (a), (d) modified from Fisher (2005).)

would translate to larger discharge from the active outlet (e.g. Mooers and Wiele, 1989).

A variety of outlets have been considered for Lake Agassiz. The oldest and most studied is the southern outlet (Figure 7.7), which is actually a complex of spillways at different elevations all with strandlines graded to them (Figure 7.7a). The Cottonwood spillway is the only tributary spillway north of the drainage divide that has a scoured subupland surface, although downstream of the drainage divide, terraces and scoured surfaces are common; with boulder lags on terraces. Streamlined hills, bars and erosional residual forms are found in all of the spillway channels and scoured surfaces, many with elongated closed depressions, some of which wrap around the stoss side of residuals (Figures 7.7a–c). The presence of the broad scoured zones and small Fish Creek spillway (Figure 7.7b), truncated by the central spillway, may represent initial meltwater channel(s) on the Big Stone Moraine. The southern outlet spillway with residuals is cut into glacial sediment, shale, granite and gneiss bedrock (Figure 7.7d). The granite south of Big Stone Lake has been sculpted by meltwater in places and palaeodischarges determined from transported boulders range between 0.1–0.36 Sv (Fisher, 2004).

The eastern outlets usually are interpreted to be the next oldest spillways, but are less well documented than the southern outlet. Downwarping the crust of the Earth to simulate isostatic loading has been modelled to identify potential outlet pathways when other geomorphic evidence such as strandlines is lacking (see Teller and Leverington,

2004, and references within). The lack of any geomorphological or sedimentological evidence for an eastern outlet to route meltwater along the Kaministikwia River system implies that such an outlet did not exist. This conclusion is especially important considering that this is the proposed drainage route for the floodwater that was to have triggered the Younger Dryas cold period (Broecker *et al.*, 1989). The absence of the hypothesised outlet has been explained by younger ice recession in the area (Lowell *et al.*, 2005), or subsequent burial by a readvance (e.g. Teller *et al.*, 2005), although documentation of a readvance is equivocal (Fisher *et al.*, 2006). Channels farther to the north and west of Lake Nipigon traditionally have been explained as outlet channels from Lake Agassiz. However, geomorphological mapping of these potential spillways is complex and their history poorly understood. Barnett (2004) identified an esker within one of these channels and explains some of the geomorphology as drainage from a localised lake rather than from Lake Agassiz.

As the ice sheet continued to recede northwards, Lake Agassiz merged with glacial Lake Ojibway in northwestern Ontario (Vincent and Hardy, 1979). The timing of this event has been estimated at ~8.0 ka ^{14}C BP (Teller and Leverington, 2004). Final drainage from Agassiz and its relationship to the 8.2 ka event has attracted considerable attention. A chronological estimate of final drainage of Agassiz using radiocarbon ages on foraminifera and shells from Hudson Bay and Hudson Strait is ~8470 BP (Barber *et al.*, 1999) about 200 years before the 8200 BP cooling recorded in a central Greenland ice core (Alley *et al.*, 1997). Subsequently, a two-stepped structure for the 8.2 ka event has been observed within marine sediments at 8490 and 8290 (Ellison *et al.*, 2006), supporting the Barber *et al.* (1999) age, but the initial meltwater pulse is not evident in the GISP2 ice-core record. The palaeohydraulics of the final drainage of Lake Agassiz were reconstructed by Clarke *et al.* (2004). Although the drainage routes and number of flood events beneath remnants of ice in Hudson Bay are uncertain, flood duration and magnitude are estimated at ~0.5 yr and ~5 Sv to drain a volume of 40 000–151 000 km^3.

A large spillway (Clearwater Lower Athabasca spillway) in northern Saskatchewan and Alberta has been interpreted as a northwestern outlet of Lake Agassiz (Smith and Fisher, 1993; Fisher and Smith, 1994). Fisher and Smith used high-elevation glaciolacustrine sediment and strandlines as evidence for Lake Agassiz extending into northwestern Saskatchewan. With deglacial chronology along the western margin of Lake Agassiz poorly known, and newly recognised moraines indicating a more northerly ice recession in northcentral Saskatchewan, the initial timing of Lake Agassiz extending into northwest Saskatchewan is uncertain (Fisher, 2005). Regardless of the water source, the Clearwater Lower Athabasca spillway is evidence for megaflooding. The spillway is 233 km long and at its head is a series of anastomosing channels (Figure 7.8) separated by erosional residual islands mostly composed of drift and Lower Cretaceous oil sand. The spillway is incised at its base into Devonian limestone or Archean-aged gneiss and granite (Tremblay, 1960). At its head the spillway is up to 12 km wide, narrowing to 2.5–5 km along most of its length; widening, then bifurcating around a moraine at its distal end (Fisher and Smith, 1993). The spillway has segments of scoured upland surfaces with streamlined residual hills and boulder lags. At the mouth of the spillway thick boulder gravels make up bedforms (Fisher, 1993; Jol *et al.*, 1996). Wood from the flood gravels (Smith and Fisher, 1993) provides a maximum age estimate for the flood at ~9900 ^{14}C BP, in good agreement with the age of two logs in sandy deltaic deposits interpreted as distal flood deposits in glacial Lake McConnell (Smith, 1994). Palaeoflow estimates, from straight reaches and largest clasts transported, yield velocities of 6.3–28.9 m s^{-1}, with 12 m s^{-1} used as a reasonable estimate giving a peak discharge of 2.16 Sv (Fisher, 1993; Fisher *et al.*, 2002). A flood duration of 1.5–3 years with increased sea-ice production in the Arctic Ocean was modelled and suggested to be the mechanism for the Preboreal Oscillation (Fisher *et al.*, 2002), a brief cooling period recorded in Greenland and northern Europe.

7.9 Southern megaflood drainage from the Great Lakes basins

Southern drainage from the Great Lakes lobes of the Laurentide Ice Sheet would have occurred from their LGM positions and subsequently as they retreated to the north. The existence of early proglacial lakes during this period is poorly known, although Lake Milwaukee occupied the Lake Michigan basin at around 16 ka ^{14}C BP (Schneider and Need, 1985). A major readvance around 15.5 ka ^{14}C BP (Larson and Schaetzl, 2001) built moraine ridges that served as dams for proglacial lakes in both the Lake Michigan and Lake Erie basins. Larson and Schaetzl (2001) provide a review of lake levels and chronology. The complex history of lake phases that postdate the 15.5 ka ^{14}C BP advance is fairly well known in comparison to proglacial lakes formed during earlier fluctuations of the ice margins. The major outlets in Great Lake basins from which megafloods were released are shown in Figure 7.9, and the evidence for these events is discussed below. Table 7.1 gives additional information on the megafloods associated with these outlets.

Lake Maumee, in the Lake Erie Basin, breached the Fort Wayne Moraine around or just before 14 ka ^{14}C BP and drained down the Wabash Valley into the Ohio and Mississippi Valleys (Eschman and Karrow, 1985; Lewis

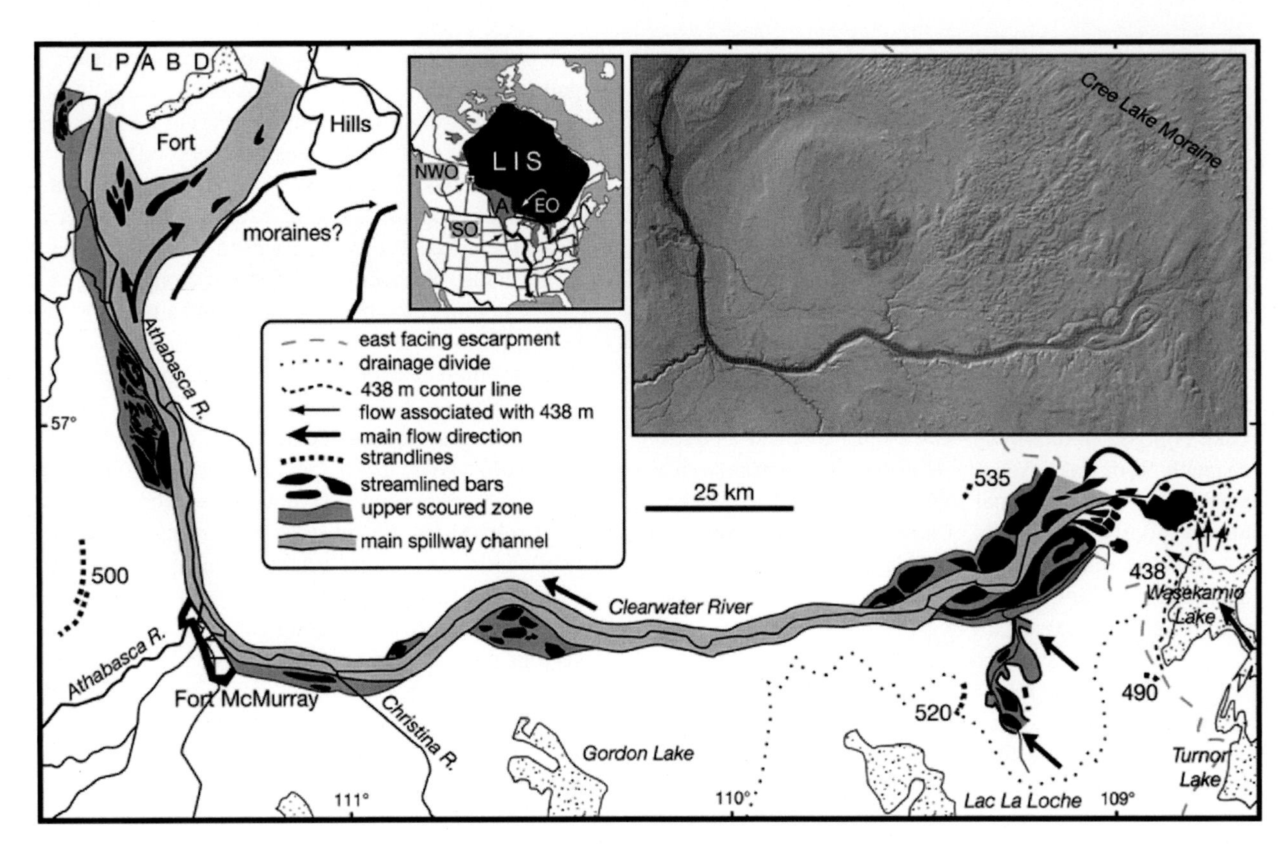

Figure 7.8. Digital elevation model and superimposed geomorphic map of the Clearwater Lower Athabasca Spillway (CLAS). LPABD: Late Pleistocene Athabasca braid delta.

et al., 1994). Fraser and Bleuer (1988) recognised two megaflood events in the valley. Although the first event was interpreted to be an outburst from stagnant ice north of the Wabash Valley, the second event, known as the Maumee Torrent, was associated with the catastrophic drainage of Lake Maumee. This flood exceeded the initial capacity of the valley and scoured a subupland outer zone surface. Within the inner channel of the valley, the effects were primarily erosional, although expansion and lee–eddy bars composed of boulder–cobble bars were deposited in the lee of resistant Silurian limestone reef obstacles on the floor of the valley. Catastrophic drainage of Lake Maumee was short-lived, ending when a resistant boulder lag developed in the outlet zone (Fraser and Bleuer, 1988).

Soon after the Maumee Torrent, ice in the Lake Michigan Basin retreated from recessional moraines fringing the southern end of the basin to form glacial Lake Chicago. Soon after 14 ka ^{14}C BP, the Chicago outlet opened, draining water southwesterly through the Sag and Des Plains channels (Plate 16) to the Illinois Valley (Figure 7.10) and then to the Mississippi River (Hansel *et al.*, 1985; Colman *et al.*, 1994). The earliest lake to drain through the outlet was the Glenwood Phase of Lake Chicago, which ended during or after the major Port Huron advance at about 13 ka ^{14}C BP. Advance of the Huron Lobe into Saginaw Bay of Michigan separated a proglacial lake in the bay (Lake Saginaw) from Lake Whittlesey, a successor to Lake Maumee in the Huron and Erie basins. As the ice began to recede from the Port Huron Moraines, a catastrophic outburst breached the divide between Lake Whittlesey and Lake Saginaw (Eschman and Karrow, 1985) and perhaps triggered the catastrophic drainage of Lake Saginaw from its outlet across the Lower Peninsula of Michigan through the Grand Valley to Lake Chicago (Kehew, 1993). Where Lake Saginaw drained across two moraine ridges to form the outlet, a resistant boulder lag developed. The Grand Valley is a typical spillway trench with a large streamlined erosional residual hill (Plate 13) and several boulder bars, one of which was deposited on a bedrock outcrop that served as a flow obstruction in the valley (Kehew, 1993).

When the megaflood arrived in glacial Lake Chicago, it formed a large delta, and must have augmented greatly the discharge from the Chicago outlet (Plate 16). Kehew (1993) argued that this influx of water would have downcut the Chicago outlet to a new level of stability (the Calumet level) controlled by a resistant boulder lag, in accordance with the outlet control model of Bretz (1951,

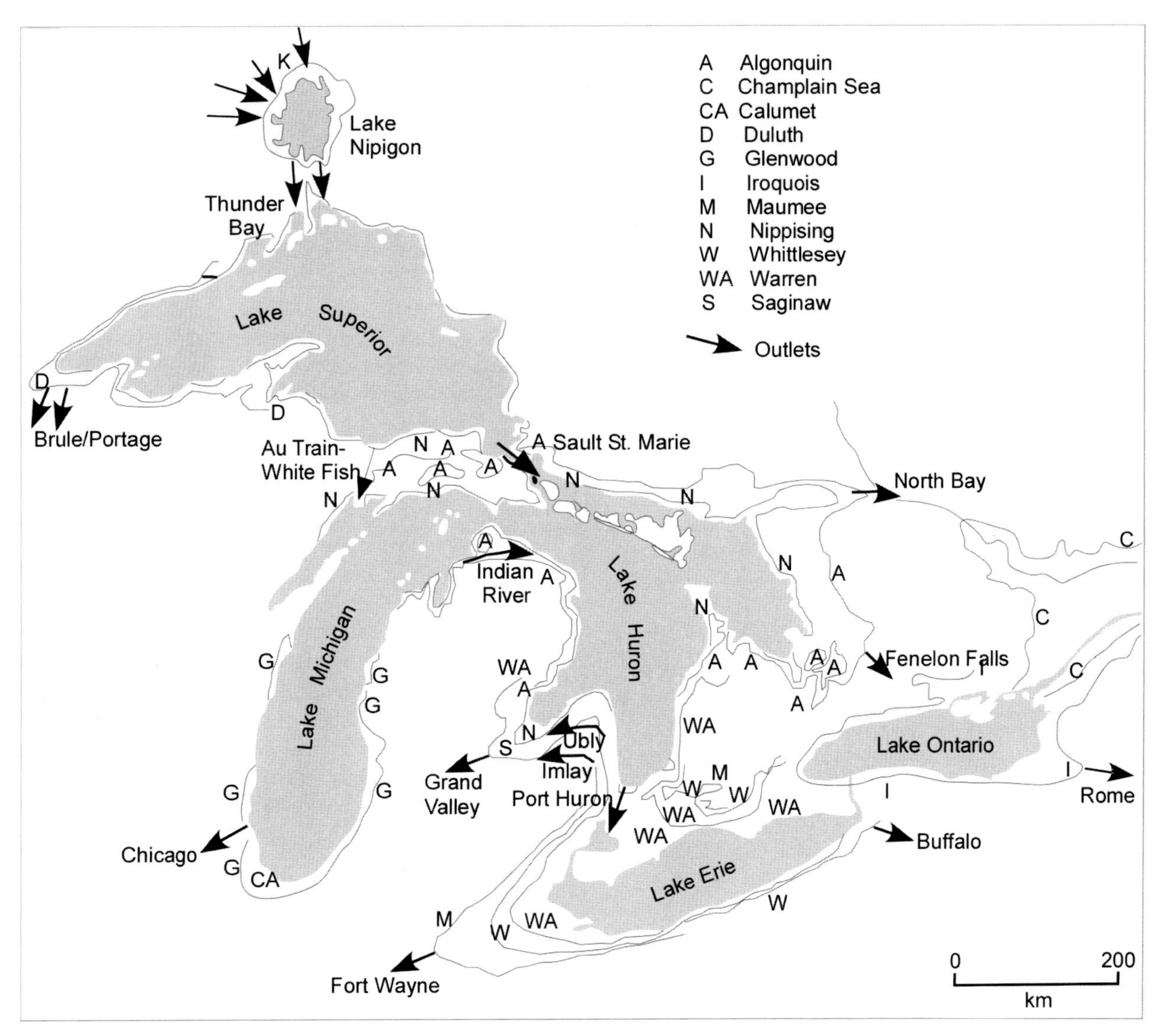

Figure 7.9. Approximate location of high shorelines recording high proglacial lake phases in the Great Lakes basins, and outlets from these basins containing landforms associated with megaflood erosion and deposition.

1955). Hansel and Mickelsen (1988) had proposed that the fall in lake level from the Glenwood to the Calumet level was due instead to a reduction in the amount of meltwater flow and runoff to the lake. In their model, the Chicago outlet was cut to its ultimate size and depth early in the history of the lake, and lake level was subsequently controlled by meltwater and runoff inputs to Lake Chicago, rather than by megaflood events.

Hajic (1990) examined the geomorphology and stratigraphy of the Illinois River valley, which begins at the junction of the Des Plains and Kanakakee Rivers downstream from the Chicago outlet. Whereas the upper Illinois Valley was incised during the last glaciation, the middle and lower valley follows the course of the ancient Mississippi River, which was diverted to its modern course by the Wisconsin advances of the Lake Michigan Lobe.

Indicators of outburst erosion and deposition occur through the entire Illinois Valley, including scoured sub-upland outer zones, inner channels, streamlined erosional residuals (Figure 7.10), and coarse gravel bars (Hajic, 1990). Sources of megafloods include: (1) catastrophic drainage from moraine-dammed lakes formed against the retreating Lake Michigan Lobe; (2) meltwater influx from the Kankakee Torrent (Ekblaw and Athy, 1925; Hajic, 1990) entering the Illinois Valley from the Kankakee River basin around 16 ka ^{14}C BP; and (3) increases in discharge through the Sag and Des Plains outlet channels responding

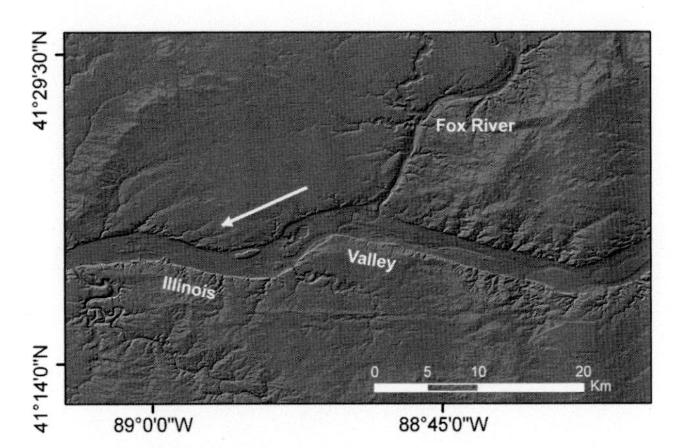

Figure 7.10. Hillshade DEM (NED 1 arc sec) of a portion of the upper Illinois Valley. The uniform, deeply incised channel typical of megaflood erosion is evident. A large streamlined erosional residual occurs in the channel near the centre of the image.

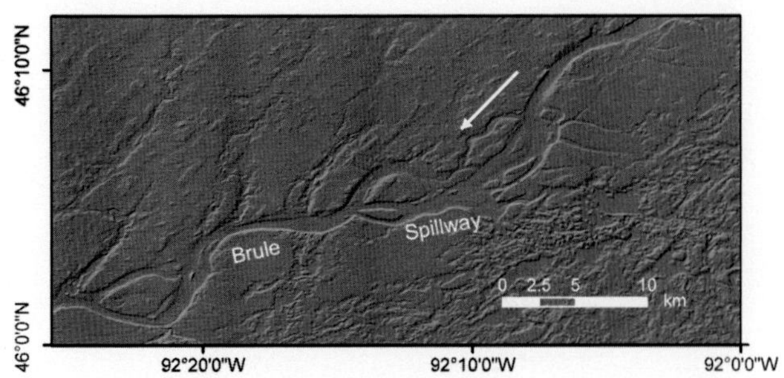

Figure 7.11. Hillshade DEM (NED 1 arc sec) of the Brule spillway (Figure 7.9), which carried catastrophic floods from Lake Duluth in the Lake Superior basin southward to the Mississippi Valley. The Portage and Brule spillways join downstream of the image to form the St Croix spillway.

to episodic megaflooding into the Lake Michigan basin. The timings and magnitudes of these events, and perhaps others, are poorly known.

Under the hypothesis that a megaflood from Lake Saginaw to Lake Chicago caused outlet incision to an elevation that would have dropped the lake to the Calumet level (189 m), sustained drainage would have characterised the outlet until the Lake Michigan Lobe retreated far enough north to open outlets to the east through the Indian River lowlands or the Straits of Mackinaw. Lake level would then have dropped below the elevation of the southern outlet. Lake Chicago returned to the Calumet level during the Two Rivers advance of the Lake Michigan Lobe at 11.7 ka ^{14}C BP (Hansel *et al.*, 1985; Colman *et al.*, 1994). The dated Calumet shoreline features were formed at this time. After retreat of the ice following the Two Rivers advance, water in the upper Great Lakes basins became confluent at about 11 ka ^{14}C BP in a large ice-dammed lake called Lake Algonquin (Figure 7.9). Although Lake Algonquin shorelines have been long recognised in northern Lake Michigan, their existence in the southern part of the basin has been contentious. Isostatic rebound modelling (Larsen, 1987; Clark *et al.*, 1994) produced a deformed Algonquin shoreline that was submerged below modern lake level in the southern part of the basin by as much as 100 m. Under this scenario, lakes in the basin in Algonquin time would be unable to drain through the Chicago outlet. To the contrary, Capps *et al.* (2007), based on previous work by Chrzastowski and Thompson (1992, 1994) and new data, summarise evidence for a post-Calumet shoreline in the southern basin that is approximately 4 m below the classical Calumet water-plane elevation and is dated at approximately 10 500 ^{14}C ka BP, an age that falls within the range of ages for Lake Algonquin. A lake at this level would have been just below the Calumet sill elevation in the southern outlet, and drainage and/or outlet downcutting could have occurred with a relatively minor rise in lake level.

Following the confluence of Lake Chicago and Lake Algonquin, a large influx of meltwater from Lake Agassiz into the Lake Michigan Basin has been suggested (e.g. Colman *et al.*, 1994). This flood, which could have raised lake level to the elevation of the Chicago outlet, has been used to explain the distinctive grey Wilmette bed in the Lake Michigan basin (Colman *et al.*, 1994). However, the source of the Wilmette beds is not discussed. Specifically, the dark grey clay making up the Wilmette bed (Wickham *et al.*, 1978) would have to have remained in suspension across the Superior basin without being deposited. With the recent questioning of a Younger Dryas age (Algonquin equivalent) opening of eastern outlets from Lake Agassiz (Lowell *et al.*, 2005; Teller *et al.*, 2005; Fisher *et al.*, 2006; Fisher and Lowell, 2006), an alternative meltwater flood or a different origin for these sediments is likely. Considerable work is needed still to unravel the deglacial events including megaflooding during ice recession in the Great Lakes region.

Early ice-dammed lakes in the Lake Superior basin, including Lake Duluth (Farrand and Drexler, 1985) drained through the Brule and Portage spillways (Figure 7.9) to the St Croix spillway system (Figure 7.11), which leads to the Mississippi Valley. The St Croix spillway contains a suite of typical erosional and depositional landforms (Figure 7.11) indicative of outburst erosion (Kehew and Lord, 1987; Johnson, 2000).

7.10 Eastern megaflood drainage from the Great Lakes basins

Although smaller than megaflood events along the southwestern margin of the Laurentide Ice Sheet, large-scale drainage events along the southeastern margin of the Laurentide Ice Sheet had a substantial impact on the development of drainage systems, landforms and the delivery of

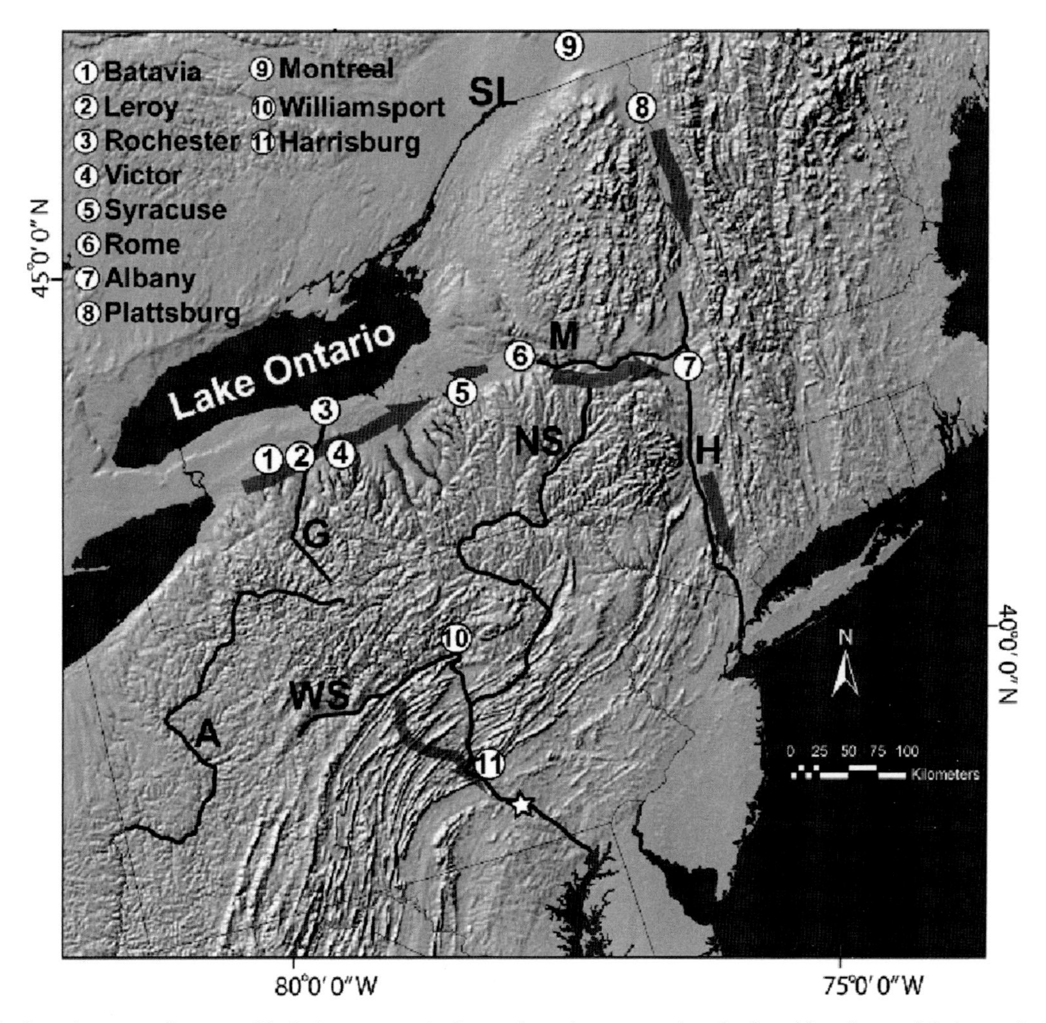

Figure 7.12. Overview map of catastrophic drainage routes in the northeast (grey arrows), and referred locations and drainage. A: Allegheny; NS: North Branch Susquehanna; M: Mohawk; H: Hudson; WS: West Branch Susquehanna; SL: St Lawrence. Star indicates location of Holtwood Gorge.

meltwater to the North and Mid Atlantic (Broecker *et al.*, 1989; Clark *et al.*, 2001; Rayburn, 2004; Rayburn *et al.*, 2005).

The majority of meltwater drainage was released from a series of ice-dammed lakes, of which glacial Lake Iroquois and its outlet spillways, the Mohawk and Hudson Valleys and the Gulf of St Lawrence (Figure 7.12) were most significant. However, evidence of very early large-scale drainage events along the southern proto-Laurentide Ice Sheet also has been found.

Magnetically reversed clays in northcentral Pennsylvania record the earliest (770–990 ka) and southernmost recognised, sizeable glacial impoundment along the West Branch of the Susquehanna River (Gardner *et al.*, 1994; Ramage *et al.*, 1998). Glacial Lake Lesley had a volume of 100 km^3, with a maximum elevation of 340 m as an early Pleistocene ice sheet impinged upon the northeastern edge of the Bald Eagle Mountain anticline near Williamsport, PA. Failure of an ice dam released catastrophic flows of meltwater over a col and into the Juniata River basin and out into the lower and middle Susquehanna River.

Subsequent field investigations in the Holtwood Gorge in southern Pennsylvania by Kochel and Parris (2000), Potter (2001) and Thompson and Sevon (2001, and references therein) identified potholes up to 8 m in depth, 5 m in width, and kilometre-long inner channels 35 m deep cut into resistant Precambrian metamorphic rocks. Further, Kochel and Parris (2000) described large, fluvially transported boulders, up to 3 m in diameter, having unique lithologies that indicate a minimum transport distance of 50 km downstream. Kochel and Parris (2000) estimate discharges on the order of 0.2 Sv would be required to transport the largest boulders and to create the macroturbulent forces necessary to cut potholes of the scale observed. Later, Wisconsin age outwash terraces and deposits in the upper and middle Susquehanna Valley buried any older flood evidence, and do not suggest that Wisconsin age meltwater flows in the northern and middle Susquehanna ever reached megaflood scale.

Although Wisconsin age ice margins also advanced into northcentral Pennsylvania, the Laurentide margin did not advance far enough south to recreate the impoundment of glacial Lake Lesley near Williamsport. However, many upland ridges in northcentral Pennsylvania did form substantial ice-dammed lakes and, although upon deglaciation some may have drained quite rapidly, the majority of these flows were not of the megaflood scale. Shaw (1989) proposed that the North Branch of the Susquehanna River may have acted as a subaerial drain for catastrophic discharges on the order of 4.8 Sv associated with the formation of drumlins in New York State, and that evidence should exist of such cataclysmic discharges. Braun (1994) refuted the proposal of Wisconsin age megafloods down the Susquehanna (Shaw, 1989) on the basis of flow restrictions created by water gaps within sandstone ridges in the Valley and Ridge province that would have created 'choke points' for catastrophic flows. Further, Braun (1994) and Braun *et al.* (2003) found an absence of slack-water megaflood deposits and an abundance of Wisconsin age deposits more indicative of normal meltwater processes. The Susquehanna Basin has long been recognised as a main outlet for glacial meltwater prior to the development of glacial Lake Iroquois and the opening of the Mohawk outlet (Muller and Prest, 1985). More work is needed to define the relationship between deglacial drainage of the Great Lakes and meltwater chronology along the North Branch of the Susquehanna River.

As ice continued to retreat northward into New York State, thick accumulations of drift formed the Valley Heads Moraines and by 14 ka ^{14}C BP (Muller and Calkin, 1993) moraines at the southern end of the Finger Lake troughs served as dams for initial proglacial ponding. Prior to impoundments created by the Valley Heads Moraines, two notable proglacial water bodies have also generally been recognised and include early glacial lakes within the Genesee River valley and glacial Lake Albany/Vermont occupying the Hudson/Champlain lowlands (Plate 17) (Connally and Sirkin, 1973; Rayburn *et al.*, 2005). In western/central New York, ponded meltwater either overflowed morainal topography southward into the Susquehanna or Allegheny drainages (Figure 7.12) or, as ice receded and exposed lower outlets, meltwater bodies coalesced and expanded along the Laurentide ice margin into the Lake Ontario basin. In the Hudson/Champlain lowlands, meltwater was conveyed southerly down the Hudson Valley to the Mid Atlantic coast as the Champlain/Hudson sub-lobe retreated. During glacial retreat a multitude of proglacial lake bodies occupied the Champlain and lower Hudson Valleys. Although the exact number and areal extents are still debated, the impoundment of proglacial lakes was the product of bedrock sills that formed thresholds by isostatic adjustment as ice-free areas rebounded (Connally and Sirkin, 1973; DeSimone and LaFleur, 1986; Stanford and Harper, 1991; Connally and Cadwell, 2002; Franzi *et al.*, 2002; Rayburn *et al.*, 2005).

West–east trending meltwater channels in central New York along the northern margin of the Allegheny Plateau have long been recognised and attributed to multiple eastward drainage events from the Great Lakes (Fairchild, 1909; Sissons, 1960; Muller, 1964; Hand and Muller, 1972; Muller and Prest, 1985). Wilson (1980) and Wilson and Muller (1981) presented erosional and sedimentologic evidence of catastrophic eastward discharges from glacial Lake Warren between ~12.6 and 12.37 ka ^{14}C BP in the Batavia–Genesee region of western New York (Plate 18). These discharges resulted as glacial ice receded from the western Allegheny Plateau and ice dams along the Onondaga scarp failed, sending discharges in excess of 0.3 Sv eastward. Wilson and Muller (1981) describe anastomosing channels, stream-eroded hills, erosional residuals and depositional features that include large-amplitude cross-beds, unstructured cobble gravel, eddy bars and scabland topography.

Another spectacular series of channels and deposits recording catastrophic eastward flows has been recognised immediately south of Syracuse, New York (Figure 7.13). Known as the Syracuse Channels (Fairchild, 1909; Muller, 1964; Hand and Muller, 1972; Thrivikramaji, 1977), these east–west channels crossing north–south-trending glacial troughs record discharges on the order of 0.2 Sv (Thrivikramaji, 1977). Although many of the channels may partially be the product of multiple glaciations and steady-state flows out of the Great Lakes, flood evidence including pendant bars, expansion bars and boulder deposits occurs in some of the Syracuse Channels. Channel incision is believed to be a result of ice marginal retreat along the escarpment and the coalescence of glacial Lake Warren II and high-level lakes in the glacial troughs south of Syracuse. Eastward outflows skirted the ice margin prior to the inception of glacial Lake Iroquois and drained out the Mohawk outlet.

Although the exact lake relationships remain conjectural (Muller and Prest, 1985) a second catastrophic eastward drainage from successors of glacial Lake Warren probably occurred through the Taylor Channel south of Leroy, New York, as ice from the Ontario Lobe made a slight readvance and blocked earlier channels. This episode of catastrophic drainage is recorded by a series of anastomosing channel complexes that terminated at cobble–gravel deltas on the western edge of glacial Lake Avon (Muller and Prest, 1985) within the Genesee Valley (Figure 7.12, Plate 18). Subsequent catastrophic outflows passed through the exposed and pre-existing Rush-Victor Channels into an

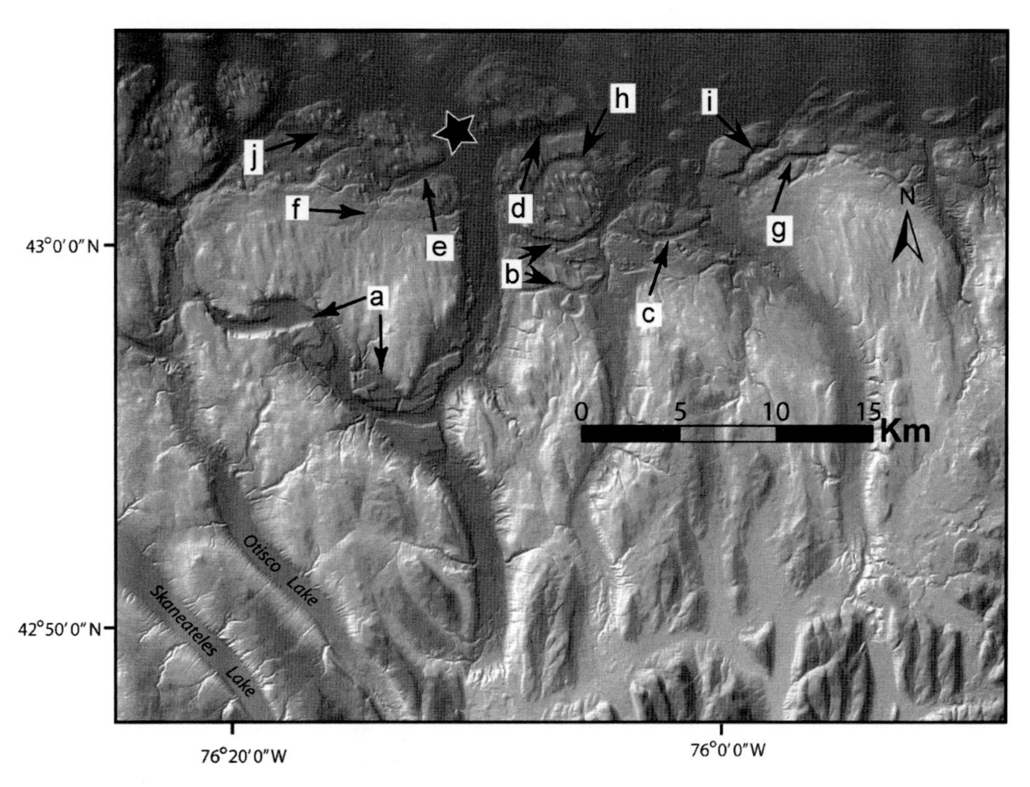

Figure 7.13. Ten-metre DEM of E–W Syracuse Channels (a–j) cut by catastrophic and non-catastrophic ice-marginal meltwater discharges conveyed from the Great Lakes. Meltwater channels were cut along the Onondaga scarp of the Allegheny Plateau that slopes northward to the Ontario Basin. Note cross-cutting relationship with N–S Finger Lakes troughs and through valleys. City of Syracuse located at star.

early phase of glacial Lake Iroquois (Muller and Prest, 1985).

The establishment of main-phase glacial Lake Iroquois occurred about 11 200 ± 190 yr ^{14}C BP (Anderson, 1988) and drained meltwater from the eastern Great Lakes through the Rome outlet into the Mohawk Valley (Figure 7.12) with an estimated average steady-state discharge of 0.6 Sv (Wall, 1995; Wall and LaFleur, 1995), into the Hudson Valley and out to the mid North Atlantic. Pre Lake Iroquois and early-phase Iroquois discharges were routed along the southern margin of the Ontario Lobe via the Mohawk Valley entering Lake Albany (Connally and Sirkin, 1973) as well as other proglacial lakes within the Hudson Valley, although few studies describe the magnitude of these flows. However, the massive steady-state discharges estimated by Wall (1995), and morphological features such as an enormous pendant bar immediately west of Schenectady (Plate 19), provide supporting evidence that substantial floods occurred through the Mohawk Valley. Perhaps the sediments attributable to these earlier floods (and hence potential evidence for these floods) are buried by more recent deposits and have yet to be found or, more likely, such deposits were probably eradicated by later floods issuing out of Lake Vermont.

At the time of the main-phase Lake Iroquois, a southern extension of the Laurentide Ice Sheet into the Adirondack uplands separated the Coveville phase of Lake Vermont in the Champlain and upper Hudson Valley (Plate 17). Stratigraphic, chronologic and morphologic evidence exist of at least three substantial proglacial flood events out of glacial Lake Vermont as it became confluent with glacial Lake Iroquois and later the Gulf of St Lawrence (Franzi *et al.*, 2002; Rayburn, 2004; Rayburn *et al.*, 2005). At about 10.9 ka ^{14}C BP (Rayburn *et al.*, 2005), the first catastrophic flood event originated at Covey Hill, Quebec, as the retreating ice margin exposed a col lower than the Lake Iroquois outlet at Rome, New York (MacClintock and Terasmae, 1960; Muller and Prest, 1985; Rayburn *et al.*, 2005). The newly exposed outlet dropped Lake Iroquois ~14.5 m to the Frontenac level (volumetric change of $570 \pm 85\,km^3$) and had an estimated late-stage discharge of between 0.8 and 0.9 Sv (Rayburn, 2004; Rayburn *et al.*, 2005). As catastrophic flow proceeded along an ice-marginal channel to Lake Vermont, it created a scoured subupland zone prior to entrenching an inner channel that completely eroded surface deposits and exposed expansive areas of Potsdam Sandstone, known locally as the 'Altona Flat Rocks' (Figure 7.14), before discharging into glacial Lake Vermont (Franzi *et al.*, 2002). As debris-laden inflows

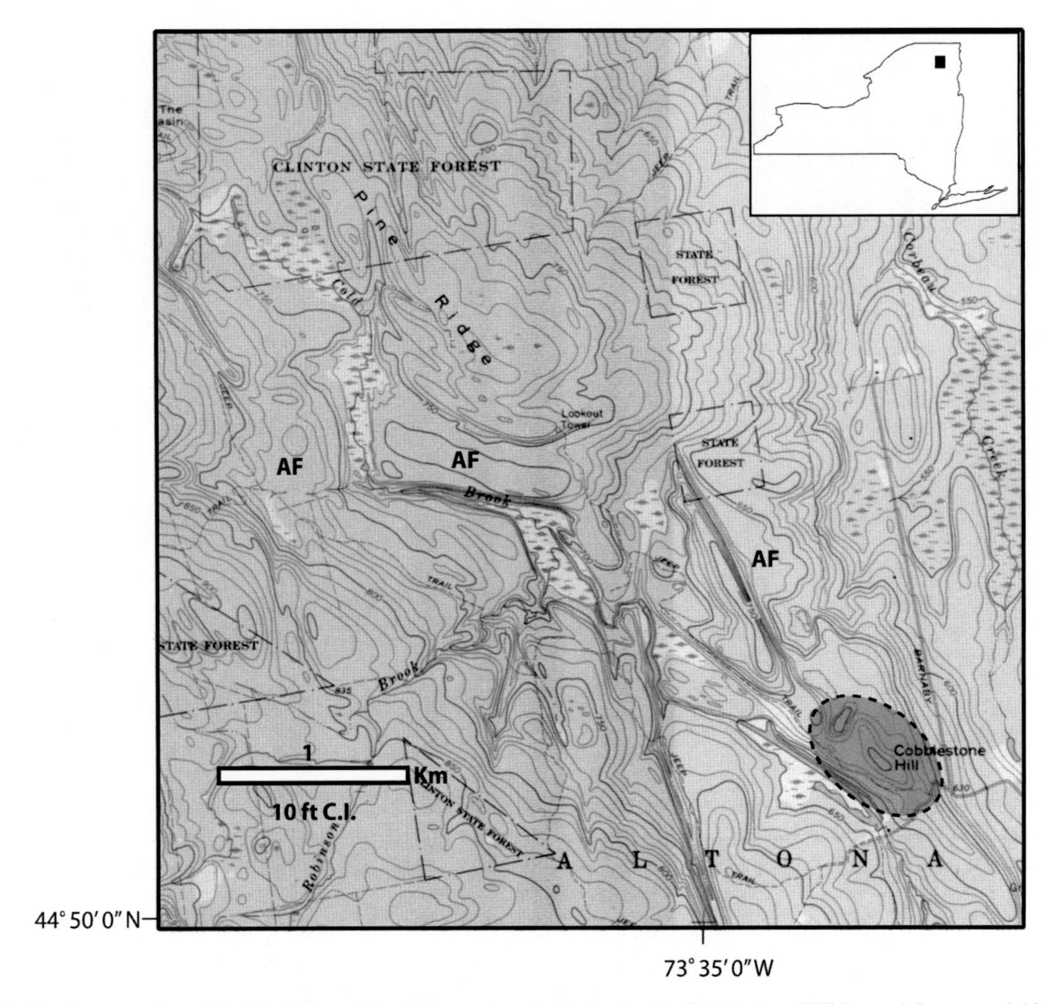

Figure 7.14. Northwest portion of the West Chazy 7.5 minute quadrangle displaying the Cobblestone Hill bar and the scoured Altona Flat Rocks (AF), as described by Franzi *et al.* (2002). Cobblestone Hill consists of boulders up to 3 m in diameter imbricated to the northwest, deposited by catastrophic discharges out of Lake Iroquois.

entered Lake Vermont, a 500 m long, 30 m tall ice-contact boulder bar deposit formed, known as Cobblestone Hill (Figure 7.14). Two clearly defined stratigraphic levels of imbricated boulders, up to 3 m in diameter, occur on Cobblestone Hill and have been correlated to the Coveville level and Upper Fort Ann levels of glacial Lake Vermont. Franzi *et al.* (2002) interpreted the upper level of boulder deposits to represent the initial influx from Lake Iroquois into Coveville-level Lake Vermont, creating a combined discharge of ~1.1 Sv down the Hudson River and into the North Atlantic (Rayburn *et al.*, 2005). As the Iroquois breakout flood inundated and overwhelmed Lake Vermont, most likely it eroded the Coveville level threshold, lowering it to the Upper Fort Ann level, which correlates with the lower elevation boulder deposits observed on Cobblestone Hill (Franzi *et al.*, 2002).

A second flood followed the initial outflows of Lake Iroquois as ice retreated north of the Adirondack Uplands and Lake Iroquois and Lake Vermont became confluent. Approximately $2500 \pm 375\,km^3$ of water must have discharged rapidly as the 76 m difference in elevation equilibrated between the Frontenac level of Lake Iroquois and the Upper Fort Ann level of Lake Vermont, resulting in a catastrophic flood pulse in the range of 1–2 Sv down the Hudson Valley (Rayburn *et al.*, 2005). This flood was nearly equivalent in size to the preceding Iroquois/Coveville Flood and led to renewed incision of the Upper Fort Ann level threshold to the Lower Fort Ann threshold (Franzi *et al.*, 2002).

Rayburn *et al.* (2005) estimate continued steady-state (non-flood) discharges of ~0.6 Sv from the Great Lakes, through the confluent proglacial lake bodies at the Lower Fort Ann level and down the Hudson Valley into the mid North Atlantic for a duration of 150–300 years. Continued ice retreat eventually opened the Gulf of St Lawrence, sending a redirected, yet comparable flood discharge (~1–2 Sv) to the North Atlantic.

The proglacial lakes within the Champlain/Hudson Valley conveyed meltwater from the Great Lakes via the

Mohawk River valley or Lake Iroquois, once it became confluent with Lake Vermont. In many ways, these proglacial lakes were more fluvial than quiescent lake bodies. Pre Lake Agassiz flood pulses that entered the system from the Great Lakes triggered failures of thresholds or confining moraine ridges and ultimately discharged and deposited flood debris onto the continental shelf and slope via the ancestral Hudson River valley (Uchupi *et al.*, 2001). Later flows from Lake Agassiz played an important role in the development of the St Lawrence drainage system (Teller, 1988).

7.11 Conclusions

Over the past several decades, the role of megafloods generated during the retreat of the Laurentide Ice Sheet has been revealed gradually. Much of our Holocene drainage network in glaciated areas is inherited from the new or modified drainage ways produced by these transient, high-magnitude events. Their geomorphological and sedimentological effects, while they may not rival the Channeled Scabland, are nevertheless spectacular. Moreover, their effects upon climate, brought on by the sudden releases of huge volumes of freshwater into the oceans may have been very significant.

Despite the progress to date, however, many questions remain unanswered. Many of the lake basins and spillways ascribed to a megaflood origin have not yet been studied in detail. Field work often has been restricted to available exposures with little subsurface investigation. With new data will come refinements and new interpretations improving our understanding of the basic hydraulic and sedimentologic processes involved. Better palaeohydraulic estimates will be hampered still by the non-steady-state erosion of spillways cut into weakly resistant surficial materials. The very existence of reconstructed megafloods in some areas is also in question, for example the eastern outlets of Lake Agassiz. Because these floods had been linked to the onset of the Younger Dryas cold event, a major palaeoclimatic triggering hypothesis is now in doubt.

Megaflood chronology is probably the largest uncertainty. Dating of predominantly erosional events that may have occurred over periods of weeks or months is extremely difficult. Proxy data from marine or lacustrine cores ultimately may be the best chance for sorting out chronological relationships, but more accurately dated ice-marginal positions would be the next logical goal. Despite these problems, it is clear that megaflooding constitutes a universal deglacial process with potentially far-reaching consequences.

Acknowledgements

Thorough reviews by Mark Johnson, Darren Sjogren and Paul Carling helped to improve the manuscript.

References

Alley, R. B., Mayewski, P. A., Sowers, T. *et al.* (1997). Holocene climatic instability: a prominent, widespread event 8200 yr ago. *Geology*, **25**, 483–486.

Anderson, T. W. (1988). Late Quaternary pollen stratigraphy of the Ottawa Valley–Lake Ontario region and its application in dating the Champlain Sea. In *The Late Quaternary Development of the Champlain Sea Basin*, ed. N. R. Gadd. Geological Association of Canada Special Paper 35, pp. 207–224.

Baker, V. R. (1973). *Paleohydrology and Sedimentology of the Lake Missoula Flooding in Eastern Washington*. Geological Society of America Special Paper 144.

Baker, V. R. (1982). *The Channels of Mars*. Austin: University of Texas Press.

Baker, V. R., Greeley, R., Komar, P. D., Swanson, D. A. and Waitt, Jr., R. B. (1987). Columbia and Snake River Plains. In *Geomorphic Systems of North America*, ed. W. L. Graf. Boulder, CO: Geological Society of America, Centennial Special Volume 2, pp. 403–468.

Baker, V. R. and Nummedal, D. (1978). *The Channeled Scabland*. Washington DC: National Aeronautics and Space Administration.

Barber, D. C., Dyke, A., Hillaire-Marcel, C. *et al.* (1999). Forcing of the cold event of 8,200 years ago by catastrophic drainage of Laurentide lakes. *Nature*, **400**, 344–348.

Barnett, P. J. (2004). *Surficial Geology Mapping and Lake Nipigon Region Geoscience Initiative Lineament Study*. Ontario Geological Survey, Open File Report 6145.053, 450–451.

Braun, D. D. (1994). Late Wisconsinan to Pre-Illinoinan glacial events in eastern Pennsylvania. In *Late Wisconsinan to Pre-Illinoinan Glacial and Periglacial Events in Eastern Pennsylvania*, ed. D. D. Braun. 57th Friends of the Pleistocene field conference guidebook, U.S. Geologic Survey Open File Report 94–434.

Braun, D. D., Pazzaglia, F. J. and Potter Jr., N. (2003). Margin of the Laurentide ice to the Atlantic Coastal Plain: Miocene-Pleistocene landscape evolution in the Central Appalachians. In *Quaternary Geology of the United States*, ed. D. J. Easterbrook. INQUA Field Guide, pp. 219–244.

Bretz, J H. (1951). The stages of Lake Chicago: their causes and correlations. *American Journal of Science*, **249**, 401–429.

Bretz, J H. (1955). *Geology of the Chicago Region: Part II, The Pleistocene*. Illinois State Geological Survey Bulletin 65, Part II.

Broecker, W. S., Kennett, J. P., Flower, B. P. *et al.* (1989). Routing of meltwater from the Laurentide Ice Sheet during the Younger Dryas cold episode. *Nature*, **341**, 318–321.

Burr, D. (2005). Clustered streamlined forms in Athabasca Valles, Mars: evidence for sediment deposition during floodwater ponding. *Geomorphology*, **69**, 242–252.

Capps, D. K., Thompson, T. A. and Booth, R. K. (2007). A Post-Calumet shoreline along southern Lake Michigan. *Journal of Paleolimnology*, **37**, 395–409.

Christiansen, E. A. (1979). The Wisconsinan deglaciation of southern Saskatchewan and adjacent areas. *Canadian Journal of Earth Sciences*, **16**, 913–938.

Chrzastowski, M. J. and Thompson, T. A. (1992). Late Wisconsinan and Holocene coastal evolution of the southern shore of Lake Michigan. In *Quaternary Coastal Systems of the United States*, eds. C. H. Fletcher and J. F. Wehmiller. SEPM Special Publication 48, 397–413.

Chrzastowski, M. J. and Thompson, T. A. (1994). Late Wisconsinan and Holocene geologic history of the Illinois-Indiana coast of Lake Michigan. *Journal of Great Lakes Research*, **20**, 9–26.

Clark, J. A., Hendriks, M., Timmermans, T. J., Struck, C. and Hilverda, K. J. (1994). Glacial isostatic deformation of the Great Lakes region. *Geological Society of America Bulletin*, **106**, 19–30.

Clark, P. U. (1994). Unstable behavior of the Laurentide Ice Sheet over deforming sediment and its implications for climate change. *Quaternary Research*, **41**, 19–25.

Clark, P. U., Marshall, S. J., Clarke, G. K. C. *et al.* (2001). Freshwater forcing of abrupt climate change during the last glaciation. *Science*, **293**, 283–287.

Clarke, G. K. C., Leverington, D. W., Teller, J. T. and Dyke, A. S. (2004). Paleohydraulics of the last outburst flood from glacial Lake Agassiz and the 8200 BP cold event. *Quaternary Science Reviews*, **23**, 389–407.

Clayton, L. and Moran, S. R. (1982). Chronology of Late Wisconsinan glaciation in middle North America. *Quaternary Science Reviews*, **1**, 55–82.

Clayton, L., Teller, J. T. and Attig, J. W. (1985). Surging of the southwestern part of the Laurentide Ice Sheet. *Boreas*, **14**, 235–241.

Colman, S. M., Clark, J. A., Clayton, L., Hansel, A. K. and Larsen, C. E. (1994). Deglaciation, lake levels, and meltwater discharge in the Lake Michigan Basin. *Quaternary Science Reviews*, **13**, 879–890.

Colman, S. M., King, J. W., Jones, G. A., Reynolds, R. L. and Bothner, M. H. (2000). Holocene and recent sediment accumulation rates in southern Lake Michigan. *Quaternary Science Reviews*, **19**, 1563–1580.

Connally, G. G. and Sirkin, L. A. (1973). Wisconsinan history of the Hudson-Champlain Lobe. In *The Wisconsinan Stage*, eds. R. F. Black, R. P. Goldthwait and H. B. Willman. Geological Society of America Memoir 136, 47–69.

Connally, G. G. and Cadwell, D. H. (2002). Glacial Lake Albany in the Champlain Valley. In *New York State Geological Association/New England Intercollegiate Geological Conference Joint, Annual Meeting Guidebook*, pp. B8, 1–26.

DeSimone, D. J. and LaFleur, R. G. (1986). Glaciolacustrine phases in the northern Hudson lowland and correlatives in western Vermont. *Northeastern Geology*, **9**, 218–229.

Dyke, A. S., Andrews, J. T., Clark, P. U. *et al.* (2002). The Laurentide and Innuitian ice sheets during the last glacial maximum. *Quaternary Science Reviews*, **21**, 9–31.

Ekblaw, G. E. and Athy, L. F. (1925). Glacial Kankakee torrent in northeastern Illinois. *Geological Society of America Bulletin*, **36**, 417–428.

Ellison, C. R. W., Chapman, M. R. and Hall, I. R. (2006). Surface and deep ocean interactions during the cold climate event 8200 years ago. *Science*, **312**, 1929–1932.

Eschman, D. F. and Karrow, P. F. (1985). Huron Basin glacial lakes: a review. In *Quaternary Evolution of the Great Lakes*, eds. P. F. Karrow and P. E. Calkin. Geological Association of Canada Special Paper 30, 79–94.

Evans, D. J. A. (2000). Quaternary geology and geomorphology of the Dinosaur Provincial Park area and surrounding plains, Alberta, Canada: the identification of former glacial lobes, drainage diversions and meltwater flood tracks. *Quaternary Science Reviews*, **19**, 931–958.

Fairchild, H. L. (1909). *Glacial Waters in Central New York*. New York State Museum Bulletin 127.

Farrand, W. R. and Drexler, C. W. (1985). Late Wisconsinan and Holocene history of the Lake Superior basin. In *Quaternary Evolution of the Great Lakes*, eds. P. F. Karrow and P. E. Calkin. Geological Association of Canada Special Paper 30, pp. 17–32.

Fisher, T. G. (1993). Glacial Lake Agassiz: The N.W. Outlet and Paleoflood Spillway, N.W. Saskatchewan and N.E. Alberta. Unpublished Ph.D. thesis, University of Calgary, Canada.

Fisher, T. G. (2003). Chronology of glacial Lake Agassiz meltwater routed to the Gulf of Mexico. *Quaternary Research*, **59**, 271–276.

Fisher, T. G. (2004). River Warren boulders: paleoflow indicators in the southern spillway of glacial Lake Agassiz. *Boreas*, **33**, 349–358.

Fisher, T. G. (2005). Strandline analysis in the southern basin of glacial Lake Agassiz, Minnesota and North and South Dakota, USA. *Geological Society of America Bulletin*, **117**, 1481–1496.

Fisher, T. G. and Lowell, T. V. (2006). Questioning the age of the Moorhead phase in the glacial Lake Agassiz basin. *Quaternary Science Reviews*, **25**, 2688–2691.

Fisher, T. G. and Smith, D. G. (1993). Exploration for Pleistocene aggregate resources using process-depositional models in the Fort McMurray region, NE Alberta, Canada. *Quaternary International*, **20**, 71–80.

Fisher, T. G. and Smith, D. G. (1994). Glacial Lake Agassiz: its northwest maximum extent and outlet in Saskatchewan (Emerson phase). *Quaternary Science Reviews*, **13**, 845–858.

Fisher, T. G., Smith, D. G. and Andrews, J. T. (2002). Preboreal oscillation caused by a glacial Lake Agassiz flood. *Quaternary Science Reviews*, **21**, 873–878.

Fisher, T. G., Lowell, T. V. and Loope, H. M. (2006). Comment on "Alternative routing of Lake Agassiz overflow during the Younger Dryas: new dates, paleotopography, and a re-evaluation" by Teller *et al.* (2005). *Quaternary Science Reviews*, **25**, 1137–1141.

Franzi, D. A., Rayburn, J. A., Yansa, C. H. and Knuepfer, P. L. K. (2002). Late glacial water bodies in the Champlain and Hudson lowlands, New York. In *New York Geological Association/New England Intercollegiate Geological*

Conference Joint Annual Meeting Guidebook, pp. A5,1–23.

Fraser, G. S. and Bleuer, N. K. (1988). Sedimentological consequences of two floods of extreme magnitude in the late Wisconsin Wabash Valley. In *Sedimentologic Consequences of Convulsive Geologic Events*, ed. H. E. Clifton. Geological Society of America Special Paper 229, pp. 111–126.

Gardner, T. W., Sasowsky, I. D. and Schmidt, V. A. (1994). Reversed polarity glacial sediments and revised glacial chronology, West Branch Susquehanna River, central Pennsylvania. *Quaternary Research*, **42**, 131–135.

Hajic, E. R. (1990). Late Pleistocene and Holocene landscape evolution, depositional subsystems, stratigraphy in the Lower Illinois River Valley and adjacent Central Mississippi River Valley. Unpublished Ph.D. Dissertation, University of Illinois, Urbana.

Hand, B. M. and Muller, E. H. (1972). Syracuse channels: evidence of a catastrophic flood. In *New York state Geological Association Guidebook*, 44th Annual Meeting, ed. J. McLelland, pp. 1–12.

Hansel, A. K. and Mickelsen, D. M. (1988). A reevaluation of the timing and causes of high lake phases in the Lake Michigan Basin. *Quaternary Research*, **29**, 113–128.

Hansel, A. K., Mickelsen, D. M., Schneider, A. F. and Larsen, C. E. (1985). Late Wisconsinan and Holocene history of the Lake Michigan Basin. In *Quaternary Evolution of the Great Lakes*, eds. P. F. Karrow and P. E. Calkin. Geological Association of Canada Special Paper 30, pp. 79–94.

Herget, J. (2005). *Reconstruction of Pleistocene Ice-Dammed Lake Outburst Floods in the Altai Mountains, Siberia*. Geological Society of America Special Paper 386.

Johnson, M. D. (2000). *Pleistocene Geology of Polk County, Wisconsin*. Wisconsin Geological and Natural History Survey Bulletin 92.

Johnson, M. D., Davis, D. M. and Pederson, J. L. (1998). Terraces of the Minnesota River valley and the character of River Warren downcutting. In *Contributions to Quaternary Studies in Minnesota*, eds. C. J. Patterson and H. E. Wright Jr. Minnesota Geological Survey Report of Investigations **49**, pp. 121–130.

Jol, H. M., Young, R., Fisher, T. G., Smith, D. G. and Meyers, R. A. (1996). Ground penetrating radar of eskers, kame terraces, and moraines: Alberta and Saskatchewan, Canada. *6th International Conference of Ground Penetrating Radar (GPR'96)*, Sendai, Japan, pp. 439–443.

Kehew, A. E. (1982). Catastrophic flood hypothesis for the origin of the Souris Spillway, Saskatchewan and North Dakota. *Geological Society of America Bulletin*, **93**, 1051–1058.

Kehew, A. E. (1993). Glacial-lake outburst erosion of the Grand Valley, Michigan and impacts on glacial lakes in the Lake Michigan basin. *Quaternary Research*, **239**, 36–44.

Kehew, A. E. and Boettger, W. M. (1986). Depositional environments of buried-valley aquifers in North Dakota. *Ground Water*, **24**, 728–735.

Kehew, A. E. and Clayton, L. (1983). Late Wisconsinan floods and development of the Souris–Pembina spillway system. In *Glacial Lake Agassiz*, eds. J. T. Teller and L. Clayton. Geological Association of Canada Special Paper 26, pp. 187–210.

Kehew, A. E. and Lord, M. L. (1986). Origin and large-scale erosional features of glacial-lake spillways in the northern Great Plains. *Geological Society of America Bulletin*, **97**, 162–177.

Kehew, A. E. and Lord, M. L. (1987). Glacial-lake outbursts along the mid-continent margins of the Laurentide Ice Sheet. In *Catastrophic Flooding*, eds. L. Mayer and D. Nash. Boston: Allen & Unwin, pp. 95–120.

Kehew, A. E. and Teller, J. T. (1994a). History of late glacial runoff along the southwestern margin of the Laurentide Ice Sheet. *Quaternary Science Reviews*, **13**, 859–877.

Kehew, A. E. and Teller, J. T. (1994b). Glacial-lake spillway incision and deposition of a coarse-grained fan near Watrous, Saskatchewan. *Canadian Journal of Earth Sciences*, **31**, 544–553.

Kehew, A. E., Beukema, S. P., Bird, B. C. and Kozlowski, A. L. (2005). Fast flow of the Lake Michigan Lobe of the Laurentide Ice Sheet: evidence from sediment-landform assemblages in southwestern Michigan, USA, *Quaternary Science Reviews*, **24**, 2335–2353.

Kochel, C. R. and Parris, A. (2000). Macroturbulent erosional and depositional evidence for large-scale Pleistocene paleofloods in the lower Susquehanna bedrock gorge near Holtwood, Pennsylvania. *Abstracts with Programs, Geological Society of America Annual Meeting 32*, p. 28.

Komar, P. D. (1983). Shapes of streamlined islands on Earth and Mars: Experiments and analysis of the minimum drag form. *Geology*, **11**, 651–654.

Kozlowski, A. L., Kehew, A. E. and Bird, B. C. (2005). Outburst flood origin of the Central Kalamazoo River Valley, Michigan, USA. *Quaternary Science Reviews*, **24**, 2354–2374.

Larsen, C. E. (1987). *Geologic History of Glacial Lake Algonquin and the Upper Great Lakes*. U.S. Geological Survey Bulletin 1801.

Larson, G. J. and Schaetzl, R. J. (2001). Origin and evolution of the Great Lakes. *Journal of Great Lakes Research*, **27**, 518–546.

Lepper, K., Fisher, T. G., Hajdas, I. and Lowell, T. V. (2007). Ages for the Big Stone moraine and the oldest beaches of glacial Lake Agassiz: implications for deglaciation chronology. *Geology*, **35**, 667–670.

Lewis, C. F. M., Moore Jr., T. C., Rea, D. K. *et al.* (1994). Lakes of the Huron Basin: their record of runoff from the Laurentide Ice Sheet. In *Late Glacial History of Large Proglacial Lakes and Meltwater Runoff Along the Laurentide Ice Sheet*, eds. J. T. Teller and A. E. Kehew. Quaternary Science Reviews 13, pp. 859–877.

Licciardi, J. M., Clark, P. U., Jenson, J. W. and MacAyeal, D. R. (1998). Deglaciation of a soft-bed Laurentide Ice Sheet. *Quaternary Science Reviews*, **17**, 427–448.

Lord, M. L. (1991). Depositional record of a glacial-lake outburst: Glacial Lake Souris, North Dakota. *Geological Society of America Bulletin*, **99**, 663–673.

Lord, M. L. and Kehew, A. E. (1987). Sedimentology and paleohydrology of glacial-lake outburst deposits in southeastern Saskatchewan and northwestern North Dakota. *Geological Society of America Bulletin*, **99**, 663–673.

Lord, M. L. and Schwartz, R. K. (2003). An array of large scale erosional bedforms: a detailed record of a glacial lake outburst flow architecture and spillway evolution. *Geological Society of America Abstracts with Programs*, **35** (6), 334.

Lowell, T. V., Fisher, T. G., Comer, G. C. *et al.* (2005). Testing the Lake Agassiz meltwater trigger for the Younger Dryas. *Eos Transactions*, **86**, 365–372.

MacClintock, P. and Terasame, J. (1960). Glacial history of Covey Hill. *Journal of Geology*, **68**, 232–241.

Maizels, J. (1997). Jökulhlaup deposits in proglacial areas. *Quaternary Science Reviews*, **16**, 793–819.

Marshall, S. J., Clarke, G. K. C., Dyke, A. S. and Fisher, D. A. (1996). Geologic and topographic controls on fast flow in the Laurentide and Cordilleran Ice Sheets. *Journal of Geophysical Research*, **101** (B8), 17827–17839.

Matsch, C. L. and Wright Jr., H. E. (1966). The southern outlet of Lake Agassiz. In *Life, Land and Water*, ed. W. J. Mayer-Oakes. Winnipeg: University of Manitoba Press, pp. 121–140.

Mooers, H. D. and Wiele, S. (1989). Glacial Lake Agassiz: glacial surges and large outflow events. *Geological Society of America Abstracts With Programs*, Annual Meeting, St. Louis, MO, A54.

Muller, E. H. (1964). Surficial geology of the Syracuse field area. *New York State Geological Association Guidebook*, 36th annual meeting, pp. 25–35.

Muller, E. H. and Calkin, P. E. (1993). Timing of Pleistocene glacial events in New York State. *Canadian Journal of Earth Sciences*, **30**, 1829–1845.

Muller, E. H. and Prest, V. K. (1985). Glacial Lakes in the Ontario Basin. In *Quaternary Evolution of the Great Lakes*, eds. P. F. Karrow and P. E. Calkin. Geological Association of Canada Special Paper 30, pp. 213–229.

O'Connor, J. E. and Baker, V. R. (1992). Magnitudes and implications of peak discharges from Glacial Lake Missoula. *Geological Society of America Bulletin*, **104**, 267–279.

Potter Jr., N. (Ed.) (2001). *The Geomorphic Evolution of the Great Valley Near Carlisle, Pennsylvania*. Southeast Friends of the Pleistocene, Annual Meeting.

Ramage, J. M., Gardner, T. W. and Sasowsky, I. D. (1998). Early Pleistocene glacial Lake Lesley, west branch Susquehanna River Valley, central Pennsylvania. *Geomorphology*, **22**, 19–37.

Rayburn, J. A. (2004). Deglaciation of the Champlain Valley New York and Vermont and its possible effects on North Atlantic climate change. Unpublished Ph.D. Dissertation, Binghamton University, Binghamton, New York.

Rayburn, J. A., Knuepfer, P. E. L. and Franzi, D. A. (2005). A series of large late Wisconsinan meltwater floods through the Champlain and Hudson Valleys, New York State, USA. *Quaternary Science Reviews*, **24**, 2410–2419.

Schneider, A. F. and Need, E. A. (1985). Lake Milwaukee: an "early" proglacial lake in the Lake Michigan Basin. In *Quaternary Evolution of the Great Lakes*, eds. P. F. Karrow and P. E. Calkin. Geological Association of Canada Special Paper 30, pp. 56–62.

Scott, J. S. (1971). *Surficial Geology of Rosetown Map-area, Saskatchewan*. Geological Survey of Canada Bulletin 190.

Sharpe, D. R. and Cowan, W. R. (1990). Moraine formation in northwestern Ontario: product of subglacial fluvial and glaciolacustrine sedimentation. *Canadian Journal of Earth Sciences*, **27**, 1478–1486.

Shaw, J. (1989). Drumlins, subglacial meltwater floods, and ocean responses. *Geology*, **17**, 853–856.

Shoemaker, E. M. (1992). Subglacial floods and the origin of low-relief ice-sheet lobes. *Journal of Glaciology*, **38**, 105–112.

Shepherd, R. G. and Schumm, S. A. (1974). Experimental study of river incision. *Geological Society of America Bulletin*, **85**, 257–268.

Sissons, J. B. (1960). Subglacial, marginal, and other glacial drainage in the Syracuse-Oneida area, New York. *Geological Society of America Bulletin*, **71**, 1575–1588.

Smith, D. G. and Fisher, T. G. (1993). Glacial Lake Agassiz: the northwestern outlet and paleoflood. *Geology*, **21**, 9–12.

Smith, D. G. (1994). Glacial Lake McConnell: paleogeography, age, duration, and associated river deltas, Mackenzie River Basin, western Canada. *Quaternary Science Reviews*, **13**, 829–843.

Stanford, S. D. and Harper, D. P. (1991). Glacial lakes of the lower Passaic, Hackensack, and lower Hudson valleys, New Jersey and New York. *Northeastern Geology*, **13**, 271–286.

Sun, C. S. and Teller, J. T. (1997). Reconstruction of glacial Lake Hind in southwestern Manitoba, Canada. *Journal of Paleolimnology*, **17**, 9–21.

Teller, J. T. (1988). Lake Agassiz and its contribution to flow through the Ottawa-St. Lawrence System. In *The Late Quaternary Development of the Champlain Sea Basin*, ed. N. R. Gadd. Geological Association of Canada Special Paper 35, pp. 281–289.

Teller, J. T. (2004). Controls, history, outbursts, and impact of large late-Quaternary proglacial lakes in North America. In *The Quaternary Period in the United States*, eds. A. R. Gillespie, S. C. Porter and B. F. Atwater. Developments in Quaternary Science 1, Elsevier, pp. 45–61.

Teller, J. T. and Leverington, D. W. (2004). Glacial Lake Agassiz: a 5000 yr history of change and its relationship to the $\partial^{18}O$ record of Greenland. *Geological Society of America Bulletin*, **116**, 729–742.

Teller, J. T. and Thorleifson, L. H. (1983). The Lake Agassiz – Lake Superior connection. In *Glacial Lake Agassiz*, eds. J. T. Teller and L. Clayton. Geological Association of Canada Special Paper 26, pp. 261–290.

Teller, J. T., Leverington, D. W. and Mann, J. D. (2002). Freshwater outbursts to the oceans from glacial Lake Agassiz and their role in climate change during the last glaciation. *Quaternary Science Reviews*, **21**, 879–887.

Teller, J. T., Boyd, M., Yang, Z., Kor, P. S. G. and Mokhtari Fard, A. (2005). Alternative routing of Lake Agassiz overflow during the Younger Dryas: new dates, paleotopography, and a reevaluation. *Quaternary Science Reviews*, **24**, 1890–1905.

Thompson Jr., G. H. and Sevon, W. D. (2001). The potholes and deeps in Holtwood gorge: an erosional enigma. In *The Geomorphic Evolution of the Great Valley Near Carlisle, Pennsylvania*, ed. N. Potter Jr. Southeast Friends of the Pleistocene Field Guide, pp. 41–64.

Thrivikramaji, K. P. (1977). Sedimentology and Paleohydraulics of Pleistocene Syracuse Channels. Unpublished Ph.D. Dissertation, Syracuse University, Syracuse, New York.

Tremblay, L. P. (1960). *Geology, La Loche, Saskatchewan. Map 10–1961*. Geological Survey of Canada, Ottawa.

Tweed, F. S. and Russell, A. J. (1999). Controls on the formation and sudden drainage of glacier-impounded lakes: implications for jökulhlaup characteristics. *Progress in Physical Geography*, **23**, 79–110.

Uchupi, E., Driscoll, N., Ballard, R. D. and Bolmer, S. T. (2001). Drainage of late Wisconsin glacial lakes and the morphology and late Quaternary stratigraphy of the New Jersey–southern New England continental shelf and slope. *Marine Geology*, **172**, 117–145.

Vincent, J. S. and Hardy, L. (1979). *The Evolution of Glacial Lakes Barlow and Ojibway, Quebec and Ontario*. Geological Survey of Canada Bulletin 316.

Wall, G. R. (1995). Post Glacial Drainage in the Mohawk River Valley with Emphasis on Paleodischarge and Paleochannel Development. Unpublished Ph.D. Dissertation, Rensselaer Polytechnic Institute, Troy, New York.

Wall, G. R. and LaFleur, R. G. (1995). The paleofluvial record of Glacial Lake Iroquois in the Eastern Mohawk valley, New York. In *New York State Geological Association Field Trip Guidebook*, eds. J. I. Garver and J. A. Smith. 67th Annual Meeting, pp. 173–203.

Waterson, N., Lowell, T. V., Fisher, T. G. and Hajdas, I. (2006). The minimum problem in dating deglaciation: a strategy from the Fort McMurray area, Akron, OH, April 20–21. *Geological Society of America Abstracts with Programs*, **38**.

Wickham, J. T., Gross, D. L., Lineback, J. A. and Thomas, R. L. (1978). *Late Quaternary Sediments of Lake Michigan*. Illinois State Geological Survey, Environmental Geology Notes 84, pp. 1–26.

Wilson, M. P. (1980). Catastrophic discharge of Lake Warren in the Batavia-Genesee region. Ph.D. Dissertation, Syracuse, New York, Syracuse University.

Wilson, M. P. and Muller, E. H. (1981). Catastrophic drainage of Lake Warren east of Batavia, N.Y. *Geological Society of America Abstracts with Programs*, **13** (7), 582.

Wolfe, B. and Teller, J. T. (1993). Sedimentologic and stratigraphic investigations of a sequence of 106 varves from glacial Lake Assinibone, Saskatchewan. *Journal of Paleolimnology*, **9**, 257–173.

8 Floods from natural rock-material dams

JIM E. O'CONNOR
and ROBIN A. BEEBEE

Summary

Breached dams formed naturally of rock or rock debris have produced many of the largest floods of Earth history. Two broad classes of natural impoundments are (1) valley-blocking accumulations of mass movements and volcaniclastic debris, and (2) closed basins rimmed by moraines, tectonic depressions, and calderas and craters formed during volcanic eruptions. Each type is restricted to particular geological and geographical environments, making their incidence non-uniform in time and space.

Floods from breached natural dams and basins result from rapid enlargement of outlets. Erosion commonly is triggered by overtopping but also by piping or mass movements within the natural dam or basin divide as the level of impounded water rises. Breaching also can be initiated by exogenous events, such as large waves caused by mass movements or ice avalanches, and upstream meteorological or dam-break floods.

The peak discharge and hydrograph of breached rock-material dams depends mainly on the impounded volume, breach geometry and breach erosion rate. For impounded water bodies that are large with respect to final breach depth, including most tectonic and volcanic basins and many ice-dammed and volcanic-dammed lakes, the peak discharge is primarily a function of final breach geometry. These floods typically last longer and attenuate less rapidly than smaller impoundments. For impoundments of smaller volume relative to final breach depth, such as most moraine-rimmed lakes and landslide and constructed dams, peak discharge is a nearly linear function of vertical breach erosion rate. Some dam-failure floods, especially in steep environments, evolve to debris flows because of downstream sediment entrainment, thus increasing peak discharge and flow volume depending on channel and valley interactions.

Floods from natural-dam failures are geomorphically important because their high flows achieve shear stresses and stream powers orders of magnitude greater than meteorological floods. Such flows can exceed critical thresholds for eroding bedrock and can transport clasts with diameters of many metres, thus forming some of the most spectacular landscapes on Earth.

8.1 Introduction

While human experience with flooding mostly is associated with rainfall or snowmelt, breached natural dams have caused most of the largest floods known on Earth (e.g. Costa and Schuster, 1988; Cenderelli, 2000; O'Connor *et al.*, 2002; O'Connor and Costa, 2004). These dam-failure floods (Table 8.1, see end of chapter) have had discharges up to $2 \times 10^7\ m^3\,s^{-1}$, exceeding those for the largest known meteorological floods by a factor of 50 (O'Connor *et al.*, 2002). Because dam-failure floods typically inundate landscapes unadjusted to such flows, they almost always produce profound and long-lasting geomorphical effects, as attested by many of the companion papers in this volume. Floods from natural-dam failures are also a significant hazard in many environments (e.g. Hewitt, 1968; Lliboutry *et al.*, 1977; Eisbacher and Clague, 1984; Clague and Evans, 1994; Risley *et al.*, 2006). Consideration of the broad-scale conditions under which natural rock-material dams form and breach on Earth provides a planetary perspective for assessing how such dams and resulting floods shape landscapes on other planetary bodies.

On Earth, several geological processes lead to impoundment of water bodies that can then rapidly drain to cause large floods (Figure 8.1; summarised by Swanson *et al.*, 1986; Costa, 1988; Costa and Schuster, 1988; Clague and Evans, 1994; O'Connor *et al.*, 2002; Carrivick and Rushmer, 2006; Korup and Tweed, 2007; among others). Two broad classes of impoundments are valley blockages and natural basins that fill with water. Valley blockage by lateral emplacement of material, such as mass movements of rock debris, volcanic products or glacial ice, can form large and deep lakes with the potential for immense downstream flooding. Overflow from tectonic basins, moraine-rimmed lakes, lakes in volcanic calderas and craters, and basins formed along isostatically depressed margins of large ice sheets can involve tremendous water volumes in both marine and non-marine environments and prompt major changes to drainage networks. Meteor impacts also produce closed basins; although no terrestrial examples are known to have filled and then breached, Coleman and Dinwiddie (2007) document an outburst flood from a breached impact crater on Mars. Water release from beneath and

Megaflooding on Earth and Mars, ed. Devon M. Burr, Paul A. Carling and Victor R. Baker.
Published by Cambridge University Press.

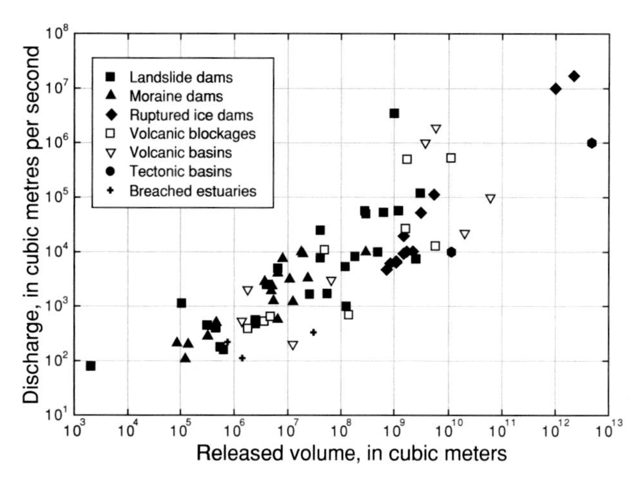

Figure 8.1. Natural dam failures, peak discharge versus volume released. Only includes ice dams that breached by overtopping (non-tunnelling). (Data from Table 8.1.)

within glaciers can also cause notable floods, including relatively small releases from alpine glaciers that nevertheless create hazards in developed mountainous areas. Even larger Holocene flows – up to 700 000 $m^3 s^{-1}$ – resulted from volcanic eruptions under ice sheets in Iceland (Waitt, 2002; Björnsson, this volume Chapter 4), as well as the 1918 Katla jökulhlaup that peaked at about 300 000 $m^3 s^{-1}$ (Tomasson, 1996). Tremendous Pleistocene floods have resulted from rapid emptying of water stored within or under continental ice sheets (e.g. Munro-Stasiuk *et al.*, this volume Chapter 6).

This chapter will describe floods from both valley-blockage and basin-fill scenarios, but will focus on breaches developed in naturally formed blockages composed of rock materials. Floods and flood features from breached ice dams are considered for comparison purposes, but are treated more comprehensively elsewhere in this volume. To aid in quantitative assessment, floods from more-studied breached constructed dams and physical model studies are included in discussion and analysis of dam-failure processes. After a general discussion of the environments leading to the formation of natural rock-material dams and basins, the controls on breaching and downstream flood characteristics are described, and how these attributes may relate to downstream flood features are considered.

8.2 Floods from breached valley blockages

Blocked valleys are the most common mechanism for producing large non-meteorological floods. While the largest such floods of the Quaternary have resulted from failure of ice dams, such as the Pleistocene Missoula Floods in the Pacific Northwest of North America and the Kuray Floods of central Asia, most historic large floods from breached valley blockages were caused by failure of landslide dams. Rarer are breaches of lava-flow dams or other volcanically emplaced materials. Failure of constructed dams can also lead to exceptional floods but these will not be considered here except in later discussion of breach and flood processes. Small floods for short distances can debouch from perched estuaries impounded by beach berms, but such floods are generally small because the gravel and sand barriers are almost always less than 3 m high (Stretch and Parkinson, 2006).

8.2.1 Landslide dams

Landslide dams (Figure 8.2) encompass a broad class of mass movements of rock, regolith and soil that block rivers or streams, typically by entering a valley from side slopes or tributary drainages. They are also the most studied type of natural dam, especially since the benchmark synthesis of natural dams by Costa and Schuster (1988) and their inventory of landslide dams (Costa and Schuster, 1991). Recent summaries have been provided by Cenderelli (2000), Korup (2002), Shang *et al.* (2003), Hewitt (2006), and Korup and Tweed (2007) that expand or elaborate on classification systems, genesis, distribution and geomorphic effects of landslide dams. Details will not be reiterated here, but to summarise: most landslide dams are triggered by precipitation, rapid snowmelt, volcanic activity and earthquakes (Costa and Schuster, 1988). Landslide dams can involve volumes exceeding 10^9 m^3 and form lakes more than 1000 m deep (Hewitt, 1998, 2006; Korup, 2002). The largest existing dam (of any kind) in the world is the Usoi landslide dam in Tajikistan, which in 1911 blocked the Bartang River valley to a depth of >550 m with a 2–2.5 × 10^9 m^3 rockslide, and is nearly twice the height of the tallest constructed dam, the 300 m high Nurek rockfill dam, also in Tajikistan. Fast-moving debris flows, rock avalanches and landslides may form dams in seconds or minutes, while more sluggish landslides and earthflows may take several years to completely block a river. Landslide dams can remain stable and unbreached, can breach after long periods, such as the 130 years between formation and breaching of Tegermach River in Kirgizstan (Costa and Schuster, 1991), or breach minutes after formation as in the case of the Pollalie Creek debris flow, which blocked the East Fork Hood River, Oregon, for only about 12 minutes before failing (Gallino and Pierson, 1985). Some landslide dams, such as the Flims rockslide on the river Rhine in Switzerland, can create multiple or complex impoundments over time, and suffer multiple episodes of partial breaching and sediment infilling (Von Poschinger, 2004; Wassmer *et al.*, 2004).

While not considered in detail here, landslides can also cause floods by mechanisms other than blocking a channel and subsequently breaching (Costa, 1991). Landslides have triggered breaches of natural and anthropogenic dams (e.g. Hermanns *et al.*, 2004), and in many mountainous environments, mass movements evolve into debris flows or sediment-laden floods (e.g. Eisbacher and Clague, 1984). In a few instances, landsliding directly into lakes has triggered large waves or expelled water, leading to downstream floods, such as for the 1983 flood of Ophir Creek, Nevada (Glancy and Bell, 2000) and, more lethally, the 1963 Mount Toc landslide into the Vajont (Vaiont) Reservoir, Italy, where a resulting 245 m wave passed over the 262 m high concrete arched dam, killing more than 2000 downvalley inhabitants (Kilburn and Petley, 2003).

Landslide dams affect all mountainous areas in the world (Costa and Schuster, 1991; Ermini and Casagli, 2003; Korup and Tweed, 2007). Hundreds have been mapped in the Himalayas (e.g. Hewitt, 1998, 2006; Shroder, 1998); Tibet and China (e.g. Hejun *et al.*, 1998; Chai *et al.*, 2000; Shang *et al.*, 2003); Europe (e.g. Eisbacher and Clague, 1984; Pirocchi, 1991; Nicoletti and Parise, 2002); Japan (Swanson *et al.*, 1986); New Zealand (Adams, 1981; Perrin and Hancox, 1992; Korup, 2002); South American Cordillera (Schuster *et al.*, 2002); and the western North America Cordillera (Costa and Schuster, 1991; Clague and Evans, 1994). Mountainous regions are particularly susceptible to landslide dams because high relief promotes mass movements, and confined valleys are readily blocked by slide debris (Costa and Schuster, 1988). Korup *et al.* (2006, 2007) determined that the largest terrestrial landslides, those with volumes greater than 10^8 m^3 and producing dams hundreds of metres tall, are concentrated in tectonically active mountain belts and volcanic arcs. More than half of these landslides occur in the steepest parts of the surrounding landscape. Additionally, certain geological environments promote landslides and landslide dams by virtue of structural orientations or weaknesses (e.g. Alden, 1928), stratigraphic combinations (e.g. Palmer, 1977; Safran *et al.*, 2006), active volcanism (e.g. Meyer *et al.*, 1986; Scott, 1989; Capra and Macías, 2002), seismicity (e.g. Davis and Karzulovíc, 1963; Nicoletti and Parise, 2002) and glacier retreat (Evans and Clague, 1994; Bovis

Figure 8.2. Bridge of the Gods landslide dam, which blocked the Columbia River, Oregon and Washington, USA, to a depth of more than 80 m at about AD 1450 before breaching cataclysmically sometime before the Lewis and Clark AD 1805 exploration of the area (O'Connor, 2004). (a) View upstream (east) toward toe of landslide, which completely filled the valley bottom. The impoundment breached at the southern distal end of the landslide, creating a new course of the Columbia River around the toe of the landslide. The breach was abrupt, resulting in bouldery flood bars deposited in the valley bottom downstream. (11 April 1928, photograph courtesy of U.S. Army Corps of Engineers.) (b) View downstream (west) of toe of Bridge of the Gods landslide and breach area. Cascade Rapids, where the Columbia River descended ~15 m over the bouldery debris, resulted from incomplete incision of the landslide dam. (9 September 1929, photograph courtesy of U.S. Army Corps of Engineers.) (c) View north-northwest across Cascade Rapids and breach area toward main body of Bridge of the Gods landslide. (Undated (but prior to 1932 inundation of this reach by Bonneville Dam) photograph courtesy of U.S. Army Corps of Engineers.)

and Jakob, 2000). The engineering community has recognised the significance of landslide dams, and impounded water and outburst floods are typically studied as hazards rather than as geomorphical agents (Whitehouse and Griffiths, 1983; Schuster and Costa, 1986; Costa and Schuster, 1988; Webby and Jennings, 1994; Jakob and Jordan, 2001; Schuster and Highland, 2001; Risley *et al.*, 2006). Recent work by geomorphologists also demonstrates the important role of landslide dams and debris flows on the evolution of fluvial systems in many mountainous environments (e.g. Hewitt, 2006; Korup, 2006; Lancaster and Grant, 2006; Hewitt *et al.*, 2008).

Volcanoes can also shed large landslides or debris avalanches, which by blocking tributary valleys set the stage for landslide dam failures. This was the case for the 1980 debris avalanche at Mount St Helens, which mobilised 2.3×10^9 m^3 of the edifice into the North Fork Toutle River valley, impounding several tributary drainages (Table 8.1) and raising the height of the natural dam bounding Spirit Lake (Voight *et al.*, 1981; Youd *et al.*, 1981). Scott (1989) documented a previous debris-avalanche damming at Mount St Helens ~2500 years ago, resulting in a lake outbreak and ensuing downstream lahar with a peak discharge exceeding 250 000 m^3 s^{-1}.

Landslide dams are subject to breaching because commonly they are composed of unconsolidated porous material and have no controlled outlet. Landslide dams that do breach, typically do so within weeks or months of formation by overtopping and rapid erosion of an outlet channel (Costa and Schuster, 1988; Ermini and Casagli, 2003). Breaching commonly is initiated by upstream meteorological flooding that rapidly increases lake levels and overflow discharges (e.g. Hancox *et al.*, 2005). Earthquakes also trigger landslide-dam failures, such as the 1786 dam-break flood on the Dadu River, China (Dai *et al.*, 2005). Rarer is failure by groundwater piping or by mass movement of the blockage itself (Costa and Schuster, 1988). Some breaches are triggered by large waves overtopping the blockage, such as the 1963 breach of the Issyk landslide dam in Kazakhstan (Gerasimov, 1965).

Geological evidence of outburst floods and debris flows from breached landslide dams increasingly is recognised in many mountainous environments. In most of these cases, bouldery fluvial or debris-flow deposits are traced to incised landslide masses filling valley bottoms (e.g. Selting and Keller, 2001; Beebee, 2003; O'Connor *et al.*, 2003). The remnant landslides and consequent flood deposits can form persistent features in valley bottoms (e.g. Schuster *et al.*, 2002; O'Connor *et al.*, 2003).

Many landslide dams do not fail cataclysmically but result in short-lived or long-lived lakes that fill with sediment before significant erosion of the blockage (Clague and Evans, 1994; Hewitt, 2006). Of the *circa* 340 landslide dams for which fate is documented in the Costa and Schuster (1991) global compilation, about 75% breached and produced downstream floods. Similarly, a compilation and analysis by Ermini and Casagli (2003) indicated that of the 282 landslide dams plotted in their Figure 8.4, 188 (67%) breached cataclysmically. In both compilations, the failure percentages may be overestimates, because landslide dams that breach cataclysmically more likely are reported in the literature than those that do not. From a more systematic survey of 232 landslide dams in New Zealand, Korup (2004) reports that about 37% apparently have failed, although this percentage may be low due to underreporting of very short-lived blockages. Predicting the stability of landslide dams is challenging because of the many internal and external factors controlling breach-triggering mechanisms and intrinsic dam stability (Costa and Schuster, 1988; Ermini and Casagli, 2003; Korup, 2004; Dunning *et al.*, 2005). Quantitative indices based on morphometric and watershed characteristics including dam height and volume, impounded water volume, watershed area and relief have been successful locally in discriminating stable from unstable dams (Ermini and Casagli, 2003; Korup, 2004) but their predictive power is low and critical values separating stability domains are apparently dependent on regional conditions (Korup, 2004).

8.2.2 Volcanogenic dams

Volcanism can directly trigger large floods and debris flows (lahars) by rapid melting of snow and ice, including the large Icelandic jökulhlaups described in this volume, Chapter 4 by Björnsson. Large snow-and-ice augmented lahars have also been documented for many of the tall stratovolcanoes of the Pacific Rim, including the November 1985 eruption of Nevado del Ruiz, Columbia, which killed more than 23 000 people in the city of Armero (Pierson *et al.*, 1990). However, volcanic activity can also indirectly produce floods by blocking drainages with primary volcanic materials such as lava and pyroclastic debris (Capra, 2007). These blockages are similar to landslide dams except for the composition of blockage material. In Table 8.1, volcanogenic landslide dams are grouped with landslides triggered by all mechanisms, although the distinction is somewhat arbitrary.

Pyroclastic-flow and lahar deposits and lava flows are all primary volcanic materials that can form blockages susceptible to breaching. Macías *et al.* (2004) summarised volcanic dams and described the 11 000 m^3 s^{-1} flood from a breached dam of pyroclastic flow material deposited in the 3 April 1982 eruption of El Chichón, Mexico. Also in Mexico, a late Pleistocene debris avalanche from Nevado de Colima volcano blocked the Naranjo River, causing a 1×10^9 m^3 lake to form and then breach, resulting in a 3.5×10^6 m^3 s^{-1} debris flow downstream (Capra and Macías, 2002). Another large lake (~1×10^9 m^3) was impounded behind a welded block-and-ash breccia that blocked the Lillooet River in British Columbia, Canada, about 2350 yr BP. This lake soon breached cataclysmically, transporting still-hot blocks up to 15 m in diameter up to 3.5 km downstream (Hickson *et al.*, 1999). Several outburst floods from breached lahars and volcanogenic debris-avalanche deposits have probably coursed down the Chakachatna River draining the Mt Spurr volcanic complex in south-central Alaska (Waythomas, 2001). Similar events have caused large floods in Japan (Kataoka *et al.*, 2008). Two floods resulted from breached pyroclastic flows and volcanogenic sediment deposits blocking the outlet of Lake Tarawera, North Island, New Zealand, including a *circa* AD 1315 flood of $5–10 \times 10^4$ m^3 s^{-1} (Table 8.1; Hodgson and Nairn, 2005). A similar flood resulted from pyroclastic flows shed during the 7.7 ka climatic eruptions of Mt Mazama, Oregon, which blocked the Williamson River (Table 8.1; Conaway, 1999). Dams of fragmental pyroclastic material are particularly susceptible to breaching because they erode quickly once overtopped.

Less easily breached are dams composed of lava flows. While many valley-filling lava flows have blocked drainages, very few are known to have breached cataclysmically. The only quantitatively well-documented floods from breached lava-flow dams are the five breakout floods in western Grand Canyon, where the Colorado River was repeatedly dammed by basalt flows cascading into the canyon from the Uinkaret volcanic field (Hamblin, 1994; Fenton *et al.*, 2002, 2004, 2006). In particular, Fenton *et al.* (2006) report a 165 ka 'hyaloclastite' dam more than 140 m high that failed, producing a peak discharge of $2.3–5.3 \times 10^5$ m^3 s^{-1} (Table 8.1), although Crow *et al.* (2008) offer alternative interpretations not requiring outburst floods from the Grand Canyon lava dams.

A unique lava-dam failure occurred at Pleistocene American Falls Lake, formed when the southward flowing Cedar Buttes basalt flow dammed the Snake River in the eastern Snake River Plain of Idaho, USA, at 72 ± 14 ka. American Falls Lake was at least 30 m deep (Scott *et al.*, 1982) and perhaps as deep as 60 m if the Cedar Buttes lava flow blocked the Snake River at an elevation near its historic grade. The lake spilled out of the lava-dammed basin on the north side of the vent, and by $\sim$16 ka gradually incised an outlet 8 to 15 m lower than the maximum blockage level. At that time, the 10^6 m^3 s^{-1} Bonneville Flood (Malde, 1968; Scott *et al.*, 1982; O'Connor, 1993) engulfed the lake, spilling out and forming large cataracts along the lake outlet channel as well as where the Cedar Buttes lava flow blocked the previous course of the Snake River to the south. The Bonneville Flood eroded the soft loess, lacustrine and fluvial sediment flanking the former Snake River route, as well as rubbly lava and hyaloclastite associated with the initial blockage, eventually draining the lake and establishing a new canyon closely following the pre-blockage route (O'Connor, 1993; unpublished mapping). Drainage of lava-dammed American Falls Lake, however, probably contributed little to the peak discharge of the Bonneville Flood.

Lava-flow dams apparently breach cataclysmically only under special conditions exemplified by the Grand Canyon case, or probably even more rarely when affected by huge upstream floods. In the Grand Canyon, basalt poured into a steep-walled canyon occupied by a large river. Consequently, lava dams were buttressed by unconsolidated talus and were composed partly of fragmental material from hydrothermal brecciation as the lava entered the river (Fenton *et al.*, 2006). Similar conditions probably enabled excavation by the Bonneville Flood of the former Snake River course. Instead of breaching cataclysmically, many lava flows entering valleys or canyons of the western US rivers formed long-lived blockages, resulting in eventual river diversion into the generally softer materials forming the palaeovalley walls (e.g. Stearns, 1931; Malde, 1982; Howard *et al.*, 1982; O'Connor *et al.*, 2003; Duffield *et al.*, 2006). Many of these lava flows travelled several kilometres down river valleys, forming massive dams of solid lava, especially where flow lobes from late in the eruption advanced down dry valleys still blocked by earlier flows.

8.3 Floods from breached basins

Breaches of divides bounding basins, such as moraine-rimmed lakes, volcanic calderas and tectonic depressions, are generally restricted to specific geological environments and climatologic situations, but floods from breached divides can be extremely large because of the potentially immense water volumes impounded in geological basins. For example, the Bonneville Flood from the tectonic basin of Great Salt Lake of the western USA, with a total volume of 4.75×10^{12} m^3, is the largest known freshwater flood volume in Earth history (Figure 8.1; O'Connor *et al.*, 2002). Several processes can form basins that can fill and then overflow. The material and geometry of the basin divide determine whether and how rapidly a breach may form, as well as the ultimate depth of incision.

8.3.1 Moraine basins

Basins formed by glacial moraines include river valleys blocked by lateral and end moraines, as well as basins formed by ice retreat from moraines deposited during times of advanced ice (Figure 8.3). The former are probably more appropriately classified as valley blockages similar to landslide dams but far more outburst floods result from the latter, which are indeed new geological basins. Floods from moraine-dammed lakes (Figure 8.3; Table 8.1) have been studied widely as an alpine hazard and geomorphological agent for several decades, recently summarised by Richardson and Reynolds (2000), Clague and Evans

Figure 8.3. Breached Neoglacial-age moraine of Cumberland Glacier and remnant Nostetuko Lake, British Columbia, Canada, after 19 July 1983 outburst flood. The lake level dropped 38.4 m, releasing 6.5×10^6 m^3 of water. (August 1984 photograph courtesy of John J. Clague.)

(2000), Cenderelli (2000) and Kattelmann (2003). Moraine dams produce persistent large-flood features because these lake releases are typically in basin headwaters and rapid breaching produces much larger flows than those resulting from even extreme meteorological events (Cenderelli and Wohl, 2003).

Large moraine-dammed lakes formed in the late Pleistocene where large valley glaciers built terminal moraines at the mouths of confined valleys. There is only one documented flood from a Pleistocene moraine dam but this was probably initiated by human interference (Carling and Glaister, 1987), although Blair (2001) and Benn *et al.* (2006) reported outburst-flood deposits that may have resulted from breached late Pleistocene moraines in the Sierra Nevada of California. In contrast, scores of large floods and debris flows resulted from rapid breaching of late Holocene moraine dams (several examples are listed in Table 8.1). Clague and Evans (2000) provide a recent summary of moraine-dammed lakes in British Columbia, Canada, and also guide readers to the extensive global literature of floods from moraine-dammed lakes.

Moraine dams form in the wake of retreating glaciers. Hence, floods from breached moraine dams are concentrated in deglaciating alpine areas where they have become a particular subject of interest because of deglaciation associated with twentieth-century warming (Lliboutry *et al.*, 1977; Liu and Sharmal, 1988; O'Connor and Costa, 1993; Clague and Evans, 1994; Kattelmann, 2003). Most recently breached moraine dams formed during the Little Ice Age of the 1700s and 1800s, the time of most extensive glaciation of the late Holocene Neoglacial period (Porter and Denton, 1967). During this period of advanced ice, the termini of valley and cirque glaciers were rimmed with moraines up to 100 m high formed of loose and unconsolidated rock debris. By the late 1900s and through most of the twentieth-century, glaciers substantially retreated from their advanced Little Ice Age positions, allowing lakes with volumes up to 1×10^8 m^3 and depths of nearly 100 m (Table 8.1; Yamada and Sharma, 1993) to form in the vacated basin between moraines and the retreating ice. Tens to hundreds of Little Ice Age moraine-dammed lakes have been mapped in glaciated regions such as Tibet (Liu and Sharmal, 1988), Nepal and Bhutan (Yamada and Sharma, 1993; Richardson and Reynolds, 2000), Peru (Lliboutry *et al.*, 1977; Reynolds, 1992), European Alps (Haeberli, 1983), southern British Columbia, Canada (Clague and Evans, 1994; McKillop and Clague, 2007a), and the Cascade Range of Oregon and Washington, USA (O'Connor *et al.*, 2001).

Many Little Ice Age moraines enclosing lakes are susceptible to breaching because of both their geotechnical characteristics and the trigger mechanisms provided by local topographic conditions. Young moraines commonly are perched on steep mountain slopes, sparsely vegetated and composed of loose and poorly sorted regolith deposited at slopes as steep as 40° – all conditions facilitating rapid erosion once overtopped. Breaching can be triggered by periods of rapid snowmelt or intense rainfall that either prompt overflow of the moraine rim or initiate outlet erosion, but the most common failure mechanism is ice-fall or rock-fall into the lake from the glacier or surrounding cirque basin, causing waves to overtop the moraine and initiate breaching (Costa and Schuster, 1988; Clague and Evans, 2000, Kershaw *et al.*, 2005). Breaching in some cases may be facilitated by lowering of the moraine crest due to melting ice cores (Reynolds, 1992). Maximum breakout volumes have approached 5×10^7 m^3 with breach depths of up to 50 m (Table 8.1).

Some moraine-rimmed basins breach in the same season of first lake appearance (e.g. O'Connor *et al.*, 2001) but most fail years or decades later depending on the time required to fill the lake basin or the incidence of breach-triggering processes such as rock-fall or ice avalanches. In general, the incidence of breaching is likely to increase as lakes get larger and then decline gradually as the moraine dams age (Clague and Evans, 2000). For example, in the Oregon Cascade Range, 13 moraine-dammed lakes formed between 1924 and 1956, by which time they had reached their maximum size. Eight of these lakes produced floods and debris flows from four complete and seven partial dam breaches between 1934 and 1987, with none since. All but one of these breaches were prior to 1971, indicating breaches are becoming rarer as the moraines age (O'Connor *et al.*, 2001). The incidence of moraine breaches in the Himalaya has not slowed, however, as lakes continue to grow and breach in the wake of retreating glaciers (Liu and Sharmal, 1988; Mool, 1995; Richardson and Reynolds, 2000; Kattelmann, 2003).

Not all moraine-dammed lakes breach cataclysmically. The large number of existing lakes impounded by Pleistocene moraines attests to the potential stability of moraine dams. Likewise, many Neoglacial moraine dams have persisted without breaching. In an inventory of moraine-dammed lakes in southern British Columbia, Canada, McKillop and Clague (2007a) found only 10 of 175 moraine-rimmed lakes had drained partly or completely. For some lakes, stability is promoted by outlet channels formed in rock or armoured by coarse material. Others are stable because low-gradient outlet channels drain the lakes (Clague and Evans, 2000). Logistic analysis of the stable and breached moraine dams in British Columbia showed that four factors were correlated with increased likelihood for breaching: large moraine height-to-width ratio, presence of ice-free moraines, large lake area, and moraines composed of sedimentary rock types (McKillop and Clague, 2007a). Moraine dams formed on stratovolcanoes may be particularly susceptible to breaching because they are generally tall and on steep slopes – in the central Oregon Cascade Range volcanoes, 8 out of 13 Neoglacial-age moraine dams have partly or completely breached (O'Connor *et al.*, 2001). This failure rate far surpasses the 6% that have failed in the mostly crystalline and sedimentary rocks of the British Columbia Coast Range (McKillop and Clague, 2007a).

8.3.2 Tectonic basins

Large tectonic basins that fill with water and subsequently breach have been the source of many great floods in Earth history (Table 8.1). Such floods can be large because of the immense volumes of water potentially held in tectonic basins. For example, the Caspian Sea – the largest terrestrial waterbody on Earth – contains 7.8×10^{13} m^3 of water and is more than 1000 m deep (although with a surface elevation of ~30 m below sea level, there is little potential energy available at present for producing megafloods). Floods from basin spills usually overwhelm pre-existing drainages, leaving pronounced flood features. Such floods commonly lower local base levels and alter hydrological pathways, initiating landscape and ecological adjustments. Consequences may include drainage integration, regional incision (e.g. House *et al.*, 2008) and changed migration pathways for aquatic species (e.g. Reheis *et al.*, 2002).

A common flood-causing scenario for tectonic basins is the filling and overtopping of a hydrologically closed basin, either because of climate change or because geological events increase the contributing drainage area or cause a sudden influx of water. Upon overtopping the basin divide, flow rapidly erodes an outlet channel, partly or completely emptying the lake. Such was the case for the Bonneville Flood of North America (Figure 8.4; Gilbert, 1890; Malde, 1968; O'Connor, 1993). Pleistocene Lake Bonneville filled the closed basin of Great Salt Lake, overtopped and eroded the alluvial fan sediment forming the drainage divide at Red Rock Pass, Idaho, and then spilled northward into the Snake River basin (Figure 8.4). About 5×10^{12} m^3 of water flowed out of the basin at a peak discharge of 10^6 m^3 s^{-1} while incising the basin divide 108 m. Similar floods elsewhere in the North American Basin and Range Province (and adjacent terrain) are indicated by incision of surface overflow channels at basin rims (e.g. Currey, 1990; Reheis *et al.*, 2002) or downstream bouldery deposits (e.g. Anderson, 1998; Reheis *et al.*, 2002). Recent work in Oregon, USA, documents two such floods from overflowing basins: (1) overflow of Pleistocene Lake Alvord into Crooked Creek of the Snake River drainage (Hemphill-Haley *et al.*, 1999; Carter *et al.*, 2006); and (2) overflow of a Pleistocene lake in the Millican basin into the Crooked River, central Oregon (Vanaman *et al.*, 2006).

The Lake Bonneville, other US Great Basin spillovers and the Oregon floods are examples of hydrologically closed basins filling during wetter climatic periods. Evidence is also emerging of closed basins breaching from sudden water inputs from upstream drainage basin changes. Floodwater from breaches of upstream ice-dammed and proglacial lakes may have entered the Pleistocene predecessor of the Caspian Sea and Pleistocene Mansi Lake, triggering erosion of their basin outlets, and resulting in large floods with volumes as great as 4.2×10^{13} m^3 (M. G. Grosswald, personal communication, 2006). Similarly, House *et al.* (2008) concluded that the lower Colorado River connected with the upper Colorado drainage at about 5 Ma as a consequence of a chain reaction of upstream basin divide breaches. Boulder deposits downstream of breached divides indicate that some of these basin breaches caused large floods. A similar sequence of events about 2 Ma may have led to integration of the Snake River into the Columbia River basin and incision of Hells Canyon and the western Snake River Plain of Idaho (Othberg, 1994; Wood and Clemens, 2002).

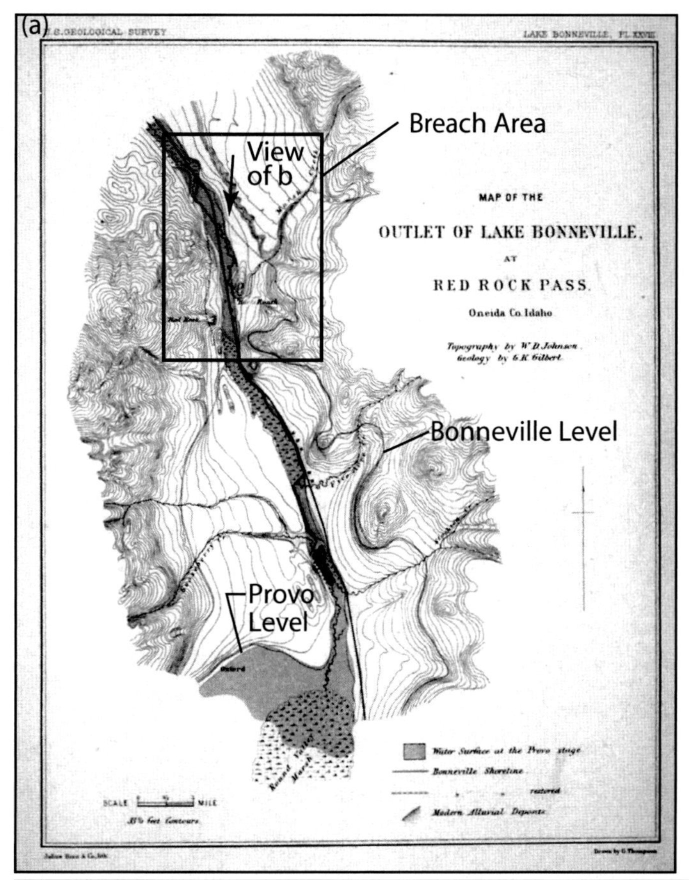

RED ROCK PASS, THE OUTLET OF LAKE BONNEVILLE, AS SEEN FROM THE NORTH.

Figure 8.4. Figures adapted from G. K. Gilbert (1890) showing conditions at pluvial Lake Bonneville outlet near Red Rock Pass, Utah, USA. (a) Map showing 1552 m elevation of maximum pluvial Lake Bonneville and subsequent 1444 m elevation of Provo shoreline after breaching and release of 4.75×10^{12} m^3 of water about 16 000 years ago. The alluvial fan of Marsh Creek formed the divide between the closed basin of Great Salt Lake and the Snake River watershed. Overtopping resulted in rapid incision of 108 m to underlying bedrock at Red Rock Pass, where downcutting ceased. (b) Sketch of view south toward Red Rock Pass from the surface of Marsh Creek fan, showing breach channel formed during incision of the fan.

Large marine floods *into* closed tectonic basins have likely resulted from eustatic sea-level rise or tectonic activity at basin margins. Hsü (1983) suggested that a desiccated Mediterranean basin was filled rapidly about 5.3 Ma by breaching at the present-day Straits of Gibraltar. Likewise, Ryan and Pitman (1999) proposed that the rising Mediterranean Sea overtopped and eroded a divide at Bosporus Strait about 7500 years ago, rapidly filling the Black Sea.

Large floods out of and into tectonic basins generally occur in semi-arid environments during times of changing hydrological conditions. Very large terrestrial freshwater floods result from sustained periods of positive water balances that allow closed basins to fill. During the Quaternary, filling episodes are associated mostly with pluvial periods during times of glacial advance (e.g. Baker, 1983; Reheis *et al.*, 2002). Likewise, marine waters are most likely to flood continental basins during times of rising sea level resulting from climate warming and melting of major ice sheets and glaciers. Large closed basins are commonly in tectonically active regions within the arid and semi-arid subtropical belts, such as the Basin and Range Province of western North America. In these areas, long periods of negative water balances allow closed basins to form without continuous overflow and drainage integration, creating the potential for large water volumes to descend over steep basin margins during singular episodes of basin filling. Marine floods into closed basins are also more likely in environments of negative water balance, where isolated basins will remain dry or contain endorheic lakes substantially below sea level, thus providing the hydraulic head for cataclysmic marine inflow (O'Connor *et al.*, 2002).

8.3.3 Volcanic basins

Large floods from basins formed by volcanic activity, in particular caldera and crater lakes, have been increasingly recognised as a hazard during the last two decades (Lockwood *et al.*, 1988; Waythomas *et al.*, 1996; Manville *et al.*, 1999, 2007; Wolfe and Begét, 2002). Calderas and volcanic craters are formed by a combination of explosive volcanism and volcanotectonic collapse (Williams, 1941). Caldera and crater lakes produce large floods because they can contain large volumes of water – for example, Lake Taupo, New Zealand, holds $\sim 6 \times 10^{10}$ m^3 (Manville *et al.*, 1999); Lake Atitlán holds $\sim 4 \times 10^{10}$ m^3 (Newhall *et al.*, 1987); and Crater Lake, Oregon, is nearly 600 m deep and contains 1.7×10^{10} m^3 (Johnson *et al.*, 1985). The largest intracaldera lake is Lake Toba in Indonesia, which contains 2.4×10^{11} m^3 of water (Chesner and Rose, 1991). While two orders of magnitude smaller than the largest tectonic basins, large caldera and crater lakes may be more likely to produce megafloods than basins of other origins because they are (1) formed essentially instantaneously without water outlets, (2) commonly rimmed by heterogeneous and structurally weak lava flows interbedded with unconsolidated pyroclastic material, and (3) located in volcanic arcs that, because of their elevated positions near continental margins, may fill rapidly from precipitation. Renewed volcanism may also promote breaching by producing displacements or waves that overtop and breach outlets (e.g. Stelling *et al.*, 2005) or by causing overtopping by rapid melting of snow and ice (e.g. Donnelly-Nolan and Nolan, 1986).

The largest well-documented megaflood from a caldera lake breakout is the post 3.4 ka flood from Aniakchak Volcano, Alaska (Plate 20; McGimsey *et al.*, 1994; Waythomas *et al.*, 1996). The caldera at Aniakchak formed from a large eruption about 3.4 ka and subsequently filled with about 3.7×10^9 m^3 of water to an average depth of $\sim$98 m. Shorelines near the elevation of the inferred low point in the rim indicate that the lake may have overflowed the basin rim for some time before rapidly breaching. Once breached, almost the entire lake emptied into the Aniakchak River valley at a peak discharge of about 10^6 m^3 s^{-1}, producing the largest known Holocene flood discharge on Earth (Waythomas *et al.*, 1996).

Several large floods from breached caldera and crater lakes have been documented in New Zealand, including the 13 March 2007 breach of the 1.8×10^6 m^3 crater lake at Mount Ruapehu (Table 8.1; Manville *et al.*, 2007). A post-1.8 ka partial breach of intracaldera Lake Taupo, on the North Island of New Zealand, emptied 2.0×10^{10} m^3 of water at a peak discharge of 1.7–3.5×10^5 m^3 s^{-1} (Manville *et al.*, 1999). The worst volcanic disaster in New Zealand resulted from an earlier breach at Mount Ruapehu on 24 December 1953, when 1.8×10^6 m^3 of water emptied rapidly from the summit crater lake (formed in 1945) into the headwaters of the Whangaehu River. Incorporating volcaniclastic debris from the steep upper slopes of the volcano, the flood transformed into a lahar with a peak discharge of $\sim 2 \times 10^3$ m^3 s^{-1} that destroyed a railroad bridge 39 km downstream, minutes before passage of a passenger train, resulting in 151 deaths (Manville, 2004).

Other documented caldera-breach floods include a 1.5 ka cataclysmic emptying of a $\sim$100 m deep lake contained in Fisher Caldera, Alaska, apparently triggered by an eruption-generated wave (Stelling *et al.*, 2005); two breaches from Okmok Caldera, Umnak Island, Alaska, within the last 1600 yr, including emptying of a 150 m deep 5.8×10^9 m^3 lake filling a caldera formed about 2000 yr BP (Wolfe and Begét, 2002; Begét *et al.*, 2005); a post-4.7 ka partial breach of Paulina Lake at Newberry Volcano, Oregon (Chitwood and Jensen, 2000; Table 8.1), a $\sim 3.3 \times 10^3$ m^3 s^{-1} late Pleistocene flood from a possibly ice-covered caldera lake on Medicine Lake Volcano, California (Donnelly-Nolan and Nolan, 1986), and a July 2002 flood of 6.5×10^7 m^3 of water from the summit caldera formed by the 1991 eruption at Pinatubo, Philippines (Lagmay *et al.*, 2007).

Like floods from other types of breached basins, crater and caldera lake megafloods are restricted geographically – obviously in this case to volcanic provinces. Larson (1989) identified 88 lakes in 75 calderas (with diameters > 2 km) in 31 volcanic regions globally. The filling and breaching of caldera and crater lakes also may be restricted to humid environments or times of local positive moisture balance because many such lakes have small drainage areas relative to their size. For example, Crater Lake formed by the 7.7 ka eruption of Mount Mazama, Oregon, and East Lake and Paulina Lake of Newberry Volcano, Oregon, all encompass 20% or more of their total drainage area, with the surface of Crater Lake making up 78% of its drainage basin (Johnson *et al.*, 1985). Resurgent volcanism can increase the likelihood of breaching by physical water displacement, formation of large waves and modification of outlet conditions.

8.4 Flood magnitude and behaviour

As a basin or river impoundment fills, a number of processes facilitating dam failure are poised to begin, any one of which may trigger downstream flooding. As water levels rise, the hydraulic gradient through the blockage increases, promoting groundwater flow, the development of piping or sapping channels and mass movements on the downstream face of the blockage. For some blockages, retrograde erosion of resulting sapping channels eventually may incise the barrier and trigger overflow. At high lake levels, rapid water rise or waves triggered by large inflows, wind and mass movements may overtop the blockage. Once sufficient discharge overtops a barrier, tractive sediment entrainment enlarges the overflow channel, increasing flow in a self-enhancing process leading to rapid breach enlargement until either the water supply diminishes or the outlet channel is armoured sufficiently to halt erosion. For some natural dam failures, breach enlargement was observed, or inferred, to proceed as a series of knickpoints migrating up the downstream blockage face or small landslides entering the outlet channel from the sides (e.g. Lee and Duncan, 1975; Plaza-Nieto and Zevallos, 1994; Dwivedi *et al.*, 2000), a process well demonstrated by experimental breaches of earthen dams (Hanson *et al.*, 2003). Once outlet geometry has stabilised, outflow will continue at a diminishing rate as the upstream impoundment level drops to the stable outlet elevation. For valley blockages, the ultimate outlet elevation may be higher, the same, or lower than the pre-impoundment blockage elevation. For breached basins, divide erosion may result in long-term integration of the basin into the hydrologic network.

Not all rock-material barriers breach cataclysmically and produce exceptional floods. The important factors controlling outburst flood likelihood and magnitude for all types of barriers are (1) the stability of the blockage or divide, (2) if breached, the speed and depth of breach growth, and volume, rate and duration of escaping water and entrained sediment, and (3) downstream water and sediment interactions that may change the volume, peak discharge, and type of flow. Each of these issues has many complicated facets, challenging geologists and engineers faced with predicting the likelihood, magnitude and downstream evolution of dam-break floods. These same factors complicate interpretation of past events from geological evidence, even for historic floods (e.g. Manville, 2004). Further analysis is hindered by the few detailed observations of dam failures. Recent summaries of approaches to assessing the likelihood and magnitude of constructed and natural dam-failure floods (some including alpine glacial dams) are provided by Clague and Evans (1994, 2000), Froehlich (1995), Walder and O'Connor (1997), Cenderelli (2000), Korup (2002), Huggel *et al.* (2002, 2003, 2004), Ermini and Casagli (2003), Wahl (2004), McKillop and Clague (2007b) and Manville *et al.* (2007). The following discussion is summarised from these works.

8.4.1 Dam failure

Rock-material dams fail by intrinsic instability, exogenous triggers, or possibly by some combination of these mechanisms. Intrinsically unstable dams fail when the impounded water either overtops and erodes an outlet, or raises piezometric gradients enough to promote piping or mass failure of the blockage. Failure at first filling – by either piping, outlet erosion or a combination of both – is particularly common for landslide dams (Costa and Schuster, 1988). Exogenic mechanisms also trigger breaches, especially for moraine impoundments for which the majority of breaches were triggered by waves from ice or rock avalanches into the lake (Table 8.1). Floods entering the impounded waterbody also have triggered natural-dam failures, mainly by initiating downcutting of existing outlet channels. In many cases it is the first seasonal flood after first overtopping (e.g. Webby and Jennings, 1994) that causes failure, but in some cases exceptional floods were required to trigger erosion of outlets that were stable for years or decades (e.g. the Gros Ventre landslide dam failure described by Alden (1928)). In rarer instances, upstream dam-break floods have triggered downstream breaches of rock-material dams, such as the upstream moraine-dam failure leading to breaching of landslide-dammed Lake Issyk (Gerasimov, 1965) and the example described previously of lava-dammed Pleistocene American Falls Lake breaching during the Bonneville Flood. Earthquakes have triggered landslide-dam and moraine-dam failures (e.g. Lliboutry *et al.*, 1977; Costa and Schuster, 1991; Dai *et al.*, 2005) and some caldera lake breaches may have been caused by intracaldera volcanic activity (e.g. Waythomas *et al.*, 1996). Human intervention also caused failure of moraine dams in Peru (Lliboutry *et al.*, 1977; Reynolds, 1992) as well as several landslide dams (Costa and

Schuster, 1991), commonly during construction of outlet works on the blockage but also by nearby construction and water diversion activities. Longer-term exogenous processes also indirectly trigger natural dam failure or basin breaching; in particular, climate change leading to higher effective moisture regimes can fill and overtop closed basins.

The likelihood of natural dams or basin rims breaching depends on geotechnical characteristics of the blockage and the potential of likely triggering processes. Important blockage characteristics are volume, height, shape, particle size distribution and layering. Likely triggering mechanisms are mass movement of the dam or barrier, piping or overtopping (either by the impoundment filling or by waves from perturbations to the impounded water). Quantitative assessments of failure likelihood are generally feasible only for large landslide dams for which several months may pass until overtopping and where downstream flooding poses significant hazards (e.g. Meyer *et al.*, 1985; Hanisch and Söder, 2000; Risley *et al.*, 2006). For many natural dams stability assessments are necessarily qualitative (e.g. Lliboutry *et al.*, 1977; Clague and Evans, 2000) because of difficulties obtaining the pertinent data in the short time available for hazard assessment. This is particularly the case for small landslide dams, which commonly fail soon after formation but still can pose major downstream flood hazards (e.g. Hancox *et al.*, 2005). Recent empirical studies of natural-dam-failure likelihood seem to offer potential for broad-scale quantification of the main breaching factors, including the dimensionless blockage index for landslide dams (Ermini and Casagli, 2003; Korup, 2004) and the logistic regression analysis for moraine dams (McKillop and Clague, 2007a). Huggel *et al.* (2004) and McKillop and Clague (2007a) have recently presented more encompassing rule-based hazard assessment procedures for moraine-dam failures and ensuing debris flows.

8.4.2 Peak discharge

Peak discharge is perhaps the most important attribute of large floods pertaining to hazards and geomorphic consequences. Consequently, estimating and predicting peak discharges and hydrographs of dam-failure floods, especially for constructed dams, has been a major focus of research in the engineering community (e.g. Ponce and Tsivoglou, 1981; MacDonald and Langridge-Monopolis, 1984; Fread, 1987; Froehlich, 1987, 1995; Singh *et al.*, 1988; Macchione and Sirangelo, 1990; Wahl, 1998, 2004). Hydrologists and geologists interested in natural-dam failures have extended this work substantially (e.g. Clague and Mathews, 1973; Evans, 1986; Costa, 1988; Costa and Schuster, 1988; Walder and Costa, 1996; Walder and O'Connor, 1997; Risley *et al.*, 2006; Davies *et al.*, 2007).

The largest impoundments generally produce the largest floods (Figure 8.5). This observation led to numerous regression equations between predictor variables, such as impoundment volume and impoundment depth, and measurements of peak discharge (Figure 8.5a), providing the simplest and most used approach to estimating peak discharge from natural dam failures (Table 8.2; Clague and Mathews, 1973; Evans, 1986; Costa and Schuster, 1988; Walder and O'Connor, 1997; Cenderelli, 2000). However, as summarised by Walder and O'Connor (1997) and Wahl (2004), regression equations as well as bounding 'envelope curves' derived from these analyses only give order-of-magnitude estimates of peak discharge. Contributing to the large uncertainty of these regression relations is that peak discharges are rarely known at the breach, other than instances for which impoundment drawdown data provide good estimates of outlet flow. Typically, the discharges incorporated into these equations are estimated by a variety of methods at various distances downstream. As the flood moves downstream beyond the breach, peak discharge may diminish due to flow-storage attenuation or increase by sediment entrainment. The primary limitation of the regression approach, however, is that other factors besides impoundment volume and depth, such as hydraulic controls and breach erosion rates, can strongly influence breach outflow (Walder and O'Connor, 1997).

The important hydraulic constraint is that the maximum possible discharge through a breach is critical flow with a specific energy equivalent to the head represented by the height of the impounded water surface above the breach bottom (Figure 8.6). Critical flow is the condition $v = (gd)^{1/2}$, where v is flow velocity, g is gravitational acceleration, and d is flow depth; and the specific energy of the flow is the sum of d and $\frac{1}{2}(v^2/g)$. This coupling between breach development (especially depth) and drawdown of the impounded waterbody forms the basis for several parameterised and physically based approaches to assessing peak discharges and outflow hydrographs from both natural-dam and constructed-dam failures (e.g. Ponce and Tsivoglou, 1981; Fread, 1987; Froehlich, 1987, 1995; Singh *et al.*, 1988; Webby and Jennings, 1994; Walder and O'Connor, 1997; Manville, 2001; and Marche *et al.*, 2006). These physically based approaches consider outflow to be critical or the similar condition of flow over a broad-crested weir. Breach growth is either parameterised by shape and time functions (e.g. the DAMBRK model of Fread (1988); Walder and O'Connor (1997)) or by physically based sediment transport and mass movement rules (e.g. the BREACH model of Fread (1987), the BEED model of Singh *et al.* (1988) and the ERODE model of Marche *et al.* (2006)). While the hydraulic characterisation of these models is probably appropriate, they are hindered by the difficulty in characterising breach growth, especially

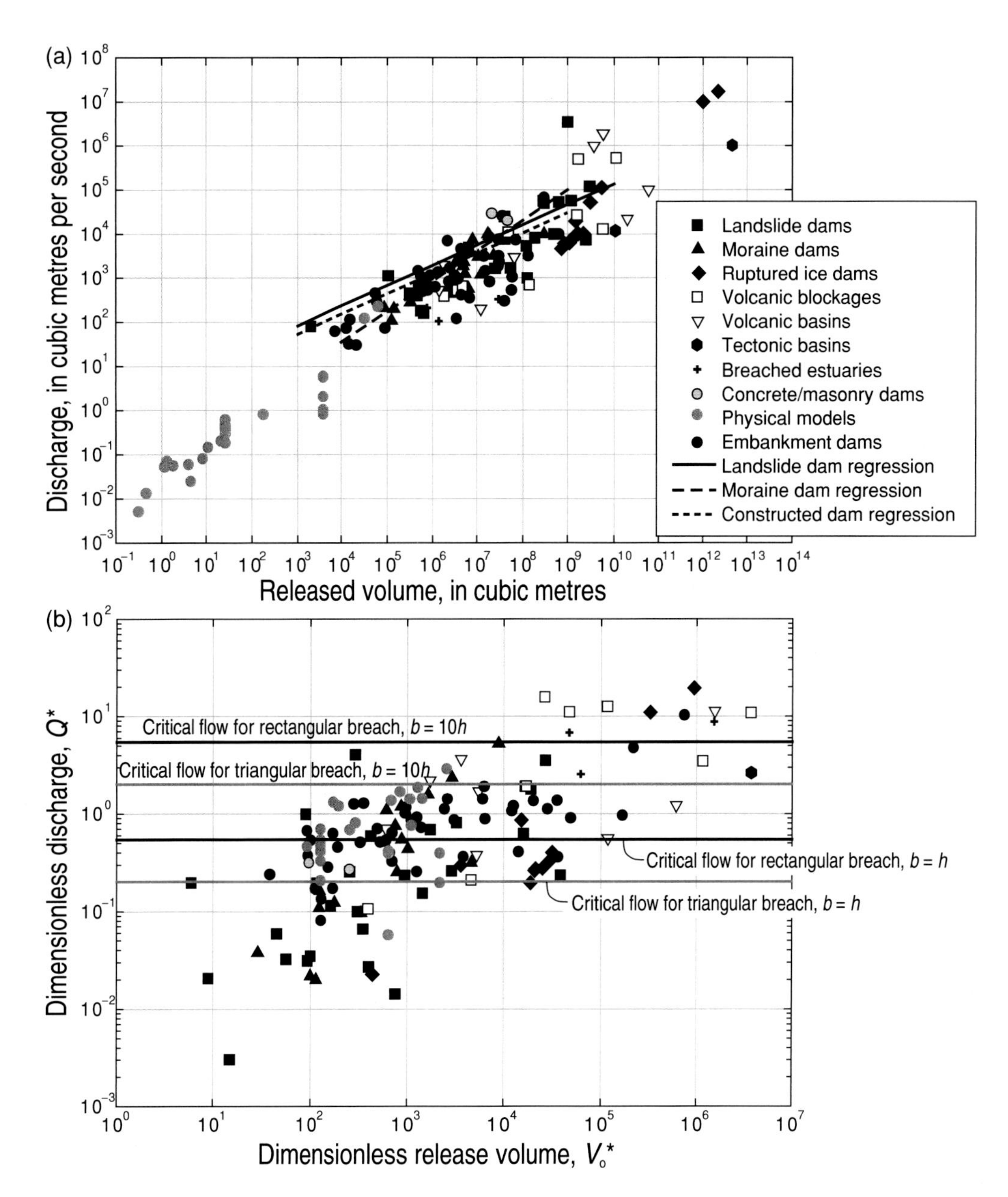

Figure 8.5. Plots of data of Table 8.1, including dam failures of all types (natural, constructed and scale models) for which reliable discharge and volume estimates are available. (a) Peak discharge, Q, versus release volume, V_o. Also shown are regression equations (Table 8.2) for various categories of dam failure over their derived ranges. (Regression equations from Cenderelli (2000) on basis of a more limited compilation than Table 8.1.) (b) Dimensionless peak discharge, $Q^* = Q/g^{1/2}h^{5/2}$, versus dimensionless volume, $V_o^* = V_o/h^3$. Also shown are maximum discharges associated with critical flow for triangular and rectangular breach cross-sections with widths equal to and a factor of 10 wider than breach depth. Other plausible breach cross-sections will have intermediate values.

the non-linearity of sediment transport, head-cut erosion, and mass movement processes enlarging the outlet (Walder and O'Connor, 1997; Wahl, 2004; Hanson *et al.*, 2005; Cencetti *et al.*, 2006). Parameterised approaches, such as Froehlich (1987), Webby and Jennings (1994) and Webby (1996), rely on dimensional analysis as a basis for empirical predictive equations of peak outflow of natural and constructed dam failures and currently are the most successful predictors of peak outflow from breached earth-fill dams (Wahl, 2004). Conditions at the breach nearly always control outflow discharge but analyses focused on outlet conditions may overestimate peak discharges if (1) downstream 'tailwater' (backwater from downstream ponding) slows outflow, or (2) the geometry of the water body inhibits

Table 8.2. Regression equations for rock-material dam failures

	Q	r^2	s	n
Landslide dams	$3.4V_o^{0.46}$	0.73	0.49	19
Moraine dams	$0.06V_o^{0.69}$	0.63	0.53	10
Constructed dams	$2.2V_o^{0.46}$	0.60	0.49	35

Note: Q, peak discharge in cubic metres per second; V_o, release volume in cubic metres; r^2, coefficient of correlation; s, standard error of the estimate; n, sample size.
Source: From Cenderelli (2000) and based on a slightly different compilation than that of Table 8.1.

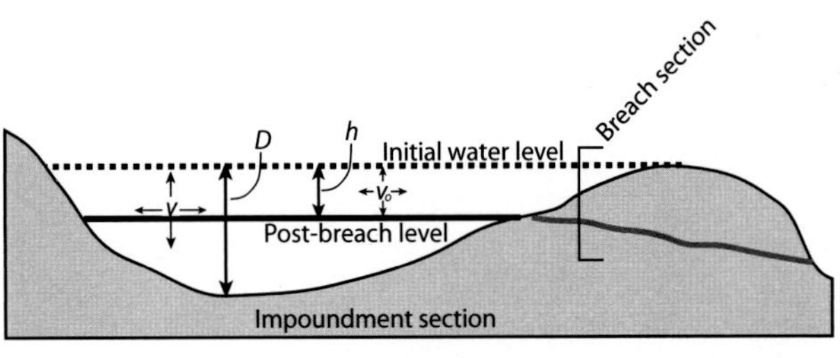

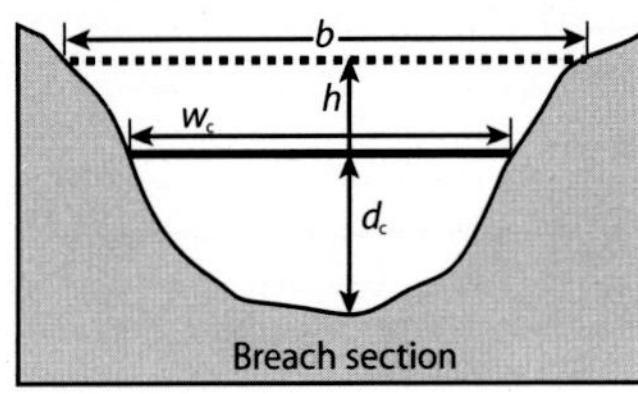

Figure 8.6. Definition diagram for impoundment and breach geometry. V is the total impoundment volume; V_o is the volume released during the breach event; D is the maximum impoundment depth, h is the change in impoundment level during the breach event, and is approximately equivalent to the maximum possible specific energy of flow through the breach; b is the breach width at the maximum impoundment elevation, w_c is the breach width at flow stage d_c associated with critical flow ($v = (gd)^{1/2}$) through the breach section with specific energy h.

movement of water to the outlet, either because of impoundment length or by internal lake constrictions limiting flow to the breach.

An important consideration, especially for breaches of natural rock-material dams, is Walder and O'Connor's (1997) distinction between 'large' and 'small' impoundments. 'Large' impoundments are those in which the breach develops fully before significant drawdown, either because the lake volume is very large relative to the breach depth or because of rapid breach erosion rates. In these cases, peak outflow is closely approximated by critical flow through the final breach geometry with a specific energy equivalent to the original water surface of the impounded water body

Table 8.3. Critical flow equations for different breach cross-section geometries

Rectangular	$Q = g^{1/2} w_c d_c^{3/2} = g^{1/2} \left(\frac{2}{3}\right)^{3/2} bh^{3/2}$
Triangular	$Q = g^{1/2} \left(\frac{1}{2}\right)^{3/2} w_c d_c^{3/2} = g^{1/2} \left(\frac{2}{3}\right)^{3/2} \left(\frac{4}{5}\right)^{5/2} bh^{3/2}$
Parabolic	$Q = g^{1/2} \left(\frac{8}{27}\right)^{3/2} w_c d_c^{3/2} = g^{1/2} \left(\frac{2}{9}\right)^{3/2} \left(\frac{3}{4}\right)^{3/2} bh^{3/2}$

Note: g, gravitational acceleration; d_c, critical depth for specific energy; h, drop in impoundment level during the release; w_c, top width of flow at critical depth d_c; b, breach width at pre-breach impoundment level (Figure 8.6).
Source: Derived from King (1954, pp. 8.8–8.11).

(unless hindered by tailwater or reservoir routing effects). For cross-section shapes where width is roughly equal to depth,

$$Q \approx g^{1/2} h^{5/2}, \qquad (8.1)$$

where Q is the peak outflow discharge, g is gravitational acceleration, and h is the difference between the surface elevation of the impounded water body and the bottom of the breach at the outflow point, generally equivalent to the drop in water surface during the course of the outflow event (Figure 8.6). For wider breaches, peak discharge will be larger by a factor approximating b/h, where b is the breach width. Table 8.3 lists specific equations for critical flow through different breach geometries. The strong dependence on breach depth clarifies why deep lakes can produce such large peak discharges. Tectonic and volcanic basins can nearly always be considered 'large' basins, and consequently their peak discharges, where independently estimated, mostly range between 1 and 20 $g^{1/2}h^{5/2}$ (Figure 8.5b).

'Small' impoundments, as classified by Walder and O'Connor (1997), are those that experience significant water-surface drawdown during breach development and breach erosion rate substantially controls peak discharge. In these cases, the peak discharge generally occurs prior to full development of the breach. This is commonly the case for landslide dams, which generally block high-relief valleys and produce deep lakes with relatively small volumes (e.g. Davies *et al.*, 2007).

The distinction between 'large' and 'small' impoundments is evident on plots of dimensionless lake volume, $V_o^* = V_o/h^3$, versus dimensionless peak discharge, $Q^* = Q/(g^{1/2}h^{5/2})$, where V_o is the outflow volume (Figure 8.5b). For $V_o^* > 10^5$, dimensionless peak discharges chiefly plot between 1 and 20, consistent with critical flow through a fully developed breach as described in Equation 8.1 and Table 8.3. Most tectonic and volcanic basins as well as several volcanic blockages are in this category, as are the large late Pleistocene Missoula and

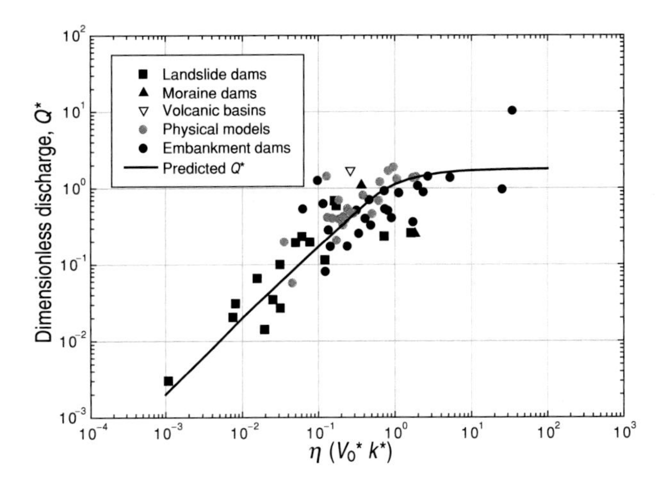

Figure 8.7. Dimensionless peak discharge , $Q^* = Q/g^{1/2}h^{5/2}$, versus the dimensionless parameter η, the product of dimensionless volume, $V_o^* = V_o/h^3$ and dimensionless erosion rate, $k^* = k/(g^{1/2}h^{1/2})$ for cases in Table 8.1 where there are reliable observations of discharge, geometry and erosion rate. Also shown is the theoretical prediction for the case of an impounded waterbody with constant-slope sides and a flat bottom being drained by a constantly growing trapezoidal breach channel with side slopes of 35°, and a bottom width that is 2.5 times the breach depth. The asymptotic relations for this prediction are given by Equations (8.5) and (8.6). The outlier Q^* value of ~10 is for the Prospect embankment dam (Table 8.1), which had a breach width about 20 times the breach depth (Wahl, 1998). (Modified from Walder and O'Connor (1997).)

Kuray ice dams. Additionally, all breaches involving V_o^* values greater than 10^4 had discharges exceeding that of critical flow for a triangular breach with $b = h$. For smaller V_o^*, particularly those less than 10^4, most dimensionless discharges are less than 1, indicating that breach erosion rate was significant factor in controlling outflow rates. This is the case for most breached landslide dams and moraine-rimmed lakes.

The distinction between 'large' and 'small' impoundments is refined by considering plausible breach erosion rates. From dimensional analysis, Walder and O'Connor (1997) define the parameter η, the product of dimensionless volume, V_o^*, and dimensionless erosion rate, k^*:

$$\eta = V_o^* k^*, \tag{8.2}$$

where $k^* = k/(g^{1/2}h^{1/2})$ and k is the vertical erosion rate during breaching. The strong relation between η and Q^* is illustrated in Figure 8.7 for the few natural and constructed dams for which breach erosion rate was independently estimated, and shows the two asymptotic functional relations between Q^* and η depending on whether $\eta < \sim 0.6$ or $\eta > \sim 1$ (Walder and O'Connor, 1997). For $\eta < \sim 0.6$ (a 'small' impoundment),

$$Q^* = \alpha \eta^{\beta}, \tag{8.3}$$

where α depends weakly on breach shape and impoundment hypsometry, and β varies slightly with impoundment hypsometry. For the 'large' impoundment situation of $\eta > \sim 1$,

$$Q^* \approx \gamma, \tag{8.4}$$

where γ, following Equation (8.1), is typically between 1 and 2, depending primarily on the ratio b/h.

Following Walder and O'Connor (1997), these relations can be cast into dimensional form and solved to form 'benchmark predictive equations' to estimate peak discharge. For a typical situation where the breach geometry is trapezoidal with 35° side slopes and ratio of the bottom width to breach depth remains a constant value of 2.5, impoundment volume above the breach bottom varies as a function of h^2, and inflow to the impoundment is small (Figure 8.7):

$$Q = 1.42(g^{1/2}h^{5/2})^{0.06}\left(\frac{kV_o}{h}\right)^{0.94}, \quad \text{for } \eta < \sim 0.6, \tag{8.5}$$

$$Q = 1.79(g^{1/2}h^{5/2}), \quad \text{for } \eta > \sim 1. \tag{8.6}$$

These equations can be modified to reflect specific conditions of lake hypsometry, breach geometry and inflow conditions, but Walder and O'Connor (1997) demonstrated that reasonable ranges of parameters for these conditions minimally affect predicted peak discharges. The coefficients in Equations (8.5) and (8.6) are slightly different than Equations (20a) and (20b) of Walder and O'Connor (1997) owing to updated numerical techniques for determining the asymptotic cases (J. S. Walder, personal communication, 2006). These relations should be universal for gravity-driven free-surface flow of viscous liquids since they are derived from only geometry and energy considerations. Breach geometry, especially width-to-depth ratios, may be systematically different in different gravitational fields because of gravitational control on mass movements, which is commonly the process by which breach channels widen.

Most landslide dams and moraines impound 'small' lakes, either defined as $V_o^* < 10^4$ (80% of breached landslide dams and 95% of moraine-dammed lakes listed in Table 8.1), or $\eta < \sim 0.6$ (85% of landslide dams with known breach rates) (Figure 8.8). However, ice dams, volcanic blockages, and tectonic and volcanic basins all have median V_o^* values approaching or surpassing 10^4; especially tectonic and volcanic basins for which V_o^* exceeds 10^5 in many cases (Figure 8.8a). For such large basins, peak outflow is primarily governed by breach geometry according to the relations in Table 8.3, although reservoir routing

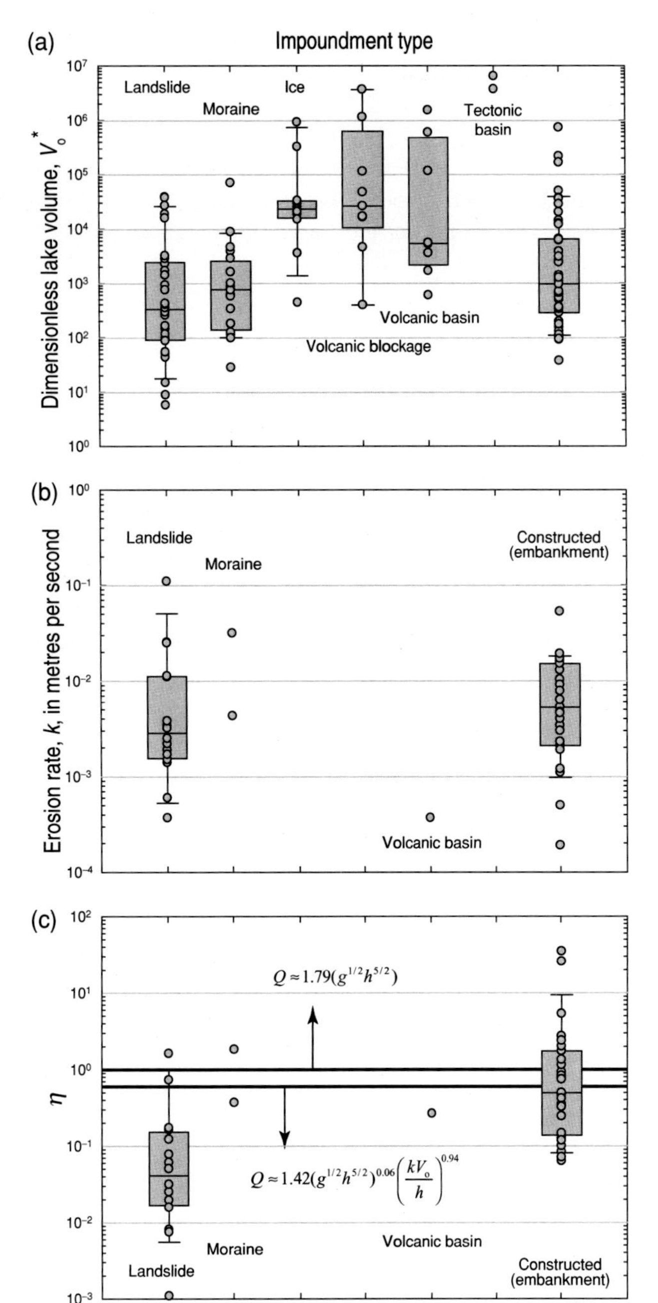

Figure 8.8. Box plots showing distribution of impoundment geometry and breaching characteristics for data of Table 8.1, categorised by impoundment type. Boxes encompass ranges between 25th and 75th percentiles. Whiskers denote 10th and 90th percentiles of data range. Solid lines within boxes are median values. Data from Table 8.1. (a) Distribution of dimensionless volume, $V_o^* = V_o/h^3$, by impoundment type. (b) Distribution of vertical erosion rate, k, for dam failures with observations of breach erosion rates. (c) Distribution of dimensionless parameter η, the product of dimensionless volume, $V_o^* = V_o/h^3$, and dimensionless erosion rate, $k^* = k/(g^{1/2}h^{1/2})$. Also shown are the fields where the asymptotic model relations (Equations (8.5) and (8.6)) are likely to be close predictors for peak discharge.

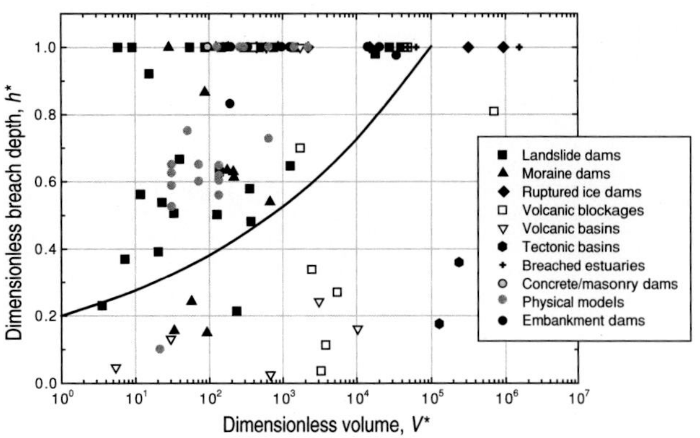

Figure 8.9. Dimensionless breach depth, $h^* = h/D$, versus dimensionless volume, $V^* = V/h^3$, for natural and constructed dam failures. The curve (Equation (8.7)) delineates an approximate minimum breach depth for breached landslide dams on the basis of data of Table 8.1.

effects or geometric controls (as well as downstream ponding) may reduce peak outflow at the breach.

In general, peak discharge depends on a combination of impoundment volume released (V_o), the drop in impounded water surface during the release (h), and the vertical breach erosion rate (k). The released volume depends on the bathymetry of the impoundment and the degree to which the blockage or basin divide is incised during outflow. For constructed embankment (earth-fill) dams, breaching nearly always results in erosion to an elevation near the previous river level, resulting in outflow of the entire impoundment volume. For natural dams and basins, breaching may erode only partly through the blockage, erode to the base of the blockage, or in much rarer cases, erode deeper than the blockage. Because of the exponential dependence of peak outflow on h, predicting discharge on the basis of the techniques outlined above 'begs the big question of "how deep [a dam or basin divide] will erode if a breach does form"' (Webby, 1996).

Less effort has been applied to predicting breach depth because most constructed dams breach to their bases, but it is an important factor in predicting downstream consequences for breached natural rock and debris dams because a high percentage erode only a fraction of the maximum impoundment depth (Table 8.1, Figure 8.9). For tectonic and volcanic basins, geological conditions typically will control breach depth. Resistant or hard units will halt erosion, generally at levels substantially above the impoundment bottom, but for landslide dams, volcaniclastic dams and moraine-dammed lakes, which are formed of unconsolidated debris, the character of the blockage materials is important and the interplay between outflow rates and breach deepening, widening and armouring will dictate the rate of breach growth and ultimate breach geometry (e.g. Clague and Evans, 1992). These interactions are parameterised by geotechnical breach erosion models such as BREACH (Fread, 1987), BEED (Singh *et al.*, 1988), ERODE (Marche *et al.*, 2006), which primarily aim to predict breach erosion rate and evolution of breach geometry, but can estimate erosion depth if the computational scheme allows for halting of vertical erosion. Although these breach erosion models are straightforward to implement, reliability is hindered by (1) uncertainty and diversity in materials forming natural blockages (e.g. Casagli *et al.*, 2003); (2) the variety and complex coupling of mass movement and fluvial erosion processes enlarging the breach channels (e.g. Walder and O'Connor, 1997); and (3) the sensitivity of predictions to the parameterisations (such as bedload transport formula) chosen to simulate these processes (e.g. Cencetti *et al.*, 2006).

In a manner conceptually similar to the Dimensionless Blockage Index of Ermini and Casagli (2003) for predicting landslide-dam stability, Webby and Jennings (1994) proposed a parametric approach to predicting breach size premised on the inference that the total volume of material eroded from a breach will be a function of the reservoir outflow volume (V_o) and the breach depth relative to the original impoundment level (h). Webby and Jennings used this to predict a breach depth of 10 to 20 m for a 50 m deep landslide-dammed lake that ultimately failed to a depth of 18.5 m. A similar approach is to consider the relation between dimensionless breach depth $h^* = h/D$, where D is the maximum depth of the impoundment (typically the water depth at the blockage for landslide dams), and the dimensionless impoundment volume, $V^* = V/D^3$, where V is the total volume of the impounded water body. Figure 8.9 shows that for nearly all values of V^*, h^* attains values ranging up to 1, indicating that erosion of the blockage to the elevation of the bottom of the impounded water body is plausible for all geometries; and that complete erosion is nearly always the case for breached ice dams and constructed dams. However, for landslide dams in particular, lower V^* result in more cases of $h^* \ll 1$. The minimum envelope line enclosing all observations of landslide-dam breaches except one is a positive function of V^* and could be used to provide a prediction of *minimum* breach depth for breaches of landslide dams such that

$$h^* > 0.2\left(V^*\right)^{0.14}, \quad \text{for } 10^0 < V^* < 10^5. \tag{8.7}$$

In dimensional terms, this equates to

$$h > 0.2V^{0.14}D^{0.58}, \quad \text{for } 10^0 < V/D^3 < 10^5. \tag{8.8}$$

These relations are consistent with the premise of Webby and Jennings (1994) that the volume eroded (as reflected in ultimate breach depth) depends partly on the volume of water available to entrain blockage materials but the strong dependence on D likely owes to the importance of

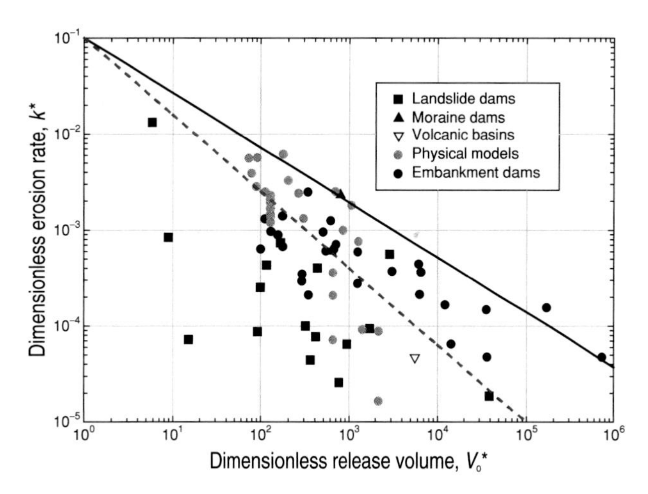

Figure 8.10. Dimensionless erosion rate, $k^* = k/(g^{1/2}h^{1/2})$, versus dimensionless volume, $V_o^* = V_o/h^3$. Also shown are approximate limiting lines for landslide dams (Equation (8.9)) and all dams composed of rock materials (Equation (8.10)). (Data from Table 8.1).

gravitational forcing on breach erosion processes of sediment entrainment and mass movement. Similarly, breach width also seems to scale with total flow volume for some types of natural and constructed dam failures (Stretch and Parkinson, 2006).

The remaining critical parameter controlling peak discharge from natural-dam failures, especially for impoundments where $\eta < \sim 0.6$, is vertical erosion rate, k. For 'small' impoundments, Q is essentially a linear function of k (Equation (8.5)) but observed vertical erosion rates for landslide dams vary 300-fold, from 3.7×10^{-4} to $1.1 \times 10^{-1}\,\text{m}\,\text{s}^{-1}$ (Figure 8.8b). A compounding challenge is that empirical data for vertical erosion rates derive from a variety of types of observations over different time periods (Table 8.1), and likely underestimate maximum instantaneous rates. Most reported k values for landslide dams, however, are between $0.001\,\text{m}\,\text{s}^{-1}$ and $0.02\,\text{m}\,\text{s}^{-1}$, which may be a reasonable range to use in conjunction with Equation (8.5) to predict peak outflows in many situations. Plausible ranges for vertical erosion rates for other types of rock debris barriers are difficult to specify because of the few observations, but may closely resemble the very similar ranges encompassed by landslide and constructed embankment dams (Figure 8.8b).

An analysis similar to that for estimating minimum breach depths shows that the *maximum* dimensionless vertical erosion rate, $k^* = k/(g^{1/2}h^{1/2})$, is inversely related to the dimensionless release volume, $V_o^* = V_o/h^3$ (Figure 8.10). For landslide dams, most observations of vertical erosion are bound by

$$k^* < 0.1\left(V_o^*\right)^{-0.57}, \quad \text{for } 10^0 < V_o^* < 10^5, \tag{8.9}$$

and for all natural and constructed rock-material dams,

$$k^* < 0.1\left(V_o^*\right)^{-0.80}, \quad \text{for } 10^0 < V_o^* < 10^6. \qquad (8.10)$$

In dimensional terms, for landslide dams,

$$k < 0.1 g^{1/2} V_o^{-0.57} h^{2.21}, \quad \text{for } 10^0 < V_o/h^3 < 10^5, \qquad (8.11)$$

and for all natural and constructed rock-material dams,

$$k < 0.1 g^{1/2} V_o^{-0.8} h^{2.90}, \quad \text{for } 10^0 < V_o/h^3 < 10^6. \qquad (8.12)$$

As with the case for predicting limits on erosion depth, the strong dependence of maximum erosion rate on h is because the potential-energy gradient across the barrier ultimately drives breach erosion processes.

8.4.3 Downstream flood behaviour

As described above, blockage stability and, if breached, the flow hydrograph at the outlet depends on many factors, chiefly the geometry of the blockage and impounded water body, ultimate breach depth, and in many cases the vertical breach erosion rate. Likewise, downstream flood behaviour depends on several factors, including the ones common to all types of floods such as the hydrograph shape and channel and valley geometries. In addition, many outsized floods from breaches of natural rock-material dams erode and deposit substantial volumes of sediment, which can significantly affect flow behaviour and geomorphological consequences.

Many breached landslide and moraine dams produce downstream debris flows, mainly by floodwaters entraining material from the breached landslide mass or moraine but also by incorporation of downstream bank and bed materials (e.g. Lliboutry *et al.*, 1977; Eisbacher and Clague, 1984; King *et al.*, 1989; Clague and Evans, 1994; Gallino and Pierson, 1985; Schuster, 2000; Huggel *et al.*, 2004; McKillop and Clague, 2007b). This has been documented best for outbursts from moraine basins, where flow bulking can increase peak discharge (comprising water and entrained debris) by an order of magnitude or more (Figure 8.11a) as well as increasing total flow volume (O'Connor *et al.*, 2001). An extreme example was the August 1985 Dig Tsho outburst in Nepal (Table 8.1), where 5.1×10^6 m^3 of water entrained $\sim 0.9 \times 10^6$ m^3 of sediment from the bounding moraine, and ultimately incorporated and redeposited $\sim 3.3 \times 10^6$ m^3 of sediment in the 40 km downstream (Vuichard and Zimmermann, 1987). Similarly, peak discharge from a breach of a moraine-rimmed lake at the foot of Collier Glacier, Oregon, grew from less than 140 ± 30 m^3 s^{-1} at the outlet to more than 500 m^3 s^{-1} 1 kilometre downstream (O'Connor *et al.*, 2001). Outbursts from crater lakes and volcanic impoundments, especially

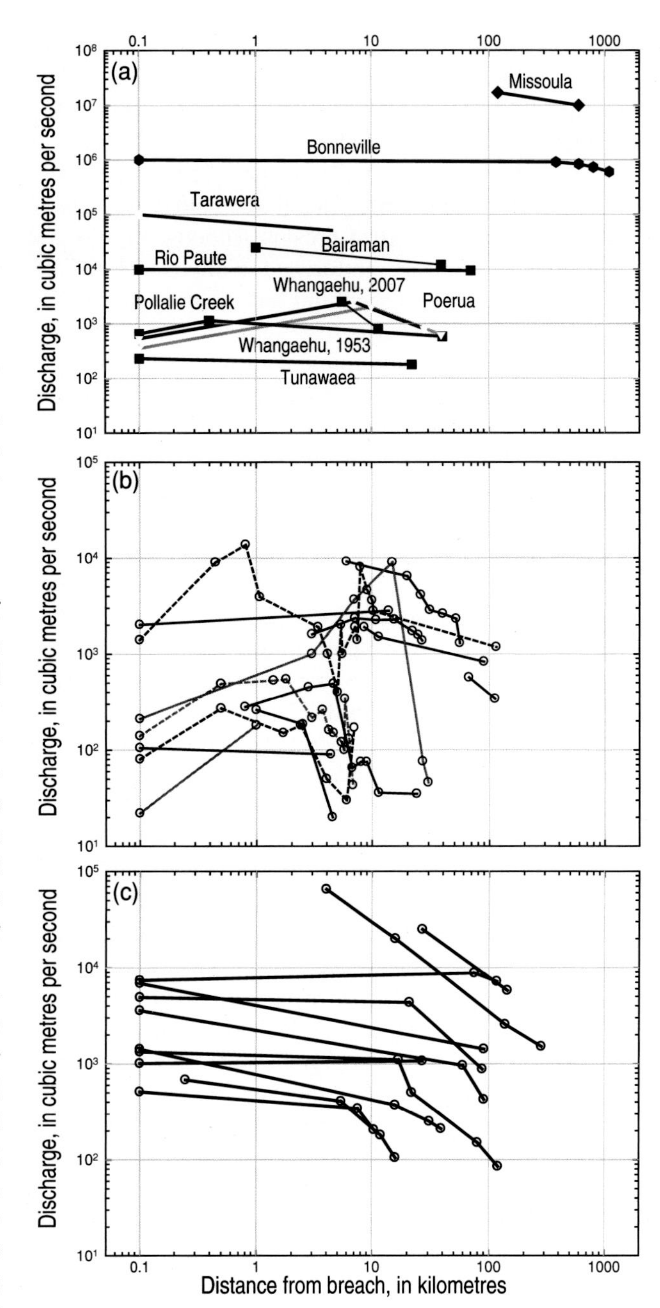

Figure 8.11. Outburst flood attenuation shown as plots of peak discharge, Q, versus distance from the breach. Estimates for breach outflow plotted at 0.1 km for portrayal on log axes. (a) Non-moraine natural dam failures. (Data from sources in Table 8.1.) (b) Floods from moraine-rimmed lakes, showing rapid downstream increases and decreases in peak discharge, chiefly owing to entrainment and deposition. (Data primarily from O'Connor *et al.* (2001).) (c) Floods from breached embankment dams. (Data chiefly from sources of Table 8.1 and Costa (1988).)

on steep stratovolcanoes or on tephra-mantled or eruption-devastated landscapes, can also bulk up into debris flows (e.g. Scott, 1988; Capra and Macías, 2002; Manville, 2004). These engorged volcanic lake breakouts have formed some of the largest and most lethal floods from natural dam breaches (e.g. Manville, 2004). Systematic analysis is difficult because of the few descriptions of flow rheology but it is clear that channel slopes greater than 0.15 are required to sustain debris flows (Clague and Evans, 1994; O'Connor *et al.*, 2001; McKillop and Clague, 2007b) except in the case of large outbursts in volcanic terrains (e.g. Scott, 1988). It is suspected that outburst floods from 'small' impoundments – with either small V_o^* or η values – are most likely to evolve into debris flows. These are typically in steep environments and involve tall dams – both conditions conducive to debris flow initiation. Plots of downstream peak-discharge changes (Figure 8.11) show that most flow bulking is within the first 10 km of the breach, consistent with entrainment requiring steep channel slopes found in basin headwaters where many landslide and moraine dams form.

Flood behaviour downstream of the breach depends on the flow rheology, outflow hydrograph and downstream channel and valley geometry. Outburst floods that evolve into debris flows near the breach, resulting in increased peak discharges and flow volumes in steep and confined channel sections, will deposit in unconfined reaches and where slopes decrease to less than ~0.15–0.2 (O'Connor *et al.*, 2001; McKillop and Clague, 2007b). Sediment deposition may sequester all or part of the initial water release, resulting in smaller downstream flood volumes compared with a clear-water flood and perhaps more rapid rates of flow attenuation (e.g. O'Connor *et al.*, 2001). In most instances, however, floods that become debris flows in their upper reaches will continue downstream on lower gradient reaches as sediment-laden water floods with total flood volumes about the same as the release volume (e.g. Manville, 2004).

Downstream attenuation of floods from breached rock-material dams will control the distribution and magnitude of resulting flood features. Attenuation characteristics depend mostly on the initial outflow hydrograph and downstream channel geometry. These are the primary inputs into flood routing models such as FLDWAV (Fread, 1993), HEC-RAS (Brunner, 2002), and the routing model applied by Risley *et al.* (2006) to hypothetical breaching of the Usoi landslide dam. In general, floods of small volume and short duration (and time-to-peak) will attenuate more rapidly due to downstream channel storage. Floods of large volume and long duration will not attenuate as rapidly because downstream channel and valley storage will be mostly filled prior to passage of the peak. The relative duration of a flood hydrograph relates directly to the quantity V_o^*. For large V_o^* values, outflow will be limited to the critical flow condition for the fully developed breach and will be sustained at near peak levels until significant drawdown of the impounded water body, leading to less rapid attenuation downstream. Additionally, a given outflow hydrograph will attenuate more rapidly downstream where there is substantial channel and valley storage in comparison to the same outflow hydrograph flowing through a confined valley or canyon environment.

Both of these latter controls can be illustrated for large outburst floods by comparing the downstream behaviour of the largest Missoula Flood ($V_o^* = 1.5 \times 10^4$) and the late Pleistocene Bonneville Flood ($V_o^* = 3.8 \times 10^6$). While the Bonneville Flood had a volume more than twice as great as maximum ice-dammed glacial Lake Missoula, its peak discharge was about 1/20 of the maximum Missoula Flood near the dam failure location. The difference in peak outflow owes to the much greater breach depth for the Missoula Flood – about 525 m compared to 108 m for the breach developed during the Bonneville Flood. The Bonneville Flood lasted for weeks or months (O'Connor, 1993), while the largest Missoula Floods lasted for only days (O'Connor and Baker, 1992). As a result, the Missoula Flood attenuated more rapidly; peak discharge diminished by more than 40% in the 600 km between the breach area and the Columbia River gorge, whereas the Bonneville Flood attenuated by a smaller percentage over twice the distance (Figure 8.11a). The difference in attenuation rate was also aided by the huge volume of storage along the Missoula Flood route. The Pasco basin, between the ice dam and Columbia River gorge, contained at maximum inundation more than 50% of the maximum volume of glacial Lake Missoula and probably substantially diminished peak discharge downstream (Denlinger and O'Connell, 2003), while the Bonneville Flood was little affected by flow storage as it coursed through the Snake River canyon.

8.5 Erosional and depositional features from natural dam failures

An outstanding aspect of extreme floods and debris flows is the size of landforms they produce and the size of material transported, as well as the anastomosing channel complexes formed where prior channels and valleys were overwhelmed by flooding (Plate 21). Such features are summarised in Chapters 2 and 3 of this volume. More specifically, Costa (1984) and (Scott, 1988) have described debris-flow and lahar landforms and deposits. Similarly, water-lain megaflood features have been described by many, including Bretz (1923, 1924, 1928, 1932), Malde (1968), Scott and Gravlee (1968), Baker (1973), Blair (1987, 2001), O'Connor (1993), Carling (1996), Cenderelli and Wohl, (1998), Benito and O'Connor (2003), Herget (2005) and

Baker (this volume Chapter 5). The scale, distribution and morphology of these features depend on flow type, magnitude and attenuation rates, as well as channel morphology and materials.

In general, the size and elevation of flood landforms scale with cross-sectional flow geometry, and clast sizes of deposited material scale with flow strength. Most assessments of megaflood features and their relation to flow are based on flow-strength measures such as shear stress and stream power. Shear stress, τ, is the tangential force applied to the bed per unit area, and for hydrostatic conditions,

$$\tau = \rho g d_f S, \qquad (8.13)$$

where ρ is the fluid density, d_f is flow depth, and S is the local flow energy gradient (which for steady and uniform flow corresponds to channel slope). Unit stream power ω (Bagnold, 1966, pp. 5–6) is the power developed (time rate of energy expenditure) per unit area of bed, and can be expressed as

$$\omega = \rho g d_f v S = \tau v, \qquad (8.14)$$

where v is flow velocity. These properties are best viewed as indices of geomorphological work, because for most natural streams only a portion of available mechanical energy is expended in accomplishing geomorphological work, with the remainder dissipated in other forms of energy loss. In addition, these formulations are for hydrostatic conditions; hence they ignore vertical and horizontal accelerations that can produce stresses on the same order as the hydrostatic forces (Iverson, 2006). Nevertheless, many researchers have shown that the volume and calibre of entrained sediment, as well as the distribution of erosional features, broadly relate to the distribution of shear stress and stream power magnitudes (e.g. Baker, 1973; O'Connor, 1993; Benito, 1997; and Cenderelli and Wohl, 2003).

Because of the large flow depths and locally steep energy gradients of floods from natural dam failures, such floods generate shear stresses and stream powers greatly exceeding those of meteorological floods (Baker and Costa, 1987). For example, the largest stream power values reconstructed from U.S. Geological Survey discharge measurements of meteorological floods – resulting from flash floods in steep basins – are on the order of $2 \times 10^4\,\mathrm{W\,m^{-2}}$ (Baker and Costa, 1987); whereas maximum stream power values exceeded $10^5\,\mathrm{W\,m^{-2}}$ for the Bonneville Flood (O'Connor, 1993), $10^5\,\mathrm{W\,m^{-2}}$ for the largest Missoula Flood (O'Connor and Waitt, 1995; Benito, 1997), and possibly $5 \times 10^6\,\mathrm{W\,m^{-2}}$ for Holocene floods from breached ice dams in the Tsangpo River gorge, Tibet (Montgomery *et al.*, 2004). Similarly, Cenderelli and Wohl (2003) concluded that floods generated by breached moraine dams produced stream power values several times greater than those from plausible meteorological floods affecting the same reaches.

Megaflood deposits commonly are distinguished by large constituent clasts (Plate 21a). Some terrestrial deposits from natural dam failures contain transported grains with intermediate diameters exceeding 10 m (Baker, 1973; Herget, 2005). Theoretical considerations and empirical studies (e.g. Baker and Ritter, 1975; Komar, 1987; Costa, 1983; Williams, 1983; O'Connor, 1993) demonstrate that tractive boulder entrainment scales with local flow strength. The empirical analyses by Williams (1983) and Costa (1983) indicate that entraining a 1 m intermediate-diameter boulder requires local stream power values of about $5 \times 10^2\,\mathrm{W\,m^{-2}}$. On the basis of reconstruction of local hydraulic conditions for the Bonneville Flood, O'Connor (1993) estimated that a 5 m boulder requires about $5 \times 10^3\,\mathrm{W\,m^{-2}}$ for transport, a value rarely attained by meteorological floods (Baker and Costa, 1987). While subject to many uncertainties (reviewed by Jacobson *et al.*, 2003), these types of flow competence relations have been used for many palaeohydraulic analyses of megafloods (e.g. Waythomas *et al.*, 1996; Manville *et al.*, 1999; Hodgson and Nairn, 2005).

Impressive erosional features are another hallmark of megafloods (Plate 21b). For many dam-breach floods, flow strength values are attained that exceed resistance thresholds of bedrock channels and valleys, leading to spectacularly eroded terrains. For large Pleistocene ice-breach and basin-breach floods in the US Pacific Northwest, threshold stream power values for eroding basalt were on the order of $10^3\,\mathrm{W\,m^{-2}}$ (Rathburn, 1989; O'Connor, 1993; Benito, 1997). Another threshold plausibly attained by many megafloods is conditions conducive to cavitation, an intensely energetic and potentially erosive process enabled by deep and high-velocity flows (Whipple *et al.*, 2000), although definitive field evidence for this process is lacking.

Flood duration can control resulting megaflood features and is commonly a differentiating factor between meteorological floods with high local stream power and shear stress values and floods resulting from breached rock-material dams. Duration controls the total energy applied to the landscape (Costa and O'Connor, 1995) and is important where landform-sculpting processes require surpassing flow-strength thresholds, such as that for bedrock erosion (Benito, 1997). Duration relates directly to dimensionless lake volume, V_o^*, hence megafloods with high V_o^* values are most likely to leave exceptional features requiring high flow strength, such as bars composed of large boulders and bedrock channels or potholes. Duration may also be a factor in deposit morphology. Large, smooth-surfaced, and well-armoured deposits probably reflect prolonged formation times. This is the case for Bonneville

Flood bars ($V_o^* = 3.8 \times 10^6$) which are densely studded with well-rounded boulders (Plate 21a) but with no preserved gravel dunes, contrasting with the seemingly less-armoured Missoula Flood ($V_o^* = 1.5 \times 10^4$) and Kuray Flood ($V_o^* = 3.6 \times 10^3$) bars that are locally covered by extensive gravel current dune fields (Baker, 1973; Carling, 1996; Herget, 2005) that perhaps owe their preservation to rapid flow diminution.

8.6 Concluding remarks

Breached natural dams have produced the largest known floods on the Earth. The largest documented floods in terms of peak discharge have been ice-dam failures during glacial epochs, but immense floods have also come from valleys temporarily blocked by landslides and primary volcanic deposits and from basins formed by tectonism, volcanism and glacial moraines (Table 8.1). The largest (by volume) freshwater flood known in Earth history was the Bonneville Flood from an overtopped tectonic basin. The largest known freshwater peak discharge of the Holocene resulted from rapid emptying of the caldera at Aniakchak Volcano (Waythomas *et al.*, 1996).

Megafloods from rock-material dams result from specific geological, topographic and climatic environments. Landslide dams and resulting floods are worldwide phenomena but are concentrated in mountainous regions where (1) relief promotes mass movements and (2) confined valleys provide effective dam sites (Costa and Schuster, 1988, 1991). Large floods and debris flows resulting directly or indirectly from volcanic processes, such as breaching of volcanically emplaced valley blockages and caldera and crater lakes, obviously are restricted to volcanic environments and are well documented in the Pacific Rim locations of Alaska, New Zealand, Mexico, Japan and the US Pacific Northwest. Floods from tectonic basins are mostly in mid-latitude semi-arid environments where epochs of greater effective moisture fill closed basins. Most breached moraine-rimmed lakes have formed and emptied in the last 150 years because of rapid alpine glacier retreat from maximum Little Ice Age positions (Clague and Evans, 2000).

The peak discharge, duration and volume of floods from breached rock-material dams are interrelated and depend chiefly on breach geometry and development rate and the size and geometry of the impounded water body. In general, the largest floods result from large water bodies breaching tall blockages, such as the Pleistocene outbursts of glacial Lake Missoula. Floods with large volumes relative to ultimate breach depth (high V_o^* values) or where breaches erode rapidly (high η values) generally will have peak discharges approximating critical flow through the final breach geometry. These discharges are generally in the range 1–20 $g^{1/2}h^{5/2}$ and hence are a strong function of breach depth. Many floods from tectonic and volcanic basins and volcanic and ice blockages are of this type, including most of the largest documented terrestrial floods. Floods from blockages where the impounded volume is small relative to breach depth ($V_o^* < 10^4$) or where erosion rates are low ($\eta \ll 1$) have peak discharges strongly dependent on breach erosion rate. This is the case for the majority of breached landslide dams and moraine-rimmed lakes (as well as most earth-fill dams).

Downstream flood behaviour, including discharge, flow type, and depositional and erosional features, depends on the breach development processes, the rate and duration of exiting flow, and interactions with the channel and valley. Breaches that result in the flow incorporating great sediment volumes, either from the blockage itself or from downstream channel boundaries, commonly evolve into debris flows (Schuster, 2000; Clague and Evans, 2000). These debris flows may have peak discharges several times the peak outflow at the breach and substantially increase the flow volume. Debris flows attenuate rapidly downstream by deposition, particularly where slopes diminish below 0.15. This behaviour is common for outbursts from moraine-rimmed and landslide-dammed impoundments. Water floods attenuate because of channel and valley storage. Large-volume floods with relatively low peak discharges (typically floods with high V_o^* values) will attenuate at a slower rate than small volume, high-peak discharge releases.

Erosional and depositional features left by large floods, including gigantic bars and eroded channel complexes, reflect the depth and breadth of inundation as well as the large forces applied by deep and high-velocity flow. Floods from natural dam failures can attain local stream power values of $\sim 10^{5-6}$ W m^{-2} – one-to-two orders of magnitude greater than those generated by the most intense meteorological floods – and are capable of eroding bedrock channels and transporting bedload clasts with diameters exceeding 10 metres. These features and deposits, whether preserved as landforms or in stratigraphic records, are the persistent legacy of great floods, the largest of which on Earth result from failure of natural dams.

Acknowledgements

Our work on natural dam failures has been aided by collaboration with John Costa, John Clague, Stephen Evans, Gordon Grant, Kyle House, Vern Manville, Joe Walder and Chris Waythomas. Reviewers Chris Waythomas, Glen Hess, Stephen Evans and Dan Cenderelli improved the manuscript. Partial support for this project was provided by the National Science Foundation award EAR-0617234.

Table 8.1. Compilation of natural, constructed and scale-model dam failures

River/Lake	Location	Date of failure	Trigger mechanism	Dam height (m) (max. impoundment depth)	Lake volume at failure (m^3)	Breach depth (m)	Volume released (m^3)	Peak discharge ($m^3 s^{-1}$)	Peak discharge determination	Breach erosion rate ($m s^{-1}$)	Breach erosion rate determination	Sources	Comments
Floods from landslide dams													
Tsatichu	Bhutan	10 Jul 2004	Overtopping?	100	5.50×10^6			5.90×10^3	Measured 35 km downstream			Dunning *et al.*, 2006	Lake volume and depth very approximate
Rio San Pedro	Chile	25 Jul 1960	Overtopping	40	2.50×10^9	40	2.50×10^9	7.45×10^3	Unknown, but complete hydrograph presented	3.70×10^{-4}	Largest dam completely cut through in 30 hours	Davis and Karzulovic (1963), Weischet (1963), Pena and Klohn (1989)	Davis and Karzulovic (1963) report new lake 26.5 m high before overtopping artificial outlet channel; blockage downstream for Lago Rinihue, but lake engulfed; V_o and V are for only the water impounded by blockage
Dadu River	China	11 Jun 1786	Overtopping	70	5.00×10^7	70	5.00×10^7					Dai *et al.* (2005)	Killed 100 000; earthquake-formed, breach triggered by aftershock; no independent assessment of Q
Dong River	China	May 1966	Overtopping	51	2.70×10^6	20	2.50×10^6	5.60×10^2	Unknown; 2 hr average	1.40×10^{-3}	Average for 4 hr duration of emptying	Li *et al.* (1986), Costa and Schuster (1991)	Complete breach in 4 hr; triggered by large rainfall inflow; breach depth estimated from Fig. 11; inflow $130 m^3 s^{-1}$
Yalong River	China	17 Jun 1967	Overtopping	175	6.80×10^8	88	6.40×10^8	5.30×10^4	Manning equation, 6 km downstream	1.90×10^{-3}	Average for 13 hr duration of emptying	Li *et al.* (1986), Chen *et al.* (1992)	Breach mostly developed in 5.5 hr, although Chen *et al.* modelling suggests full breach development in 2.5 hr
Rio Toro	Costa Rica	13 Jul 1992	Piping	52	5.00×10^5	12	4.50×10^5	4.00×10^2	Unknown			Mora *et al.* (1993)	Breached in a series of events; recorded data for 13 July event; breach drop estimated from Fig. 8.5; volume released estimated from given lake dimensions
Rio Paute	Ecuador	10 May 1993	Overtopping	83	2.10×10^8	40	1.85×10^8	8.25×10^3	Reservoir outflow; 5 min average	1.11×10^{-2}	For 6 hr, 'greater volume having been discharged'	Plaza Nieto and Zevallos, (1994), Canuti *et al.* (1994)	Partly manipulated overflow; intense breaching followed retrograde erosion of landslide
Rio Pisque	Ecuador	26 Jan 1990	Overtopping	45	3.60×10^6	30	2.50×10^6	4.80×10^2	Unknown	1.50×10^{-3}	30 m over 5.5 hr	Plaza-Nieto *et al.* (1990), Zevallos-Moreno (1993 written communication)	Breach rate $0.0024 m s^{-1}$ for 3.5 hr of most intense erosion; failed by headward erosion of overtopped constructed outlet channel; flood data for 'main flood'

Birehi Ganga River	India	25 Aug 1894	Overtopping	237	4.40×10^{8}	120	2.85×10^{8}	5.67×10^{4}	Reservoir outflow; 1 hr average	2.54×10^{-2}	90 m in 1 hr	Lubbock (1894), Strachey (1894), Malde (1968), Costa (1988), Costa and Schuster (1991)	Landslide volume from dimensions in Strachey (1894)
Issyk River	Kirgizstan	8 Jul 1963	Overtopping	55	1.26×10^{8}	55	1.26×10^{8}	1.00×10^{3}	Unknown	6.00×10^{-4}	Average for 24 hr duration	Gerasimov (1965), Schuster *et al.* (2002)	Retrograde erosion, triggered by overtopping by flood wave from upstream breached moraine dam
Tegermach River	Kirgizstan	18 Jun 1966	Piping	90	6.60×10^{6}	90	6.50×10^{6}	4.96×10^{3}	Chezy equation, unknown distance downstream	2.50×10^{-2}	Average for 1 hr duration	Glazyrin and Reyzvikh (1968), Costa and Schuster (1991)	Failed by piping, with development downstream sapping channels
Naranjo River (Colima)	Mexico	18.5 ka	Overtopping?	150	1.00×10^{9}	150	1.00×10^{9}	3.50×10^{6}	Superelevation of debris flow, 10 km downstream			Capra and Macias (2002)	Volcanic debris avalanche, filled canyon for 25 km; flow eroded $1.5 \cdot 10^{6}$ m^3 of material from debris avalanche
Poerua River	New Zealand	12 Oct 1999	Overtoppping	80	6.00×10^{6}	45	4.00×10^{6}	2.50×10^{3}	Multiple; conveyance, 1 d modelling, 5.5 km downstream, perhaps some bulking up			Hancox *et al.* (2005)	Landslide volume 10–15 million; Lake volume 5–7 million m^3
Tunawaea Stream	New Zealand	22 Jul 1992	Overtopping	50	9.00×10^{5}	18.5	6.30×10^{5}	1.60×10^{2}	Dam-break modelling fitted to gauge record	3.40×10^{-3}	18.5 m over 1.5 hrs; fitted modelling	Jennings *et al.* (1993), Webby and Jennings (1994)	Discharge for breakout only, river flow subtracted
Bairaman River	Papua New Guinea	12 Sep 1986	Overtopping	130	5.00×10^{7}	70	4.00×10^{7}	2.50×10^{4}	Conveyance, 1 km downstream	1.13×10^{-2}	70 m over 103 min	King *et al.* (1989)	Landslide volume eroded twice total water volume; human-induced overtopping
Indus River	Pakistan	Jun 1841	Overtopping	150	1.20×10^{9}	150	1.20×10^{9}	5.70×10^{4}	Estimate by Hewitt	1.70×10^{-3}	24 hr average	Hewitt (1968), Shroder *et al.* (1991)	Shroder *et al.* report 540 000 $m^3\,s^{-1}$, but value implausible; Hewitt reported drained in 24 hr; overtopping triggered by landslide generated wave
Rio Mantaro	Peru	28 Oct 1945	Overtopping	96.3	3.15×10^{8}	55.8	3.00×10^{8}	5.00×10^{4}	Manning equation applied by authors, 90 km downstream, based on data in Snow	2.20×10^{-3}	7 hr average	Snow (1964)	Lake volumes calculated from Snow (1964) measurements
Rio Mantaro	Peru	6 Jun 1974	Overtopping	170	6.70×10^{8}	107	5.00×10^{8}	1.00×10^{4}	Reservoir outflow	2.50×10^{-3}	12 hr average	Lee and Duncan (1975)	

(*cont.*)

Table 8.1. (cont.)

River/Lake	Location	Date of failure	Trigger mechanism	Dam height (m) (max. impoundment depth)	Lake volume at failure (m^3)	Breach depth (m)	Volume released (m^3)	Peak discharge ($m^3 s^{-1}$)	Peak discharge determination	Breach erosion rate ($m s^{-1}$)	Breach erosion rate determination	Sources	Comments
Ching-Shui	Taiwan	18 May 1951	Overtopping	217	1.57×10^8	200	1.20×10^8	5.36×10^3	'average discharge' drawdown?	3.20×10^{-3}	6 hr 70 m lake drawdown	Chang (1984), Hung (2000)	Also reported as Chin-Shui-Chi River; dam formed from two failures; breach depth from inspection of Figs. 3 and 4 of Hung (2000)
Ching-Shui	Taiwan	24 Aug 1979	Overtopping	90	4.00×10^7	90	4.00×10^7	7.78×10^3	Slope-area, unknown location			Chang (1984), Taiwan Council for Economic Development, 1993 written comm. to J. E. Costa	Also reported as Chin-Shui-Chi River
Yigong River	Tibet	10 Jun 2000	Overtopping	55.1	3.00×10^9	54	3.00×10^9	1.20×10^5	Gauged, 17 km downstream			Shang *et al.* (2003)	
Trib. of Granite Creek	USA, Alaska	10 Aug 1971	Overtopping	26	2.52×10^7	26	2.52×10^7	1.66×10^3	Slope-area, 11 km downstream			Lamke (1972), Costa (1988)	Lake volume estimated from reported dimensions
Foreman Creek	USA, California	4 Jan 1982	Unknown	7	2.00×10^3	7	2.00×10^3	8.00×10	Gauged?	1.10×10^{-1}	Average for 1 min period	Weber *et al.* (1986)	
Navarro River	USA, California	23 Mar 1995	Unknown	8.5	7.70×10^5	5.5	5.43×10^5	1.80×10^2	Gauge, ~20 km downstream			Sowma-Bawcom (1996)	Breach depth estimated from reported post-breach lake data
Polallie Creek	USA, Oregon	25 Dec 1980	Overtopping	10.6	1.05×10^5	10.6	1.05×10^5	1.13×10^3	Mean of slope-conveyance measurement range			Gallino and Pierson (1985)	Total debris flow discharge (850–1400 cm) from downstream (0.5 km) slope-conveyance
N. Fork Toutle River	USA, Washington	27 Aug 1980	Overtopping	9	3.50×10^5	9	3.10×10^5	4.50×10^2	Visual estimate at breach	3.80×10^{-3}	9 m over 40 min duration	Jennings *et al.* (1981), Costa and Schuster (1991)	
N. Fork Toutle River	USA, Washington	20 Jan 1992	Overtopping	4.5	2.50×10^6	4.5	2.47×10^6	4.77×10^2	Unknown			Costa and Schuster (1991)	Jackson Lake; triggered by inflow of North Fork Toutle River
Gros Ventre River	USA, Wyoming	18 May 1927	Overtopping	70	8.00×10^7	15	5.40×10^7	1.70×10^3	Gauged, 240 km downstream			Malde (1968)	
Floods from moraine dams													
Lugge Tsho	Bhutan	7 Oct 1994	Unknown	147	1.06×10^8	23	4.80×10^7					Richardson and Reynolds (2000), Watanbe and Rothacher (1996), Yamada *et al.* (2004)	
Lake Nostetuko	Canada, British Columbia	19 Jul 1983	Unknown			38.4	6.50×10^6	5.70×10^2	Gauged, 67 km downstream			Blown and Church (1985), Evans (1987), Clague and Evans (1994)	4 m wave triggered by ice avalanche

Queen Bess Lake	Canada, British Columbia	12 Aug 1997	Ice avalanche wave	60	2.00×10^7	9	6.50×10^6	4.00×10^3	Approx value from many methods at 1 km			Kershaw *et al.* (2005)	Total lake volume estimated from inspection of Fig. 3; 3500–4500 estimate for peak flow through breach (after overtopping wave)
Keppel Cove Tarn	UK	29 Oct 1927	Overtopping	10	1.24×10^5	10	1.24×10^5	1.08×10^2	Step-backwater; 0.1 km downstream			Carling and Glaister (1987)	Overtopping triggered by precipitation
Moraine Lake No. 13; Almatinka River	Kazahkstan	3 Aug 1977	Piping?	8.2	9.64×10^4	5.2	8.64×10^4	2.10×10^2	At breach; reservoir outflow?	4.30×10^{-3}	Average for 20 min duration	Yesenov and Degovets (1979)	Sapping and retrograde erosion; evolved into 11 000 $m^3 s^{-1}$ debris flow; breach data from Fig. 1 of Yesenov and Degovets
Dig Tsho	Nepal	4 Aug 1985	Ice avalanche wave	18	5.10×10^6	18	5.10×10^6	2.35×10^3	Step-backwater, 4 km downstream			Galay (1985), Vuichard and Zimmermann (1987), Cenderelli and Wohl (2003)	
Dudh Khosi Valley	Nepal	2 Sep 1977	Ice core melting?	30	4.90×10^6	30	4.90×10^6	1.90×10^3	Step-backwater, 8.6 km downstream			Buchroithner *et al.* (1982) Fushimi *et al.* (1985), Cenderelli and Wohl (2003)	
Tam Pokhari Glacier Lake	Nepal	3 Sep 1998	Ice avalanche wave	60	1.90×10^7	52	1.77×10^7	1.00×10^4	Gauged, 65 km downstream; minimum estimate because water above gauge			Dwivedi *et al.* (2000), Dwivedi (2006 written communication)	Breach enlargement by retrograde erosion
Laguna Jancaruish; Quebrada Los Cedros	Peru	20 Oct 1950	Ice avalanche wave	21	8.00×10^6	21	8.00×10^6	7.50×10^3	Downstream estimate, unknown method			Lliboutry *et al.* (1977)	Destroyed while outlet works under construction; assumed that the entire lake emptied
Midui Lake	Tibet	15 Jul 1988	Ice avalanche wave	31	6.40×10^6	19	5.40×10^6	1.25×10^3	Gauged? 1.8 km downstream	3.17×10^{-2}	'Bursting duration is up to 10 min'	Li and You (1992), Ding and Liu (1992)	
Boqu River	Tibet	11 Jul 1981	Piping	50		32	1.90×10^7	9.17×10^3	Manning equation; 6 km downstream			Xu (1988), Xu and Feng (1994)	Evolved into debris flow; maximum lake depth estimated from Fig. 3 of Xu and Feng
Demenhai Lake	Tibet	26 Sep 1964	Ice avalanche; 10 m wave	27	4.16×10^6	17	3.70×10^6	2.81×10^3	Debris flow estimate; 14 km downstream			Lü and Li (1986)	
Gelhapuco	Tibet	21 Sep 1964	Ice avalanche wave	41		41	2.34×10^7	3.26×10^3	Debris flow estimate; 30 km downstream			Xu and Feng (1994)	

(cont.)

Table 8.1. (cont.)

River/Lake	Location	Date of failure	Trigger mechanism	Dam height (m) (max. impoundment depth)	Lake volume at failure (m^3)	Breach depth (m)	Volume released (m^3)	Peak discharge ($m^3 s^{-1}$)	Peak discharge determination	Breach erosion rate ($m s^{-1}$)	Breach erosion rate determination	Sources	Comments
Qunbixiama-Cho	Tibet	10 Jul 1940	Ice avalanche wave	50		50	1.24×10^7	1.20×10^3	Unknown determination; 50 km downstream			Xu and Feng (1994)	
Sangwang-Cho	Tibet	10 Jul 1954	Ice avalanche wave	40		40	3.00×10^8	1.00×10^4	Unknown determination; 120 km downstream			Xu and Feng (1994)	
Longda-Cho	Tibet	25 Aug 1964	Unknown	22		22	1.08×10^7	3.10×10^3	Debris flow estimate; unknown location			Xu and Feng (1994)	
Diller Glacier; Wychus Creek	USA, Oregon	7 Sep 1970	Overtopping by excess precipitation	22.4	3.20×10^5	22.4	3.20×10^5	2.80×10^2	Debris flow estimate; 0.8 km downstream			O'Connor *et al.* (2001)	Debris flow
Collier Glacier; White Branch	USA, Oregon	Jul 1942	Overtopping by excess meltwater?	10	6.70×10^5	5.4	4.60×10^5	5.00×10^2	Debris flow estimates; 0.5 km and 1.4 km downstream			O'Connor *et al.* (2001)	Debris flow
East Bend Glacier; Crater Creek	USA, Oregon	7 Oct 1966	Ice avalanche?	18.1	3.40×10^5	4.4	1.40×10^5	2.00×10^2	Debris flow estimates; 0.5 km and 1.7 km downstream			O'Connor *et al.* (2001)	Debris flow (lake volume reported in O'Connor *et al.*, 2001, Table 8.2 incorrect)
Floods from breached ice dams													
('non-tunnel' events of Walder and Costa, 1996; J. S. Walder, personal communication)													
Chong Khumdan Glacier, Upper Shyok River	Pakistan	15 Aug 1929	Subglacial tunnelling, then collapse	150	1.50×10^9	150	1.50×10^9	1.95×10^4	Gauged, unknown distance downstream			Hewitt (1968, 1982)	
Altai Mountains; River Ob	Russia	Pleistocene	Unknown	650	6.07×10^{11}	650	1.00×10^{12}	1.00×10^7	Step-backwater			Herget (2005)	Supercedes Baker *et al.* (1993) Kuray flood analysis
Lake George	USA, Alaska	13 Jul 1958	Overtopping	48.8	2.20×10^9	48.8	2.20×10^9	1.02×10^4	Gauged, 27 km downstream			Hulsing (1981), Lipscomb (1989)	Maximum lake level measured on 13 July; peak flow 18 July
Lake George	USA, Alaska	26 Jun 1959	Overtopping	35.1	1.10×10^9	35.1	1.10×10^9	6.32×10^3	Gauged, 27 km downstream			Hulsing (1981), Lipscomb (1989)	
Lake George	USA, Alaska	12 Jul 1960	Overtopping	41.2	1.50×10^9	41.2	1.50×10^9	9.29×10^3	Gauged, 27 km downstream			Hulsing (1981), Lipscomb (1989)	
Lake George	USA, Alaska	20 Jul 1961	Overtopping	43.3	1.70×10^9	43.3	1.70×10^9	1.01×10^4	Gauged, 27 km downstream			Hulsing (1981), Lipscomb (1989)	
Lake George	USA, Alaska	26 Jun 1962	Overtopping	29.3	7.40×10^8	29.3	7.40×10^8	4.67×10^3	Gauged, 27 km downstream			Hulsing (1981), Lipscomb (1989)	

Lake George	USA, Alaska	26 Jun 1964	Overtopping	29.9	8.60×10^8	29.9	8.60×10^8	6.12×10^3	Gauged, 27 km downstream	Hulsing (1981), Lipscomb (1989)	
Lake George	USA, Alaska	8 Jul 1965	Overtopping	32	1.10×10^9	32	1.10×10^9	6.69×10^3	Gauged, 27 km downstream	Hulsing (1981), Lipscomb (1989)	
Russell Fiord	USA, Alaska	8 Oct 1986	Unknown	25.5	5.40×10^9	25.5	5.40×10^9	1.13×10^5	Reservoir outflow, 1 hr average	Mayo (1989), Trabant *et al.* (2003)	Peak discharge 6 hr after erosion begins
Russell Fiord	USA, Alaska	13 Aug 2002	Overtopping	14.9	3.10×10^9	14.9	3.10×10^9	5.20×10^4	Reservoir drawdown, 0.5 hr average	Trabant *et al.* (2003)	Peak discharge 21 hr after initial erosion of breach
Missoula Flood	USA, Idaho	Pleistocene	Unknown	525	2.20×10^{12}	525	2.20×10^{12}	1.70×10^7	Step-backwater	O'Connor and Baker (1992)	
Volcanogenic valley blockages											
Tadami River (Numaza-wako)	Japan	5 ka	Overtopping	100	1.70×10^9	70	1.60×10^9	2.70×10^4	Mean of 5 competence estimates 11.4–17.8 km downstream	Kataoka *et al.* (2008)	15 km of valley filled by ignimbrite
Magdalena River (El Chichon)	Mexico	26 May 1982	Overtopping	10	4.80×10^7	10	4.80×10^7	1.10×10^4	Calibrated dam-break modelling	Macias *et al.* (2004)	Dammed 10 m above previous level by hot pyroclastic flow; lake emptied in 1–2 hr
Tarawera River	New Zealand	AD 1315	Unknown	118	4.00×10^9	40	1.70×10^9	5.00×10^5	Competence	Hodgson and Nairn (2005)	Dammed 30 m above previous level by pyroclastic flow; present lake depth from http://www.ebop.govt.nz/Water/Lakes/Lake-Statistical-Information.asp
Tarawera River	New Zealand	1 Nov 1904	Unknown	91	2.44×10^9	3.35	1.40×10^8	7.00×10^2	Gauged	Hodgson and Nairn (2005)	Dammed 12.8 m by reworked pyroclastic material; released volume estimated by multiplying reported lake area by lake fall
Marella River (Mapanuepe Lake)	Philippines	25 Aug 1991	Overtopping	15		6	3.60×10^6	5.30×10^2	Measured, unspecified location	Umbal and Rodolfo, 1996	Dammed by lahars generated from pyroclastic flows
Marella River (Mapanuepe Lake)	Philippines	21 Sep 1991	Overtopping	24	7.50×10^7	6.5	4.70×10^6	6.50×10^2	Measured, unspecified location	Umbal and Rodolfo, 1996	Dammed by lahars generated from pyroclastic flows
Marella River (Mapanuepe Lake)	Philippines	12 Oct 1991	Overtopping	22	4.00×10^7	2.5	1.80×10^6	3.90×10^2	Measured, unspecified location	Umbal and Rodolfo, 1996	Dammed by lahars generated from pyroclastic flows

(*cont.*)

Table 8.1. (cont.)

River/Lake	Location	Date of failure	Trigger mechanism	Dam height (m) (max. impoundment depth)	Lake volume at failure (m^3)	Breach depth (m)	Volume released (m^3)	Peak discharge ($m^3 s^{-1}$)	Peak discharge determination	Breach erosion rate ($m s^{-1}$)	Breach erosion rate determination	Sources	Comments
Colorado River	USA, Colorado	165 ka	Unknown	302	1.10×10^{10}	302	1.10×10^{10}	5.30×10^{5}	Step-backwater			Fenton *et al.* (2006)	Dammed by lava flow; dam height 140–366 m, flood data from 'scenario A' of Fenton *et al.*
Williamson River (Mazama)	USA, Oregon	~7.6 ka	Unknown	21	6.50×10^{9}	17	5.70×10^{9}	1.30×10^{4}	Competence			Conaway (1999)	Existing lake further dammed 12 m by pyroclastic flow; lake eroded divide below previous outlet
Breached estuaries													
Mhlanga	South Africa			2.5	7.5×10^{5}	2.5	7.50×10^{5}	2.10×10^{2}	Water-level recording		Breach width 30 m	Parkinson and Stretch (2007) (Table 8.2)	
Wamberal	South Africa			2.8	1.375×10^{6}	2.8	1.38×10^{6}	1.05×10^{2}	Water-level recording			Parkinson and Stretch (2007) (Table 8.2)	Artificially breached
Bot	South Africa			2.7	3.00×10^{7}	2.7	3.00×10^{7}	3.30×10^{2}				Parkinson and Stretch (2007) (Table 8.2)	
Floods from tectonic basins													
Lake Bonneville	USA, Idaho	Pleistocene	Overtopping or sapping	300	6.50×10^{12}	108	4.75×10^{12}	1.00×10^{6}	Step backwater			Jarrett and Malde (1987), O'Connor (1993)	Bonneville volume and depth from Fig. 16 of Currey (1990)
Crooked Creek	USA, Oregon	Pleistocene	Unknown	68	4.00×10^{10}	12	1.13×10^{10}	1.00×10^{4}	Critical flow at outlet			Carter *et al.* (2006)	Spillover of closed tectonic basin; total lake volume estimated from dimensions and reported spill volume
Floods from calderas and crater lakes													
Waikato River; Lake Taupo	New Zealand	~1.8 ka	Unknown	198	8.00×10^{10}	32	1.99×10^{10}	2.21×10^{4}	Competence, 14.3 km downstream			Manville *et al.* (1999)	
Waikato River; Lake Taupo	New Zealand	26.5 ka	Unknown	330	1.10×10^{11}	75	6.00×10^{10}	1.00×10^{5}	Competence, 12 km downtream			Manville and Wilson (2004), Manville *et al.* (2007)	Lake volumes estimated from data reported in Manville *et al.* (2007)
Whangaehu River, Mount Ruapehu, Crater Lake	New Zealand	24 Dec 1953	Unknown			7.9	1.80×10^{6}	2.00×10^{3}	Calibrated dam-break modelling			Manville (2004)	
Whangaehu River, Mount Ruapehu, Crater Lake	New Zealand	28 Mar 2007	Piping?	134	1.30×10^{7}	6.3	1.40×10^{6}	5.30×10^{2}	Lake drawdown at outlet	3.70×10^{-4}	Lake emptied in less than 90 min	Manville *et al.* (2007), Manville and Cronin (2007), Manville pers. comm., 11 Feb 2008	Crater lake breach, tephra barrier, eroded to lava-rock sill

Pinatubo Caldera	Philippines	10 Jul 2002	Overtopping	175	1.61×10^8	23	6.50×10^7	3.00×10^3	Unknown			Lagmay *et al.* (2007), Antonia *et al.* (2003)	
Aniakchak River	USA, Alaska	Post 3.4 ka	Unknown	183	3.70×10^9	183	3.70×10^9	1.00×10^6	Step backwater at 1 km			Waythomas *et al.* (1996)	
Crater Creek, Okmok Caldera, Umnak Island	USA, Alaska	1.5 ka–1.0 ka	Unknown	150	5.80×10^9	150	5.80×10^9	1.90×10^6	Unknown			Wolfe and Begét (2002)	'Maximum possible discharge' perhaps from critical flow calculation at breach
Crater Creek, Okmok Caldera, Umnak Island	USA, Alaska	AD 1817	Unknown	8				2.00×10^4	Competence, 5 km downstream			Wolfe and Begét (2002)	
Paulina Creek; Paulina Lake, Newberry Caldera	USA, Oregon	After AD 1680, before AD 1730	Unknown	78	3.20×10^8	2	1.24×10^7	2.00×10^2	Critical flow (110–280 $m^3 s^{-1}$) at downstream falls			Chitwood and Jensen (2000)	Release volume estimated from lake area; present lake depth and volume from Johnson *et al.* (1985)
Floods from constructed dams (earthen and rockfill)													
Oros	Brazil	26 Mar 1960	Overtopping	35.8	6.60×10^8	35.8	6.60×10^8	9.63×10^3	Unknown	1.20×10^{-3}	Formation time 8.5 hr	MacDonald and Langridge-Monopolis (1984), Costa (1988), Froehlich (1995), Wahl (1998)	
Salles Oliveira	Brazil	1977	Overtopping	38.4	7.15×10^7	38.4	7.15×10^7	7.20×10^3	Unknown	5.30×10^{-3}	Failure time 2 hr	Wahl (1998)	
Lake Ha! Ha!	Canada, Quebec	19 Jul 1996	Overtopping			10.6	6.00×10^7	1.02×10^3	Lake drawdown, includes 160 inflow			Capart *et al.* (2007)	Earthen dike, increasing level of lake
Bradfield Dam	UK	3 Nov 1864	Piping			29	3.20×10^6	1.13×10^3	Unknown	1.61×10^{-2}	Failure in less than 30 min	Costa (1988), Macchione and Sirangelo (1990), Wahl (1998)	
Eigiau	UK	1925	Unknown			10.5	4.50×10^6	4.00×10^2	Unknown ('surveyed')			Macchione and Sirangelo (1990)	
Ashalim Dam	Israel	22 Dec 1993	Piping?	8	5×10^5	8	5.00×10^5	6.50×10^2	600–700 slope-area, 1 km downstream			Greenbaum (2007)	Earthen dam, 14 m high, overtopping unlikely
Butler	USA, Arizona	1982	Overtopping			7.16	2.38×10^6	8.10×10^2	Slope-area, 600 m downstream			Froehlich (1995)	

(cont.)

Table 8.1. (cont.)

River/Lake	Location	Date of failure	Trigger mechanism	Dam height (m) (max. impoundment depth)	Lake volume at failure (m^3)	Breach depth (m)	Volume released (m^3)	Peak discharge ($m^3 s^{-1}$)	Peak discharge determination	Breach erosion rate ($m s^{-1}$)	Breach erosion rate determination	Sources	Comments
Baldwin Hills	USA, California	23 Dec 1963	Subsidence			12.2	9.10×10^5	1.13×10^3	Reservoir outflow, 15 min average	1.03×10^{-2}	Formation time 0.33 hr	MacDonald and Langridge-Monopolis (1984), Costa (1988), Macchione and Sirangelo (1990), Froehlich (1995), Wahl (1998)	
Davis Reservoir	USA, California	18 Jun 1914	Piping	11.9	5.80×10^7	11.6	5.80×10^7	5.10×10^2	Unknown	5.00×10^{-4}	Maximum development 7 hr	Ponce (1982), MacDonald and Langridge-Monopolis (1984), Wahl (1998)	
Hell Hole	USA, California	23 Dec 1964	Piping			35.1	3.06×10^7	7.36×10^3	Reservoir outflow, 1 hr average	1.30×10^{-2}	Formation time 0.75 hr	Scott and Gravlee (1968), MacDonald and Langridge-Monopolis (1984), Froehlich (1995), Wahl (1998)	
Lake Frances	USA, California	1899	Piping			14	7.89×10^5			3.40×10^{-3}	Failure time 1 hr	MacDonald and Langridge-Monopolis (1984), Wahl (1998)	
Pud-dingstone	USA, California	7 Apr 1926	Overtopping	15.2	6.17×10^5	15.2	6.17×10^5	4.80×10^2	Reservoir outflow, 15 min average	1.69×10^{-2}	Formation time 0.25 hr	Ponce (1982), Froehlich (1995), Wahl (1998)	
Apishapa	USA, Colorado	22 Aug 1923	Piping			28	2.22×10^6	6.85×10^3	Reservoir outflow, 15 min average	1.04×10^{-2}	Formation time 0.75 hr	MacDonald and Langridge-Monopolis (1984), Costa (1988), Macchione and Sirangelo (1990), Froehlich (1995), Wahl (1998)	
Castlewood	USA, Colorado	3 Aug 1933	Overtopping	21.6	6.17×10^6	21.6	6.17×10^6	3.57×10^3	Reservoir outflow, 15 min average	1.80×10^{-2}	Maximum development 0.33 hr	MacDonald and Langridge-Monopolis (1984), Costa (1988), Froehlich (1995), Wahl (1998)	

Ireland No. 5	USA, Colorado	1984	Piping			3.81	1.60×10^4	1.10×10^2	Slope-area	2.10×10^{-3}	Formation time 0.5 hr	Froehlich (1995)	
Lawn Lake	USA, Colorado	15 Jul 1982	Piping			6.71	7.98×10^5	5.10×10^2	Calibrated dam-break model			Jarrett and Costa (1986), Wahl (1998)	
Lily Lake	USA, Colorado	1951	Piping			3.35	9.25×10^4	7.10×10	Slope-area			Froehlich (1995), Wahl (1998)	
Lower Latham	USA, Colorado	1973	Piping			5.79	7.08×10^6	3.40×10^2	Slope-area	1.10×10^{-3}	Formation time 1.5 hr	Froehlich (1995), Wahl (1998)	
Prospect	USA, Colorado	1980	Piping			1.68	3.54×10^6	1.16×10^2	Resevoir outflow	1.90×10^{-4}	Formation time 2.5 hr	Froehlich (1995), Wahl (1998)	Breach width 88.4 m
Schaeffer	USA, Colorado	5 Jun 1921	Overtopping	30.5	4.44×10^6	30.5	4.44×10^6	4.50×10^3	Slope-area; 13 km downstream	1.52×10^{-2}	Completely drained in 0.5 hr	Costa (1988)	
Martin Cooling Pond Dike	USA, Florida	1979	Foundation defect			8.53	1.36×10^8	3.12×10^3	Unknown			Wahl (1998)	
Kelly Barnes Lake	USA, Georgia	6 Nov 1977	Piping			11.3	7.77×10^5	6.80×10^2	Slope-area; 250 m downstream	6.30×10^{-3}	Failure time 0.5 hr	Ponce (1982), MacDonald and Langridge-Monopolis (1984), Costa (1988), Froehlich (1995), Wahl (1998)	
Sinker Creek	USA, Idaho	19 Jun 1943	Mass movement ('seepage slide')	21.3	3.30×10^6	21.3	3.30×10^6			3.00×10^{-3}	Failure time 2 hr	MacDonald and Langridge-Monopolis (1984), Wahl (1998)	
Teton Dam	USA, Idaho	5 Jun 1976	Piping		3.10×10^8	77.4	3.10×10^8	6.51×10^4	Slope-area; 4 km downstream	1.72×10^{-2}	Formation time 1.25 hr	MacDonald and Langridge-Monopolis (1984), Costa (1988), Froehlich (1995)	
Mill River	USA, Massachusetts	16 May 1874	Unknown			13.1	2.50×10^6	1.65×10^3	Unknown			Costa (1988)	
French Landing	USA, Michigan	1925	Piping			8.53	3.87×10^6	9.29×10^2	Reservoir outflow; 1 hr average	4.00×10^{-3}	Formation time 0.58 hr	MacDonald and Langridge-Monopolis (1984), Costa (1988), Froehlich (1995), Wahl (1998)	
Fred Burr	USA, Montana	1948	Piping			10.2	7.50×10^5	6.54×10^2	Slope-area; unknown distance downstream			Costa (1988) Froehlich (1995), Wahl (1998)	

(cont.)

Table 8.1. (cont.)

River/Lake	Location	Date of failure	Trigger mechanism	Dam height (m) (max. impoundment depth)	Lake volume at failure (m^3)	Breach depth (m)	Volume released (m^3)	Peak discharge ($m^3 s^{-1}$)	Peak discharge determination	Breach erosion rate ($m s^{-1}$)	Breach erosion rate determination	Sources	Comments
Frenchman Creek	USA, Montana	5 Apr 1952	Piping			10.8	1.60×10^7	1.42×10^3	Unknown			MacDonald and Langridge-Monopolis (1984), Costa (1988), Froehlich (1995), Wahl (1998)	
Lower Two Medicine	USA, Montana	8 Jun 1964	Piping	11.3	2.96×10^7	11.3	2.96×10^7	1.80×10^3	Slope-area; 4 km downstream			MacDonald and Langridge-Monopolis (1984), Costa (1988), Froehlich (1995), Wahl (1998)	
Swift	USA, Montana	8 Jun 1964	Overtopping	57.6	3.70×10^7	47.9	3.70×10^7	2.50×10^4	Slope-area; 27 km downstream	5.32×10^{-2}	Maximum development time 0.5 hr	MacDonald and Langridge-Monopolis (1984), Costa (1988), Wahl (1998)	
Lake Avalon	USA, New Mexico	1 Oct 1904	Piping			13.7	3.15×10^7	2.30×10^3	Unknown	1.90×10^{-3}	Failure time 2 hr	Ponce (1982), Macchione and Sirangelo (1990), Wahl (1998)	
Elk City	USA, Oklahoma	1 May 1936	Overtopping	9.44	1.18×10^6	9.44	1.18×10^6	6.10×10^2	Unknown			Ponce (1982), Wahl (1998)	
Porter Hill	USA, Oregon	27 Feb 1993	Mass movement			2.5	1.50×10^4	3.10×10	Manning equation; 150 m downstream			Costa and O'Connor (1995)	
Lake Latonka	USA, Pennsylvania	27 Oct 1966	Mass movement			6.25	4.09×10^7	2.90×10^2	'surveyed'	1.20×10^{-3}	Failure time 3 hr	Macchione and Sirangelo (1990), Ponce (1982), Wahl (1998)	
Laurel Run	USA, Pennsylvania	20 Jul 1977	Overtopping	14.1	5.55×10^5	14.1	5.55×10^5	1.05×10^3	Slope-area 1.6 km downstream			MacDonald and Langridge-Monopolis (1984), Costa (1988), Froehlich (1995), Wahl (1998)	
North Branch Trib.	USA, Pennsylvania	20 Jul 1977	Overtopping	5.5	2.20×10^4	5.5	2.20×10^4	2.94×10	Slope-area			MacDonald and Langridge-Monopolis (1984), Costa (1988), Wahl (1998)	

Otto Run	USA, Pennsylvania	20 Jul 1977	Overtopping			5.8	7.40×10^3	6.00×10	Slope-area			MacDonald and Langridge-Monopolis (1984), Costa (1988), Wahl (1998)	
Sandy Run	USA, Pennsylvania	1977	Unknown			8.5	5.67×10^4	4.35×10^2	Unknown			MacDonald and Langridge-Monopolis (1984), Wahl (1998)	
South Fork (Johnstown)	USA, Pennsylvania	31 May 1889	Overtopping	24.6	1.89×10^7	24.6	1.89×10^7	8.50×10^3	Reservoir outflow, 30 min average	9.10×10^{-3}	Reservoir emptied in 45 min (McCullough, p. 102)	McCullough 1968, MacDonald and Langridge-Monopolis (1984), Costa (1988), Froehlich (1995)	
South Fork Tributary	USA, Pennsylvania	1977	Overtopping, from upstream dam failure		3.70×10^3	1.83	1.22×10^2	1.22×10^2	Slope-area			Wahl (1998)	Very large inflow from upstream dam failure
Goose Creek	USA, South Carolina	16 Jul 1916	Overtopping			1.37	1.06×10^7	5.65×10^2	Unknown		Failure time 0.5 hr	Ponce (1982), MacDonald and Langridge-Monopolis (1984), Wahl (1998)	Reported discharge seems implausible (Wahl, 1998)
DMAD	USA, Utah	6 Jun 1983	Unknown			8.8	1.97×10^7	7.93×10^2	Reservoir outflow			Costa (1988)	
Hatchtown	USA, Utah	25 May 1914	Piping			16.8	1.48×10^7	3.08×10^3	Reservoir outflow; 1 hr average	4.70×10^{-3}	Formation time 1 hr	MacDonald and Langridge-Monopolis (1984), Costa (1988), Froehlich (1995), Wahl (1998)	Discharge from Froehlich (1995)
Little Deer Creek	USA, Utah	16 Jun 1963	Piping			22.9	1.36×10^6	1.33×10^3	Slope-area	1.93×10^{-2}	Formation time 0.33 hr	MacDonald and Langridge-Monopolis (1984), Costa (1988), Froehlich (1995), Wahl (1998)	
Quail Creek	USA, Utah	1 Jan 1989	Piping			16.7	3.08×10^7	3.11×10^3	Reservoir outflow, 15 min average	4.60×10^{-3}	Formation time 1 hr	O'Neill and Gourley (1991, U.S. Geological Survey, unpublished data), Froehlich (1995), Wahl (1998)	

(cont.)

Table 8.1. (cont.)

River/Lake	Location	Date of failure	Trigger mechanism	Dam height (m) (max. impoundment depth)	Lake volume at failure (m^3)	Breach depth (m)	Volume released (m^3)	Peak discharge ($m^3 s^{-1}$)	Peak discharge determination	Breach erosion rate ($m s^{-1}$)	Breach erosion rate determination	Sources	Comments
Buffalo Creek	USA, Virginia	26 Feb 1972	Seepage	14		14	4.84×10^5	1.42×10^3	Slope-area	7.80×10^{-3}	Failure time 0.5 hr	MacDonald and Langridge-Monopolis (1984), Costa (1988), Wahl (1998)	Coal waste embankment
Centralia, Res. No. 3	USA, Washington	5 Oct 1991	Unknown	5.2	1.33×10^4	5.2	1.33×10^4	7.10×10	Slope-area			Costa (1994)	
Unnamed	USA, Washington	18 Apr 2000	Unknown	7.5		5.5	6.06×10^4	2.80×10^2	Manning equation			Carson (2001, 2001 written commun.)	
Wheatland no. 1	USA, Wyoming	8 Jul 1969	Mass movment; piping			12.2	1.16×10^7			2.30×10^{-3}	Formation time 1.5 hr	Ponce (1982), MacDonald and Langridge-Monopolis (1984), Wahl (1998)	
Floods from constructed dams (masonry and concrete)													
Malpasset	France	2 Jan 1959	Unknown	61	2.20×10^7	61	2.20×10^7	2.83×10^4	Unknown			Costa (1988)	Concrete arch
St Francis	USA, California	12 Mar 1928	Mass movement?			56.4	4.71×10^7	1.98×10^4	Slope-conveyance			Costa (1988), Rogers and McMahon (1993)	Concrete gravity; discharge is upper end of reported range
Floods from constructed dams (physical simulation of natural and constructed dam breaches)													
Test 3	Austria	1982	Overtopping	0.36	2.4	0.27	1.82	5.40×10^{-2}	Reservoir outflow	9.10×10^{-3}	Measured	Simmler and Samet (1982)	Breach experiment, homogenous dam
Test 9	Austria	1982	Overtopping	0.3	1.97	0.195	1.32	6.80×10^{-2}	Reservoir outflow	8.40×10^{-3}	Measured	Simmler and Samet (1982)	Breach experiment, central core
Test 10	Austria	1982	Overtopping	0.3	1.97	0.18	1.18	5.10×10^{-2}	Reservoir outflow	4.30×10^{-3}	Measured	Simmler and Samet (1982)	Breach experiment, upstream lining
FAFUM	Germany		Overtopping	0.22	6.95	0.16	4.55	2.40×10^{-2}	Calibrated weir			Bechteler and Kulisch (1994)	Sand
IMPACT Lab. Trial; Test 5; Series 1	UK	2002	Overtopping	0.5	1.80×10^2	0.5	1.80×10^2	7.80×10^{-1}	measured	2.00×10^{-4}	Measured	Breach Formation (WP2) Technical Report, Dec 2004 (www.IMPACT-project.net), accessed 13 Feb 2008	One representative experiment; Series 1, test 5; intended to be equivalent to Field test #2
Poerua prototype	New Zealand		Overtopping	0.65	0.7	<0.35	0.475	1.30×10^{-2}	Measured		Breach development time of 0.25 hr	Davies *et al.* (2007)	Data from trial with Poerua prototype geometry

IMPACT Field test 2	Norway	1 Oct 02	Overtopping	5	33000	5	3.30×10^4	1.18×10^2	Measured	1.67×10^{-2}	Vertical erosion complete in 5 min	Vaskinn *et al.* (2004), Breach Formation (WP2) Technical Report, Dec 2004 (www.IMPACT-project.net), accessed 13 Feb 2008	Breach experiment, homogeneous gravel dam
IMPACT Field test 3	Norway	21 Aug 03	Overtopping	6	65000	6	6.50×10^4	2.20×10^2	Measured	1.00×10^{-2}	Breach developed in 10 min	See test 2	Rockfill with moraine core
Laboratory Sim. 1	Peru	1974	Overtopping	0.68	42.9	0.44	3.20×10		Reservoir outflow			Lee and Duncan (1975)	1:250 model of 1974 Manntaro River landslide dam; Lee and Duncan Table 8.2 values scaled; only physical dimension data (volume, breach depth) used in analysis, since scaled discharge values seem unreasonable
Laboratory Sim. 2	Peru	1974	Overtopping	0.68	42.9	0.41	3.00×10		Reservoir outflow			See Sim. 1	As for Sim. 1
Laboratory Sim. 3	Peru	1974	Overtopping	0.68	42.9	0.38	3.00×10		Reservoir outflow			See Sim. 1	As for Sim. 1
Laboratory Sim. 4	Peru	1974	Overtopping	0.68	42.9	0.42	3.20×10		Reservoir outflow			See Sim. 1	As for Sim. 1
CEHIDRO Lab	Portugal	2001	Overtopping	0.5	2.7	0.05	3.25×10^{-1}	5.00×10^{-3}	Measured			Franca and Almeida (2002)	Data from Fig. 5 example
A-1	Thailand		Overtopping	0.6	28	0.6	2.80×10	6.00×10^{-1}	Measured	3.50×10^{-3}	Breach deformation time	Chinnarasri *et al.* (2004)	Homogeneous earthfill
A-2	Thailand		Overtopping	0.6	28	0.6	2.80×10	2.84×10^{-1}	Measured	4.00×10^{-3}	Breach deformation time	Chinnarasri *et al.* (2004)	Homogeneous earthfill
A-3	Thailand		Overtopping	0.6	28	0.6	2.80×10	4.63×10^{-1}	Measured	4.60×10^{-3}	Breach deformation time	Chinnarasri *et al.* (2004)	Homogeneous earthfill
B-1	Thailand		Overtopping	0.6	28	0.6	2.80×10	1.78×10^{-1}	Measured	3.30×10^{-3}	Breach deformation time	Chinnarasri *et al.* (2004)	Homogeneous earthfill, finer
B-2	Thailand		Overtopping	0.6	28	0.6	2.80×10	4.04×10^{-1}	Measured	5.50×10^{-3}	Breach deformation time	Chinnarasri *et al.* (2004)	Homogeneous earthfill, finer
B-3	Thailand		Overtopping	0.6	28	0.6	2.80×10	3.63×10^{-1}	Measured	4.00×10^{-3}	Breach deformation time	Chinnarasri *et al.* (2004)	Homogeneous earthfill, finer
C-1	Thailand		Overtopping	0.6	28	0.6	2.80×10	3.47×10^{-1}	Measured	2.90×10^{-3}	Breach deformation time	Chinnarasri *et al.* (2004)	Homogeneous earthfill, clay lining
C-2	Thailand		Overtopping	0.6	28	0.6	2.80×10	4.40×10^{-1}	Measured	4.60×10^{-3}	Breach deformation time	Chinnarasri *et al.* (2004)	Homogeneous earthfill, clay lining
C-3	Thailand		Overtopping	0.6	28	0.6	2.80×10	4.05×10^{-1}	Measured	5.00×10^{-3}	Breach deformation time	Chinnarasri *et al.* (2004)	Homogeneous earthfill, clay lining

(cont.)

Table 8.1. (cont.)

River/Lake	Location	Date of failure	Trigger mechanism	Dam height (m) (max. impoundment depth)	Lake volume at failure (m^3)	Breach depth (m)	Volume released (m^3)	Peak discharge ($m^3 s^{-1}$)	Peak discharge determination	Breach erosion rate ($m s^{-1}$)	Breach erosion rate determination	Sources	Comments
Run 1	Thailand		Overtopping	0.8	16	0.52	1.04×10			1.25×10^{-2}	Measured maximum instantaneous rate vertical lowering	Chinnarasri *et al.* (2003)	
Run 2	Thailand		Overtopping	0.8	16	0.5	1.00×10			8.50×10^{-3}	Measured maximum instantaneous rate vertical lowering	Chinnarasri *et al.* (2003)	
Run 3	Thailand		Overtopping	0.8	16	0.47	9.40			6.00×10^{-3}	Measured maximum instantaneous rate vertical lowering	Chinnarasri *et al.* (2003)	
Run 5	Thailand		Overtopping	0.8	16	0.42	8.40			5.00×10^{-3}	Measured maximum instantaneous rate vertical lowering	Chinnarasri *et al.* (2003)	
ARS Embankment 1, Soil 1	USA, Oklahoma	2000?	Overtopping	1.83	4000	1.83	4.00×10^3	5.50	Measured	1.50×10^{-3}	'Formation stage'	Hanson *et al.* (2003)	Sandy soil
ARS Embankment 1, Soil 2	USA, Oklahoma	2000?	Overtopping	1.83	4000	1.83	4.00×10^3	8.00×10^{-1}	Measured	3.00×10^{-4}	'Formation stage'	Hanson *et al.* (2003)	Fine soil
ARS Embankment 2, Soil 1	USA, Oklahoma	2000?	Overtopping	1.22	4000	1.22	4.00×10^3	2.00	Measured	3.00×10^{-4}	'Formation stage'	Hanson *et al.* (2003)	Sandy soil
ARS Embankment 2, Soil 2	USA, Oklahoma	2000?	Overtopping	1.22	4000	1.22	4.00×10^3	1.00	Measured	5.65×10^{-5}	'Formation stage'	Hanson *et al.* (2003)	
ARS FS#1	USA, Oklahoma	1999	Overtopping	1.83	4000	1.83	4.00×10^3	5.80	Measured	8.70×10^{-4}	'Formation stage'	Hahn *et al.* (2000)	

Source: Updated from Walder and O'Connor (1997).

References

Adams, J. E. (1981). Earthquake-dammed lakes in New Zealand. *Geology*, **9**, 215–219.

Alden, W. C. (1928). Landslide and flood at Gros Ventre, Wyoming. *Transactions of the American Institute of Mining and Metallurgical Engineers*, **76**, 347–360.

Anderson, D. E. (1998). Late Quaternary paleohydrology, lacustrine stratigraphy, fluvial geomorphology and modern hydroclimatology of the Armargosa River/Death Valley hydrologic system, California and Nevada. Unpublished Ph. D. thesis, University of California, Riverside.

Antonia, M., Bornas, V. and Newhall, C. G. (2003). The 10 July 2002 caldera-rim breach and breakout lahar from Mt. Pinatubo, Philippines: Results and lessons from attempting and artificial breach. *Programme and abstracts of the Volcanological Society of Japan*, 2003, p. 91.

Bagnold, R. A. (1966). *An Approach to the Sediment Transport Problem from General Physics*. U.S. Geological Survey Professional Paper 422-I.

Baker, V. R. (1973). *Paleohydrology and Sedimentology of Lake Missoula Flooding in Eastern Washington*. Geological Society of America Special Paper 144.

Baker, V. R. (1983). Late Pleistocene fluvial systems. In *The Late Pleistocene*, ed. S. C. Porter; Volume 1 of *Late-Quaternary Environments of the United States*, ed. H. E. Wright, Jr. Minneapolis, MN: University of Minnesota Press, pp. 115–129.

Baker, V. R. and Costa, J. E. (1987). Flood power. In *Catastrophic Flooding*, eds. L. Mayer and D. Nash. Boston, MA: Allen and Unwin, pp. 1–24.

Baker, V. R. and Ritter, D. F. (1975). Competence of rivers to transport coarse bed load material. *Geological Society of America Bulletin*, **86**, 975–978.

Baker, V. R., Benito, G. and Rudoy, A. N. (1993). Paleohydrology of late Pleistocene superflooding, Altay Mountains, Siberia. *Science*, **259**, 348–350.

Bechteler, W. and Kulisch, H. (1994). Physical 3D-simulations of erosion-caused dam-breaks. *International Workshop of Floods and Inundations Related to Large Earth Movements*, Trento, Italy, 4–7 Oct., 1994, unpaginated.

Beebee, R. A. (2003). Snowmelt hydrology, paleohydrology, and landslide dams in the Deschutes River basin, Oregon. Unpublished Ph.D. thesis, University of Oregon, Eugene, Oregon.

Begét, J. E., Larsen, J. F., Neal, C. A., Nye, C. J. and Schaefer, J. R. (2005). *Preliminary Volcano-hazard Assessment for Okmok Volcano, Umnak Island Alaska. Report of Investigations 2004–3*. Alaska Department of Natural Resources Division of Geological and Geophysical Surveys.

Benito, G. (1997). Energy expenditure and geomorphic work of the cataclysmic Missoula flooding in the Columbia River Gorge, USA. *Earth Surface Processes and Landforms*, **22**, 457–472.

Benito, G. and O'Connor, J. E. (2003). Number and size of last-glacial Missoula floods in the Columbia River valley between the Pasco Basin, Washington, and Portland, Oregon. *Geological Society of America Bulletin*, **115**, 624–638.

Benn, D. I., Owen, L. A., Finkel, R. C. and Clemmens, S. (2006). Pleistocene lake outburst floods and fan formation along the eastern Sierra Nevada, California: implications for the interpretation of intermontane lacustrine records. *Quaternary Science Reviews*, **25**, 2729–2748.

Blair, T. C. (1987). Sedimentary processes, vertical stratification sequences, and geomorphology of the Roaring River alluvial fan, Rocky Mountain National Park, Colorado. *Journal of Sedimentary Petrology*, **57**, 1–18.

Blair, T. C. (2001). Outburst flood sedimentation of the proglacial Tuttle Canyon alluvial fan, Owens Valley, California, U. S.A. *Journal of Sedimentary Research*, **71**, 657–679.

Blown, I. and Church, M. (1985). Catastrophic lake drainage within the Homathko River basin, British Columbia. *Canadian Geotechnical Journal*, **22**, 551–563.

Bovis, M. J. and Jakob, M. (2000). The July 29, 1998 debris flow and landslide dam at Capricorn Creek, Mount Meager volcanic complex, southern Coast Mountains, British Columbia. *Canadian Journal of Earth Sciences*, **27**, 1321–1334.

Bretz, J H. (1923). The Channeled Scabland of the Columbia Plateau. *Journal of Geology*, **31**, 617–649.

Bretz, J H. (1924). The Dalles type of river channel. *Journal of Geology*, **32**, 139–149.

Bretz, J H. (1928). Bars of the Channeled Scabland. *Geological Society of America Bulletin*, **39**, 643–702.

Bretz, J H. (1932). *The Grand Coulee*. American Geographical Society Special Publication 15, New York: American Geographical Society.

Brunner, G. W. (2002). *HEC-RAS River Analysis System User's Manual*, U.S. Army Corps of Engineers, Hydrologic Engineering Center.

Buchroithner, M. F., Jentsch, G. and Wanivenhaus, B. (1982). Monitoring of recent geological events in the Khumbu area (Himalaya, Nepal) by digital processing of Landsat MSS data. *Rock Mechanics*, **15**, 181–197.

Canuti, P., Frassoni, A. and Natale, L. (1994). Failure of the Rio Paute landslide dam. *Landslide News*, **8**, 6–7.

Capart, H., Spinewine, B., Young, D. L. *et al.* (2007). The 1996 Lake Ha! Ha! breakout flood, Québec: Test data for geomorphic flood routing methods. *Journal of Hydraulic Research*, **45** (extra issue), 97–109.

Capra, L. (2007). Volcanic natural dams: identification, stability, and secondary effects. *Natural Hazards*, **43**, 45–61.

Capra, L. and Macías, J. L. (2002). The cohesive Naranjo debris-flow deposit (10 km^3): a dam breakout flow derived from the Pleistocene debris-avalanche deposit of Nevado de Colima Volcano (México). *Journal of Volcanology and Geothermal Research*, **117**, 213–235.

Carling, P. A. (1996). Morphology, sedimentology and palaeohydraulic significance of large gravel dunes, Altai Mountains, Siberia. *Sedimentology*, **43**, 647–664.

Carling, P. A. and Glaister, M. S. (1987). Reconstruction of a flood resulting from a moraine-dam failure using geomorphological evidence and dam-break modeling. In *Catastrophic Flooding*, eds. L. Mayer and D. Nash. Boston, MA: Allen and Unwin, pp. 181–200.

Carrivick, J. L. and Rushmer, L. (2006). Understanding high-magnitude outburst floods. *Geology Today*, **22**, 60–65.

Carson, R. J. (2001). Dam failure in southeastern Washington. *Geological Society of America Abstracts with Program*, **33** (6), p. 284.

Carter, D. T., Ely, L. L., O'Connor, J. E. and Fenton, C. R. (2006). Late Pleistocene outburst flooding from pluvial Lake Alvord into the Owyhee River, Oregon. *Geomorphology*, **75**, 346–367.

Casagli, N., Ermini, L. and Rosati, G. (2003). Determining grain size distribution of the material composing landslide dams in the Northern Apennines: sampling and processing methods. *Engineering Geology*, **69**, 83–97.

Cencetti, C., Fredduzzi, A., Marchesini, I., Naccini, M. and Tacconi, P. (2006). Some considerations about the simulation of breach channel erosion on landslide dams. *Computational Geosciences*, **10**, 201–219.

Cenderelli, D. A. (2000). Floods from natural and artificial dam failures. In *Inland Flood Hazards: Human Riparian, and Aquatic Communities*, ed. E. E. Wohl. New York: Cambridge University Press.

Cenderelli, D. A. and Wohl, E. E. (1998). Sedimentology and clast orientation of deposits produced by glacial-lake outburst floods in the Mount Everest Region, Nepal. In *Geomorphological Hazards in High Mountain Areas*, eds. J. Kalvoda and C. L. Rosenfield. Dordrecht, the Netherlands: Kluwer Academic Publishers, pp. 1–26.

Cenderelli, D. A. and Wohl, E. E. (2003). Flow hydraulics and geomorphic effects of glacial-lake outburst floods in the Mount Everest Region, Nepal. *Earth Surface Processes and Landforms*, **28**, 385–407.

Chai, H. J., Liu, H. C., Zhang, Z. Y. and Wu, Z. W. (2000). The distribution, causes and effects of damming landslides in China. *Journal of the Chengdu Institute of Technology*, **27**, 1–19.

Chang, S. C. (1984). Tsao-Ling landslide and its effect on a reservoir project. *Proceedings of the IVth International Symposium on Landslides*, 16–21 September, Toronto, 469–473.

Chen, Y. J., Zhou, F., Feng, Y. and Xia, Y. C. (1992). Breach of a naturally embanked dam on Yalong River. *Canadian Journal of Civil Engineering*, **19**, 811–818.

Chesner, C. A. and Rose, W. I. (1991). Stratigraphy of the Toba Tuffs and the evolution of the Toba Caldera Complex, Sumatra, Indonesia. *Bulletin of Volcanology*, **53**, 343–356.

Chinnarasri, C., Tingsanchali, T., Weesakul, S. and Wongwises, S. (2003). Flow patterns and damage of dike overtopping. *International Journal of Sediment Research*, **18**, 301–309.

Chinnarasri, C., Jirakitlerd, S. and Wongwises, S. (2004). Embankment dam breach and its outflow characteristics. *Civil Engineering and Environmental Systems*, **21**, 247–264.

Chitwood, L. A. and Jensen, R. A. (2000). Large prehistoric flood along Paulina Creek. In *What's New at Newberry Volcano, Oregon*, eds. R. A. Jensen and L. A. Chitwood. Bend, OR: U.S.D.A. Forest Service, pp. 31–40.

Clague, J. J. and Evans, S. G. (1992). A self-arresting moraine dam failure, St. Elias Mountains, British Columbia. *Current Research, Part A, Geological Survey of Canada, Paper 92–1A*, pp. 185–188.

Clague, J. J. and Evans, S. G. (1994). *Formation and Failure of Natural Dams in the Canadian Cordillera*. Geological Survey of Canada Bulletin 464.

Clague, J. J. and Evans, S. G. (2000). A review of catastrophic drainage of moraine-dammed lakes in British Columbia. *Quaternary Science Reviews*, **19**, 1763–1783.

Clague, J. J. and Mathews, W. H. (1973). The magnitude of jökulhlaups. *Journal of Glaciology*, **12**, 501–504.

Coleman, N. M. and Dinwiddie, C. L. (2007). Hydrologic analysis of the birth of Elaver Vallis, Mars by catastrophic drainage of a lake in Morella Crater. *Seventh International Conference on Mars*, Abstract 3334. Pasadena, CA, July 9–13, 2007.

Conaway, J. S. (1999). Hydrogeology and paleohydrology of the Williamson River basin, Klamath County, Oregon. Unpublished M.S. thesis, Portland State University, Portland, Oregon.

Costa, J. E. (1983). Paleohydraulic reconstruction of flash-flood peaks from boulder deposits in the Colorado Front Range. *Geological Society of America Bulletin*, **94**, 986–1004.

Costa, J. E. (1984). Physical geomorphology of debris flows. In *Developments and Applications of Geomorphology*, eds. J. E. Costa and P. J. Fleisher. Berlin: Springer-Verlag, pp. 268–317.

Costa, J. E. (1988). Floods from dam failures. In *Flood Geomorphology*, eds. V. R. Baker, R. C. Kochel and P. C. Patton. New York: John Wiley and Sons, pp. 439–469.

Costa, J. E. (1991). Nature, mechanics, and mitigation of the Val Pola Landslide, Valtellina, Italy, 1987–1988. *Zeitschrift fur Geomorphologie*, **35**, pp. 15–38.

Costa, J. E. (1994). *Multiple Flow Processes Accompanying a Dam-break Flood in a Small Upland Watershed, Centralia, Washington*. Water-Resources Investigations Report 94–4026. U.S. Geological Survey.

Costa, J. E. and O'Connor, J. E. (1995). Geomorphically effective floods. In *Natural and Anthropogenic Influences in Fluvial Geomorphology*. American Geophysical Union Monograph 89, eds. J. E. Costa, A. J. Miller, K. W. Potter and P. R. Wilcock, pp. 45–56.

Costa, J. E. and Schuster, R. L. (1988). The formation and failure of natural dams. *Geological Society of America Bulletin*, **100**, 1054–1068.

Costa, J. E. and Schuster, R. L. (1991). *Documented Historical Landslide Dams from Around the World*. U.S. Geological Survey Open-File Report 91–239.

Crow, R., Karlstrom, K. E., McIntosh, W., Peters, L. and Dunbar, N. (2008). History of Quaternary volcanism and lava dams in western Grand Canyon based on lidar analysis, $^{40}Ar/^{39}Ar$ dating, and field studies: implications for flow stratigraphy, timing of volcanic events, and lava dams. *Geosphere*, **4**, 183–206.

Currey, D. R. (1990). Quaternary palaeolakes in the evolution of semidesert basins, with special emphasis on Lake Bonneville and the Great Basin, U.S.A. *Palaeogeography, Palaeoclimatology, and Palaeoecology*, **76**, 189–214.

Dai, F. C., Lee, C. F., Deng, J. H. and Tham L. G. (2005). The 1786 earthquake-triggered landslide dam and subsequent dam-break flood on the Dadu River, southwestern China. *Geomorphology*, **65**, 205–221.

Davies, R. R., Manville, V., Kunz, M. and Donadini, L. (2007). Landslide dambreak flood magnitudes: case study. *Journal of Hydraulic Engineering*, **133**, 713–720.

Davis, S. N. and Karzulovíc, J. (1963). Landslides at Lago Riñihue. *Bulletin of the Seismological Society of America*, **53**, 1403–1414.

Denlinger, R. P. and O'Connell, D. (2003). Two dimensional flow constraints on catastrophic outflow of glacial Lake Missoula over three dimensional terrain. In *Proceedings of Third International Paleoflood Workshop*, Hood River, Oregon, p. 17.

Ding Y. and Liu J. (1992). Glacier lake outburst flood disasters in China. *Annals of Glaciology*, **16**, 180–184.

Donnelly-Nolan, J. M. and K. M. Nolan (1986). Catastrophic flooding and eruption of ash-flow tuff at Medicine Lake volcano, California. *Geology*, **14**, 875–878.

Duffield, W., Riggs, N., Kaufman, D. *et al.* (2006). Multiple constraints on the age of a Pleistocene lava dam across the Little Colorado River, Arizona. *Geological Society of America Bulletin*, **118**, 431–429.

Dunning, S. A., Petley, D. N. and Strom, A. L. (2005). The morphologies and sedimentology of valley confined rock-avalanche deposits and their effect on potential dam hazard. In *Proceeding of the International Conference on Landslide Risk Management*, eds. O. Hungr, R. Fell, R. Couture and E. Eberhardt. London: Balkema, pp. 691–704.

Dwivedi, S. K., Archarya, M. D. and Simard, D. (2000). The Tam Pokhari Glacier Lake outburst flood of 3 September 1998. *Journal of Nepal Geological Society*, **22**, 539–546.

Eisbacher, G. H. and Clague, J. J. (1984). *Destructive Mass Movements in High Mountains: Hazard and Management.* Geological Survey of Canada Paper 84–16.

Ermini, L. and Casagli, N. (2003). Prediction of the behavior of landslide dams using a geomorphological dimensionless index. *Earth Surface Processes and Landforms*, **28**, 31–47.

Evans, S. G. (1986). The maximum discharge of outburst floods caused by the breaching of man-made and natural dams. *Canadian Geotechnical Journal*, **23**, 385–387.

Evans, S. G (1987). The breaching of moraine-dammed lakes in the southern Canadian Cordillera. In *Proceedings of the International Symposium on Engineering Geological Environment in Mountainous Areas*, Vol. 2, Beijing, pp. 141–150.

Evans, S. G. and Clague, J. J. (1994). Recent climatic change and catastrophic geomorphic processes in mountain environments. *Geomorphology*, **10**, 107–128.

Fenton, C. R., Webb, R. H., Cerling, T. E., Poreda, R. J. and Nash, B. P. (2002). Cosmogenic ^{3}He ages and geochemical discrimination of lava-dam outburst-flood deposits in western Grand Canyon, Arizona. In *Ancient Floods, Modern Hazards, Principles and Applications of Paleoflood Hydrology.* American Geophysical Union Water Science and Application Series 4, eds. P. K. House, R. H. Webb and D. R. Levish, pp. 191–215.

Fenton, C. R., Poreda, R. J., Nash, B. P., Webb, R. H. and Cerling, T. E. (2004). Geochemical discrimination of five Pleistocene lava-dam outburst-flood deposits, Grand Canyon. *Journal of Geology*, **112**, 91–110.

Fenton, C. R., Webb, R. H. and Cerling, T. E. (2006). Peak discharge of a Pleistocene lava-dam outburst flood in Grand Canyon, Arizona, USA. *Quaternary Research*, **65**, 324–335.

Franca, M. J. and Almeida, A. B. (2002). Experimental tests on Rockfill dam breaching process. In *Proceedings of IAHR–International Symposium on Hydraulic and Hydrologic Aspects of Reliability and Safety Assessment of Hydraulic Structure*, St. Petersburg, unpaginated.

Fread, D. L. (1987). *BREACH: An Erosion Model for Earthen Dam Failures.* Silver Spring, MD: Hydrologic Research Laboratory, National Weather Service.

Fread, D. L. (1988). *The NWS DAMBRK Model: Theoretical Background/User Documentation.* Silver Spring, MD: Hydrologic Research Laboratory, National Weather Service.

Fread, D. L. (1993). NWS FLDWAV model: the replacement of DAMBRK for dam-break flood prediction. *Proceedings of the 10th Annual Conference of the Association of State Dam Safety Officials, Inc.* Kansas City, Missouri, pp. 177–184.

Froehlich, D. C. (1987). Embankment-dam breach parameters. In *Proceedings of the 1987 National Conference on Hydraulic Engineering*, ed. R. M. Ragan. New York: American Society of Civil Engineers, pp. 570–575.

Froehlich, D. C. (1995). Peak outflow from breached embankment dam. *Journal of Water Resources Planning and Management*, **121**, 90–97.

Fushimi, H., Ikegami, K., Higuchi, K. and Shankar, K. (1985). Nepal case study: catastrophic floods. In *Techniques for Prediction of Runoff from Glacierized Areas.* International Association of Hydrological Sciences (IAHS-AISH) Publication No. 149, ed. G. J. Young, pp. 125–130.

Galay, V. J. (1985). Hindu Kush-Himalayan erosion and sedimentation in relation to dams. In *International Workshop on Water Management in the Hindu Kush-Himalaya region*, Chengu, China, 1985. Katmandu, Nepal: International Centre for Integrated Mountain Development (ICIMOD).

Gallino, G. L. and Pierson, T. C. (1985). *Polallie Creek Debris Flow and Subsequent Dam-break Flood of 1980, East Fork Hood River Basin, Oregon.* U.S. Geological Survey Water Supply Paper 2273.

Gerasimov, V. A. (1965). Issykskaia katastrofa 1963 g. i otrazhenie ee in geomorfoogii doliny r. Issyk. [The Issyk catastrophe in 1963 and its effect on geomorphology of the

Issyk River valley.] *Akademiia Nauk SSSR, Izvestiia Vsesoiuznogo, Geograficheskogo Obshchestva*, **97–6**, 541–547. (In Russian; author in possession of translation.)

Gilbert, G. K. (1890). *Lake Bonneville*, U.S. Geological Survey Monograph 1.

Glancy, P. A. and Bell, J. W. (2000). *Landslide-induced Flooding at Ophir Creek, Washoe County, Western Nevada, May 30, 1983*. U.S. Geological Society Professional Paper 1617.

Glazyrin, G. Y. and Reyzvikh, V. N. (1968). Computation of the flow hydrograph for the breach of landslide lakes. *Soviet Hydrology, Selected Papers* (published by the American Geophysical Union), **5**, 492–496.

Greenbaum, N. (2007) Assessment of dam failure flood and a natural, high-magnitude flood in a hyperarid region using paleoflood hydrology, Nahal Ashalim catchment, Dead Sea, Israel. *Water Resources Research*, **43**, doi:10.1029/2006WR004956.

Haeberli, W. (1983). Frequency and characteristics of glacier floods in the Swiss Alps. *Annals of Glaciology*, **4**, 85–90.

Hahn, W., Hanson, G. J. and Cook, K. R. (2000). Breach morphology observations of embankment overtopping test. In *Proceedings of the 2000 Joint Conference on Water Resources Engineering and Water Resources Planning & Management National Conference on Hydraulic Engineering*, eds. R. H. Hotchkiss and M. Glade. Reston, VI: American Society of Civil Engineers.

Hamblin, W. K., (1994). *Late Cenozoic Lava Dams in the Western Grand Canyon*. Geological Society of America Memoir 183.

Hancox, G.T., McSaveney, M. J., Manville, V. R. and Davies, T. R. (2005). The October 1999 Mt. Adams rock avalanche and subsequent landslide dam-break flood and effects in Poerua River, Westland, New Zealand. *New Zealand Journal of Geology and Geophysics*, **48**, 683–705.

Hanisch, J. and Söder, C. O. (2000). Geotechnical assessment of the Usoi landslide dam and the right bank of Lake Sarez. In *Usoi Landslide Dam and Lake Sarez. An Assessment of Hazard and Risk in the Pamir Mountains, Tajikistan,. International Strategy for Disaster Reduction Prevention Series No. 1*, eds. D. Alford and R. L. Schuster. Geneva: United Nations, pp. 23–42.

Hanson, G. J., Cook, K. R., Hahn, W. and Britton, S. L. (2003). Observed erosion processes during embankment overtopping test. *American Society of Agricultural Engineers Meeting Paper 032066*. St. Joseph, MI: American Society of Agricultural Engineers.

Hanson, G. J., Cook, K. R. and Hunt, S. L. (2005). Physical modeling of overtopping erosion and breach formation of cohesive embankments. *Transactions of the American Society of Agricultural Engineers*, **48**, 1783–1794.

Hejun, C., Hanchao, L. and Zhuoyuan, A. (1998). Study on the categories of landslide-damming of rivers and their characteristics. *Journal of the Chengdu Institute of Technology*, **25**, 411–416.

Hemphill-Haley, M. A., Lindberg, D. A. and Reheis, M. R. (1999). Lake Alvord and Lake Coyote: a hypothesized flood. In *Quaternary Geology of the Northern Quinn River and Alvord Valleys, Southeastern Oregon. 1999 Friends of the Pleistocene Pacific Cell Field Trip Guidebook*, ed. C. Narwold. Arcata, CA: Humboldt State University, pp. A21–A27.

Herget, J. (2005). *Reconstruction of Pleistocene Ice-dammed Lake Outburst Floods in the Altai Mountains, Siberia*. Geological Society of America Special Paper 386.

Hermanns, R. L., Niedermann, S., Ivy-Ochs, S. and Kubik, P. W. (2004). Rock avalanching into a landslide-dammed lake causing multiple dam failure in Las Conchas valley (NW Argentina): evidence from surface exposure dating and stratigraphic analyses. *Landslides*, **1**, 113–122.

Hewitt, K. (1968). Record of natural damming and related floods in the Upper Indus Basin. *Indus*, **10** (3), 11–19.

Hewitt, K. (1982). Natural dams and outburst floods of the Karakoram Himalaya. In *Hydrological Aspects of Alpine and High-Mountain Areas*. International Association of Hydrological Sciences Publication 138, pp. 259–269.

Hewitt, K. (1998). Catastrophic landslides and their effects on the Upper Indus streams, Karakoram Himalaya, northern Pakistan. *Geomorphology*, **26**, 47–80.

Hewitt, K. (2006). Disturbance regime landscapes: mountain drainage systems interrupted by large rockslides. *Progress in Physical Geography*, **30**, 365–393.

Hewitt, K., Clague, J. J. and Orwin, J. F. (2008). Legacies of catastrophic rock slope failures in mountain landscapes. *Earth Science Reviews*, **87**, 1–38.

Hickson, C. J., Russel, J. K. and Stasiuk, M. V. (1999). Volcanology of the 2350 B.P. eruption of Mount Meager Volcanic Complex, British Columbia, Canada: implications for hazards from eruptions in topographically complex terrain. *Bulletin of Volcanology*, **60**, 489–507.

Hodgson, K. A. and Nairn, I. A. (2005). The c. AD 1315 syn-eruption and AD 1904 post-eruption breakout floods from Lake Tarawera, Haroharo caldera, North Island, New Zealand. *New Zealand Journal of Geology and Geophysics*, **48**, 491–506.

House, P. K., Pearthree, P. A., and Perkins, M. D. (2008). Stratigraphic evidence for the role of lake-spillover in the birth of the lower Colorado River in southern Nevada and western Arizona. In *Geologic and Biologic Evolution of the Southwest*, Geological Society of America Special Paper 439, eds. M. C. Reheis, R., Hershler and D. M. Miller, pp. 333–351.

Howard, K. A., Shervais, J. W. and McKee, E. H. (1982). Canyon-filling lavas and lava dams on the Boise River, Idaho, and their significance for evaluating downcutting during the last two million years. In *Cenozoic Geology of Idaho*, eds. B. Bonnichsen and R. M. Breckenridge. Moscow, ID: Idaho Bureau of Mines and Geology, pp. 629–644.

Hsü, K. J. (1983). *The Mediterranean was a Desert: a Voyage of the Glomar Challenger*. Princeton, NJ: Princeton University Press.

Huggel, C., Kääb, A., Haeberli, W., Teysseire, P. and Paul, F. (2002). Remote sensing based assessment of hazards from glacier lake outbursts: a case study in the Swiss Alps. *Canadian Geotechnical Journal*, **39**, 316–330.

Huggel, C., Kääb, A., Haeberli, W. and Krummenacher, B. (2003). Regional-scale GIS-models for assessment of hazards from glacier lake outbursts: evaluation and application in the Swiss Alps. *Natural Hazards and Earth System Sciences*, **3**, 647–662.

Huggel, C., Haeberli, W., Kääb, A., Bieri, D. and Richardson, S. (2004). An assessment procedure for glacial hazards in the Swiss Alps. *Canadian Geotechnical Journal*, **41**, 1068–1083.

Hulsing, H. (1981). *The Breakout of Alaska's Lake George*. U.S. Geological Survey.

Hung, J.-J. (2000). Chi-Chi earthquake induced landslides in Taiwan. *Earthquake Engineering and Engineering Seismology*, **2** (2), 25–33.

Iverson, R. M. (2006). Langbein lecture: Shallow flows that shape Earth's surface. *EOS Transactions, American Geophysical Union*, **87** (52), Abstract H22A-01.

Jacobson, R. B., O'Connor, J. E., and Oguchi, T. (2003). Surficial geologic tools in fluvial geomorphology. In *Tools in Fluvial Geomorphology*, eds. G. M. Kondolf and H. Piégay. Chichester, UK: John Wiley and Sons, pp. 25–57.

Jakob, M. and Jordan, P. (2001). Design flood estimates in mountain streams: the need for a geomorphic approach. *Canadian Journal of Civil Engineering*, **28**, 425–439.

Jarrett, R. D. and Costa, J. E. (1986). *Hydrology, Geomorphology, and Dam-break Modeling of the July 15, 1982, Lawn Lake Dam and Cascade Dam Failures, Larimer County, Colorado*. U.S. Geological Survey Professional Paper 1369.

Jarrett, R. D. and Malde, H. E. (1987). Paleodischarge of the late Pleistocene Bonneville Flood, Snake River, Idaho, computed from new evidence. *Geological Society of America Bulletin*, **99**, 127–134.

Jennings, M. E., Schneider, V. R. and Smith P. E. (1981). The 1980 eruptions of Mount St. Helens, Washington, Computer assessments of potential flood hazards from breaching of two debris dams, Toutle River and Cowlitz River systems. In *The 1980 Eruptions of Mount St. Helens Washington*, U.S. Geological Survey Professional Paper 1250, eds. P. W. Lipman and D. R. Mullineaux, pp. 829–836.

Jennings, D. N., Webby, M. G. and Parkin, D. T. (1993). Tunawaea landslide dam, King Country, New Zealand. *Landslide News*, **7**, 25–27.

Johnson, D. M., Petersen, R. R., Lycan, D. R. *et al.* (1985). *Atlas of Oregon Lakes*. Corvallis, OR: Oregon State University Press.

Kataoka, K. S., Urabe, A., Manville, V. and Kajiyama, A. (2008). Breakout flood from an ignimbrite-dammed valley after the 5 ka Numazawako eruption, northeast Japan. *Geological Society of America Bulletin*, **120**, 1233–1247.

Kattelmann, R. (2003). Glacial lake outburst floods in the Nepal Himalaya: A manageable hazard? *Natural Hazards*, **28**, 145–154.

Kershaw, J. A., Clague, J. J. and Evans, S. E. (2005). Geomorphic and sedimentological signature of a two-phase outburst flood from moraine-dammed Queen Bess Lake, British Columbia, Canada. *Earth Surface Processes and Landforms*, **30**, 1–25.

Kilburn, C. R. J. and Petley, D. N. (2003). Forecasting giant, catastrophic slope collapse: lessons from Vajont, Northern Italy. *Geomorphology*, **54**, 21–32.

King, H. W. (1954). *Handbook of Hydraulics*, 4th edn. New York: McGraw Hill.

King, J., Loveday, I. and Schuster, R. L. (1989). The 1985 Bairaman landslide dam and resulting debris flow, Papua New Guinea. *Quarterly Journal of Engineering Geology, London*, **22**, 257–270.

Komar, P. D. (1987). Selective gravel entrainment and the empirical evaluation of flow competence. *Sedimentology*, **34**, 1165–1176.

Korup, O. (2002). Recent research on landslide dams: a literature review with special attention to New Zealand. *Progress in Physical Geography*, **26**, 206–235.

Korup, O. (2004). Geomorphometric characteristics of New Zealand landslide dams. *Engineering Geology*, **73**, 13–35.

Korup, O. (2006). Rock-slope failure and the river long profile. *Geology*, **34**, 45–48.

Korup, O. and Tweed, F. (2007). Ice, moraine, and landslide dams in mountainous terrain. *Quaternary Science Reviews*, **26**, 3406–3422.

Korup, O., Strom, A. L. and Weidinger, J. T. (2006). Fluvial response to large rock-slope failures: examples from the Himalayas, the Tien Shan, and the Southern Alps in New Zealand. *Geomorphology*, **78**, 3–21.

Korup, O., Clague, J. J., Hermanns, R. L. *et al.* (2007). Giant landslides, topography, and erosion. *Earth and Planetary Science Letters*, **261**, 578–579.

Lagmay, A. M. F., Rodolfo, K. S., Siringan, F. P. *et al.* (2007). Geology and hazard implications of the Maraunot notch in the Pinatubo Calera, Philippines. *Bulletin of Volcanology*, **69**, 797–809.

Lamke, R. D. (1972). *Floods of the Summer of 1971 in Southcentral Alaska*. U.S. Geological Survey Open-file Report.

Lancaster, S. T. and Grant, G. E. (2006). Debris dams and the relief of headwater streams. *Geomorphology*, **82**, 84–97, doi:10.1016/j.geomorph.2005.08.020.

Larson, G. L. (1989). Geographical distribution, morphology and water quality of caldera lakes: a review. *Hydrobiologia*, **171**, 23–32.

Lee, K. L. and Duncan, J. M. (1975). *Landslide of April 25, 1974, on the Mantaro River, Peru*. Washington, DC: Committee on Natural Disasters, Commission on Sociotechnical Systems, National Research Council.

Li, D. and You, Y. (1992). Bursting of the Midui Moraine Lake in Bomi, Xizang. *Mountain Research* [China], **10**–4, 219–224. [In Chinese, English summary.]

Li, T., Schuster, R. L. and Wu, J. (1986). Landslide dams in south-central China. In *Landslide Dams: Process, Risk, and Mitigation*. American Civil Engineers Special Publication No. 3, ed. R. L. Schuster, pp. 146–162.

Lipscomb, S. W. (1989). *Flow and Hydraulic Characteristics of the Knik-Matanuska River Estuary, Cook Inlet, Southcentral Alaska*. U.S. Geological Survey Water-Resources Investigations Report 89–4064.

Liu, C. and Sharmal, C. K. (Eds.) (1988). *Report on First Expedition to Glaciers and Glacier Lakes in the Pumqu (Arun) and Poiqu (Bhote-Sun Kosi) River Basins, Xizang (Tibet), China.* Beijing: Science Press.

Lliboutry, L., Arnao, V. M., Pautre, A. and Schneider, B. (1977). Glaciological problems set by the control of dangerous lakes in Cordillera Blanca, Peru: I. Historical failures of morainic dams, their causes and prevention. *Journal of Glaciology*, **18**, 239–254.

Lockwood, J. P., Costa, J. E., Tuttle, M. L., Nni, J. and Tebor, S. G. (1988). The potential for catastrophic dam failure at Lake Nyos Maar, Cameroon. *Bulletin of Volcanology*, **50**, 340–349.

Lü, R. and Li, D. (1986). Debris flow induced by ice lake burst in the Tangbulang Gully, Gonbujiangda, Xizang (Tibet). *Journal of Glaciology and Geocryology* [China], **8**–1, 61–70. (In Chinese, English summary.)

Lubbock, G. (1894). The Gohna lake. *Geographical Journal*, **4**, 457.

Macchione, F. and Sirangelo, B. (1990). Floods resulting from progressively breached dams. In *Proceedings of Lausanne Symposia, August 1990, Hydrology in Mountainous Regions.* International Association of Hydrological Sciences (IAHS-AISH) Publication No. 194, pp. 325–332.

MacDonald, T. C. and Langridge-Monopolis, J. (1984). Breaching characteristics of dam failures. *Journal of Hydraulic Engineering*, **110**, 567–586.

Macías, J. L., Capra, L., Scott, K. M. *et al.* (2004). The 26 May 1982 breakout flows derived from failure of a volcanic dam at El Chichón, Chiapas, Mexico. *Geological Society of America Bulletin*, **116**, 233–246.

Malde, H. E. (1968). *The Catastrophic Late Pleistocene Bonneville Flood in the Snake River Plain, Idaho.* U.S. Geological Survey Professional Paper 596.

Malde, H. E. (1982). The Yahoo Clay, a lacustrine unit impounded by the McKinney Basalt in the Snake River canyon near Bliss, Idaho. In *Cenozoic Geology of Idaho*, eds. B. Bonnichsen and R. M. Breckenridge. Moscow, ID: Idaho Bureau of Mines and Geology, pp. 617–628.

Manville, V. (2001). *Techniques for Evaluating the Size of Potential Dam-break Floods from Natural Dams.* Science Report no. 2001/28, Institute of Geological and Nuclear Sciences, Wellington, New Zealand.

Manville, V. (2004). Paleohydraulic analysis of the 1953 Tangwai lahar: New Zealand's worst volcanic disaster. *Acta Vulcanologica.* **16** (1–2), unpaginated.

Manville, V. and Cronin, S. J. (2007). Breakout lahar from New Zealand's Crater Lake. *EOS, Transactions, American Geophysical Union*, **88**, 442–443.

Manville, V. and Wilson, C. J. N. (2004). The 26.5 ka Oruanui eruption, New Zealand: a review of the roles of volcanism and climate in the post-eruptive sedimentary response. *New Zealand Journal of Geology and Geophysics*, **47**, 525–547.

Manville, V., White, J. D. L., Houghton B. F. and Wilson, C. J. N. (1999). Paleohydrology and sedimentology of a post-1.8 ka breakout flood from intracaldera Lake Taupo, North Island, New Zealand. *Geological Society of America Bulletin*, **111**, 1435–1447.

Manville, V., Hodgson, K. A. and Nairn, I. A. (2007). A review of break-out floods from volcanogenic lakes in New Zealand. *New Zealand Journal of Geology and Geophysics*, **50**, 131–150.

Marche, C., Mahdi, T. and Quach, T. (2006). Erode: une methode fiable pour etablir l'hydrogramme de rupture potentielle par surverse de chaque digue en terre. *Transactions of the International Congress on Large Dams*, **22** (3), 337–360.

Mayo, L. R. (1989). Advance of Hubbard Glacier and 1986 outburst of Russell Fiord, Alaska, U.S.A. *Journal of Glaciology*, **13**, 189–194.

McCullough, D. G. (1968). *The Johnstown Flood.* New York: Simon and Schuster.

McGimsey, R. G., Waythomas, C. F. and Neal, C. A. (1994). High stand and catastrophic draining of intracaldera Surprise Lake, Aniakchak Volcano, Alaska. In *Geologic Studies in Alaska by the U.S.A. Geological Survey, 1993.* U.S. Geological Survey Bulletin 2107, eds. A. B. Till and T. E. Moore, pp. 59–71.

McKillop, R. J. and Clague, J. J. (2007a). Statistical, remote sensing-based approach for estimating the probability of catastrophic drainage from moraine-dammed lakes in southwestern British Columbia. *Global and Planetary Change*, **56** 153–171; doi:10.1016/j.gloplacha.2006.07.004.

McKillop, R. J. and Clague, J. J. (2007b). A procedure for making objective preliminary assessments of outburst flood hazard from moraine-dammed lakes in southwestern British Columbia, *Natural Hazards*, **41**, 131–157, doi:10.1007/s11069–006–9028–7.

Meyer, W., Sabol, M. A., Glicken, H. X. and Voight, B. (1985). *The Effects of Ground Water, Slope Stability and Seismic Hazards on the Stability of the South Fork Castle Creek Blockage in the Mount St. Helens Area, Washington.* U.S. Geological Survey Professional Paper 1345.

Meyer, W., Sabol, M. A. and Schuster, R. L. (1986). Landslide dammed lakes at Mount St. Helens, Washington. In *Landslide Dams: Process, Risk, and Mitigation.* American Civil Engineers Special Publication 3, ed. R. L. Schuster, pp. 21–41.

Montgomery, D. R., Hallet, B., Liu, Y. *et al.* (2004). Evidence for Holocene megafloods down the Tsangpo River gorge, southeastern Tibet. *Quaternary Research*, **62**, 201–207.

Mool, P. K. (1995). Glacier lake outburst floods in Nepal. *Journal of Nepal Geological Society*, **11**, 273–280.

Mora, S., Madrigal, C., Estrada J. and Schuster, R. L. (1993). The 1992 Rio Toro Landslide Dam, Costa Rica. *Landslide News*, **7**, 19–22.

Newhall, C. G., Paull, C. K., Bradbury, J. P., *et al.* (1987). Recent geologic history of Lake Atitlán, a caldera lake in western Guatemala. *Journal of Volcanology and Geothermal Research*, **33**, 81–107.

Nicoletti, P. G. and Parise, M. (2002). Seven landslide dams of old seismic origin in southeastern Sicily (Italy). *Geomorphology*, **46**, 203–222.

O'Connor, J. E. (1993). *Hydrology, Hydraulics, and Geomorphology of the Bonneville Flood.* Geological Society of America Special Paper 274.

O'Connor, J. E. (2004). The evolving landscape of the Columbia River Gorge: Lewis and Clark and cataclysms on the Columbia. *Oregon Historical Quarterly*, **105**, 390–421.

O'Connor, J. E. and Baker, V. R. (1992). Magnitudes and implications of peak discharges from glacial Lake Missoula. *Geological Society of America Bulletin*, **104**, 267–279.

O'Connor, J. E. and Costa, J. E. (1993). Geologic and hydrologic hazards in glacierized basins resulting from 19th and 20th century global warming. *Natural Hazards*, **8**, 121–140.

O'Connor, J. E. and Costa, J. E. (2004). *The World's Largest Floods, Past and Present: Their Causes and Magnitudes.* U.S. Geological Survey Circular 1254.

O'Connor, J. E. and Waitt, R. B. (1995). Beyond the Channeled Scabland: A field trip guide to Missoula Flood features in the Columbia, Yakima, and Walla Walla Valleys of Washington and Oregon. *Oregon Geology*, **57**, Part I pp. 51–60, Part II pp. 75–86, Part III pp. 99–115.

O'Connor, J. E., Hardison III, J. H. and Costa, J. E. (2001). *Debris Flows from Failures of Neoglacial Moraine Dams in the Three Sisters and Mt. Jefferson Wilderness areas, Oregon.* U.S. Geological Survey Professional Paper 1608.

O'Connor, J. E., Grant, G. E. and Costa, J. E. (2002). The geology and geography of floods. In *Ancient Floods, Modern Hazards, Principles and Applications of Paleoflood Hydrology*. American Geophysical Union Water Science and Application Series 4, eds. P. K. House, R. H. Webb and D. R. Levish, pp. 191–215.

O'Connor, J. E., Curran, J. H., Beebee, R. A., Grant, G. E. and Sarna-Wojcicki, A. (2003). Quaternary geology and geomorphology of the lower Deschutes River canyon, Oregon. In *A Peculiar River: Geology, Geomorphology, and Hydrology of the Deschutes River, Oregon.* American Geophysical Union Water Science and Application Series No. 7, eds. J. E. O'Connor and G. E. Grant, pp. 73–94.

O'Neill, A. L. and Gourley, C. (1991). Geologic perspectives and cause of the Quail Creek dike failure. *Bulletin of the Association of Engineering Geologists*, **28**, 127–145.

Othberg, K. L. (1994). *Geology and Geomorphology of the Boise Valley and Adjoining Areas, Western Snake River Plain, Idaho.* Bulletin 29, Idaho Geological Survey. Moscow, ID: Idaho Geological Survey Press.

Palmer, L. (1977). Large landslides of the Columbia River Gorge, Oregon and Washington. In *Geological Society of America Reviews in Engineering Geology, Volume III.* Boulder, CO: Geological Society of America, pp. 69–84.

Parkinson, M. and Stretch, D. (2007). Breaching timescales and peak outflows for perched, temporary open estuaries, *Coastal Engineering Journal*, **49**, 267–290.

Pena, H. and Klohn, W. (1989). Non-meteorological flood disasters in Chile. In *Hydrology of Disasters*, eds. Ö. Starosolszky and O. M. Melder. London: James and James, pp. 243–258.

Perrin, N. D. and Hancox, G. T. (1992). Landslide dammed lakes in New Zealand: preliminary studies on the distributions, causes, and effects. In *Landslides: Vol. 2 Proceedings of the 6th International Symposium on Landslides*, ed. D. H. Bell. Christchurch, New Zealand: A. A. Balkema, pp. 1457–1466.

Pierson, T. C., Janda, R. J., Thouret, J.-C. and Borrero, C. A. (1990). Perturbation and melting of snow and ice by the 13 November 1985 eruption of Nevado del Ruiz, Colombia, and consequent mobilization, flow, and deposition of lahars. *Journal of Volcanology and Geothermal Research*, **41**, 17–66.

Pirocchi, A. (1991). *Landslide Dammed Lakes in the Alps: Typology and Evolution.* Unpublished doctoral thesis, Pavia, Italy (in Italian).

Plaza-Nieto, G. and Zevallos, O. (1994). The 1993 La Josefina rockslide and Rio Paute landslide dam, Ecuador. *Landslide News*, **8**, 4–6.

Plaza-Nieto, G., Yepes, H. and Schuster, R. L. (1990). Landslide dam on the Pisque River, northern Ecuador. *Landslide News*, **4**, 2–4.

Ponce, V. M. (1982). *Documented Cases of Earth Dam Breaches.* San Diego State University Civil Engineering Series No. 82149.

Ponce, V. M. and Tsivoglou, A. M. (1981). Modeling gradual dam breaches. *Journal of the Hydraulics Division, Proceedings of the American Society of Civil Engineers*, **107**, 829–838.

Porter, S. C. and Denton, G. H. (1967). Chronology of neoglaciation in the North American Cordillera. *American Journal of Science*, **265**, 177–210.

Rathburn, S. L. (1989). Pleistocene glacial outburst flooding along the Big Lost River, east-central Idaho. Unpublished M. S. thesis, University of Arizona, Tucson, Arizona.

Reheis, M. C., Sarna-Wojcicki, A. M., Reynolds, R. L., Repenning, C. A. and Mifflin, M. D. (2002). Pliocene to Middle Pleistocene lakes in the western Great Basin: ages and connections. In *Great Basin Aquatic Systems History*, eds. R. Hershler, D. B. Madsen and D. R. Currey. Washington DC: Smithsonian Institution Press, pp. 53–108.

Reynolds, J. J. (1992). The identification and mitigation of glacier-related hazards: examples from the Cordillera Blanca, Peru. In *Geohazards: Natural and Man-made*, eds. G. J. H. McCall, D. J. C. Laming and S. C. Scott. London: Chapman and Hall, pp. 143–157.

Richardson, S. D. and Reynolds, J. M. (2000). An overview of glacial hazards in the Himalayas. *Quaternary International*, **65–66**, pp. 31–47.

Risley, J. C., Walder, J. S. and Denlinger, R. P. (2006). Usoi Dam wave overtopping and flood routing in the Bartang and Panj Rivers, Tajikistan. *Natural Hazards*, **38**, 375–390.

Rogers, J. D. and McMahon, D. J. (1993). Reassessment of the St. Francis Dam Failure. In *Proceedings of Dam Safety '93, 10th Annual ASDSO Conference*, Kansas City, Missouri, September 26–29, pp. 333–339.

Ryan, W. and Pitman, W. (1999). *Noah's Flood: The New Scientific Discoveries about the Event that Changed History.* New York: Simon and Schuster.

Safran, E. B., Anderson, S. W., Ely, L. *et al.* (2006). Controls on large landslide distribution in the southeastern interior

Columbia River basin. *EOS Transactions, American Geophysical Union*, **87** (52), Abstract H53B-0623.

Schuster, R. L. (2000). Outburst debris flows from failure of natural dams. In *Proceedings 2nd International Conference on Debris Flow Hazard Mitigation*, 16–20 August, Taipei, pp. 29–42.

Schuster, R. L. and Costa, J. E. (1986). A perspective on landslide dams. In *Landslide Dams: Process, Risk, and Mitigation*. American Civil Engineers Special Publication 3, ed. R. L. Schuster, pp. 1–20.

Schuster, R. L. and Highland, L. M. (2001). *Socioeconomic and Environmental Impacts of Landslides in the Western Hemisphere*. U.S. Geological Survey Open-file Report 01–0276.

Schuster, R. L., Salcedo, D. A. and Valenzuela, L. (2002). Overview of catastrophic landslides of South America in the twentieth century. In *Catastrophic Landslides: Effects, Occurrence, and Mechanisms*. Geological Society of America Reviews in Engineering Geology Volume XV, eds. S. G. Evans and J. V. DeGraff, pp. 1–34.

Scott, K. M. (1988). *Origin, Behavior, and Sedimentology of Lahars and Lahar-runout Flows in the Toutle-Cowlitz River System*. U.S. Geological Survey Professional Paper 1447-A.

Scott, K. M. (1989). *Magnitude and Frequency of Lahars and Lahar-runout Flows in the Toutle-Cowlitz River System*. U.S. Geological Survey Professional Paper 1447-B.

Scott, K. M. and Gravlee, G. C. (1968). *Flood Surge on the Rubicon River, California: Hydrology, Hydraulics and Boulder Transport*. U.S. Geological Survey Professional Paper 422-M.

Scott, W. E., Pierce, K. L., Bradbury, J. P. and Forester R. M. (1982). Revised Quaternary stratigraphy and chronology in the American Falls area, southeastern Idaho. In *Cenozoic Geology of Idaho*, eds. B. Bonnichsen and R. M. Breckenridge. Moscow, ID: Idaho Bureau of Mines and Geology, pp. 581–595.

Selting, A. J. and Keller, E. A. (2001). The Mission debris flow: an example of a prehistoric landslide dam failure, Santa Barbara, California. *Geological Society of America Abstracts with Programs*, **33** (6).

Shang, Y., Yang, Z., Li, L. *et al.* (2003). A super-large landslide in Tibet in 2000: background, occurrence, disaster, and origin. *Geomorphology*, **54**, 225–243.

Shroder Jr., J.F. (1998). Slope failure and denudation in the western Himalaya. *Geomorphology*, **26**, 81–105.

Shroder Jr., J. F., Cornwell K. and Khan, M. S. (1991). Catastrophic breakout floods in the western Himalaya, Pakistan. *Geological Society of America Abstracts with Program*, **23** (5), 87.

Simmler, H. and Samet, L. (1982). Dam failure from overtopping studied on a hydraulic model. In *Fourteenth International Congress on Large Dams, Rio de Janeiro, Brazil, 3–7 May, 1982*, 427–446.

Singh, V. P., Scarlatos, P. D., Collins, J. G. and Jourdan, M. R. (1988). Breach erosion of earthfill dams (BEED) model. *Natural Hazards*, **1**, 161–180.

Snow, D. T. (1964). Landslide of Cerro condor-Sencca, Department of Ayacucho, Peru. In *Engineering Geology Case Histories*. Geological Society of America No. 5, ed. G. A. Kiersch, pp. 1–5.

Sowma-Bawcom, J. A. (1996). Breached landslide dam on the Navarro River, Mendocino County, California. *California Geology*, September/October 1996, pp. 120–127.

Stearns, H. T. (1931). *Geology and Water Resources of the Middle Deschutes River Basin, Oregon*. U.S. Geological Survey Water Supply Paper 637-D, pp. 125–220.

Stelling, P., Gardner, J. and Beget, J. E. (2005). Eruptive history of Fisher Caldera, Alaska, USA. *Journal of Volcanology and Geothermal Research*, **139**, 163–183.

Strachey, R. (1894). The landslip at Gohna, in British Garwhal. *Geographical Journal*, **4**, 172–170.

Stretch, K. and Parkinson, M. (2006). The breaching of sand barriers at perched, temporary open/closed estuaries: a model study. *Coastal Engineering Journal*, **48** (1), 13–30.

Swanson, F. J., Oyagi, N. and Tominaga, M. (1986). Landslide dams in Japan. In *Landslide Dams: Process, Risk, and Mitigation*. American Civil Engineers Special Publication No. 3, ed. R. L. Schuster, pp. 131–145.

Tomasson, H. (1996). The jökulhlaup from Katla in 1918. *Annals of Glaciology*, **22**, 249–254.

Trabant, D. C., March, R. S. and Thomas, D. S. (2003). *Hubbard Glacier, Alaska: Growing and Advancing In Spite of Global Climate Change and the 1986 and 2002 Russell Lake Outburst Floods*. U.S. Geological Survey Fact Sheet 001–03.

Umbal J. V. and Rodolfo, K. S. (1996). The 1991 lahars of southwestern Mount Pinatubo and evolution of the lahar-dammed Mapanuepe Lake, In *Fire and Mud: Eruptions and Lahars of Mount Pinatubo*, eds. C. G. Newhall and R. S. Punongbayan, Seattle: University of Washington Press, pp. 951–970.

Vanaman, K. M., O'Connor, J. E. and Riggs, N. (2006). Pleistocene history of Lake Millican, central Oregon. *Geological Society of America Abstracts with Programs*, **38** (7), p. 71.

Vaskinn, K. A., Lovoll, A., Hoeg, K. *et al.* (2004). Physical modeling of breach formation – large scale field tests. In *Dam Safety 2004: Proceedings of the Association State Dam Safety Officials*, Phoenix, Arizona, September 2004.

Voight, B., Glicken, H., Janda, R. J. and Douglass, P. M. (1981). Catastrophic rockslide avalanche of May 18. In *The 1980 Eruptions of Mount St. Helens Washington*. U.S. Geological Survey Professional Paper 1250, eds. P. W. Lipman and D. R. Mullineaux, pp. 821–828.

Von Poschinger, A. (2004). The Flims rockslide dam. In *Security of Natural and Artificial Rockslide Dams: Extended Abstracts Volume, NATO Advanced Research Workshop*, eds. K. Abdrakhmatov, S. G. Evans, R. Hermanns G. Scarascia Mugnozza and A. L. Strom. Bishkek, Kyrgyzstan. pp. 141–144

Vuichard, D. and Zimmermann, M. (1987). The 1985 catastrophic drainage of a moraine-dammed lake, Khumbu Himal:

cause and consequences. *Mountain Research and Development*, **7**, 91–110.

Wahl, T. L. (1998). *Prediction of Embankment Dam Breach Parameters: A Literature Review and Needs Assessment.* Dam Safety Research Report DSO-98–004. U.S. Department of the Interior, Bureau of Reclamation, Dam Safety Office.

Wahl, T. L. (2004). Uncertainty of predictions of embankment dam breach parameters. *Journal of Hydraulic Engineering*, **130**, 389–397.

Waitt, R. B. (2002). Great Holocene floods along Jökulsá á Fjöllum, North Iceland. In *Flood and Megaflood Processes, Recent and Ancient Examples*. International Association Sedimentologists Special Publication 32, eds. I. P. Martini, V. R. Baker and G. Garzón, pp. 37–51.

Walder, J. S. and Costa, J. E. (1996). Outburst floods from glacier-dammed lakes: the effect of mode of lake drainage on flood magnitude. *Earth Surface Processes and Landforms*, **21**, 701–723.

Walder, J. S. and O'Connor J. E. (1997). Methods for predicting peak discharge of floods caused by failure of natural and constructed earthen dams. *Water Resources Research*, **33**, 2337–2348.

Wassmer, P., Schneider, J. L., Pollet, N. and Schmitter-Voirin, C. (2004). Effects of the internal structure of a rock-avalanche dam on the drainage mechanism of its impoundment: Flims sturzstrom and Ilanz paleo-lake, Swiss Alps. *Geomorphology*, **61**, 3–17.

Watanabe, T. and Rothacher, D. (1996). The 1994 Lugge Tsho glacial lake outburst flood, Bhutan Himalaya. *Mountain Research and Development*, **16**, 77–81.

Waythomas, C. F. (2001). Formation and failure of volcanic debris dams in the Chakachatna River valley associated with eruptions of the Spurr volcanic complex, Alaska. *Geomorphology*, **39**, 111–129.

Waythomas, C. F., Walder, J. S., McGimsey, R. G. and Neal, C. A. (1996). A catastrophic flood caused by drainage of a caldera lake at Aniakchak volcano, Alaska, and implications for volcanic-hazards assessment. *Geological Society of America Bulletin*, **108**, 861–871.

Webby, M. G. (1996). Discussion [of Froehlich (1995)]. *Journal of Water Resources Planning and Management*, **122**, 316–317.

Webby, M. G. and Jennings, D. N. (1994). Analysis of dam-break flood caused by failure of Tunawaea landslide dam. *Proceedings of International Conference on Hydraulics in Civil Engineering 1994*. Queensland, Australia: University of Brisbane, pp. 163–168.

Weber, G. E., Nielsen, H. P. and Kraeger, B. K. (1986). The Foreman Creek flood: failure of a landslide formed debris dam during the 1–4–82 storm, Sand Cruz Co., CA. *Geological Society of America Abstracts with Programs*, **18** (2), p. 196.

Weischet, W. (1963). Further observations of geologic and geomorphic changes resulting from the catastrophic earthquake of May 1960, in Chile. *Bulletin of the Seismological Society of America*, **53**, 1237–1257.

Whipple, K. X., Hancock, G. S. and Anderson, R. S. (2000). River incision into bedrock: mechanics and relative efficacy of plucking, abrasion, and cavitation. *Geological Society of America Bulletin*, **112**, 490–503.

Whitehouse, I. E. and Griffiths, G. A. (1983). Frequency and hazard of large rock avalanches in the central Southern Alps. *Geology*, **11**, 331–334.

Williams, H. (1941). Calderas and their origins. *University of California Bulletin of Geological Sciences*, **25**, 239–346.

Williams, G. P. (1983). Paleohydrological methods and some examples from Swedish fluvial environments, I: Cobble and boulder deposits. *Geografiska Annaler*, **65A**, 227–243.

Wolfe, B. A. and Begét, J. E. (2002). Destruction of an Aleut village by a catastrophic flood release from Okmok Caldera, Umnak Island, Alaska. *Geological Society of America Abstracts with Programs*, **34** (6), p. 126.

Wood, S. H. and Clemens, D. M. (2002). Geologic and tectonic history of the western Snake River Plain, Idaho and Oregon. In *Tectonic and Magmatic Evolution of the Snake River Plain Volcanic Province*. Idaho Geological Survey Bulletin 30, eds. B. Bonnichsen, C. M. White and M. McCurry. Moscow, ID: Idaho Geological Survey Press, pp. 69–103.

Xu, D. (1988). Characteristics of debris flow caused by outburst of glacial lakes on the Boqu River in Xizang, China. *Geojournal*, **17**, 569–580.

Xu, D. and Feng, Q. (1994). Dangerous glacier lakes and their outburst features in the Tibetan Himalayas. *Bulletin of Glacier Research*, **12**, 1–8.

Yamada, T. and Sharma, C. K. (1993). Glacier lakes and outburst floods in the Nepal Himalaya. In *Snow and Glacier Hydrology, Proceedings of the Katmandu Symposium, November, 1992*, International Association of Hydrological Sciences Publication No. 218, pp. 319–330.

Yamada T., Naito, N., Kohshima, S. *et al.* (2004). Outline of 2002 research activities on glaciers and glacier lakes in Lunana region, Bhutan Himalayas. *Bulletin of Glaciological Research*, **21**, 79–90.

Yesenov, U. Y. and Degovets, A. S. (1979). Catastrophic mudflow on the Bol'shaya Almatinka River in 1977. *Soviet Hydrology, Selected Papers* (published by the American Geophysical Union), **18**, 158–160.

Youd, T. L., Wilson, R. C. and Schuster, R. L. (1981). Stability of blockage in North Fork Toutle River. In *The 1980 Eruptions of Mount St. Helens Washington*. U.S. Geological Survey Professional Paper 1250, eds. P. W. Lipman and D. R. Mullineaux, pp. 821–828.

9

Surface morphology and origin of outflow channels in the Valles Marineris region

NEIL M. COLEMAN
and VICTOR R. BAKER

Summary

The outflow channels that emptied into Chryse Planitia provide the best evidence that great quantities of water once flowed on the Martian surface. Some channels were created when the cryosphere ruptured and groundwater discharged from chaos or from cavi along major fault zones. Some chaos formed on channel floors when fluvial erosion thinned the cryosphere, leading to catastrophic breakout of confined groundwater. These chaos can be used to estimate the cryosphere thickness, crustal heat flux and climate trends. At Iamuna Chaos the cryosphere was 700–1000 m thick when Ravi Vallis formed, indicating a cold, long-term climate similar to present-day Mars. The discovery of outflow channels at elevations >2500 m in Ophir Planum shows that Hesperian recharge likely occurred in upslope areas to the west (e.g. Sinai Planum, Tharsis highlands, Syria Planum). The larger circum-Chryse channels were carved by floods that issued directly from the ancestral canyons, which likely were smaller and less interconnected than today. A plausible mechanism for water release was catastrophic drainage of chasm lakes caused by the collapse of topographic barriers or ice-debris dams. We report evidence that a megaflood filled Capri Chasma and overtopped its eastern rim, carving two crossover channels and spectacular dry falls cataracts. This flooding may represent an initial outpouring of canyon lakes via a gateway in eastern Coprates Chasma. The recent discovery of hematite and abundant hydrated sulphates in the Valles Marineris canyons provides compelling evidence of a water-rich history. Ultimately, the interconnection and draining of lakes in the ancestral Valles Marineris produced floods that carved the Simud-Tiu channels, permanently lowered regional groundwater levels, and ended most fluvial activity. If a frozen ocean existed when the northernmost Simud-Tiu channels formed, its surface would have been lower than an elevation of –3760 m.

9.1 Introduction

This chapter examines the morphology and origins of the circum-Chryse outflow channels (except for Mawrth Vallis, an older channel not reviewed here). These large channels provide the best evidence that great quantities of water once flowed at the surface of Mars (see Figure 9.1 and online regional maps at http://planetarynames.wr.usgs.gov/mgrid_mola.html). They also provide indirect evidence that ice-covered lakes were present in at least some of the ancestral canyons of the Valles Marineris. Since their discovery in Mariner 9 images these channels have inspired a wealth of ideas and scientific debates about their origin. Much of the disagreement about the outflow channels and the fluid that carved them stems from varying assumptions about subsurface conditions and the prevailing climate when they formed. Researchers have variously concluded they were carved by wind, rivers of liquid carbon dioxide, lava flows, carbon dioxide-charged debris avalanches, glaciers, aqueous floods or combinations of these. See Carr (1996) and Coleman (2005) for discussion about alternative processes. Some outflow channels are known to have formed in recent geologic time when the climate and atmosphere would have been substantially the same as today (Burr *et al.*, 2002). Alternatives to an aqueous flood origin do not provide consistency or coherence in explaining the whole assemblage of landforms associated with the channels (Carr, 1996; Baker, 2001, 2006). The largest channels issued from the Valles Marineris. The recent discovery of hematite and abundant hydrated sulphates in these chasms provides compelling evidence of a water-rich history (Gendrin *et al.*, 2005; 2006; Bishop *et al.*, 2007; Glotch and Rogers, 2007; Kundson *et al.*, 2007; Mangold *et al.*, 2007; Roach *et al.*, 2007). Based on this considerable body of evidence, and consistent with the megaflood theme of this volume, we adopt the prevailing view that the outflow channels were carved by aqueous floods with varying degrees of entrained ice and debris. Volcanism, impact cratering and other physical processes, possibly including ice-related processes (Lucchitta, 1982; Costard and Baker, 2001), have subsequently modified the channel floors over billions of years.

Numerous researchers have studied the circum-Chryse channels. Example works include Moore *et al.* (1995), who suggested that these outflow channels were produced by hydrologic cycling involving groundwater movement from a surface water body in Chryse Planitia. At that time Chryse was thought to be a closed basin – laser

Megaflooding on Earth and Mars, ed. Devon M. Burr, Paul A. Carling and Victor R. Baker.
Published by Cambridge University Press.

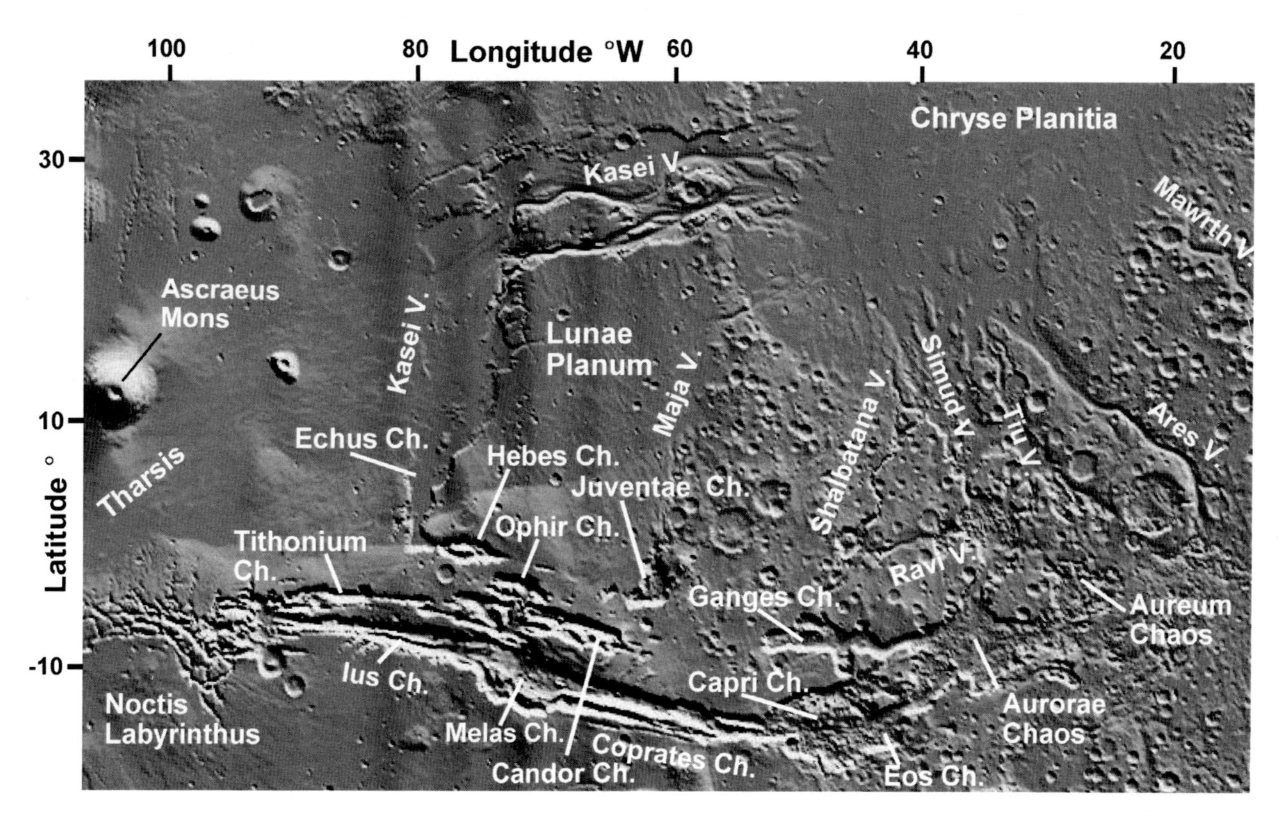

Figure 9.1. Regional view of the Valles Marineris and the circum-Chryse outflow channels. (Credit: MOLA shaded relief image, NASA Goddard.)

altimetry data confirm it is not. Tanaka (1997) mapped the surface geology of Chryse Planitia and described five major sedimentary units of Hesperian to Early Amazonian age. Nelson and Greeley (1999) mapped the geology of Xanthe Terra and interpreted the outflow channel history, with emphasis on Simud, Tiu, Ares, Mawrth and Shalbatana Valles. They proposed that three main outflow episodes occurred in Xanthe Terra: (1) broad sheetwash during the Early Hesperian Epoch, originating from the eastern Valles Marineris and chaos; (2) early channel formation; and (3) subsequent flooding that deepened the channels to their present forms.

At the kilometre scale, surface morphology provides important clues about the processes that formed the channels, but such a focused view cannot reveal the sources of the erosive liquids. The key to understanding the origin of the outflow channels is to look at much larger scales, on the order of 100–1000 km, where we can see the regional relationships from fluid sources in canyons and chaos, along channel reaches, to distal ponding areas where lakes or seas may have formed. Ivanov and Head (2001) used high-resolution MOLA grids to study the topography of Chryse Planitia and the largest outflow channels that enter this lowland (i.e., Kasei, Maja, Simud, Tiu, Ares and Mawrth Valles). They concluded that the Simud-Tiu channels appear to be younger than other circum-Chryse channels. They also examined the elevations of channel termini (i.e., locations where distinctive channel morphology changes abruptly) and compared these to elevations of hypothesised ancient shorelines. These studies found that the termini have elevations close to those of the Contact 2 shoreline of Parker *et al.* (1993) and tentatively concluded that the channels flowed into a pre-existing standing body of water. If this interpretation is correct, it means that an ocean may have existed in the northern plains prior to the formation of the circum-Chryse outflow channels rather than being caused by them. One problem for a Contact 2 ocean is that erosional features can be traced far from channel mouths into the northern lowlands, at elevations below the Contact 2 elevation of –3760 m, and no deltas are observed at the mouths of the outflow channels where the coarsest flood debris would have accumulated. Tanaka (1999) and Ivanov and Head (2001) suggested the discharges may have continued into the northern plains as density currents (hyperpycnal flows), forming submarine channels and depositing turbidites. However, it is unclear how density currents could have occurred because a shallow ocean in the northern plains likely would have frozen over its full depth, unless the water had extreme salinity with mixed salts. During the Hesperian Period, the cryosphere thickness at one location near the equator has been estimated at 700–1000 m

(Coleman, 2005). Water bodies in the northern plains would have frozen over much greater depths due to the reduced solar warmth at high latitudes and the high thermal conductivity of water.

9.2 Outflow channels that emerged from the chasmata or other large basins

The history of the circum-Chryse outflow channels is closely interwoven with the genesis of the canyons. Landscape evolution has created obstacles to reconstructing the paleohydrology of the channels. Simud, Tiu, Maja and Kasei Valles have their most distant source regions within the canyons and were created during the Hesperian Period when these chasms were actively growing. Ares Vallis appears linked to early overflows from Argyre Planitia. We cannot know the size and shape of the Valles Marineris at the time the outflow channels formed, but it is reasonable to conclude the ancestral canyons were narrower, smaller and more isolated from each other. If ice-covered lakes existed in some of the canyons, much less water would have been required to fill them compared with the geometry of the basins that we see today. Present-day examples of chasms that may resemble the early, ancestral canyons include the fully enclosed Hebes Chasma, depressions east of Tithonium Chasma, and a series of narrow canyons south of Coprates Chasma (see http://planetarynames.wr.usgs.gov/images/mc18_mola.pdf).

Rodriguez *et al.* (2005) studied the geological and hydrological history of Aureum and Hydaspis chaos and adjacent highland areas. They did not consider water storage in canyons, but instead proposed that the circum-Chryse outflow channels were excavated primarily by catastrophic and non-catastrophic releases of pressurised groundwater from extensive cavern systems. Rodriguez *et al.* (2005) further suggested that water release and collapse of caverns generated chaotic terrains, and progressive subsidence along with magmatic activity could have reactivated the outflow activity.

Tanaka (1997) re-examined Viking images of Chryse Planitia and surrounding areas and mapped five major sedimentary deposits of Hesperian to Early Amazonian age. He concluded that the Valles Marineris canyons and nearby chaotic terrain at the heads of the outflow channels may have been formed largely by huge mass flows. In this hypothesis, liquefaction of crustal material would have been triggered by seismic activity and fluid expulsion, leading to flow of material from interconnected canyons, chaos and Chryse channels. Tanaka (1997) further suggested that hyperconcentrated floods and mass flows may have been the predominant processes that eroded these outflow channels. Likewise, Rodriguez *et al.* (2006) concluded that multistage debris flows produced extensive sedimentary deposits in the lower channel floors of Simud-Tiu Valles. Schultz (1998) proposed the Valles Marineris formed in three stages: (1) Late Noachian to Early Hesperian dyke emplacements radial to Syria Planum; (b) localised subsidence after the Early Hesperian Epoch that formed ancestral canyons; and (3) regional normal faulting that overprinted the ancestral basins to form structural troughs, primarily during the Amazonian Period. Consistent with Schultz (1998) and Lucchitta *et al.* (1994), we consider that the Valles Marineris canyons were mainly produced by regional normal faulting, with a combination of secondary processes playing significant roles (i.e., landslides, disruption of terrain by the loss of ground ice, fluvial erosion with debris transport, volcanism, impact events and possible glacial activity).

The present-day canyon walls and floors have been altered since the time of the fluvial events by continued growth of the canyons, landslides, cratering events, possible volcanic activity on the canyon floors, and the creation of vast expanses of chaotic terrain in the eastern Valles Marineris. This extensive modification of the canyon walls likely obliterated most evidence of palaeolake shorelines. However, evidence has been preserved that a lake filled and overflowed Capri Chasma (Coleman *et al.*, 2007a). The existence of interior layered deposits (ILD) in most of the canyons provides further evidence of palaeolakes. Lucchitta *et al.* (1992) reviewed five hypotheses for their origin: that the deposits are (1) erosional remnants of the material in the canyon walls; (2) aeolian; (3) landslide material; (4) lake deposits; or (5) volcanic material. Lucchitta *et al.* (1992) favoured lacustrine or volcanic processes for ILD formation, and Chapman and Tanaka (2001) suggest that the ILDs may have been formed by sub-ice volcanic eruptions. The availability of high-resolution imagery and data on thermal properties and surface compositions reveals that ILDs contain little or no olivine (a mineral that weathers quickly), are fine grained and easily eroded, and contain layers enriched in hematite and hydrated sulphates (Gendrin *et al.*, 2005, 2006; Bishop *et al.*, 2007; Glotch and Rogers, 2007; Kundson *et al.*, 2007; Mangold *et al.*, 2007; Roach *et al.*, 2007).

In the present epoch, the main canyons of the Valles Marineris are highly interconnected from their beginnings high on the Tharsis plateau to the distant terminus of channels in the lowlands of Chryse Planitia (Figure 9.1). The canyons begin on the Tharsis plateau in a 700 km wide morass of structurally controlled valleys named Noctis Labyrinthus (see map at http://planetarynames.wr.usgs.gov/images/mc17_mola.pdf). This Hesperian structural complex provided a network of surface and underground pathways that would have enhanced drainage from elevated terrain in central Tharsis. The underlying structural complex is probably larger than mapping indicates because Amazonian and upper Hesperian units appear to have overlapped and concealed much of its

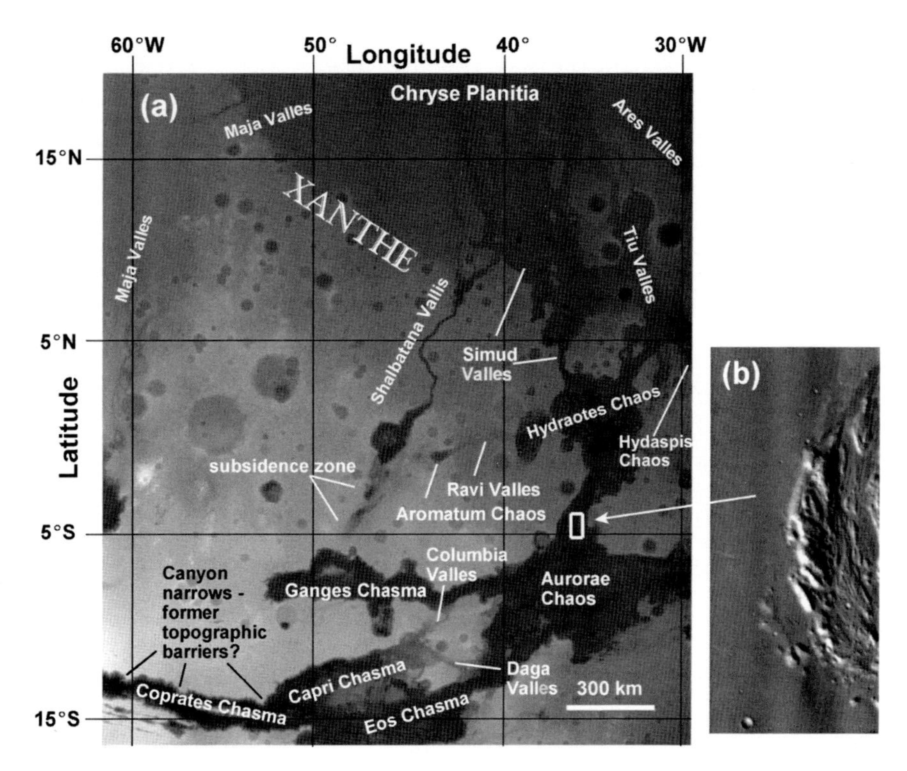

Figure 9.2. (a) Chasmata, outflow channels and chaotic terrain of the eastern Valles Marineris. Base is MOLA MEGDR image file at 128 pixels/degree. (After Coleman *et al.* (2007a).) (b) High-standing, water-carved features in the valley that connects Aurorae Chaos to Hydraotes Chaos. (Subframe of THEMIS image I08577010.)

surface expression. East of Noctis Labyrinthus lie the canyons of Tithonium Chasma and Ius Chasma. Ius runs more than 700 km from Noctis to the depths of Melas Chasma. North of Melas Chasma are Candor Chasma and Ophir Chasma. These canyons are now interconnected and also form a continuous canyon system eastward via Coprates Chasma to Capri Chasma, Eos Chasma and the vast lowlands of Aurorae Chaos. A single broad valley opens northward from Aurorae Chaos to Hydraotes Chaos, where the channels of Simud and Tiu Valles begin.

9.2.1 Simud and Tiu Valles

These channels both begin as canyons at the northern margin of Hydraotes Chaos (Figure 9.2). Two canyons feed into Tiu Valles, the larger of which is 60 km wide and >2200 m deep. Simud Valles start in a canyon of similar depth that is 25 km wide. Both of the larger canyons have floor elevations of –4 km, with Tiu Valles being a few tens of metres deeper. There is no cross-cutting evidence farther north to indicate that one channel system is older than the other. Simud and Tiu Valles did not originate as flows solely from Hydraotes Chaos because there is evidence of extensive flow in the valley that leads northward into Hydraotes Chaos from Aurorae Chaos. A high 'island' of water-eroded features stands more than 1100 m above the valley floor (Figure 9.2b). These erosional landforms were left high and dry after the adjacent channel was deepened by continued flooding or multiple episodes of flooding. South of Hydraotes Chaos lies the vast lowland of Aurorae Chaos. To the west a broad canyon connects this lowland to Ganges Chasma (Figure 9.2), while eastward occur the large basins of Aureum, Arsinoes and Pyrrhae Chaos (see http://planetarynames.wr.usgs.gov/images/mc19_mola.pdf). These features may have been formed by the loss of underground volatiles and lateral drainage of groundwater into Aurorae Chaos. At its southwest corner, Aurorae Chaos is connected by a broad valley to Eos Chasma and Capri Chasma.

In the eastern Valles Marineris, at the transition between canyons and chaotic lowlands, there is clear evidence that floods spilled over from Capri Chasma into ancestral Ganges Chasma and southernmost Aurorae Chaos (Figures 9.2 and 9.3). Two large channel systems known as Columbia and Daga Valles exist at the eastern margin of Capri Chasma (Figure 9.3). At first glance, these channels appear to be relict 'hanging' valleys that were carved by floods before the canyons were born. Closer inspection reveals a different story. Both channel systems have similar high-water marks, were eroded to great depths, and are hanging only at their downstream termini (Coleman *et al.*, 2007a). Using MOLA data we determined the elevations of

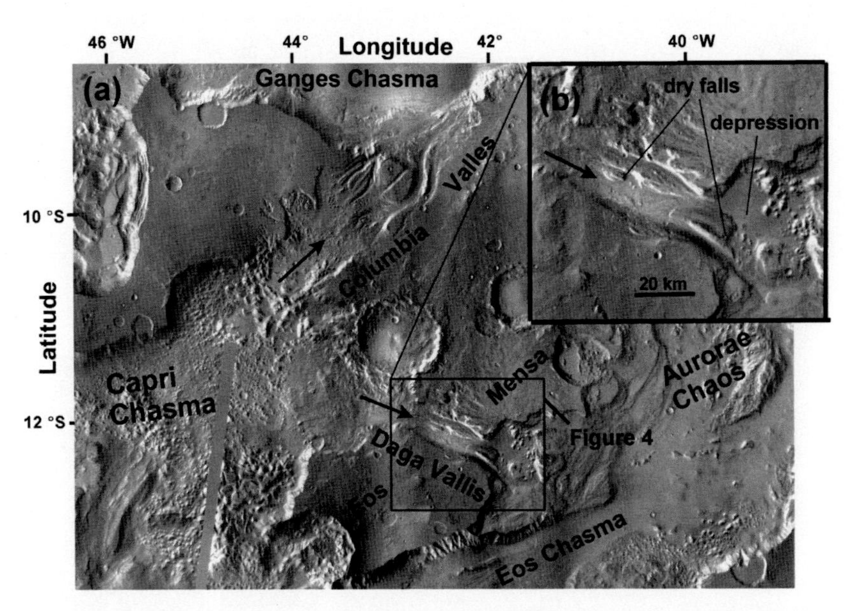

Figure 9.3. (a) View of eastern Capri Chasma and crossover channels to Ganges Chasma (Columbia Valles) and Aurorae Chaos (Daga Vallis). Arrows show flow directions (after Coleman *et al.*, 2007a). (b) Close-up of Daga Vallis. (Image credit: THEMIS Daytime IR Mosaic (Christensen *et al.*, 2006).)

eroded channel margins (rather than channel floors) to estimate the highest elevation to which floodwaters rose. The spillover to Ganges Chasma (Columbia Valles) occurred at an elevation of ~1100–1150 m and the overflow into Eos Mensa (Daga Vallis) occurred at ~1150 m. Therefore, the estimated elevation of the palaeolake surface in Capri Chasma was ~1150 m (Coleman *et al.*, 2007a). This is a minimum elevation because continued subsidence in the Valles Marineris region may have lowered the terrain over billions of years.

Daga Vallis is deeper than Columbia Valles. Subtracting the elevation of the overflow on the floor of western Daga Vallis (i.e. −65 m) from the initial flood level (1150 m) reveals that a lake ~1215 m deep drained through the spillover channels (Coleman *et al.*, 2007a). The thalweg slope of Daga Vallis is large, ~0.02, which even in the lower gravity of Mars would have induced high floodwater velocities. Coleman *et al.* (2007a) estimate a peak discharge rate of 1–6 × 10^8 m^3 s^{-1}. Daga Vallis terminates as a spectacular 500 m high cataract of a dry falls. The cataract headwall forms the western margin of a closed depression 25 km wide (centred at 12.26° S, 41.80° W) that likely was formed by a combination of plunge-pool erosion below the falls and large-scale genesis of chaotic terrain in the northern part of the depression. The depression probably was not a crater because its rim would have diverted flow around it. This depression acted as an enormous 'stilling basin' that dissipated the flow power. North of the cataract an inner channel of Daga Vallis leads down into the deepest parts of the depression. Sediments deposited by late-stage waning flows likely filled in the deepest parts of the plunge pool at the base of the cataract. Two outlets were eroded at the southern rim of the depression.

What were the sources of the floodwaters that debouched from Capri Chasma? This canyon and Eos Chasma to the south both contain chaotic terrain that may have been formed by groundwater breakouts. But we do not know the ancestral size of these canyons and whether chaotic terrain had formed within them at that time. An alternative explanation is that a partial juncture between ancestral Capri Chasma and Coprates Chasma may have been breached, catastrophically draining canyon lakes to the west in Coprates Chasma and possibly other interconnected canyons (i.e., Melas, Candor and Ophir Chasmata). The existence of two large spillover channels at approximately the same elevation indicates that the channels formed at the same time in response to a massive influx of water to ancestral Capri Chasma (Coleman *et al.*, 2007a). If the palaeolake had filled slowly, the overflow of the canyon lake likely would have eroded a single channel because erosion would have begun and concentrated at one place. Estimated peak discharge rates of $>10^8$ m^3 s^{-1} for Daga Vallis exceed realistic estimates for most outflow channels but are entirely consistent with catastrophic release from a canyon lake. As a historical note, the presence of multiple spillways at the same elevation provided key support for J Harlen Bretz in the great 'Channeled Scabland' debate (Baker, 1981).

Several locations in eastern Coprates Chasma are relatively narrow today (Figure 9.2) and may indicate where junctures could have been created by the collapse of canyon walls, releasing lakewaters that had been ponded in interconnected canyons west of Capri Chasma. A cascade of megafloods would have been possible, analogous to those that carved the Channeled Scabland in Washington State. Relatively narrow floodways could have been susceptible to blockage by debris and ice, making it possible that multiple episodes of flooding issued from Coprates Chasma into ancestral Capri Chasma. Indeed, if a new gateway in eastern Coprates Chasma was the primary source for the floods that issued from Capri Chasma, the hydrographs of those floods may have resembled those hypothesised for the Missoula palaeofloods, in which floodwaters rose slowly to a crest, peaked and then fell rapidly. Baker (1981) reported field evidence that the Missoula floods receded rapidly because bedforms such as giant current ripples were not eroded away. Therefore, unlike outflow channels sourced by groundwater (e.g. Ravi, Allegheny and Walla Walla Valles), those at the eastern end of Capri Chasma could have briefly carried flows at nearly their full depth.

If Daga and Columbia Valles were indeed formed at the same time by a massive influx of floodwater that overtopped Capri Chasma, then the valley that now connects Eos Chasma and Aurorae Chaos did not yet exist, otherwise the flood would have passed through it instead of carving new channels. It is quite possible that a topographic barrier existed between the ancestral Capri and Eos Chasmata at the time of the flooding, preventing bypass flow to the south. The termini of both Daga and Columbia Valles are hanging valleys, indicating that back wearing from subsequent canyon growth occurred in the outlet from Ganges Chasma and at the juncture between Eos Chasma and Aurorae Chaos (Figures 9.2 and 9.3). Our interpretation is consistent with geochemical evidence for past aqueous activity in the Valles Marineris (Gendrin *et al.*, 2005, 2006; Bishop *et al.*, 2007; Glotch and Rogers, 2007; Kundson *et al.*, 2007; Mangold *et al.*, 2007; Roach *et al.*, 2007).

Our interpretation adds important details to the work of Tanaka (1999), who proposed a debris-flow mechanism for emplacement of the Simud-Tiu deposits that extend more than 2000 km from Hydraotes Chaos northward to Acidalia Planitia. He proposed that catastrophic flooding during the Late Hesperian Epoch carved deep canyons in Xanthe Terra. After multiple episodes of erosion and deposition, fluidised rock material was excavated from the lower parts of highlands exposed along Simud and Tiu Valles. Chaotic and hummocky terrains developed over hundreds of thousands of square kilometres, perhaps initiated by intense seismic activity from a large impact event. Tanaka (1999) proposed that fluidised material slid out from the base of chaotic terrains, coalesced into a single flow unit, and then flowed northward thousands of kilometres into Chryse Planitia and the northern plains.

We examined landforms associated with Tiu-Simud Valles at low elevations in Chryse Planitia to gain insights about a possible northern ocean. MOLA data reveal that the Tiu-Simud channels persist northward into the area bounded by latitude 24–30° N and longitude 28–38° W. The channels range in depth from 150 to 200 m, with floor elevations between –3950 and –4030 m. These elevations are 190 to 270 m below the hypothesised Contact 2 shoreline. This fact suggests that at the time of the initial flows in these distal channels there was no water body in the northern plains with a surface elevation higher than –4 km. Continuing flows in Simud-Tiu Valles may have created a sea that rose above this level. It is also possible that a frozen body of water derived from previous flows may have existed north of latitude 30° N at elevations where the channels disappear. Our interpretation parallels that of Carr and Head (2003) who also found no support for higher shorelines but concluded that the Vastitas Borealis Formation (VBF) may consist of sublimation residue from ponded flood effluents (Kreslavsky and Head, 2002). Carr and Head (2003) report an average elevation for the entire VBF of -3903 ± 393 m. The disappearance of the distal Simud-Tiu channels could therefore be explained by the presence of a frozen sea with a surface in this elevation range.

South of ~17° N some of the channel floors of Simud-Tiu Valles display a reversed gradient, i.e. the terrain slopes gently upward in the downstream (northerly) direction. These reversed gradients are very small. Examples: (1) a 150 km long reach of Simud Valles (centred at 16.85° N, 37.92° W) has a reversed slope of 0.0003; (2) a 72 km reach of Simud Valles (centred at 16.27° N, 41.19° W) has a reversed slope of 0.0008. These gentle reversed slopes may simply have been caused by reduced fluvial erosion rates near the channel mouths. Greater channel depths south of these areas were apparently caused by terrain collapse, generating a characteristic knobby terrain. We estimated the preflood regional slope from Columbia Valles to Chryse Planitia by measuring rim elevations along the western margin of Simud Valles and its headwater canyons. The elevation gradually decreases 4300 m over a distance of 1850 km. The initial overland flows in Simud Valles therefore had a mean energy slope of 0.002. No gradient reversal is seen in this preflood topography, which suggests that the reversed slopes on parts of the channel floors do not have a regional cause such as crustal upwarping.

The channel floor elevations near the mouths of Simud-Tiu Valles occur within a range of –3680 to –3770 m. Most of the Valles Marineris canyons have low points below this elevation range. For example, Ophir, Candor, Melas and Coprates Chasmata have deep points lower than –4600 m. This evidence means that residual lakes up to 1 km deep could have remained in the Valles Marineris even after the canyon lakes drained. Alternatively, significant subsidence of the canyon floors may have occurred during

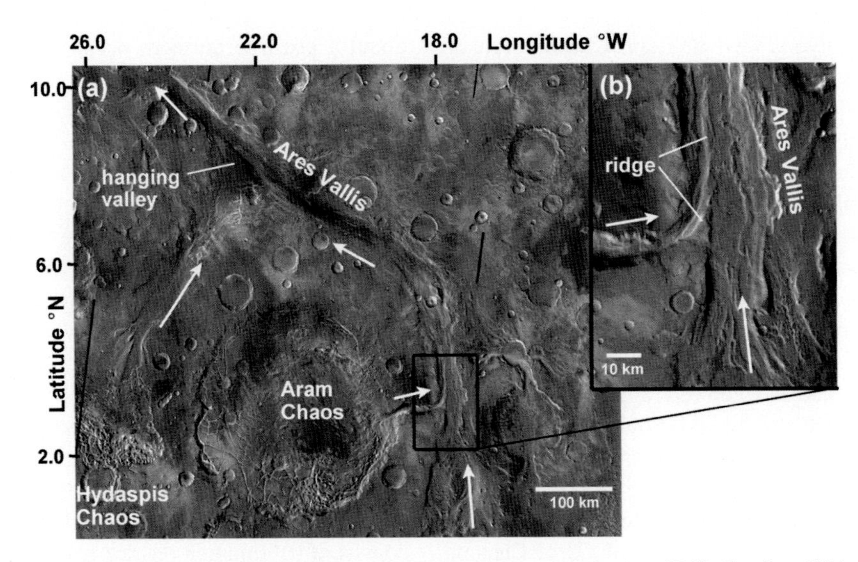

Figure 9.4. (a) Locations of Aram Chaos and eastern Hydaspis Chaos, two source areas for Ares Vallis flooding. White arrows show flow directions. Other sources (e.g. Iani and Margaritifer Chaos) lie south of this image. Subfigure (b) shows where channel that drained Aram Chaos intersects Ares Vallis. Note ridge that separates channels on Ares floor. (Image credit: THEMIS Daytime IR Mosaic (Christensen *et al.*, 2006).)

the Amazonian Period long after the channels formed. Terrain collapse could likewise have contributed to the reversed hydraulic gradient near the mouths of Simud and Tiu Valles and may also have obliterated evidence of inner channels in these broad valleys.

The channels of Simud-Tiu Valles were carved by the last large floods that emptied into Chryse Planitia. This supposition is logical in the context of regional hydrology because the draining of the main canyons of the Valles Marineris would have rapidly and permanently lowered the groundwater elevations in adjacent regions, thus preventing the recurrence of large outflows in other channels. The rapid decline of lake levels would have created groundwater overpressures beneath the basins, which would have destabilised canyon walls and adjacent highlands and triggered landslides, new chaos formation, and groundwater outflows, adding to the flood discharges. Fluvial incision of channels east of Capri Chasma and Eos Chasma may have penetrated the cryosphere and triggered additional chaos formation and groundwater outflow (see Coleman, 2005). These processes may explain the existence of extensive chaotic terrain in the eastern Valles Marineris lowlands (e.g. Hydraotes and Aurorae Chaos) (Coleman *et al.*, 2007a). The final flows apparently discharged from western Hydaspis Chaos (Figure 9.2), carving an erosional scarp 10–20 m high that crosses most of the Tiu channel. These Hydaspis flows travelled northward along the existing Tiu channel. The rim around Hydaspis Chaos has elevations less than –1500 m, indicating the outflow elevation.

9.2.2 Ares Vallis

Ares Vallis is older than the Simud-Tiu channels, and had multiple sources that included discrete features such as Aram Chaos in the eastern Valles Marineris (Figure 9.4). We include Ares in this section because its origin appears linked to early overflows from Argyre Planitia, a large impact basin in the southern hemisphere. Ares Vallis is the final drainage channel for what may be the longest known surface-water pathway in the Solar System (Parker *et al.*, 2000; Clifford and Parker, 2001), extending from the south polar region to the northern plains (see online maps (e.g., MC-26, MC-19 and MC-11) at http://planetarynames.wr.usgs.gov/mgrid_mola.html). This extended flow system began as channels that drained the south polar region then flowed northward to form a sea in Argyre Planitia. The creation of Uzboi Vallis indicates that Argyre eventually overflowed its northeastern rim near 35° S, 36° W. Northward the channels are extensively overprinted with large craters (Hale, Bond and Holden), which suggests the Argyre overflow occurred in the Late Noachian Epoch. Other segments of this channel system include Ladon Valles and a series of depressions (including Margaritifer Chaos) that continue northward to the equator where the main channel of Ares Vallis emerges from Iani Chaos.

The Ares channels incise Noachian terrain everywhere except near the channel mouth, where Hesperian ridged plains material is exposed along the eastern banks (Scott and Tanaka, 1986). Rogers *et al.* (2005) used THEMIS and TES data to study the mineralogy of ancient crust exposed in Ares Vallis bedrock. They concluded that the region experienced one or more stages of olivine-enriched magmatism during formation of the upper crust. They also concluded that the exposure of materials >3 Gyr old that contain ~25% olivine indicates that chemical weathering has been limited in the region. We note an alternative explanation for the presence of olivine at the Martian surface. Over billions of years, very low rates of physical erosion may suffice to continually expose fresh rock surfaces.

The presence of hanging valleys and evidence of multiple discharge events reveals a complex history of palaeofloods in Ares Vallis. In addition to the early overflow of Argyre, subsequent floodwaters issued from Margaritifer, Iani, Aram and eastern Hydaspis Chaos. Using MOLA data we can examine the relative ages of the various tributary channels. The unnamed channel that begins at eastern Hydaspis Chaos (Figure 9.4, lower left) forms a hanging valley where it intersects Ares north of Aram Chaos. As discussed previously, the youngest flows in Tiu Valles emanated from western Hydaspis Chaos (see map at http://planetarynames.wr.usgs.gov/images/mc11_mola.pdf). Because the Tiu channels cross-cut the mouth of Ares Vallis, the discharges from western Hydaspis are younger than earlier flows from eastern Hydaspis that flowed both northeast to Ares and northwest toward Tiu Valles. Aram Chaos itself occupies a large unnamed crater (Figure 9.4). An unnamed channel incises the eastern rim of this crater, showing that a palaeolake filled the crater and drained into Ares Vallis after the rim was breached. MOLA data do not have enough resolution to reveal cross-cutting relations between this unnamed channel and the adjacent floor of Ares Vallis that was eroded by flows from Iani Chaos. Figure 9.4b shows a close-up of this channel intersection, revealing parallel channels divided by a ridge. There is no evidence that Ares cross-cuts the channel that drained Aram Chaos. The channels may have flowed at the same time.

Komatsu and Baker (1997) analysed discharges in a deep, constricted reach of Ares Vallis. Assuming bankfull flows, which may not be realistic (see Wilson *et al.*, 2004), they estimated the range of possible maximum discharge from 10^8 to $10^9\,m^3\,s^{-1}$. We have re-evaluated these early calculations using MOLA data for the channel floor elevations. The Ares channel between latitude 7° N and 10° N displays an inconsistent energy slope and probably has been altered by post-fluvial processes that changed the floor elevation. There is a 20 km long reach between latitude 9.2° N and 9.7° N with a consistent energy slope of 0.0017 that may represent a reach where the channel floor has been less altered. We estimate peak flow by assuming the channel carried flooding to only half its present depth. This depth restriction is consistent with Wilson *et al.* (2004) who argued that the magnitude of palaeoflooding in most Martian channels would be overestimated by assuming bankfull flow in the channels observed today. Given a channel width of 21.5 km, depth of 1600 m, mean flow depth of 600 m (maximum depth of 800 m), energy slopes ranging from 0.00017 to 0.0017, and Manning n values of 0.04 to 0.06, we estimate velocities of 10 to 45 $m\,s^{-1}$ and peak discharge rates of 1–6 $\times\ 10^8\,m^3\,s^{-1}$.

In 1997, Mars Pathfinder landed at the mouth of Ares Vallis near MOLA coordinates 19.10° N, 33.24° W. Tanaka (1999) observed that boulders up to 1 m in diameter at the Pathfinder site are too large to have been transported by floods where the channel system has broadened out to hundreds of kilometres in width. Tanaka (1999) presented debris flows as an alternative way to transport boulders. Pathfinder landed on the margin of a floodplain where Tiu Valles cross-cut the mouth of Ares Vallis. Here the Tiu Valles channel is 70 km wide with channel depths up to 150 m, and can be traced N-NE of Pathfinder more than 650 km into Chryse Planitia. North of Pathfinder (at 20.2° N, 33.8° W) stand water-carved landforms 200 m higher than the channel floor. These landforms, along with the length and depth of the Tiu channel and the truncation of the mouth of Ares Vallis, suggest that fluvial erosion was more effective than debris deposition in this reach of Tiu Valles. Another possibility is that deeper channels may have been present that were partly filled by late-stage hyperconcentrated flows. Additional research is needed regarding the flow conditions that created the Ares and Simud-Tiu channel systems and the surface deposits around Mars Pathfinder.

9.2.3 Kasei Valles

Kasei Valles represent a canyon system and network of outflow channels more than 3500 km long. This drainage basin begins on the equator west of Hebes Chasma as a box canyon in southernmost Echus Chasma (Figure 9.1). East of Echus Chasma is Lunae Planum, a plateau surfaced by ridged plains material of Early Hesperian age. To the west rises the Tharsis Plateau, veneered with Amazonian-age volcanics. Scott and Tanaka (1986) and Tanaka (1997) interpreted that Echus Chasma was invaded by flood basalts of middle-Amazonian age, specifically Member 5 of the Tharsis Montes formation. This geologic unit forms a halo around the three aligned Tharsis Montes volcanoes (Arsia, Pavonis and Ascraeus Mons) and terminates eastward where it embays Echus Chasma.

Carr (1981) was the first to note that young lava flows from Tharsis may have partly buried the western margin of Echus Chasma. THEMIS images show overlapping lava flows along the western margin of Kasei Valles (Figure 9.5b). THEMIS images and MOLA data provide evidence that lavas poured into Kasei Valles over a swath >1000 km wide from latitude 5.5° N to 23.5° N. These lava-covered surfaces of Amazonian age are distinctive because of their smoothness and paucity of craters compared to Hesperian terrain. Future research will be needed to better estimate how much of the source area for Kasei Valles became infilled and concealed by the extensive Amazonian volcanics.

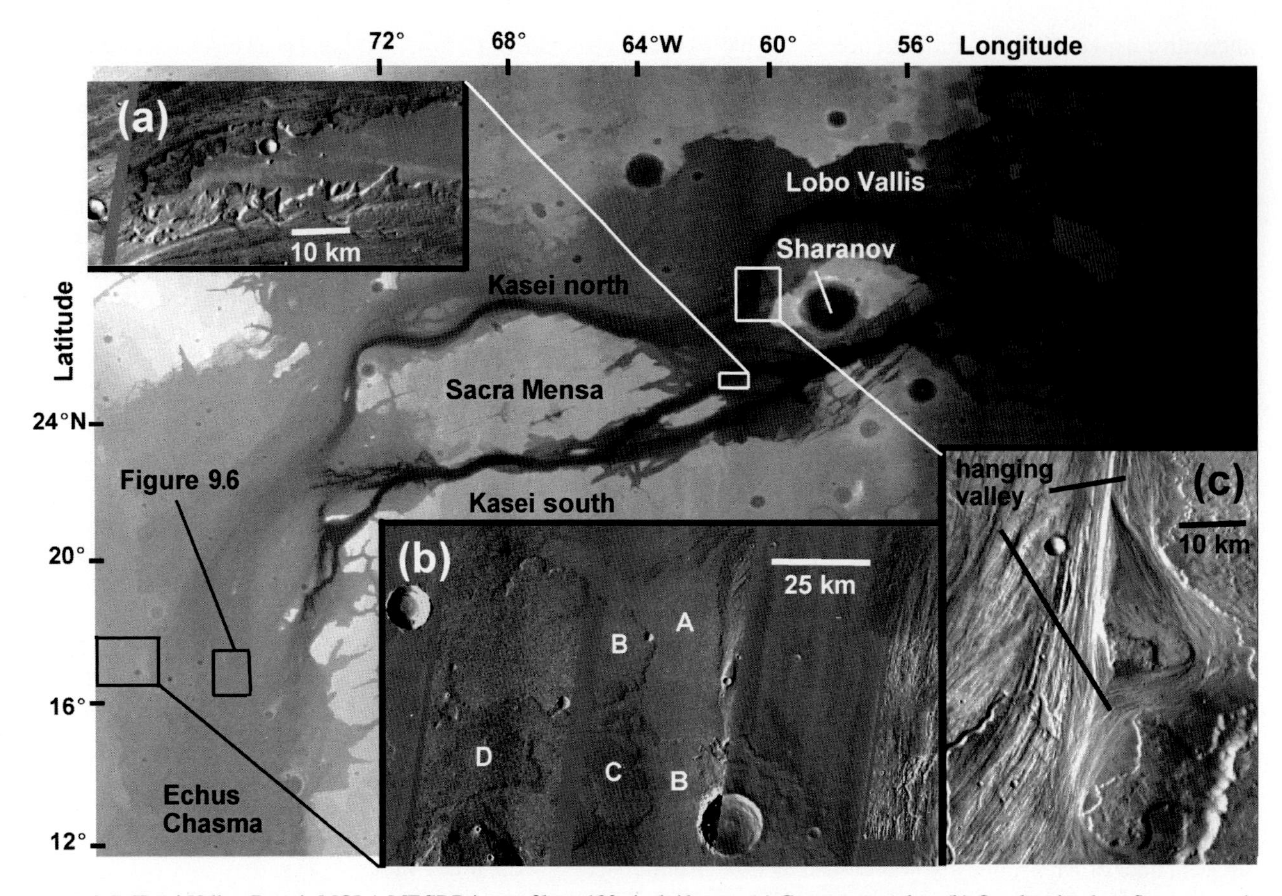

Figure 9.5. Kasei Valles. Base is MOLA MEGDR image file at 128 pixels/degree. (a) Cataract complex. (b) Overlapping lava flows entered the channel from the west. Oldest flow is A and youngest is D. Right side of image is flood scoured and free of lava. Crater at bottom postdates the flooding – its western ejecta blanket was covered by lava unit B. (c) Valley carved by early overland flooding that was left hanging on both ends when adjacent channel became deeply incised. (Image credit for subfigures (a), (b) and (c): THEMIS Daytime IR Mosaic (Christensen *et al.*, 2006).)

Lava flows travelled northward on some of the channel floors. There is no evidence that they were involved in eroding the channels because they partly infill them and form features of positive relief. Figure 9.6 shows a lava flow front within a channel at the northern end of Echus Chasma. The surface of the lava flow dips gently to the north, becoming steeper near the flow front where the elevation drops 20 m over a distance of 1 km or less. In most places on the channel floor the lavas were thick enough to conceal the flood-scoured terrain, however some high-standing islands of water-carved terrain poke through the lava fields. Material north of the lava terminus smoothly embays the scoured terrain, and MOLA elevations vary <3 m over 30 km. 'Ghosts' of longitudinal ridges can be seen through this surface. The smooth material may have been deposited as sediments during the waning stage of the final flood in this channel or, alternatively, the sediments may have been deposited at the time of the lava invasion by mudflows spawned by the large-scale melting of ground and surface ice.

The morphology of Kasei Valles varies along its course. Amazonian lava flows with lobate margins dominate Echus Chasma in the southern part of the drainage basin. Water-carved features appear beyond the lava flows north of 15° N, where many hundreds of longitudinal ridges cover the channel floors (Figure 9.6). These ridges likely were produced by hydrodynamic erosion during catastrophic scabland flooding that deeply eroded and scoured the surface. The scabland attains a spectacular width of 300 km at latitude 16.5° N, but this is a minimum width because basalt flows overlap the western margin. A distinctive streamlined 'island' was eroded in the lee of a crater at 15.2° N, 75.3° W. Here the channel system diverges into two main channels that continue northward on subparallel paths. IAU nomenclature does not differentiate these two channels, so here we refer to them as the Kasei north and south channels (Figure 9.5). Near longitude 73° W both channels turn eastward where they gradually descend to Chryse Planitia (Figures 9.1 and 9.5) through a complex of anastomosing valleys. The Kasei south channel branches at

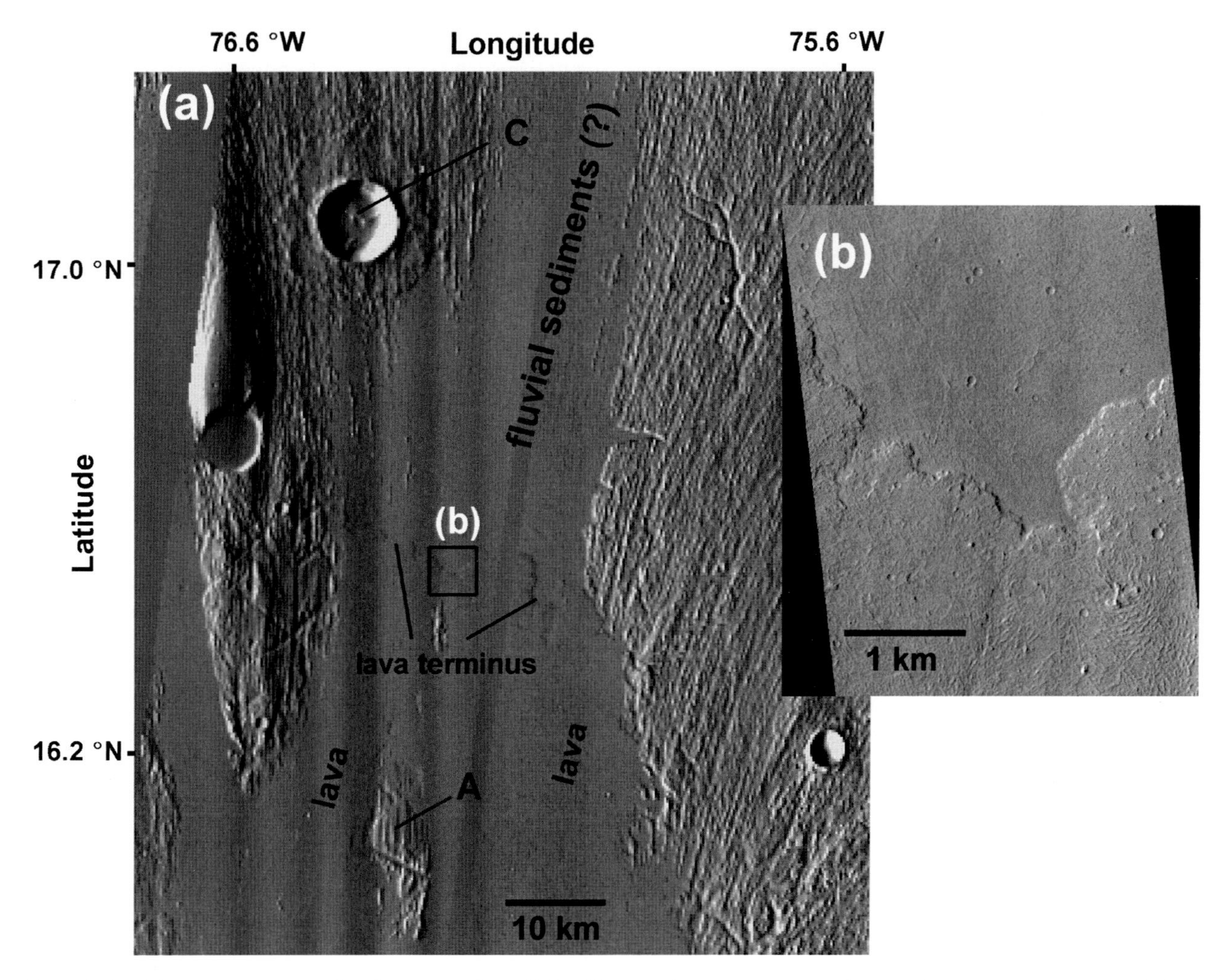

Figure 9.6. (a) Scoured terrain carved by flooding in Kasei Valles. Flow was from bottom to top of image. Amazonian-aged lava invaded channel from the south. See lobate terminus at centre and MOC image in subfigure (b). Near lava terminus the elevation drops ~20 m over 1 km. Material north of lava flow smoothly embays the scoured terrain, and elevations vary <3 m over 30 km. This smooth material may consist of sediments deposited during waning stage of final flood, or it may instead have been deposited at the time of lava invasion by mudflows spawned by melting of ground or surface ice. 'Island' of scoured terrain at A stands 50 m above the adjacent lava fields. Crater at C postdates Kasei flooding. Its lobate ejecta blanket was too thin to conceal underlying longitudinal ridges. (Image credit: (a) THEMIS daytime IR mosaic (Christensen *et al.*, 2006); (b) MOC image R2301324, Malin Space Science Systems.)

the southeastern side of Sacra Mensa (Figure 9.5). An elongated cataract complex more than 30 km long is preserved on the channel floor (Figure 9.5a). MOLA data reveal the headwall of this spectacular dry falls to be ~250 m high. This cataract complex demonstrates that flooding was the last major event in this channel reach, and that the channel floor was not formed or modified by valley glaciation.

As shown in Figure 9.5, the Kasei north channel divides around a highland plateau topped by Sharanov Crater. The branch north of Sharanov is called Lobo Vallis and it proceeds eastward where it further divides. The other branch of the north channel carved a scabland to the southeast. The Kasei south channel cross-cuts this scabland and therefore is younger. Within the north channel is a small inner channel that is not cross-cut where it intersects the south channel at 25.42° N, 59.03° W. At the mouth of the inner channel MOLA data reveal a lobe of material superimposed on the floor of the Kasei south channel, which indicates that flow occurred in the inner channel after flow ceased in the south channel. East of the sediment lobe is a 50 m high fault scarp that trends north–south across the entire south channel. The fault movement postdated the fluvial channel because the scarp itself is not incised.

Several hanging valleys exist in the region west of Sharanov Crater. Figure 9.5c shows a meander bend that stands 450 m higher than the adjacent channel. East of

Sharanov the north and south Kasei channels branch extensively forming a channelled scabland more than 350 km wide. This scabland disappears on the Chryse plains east of longitude 45° W where the surface was eroded by the younger Simud-Tiu flooding.

Robinson and Tanaka (1990) used a stereomodel of Viking images to perform flow calculations for late-stage flooding in the north channel of Kasei Valles. They concluded that groundwater breakouts alone would have been insufficient to generate the estimated discharge rates and suggested that water had been ponded and catastrophically released from the box canyon at the southern terminus of Echus Chasma. The work of Robinson and Tanaka (1990) can now be updated using high-resolution MOLA data and THEMIS images. Williams *et al.* (2000) concluded that Kasei's south channel was younger than the northern channel. They used MOC images and early MOLA profiles to evaluate the flow rates and duration in Kasei Valles. Based on cross-cutting relations they concluded that the channel network of Nilus Mensae was developed by sapping processes during a minimum interval of 50 000 years between the final flow events in the north and south channels. Williams *et al.* (2000) estimated flow discharges of $\leq 2 \times 10^7\,m^3\,s^{-1}$ and $\leq 0.06 \times 10^7\,m^3\,s^{-1}$ for the north and south channels, respectively. Their discharge estimate for the final flood in northern Kasei Vallis was orders of magnitude less than calculated by Robinson and Tanaka (1990). Williams *et al.* (2000) interpret that Robinson and Tanaka (1990) overestimated the channel width and flow cross-section.

Williams and Malin (2004) reported evidence for late-stage fluvial activity. They analysed two inner channels in the south branch of Kasei Valles. Based on the existence of a concave-upward channel profile, perched boulders on channel banks, and arcuate alcoves that appear to have formed by undermining, they interpreted that a Newtonian fluid such as water carved the inner channels. Williams and Malin (2004) suggested that a platy surface texture on the floors of the inner channels was the remnant of a mudflow produced during the waning stages of flooding in the south channel. In general, we have found that the Kasei channel floors east of longitude 70° W have flat profiles and smooth surface textures that would indeed be consistent with mudflow deposits. The channel margins above the smooth floors typically display longitudinal ridges, which likely formed during an earlier phase of strong hydrodynamic erosion. The smooth channel floors may have formed during a later fluvial event, or during the waning phase of a large flood characterised by a continuum of flow conditions ranging from strongly erosive to lower-energy (depositional) regimes. Local resurfacing by lava flows or aeolian deposits may also have occurred.

For Kasei Valles we propose a complex history of initial scabland flooding with intense hydrodynamic erosion and the incision of deep channels with longitudinal ridges, followed by a later stage or episode of mudflows that partly infilled the channels and flattened their surface profiles. It is unclear to what extent volcanism played a direct role in the Kasei flooding. However, given the long history of volcanism in Tharsis, extending from the Noachian to the Amazonian Periods, multiple releases of water would have been possible through the interaction of flood basalts with surface ice deposits. Ice may have preferentially accumulated at low latitudes during times of high obliquity (Laskar *et al.*, 2004; Head *et al.*, 2005; Forget *et al.*, 2006). The effusion of flood basalts in eastern Tharsis and the invasion of Echus Chasma and the western Kasei channels by lavas could have triggered multiple episodes of flooding.

9.2.4 Maja Valles

Maja Valles have an overall length of almost 1700 km (Figure 9.2). They begin at the northern end of Juventae Chasma where the canyon terminates as a blunt trough, divide into multiple channels, and traverse 1300 km of highland terrain. At the highland margin, the channels converge into a single valley that descends through a gorge onto the plains of Chryse Planitia at 17.88° N, 53.80° W. There the channel divides, with the main channel continuing northeastward and another trending due north where it cross-cuts the older Maumee Valles, which were also formed by highland runoff. Distinct Maja channels can be traced more than 360 km from the highland rim eastward into Chryse. The deepest channel floor has a base elevation of –3650 m at 20.798° N, 48.309° W. This channel terminates abruptly at a north–south ridge ~120 m high that forms the western margin of a field of polygonally fractured ground centred at 21.10° N, 47.44° W (Coleman and Baker, 2007). Viking I landed on or near the same ridge in an area between the distal channels of Kasei Valles and Maja Valles.

Although Juventae Chasma is not connected to a long chain of canyons that could have provided multiple episodes of floodwaters, there are large areas of collapsed terrain located south and west of Juventae that may have formed in response to large-scale groundwater migration northward from a possible palaeolake in ancestral Candor Chasma (Coleman and Baker, 2007). These collapsed areas can be seen at http://planetarynames.wr.usgs.gov/images/mc18_mola.pdf.

Chapman *et al.* (2003) examined Juventae Chasma, its ILDs, and the Maja Valles system and concluded that if the ILDs represent sub-ice volcanoes, then at least two periods of flooding and ILD formation occurred. They observed depositional bars with megaripples (MOC image M15–00976) on the floor of the western channel of Maja

Plate 1. The Potholes cataract complex with two large and distinctive alcoves formed by Missoula-flood flow from the Quincy Basin (right side of the image) towards the valley of the modern Columbia River (left side of the image). The black (and one blue) water bodies downstream of the cataract are plunge pools and the black water bodies upstream occupy linear flood-scours in the bedrock. Horizontal field of view is *c.* 12 km.

Plate 2. The Upper Grand Coulee (the artificial Banks Lake top right), with linear flood-scoured grooves leading from the area now occupied by the lake southeast towards the Dry Falls (centre) with flood-scoured plunge pools and a scoured extension in bedrock to the west near Deep Lake (far left). Horizontal field of view is *c.* 16 km.

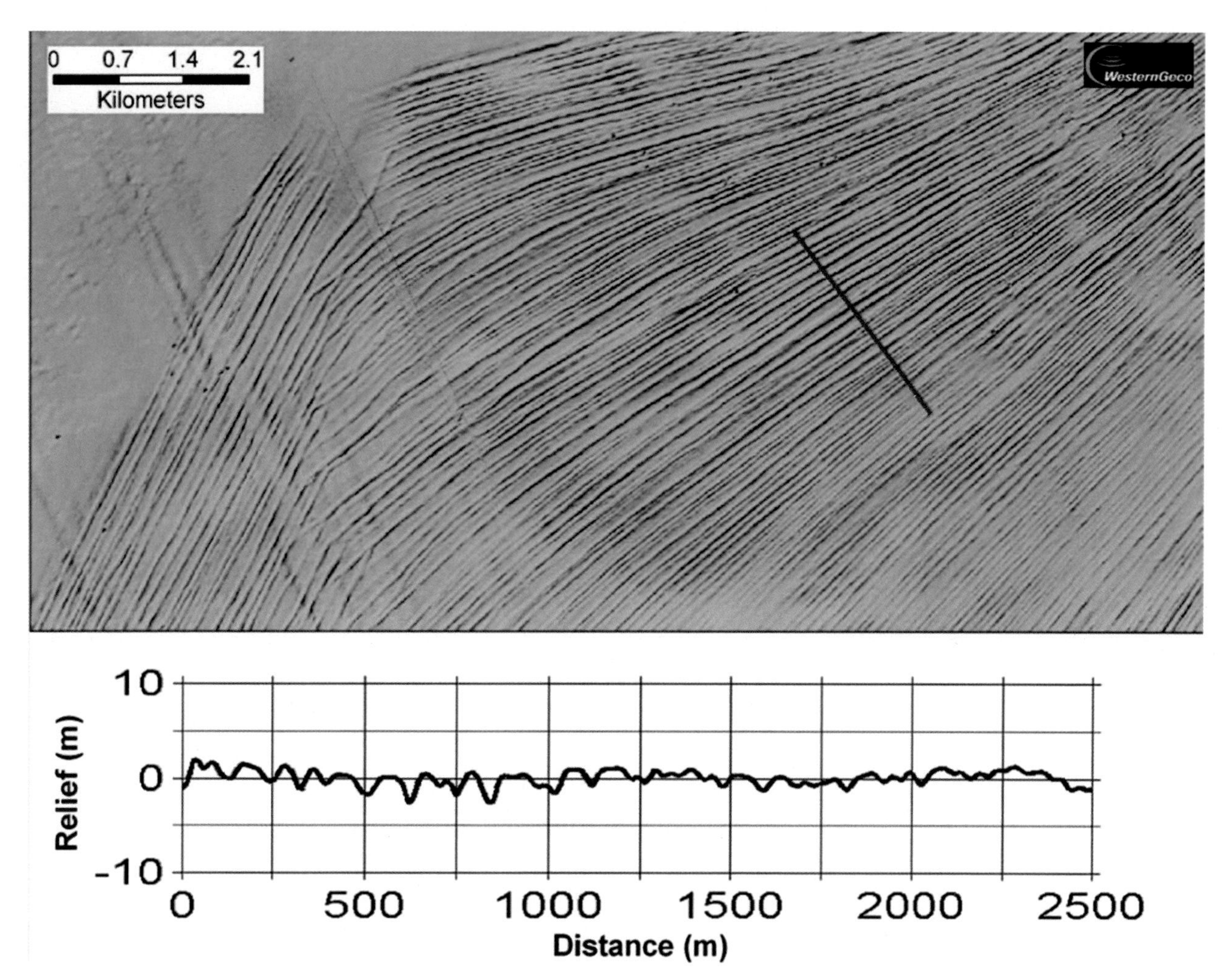

Plate 3. Furrows on the continental rise seaward of the Sigbee Escarpment, northwest Gulf of Mexico. (Redrawn from Bean (2003).)

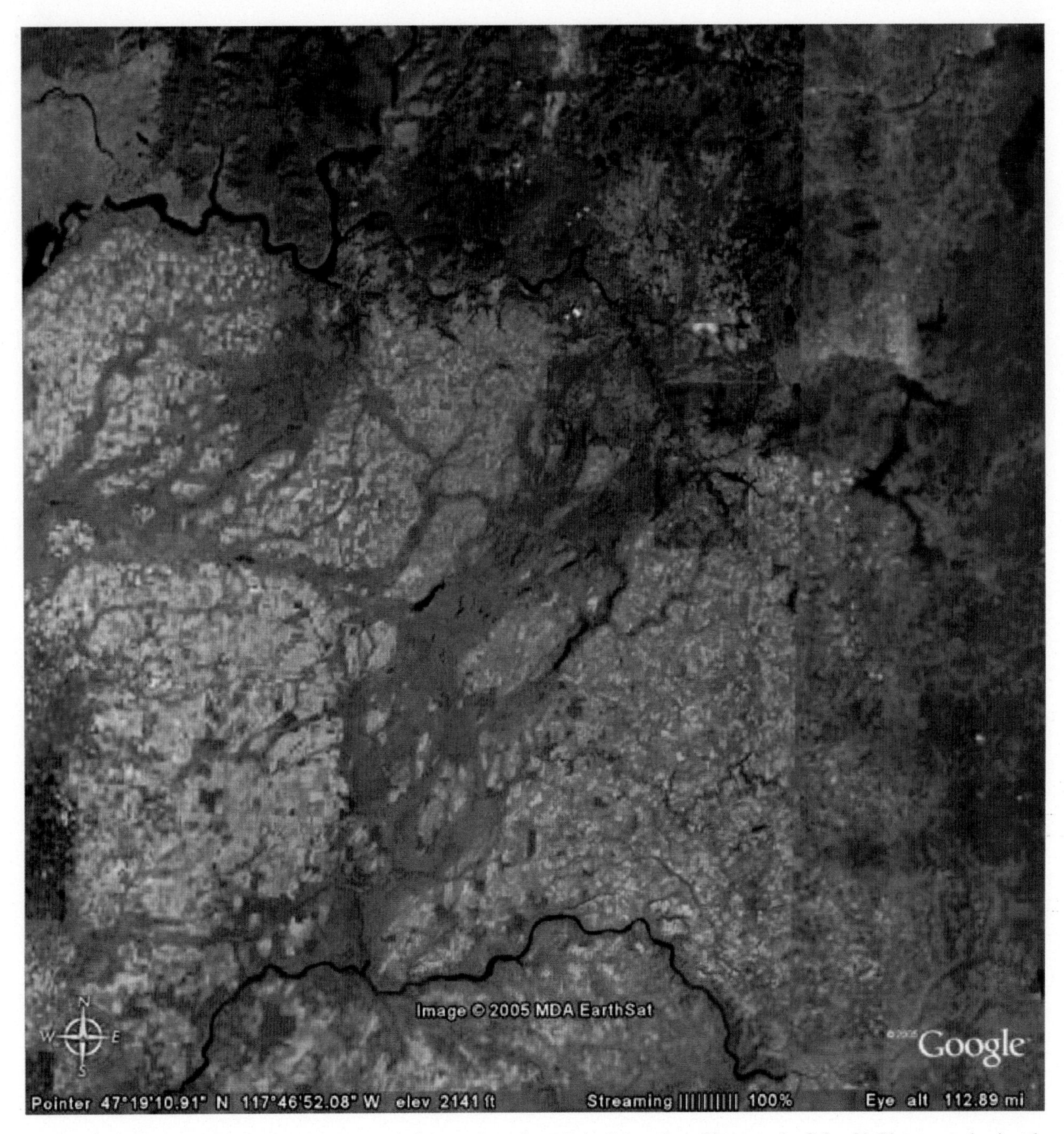

Plate 4. Anastomosed channel patterns appear as a dark network cut through the light-coloured loess on the Columbia Plateau, predominately within Washington State. Horizontal field of view is *c.* 200 km.

Plate 5. Putative erosional antidunes (largest to the left) can be seen above the line of trees and formed in the lee of the centrally placed hill that was submerged by waters draining from glacial Lake Chuja, Altai Mountains, Siberia, in the Late Quaternary. Field of view is approximately one kilometre.

Plate 6. Oblique aerial view of a portion of the Four Thousand Islands reach of the River Mekong, looking upstream (channel width is several kilometres).

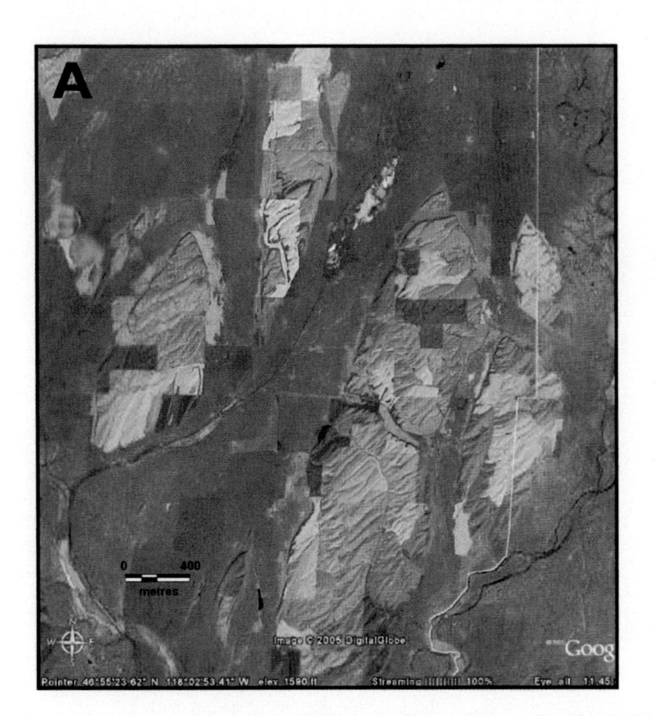

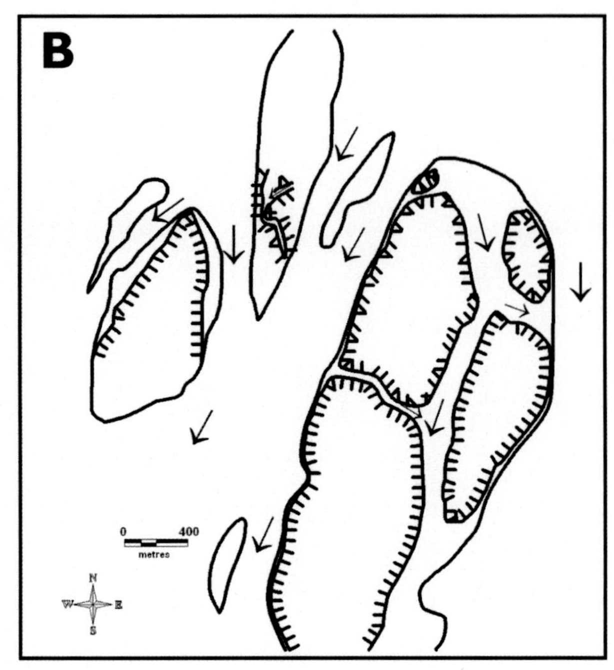

Plate 7. (A) Streamlined loess hills in the Cheny-Palouse Scabland. (B) Interpretation of image showing extent of residual hills, scarp-lines and inferred directions of palaeoflow.

Plate 8. Oblique view of an upstream scour hollow – dark green patch upstream of the obstacle posed by a bedrock hill near Chagan-Uzun in the western Chuja Basin, Altai Mountains.

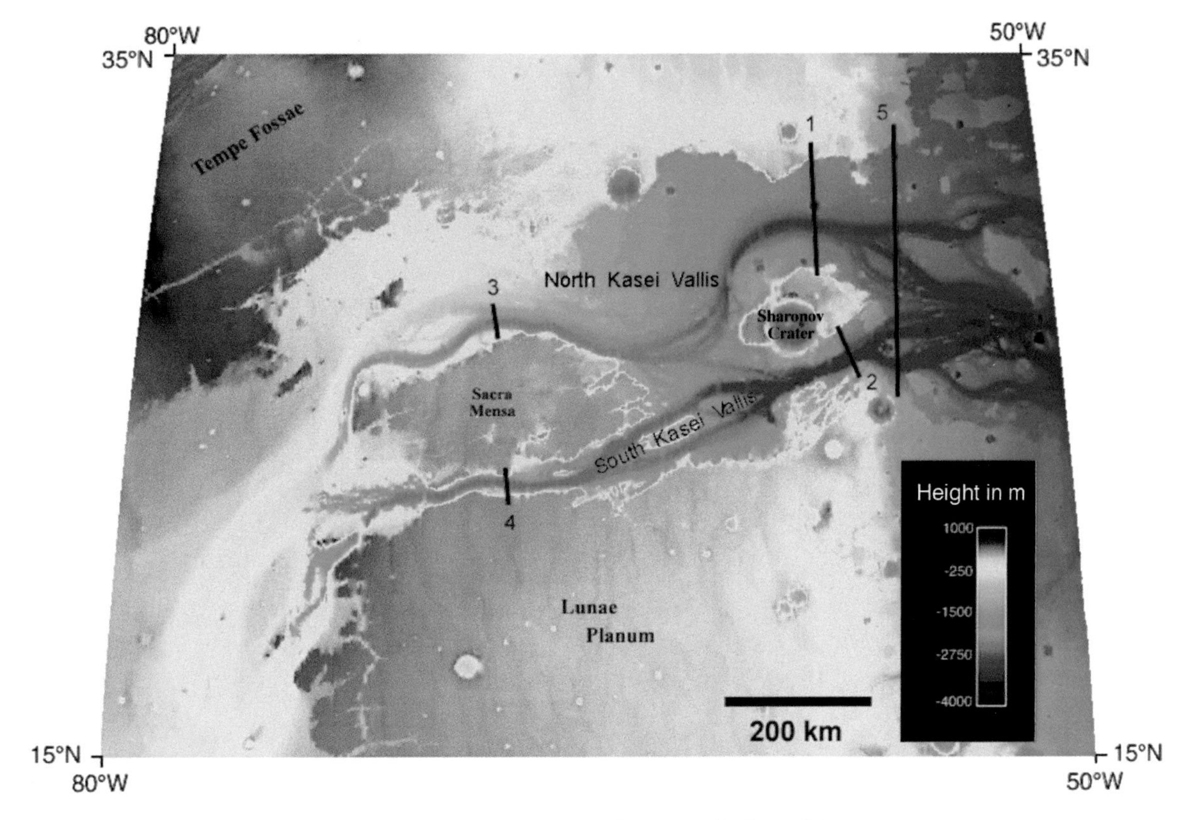

Plate 9. MOLA DTM of Kasei Valles (black lines indicate the cross-profiles shown in Figure 2.4).

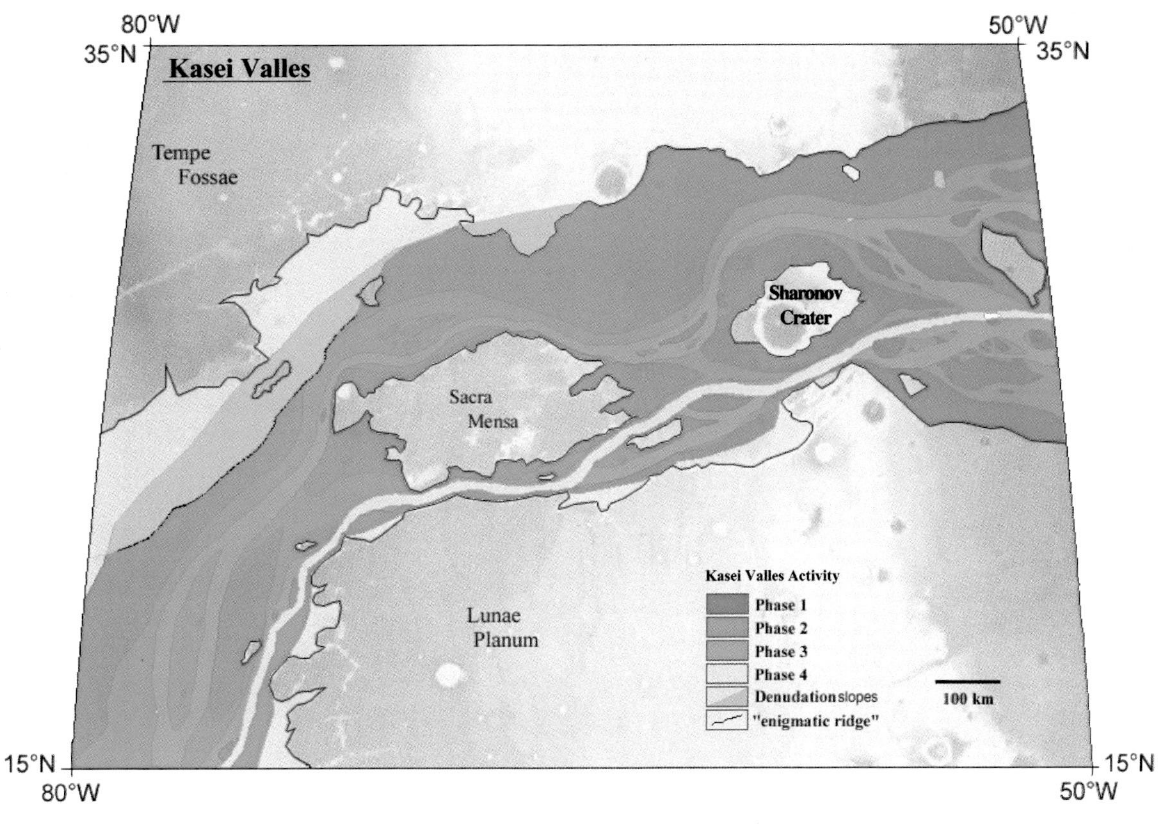

Plate 10. Phases of incision (terraces) within Kasei Valles (see Lanz, 2004).

Plate 11. View from top of a giant bar (formed in a valley wall embayment downstream of the ridge to the right) across the Katun River (flow right to left), Altai Mountains, Siberia. The Katun River is incised within Holocene terraces (middle distance). In the background is Log Korkobi bar (Carling *et al.*, 2002) set against the mountain wall. Note that the surface of the bar is not flat, as would be the case if it were a terrace, and the bar top slopes back into the tributary valley. The tributary stream was cut through the bar and the terraces and is now graded to the level of the modern Katun River. Run-up deposits are present below point A.

Plate 12. Oblique aerial view of *c*. 5 m high standing waves during turbid flood in the Kuiseb River, Namibia. Flow towards right. The sand dunes are 150 m high and provide scale.

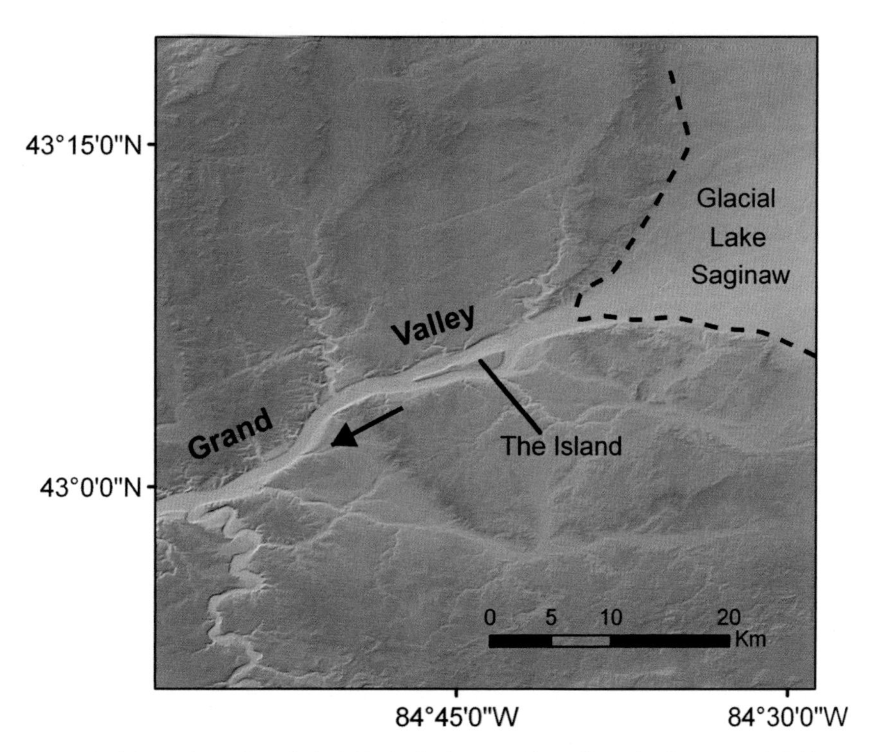

Plate 13. DEM (NED 1 arc sec) of the outlet region of glacial Lake Saginaw and its spillway leading across Michigan to glacial Lake Chicago. Note large streamlined erosional residual form in spillway ('the Island').

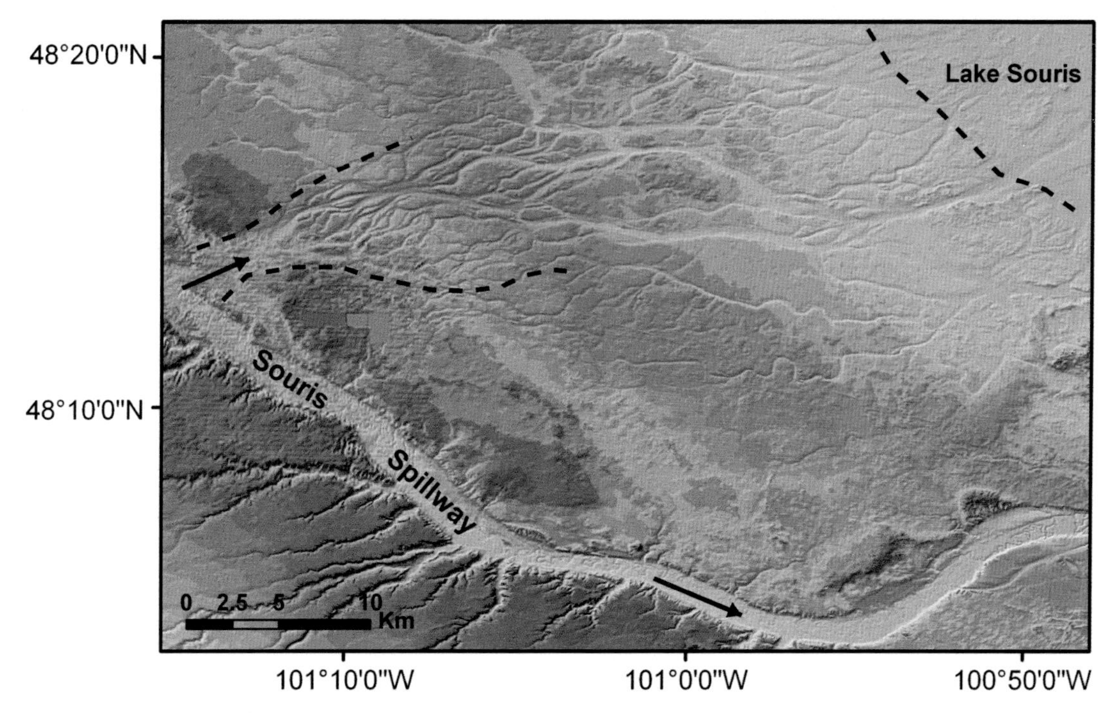

Plate 14. DEM (NED 1 arc sec) showing Souris spillway and anastomosing channels formed by the overflow of the glacial Lake Regina outburst flows from the Souris spillway to glacial Lake Souris. Location of scene shown on Figure 7.6.

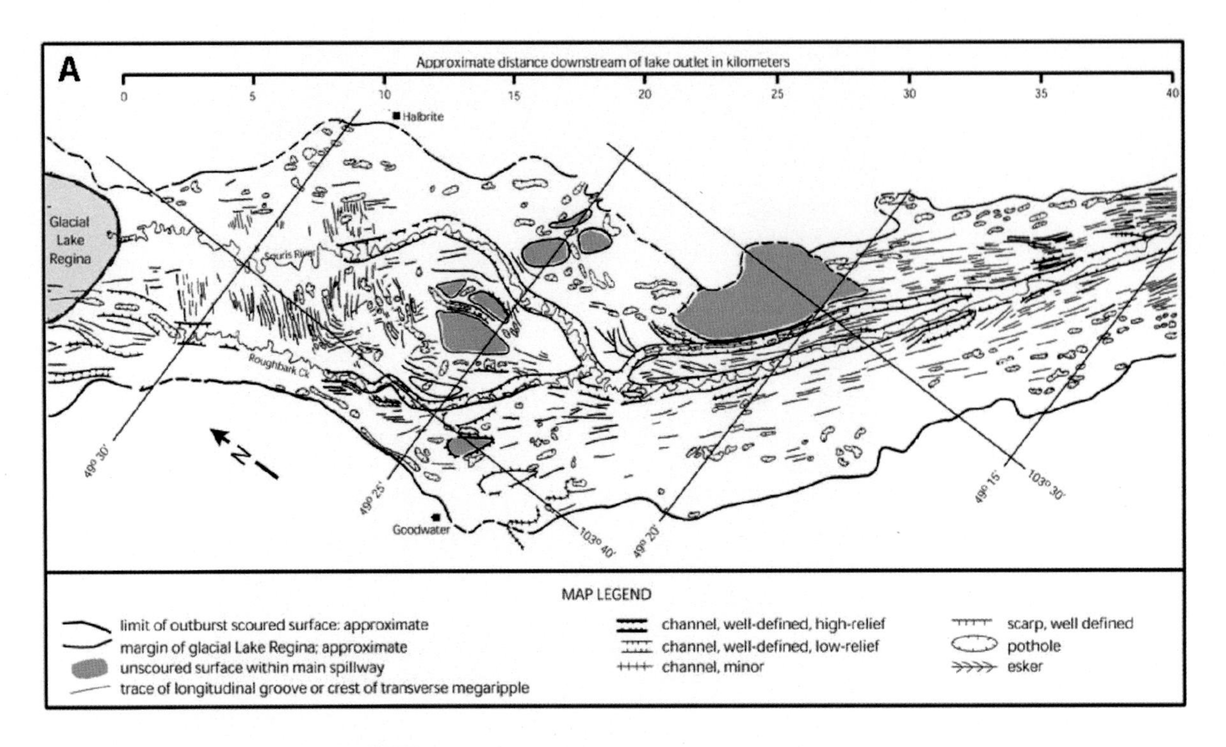

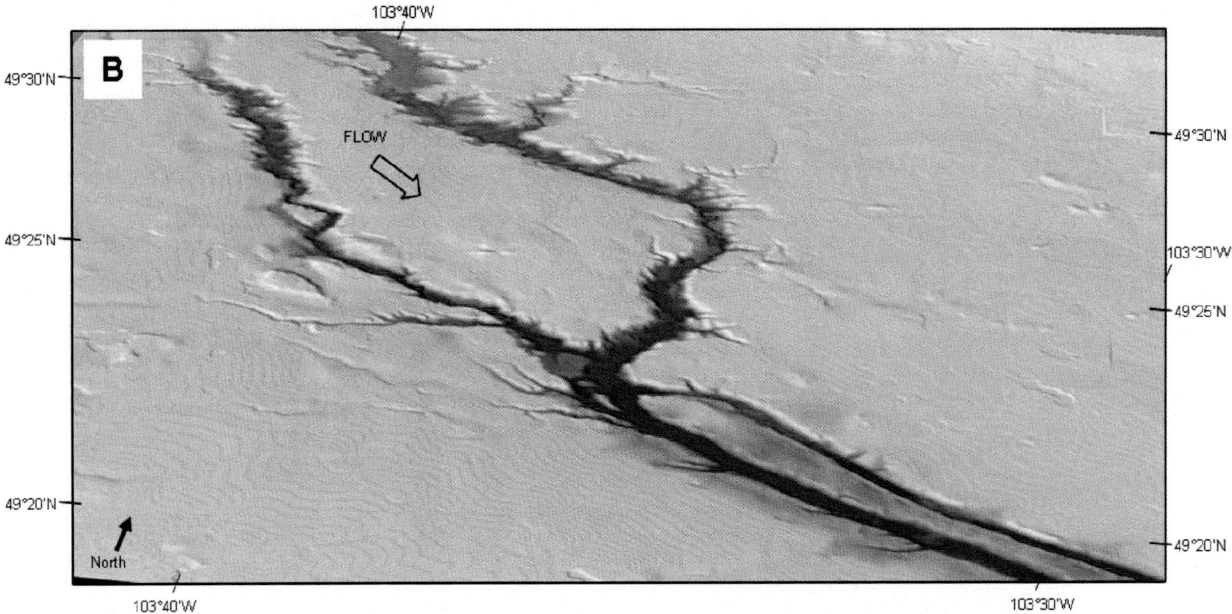

Plate 15. (A) Map showing distribution of bedforms and major spillway features of the Souris spillway immediately downstream of the outlet of glacial Lake Regina (Fig. 7.6). Note that transverse forms are limited to upper 12 km of spillway and longitudinal forms dominant downstream. All distinct transverse and longitudinal forms observed are shown; potholes that are large (~100s metres) are shown but most are too small to be mapped. (B) Oblique view image of map area (A) to illustrate geomorphic form. (Data from GeoBase®.)

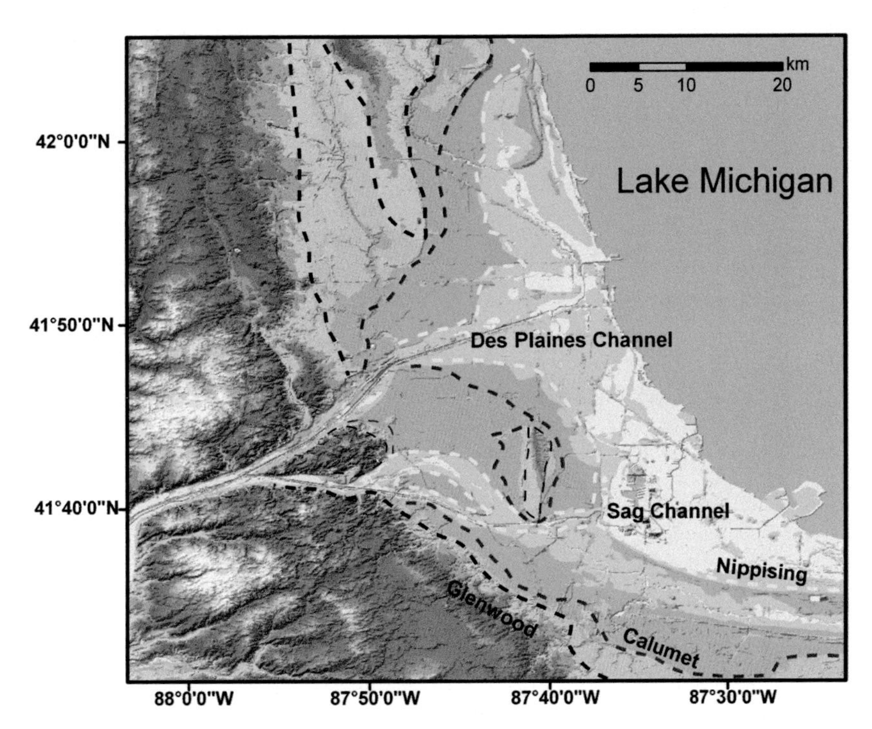

Plate 16. Hillshade DEM (NED 1 arc sec) of the Chicago outlet. Dashed lines show the approximate shoreline levels associated with the Glenwood, Calumet and Nippising phases.

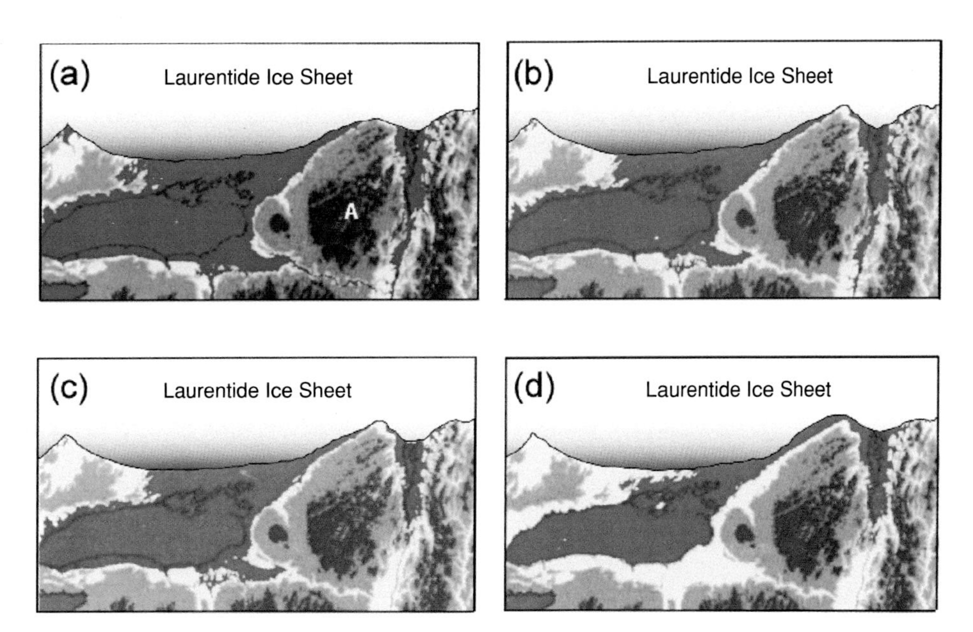

Plate 17. Reconstructed proglacial lakes within the Ontario Basin and Champlain Lowlands from transverse mercator projected 80 m Digital Elevation Models (DEM). Lake Ontario Basin, outlined in black, and Adirondack Mountains. (A) prevents confluence of water bodies. (a) Main phase Lake Iroquois with drainage out the Mohawk River and the Coveville level of Lake Vermont. (b) Flood I: ice retreat exposed a col at Covey Hill, Quebec, lower than the Rome outlet to the Mohawk. The new outlet dropped Lake Iroquois to the Frontenac level releasing catastrophic flows into the Coveville level of Lake Vermont. (c) Flood II: breakout of Frontenac level Lake Iroquois into the Coveville level created the threshold of the Upper Fort Ann level of Lake Vermont. (d) Flood III: continued ice retreat led to the confluence of Lake Iroquois and Lake Vermont creating further erosion and the development of the Lower Fort Ann level of Lake Vermont. Continued ice retreat opened the Gulf of St Lawrence for later floods from Lake Agassiz to the North Atlantic. (Modified from Rayburn *et al.* (2005).)

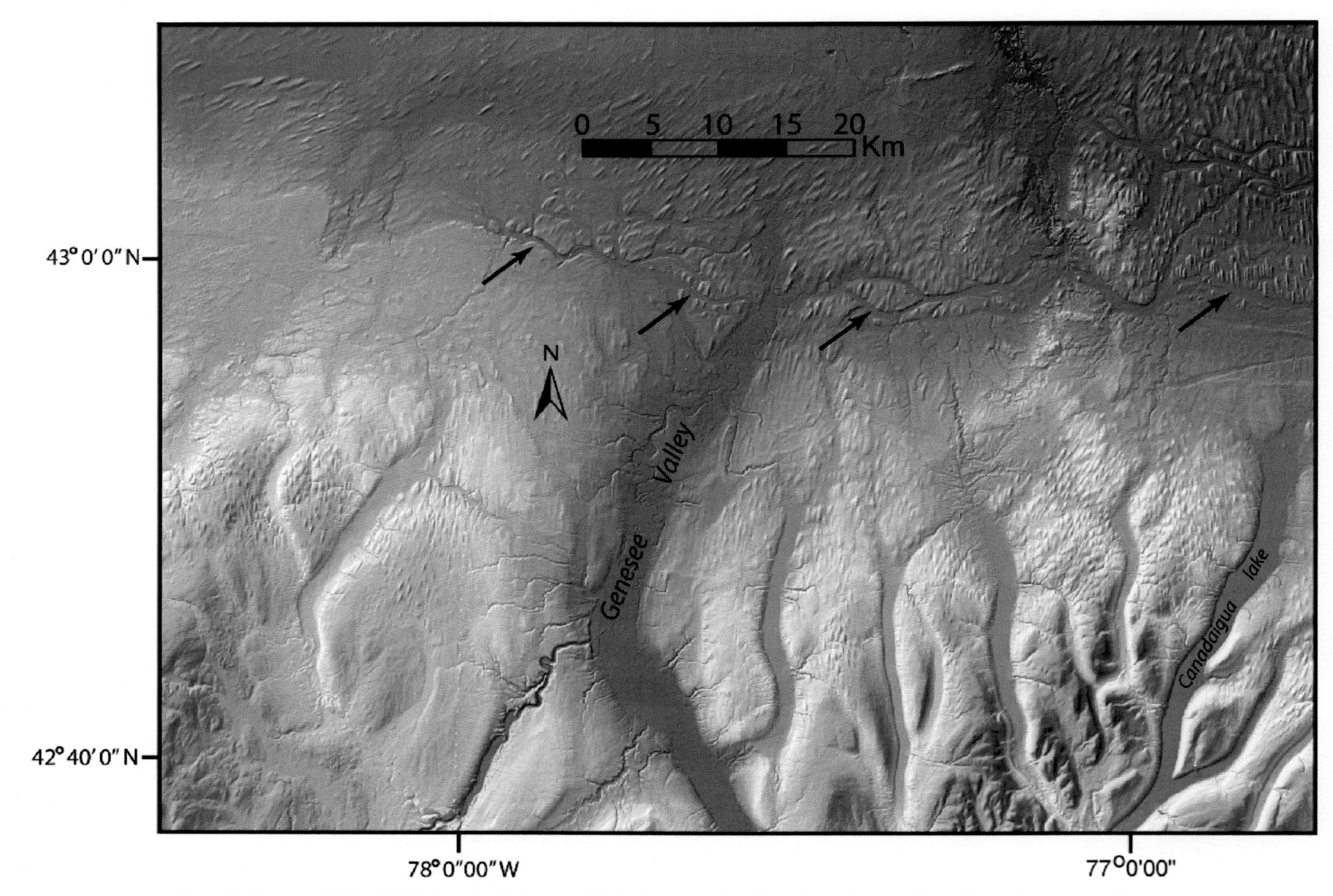

Plate 18. Ten-metre DEM of E–W channels (black arrows) in the Batavia region. Channels are interpreted as spillways from catastrophic drainage events issuing out of the Great Lakes. Wilson (1980) evaluated cobble–gravel delta deposits, pendant bars, expansion bars and scabland topography where glacial deposits were stripped off by catastrophic discharges. Note cross-cutting relationships with N–S Finger Lakes troughs and through valleys and abundant drumlin swarms.

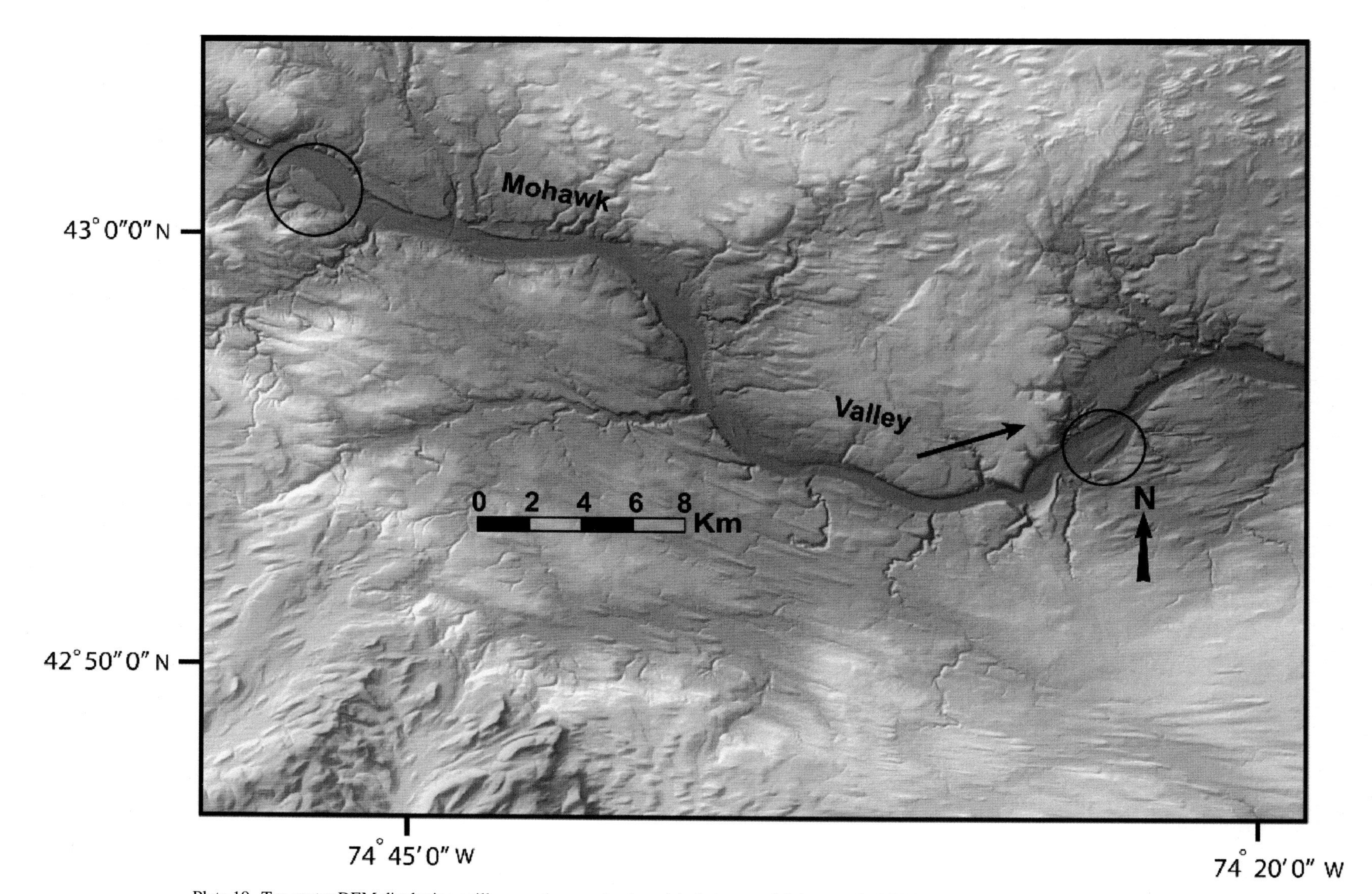

Plate 19. Ten-metre DEM displaying spillway and expansion bars (circles) west of Schenectady, NY. Bars are most likely the result of catastrophic discharges from Lake Iroquois down the Mohawk similar to those described by Wall (1995).

Plate 20. Aniakchak Caldera, Alaska, USA, breached sometime after 3400 years ago, producing the largest known terrestrial flood of the Holocene (Waythomas *et al.*, 1996). (a) Eastward aerial view. Formed during a catastrophic eruption about 3400 years ago, it is about 10 km across and averages 500 m deep. After formation, the caldera filled to a maximum depth of 183 m before breaching through a low point on the western caldera rim, incising 'The Gates' and releasing 3.7×10^9 m^3 of water at a peak discharge of 10^6 m^3/s. (Photograph by M. Williams, National Park Service, 1977.) (b) Eastward view of 'The Gates', the ~200 m deep breach formed in the rim of Aniakchak Caldera. (3 July 1992, photograph by C. A. Neal, U.S. Geological Survey.)

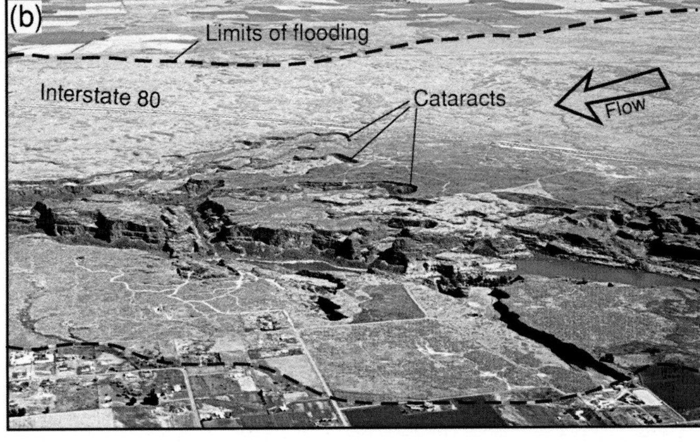

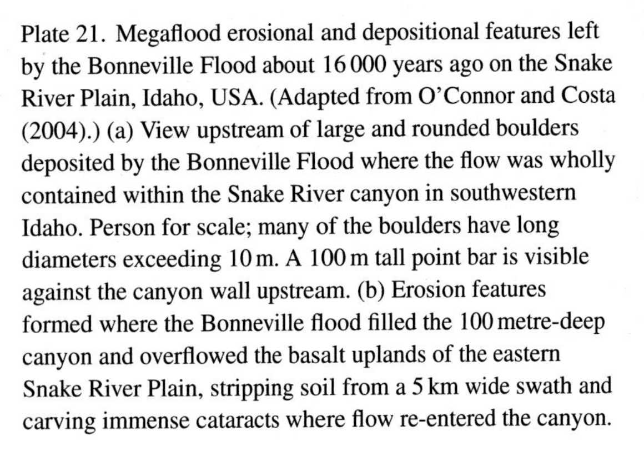

Plate 21. Megaflood erosional and depositional features left by the Bonneville Flood about 16 000 years ago on the Snake River Plain, Idaho, USA. (Adapted from O'Connor and Costa (2004).) (a) View upstream of large and rounded boulders deposited by the Bonneville Flood where the flow was wholly contained within the Snake River canyon in southwestern Idaho. Person for scale; many of the boulders have long diameters exceeding 10 m. A 100 m tall point bar is visible against the canyon wall upstream. (b) Erosion features formed where the Bonneville flood filled the 100 metre-deep canyon and overflowed the basalt uplands of the eastern Snake River Plain, stripping soil from a 5 km wide swath and carving immense cataracts where flow re-entered the canyon.

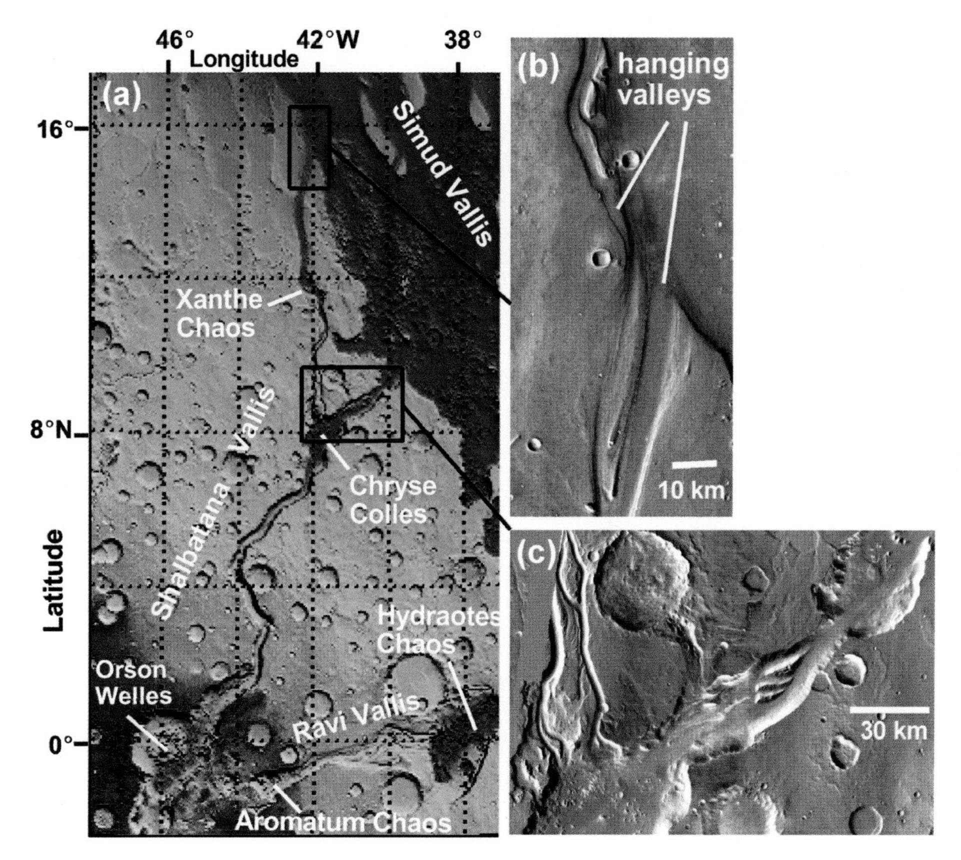

Plate 22. (a) MOLA elevation map of Shalbatana Vallis and Ravi Vallis. (b) Northern terminus of Shalbatana Vallis is cross-cut by Simud Valles. (c) Close-up view of Chryse Colles, a depression where Shalbatana divides into two channel systems. (Credit for subfigures (b) and (c): THEMIS daytime IR mosaic (Christensen *et al.*, 2006).)

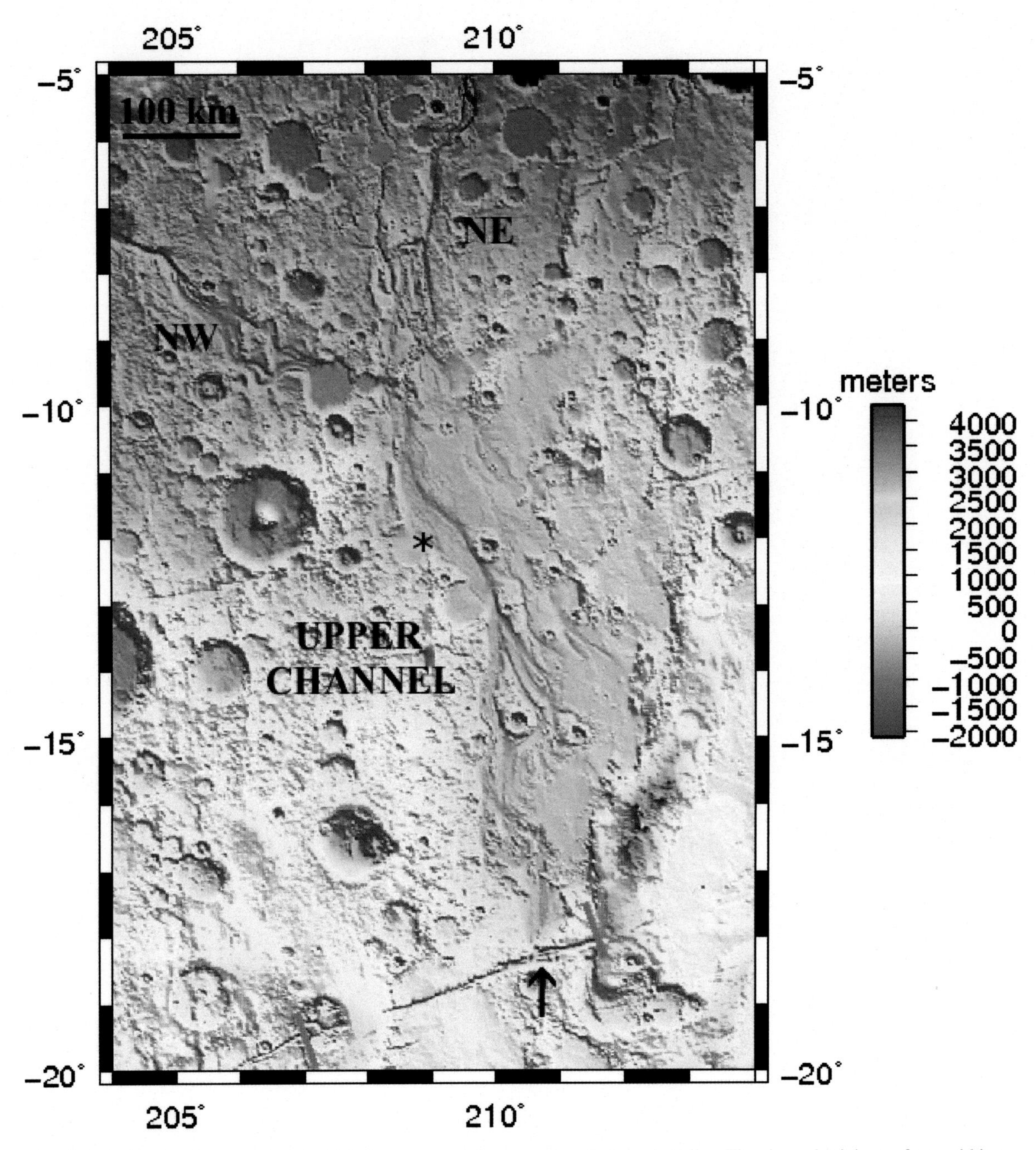

Plate 23. MOLA shaded relief overlain with MOLA colour-coded topography of Mangala Valles. The channel originates from within a segment of the fissure that has widened into a trough (grey brackets), and passes through a notch (vertical black arrow) in a degraded crater rim on the north side of Mangala Fossa. The upper channel stretches approximately 550 km northward (downstream) from the channel's origin, where it diverges into a northwestern (NW) and a northeastern (NE) branch. Figure 10.1 is located towards the north end of the upper channel (as denoted by black asterisk).

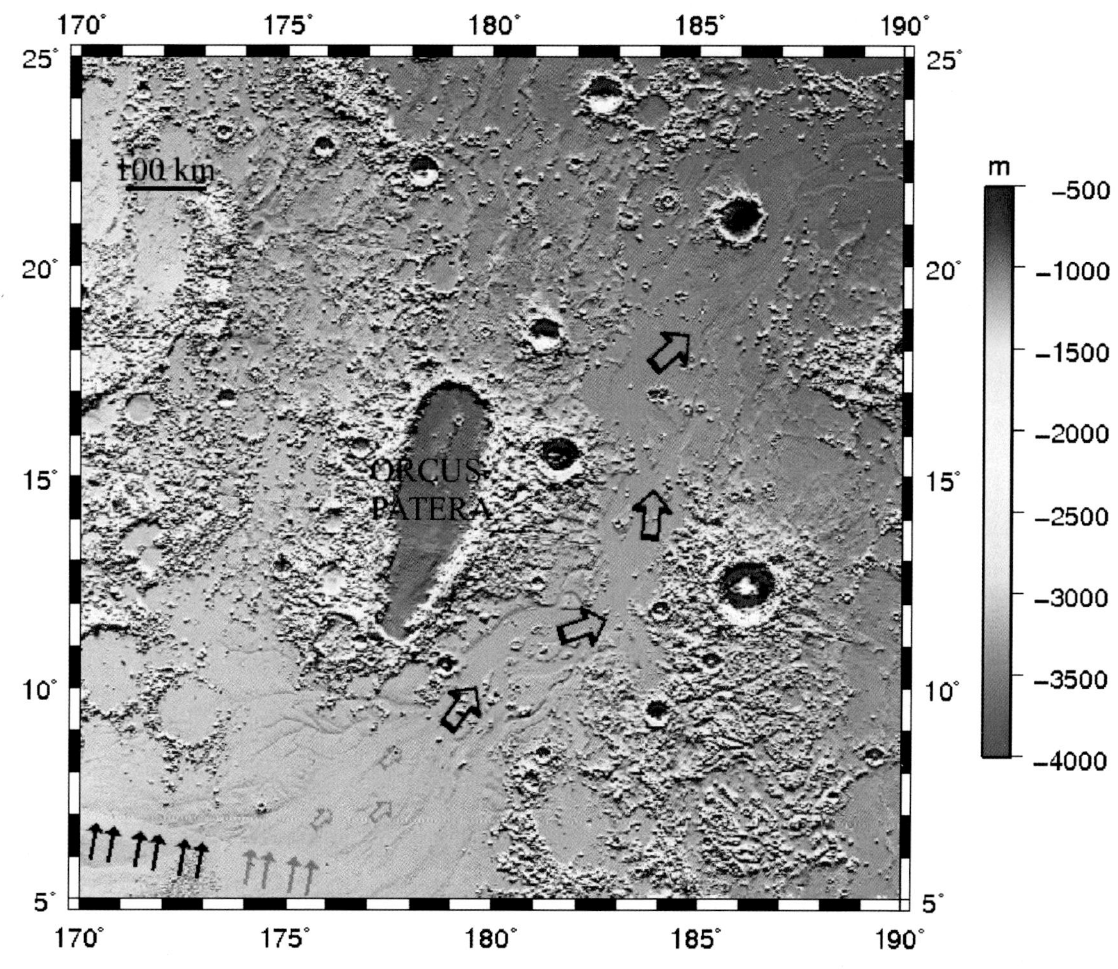

Plate 24. MOLA shaded relief overlain with MOLA colour-coded topography of Marte Valles. The floodwater is hypothesised to have come from the eastern-most segment of a Cerberus Fossae fissure (black solid arrows) that has been buried by lava flows (grey solid arrows.) Very shallow (buried) channels stretch from the buried fissure segment northeastward to Marte Vallis (grey open arrows), which itself stretches northeastward to Amazonis Planitia (black open arrows).

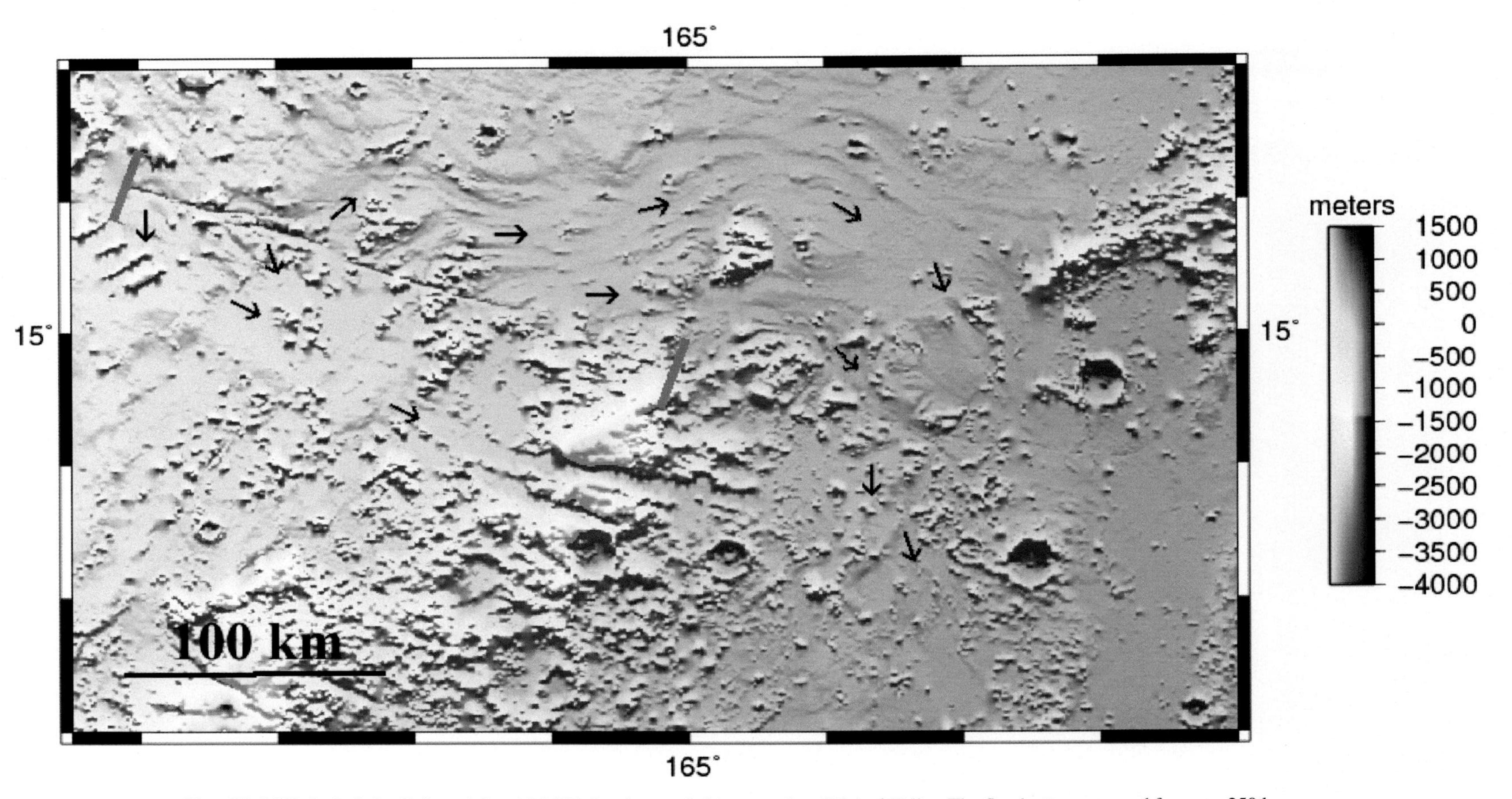

Plate 25. MOLA shaded relief overlain with MOLA colour-coded topography of Grjotá Valles. The floodwaters emerged from a ~250 km long segment of the Cerberus Fossae fissure (bracketed in grey), of which ~150 km show evidence for water release. Water spilled over both to the north and to the south, flowing generally eastward through knobs of outlying southern highland terrain before turning southward toward the Cerberus plains.

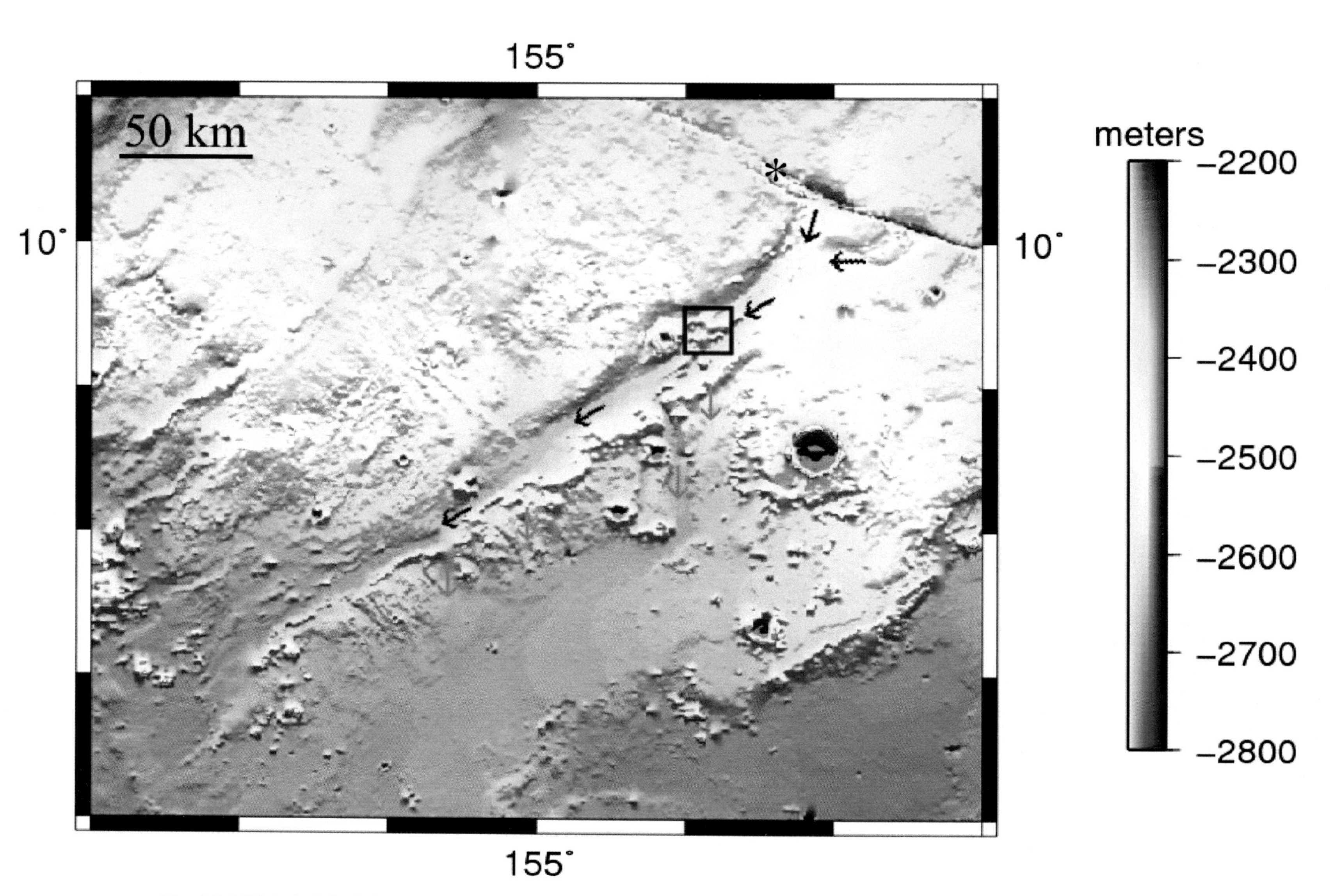

Plate 26. MOLA shaded relief overlain with MOLA colour-coded topography of Athabasca Valles. The floodwaters emerged from two locations along the Cerberus Fossae fissure and flowed southwestwardly (black arrows) to the western Cerberus plains. The channel is bordered on its southern side by a wrinkle ridge; the floodwaters breached the wrinkle ridge, producing distributary channels (grey arrows). The asterisk denotes the location of Figure 10.3, and the black box denotes the location of Figure 10.4.

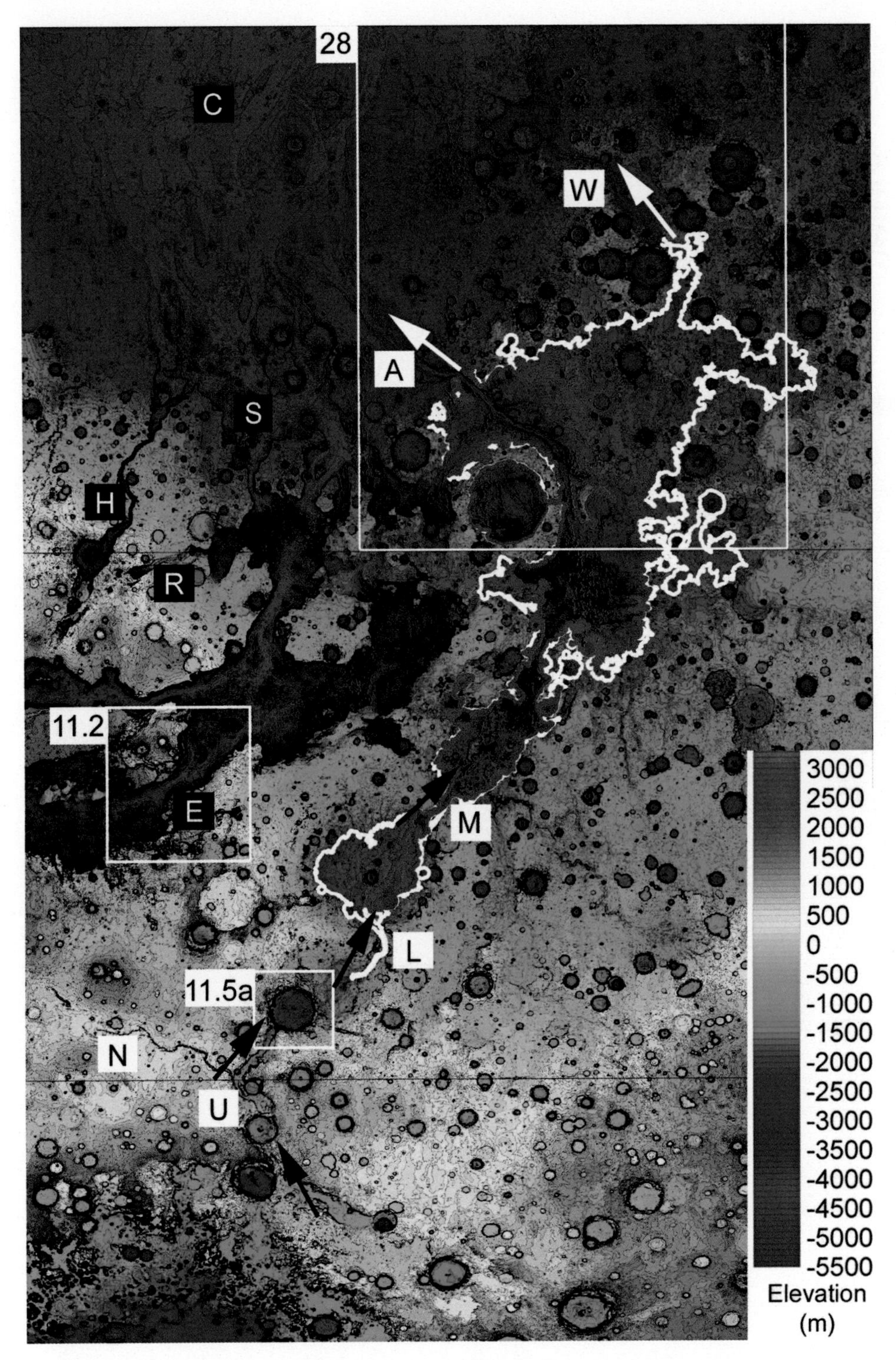

Plate 27. Chryse Trough region in 64 pixel/degree gridded MOLA topography, cylindrical projection, bounded by 30° N, 45° S, 0° E and 50° W. Straight black lines represent the equator and 30° S (30° of latitude is 1778 km); and white outlines give the locations of Plate 28, and Figures 11.2 and 11.5a. Black arrows indicate flow direction from Argyre through the ULM system; the coarse white and red outlines are the –1880 and –2000 m contours within the Margaritifer basin, respectively; and the white arrows indicate outflows from that basin through Mawrth (W) and Ares (A) Valles. Labelled are Uzboi Vallis (U), Ladon Valles (L), Morava Valles (M) and the Nirgal Vallis (N) tributary to Uzboi. Contrasting letters indicate regional features: Eos Chasma (E), Ravi Vallis (R), Shalbatana Vallis (H), Simud Valles (S) and the Chryse basin (C).

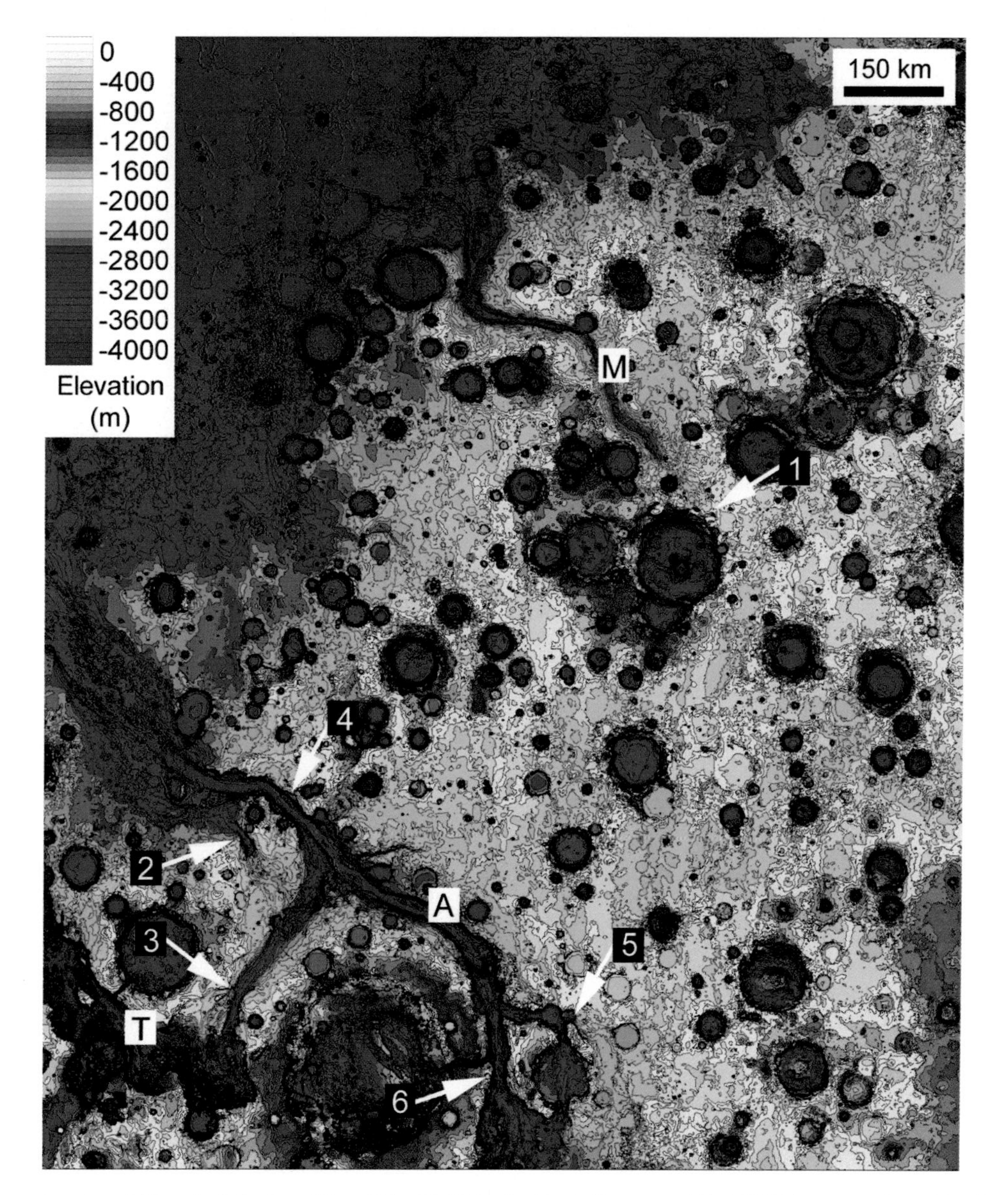

Plate 28. The head region of Mawrth (M), Ares (A) and Tiu (T) Valles in 64 pixel/degree gridded MOLA topography, cylindrical projection, bounded by 30° N, 0° N, 5° W and 30° W, 100 m contour interval. Margaritifer basin overflow points (elevations given prior to incision) include the outlet to Mawrth at –1880 m (1), the western outlet to Ares at –2000 m (2), the inlet from Tiu to Ares at –1800 m (3), the main outlet to Ares at –2000 m (4), and two other points at –2000 m (5, 6) where Ares overflowed a second divide headward in the basin. The southern extension of the basin can be seen in Plate 27. Aram Chaos, the highly degraded impact crater under the label '6', is ~280 km in diameter. No water source is evident for Mawrth Vallis without an overflow of the Margaritifer basin to the south.

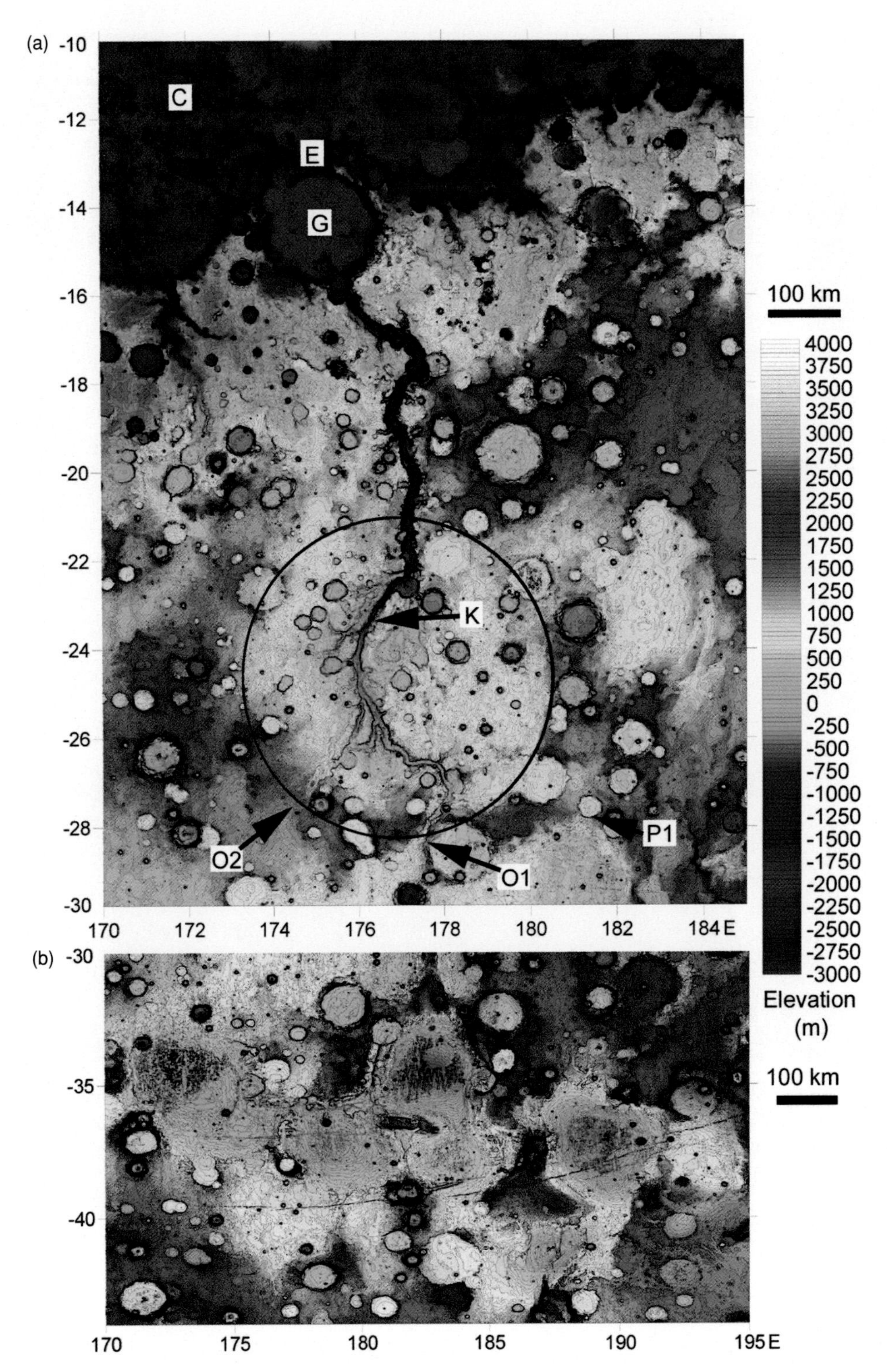

Plate 29. (a) Ma'adim Vallis, showing the intermediate basin in a black outline, the main (O1) and western (O2) outlets of the Eridania basin, an unused pass (P1) that was formerly blocked by a crater rim, the location of the knickpoint (K), Gusev crater (G), the outlet breach to Gusev (E), and chaotic terrain north of the crater. (b) The eastern part of the Eridania basin at the head of Ma'adim Vallis. Note the high concavity of the sub-basin floors relative to other crater floors in the area. Both are 128 pixel/degree gridded MOLA topography, 50 m contour interval. (Part (a) is from Irwin *et al.* (2004), Copyright 2004 American Geophysical Union, reproduced/modified by permission of American Geophysical Union.)

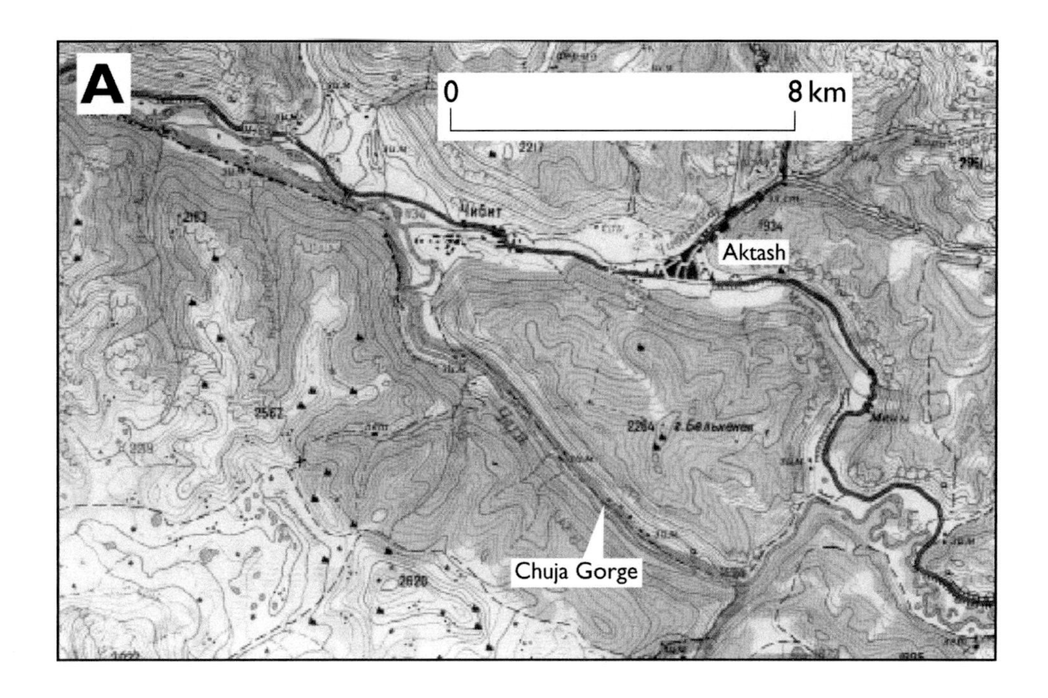

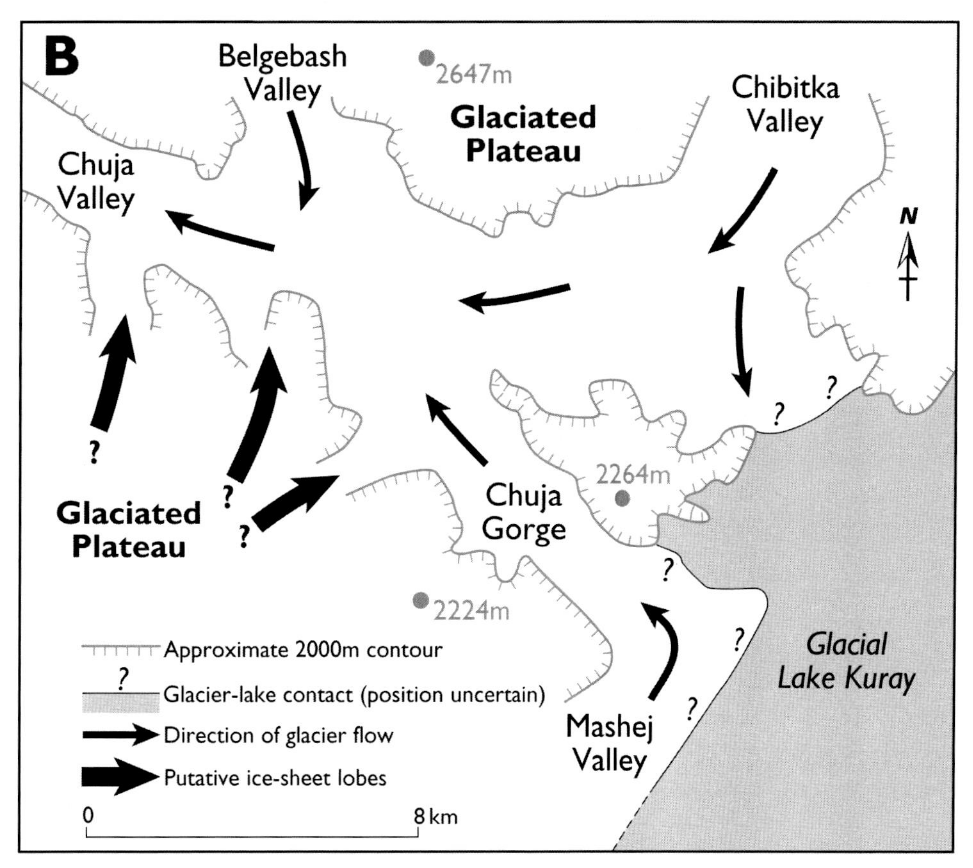

Plate 30. (A) Map of the modern geography of the region around the town of Aktash in the vicinity of the former ice dam. (Reproduced from 1:200 000 scale map produced by the Federal Service of Geodesy and Cartography, Russia, 1992.) (B) Schematic cartoon of the probable configuration of the ice dam in the same location as A showing valleys filled with glacier ice to an altitude of 2100 m at times of maximum lake levels (see text for details).

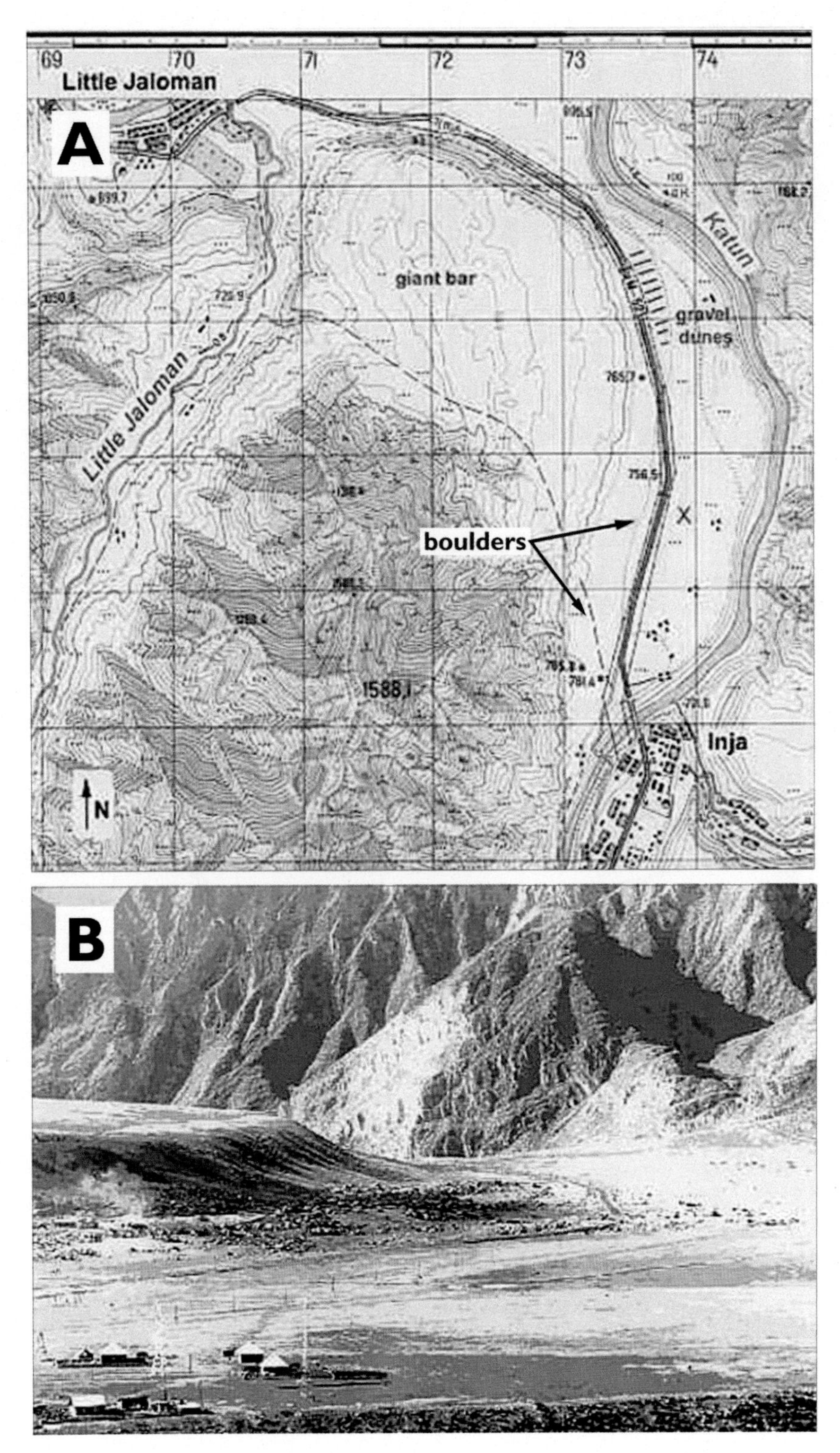

Plate 31. (A) Map of the modern geography of the region of the Katun River valley between the villages of Little Jaloman and Inja in the vicinity of the Little Jaloman giant bar. (Reproduced from 1:200 000 scale map produced by the Federal Service of Geodesy and Cartography, Russia, 1992.) (B) View to the northwest from Inja bar across Katun River valley towards the Little Jaloman bar showing concentrations of boulders lags on the eroded flank of the bar and a boulder field strewn across the terrace surface (see arrows in panel A).

Plate 32. Features of major bars of the Katun Valley: (A) The surface of the Little Jaloman bar is inclined up the tributary valley of the Little Jaloman River with the steep bar front (to right of image) facing the Katun valley (the black arrow points up-tributary). The Little Jaloman River, flowing left to right, cuts the bar into two at point A. Excellent sediment exposures are located opposite A on the slope hidden from view below the white arrow. (B) View of 70 m high section of the bar just within the mouth of the Little Jaloman valley (section is below white arrow in panel a). Major bed sets (white arrow) are coarsest and thickest (10 m to 20 m) at the base of the bar and thin and fine upwards and into the tributary valley (to right). Also shown are examples of a silt layer and diamicton. (C) Exposure of the Inja bar just within the mouth of the Inja valley. The succession (*c.* 50 m) is primarily granule gravel (Ggl/b) with discontinuous single-clast thick, cobble to small-boulder lags along certain bedding planes. These cobble horizons suggests some rough separation of the deposits into repetitive units composed of thick Ggl alternating with thinner Ggl/b. A large rounded pebble pod crops-out bottom right with figure for scale to the left (see also Plate 37B).

Plate 33. Principal facies of the exposed sequence of the Komdodj bar. (A) The lower part of the sequence consists of a basal facies (Ggl) that rests conformably over a silt layer that in turn caps the large sandy clinostratifications and grades upwards into a coarser laminated unit (Gd/cl). Note: The basal contact between downstream-dipping clinoforms and finer gravels above is similar to that shown in Plate 35 for the Kezek–Jala bar, except that here the pre-bar sequence was formed in a more sheltered setting, hence the sediments are finer and the contact is mainly conformable with only local erosion of the contact. (B) The upper part of sequence in the Komdodj bar is locally massive or parallel bedded, but in this photograph it is cross-bedded to locally massive gravel (Gdt) unconformably overlaying laminated gravel (Ggl or Gd/cl). Figure for scale (1.9 m). Note the direction of the cross-bedding in panel B is down the Katun, whereas some directional structures (not shown) in Ggl facies in panel A indicate flow up the Katun River.

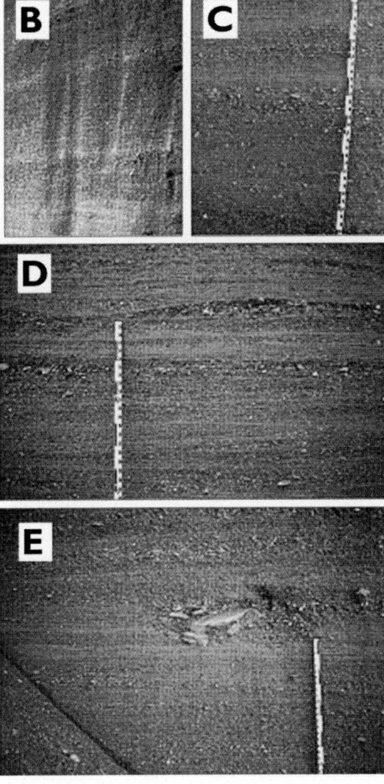

Plate 34. Facies of the giant bars of the Katun River valley. (A) Thick sequence of the Komdodj bar with laminated gravel (Ggl, Gd/cl) overlying silt and very fine sand (Fm/Fr) and in turn capped by coarse gravel in large clinostratifications (Gbp) (section is 50 to 70 m high). (B–E) (Scale bar is graduated in cm and 5 cm intervals) Principal characteristics of the fine-grained laminated gravel (Ggl) include parallel laminations (B, C), clusters of pebble gravel along some bedding planes and associated undulated laminae (D), and (E) local small scour with flat imbricated pebble indicating current flow up a valley tributary and away from the main Katun River valley. Scale bar in each image has major graduations of 10 cm.

Plate 35. Sediment exposures along the road (km 682) at the Kezek-Jala bar. (A) Thick sequences of horizontally bedded fine-grained gravel constitute the upper part of the exposure, and contain frequent outsized clasts. (B) The basal tangential major clinoforms (lower half of image) consist of coarse gravel (Gc/bp) and dip down the Katun valley. Note the uncomformable contact between the cobble gravel and the horizontally bedded fine-grained gravels (Ggl) above. (C) Erosional remnants of silt and massive-to-plane-bedded sand (Sm/l, Table 13.1) occur along the unconformable contact. Scale bar bottom right is 1 m in length.

Plate 36. Features of the Log Korkobi bar, see Figure 13.3 for location. (A) Geomorphologic-geologic setting of the bar downstream from a resistant bedrock ridge on the true right side of the valley. Limited 3 m high exposures occur along the track (arrowed). In the background and to the right is the Little Jaloman village below the Little Jaloman bar. (B) Large 1.5 m size angular boulders eroded from the upstream bedrock ridge and dumped in the bar (track section arrowed in A). (C) Imbrication clusters in the finer deposits of the exposure shown in B, indicating a current flowing counter to the direction of the Katun River, thus attesting to the presence of a gyre downstream from the bedrock ridge. Scale bar is graduated in cm and 5 cm intervals.

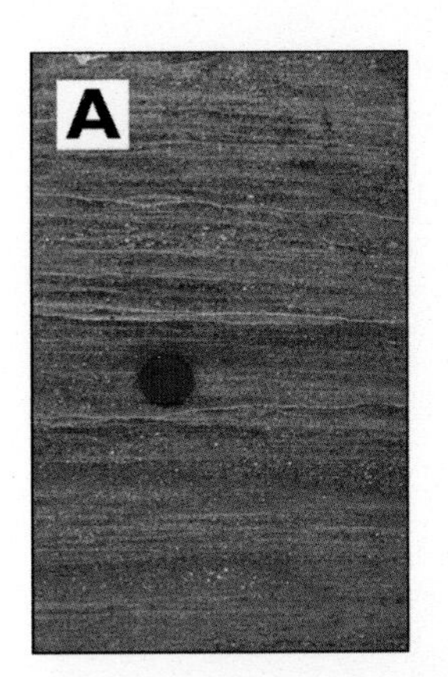

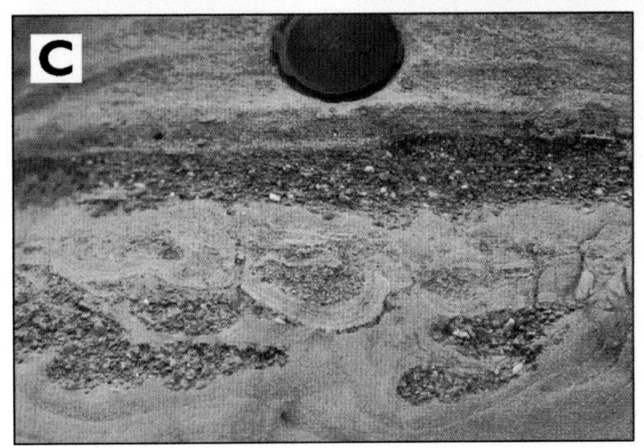

Plate 37. (A) Thin, wavy lamination of coarse sand and granules in giant bar deposits (lens cap is 50 mm in diameter). (B) Large gravel pod within horizontally bedded fine gravels of the Inja bar, in the entrance to the tributary Inja River valley near Inja village. Person 1.9 m tall provides scale. Cobble pod is the same one as shown in Plate 32C. (C) Fine silt beds convoluted with superimposed fine gravel due to loading of the silt layers by the deposited gravels. Lens cap is 50 mm in diameter. (D) Obique view looking downwards onto exposed coarse, red diamicton. Notebook is 21.5 cm long.

Plate 38. Geomorphologic setting and sediments in the tributary valley behind the Inja bar and near the village of Injuska. (A) Backside of the Inja bar showing the level of the main Inja bar, the limnic fill dissected by incision by the Inja River (with recent rill erosion). The Katun River is flowing left to right beyond the giant bar and the tributary Inja River is flowing right to left in the defile (just above figure's head) across the image. (B) View looking up the Inja River tributary valley from the Inja bar top. The village of Injuska is situated just to the left and below the left-hand dot in the line of dots marking an end moraine of a glacier that fed fine sediments to a small lake impounded behind the giant bar. (C) Laminated gravel overlain (and interdigitating elsewhere) with silty glacio-lacustrine thin laminations (Lac). (D) Half-metre high exposure of finely laminated lacustrine unit (0.5 m long pick for scale) intercalated with flood gravels above. Lower end of pick is located at the interface between the lacustrine deposits and flood gravels below. Location: Injuska village. (E) The Little Jaloman valley, in common with most of the valleys, does not contain glacio-lacustrine deposits, rather the most up-valley, laminated gravel (Ggl) of the bar has been variously eroded and capped by gravel (Ggt) transported by the Little Jaloman River.

Plate 39. (A) An example of the upper succession of the Little Jaloman bar shows laminated fine gravel capped by pebble gravels and thin silt layers. These cycles are similar to those lower in the succession except that the lower examples (e.g. Plate 32C) are coarser and are thicker overall. Scale bar is graduated in cm and 5 cm. See also Plate 38D. (B) Section inclined at 45° away from observer consists of thinly bedded sequence of granule-gravels in upper part of Little Jaloman bar. Length of inclined section to just above letter 'B' is about 6 m.

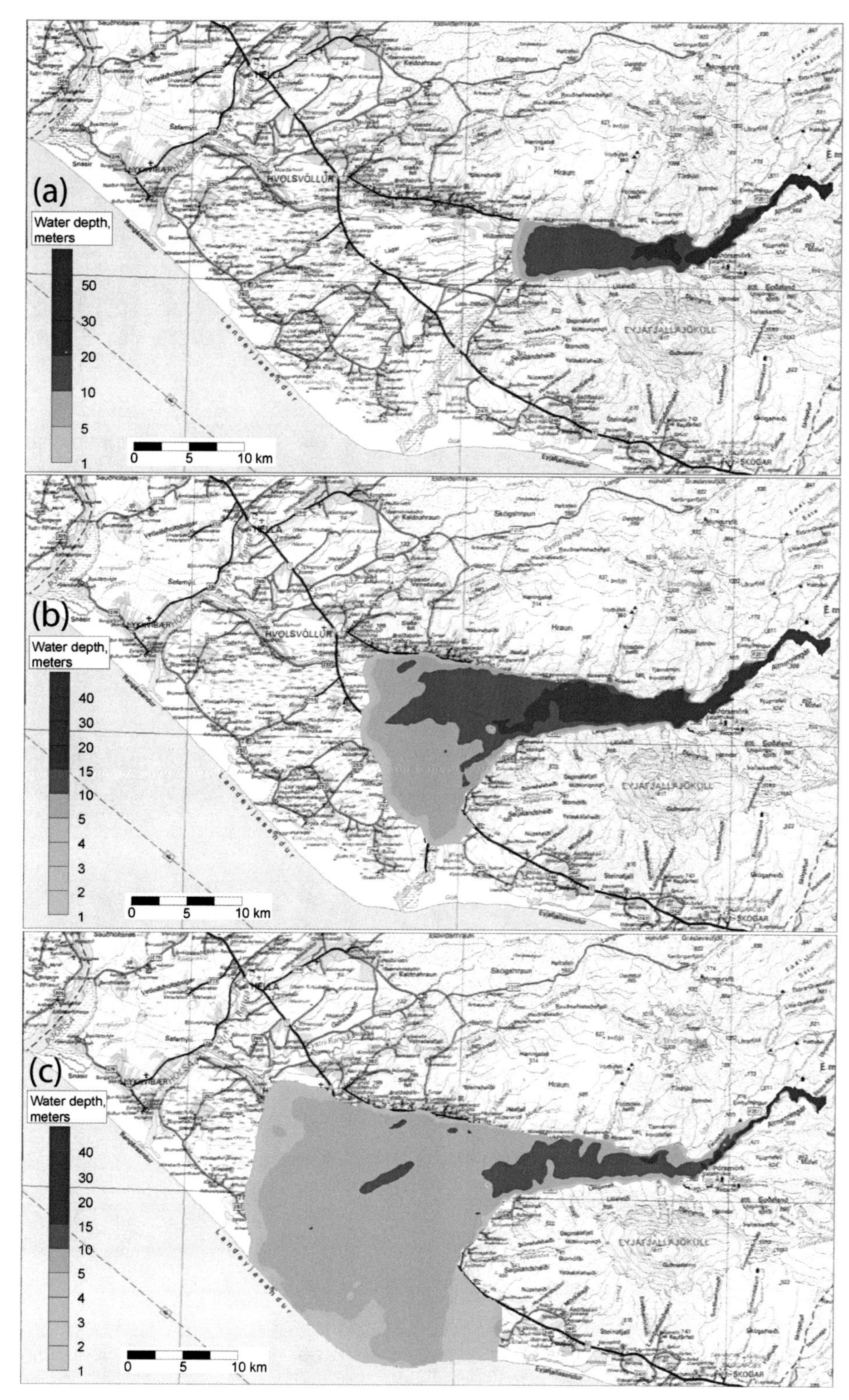

Plate 40. Depth of a simulated jökulhlaup to the west from Katla (a) two and a half, (b) four and (c) six hours after the assumed start of the flood.

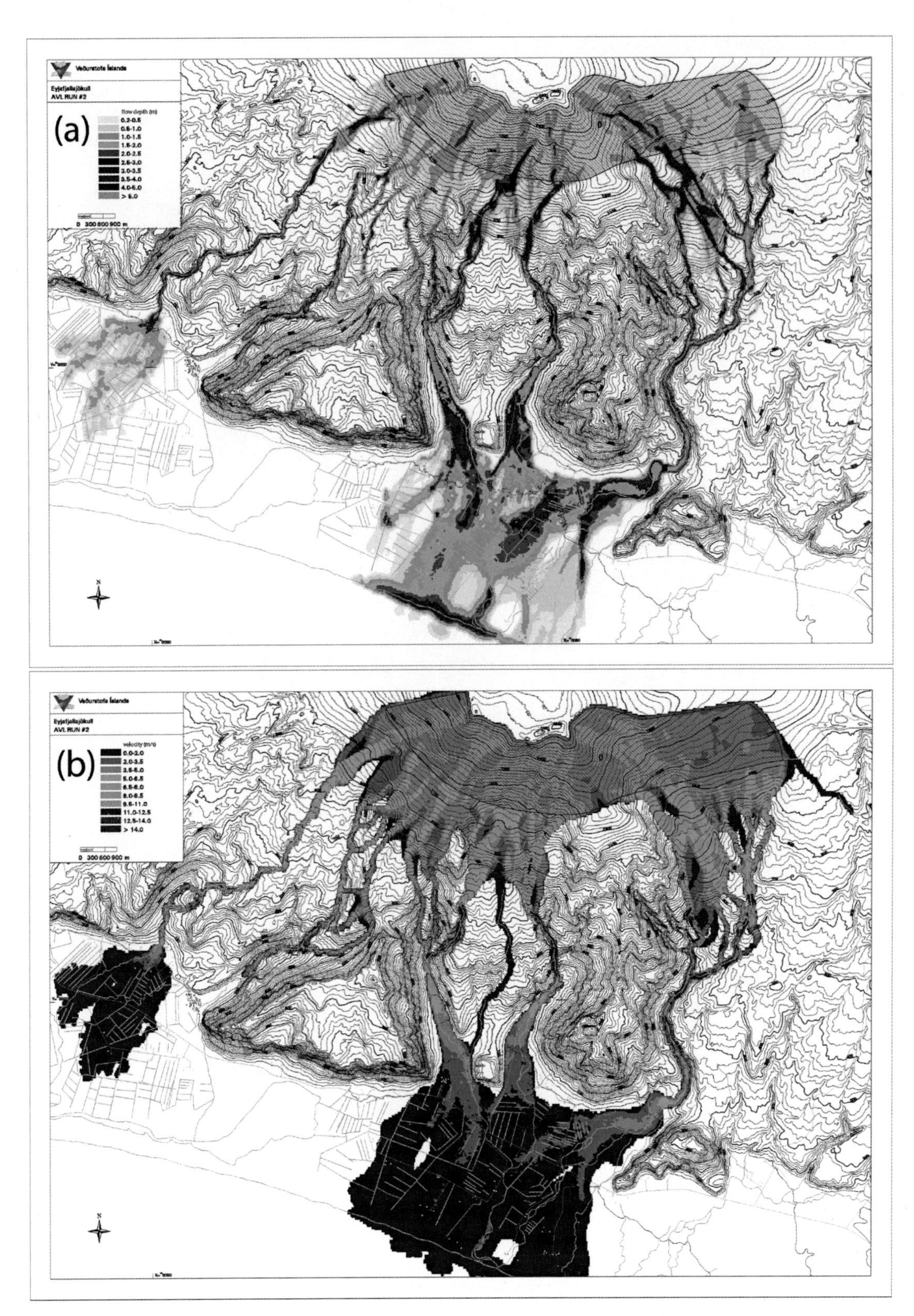

Plate 41. Simulation of a jökulhlaup to the south from Eyjafjallajökull. (a) Maximum flow depth. (b) Maximum flow speed. (See text for explanations.)

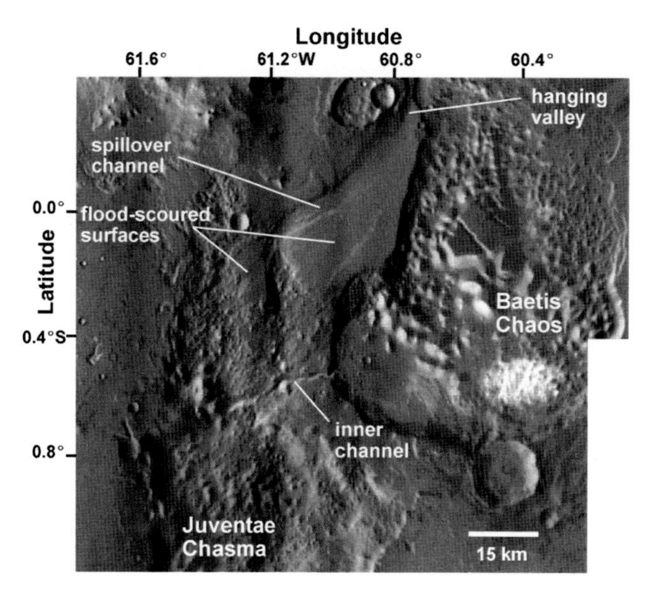

Figure 9.7. Chasm and chaos source areas for Maja Valles. Flood-scoured surfaces mark minimum elevations for the initial overflows (i.e. 1180 m) (After Coleman and Baker (2007).) (Image credit: THEMIS Daytime IR Mosaic (Christensen *et al.*, 2006).)

Valles, demonstrating that catastrophic floods carved the Maja channels. Other processes such as glacial erosion or debris flows do not produce megaripples. Catling *et al.* (2006) described various hypotheses for the origin of light-toned layered deposits in the Valles Marineris. They closely examined Juventae Chasma and interpreted its deposits to be sulphate-rich sedimentary rocks that formed either as evaporites from a large sea or that they accumulated by dry deposition of volcanic sulphate aerosols in association with snow or ice at low latitudes during times of high obliquity. Catling *et al.* (2006) interpreted that at least some of the layered material was deposited long before the adjacent Hesperian plateau basalts. They dismissed a volcanic origin based on the abundance of gypsum as inferred from OMEGA spectroscopy data. Bishop *et al.* (2007) found a concentration of hydrated sulphates in southwestern Juventae Chasma, but also reported the presence of hydrated material in places across the interior of the chasm.

Catling *et al.* (2006) identified terraces of possible fluvial origin high on the western margin of Juventae Chasma that they interpreted as having formed before or contemporaneous with the chasma itself. Consistent with Chapman *et al.* (2003), we find clear morphological evidence that a lake existed in ancestral Juventae Chasma and that the water eventually overflowed or breached the canyon wall at its northern terminus. The overflow level from the canyon can be estimated using the elevations of water-eroded landforms on the plateau west of Baetis Chaos (Figure 9.7) (Coleman and Baker, 2007). This plateau has a wide flood-scoured surface and a crossover channel eroded ~100 m deep, the margins of which indicate an elevation of 1150 to 1180 m for the initial overflow flooding from Juventae Chasma. This overflow elevation range also defines the minimum potentiometric surface for groundwater that discharged into and overflowed Juventae Chasma.

A broad valley with a small inner channel leads down from the northern end of Juventae Chasma into Baetis Chaos (Figure 9.7). This inner channel has a floor elevation of approximately −100 m where it exits Juventae Chasma. Subtracting this from the initial overflow elevation of 1150 to 1180 m reveals that a water column 1250 to 1280 m deep in Juventae Chasma was drained by the flooding. The present-day canyon floor has a low point at −4400 m. If any part of the ancestral canyon was as deep as it is now, a residual lake with a maximum depth of ~4300 m would have remained after the flooding.

The genesis of Baetis Chaos (Figure 9.7) may have been triggered by deep fluvial incision during the catastrophic flooding from Juventae Chasma, creating incipient conditions for groundwater breakout (see Coleman, 2005). The 15 km wide channel of Maja Valles emerges from the northern end of Baetis Chaos. Floodwaters also entered Chia Crater to form a lake that overflowed its northern rim. Just west of Chia Crater the main channel splits into two valleys (see http://planetarynames.wr.usgs.gov/images/mc10_mola.pdf) that continue northward for hundreds of kilometres. The western channel is a 'hanging' valley where it diverges from the main channel, which is 350 m deeper. The depth of the main channel indicates that it incised more quickly and eventually captured all of the flow, causing flooding to cease in the western channel. The megaripples found by Chapman *et al.* (2003) occur along the western channel and likely were formed by early overland flows.

Northward, broad scabland flooding is indicated along the central reaches of Maja Valles by the presence of streamlined 'islands' that formed on the channel floor between latitudes 9.2 and 11.6° N (see http://planetarynames.wr.usgs.gov/images/mc10_mola.pdf). In this latitude band, the terrain was eroded over a width of ~90 km indicating that the flows consisted of broad, relatively shallow scabland flooding. A divergence of flow occurred at 12° N, 58° W, where the main channel branched into two large channels. Ister Chaos and other chaotic terrain occur on the floor of the eastern channel, which forms hanging valleys at the divergence point, and also further north where the two channels recombine before descending to the floor of Chryse Planitia. These cross-cutting relations show that flow continued in the western channel longer than in the eastern channel.

Chapman *et al.* (2003) used MOLA data to construct a topographic profile across the channels in the area southwest of Chia Crater where overland flooding was relatively narrow and the channels are well defined. Using floodwater depths of 440 m and 590 m they estimated maximum discharges of 2×10^8 to 4×10^8 $m^3\,s^{-1}$, but also concluded that discharges up to 10^6 $m^3\,s^{-1}$ may be more realistic given that bankfull floods rarely fill valleys (see Wilson *et al.*, 2004).

9.3 Outflow channels that emerged from discrete chaotic terrain or fault zones

In this section we discuss circum-Chryse outflow channels that did not originate at the mouths of canyons. Ravi, Shalbatana, Allegheny, Walla Walla and Elaver Valles issued from chaotic terrain or fault zones represented at the surface as cavi. Distant canyon lakes and elevated groundwater pressures may nonetheless have contributed to the formation of these channels. For example, Carr (1995) described a 40 km wide depression that leads northward from Ganges Chasma to the origin of Shalbatana Vallis. He described this depression as evidence for channelised subsurface flow of water, and suggested that a lake in Ganges Chasma may have drained northward to the source of Shalbatana Vallis. For this interpretation to be correct, the outlet on the eastern side of present-day Ganges Chasma (Figure 9.2) could not have existed at the time of the Shalbatana flood. Gridded MOLA data reveal that the depression leading north from Ganges Chasma broadens and extends eastward to another channel, Ravi Vallis, and therefore both channels may partly owe their origin to lateral recharge from a distant canyon lake.

9.3.1 Ravi Vallis

Ravi Vallis begins near the equator at the eastern margin of Aromatum Chaos (Figures 9.2 and 9.8). East of the chaos the main channel diverges and converges in an anastomosing pattern. Multiple channels are present, but two are larger than the others and here we refer to them as the northern and southern channels. The northern channel became the deepest because it ultimately captured all of the discharge in the system, focusing erosion within its reaches. Eastward, the southern channel terminates as a hanging valley where it rejoins the northern channel. Therefore, flows had ceased in the southern channel while flow continued in the northern channel.

Before the flooding, groundwater in Xanthe Terra would have been confined beneath a thick cryosphere, with a potentiometric surface higher than the land surface. Eventually, at the present location of Aromatum Chaos, the confining layer was ruptured, probably by a tectonic event. Disruption by a geothermal or impact event may also have been possible. Groundwater would have rushed upward under high artesian pressure from beneath the cryosphere. This catastrophic upwelling would have undermined and transported near-surface geological materials as the water flowed torrentially downslope, rapidly eroding the terrain. The flow readily would have maintained itself because refreezing of the cryosphere would have been inhibited by the high flow rate, high heat capacity of water, relatively warm groundwater temperatures, and hydrodynamic erosion of the outflow zone (Coleman, 2005).

Just east of Aromatum Chaos, Ravi Vallis is 23 km wide and the channel floor is covered with longitudinal ridges (Figure 9.8b). The surface morphology shows that hydrodynamic erosion dominated in this narrowest part of the channel where flow depths and erosive power were greatest. Further downstream, the presence of streamlined islands and erosional scarps shows that the channel complex is 45 km wide and may attain a width of 70 km (Coleman, 2005). The great width of the early overland flooding indicates that the initial flow was shallow scabland flooding and that a deep preflood valley did not exist. Based on average flow depths of 30–100 m, mean velocities of 6–20 $m\,s^{-1}$ and discharges of 0.4–5 $\times$ 10^7 $m^3\,s^{-1}$ have been estimated for the initial overland flows in Ravi Vallis (Coleman, 2005). Leask *et al.* (2006) also analysed Ravi Vallis flooding. Using water depths of 50–150 m, they estimated flow speeds of $\sim$10–25 $m\,s^{-1}$, a maximum discharge rate of $\sim 3 \times 10^7$ $m^3\,s^{-1}$ just after the start of the flood, and $< 1 \times 10^7$ $m^3\,s^{-1}$ in the late stages. They calculated a flood duration of two to ten weeks and a minimum total water volume of 1.1–6.5 $\times$ 10^4 km^3.

Iamuna and Oxia Chaos (Figure 9.8) were spawned on the floor of the northernmost channel in its deepest reaches (Coleman, 2005), resulting in new outbreaks of groundwater. A smaller unnamed chaos exists near the terminus of the southern channel. The formation of chaos on the channel floor provides evidence that the flooding lasted long enough to thin the cryosphere by fluvial incision, permitting secondary breakouts of groundwater at multiple locations. These flows from Iamuna and Oxia Chaos, like the outflows from Aromatum Chaos, may have eventually reached Chryse Planitia via lowlands to the east. The Ravi Vallis channels are abruptly truncated 220 km from their source by scarps at the steep, faulted margin of Hydraotes Chaos. The chaos therefore postdated all features associated with Ravi Vallis, and obliterated any evidence of cross-cutting relations (relative age) between Ravi and Simud Valles.

Confined groundwater was the apparent source for the initial outflows that carved Ravi Vallis. Water released from storage in a confined aquifer represents only the secondary effects caused by changes in the fluid pressure (i.e., aquifer compaction and water expansion). As the flood progressed, the aquifer fluid pressures decreased and flow conditions eventually would have transitioned from confined to unconfined. The unconfined dewatering of an aquifer is more protracted because releases from

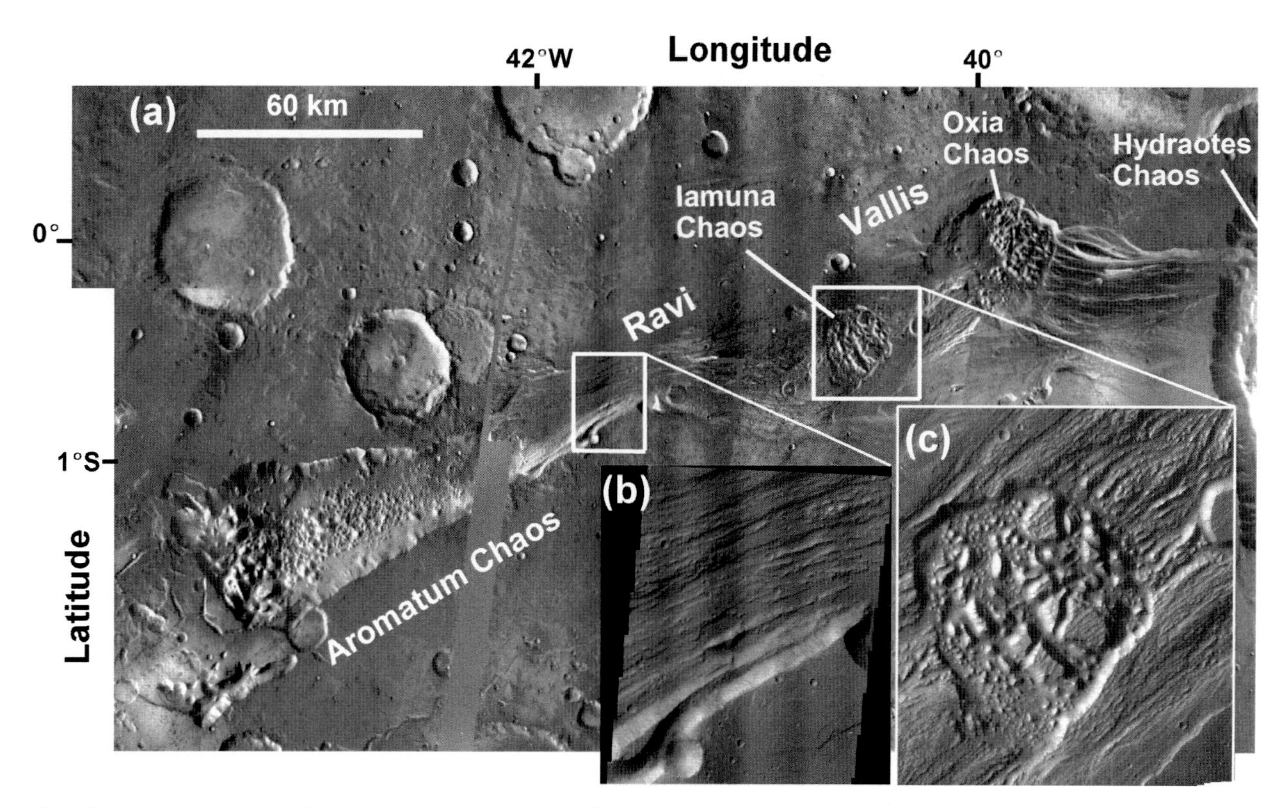

Figure 9.8. (a) Location map of Ravi Vallis and associated chaotic terrain. (After Coleman (2005).) (b) Longitudinal ridges carved by hydrodynamic erosion on the floor of Ravi Vallis. (c) Close-up view of Iamuna Chaos. Longitudinal ridges have been preserved on top of the interior chaos blocks. (Image credit: Christensen *et al.* (2006); subfigures (a) and (c) are from THEMIS daytime IR mosaic; (b) is from THEMIS image V03123001.)

storage represent actual dewatering of the pores and fractures (Freeze and Cherry, 1979). The hydrograph for flow released from a confined aquifer should display initial rapid release of water, resulting in broad scabland erosion. We interpret that subsequent unconfined drainage, augmented by distant lateral recharge from an ice-covered lake in ancestral Ganges Chasma, would have produced a long recession of flood stage that would have concentrated erosion in the deepest parts of Ravi Vallis, creating an inner channel. Therefore, it is unlikely that floodwaters ever filled Ravi Vallis to its full depth (Coleman, 2005). Calculations that assume bankfull flow will overestimate flow velocities and discharge rates for this and other palaeofloods that were spawned by groundwater breakouts (Wilson *et al.*, 2004).

Groundwater breakout mechanism for Iamuna Chaos

The floodwaters that carved Ravi Vallis incised the crust more than 700 m, thinned the cryosphere, and triggered the release of groundwater to form secondary chaos zones on the channel floors (Coleman, 2005). The inner channel within the northern channel of Ravi Vallis leads directly into the deepest part of Iamuna Chaos, which appears to have been the point of initial groundwater outbreak (Figure 9.8c). The critical condition for groundwater breakout occurred as the floodwater discharge rate waned and the inner channel reached its maximum depth. So long as water continued flowing in the channel, the overpressure in the underlying confined aquifer would have been partly offset by the floodwater depth. As the flood stage declined, the overpressure from below would have increased until it exceeded the overburden pressure and rock strength and broke through the channel floor, initiating a new groundwater outbreak (Coleman, 2005). With the pressures in the confined aquifer being relieved downstream at Iamuna Chaos, discharge from Aromatum Chaos would have diminished and stopped. The new flows from Iamuna Chaos would have further eroded the northern channel, incising a new inner channel that spawned another secondary outbreak at the site of Oxia Chaos. Discharges from Oxia Chaos would similarly have terminated the flow from Iamuna Chaos. Based on the depth of fluvial incision at Iamuna Chaos, and an analysis of overpressures beneath the channel floor, the cryosphere was 700–1000 m thick at the time of the Ravi Vallis flood. This cryosphere thickness was used to estimate a crustal heat flux of $>50\,\mathrm{mW\,m^{-2}}$ at this location at the time of the flood (Coleman, 2005). The cryosphere thickness at this equatorial location indicates a cold, long-term climate trend similar to that of present-day Mars.

We cannot rule out the possibility that Iamuna and Oxia Chaos formed much later than the Ravi Vallis flooding. However, over time a thick cryosphere would have reformed beneath the channel floor, sealing the top of the underlying aquifer and preventing renewed groundwater outbreaks. On Earth, permafrost layers that are 100–1000 m thick can develop or decay in only 10^4–10^5 years (Lunardini, 1995). Therefore, it is likely that Iamuna and Oxia Chaos formed during the final stages of the Ravi Vallis flooding, or soon thereafter.

Komatsu *et al.* (2000) suggested that dissociation of carbon dioxide ice or clathrate could be a triggering mechanism for chaos formation. However, the initial accumulation of these materials in the subsurface would have been difficult given a higher crustal heat flux in the Hesperian Period and because the thermal conductivity of carbon dioxide ice and clathrate is significantly lower than that of water ice (Urquhart and Gulick, 2003).

9.3.2 Shalbatana Vallis

Shalbatana Vallis is northwest of Ravi Vallis and has its origin in a northeast-trending canyon that is 225 km long (Plate 22). This canyon may be the northernmost part of a long linear depression that extends southward to Ganges Chasma. Shalbatana Vallis appears linked to Ganges Chasma in the same manner as Ravi Vallis. The southern part of the depression includes the interior of Orson Welles, a crater that was ~130 km wide. We cannot know how large this depression was when Shalbatana was formed. It likely contained a sizeable ice-covered lake that would have been catastrophically drained when water overflowed the rim of the canyon at an elevation of –300 to –350 m and carved an outlet to the northeast. The Shalbatana channel begins at this outlet at 1.75° N, 43.55° W. Just north of its origin the channel is more than 2 km deep and ~18 km wide, and it continues as a single deep channel over the southern half of its course.

Cabrol *et al.* (1997) theorised that the chaos that formed along Shalbatana Vallis were generated by magma intrusion of the points of intersection between Tharsis radial and concentric faults. They further proposed that Shalbatana Vallis was carved by water sources from chaos and a lake in Ganges Chasma. Similarly, Rodriguez *et al.* (2003) hypothesised that the subsidence zone north of Ganges Chasma was produced by the collapse of a complex system of chains of cavernous bodies that were created by the intrusion of dykes into ground ice. They proposed that the water that formed Shalbatana and Ravi Valles was derived from three sources: melting of the cryosphere by intrusive magmatism, drainage from a palaeolake in Ganges Chasma, and catastrophic outflows from confined aquifers. One complication for models that invoke igneous intrusions to create Shalbatana Vallis (or Ravi Vallis) is that no surface volcanics that correspond to the Hesperian age of these floods have been mapped in the source regions for these channels. Additional research is needed to assess whether dyke intrusions played a role in forming these channels. For example, mapping by Langlais *et al.* (2004) indicates the presence of a linear magnetic anomaly over Ganges Chasma that extends northward toward Shalbatana Vallis.

The flooding that created Shalbatana Vallis (and Ravi Vallis) occurred in the Middle Hesperian to Late Hesperian Epochs because the channel eroded ridged plains material of Early Hesperian age (Scott and Tanaka, 1986). The channel also incised Hesperian terrain mapped as Chryse Unit 1 by Tanaka (1997). The overall length of Shalbatana Vallis is ~880 km from its canyon source to the terminus of the northernmost channel. A depression named Chryse Colles occurs as a broadening of the valley floor 430 km along the channel where it divides into two valley systems (Plate 22c). The colles is ~2100 m deep and has a relatively flat floor with an elevation of –3800 m, similar to the Simud channel floor to the east. This colles may originally have been a chaos with interior blocks that became degraded by continuing flow in Shalbatana Vallis or by post-fluvial processes such as the cyclic formation and loss of ground ice. From Chryse Colles the main channel continues 120 km to the northeast where it descends to the broad lowland of Simud Valles and terminates in a region of knobs and eroded mesas. Nelson and Greeley (1999) reported that a lobe of Shalbatana sediments was deposited on the Simud floor, which would suggest a late stage of Shalbatana flooding. We find no evidence of this lobe in THEMIS images or MOLA data, nor is there evidence of an inner channel that would have been carved by late Shalbatana flows on the Simud valley floor. Cross-cutting relations between the channels may have been concealed by the post-fluvial degradation and collapse of mesas within Simud Vallis, forming the knobby terrain.

A second Shalbatana channel more than 450 km long trends due north from Chryse Colles (Plate 22a). This northern channel is 'hanging' at both ends, which shows that the channel that exited the northeast corner of Chryse Colles flowed for a longer time, became more deeply incised, and eventually captured all of the flow. Flow then apparently ceased in the northern channel. Xanthe Chaos, which has a low point at an elevation of –4052 m (11.532° N, 41.882° W), formed within the northern channel. We propose that this chaos formed within the Shalbatana channel when the palaeoflood deeply incised the cryosphere, creating conditions for breakout of confined groundwater (see Coleman, 2005). The maximum depth of the chaos (~1200 m) may approximate the cryosphere

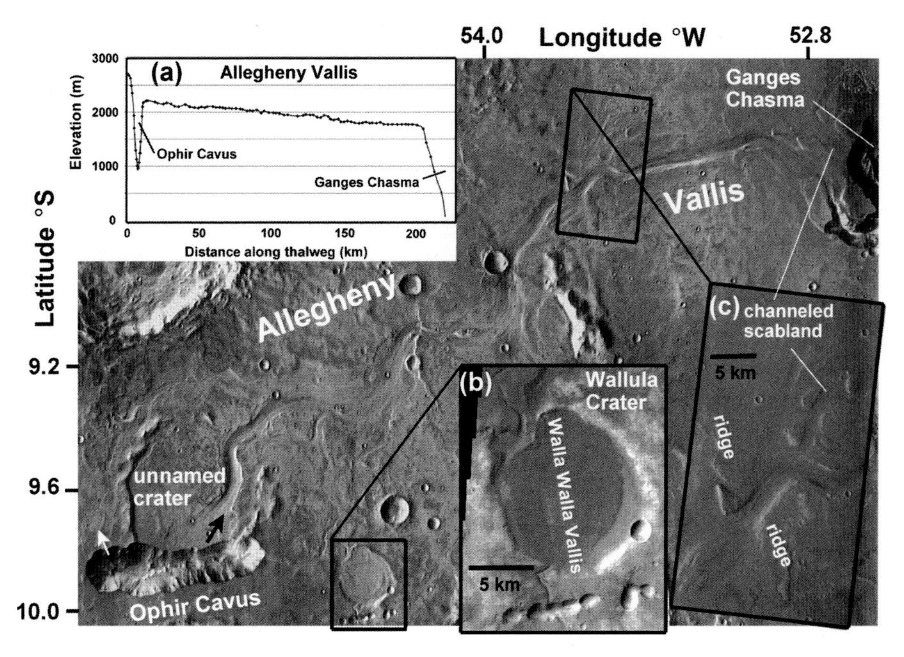

Figure 9.9. View of Allegheny Vallis from its source at Ophir Cavus to its terminus at Ganges Chasma. White arrow shows where water initially crested the NW rim of the cavus at an elevation >2500 m. Black arrow shows where discharge from cavus entered and filled unnamed crater. (Base image credit: THEMIS daytime IR mosaic (Christensen *et al.*, 2006).) (a) Plot of channel thalweg. (b) Close-up of Wallula Crater and the pit chain that sourced Walla Walla Vallis. (THEMIS image V10176001.) (c) Channeled scabland carved by release of floodwaters that ponded behind a ridge. (THEMIS image V06781022.)

thickness at the time of the flooding. The terminus of the northern channel forms a 'hanging' valley with a floor elevation of –3623 m that was cross-cut by Simud Valles (Plate 22b). The floor of Simud is more than 150 m deeper than the hanging valley. We conclude that the final flow in Simud Valles occurred after the carving of the northern Shalbatana channel.

9.3.3 Allegheny Vallis and Walla Walla Vallis

Allegheny Vallis is a 190 km long outflow channel in Xanthe Terra that ends at the western margin of Ganges Chasma (Figure 9.9). From the source to the terminus the channel floor gradually drops in elevation by ~435 m, indicating an average energy slope of 0.0023. A plot of the channel thalweg is shown in Figure 9.9a. The channel was carved by floodwaters that erupted from Ophir Cavus, a pit 2000 m deep and 38 km long. This cavus is located along the central part of Ophir Catenae, a 600 km long alignment of cavi (see http://planetarynames.wr.usgs.gov/images/mc18_mola.pdf) that extends from Candor Chasma to south of Ganges Chasma (Coleman *et al.*, 2007b). The catenae are likely the surface expression of a long, en echelon, dilational fault system. We theorise that a tectonic event created these faults, cracked the cryosphere, and dramatically released confined groundwater onto the surface of Mars. Two other channels, Walla Walla Vallis and Elaver Vallis, also were spawned along the trend of Ophir Catenae. Elaver Vallis eroded terrain of Lower Hesperian age (Witbeck *et al.*, 1991). Therefore, if all three channels formed around the same time, then the floods occurred in the Middle to Late Hesperian Epochs.

Ophir Cavus developed at the southern margin of a large, eroded, unnamed crater (Figure 9.9). Abrupt truncation of Allegheny Vallis at the northern rim of the cavus suggests that the depression grew to its present size sometime after the fluvial event. MOLA data indicate that the initial floodwaters erupted onto the plains at an elevation >2500 m (see white arrow in Figure 9.9). Flow from the cavus (black arrow in Figure 9.9) filled an unnamed crater and created a transient lake (Coleman *et al.*, 2007b). Eventually the northeastern rim was overtopped or breached and the lake drained, eroding a channel complex to the northeast. The flow divided and converged, carving a mid-channel 'island' and a series of terraces. The floodwaters then eroded a series of sweeping meanders to the north. Two ridge obstructions apparently existed along the path of the flood. The northernmost obstruction was larger and caused a transient lake to form. The lakewaters rose until the obstruction was surpassed, catastrophically draining the lake and carving a scabland to the northeast (Figure 9.9c). Eventually one channel grew deep enough to capture all the flow. The scabland then drained and the main channel was eroded across the plains to the western margin of Ganges Chasma. The channel ends as a broad, branching

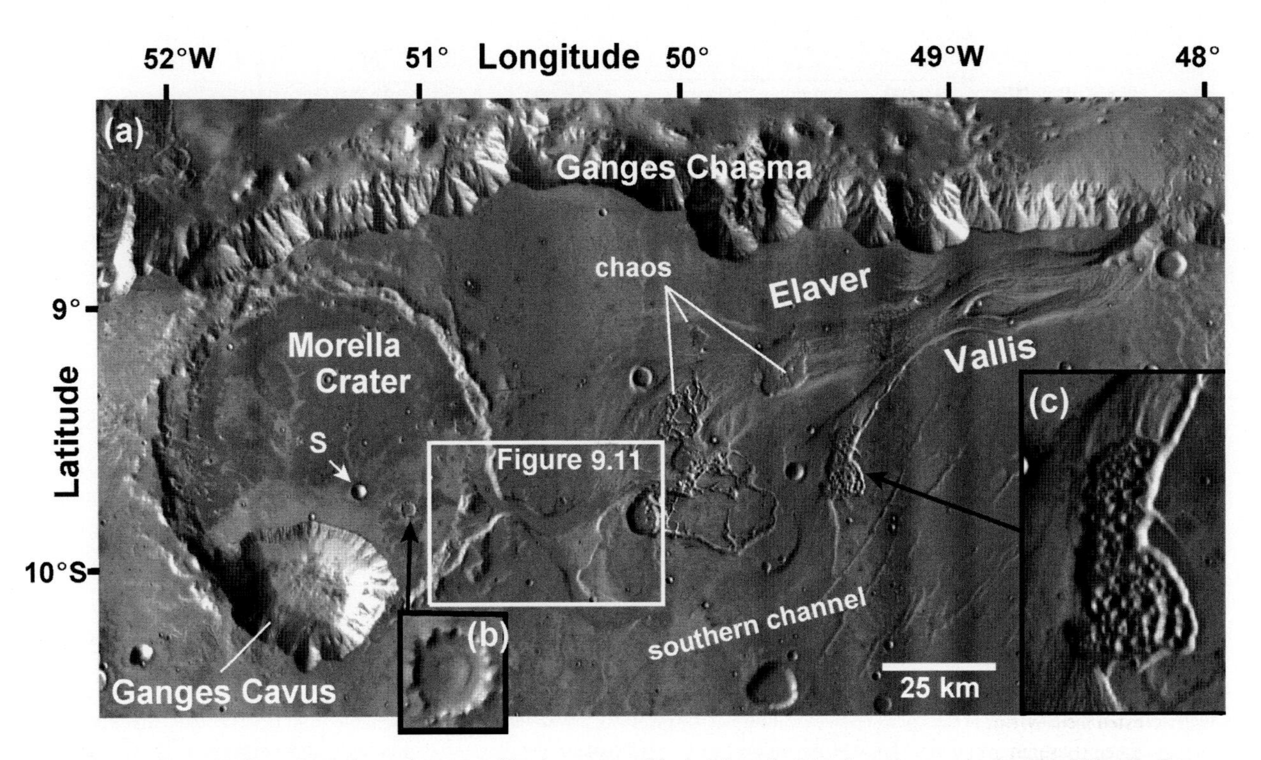

Figure 9.10. (a) Elaver Vallis, a channel complex created by the catastrophic draining of a lake through a gap in the rim of Morella Crater. (After Coleman and Dinwiddie (2007).) The channels abruptly terminate (upper right) at the southern margin of Ganges Chasma. S is Somerset Crater. The white box shows the location of Figure 9.11. (b) Johnstown Crater. (c) Unnamed chaos that formed on southern channel floor. (Credit for images: THEMIS daytime IR mosaic (Christensen *et al.*, 2006).)

scabland that terminates abruptly along the canyon rim. The morphology indicates that the rim of Ganges Chasma continued to grow westward after the flooding had ended (Coleman *et al.*, 2007b).

Walla Walla Vallis was carved by groundwater discharges from an unnamed pit chain (Figure 9.9b). MOLA pass 10452 crossed over this area and revealed that one of the pits is more than 130 m deep. The terrain north of the pit chain stands higher than 2520 m, indicating that the confined aquifer had a potentiometric surface at or above this elevation. The Walla Walla floodwaters carved a small channel that leads down into Wallula Crater, a subdued Noachian crater 12 km in diameter with a base elevation of 2400 m. A north-trending erosional channel across Wallula Crater is faintly visible in Figure 9.9b but is too shallow or narrow to be seen in MOLA data. Ponded water may have accumulated in Wallula Crater until the water breached or overtopped its northwestern rim. Beyond this overflow point Walla Walla Vallis splits into two channels. THEMIS images indicate that at least one of the branches merges northward with Allegheny Vallis, suggesting that the Walla Walla floodwaters contributed to the discharges in Allegheny Vallis. Additional research and high-resolution images will be needed to confirm the superposition relationship between these channels.

The Walla Walla flood eroded Noachian terrain of the cratered unit of the Plateau Sequence (unit Npl_1 of Witbeck *et al.*, 1991). The source of Walla Walla Vallis is only 25 km from the source of Allegheny Vallis. Both floods erupted from segments of Ophir Catenae, which indicates they may have been concurrent in time, or nearly so. If the floods were concurrent, the Walla Walla flooding would have ceased quickly because Allegheny Vallis eroded a much larger channel with a thalweg lower in elevation than the floor of Wallula Crater. The outflow at Allegheny Vallis would therefore have pirated much of the regional discharge, depressurising the confined aquifer system and terminating the flow in Walla Walla Vallis.

9.3.4 Elaver Vallis

Elaver Vallis is a 160 km long channel system that begins at a notch in the rim of 80 km wide Morella Crater (Figures 9.10 and 9.11). The morphology of the landforms indicates that a water body existed inside Morella, and this makes Elaver Vallis unique because it was the only circum-Chryse outflow channel created by the catastrophic drainage of a crater lake. The channel system incises Early Hesperian strata (unit Hpl_3) of the Plateau Sequence (Scott and Tanaka 1986; Witbeck *et al.*, 1991); therefore, the flooding likely occurred in the Middle to Late Hesperian Epochs. Extensive cratering on the channel floors rules out an Amazonian age.

There is only one outlet from Morella Crater, and no channels enter it, so groundwater must have

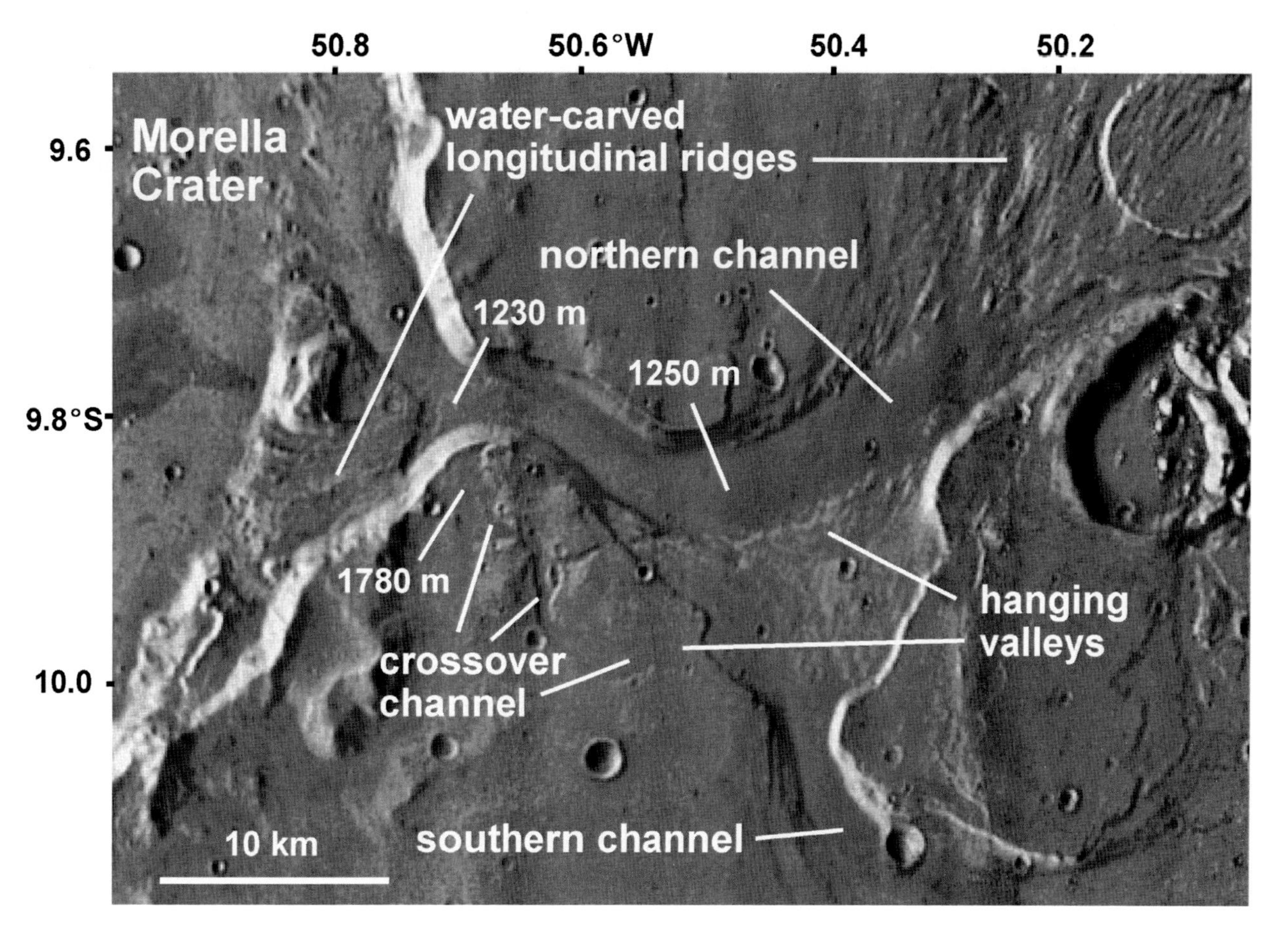

Figure 9.11. The source of Elaver Vallis is a water gap where floodwaters breached the rim of Morella Crater. Flow in the channel was from left to right. The elevation of a smaller crossover channel (1780 m) marks the level at which the rising lake overtopped the crater rim. The high point on the floor of Elaver Vallis (1250 m) marks the elevation of the residual lake surface after the crater lake drained. (Credit for image: THEMIS daytime IR mosaic (Christensen *et al.*, 2006).)

discharged from Ganges Cavus, a 3 km deep pit in Morella (Figure 9.10). We interpret this easternmost cavus of Ophir Catenae (http://planetarynames.wr.usgs.gov/images/mc18_mola.pdf) as a subsidence feature produced mainly by dilatant faulting. The inception of Ganges Cavus ruptured the cryosphere, permitting confined groundwater to rush upward into the crater, which caused additional undermining and erosion in the cavus. Water discharged onto the adjacent crater floor at an elevation of 1080 m. The crater acted, in effect, like an enormous standpipe that slowly filled with groundwater. The initial breakout at Ganges Cavus was controlled by massive groundwater overpressure that ruptured the cryosphere and destroyed the aquifer at the outflow zone. This initial flow was unconstrained by aquifer properties and cannot be analysed using typical aquifer analysis methods. However, subsequent discharges can be analysed because they represent flow from intact aquifers around Ganges Cavus (Coleman and Dinwiddie, 2007).

The palaeolake surface would have risen until the crater wall was breached, or until the pressure of the lake water column equalled the pressure in the subcryosphere aquifer system. Growth of ice on the crater palaeolake would have been minimised by the continuous influx of relatively warm groundwater. The water never reached the height of the groundwater potentiometric surface. Overtopping of the crater rim verifies that groundwater inflow continued after the crater holding capacity was reached, leading to catastrophic flooding that eroded a water gap (Figure 9.11). The lake ultimately rose to an elevation of 1780 m because a small channel at this elevation was formed and preserved on the crater rim as a hanging valley while a deeper channel eroded at the water gap. Subtracting the channel floor elevation just east of the water gap (i.e. 1250 m) from the 1780 m high-water mark (Figure 9.11) shows that, at a minimum, a 530 m deep water column catastrophically drained through the breach in Morella Crater (Coleman *et al.*, 2007b). Given that the crater has an average inner diameter of 70 km, the minimum volume drained from the lake was 2.0×10^{12} m^3 (Coleman and Dinwiddie, 2007). This volume is so large that the Mississippi River at high flood stage ($\sim$30 000 m^3 s^{-1}) would require >770 days to fill it. This volume is well defined because most of the floor of Morella Crater lies below 1250 m, meaning a residual lake 170 m deep (excluding cavus depth) remained after the flood. Using aquifer parameters that provide an approximate upper value for groundwater inflow rates, Coleman and Dinwiddie (2007) estimated that more than eight years were needed to initially fill Morella. Their analysis assumed that Ganges Cavus, which eventually grew to a volume of $\sim 1.8 \times 10^{12}$ m^3, was the outflow source and that it formed and filled with water relatively quickly (~1 month) after the initial rupture of the cryosphere. Otherwise, the time to fill Morella Crater could have been longer than eight years.

The rim of Morella Crater served as a natural dam for the interior lake. Failure of this rock-material dam catastrophically drained the palaeolake and incised Elaver Vallis. Similarly, the failure of natural dams has produced many of the largest floods in Earth's history. We refer readers to O'Connor and Beebee (this volume Chapter 8) for an extensive discussion of terrestrial floods caused by failure of natural rock-material dams. They also provide analytical methods for estimating the magnitude and behaviour of these megafloods. Using the methods of Walder and O'Connor (1997), Coleman and Dinwiddie (2007) estimated peak flows through the Morella Crater water gap ranging from $\sim 1 \times 10^6$ m^3 s^{-1} to an upper limit of $\sim 3 \times 10^7$ m^3 s^{-1} for a scenario where the entire water gap formed rapidly. During the flood, as the lake level dropped 530 m, the overlying pressure at Ganges Cavus would have diminished by 2 MPa, causing the groundwater outflow rate to increase (Coleman and Dinwiddie, 2007). This water flowed up into the water column of the residual lake, which explains why the discharge did not erode a channel across the floor of Morella Crater toward the outlet. Subsequent outflow would have continued at declining rates until the potentiometric surface fell below the channel floor elevation of 1250 m (Figure 9.11).

The floor of Morella Crater is relatively flat and smooth, consistent with lacustrine deposition. Elevations vary less than ~90 m along a 40 km profile from east to west across the crater's centre. Two small craters occur on Morella's floor northeast of Ganges Cavus. Somerset Crater has a continuous rim and is deeper than Johnstown Crater, which has a very different, 'scalloped' appearance. The rim of Johnstown is highly degraded with numerous arcuate embayments (Figure 9.10b). Several of these cove-like openings incise the crater rim to the floor elevation of Morella Crater. Based on its greater depth and minimal erosion, Somerset Crater postdates the lacustrine event in Morella Crater. Degradation by rim erosion and infilling shows that Johnstown Crater likely existed before the Elaver Vallis flooding. Although Morella is relatively flat, there is evidence of erosion on the crater floor. Dendritic patterns suggest drainage of water from the northern half of the crater toward Ganges Cavus. In daytime infrared images (Figure 9.10) the dendritic patterns are lighter than their surroundings. Gridded MOLA data reveal these to be shallow branching valleys 20–40 m deep. The valleys are at or below the base level of Morella's water gap (1250 m) and would have remained inundated by the residual crater lake. Therefore, the valleys probably were eroded after the carving of Elaver Vallis. One possibility is that basal or differential melting of the frozen residual lake may have generated enough liquid water to incise these valleys.

Turning to the channel system itself, east of Morella Crater the initial flows in Elaver Vallis produced broad scabland flooding more than 65 km wide (Figure 9.10). Two main channels were produced, one branch to the north and one to the south that forms a long meander with hanging valleys at both ends. Just east of the Morella water gap, where the two channels diverged, the southern branch is a 'hanging' valley 200 m higher than the northern channel (Figure 9.11). The presence of this hanging valley confirms that flow persisted in the northern branch after flow ceased in the southern channel. At least four chaos zones appear on the channel floors. The chaos in the southern channel (Figure 9.10c) has a floor elevation ~420 m deeper than the high ground on its eastern side. The easternmost chaos in the northern channel is ~500 m deeper than the terrain that flanks the channel. Elaver Vallis abruptly terminates at a canyon on the southern rim of Ganges Chasma, showing that Ganges continued to grow southward after the fluvial episode ended.

Analysis of Elaver, Allegheny and Walla Walla Valles reveals important details about hydrologic conditions during the Late Hesperian Epoch. These channels were sourced by aquifers with potentiometric surface elevations too high to be explained by distant polar recharge and discharge from a globally connected aquifer system (see Carr, 2002). The Tharsis plateau (Figure 9.1) stands much higher (>5000 m) than the channel outflows and is one likely source region for Hesperian groundwater recharge because of its long-term, extensive volcanic activity. Sinai Planum and Syria Planum were also possible recharge areas. Could the elevations of the channel source areas have changed over time? Extensional tectonics in the Valles Marineris seem more likely to have lowered the channels than to have raised them. Therefore, if the channel source elevations have not changed appreciably, or have lowered since the Hesperian Period, or if relative elevation differences have been preserved, then the outflows must have derived from regional recharge at higher elevations (Coleman *et al.*, 2007b).

If the ancestral Valles Marineris canyons were deep enough, rising hydraulic pressures of the confined regional groundwater system coupled with tectonic stimuli would have created conditions favourable for groundwater breakouts and the growth of deep, ice-covered lakes. The floors and lower walls of the Valles Marineris canyons would have been weak points because of their minimal overburden and low elevation with respect to the elevated potentiometric surface. It is not possible to know the size and shape of the canyons when Allegheny and Walla Walla Valles formed. However, it is reasonable to conclude the ancestral canyons would have been narrower, smaller in volume, and more isolated, analogous to present-day Hebes Chasma. Using estimates for the strength of a basaltic rock mass (Schultz, 1995), analysis of possible fluid pressures beneath the canyon floors shows that formation of lakes in the ancestral canyons may have been inevitable as a consequence of rising groundwater potentiometric levels (Coleman *et al.*, 2007b). Conditions appear to have been so favourable for groundwater breakouts that ice-covered lakes probably already existed in the canyons when the high channels formed. In particular, a high-standing, ice-covered lake likely existed in eastern Candor Chasma because this canyon is intersected by the Ophir Catenae fault system from which Allegheny Vallis and Walla Walla Vallis originated.

What are plausible sources of groundwater recharge at low latitudes? Water ice could have accumulated preferentially at low latitudes during times of high obliquity (Forget *et al.*, 2006). The effusion of Hesperian flood basalts in Tharsis and near the Valles Marineris would have dramatically melted these ice deposits, yielding large volumes of liquid water. If a thick regional cryosphere were present, it is unclear how melt water could have penetrated the surface to recharge the aquifers. High crustal heat fluxes in Tharsis may have greatly thinned the cryosphere. Another possibility is that the ancestral Noctis Labyrinthus canyons and their underlying faults could have effectively channeled surface water and groundwater eastward from central Tharsis to the main canyons of the Valles Marineris (Coleman *et al.*, 2007b). Lake waters would have gradually accumulated beneath ice covers. Eventually the cryosphere beneath the canyon floors would disappear under these conditions, linking the canyon lakes to the groundwater system. In this alternative recharge model, much of the regional recharge would have occurred beneath ice-covered canyon lakes rather than on the Tharsis plateau itself. This interpretation may reconcile the presence of a thick Hesperian cryosphere with the evidence for groundwater as a source for some of the circum-Chryse outflow channels.

9.4 Conclusions

Unravelling the history of the outflow channels is the key to understanding the surface-water and groundwater hydrology of Mars. The origin of the circum-Chryse outflow channels is closely linked to the birth and evolution of the Valles Marineris. We cannot know the size and shape of the chasms when the channels formed, but those ancestral chasms were probably narrower, smaller and more isolated from each other. Groundwater recharge in uplands to the west likely caused the high groundwater potentiometric surface in Ophir Planum that led to the outflows at Allegheny, Walla Walla and Elaver Valles. Ganges Chasma as seen today probably did not exist at that time, otherwise the groundwater breakouts would have occurred in its depths instead of on the adjacent plateaus. We can also envision a time when ice-covered lakes were present in at least some (if not most) of the western Valles Marineris. These isolated canyons eventually became interconnected and drained eastward, perhaps causing a succession of megafloods that created the chaotic lowlands of the eastern Valles Marineris. Ultimately, the draining of the Valles Marineris would have rapidly and permanently lowered the groundwater elevations in adjacent regions, preventing the recurrence of large outflows in other channels.

Catastrophic decline of lake levels would have created groundwater overpressures beneath the chasms, which would have destabilised canyon walls and triggered landslides, new chaos formation and groundwater outflows, enhancing the flood discharges.

Acknowledgements

We appreciate the thoughtful reviews provided by Baerbel Lucchitta, Goro Komatsu, J. Alexis Rodriguez Palmero and Paul Carling.

References

Baker, V. R. (Ed.) (1981). *Catastrophic Flooding: The Origin of the Channeled Scabland.* Stroudsburg, PA: Dowden, Hutchinson & Ross, Inc.

Baker, V. R. (2001). Water and the martian landscape. *Nature*, **412**, 228–236.

Baker, V. R. (2006). Geomorphological evidence for water on Mars. *Elements*, **2**, 139–143.

Bishop, J. L., Noe Dobrea, E., Murchie, S. L. *et al.* and the MRO CRISM Team (2007). Sulfates and mafic minerals in Juventae Chasma as seen by CRISM in coordination with OMEGA, HIRISE, and context images. In *Seventh International Conference on Mars*, Abstract 3350, Lunar and Planetary Institute, Houston (CD-ROM).

Burr, D. M., Grier, J. A., McEwen, A. S. and Keszthelyi, L. P. (2002). Repeated aqueous flooding from the Cerberus Fossae: evidence for very recently extant, deep groundwater on Mars. *Icarus*, **159**, 53–73.

Cabrol, N. A., Grin, E. A. and Dawidowicz, G. (1997). A model of outflow generation by hydrothermal underpressure drainage in volcano-tectonic environment, Shalbatana Vallis (Mars). *Icarus*, **125**, 455–464.

Carr, M. H. (1981). *The Surface of Mars.* New Haven: Yale University Press.

Carr, M. H. (1995). The Martian drainage system and the origin of valley networks and fretted channels. *Journal of Geophysical Research*, **100**, 7479–7507.

Carr, M. H. (1996). *Water on Mars*. New York: Oxford University Press.

Carr, M. H. (2002). Elevations of water-worn features on Mars: implications for circulation of groundwater. *Journal of Geophysical Research*, **107**, 5131, doi:10.1029/2002JE001845.

Carr, M. H. and Head, J. W. (2003). Oceans on Mars: an assessment of the observational evidence and possible fate. *Journal of Geophysical Research*, **108**, 5042, doi:10.1029/2002JE001963.

Catling, D. C., Wood, S. E., Leovy, C. *et al.* (2006). Light-toned layered deposits in Juventae Chasma, Mars. *Icarus*, **181**, 26–51.

Chapman, M. G. and Tanaka, K. L. (2001). Interior trough deposits on Mars: subice volcanoes? *Journal of Geophysical Research*, **106** (E5), 10,087–10,100.

Chapman, M. G., Gudmundsson, M. T., Russell, A. J. and Hare, T. M. (2003). Possible Juventae Chasma subice volcanic eruptions and Maja Valles ice outburst floods on Mars: implications of Mars Global Surveyor crater densities, geomorphology, and topography. *Journal of Geophysical Research*, **108** (E10), doi:10.1029.2002JE002009.

Christensen, P. R., Gorelick, N. S., Mehall, G. L. and Murray, K. C. (2006). *THEMIS Public Data Releases*, Planetary Data System node, Arizona State University, http://themis-data.asu.edu.

Clifford, S. M. and Parker, T. J. (2001). The evolution of the Martian hydrosphere: implications for the fate of a primordial ocean and the current state of the northern plains, *Icarus*, **154**, 40–79.

Coleman, N. (2005). Martian megaflood triggered chaos formation, revealing groundwater depth, cryosphere thickness, and crustal heat flux. *Journal of Geophysical Research*, **110**, doi:10.1029/2005JE002419.

Coleman, N. M. and Baker, V. R. (2007). Evidence that a paleolake overflowed the rim of Juventae Chasma, Mars. In *Lunar and Planetary Science Conference* (CD-ROM), XXXVIII, Abstract 1046.

Coleman, N. M. and Dinwiddie, C. L. (2007). Hydrologic analysis of the birth of Elaver Vallis, Mars by catastrophic drainage of a lake in Morella Crater. In *Seventh International Conference on Mars*, Abstract 3107, Lunar and Planetary Institute, Houston (CD-ROM).

Coleman, N. M., Dinwiddie, C. L. and Baker, V. R. (2007a). Evidence that floodwaters filled and overflowed Capri Chasma, Mars. *Geophysical Research Letters*, **34**, L07201, doi:10.1029/2006GL028872.

Coleman, N. M., Dinwiddie, C. L. and Casteel, K. (2007b). High outflow channels on Mars indicate Hesperian recharge at low latitudes and the presence of canyon lakes. *Icarus*, **189**, doi:10.1016/j.icarus.2007.01.020, 344–361.

Costard, F. and Baker, V. R. (2001). Thermokarst landforms and processes in Ares Vallis, Mars. *Geomorphology*, **37**, 289–301.

Forget, F., Haberle, R. M., Montmessin, F., Levrard, B. and Head, J. W. (2006). Formation of glaciers on Mars by atmospheric precipitation at high obliquity. *Science*, **311**, 368–371.

Freeze, R. A. and Cherry, J. A. (1979). *Groundwater*. Upper Saddle River, NJ: Prentice-Hall.

Gendrin, A., Mangold, N., Bibring, J.-P. *et al.* (2005). Sulfates in Martian layered terrains: the OMEGA/Mars Express view. *Science*, **307**, 1587–1591.

Gendrin, A., Bibring, J.-P., Quantin, C. *et al.* and the OMEGA team (2006). Two years of sulfate mapping in Valles Marineris and Terra Meridiani as seen by OMEGA/Mars Express. In *Lunar and Planetary Science Conference* (CD-ROM), XXXVII, Abstract 1872.

Glotch, T. D. and Rogers, A. D. (2007). Evidence for aqueous deposition of hematite- and sulfate-rich light-toned deposits in Aureum and Iani Chaos, Mars. *Journal of Geophysical Research*, **112**, doi:10.1029/2006JE002863.

Head, J. W., Neukum, G., Jaumann, R. *et al.* and the HRSC Co-Investigator Team (2005). Tropical to mid-latitude snow and ice accumulation, flow and glaciation on Mars. *Nature*, **434**, 346–351.

Ivanov, M. A. and Head, J. W. (2001). Chryse Planitia, Mars: topographic configuration, outflow channel continuity and sequence, and tests for hypothesized ancient bodies of water using Mars Orbiter Laser Altimeter (MOLA) data. *Journal of Geophysical Research*, **106** (E2), 3275–3295.

Kundson, A. T., Arvidson, R. E., Christensen, P. R. *et al.* and the CRISM Science Team (2007). Aqueous geology in Valles Marineris: new insights in the relationship of hematite and sulfates from CRISM and HIRISE. In *Seventh International Conference on Mars*, Abstract 3370, Lunar and Planetary Institute, Houston (CD-ROM).

Komatsu, G. and Baker, V. R. (1997). Paleohydrology and flood geomorphology of Ares Vallis. *Journal of Geophysical Research*, **102** (E2), 4151–4160.

Komatsu, G., Kargel, J., Baker, V., Strom, R., Ori, G., Mosangini, C. and Tanaka, K. (2000). A chaotic terrain formation hypothesis: explosive outgas and outflow by dissociation of clathrate on Mars. In *Lunar and Planetary Science Conference* (CD-ROM), XXXI, Abstract 1434.

Kreslavsky, M. A. and Head, J. W. (2002). Fate of outflow channel effluents in the northern lowlands of Mars: the Vastitas Borealis Formation as a sublimation residue from frozen ponded bodies of water. *Journal of Geophysical Research*, **107**, doi:10.1029/2001JE001831.

Langlais, B., Purucker, M. E. and Mandea, M. (2004). Crustal magnetic field of Mars. *Journal of Geophysical Research*, **109**, doi:10.1029/2003JE002048.

Laskar, J., Correia, A. C. M., Gastineau, M. *et al.* (2004). Long term evolution and chaotic diffusion of the insolation quantities of Mars. *Icarus*, **170**, 343–364.

Leask, H. J., Wilson, L. and Mitchell, K. L. (2006). Formation of Ravi Vallis outflow channel, Mars: morphological development, water discharge, and duration estimates. *Journal of Geophysical Research*, **111**, doi:10.1029/2005JE002550.

Lucchitta, B. K. (1982). Ice sculpture in the Martian outflow channels. *Journal of Geophysical Research*, **87**, 9951–9973.

Lucchitta, B. K., McEwen, A. S., Clow, G. D. *et al.* (1992). The canyon system on Mars. In *Mars*, eds. H. H. Kieffer *et al.* Tucson: University of Arizona Press, pp. 453–492.

Lucchitta, B. K., Isbell, N. K. and Howington-Kraus, A. (1994). Topography of Valles Marineris: implications for erosional and structural history. *Journal of Geophysical Research*, **99** (E2), 3783–3798.

Lunardini, V. J. (1995). *Permafrost Formation Time*, CRREL Report 95–8, U.S. Army Cold Regions Research and Engineering Laboratory, Hanover, NH.

Mangold, N., Gendrin, A., Quantin, C. *et al.* and the OMEGA and HRSC Co-Investigator Team (2007). An overview of the sulfates detected in the equatorial regions by the OMEGA/MEX spectrometer. In *Seventh International Conference on Mars*, Abstract 3141, Lunar and Planetary Institute, Houston (CD-ROM).

Moore, J. M., Clow, G. D., Davis, W. L. *et al.* (1995). The circum-Chryse region as a possible example of a hydrologic cycle on Mars: geologic observations and theoretical evaluation. *Journal of Geophysical Research*, **100** (E3), 5433–5447.

Nelson, D. M. and Greeley, R. (1999). Geology of Xanthe Terra outflow channels and the Mars Pathfinder landing site. *Journal of Geophysical Research*, **104** (E4), 8653–8669.

Parker, T. J., Clifford, S. M. and Banerdt, W. B. (2000). Argyre Planitia and the Mars global hydrologic cycle. In *Lunar and Planetary Science Conference* (CD-ROM), XXXI, Abstract 2033.

Parker, T. J., Gorsline, D. S., Saunders, R. S., Pieri, D. C. and Schneeberger, D. M. (1993). Coastal geomorphology of the Martian northern plains. *Journal of Geophysical Research*, **98** (E6), 11,061–11,078.

Roach, L. H., Mustard, J. F., Murchie, S. L. *et al.* and the CRISM Science Team (2007). Magnesium and iron sulfate variety and distribution in east Candor and Capri Chasma, Valles Marineris. In *Seventh International Conference on Mars*, Abstract 3223, Lunar and Planetary Institute, Houston (CD-ROM).

Robinson, M. S. and Tanaka, K. L. (1990). Magnitude of a catastrophic flood event in Kasei Valles, Mars. *Geology*, **18**, 902–905.

Rodriguez, J. A., Sasaki, S. and Miyamoto, H. (2003). Nature and hydrological relevance of the Shalbatana complex underground cavernous system. *Geophysical Research Letters*, **30**, 1304, doi:10.1029/2002GL016547.

Rodriguez, J. A., Sasaki, S., Kuzmin, R. O. *et al.* (2005). Outflow channel sources, reactivation, and chaos formation, Xanthe Terra, Mars. *Icarus*, **175**, 36–57.

Rodriguez, J. A. P., Tanaka, K. L., Miyamoto, H. and Sasaki, S. (2006). Nature and characteristics of the flows that carved the Simud and Tiu outflow channels, Mars. *Geophysical Research Letters*, **33**, L08S04, doi:10.1029/2005GL024320.

Rogers, A. D., Christensen, P. R. and Bandfield, J. L. (2005). Compositional heterogeneity of the ancient Martian crust: analysis of Ares Vallis bedrock with THEMIS and TES data. *Journal of Geophysical Research*, **110**, doi:10.1029.2005JE002399.

Schultz, R. (1995). Limits on strength and deformation properties of jointed basaltic rock masses. *Rock Mechanics and Rock Engineering*, **28** (1), 1–15.

Schultz, R. (1998). Multiple-process origin of Valles Marineris basins and troughs, Mars. *Planetary Space Science*, **46** (6/7), 827–834.

Scott, D. H. and Tanaka, K. L. (1986). *Geologic Map of the Western Equatorial Region of Mars*. U.S. Geological Survey Miscellaneous Investigations Series, Map I-1802-A.

Tanaka, K. (1997). Sedimentary history and mass-flow structures of Chryse and Acidalia Planitiae, Mars. *Journal of Geophysical Research*, **102** (E2), 4131–4149.

Tanaka, K. (1999). Debris-flow origin for the Simud/Tiu deposit on Mars. *Journal of Geophysical Research*, **104** (E4), 8637–8652.

Urquhart, M. and Gulick, V. (2003). Plausibility of the "White Mars" hypothesis based upon the thermal nature of the Martian subsurface. *Geophysical Research Letters*, **30**, 1622, doi:10.1029/2002GL016158.

Walder, J. and O'Connor, J. (1997). Methods for predicting peak discharge of floods caused by failure of natural and constructed earthen dams. *Water Resources Research*, **33**, 2337–2348.

Williams, R. M. and Malin, M. C. (2004). Evidence for late stage fluvial activity in Kasei Valles, Mars. *Journal of Geophysical Research*, **109**, doi:10.1029/2003JE002178.

Williams, R. M., Phillips, R. J. and Malin, M. C. (2000). Flow rates and duration within Kasei Valles, Mars: implications for the formation of a martian ocean. *Geophysical Research Letters*, **27**, 1304, 1073–1076.

Wilson, L., Ghatan, G. J., Head, J. W. and Mitchell, K. L. (2004). Mars outflow channels: a reappraisal of the estimation of water flow velocities from water depths, regional slopes, and channel floor properties. *Journal of Geophysical Research*, **109**, doi:10.1029/2004JE002281.

Witbeck, N. E., Tanaka, K. L. and Scott, D. H. (1991). *Geologic Map of the Valles Marineris Region, Mars*. U.S. Geological Survey Miscellaneous Investigations Series, Map I-2010.

10

Floods from fossae: a review of Amazonian-aged extensional–tectonic megaflood channels on Mars

DEVON M. BURR, LIONEL WILSON and ALISTAIR S. BARGERY

Summary

The four youngest megaflood channels on Mars – Mangala Valles, Marte Vallis, Grjotá Valles and Athabasca Valles – date to the Amazonian Period and originate at fissures. The channels show common in-channel morphological indications of flood activity (streamlined forms, longitudinal lineations, scour), as well as evidence for volcanic, tectonic, sedimentary and/or glacial/ground ice processes. The fissure sources and channel termini have varied expressions, suggesting various triggering mechanisms and fates for the floodwaters. Possible triggering mechanisms include magmatic processes (dyke intrusion), tectonic processes (extensional faulting) and a combination of both types of processes. Surface morphology suggests that each of these mechanisms may have operated at different times and locations. Upon reaching the surface, the water likely would have fountained at least a few tens of metres above the surface, producing some water and/or ice droplets at the fountain margins. The likely sources of the floodwater are subsurface aquifers of a few kilometres' thickness and a few tens of degrees Celsius in temperature.

10.1 Introduction

Megaflooding on Mars has varied in origin and amount throughout the history of the planet. During the Noachian Period, the most ancient period, flooding originated from crater basins (Irwin and Grant, this volume Chapter 11). During the Early Hesperian Epoch, megafloods originated at chaos terrain often set within Valles Marineris chasmata (Coleman and Baker, this volume Chapter 9). During the Amazonian Period, the most recent period, megaflooding originated from fossae produced by extensional tectonism.

This chapter provides a review of the four megaflood channels originating at fossae that experienced flow during the Amazonian Period. These megaflood channels are the largest examples of a suite of aqueous flow channels that originate at volcanotectonic fissures. Granicus Valles originate from the Elysium Fossae on the flanks of Elysium Mons (Mouginis-Mark *et al.*, 1984). Other smaller channels originate at the Ceraunius and Olympica Fossae in the northwest Tharsis region (Mouginis-Mark, 1990), and from unnamed fractures on Ascraeus Mons and the Olympus Mons aureole (Mouginis-Mark and Christensen, 2005). These channels show that aqueous flows from fossae have occurred throughout volcanic provinces on Mars. The focus of this chapter is megaflood flow; thus, this review focuses on the four largest known and best investigated to date of these channels. These four channels are very roughly an order of magnitude smaller in each dimension than the largest flood channels, the Hesperian-aged circum-Chryse channels (Coleman and Baker, this volume Chapter 9), but still discharged geomorphically significant volumes of water.

Because the youth of these megaflood channels is noteworthy, this chapter starts with a brief overview of age-dating of young features on Mars before then giving the inferred ages of these tectonic megaflood channels. Next, the morphology of each channel is reviewed, including the source morphology, in-channel bedforms, any inferred flow characteristics, and possible floodwater sinks. Then a synopsis is provided of the possible mechanisms that may have triggered groundwater release from fossae and led to surface flood flow. Finally, some thermal and mechanical aspects of the inferred groundwater movement to and within the source fossae are discussed.

10.2 Ages

10.2.1 Issues related to dating of Amazonian-aged channels

The most recent period on Mars is the Amazonian (Tanaka, 1986), lasting from ~3.0 Ga to the present (Hartmann and Neukum, 2001). The absolute dating of the surface features on Mars is accomplished by comparison of actual impact crater size–frequency distributions with model size–frequency distributions for fixed ages, referred to as 'isochrons' (e.g. Hartmann and Neukum, 2001). Crater counts in support of dating a flood in a channel are performed on terrain in the channel interpreted to have been created or modified by the floodwaters, e.g. flood-scoured channel floors. The actual size–frequency distributions of impact craters on channel floors indicate the age of only the most recent, geomorphically effective flood. This supposition is founded upon the fact that floods are 'self-censoring',

Megaflooding on Earth and Mars, ed. Devon M. Burr, Paul A. Carling and Victor R. Baker.
Published by Cambridge University Press.

i.e., more recent, larger floods have been shown to erase the sedimentological evidence of previous smaller floods (e.g. House *et al.*, 2001). Likewise on Mars, previous smaller floods would not be visible in the channel floor cratering record, although the existence and ages of previous floods may be indicated by flood terraces (Berman and Hartmann 2002, Burr *et al.*, 2002b). Ideally, to derive the age of a flood from crater counts, a flood should have erased all craters completely but this may not always be the case (see e.g. Figure 13 of Burr *et al.*, 2002b).

A second issue with regard to age-dating of young Martian flood channels is the uncertain effect of secondary craters. Secondary craters are those formed by impact of material thrown out from the surface of a planet by a (larger) primary impact. Because secondaries are smaller than primaries, they are erased more quickly; thus, they affect the age-dating primarily of the young surfaces of Mars, where they are subject to the least amount of erasure and where larger craters are not present for use in age-dating. The discovery of efficient secondary production in young lava flows on Mars (McEwen *et al.*, 2005) raises the question of whether Martian isochrons, derived from and tied to radiometrically dated lunar samples, properly account for secondary crater production (see McEwen and Bierhaus, 2006, for a review). Because of significant secondary production, dates of the inferred youngest flood channel have varied by up to two orders of magnitude (compare the ages of Athabasca Valles in Burr *et al.*, 2002b and McEwen *et al.*, 2005). However, even with this variance, this channel still retains an age-date of late Amazonian. Older surfaces should have still less error and, overall, age-dating of features to within the Amazonian Period appears robust (Hartmann, 2005). Each of the four fissure-headed channels discussed below has been inferred to have seen flooding during the Amazonian Period (see next section, 'Channel ages').

A final point about the ages of these channels is the possibility of channel exhumation. The documentation of extensive layered terrain on Mars (Malin and Edgett, 2000) and of exhumation of inverted fluvial channels from within the rock record (Williams and Edgett, 2005; Williams *et al.*, 2005) suggest that Martian outflow channels also could be exhumed. Extensive burial and partial exhumation of one fissure-headed channel (Athabasca Valles) has been argued on the basis of inferred scour upslope of its source (Edgett and Malin, 2003). In such a case, the derived model crater-count age for the channel floor would reflect an exposure age, not a formation age. However, in this particular case, the scour has an alternative interpretation, namely, as an effect of floodwater gushing upslope from the fissure due to the force of the water eruption (Burr *et al.*, 2002b; Head *et al.*, 2003; Manga, 2004). Evidence for the proposed exhumation is not apparent within this channel and no evidence for exhumation has been deduced for other Amazonian-aged fissure-headed channels.

In the light of these issues regarding the age-dating of young channels, the ages of each of the four largest tectonic outflow channels are now discussed.

10.2.2 Channel ages

Mangala Valles The oldest and largest tectonic outflow channel is Mangala Valles (Plate 23). The most upslope southern end of Mangala Valles coincides with the Memnonia Fossae, flat-floored features inferred to be a graben system formed at the coincidence of dyke emplacement and normal faulting (see Hanna and Phillips, 2006). Flooding down Mangala is inferred to have coincided with faulting on the Memnonia Fossae on the basis of crater counts and stratigraphic relationships (Tanaka and Chapman, 1990) as mapped at 1:500 000-scale on Viking images (Chapman and Tanaka, 1993; Craddock and Greeley, 1994; Zimbelman *et al.*, 1994). Two periods of coeval faulting and flooding originally were inferred from Viking data, the first dated as late Hesperian age and the second originally dated at the Hesperian–Amazonian boundary (Tanaka and Chapman, 1990). Subsequent analysis on improved images interpreted the youngest materials in Mangala Valles as Amazonian in age, based on superposition relationships and lack of craters greater than 2 km in diameter (Zimbelman *et al.*, 1992).

Marte Vallis Marte Vallis (Plate 24) is the largest and oldest of the three Amazonian-aged outflow channels that surround the Cerberus plains, where it is located at the eastern end of the plain (Burr *et al.*, 2002b). The channel originally was interpreted as being continuous with one or both of the other two Cerberus plains channels (Tanaka and Scott, 1986; Berman and Hartmann, 2002, Figure 1; Plescia, 2003). However, topography from the Mars Orbiter Laser Altimeter (MOLA; Smith *et al.*, 1998) and crater counts on the channel floor both suggest that the three Cerberus channels are spatially and temporally distinct (Burr *et al.*, 2002b; Berman and Hartmann, 2002). Marte Vallis has been dated by counting craters on Mars Orbiter Camera (MOC) images (Malin and Edgett, 2001) of both the channel floor, inferred to date to the last flood, and the dark lobate material that embays much of the channel, inferred to date since the last flood. These crater counts produced similar ages spanning a few to ~200 Ma (Berman *et al.*, 2001; Burr *et al.*, 2002b; Berman and Hartmann, 2002).

Grjotá Valles A previously unnamed channel to the north of the Cerberus plains (Plate 25) (Burr *et al.*, 2002b) has been named Grjotá Valles (Plescia, 2003). Age-dating of the Grjotá Valles floor is affected by its weathered character, its indistinct flood boundaries, and the presence of some aeolian dunes (Burr *et al.*, 2002b). Despite these factors, crater counts on MOC images of inferred flood-formed surfaces – i.e., longitudinally grooved surfaces adjacent to streamlined forms – fall along the isochrons of Hartmann (1999) and give an age of ~10–40 Ma (Burr *et al.*, 2002b).

Athabasca Valles Athabasca Valles (Plate 26) are at the western end of the Cerberus plains. They have the most pristine geomorphology of all the Amazonian-aged, fissure-headed channels, and crater counts on the channel floor have yielded the youngest model ages. Crater counts on MOC images of the longitudinally grooved terrain in the topographic channel give an age of 2–8 Ma (Burr *et al.*, 2002b) based on comparison with the Hartmann (1999) isochron model. Other crater counts within the topographic channel give an age of 3 Ma (Werner *et al.*, 2003) based on the Neukum and Ivanov (1994, described by Hartmann and Neukum, 2001) model. This 3 Ma unit is not characterised morphologically but is attributed to volcanism; the most recent flood erosion in that study is dated to 1.6 Ga, with a possible last fluvial event at 30 Ma (Werner *et al.*, 2003). These geological units of Werner and colleagues are defined on the basis of the match of crater counts with model isochrons, not with morphological characterisation, nor are the locations of the areas counted described; together this makes it difficult to independently evaluate the origin of the units (see also McEwen *et al.*, 2005, p. 376). Widespread pitted mounds in the channel have been interpreted as hydrovolcanic rootless cones, which form by lava overrunning water-rich or ice-rich ground (Greeley and Fagents, 2001; Lanagan *et al.*, 2001; Lanagan, 2004; Jaeger *et al.*, 2007). Because water is unstable near the surface of Mars in the present climate, the presence of rootless cones implies that lava emplacement was preceded closely by flooding (Lanagan *et al.*, 2001; Lanagan, 2004). The age of 3 Ma is derived by Werner *et al.* (2003) from a MOC image that shows features of the type interpreted by Lanagan *et al.* (2001) and Lanagan (2004) to be rootless cones. On this basis, an age of 3 Ma, if correctly attributed to lava, would still imply flooding of a similarly young age. Subsequent investigation with Thermal Emission Imaging Spectrometer (THEMIS) infrared (IR) images (Christensen *et al.*, 2003) and crater counts on MOC images showed that ~80% of the craters on the Athabasca Valles grooved channel floor are secondaries (McEwen *et al.*, 2005). Because this efficient secondary production called into question the use of isochrons for young surfaces, the channel was re-dated using statistical methods, with an age of 1.5–200 Ma (McEwen *et al.*, 2005). The possible exhumation of Athabasca Valles is discussed in the previous section.

10.3 Morphology

As shown above, all four known outflow channels that experienced megaflooding in the Amazonian Period have tectonic sources, either fissures or graben. The channels also all show common in-channel morphological indicators of flooding, including streamlined forms, scour and longitudinal lineations. However, the morphologies, both at their sources and at their termini, vary. Channel location, source and channel morphology, and possible floodwater sinks are discussed below.

10.3.1 Mangala Valles

Centred near 15° S 210° E, Mangala Valles (Plate 23) are located in the southern highlands just south of the hemispheric dichotomy boundary of Mars. They are situated to the west of the Tharsis rise and to the south of Amazonis Planitia, and begin at an elevation of about 0 m. From this origination point at Mangala Fossa, one of the Memnonia Fossae, the main channel stretches ~850 km northward to its dual outlets at the south side of the eastern Medusae Fossae Formation (Plate 23). The upper (southern-most) ~550 km of the channel is a single reach situated between north–south trending fault blocks (Zimbelman 1989). The most proximal ~150 km show primarily scoured terrain with a few smaller streamlined forms, and the next ~400 km show less scour with much larger streamlined forms behind impact craters or other obstacles. The upper reach expands rapidly from 5.5 km through a notch in an eroded impact crater north of Mangala Fossa to ~50 km in width within a few kilometres downstream. Channel depth varies from a few tens of metres to a few hundred metres, and the overall slope of the upper reach is ~0.0005 m m^{-1} (~0.03°). Downslope (northward) of this upper reach, the channel diverges into two separate branches. These two divergent branches are as narrow as approximately 10 km and as deep as 1000 m in some locations. The northwestward branch has a number of in-filled impact craters along its path, whereas the north-northeastward branch shows a more anastomosing plan-view form.

The channel system was inferred to have been formed by catastrophic flooding on the basis of streamlined bars within the channel (see Carling *et al.*, this volume Chapter 3) and braided plan-view morphology (Milton, 1973; Sharp and Malin, 1975). The oblique orientation of the most proximal streamlined forms (Tanaka and Chapman 1990) originally suggested that water may have been released from more than one location (Zimbelman *et al.*, 1992). The most recent analysis, using THEMIS and MOLA data (Ghatan *et al.*, 2005), suggests that the groundwater filled and overflowed the source trough, generating oblique overland flow, and that this sheet flow then coalesced into channelised flow through the 5.5 km wide notch to the north side of the source trough. The time scale for the initial trough filling may have been from as little as ~2 hours (Leask *et al.*, 2007) up to 10^7 hours (Ghatan *et al.*, 2005), indicating the considerable uncertainty associated with these estimates.

The context and morphology of Mangala Valles suggest a complex history providing multiple hypotheses. Early interpretations of Mangala Valles as an outflow channel suggested the origin to have been breaching of a large surface reservoir or lake (Sharp and Malin, 1975). In subsequent analysis, the origination of the channel at a fissure or graben radial to the Tharsis rise was interpreted as indicating that water release resulted from fault-induced cracking into a perched aquifer (Tanaka and Chapman, 1990). To other workers, the proximity of the channel to the Tharsis rise suggested floodwater generation by Tharsis magmatic melting of near-surface ground ice, which would have then migrated in the subsurface to the tectonic fissures to be released to flow down the channel (Zimbelman *et al.*, 1992). Under this scenario, water release likely would have been artesian (Zimbelman *et al.*, 1992). The most recent analyses of the Mangala Vallis source region by Ghatan *et al.* (2005) and Leask *et al.* (2006) also inferred tectonic tapping of a pressurised aquifer, with Tharsis being the most likely source of the groundwater. This most recent hypothesis builds on the global hydrosphere model of Clifford (1993) and Clifford and Parker (2001), in which the crust of the planet is divided into the cryosphere (the uppermost portion that remains below the water freezing temperature) and the hydrosphere (a sub-cryospheric zone where liquid water accumulates), and on the interpretation of the graben as reflecting subsurface dykes (Wilson and Head, 2002). In this hypothesis, dyke emplacement resulted in graben formation and cracking of the cryosphere, which released pressurised groundwater to the surface; the minimum duration of the flooding is estimated to have been 1–3 months (Ghatan *et al.*, 2005; Leask *et al.*, 2007). The interpretation of sub-parallel symmetric ridges around the eastern end of Mangala Fossa as dunes emplaced by a phreatomagmatic eruption plume resulting from dyke emplacement, cryosphere cracking and magma–groundwater mixing (Wilson and Head, 2004) is consistent with this hypothesis.

As an alternative to the idea of a pre-pressurized aquifer due to cryosphere growth, the high aquifer pore pressures responsible for the flooding at Mangala Valles may have been a direct result of the stress release during the tectonic event forming the graben (Hanna and Phillips, 2006). The presence of chaos within a crater near the Mangala Valles source graben may be a consequence of elevated palaeopore pressures and thus provide circumstantial evidence consistent with this model (Hanna and Phillips, 2006).

In addition to catastrophic aqueous flow, phreatomagmatism and tectonism, the channel also shows evidence of ice processes. The pitted proximal channel floor (Figure 10.1) has been interpreted to be the result of ice blocks stranded during early stage sheet flow of meltwater and ice, possibly in association with temporary ice dams (Zimbelman *et al.*, 1992). Lobate ridges deposited on the Mangala Fossa margins adjacent to the Mangala Valles channel origin (Figure 10.2) have been interpreted to be glacial deposits, with the glaciation being the consequence of floodwater release (Head *et al.*, 2004). A candidate opportunity for ice formation is the cooling of the thin sheet of water that would have spread across the floor of the Mangala Fossa graben in the first stages of water release from graben-bounding fractures (Leask *et al.*, 2007). Additional surface ice formation could have occurred during the filling of the graben if the temperature of the released water were less than 4 °C; the fact that water has its maximum density at this temperature suppresses convection as the water is cooled from above by evaporation at its surface (Leask *et al.*, 2007).

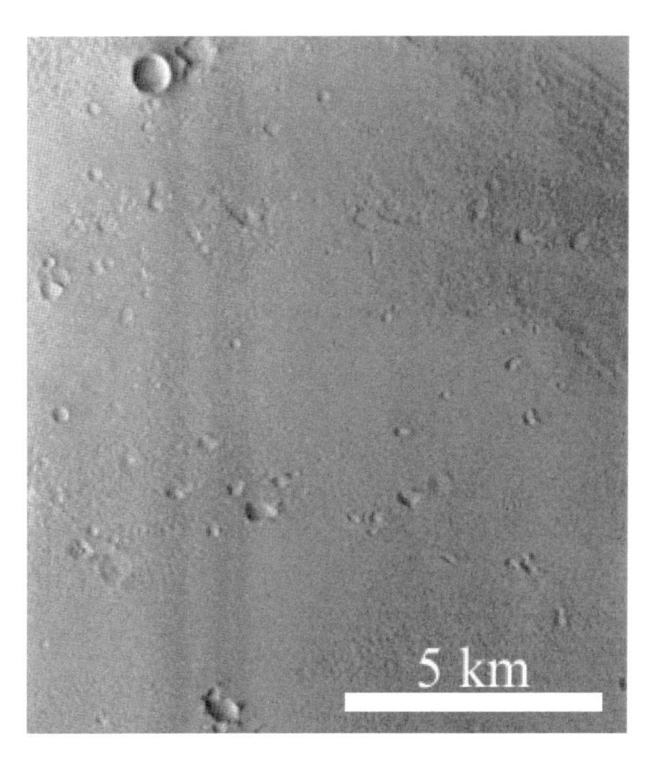

Figure 10.1. Portion of THEMIS visible image V23283003 located near 12.2° S 208.7° E showing pitted terrain in Mangala Valles, where irregular pits were interpreted to be the effects of ice blocks deposited by the flood flow (Zimbelman *et al.*, 1992). See Plate 23 for locations. North is up, illumination is from the left.

The terminal pathways and eventual sinks for Mangala Valles floodwaters are ambiguous. According to the interpretation of Ghatan *et al.* (2005), the topographic data suggest that the eastern branch formed first but was later pirated by the western branch, whereas previous crater statistics support the opposite sequence (Chapman and Tanaka, 1993). Ghatan *et al.* (2005) suggest that, as flooding subsided, residual water froze and

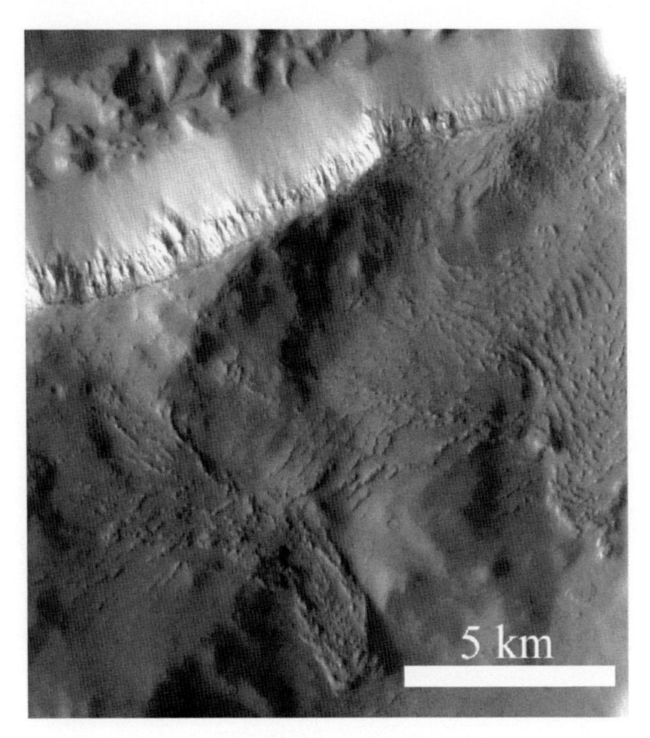

Figure 10.2. Portion of THEMIS visible image V06597003 located near 18° S 211.5° E, showing ridges around the source region of Mangala Valles, interpreted to be glacial deposits (Head *et al.*, 2004). North is up, illumination is from the left.

sublimated, leaving a residue in the deepest parts of the channel.

Additional sinks for the floodwater are suggested by morphology and may have been different for the northeastern and northwestern branches. The northeastern branch terminates at the southern side of a broadly level plains unit surrounding Medusae Fossae Formation deposits (Tanaka and Chapman, 1990). Approximately 300 kilometres to the north, a group of streamlined forms with impact craters or knobs at their southern ends is apparent in images and MOLA topography (centred near 1° N 207° E). The kilometre-scale size and large width:length ratio of these forms distinguish them from the aeolian streamlining (yardangs) that overprints this region and suggest their formation by floodwater flow. The forms are large in width relative to the northeastern branch but also shallower. A possible scenario for their formation is that upon exiting the deep and narrow northeastern branch, the Mangala Valles floodwaters spread out and became more quiescent, depositing their sediment load in the lee of flow obstacles.

The northwestern branch terminus is largely embayed by Medusae Fossae Formation deposits. To the west of those deposits is a 75 km diameter impact crater with a terraced fan sourced by a short broad valley. This fan is similar to a terraced fan in Coprates Chasma that is interpreted as representing erosion and redistribution of fan material during lake-level drops (Weitz *et al.*, 2006; see also Di Achille *et al.*, 2006). Thus, this fan-shaped deposit may indicate ponding and episodic recession of Mangala Valles floodwaters within this crater. However, any path between the outlet of the northwestern branch terminus and this impact crater is largely covered by the Amazonian-aged Medusae Fossae Formation.

In summary, morphology and modelling of Mangala Valles suggest a tectonovolcanic release of pressurised groundwater resulting in a catastrophic water flood and possible glaciation during the late Hesperian to early Amazonian Epochs.

10.3.2 Marte Vallis

In its current topographic expression, Marte Vallis originates near 180° E, 10° N at the eastern side of the Cerberus plains at an elevation of approximately −3100 m. Situated between the Orcus Patera structure to the northwest and southern highland outliers to the southeast, Marte Vallis stretches first ~250 km northeastward and then ~750 km north-northeastward to Amazonis Planitia (Plate 24). The channel system is consistently broad, spanning several tens of kilometres in width, and a few tens of metres deep, with an overall slope of ~0.0002 m/m (0.013°). It is embayed by a dark lobate material interpreted as lava (Plescia, 1990). This dark embaying material surrounds a number of brighter streamlined forms. The low, contrasting albedo of this embaying material caused the channel's streamlined forms and anastomosing morphology to be readily visible in Viking images, leading to its early identification as an outflow channel (Tanaka and Scott, 1986). Multiple levels along the sides of Marte Vallis are interpreted as terraces (e.g. Berman and Hartmann, 2002), which may reflect multiple flood events (Burr *et al.*, 2002b). Pirating of anastomosing channels within Marte Vallis as inferred from MOC images also suggests at least two episodes of flooding (Fuller and Head, 2002). The (post-embayment) depth of the channel is only on average a few tens of metres.

Marte Vallis may have had its source in the other (more western) Cerberus flood channels (Tanaka and Scott, 1986; Berman and Hartmann, 2002; Plescia, 2003), although MOLA topography and crater counts on MOC images indicate that the three Cerberus channels – Marte, Grjótá and Athabasca – are distinct (Burr *et al.*, 2002b). To the southwest (i.e., upslope) of the first topographic expression of Marte Vallis, MOLA topography of the eastern Cerberus plains shows broad, shallow channels leading from a subdued linear depression, which is collinear from the Cerberus Fossae (Plate 24). The broad, shallow channels are not visible on the southern side of the linear depression. This topography is the basis for the hypothesis that Marte Vallis originated at a section of the Cerberus Fossae graben that is now buried by subsequent lava flows (Burr *et al.*, 2002b). The Cerberus plains are interpreted as being filled with lava from the Fossae and other volcanic vents (Plescia, 1990, 2003; Sakimoto *et al.*, 2001). The broad, shallow channels stretching from the subdued linear depression (interpreted as a buried fissure segment) are hypothesised to have conveyed the floodwater from the now-buried fissure to the topographic channel that remains visible today, and then to have been buried by lava (Burr *et al.*, 2002b). Another possible source for the Marte Vallis floodwaters is breaching of a lake (Scott and Chapman, 1991) empounded behind a debris dam, ice dam or wrinkle ridge (Moller *et al.*, 2001). Evidence for this lake is lacking but may be buried beneath subsequent lava. The presence of rootless cones in Amazonis Planitia (Lanagan *et al.*, 2001; Greeley and Fagents, 2001) supports the idea derived from the channel's larger scale morphology (Tanaka and Scott, 1986; Fuller and Head, 2002) that Amazonis Planitia was the sink for the Marte Vallis water flows.

10.3.3 Grjotá Valles

Grjotá Valles are the least well-defined of the tectonic outflow channels. Located in remnants of southern highland terrain north of the Cerberus plains, they are oriented east–west and centred near 165° E, 15° N (Plate 25). The channel originates at the most northern major Cerberus fossa as indicated by local scour, orientation of nearby flood-formed features (scattered mesoscale streamlined forms and longitudinal lineations), and MOLA topography (Burr *et al.*, 2002b; Plescia, 2003). Recent mapping using THEMIS and MOC images shows that the origin of the floodwater was distributed discontinuously over a 250 km stretch of fissure as measured end to end; the summed lengths within this stretch from which water actually emerged total ~150 km (Burr and Parker, 2006). The elevation of this distributed source ranges from approximately −2100 at the eastern end to −2400 m at the western end. The water flowed both to the north and to the south of the fissure, anastomosing through the remnant highland terrain knobs in which it originated, thus making a conclusive determination of flow cross-sectional area difficult. The better-defined, northern branch is up to several tens of kilometres in width and of order ten metres in depth on average, although a small number of short segments are up to a few tens of metres deep. The overall slope of the channel is ~0.0006 m/m (0.035°).

This broadly distributed plan-view morphology and small channel depth may be due to the lack of any significant surface topography controlling channel development (Burr *et al.*, 2002a; Burr and Parker, 2006). Mangala Valles are located between north–south trending fault blocks, Marte Vallis sits between Orcus Patera and southern highland outliers, and Athabasca Valles are bounded by an Elysium Mons wrinkle ridge (see next section). In contrast, Grjotá Valles have no such apparent topographic structures to confine surface flow. The distributed nature of the outflow at the fissure-source of the channel may be another factor in this distributed plan-view morphology (Burr and Parker, 2006).

The sink for the floodwaters is uncertain. Some of the floodwater appears to have flowed southward toward the Cerberus plains (Burr *et al.*, 2002b; Berman and Hartmann, 2002; Plescia, 2003) but the termination of this southward flow is poorly defined (Burr and Parker, 2006). Inferred rootless cones in the channel imply that some of the water infiltrated (Burr and Parker, 2006). However, infiltration may have been minimised by freezing (see Clifford and Parker, 2001; see Burr *et al.*, 2002b for discussion) and the inferred rootless cones are limited in extent and open to other possible interpretations (Burr *et al.*, 2008). Rootless cones in the far western Amazonis Planitia (Lanagan *et al.*, 2001; Lanagan, 2004; Greeley and Fagents, 2001), which is to the northeast of Grjotá Valles, raise the possibility that some floodwater may have gone to the northeast, and streamlined forms hint at flow to the northeast (Burr and Parker, 2006). However, the continuation of that flow as far as Amazonis Planitia is not discernible in available data (Burr and Parker, 2006). Previous suggestions for the lack of an obvious sink – embayment by lava, blanketing by tephra, aeolian erosion/deposition, mass wasting and poor data coverage (Burr *et al.*, 2002b) – now appear unlikely, given the improved extent and quality of MOLA and THEMIS data used in recent mapping (Burr and Parker, 2006). This mapping, in conjunction with recent modelling (Bargery and Wilson, 2006), suggests that the floodwaters of Grjotá, with their shallow depth and widely distributed nature, either evaporated or froze and sublimated during and/or after flow. On this basis, the sink for the floodwaters is hypothesised to have been primarily the atmosphere (Burr and Parker, 2006).

10.3.4 Athabasca Valles

Athabasca Valles have the most pristine morphology of any outflow channel system on Mars. The main topographic channel originates at two locations along the southernmost major member of the Cerberus Fossae near 10.2° N 157.2° E (Plate 26). The easternmost of these linear sources is ~18 km in length along the fossae and the western source is ~22 long, with elevations of approximately −2400 m. The channels emanating from these two sources come together ~20 km to the south of the fossae to create a single channel. About 40 km below this confluence, the Athabasca Valles become distributary, with smaller channels breaching through the wrinkle ridge on the southern

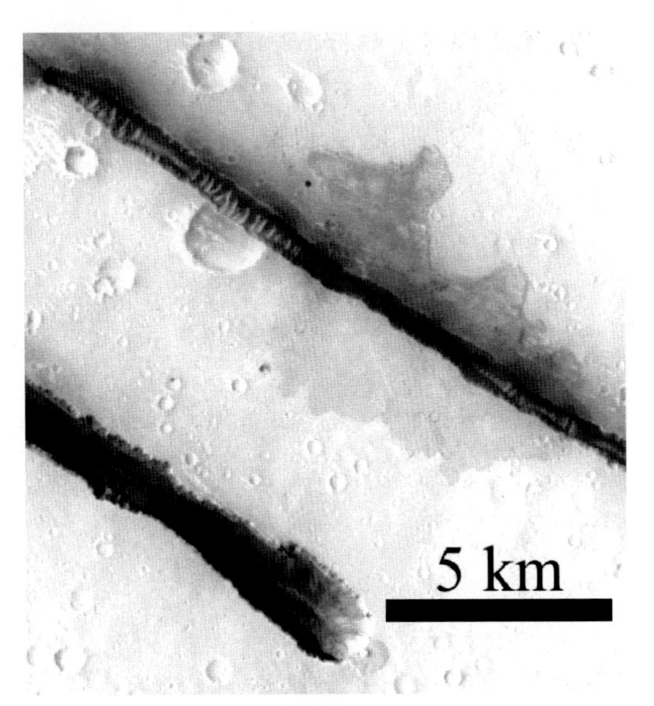

Figure 10.3. Portion of MOC image M02-01973 located near 10.4° N 156.2° E, showing a segment of the Cerberus Fossae. The fossae are sharp edged, indicating recent tectonism and/or melting of ice-rich ground. The dark lobate features to the north, south and at the southeastern tip are interpreted as spatter-fed lava. See Plate 26 for location. North is up, illumination is from the left.

side of the channel and stretching southward to the Cerberus plains. Overall, the channel is ~350 km in length, up to 100 m in depth, with a slope of ~0.0005 m/m (0.03°). The topographic channel varies from ~15 km to ~25 km in width, although the floodwaters may have locally overflowed it (Keszthelyi *et al.*, 2004b).

The Athabasca Valles system shows evidence for volcanism, tectonism, sedimentation during temporary ponding, glacial processes and ground ice. Volcanic processes are indicated by dark, lobate features surrounding the Cerberus Fossae (Figure 10.3) that have been interpreted as (spatter-fed) lava flows (Burr *et al.*, 2002a; Head *et al.*, 2003). Platy-ridged terrain on the edge of the channel has been interpreted as analogous to rafted lava plates in Iceland (Keszthelyi *et al.*, 2000, 2004a). The hummocky or polygonal material in the channel may also be lava (Burr *et al.*, 2002b; Jaeger *et al.*, 2007). Various pitted mounds in the channel have been interpreted as being composed of lava, including hypothesised rootless cones (Lanagan *et al.*, 2001; Jaeger *et al.*, 2007) and/or basaltic ring structures (Jaeger *et al.*, 2003, 2005).

Similar features also have been hypothesised to be the result of glacial and ground ice processes. A localised set of ring features has been hypothesised to be kettle holes formed by deposition of sediment-rich ice blocks (Gaidos and Marion, 2003; see also Burr *et al.*, 2005). Pitted mounds and cones have also been hypothesised to be collapsed pingos (Burr *et al.*, 2005; Page and Murray, 2006). Some areas of patterned ground in the channel have been hypothesised to be thermal contraction polygons (Burr *et al.*, 2005) or hyperconcentrated flow deposits (Rice *et al.*, 2002).

Recent tectonic activity in the Athabasca Valles region is indicated by small in-channel fissures and fissure morphology. The fissures at which the flood channel originates have sharp edges (such as are shown in Figure 10.3), which has been interpreted as evidence of post-lava flow tectonism (Berman and Hartmann, 2002). Alternatively, this sharp-edge morphology may be a result of melting due to dyke intrusion into ice-rich ground (Head *et al.*, 2003). Smaller, open, sharp-edged fissures also cut across the channel floor. The presence of such fissures in a streamlined form (Burr *et al.*, 2005) indicates that they formed more recently than the flooding that created the streamlined forms. Some of these small fissures are filled, possibly with lava and/or with flood sediments, whereas some are open (Keszthelyi *et al.*, 2004b; Burr *et al.*, 2005). This variation suggests that tectonic activity is interleaved with volcanic and/or flood events.

The pristine morphology of the Athabasca Valles has provided some indication of the floodwater flow characteristics. A set of transverse linear forms in the channel were analysed and shown to have dune morphology (Burr *et al.*, 2004). Inferred to be flood-formed dunes, they indicate that the floodwater flow was subcritical at the location and time of their formation (Burr *et al.*, 2004). Temporary in-channel floodwater ponding due to hydraulic damming and associated sedimentation has been hypothesised on the basis of a cluster of streamlined forms (Figure 10.4) (Burr, 2005). According to this hypothesis, a crater and its ejecta hydraulically dammed the floodwaters and produced deposition of flood sediments, which were streamlined behind in-channel obstacles during ponded water outflow. The layering of these streamlined forms (Figure 10.4) may thus be made up of layered flood sediments, likely interleaved with lava flows, volcanic ash deposits and dust (Burr, 2005).

Morphology shows that at the distal end of the channel, Athabasca Valles floodwaters flowed to both the eastern and western Cerberus plains. A shallow spillway, Lethe Vallis, is seen in MOLA topography and more recent visible wavelength images, indicating that some water flowed eastward (Plescia, 2003). The past presence of a lake in the western Cerberus plains is suggested on the basis of MOLA topography, which shows a shallow basin, and MOC images showing scarps and benches, which are interpreted to be shorelines (Lanagan and McEwen, 2003; Lanagan, 2004). The interpretation of pitted mounds

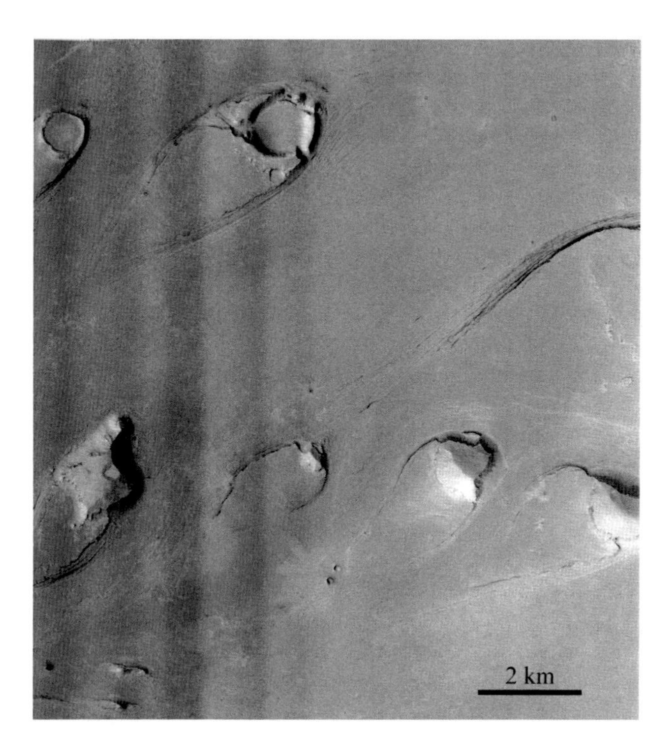

Figure 10.4. Mosaic of MOC narrow-angle images showing a cluster of streamlined forms in Athabasca Valles, inferred to have been formed by hydraulic damming/ponding, surrounded by longitudinal lineations. Scattered small impact craters with bright ejecta are secondaries from Zunil (McEwen *et al.*, 2005). See Plate 26 for location of mosaic. North is up, illumination is from the left. (Image credit: Malin Space Science Systems.)

as rootless cones both within this basin and at the mouth of Athabasca Valles also supports the hypothesis of distal floodwater ponding and suggests that the infiltration of floodwaters was quickly followed by lava flows (Lanagan *et al.*, 2001). The morphology of the material at the channel mouth on which the inferred cones sit has been interpreted as lava (Keszthelyi *et al.*, 2000, 2004a; see also Jaeger *et al.*, 2007), which would have covered any flood deposits and is consistent with the hypothesis of rootless cones. Alternatively, the same platy-ridged material in the western Cerberus plains has also been interpreted as the remnants of an ice-covered lake left by floodwaters from the Cerberus Fossae (Murray *et al.*, 2005; Kossacki *et al.*, 2006), and pitted mounds have been interpreted as pingos. For either interpretation, the Athabasca Valles sink would be primarily the western Cerberus plains, with smaller proportions of floodwater flowing to the east.

In summary, the morphology of the four Amazonian-aged, tectonic outflow channels suggests several similar processes in addition to floodwater flow. To some extent, the number of additional processes may be partially a function of the amount of analysis a channel has received; Grjotá Valles may show the fewest additional processes because they have received little analysis to date. Marte Vallis is embayed by lava flows, which may have covered post-flood modification. Table 10.1 summarises the processes that are interpreted variously to have occurred in each channel.

10.4 Mechanisms triggering water release

Possible water release mechanisms for floods from fissures can be classified as volcanic, tectonic or a combination of both types.

10.4.1 Volcanic

The geometries of the sources of tectonic outflow channels suggest that if a volcanic mechanism is involved in causing catastrophic groundwater release, it involves subsurface dyke emplacement (Wilson and Head, 2002). Chaotic terrain has been interpreted elsewhere to be the result of sill intrusion and resultant cryosphere disruption (Leask *et al.*, 2007) but is not evident at the source for any of these channels. (A region of chaotic terrain is located along the Mangala Valles source graben near 19.3° N, 207.3° E (Hanna and Phillips, 2006) but this chaos is ~200 km west of the channel and confined within a crater.) The dyke-induced mechanism relies upon the presence of a sub-cryosphere aquifer system that already has been pressurised, as predicted by Carr (1979) and Clifford (1993), a consequence of the growth of a global subsurface cryosphere over geological time. As detailed in Head *et al.* (2003) for Athabasca Valles, this proposed mechanism entails dyke emplacement resulting in cracking of this cryosphere. In addition to causing local surface eruptions producing spatter-fed flows or phreatomagmatic deposits, this cracking provides a pathway for groundwater to rise from the pressurised aquifer to the surface. Water fountains are hypothesised to have resulted at the surface with the water then draining downslope. Subsequent collapse of the ice-rich soil adjacent to the dyke as a wave of heating spreads away from it could have produced the sharp-edged morphology of the fissures. The interpretation of geomorphic features associated with the Athabasca Valles source region as spatter-fed flows, collapse pits and exposed dyke tops supports this hypothesised mechanism (Head *et al.*, 2003), as does the presence of possible phreatomagmatic deposits (Wilson and Head, 2004) and exposed dyke tops (Leask *et al.*, 2007) linked to the Mangala Valles source graben. The use of aquifer permeabilities equal to those of young basalt aquifers on Earth can produce the amplitudes and durations of the discharges estimated from surface topography at Athabasca Valles (Manga, 2004). However, it is not clear that this is the case at Mangala Valles; the problem is the length of time for which the high discharge rate appears to have been maintained, rather than the size of the initial water volume flux (Leask *et al.*, 2007).

Table 10.1. Morphological processes in fossae-fed channels

	Mangala	Marte	Grjotá	Athabasca
Approximate location	15° S 210° E (western Tharsis flank)	10° N 180° E (east of Cerberus plains)	15° N 165° E (north of Cerberus plains)	10° N 155° E (west of Cerberus plains)
Approximate elevation at source	0 m	−3100 m	−2100 to −2400 m	−2400 m
Source morphology	Widened fossa segment, with overflow through notch in eroded impact crater rim	Buried by Cerberus plains lava	Distributed along ~250 km of fossa	Dual source locations on fossa
Possible sink(s)	NW branch: ponding in crater w/terraced fan NE branch: broad flow northward	Ponding, freezing/ sublimation and infiltration in Amazonis Planitia	Infiltration combined with some freezing/sublimation	Mostly ponding in western Cerberus plains, freezing/ sublimation, infiltration
Process	*Inferred geologic evidence*			
Volcanism	Dyke emplacement,[a] phreatomagmatic dunes[b]	Embaying lava flows,[i] lava burial of source[j]	Embaying lava flows[l]	Lava flows at source,[j,m] dyke emplacement,[n] rootless cones,[o,p] basaltic ring structures[q,r]
Tectonics	Rotated fault block,[c] graben formation[d,e]			Sharp-edged fissures[s,n]
Glaciation, ground ice	Kettle holes,[f] moraines and other glacial deposits[g]			Kettle holes,[t] ice wedge polygons,[u] pingos[u]
Flood sedimentation	Unit Amch[h]	In Amazonis Planitia sink[k]		Dunes,[v] streamlined forms,[m,j] hyperconcentrated flow deposits[w]

Sources: [a] Wilson and Head (2002)
[b] Wilson and Head (2004)
[c] Zimbelman (1989)
[d] Tanaka and Chapman (1990)
[e] Craddock and Greeley (1994)
[f] Zimbelman *et al.* (1992)
[g] Head *et al.* (2004)
[h] Ghatan *et al.* (2005)
[i] Plescia (1990)
[j] Burr *et al.* (2002b)
[k] Fuller and Head (2002)
[l] Burr and Parker (2006)
[m] Burr *et al.* (2002a)
[n] Head *et al.* (2003)
[o] Lanagan *et al.* (2001)
[p] Greeley and Fagents (2001)
[q] Jaeger *et al.* (2003)
[r] Jaeger *et al.* (2005)
[s] Berman and Hartmann (2002)
[t] Gaidos and Marion (2003)
[u] Burr *et al.* (2005a)
[v] Burr *et al.* (2004)
[w] Rice *et al.* (2002)

Repeated short-duration high-discharge floods (e.g. Hanna and Phillips, 2006) can explain both the high discharges and the large volumes of water required.

10.4.2 Tectonic

Tectonic mechanisms for catastrophic groundwater release involve faulting, but do not require intrusive volcanism to induce the faulting. A correlation between faulting and floodwater release was first proposed for Mangala Valles on the basis of the similar crater-count model ages of the flood channel and the associated faults (Tanaka and Chapman, 1990). The mechanism entailed breaching of pressurised perched aquifers by the faulting. It was suggested that the elevated water pressures might have been produced by groundwater circulation induced by Tharsis magmatism, Tharsis-centred tectonic uplift, or compaction of saturated sediments by the load imposed by lava flow emplacement.

A more recent hypothesis invokes dyke-induced extensional tectonism as the causal mechanism of all aspects of the process (Hanna and Phillips, 2006). Based on a recent model of the Martian crust (Hanna and Phillips, 2005), this mechanism involves a tectonic pressurisation of the aquifers as a result of the release of the extensional stresses in the crust, with excess pore pressures up to 10 MPa being produced. This extensional pressurisation results in a discharge of water to the surface through the tectonically generated fault. Although other factors, such as perched aquifers, volcano–ice interactions, or phreatomagmatic aquifer pressurisation, may also be involved in particular cases, they are not necessary; Hanna and Phillips (2006) show that extensional tectonism alone is sufficient to produce groundwater discharges and flood volumes equal to those modelled for surficial flow in Athabasca and Mangala Valles.

10.4.3 Combined factors

The catastrophic release of groundwater may require both volcanic and tectonic processes. The model of Zimbelman *et al.* (1992) for Mangala Valles proposed that groundwater was generated through melting of ground ice as the geotherm was steepened by enhanced regional volcanic activity during Tharsis formation, and subsequently this groundwater was released through fractures associated with the Mangala Fossa graben where it intersects a north–south trending fault. However, the melting of ground ice in this way may not have been able to provide a great enough water volume or discharge to have carved the Mangala Valles (McKenzie and Nimmo, 1999; Ghatan *et al.*, 2005), so this scenario may not be appropriate for that particular channel system (see Ghatan *et al.*, 2005, for discussion). Nevertheless, the very recent tectonic and volcanic activity that has been documented in Athabasca Valles suggests that a combination of the two processes may well be possible in other situations.

As an example, in a more recent model of the Mangala system, Wilson and Head (2002) suggested that the Mangala Fossa graben was produced by faulting above one of a series of dykes (the Memnonia Fossae) propagating laterally from a mantle source beneath Arsia Mons. The locations and radial orientations of these dykes, and others inferred to be responsible for the graben of the nearby Sirenum, Icaria, Thaumasia and Claritas Fossae, are controlled by the strength and orientation of the regional tectonic stress field. The widths and depths of the graben are determined by a combination of the width of the dyke as its upper tip approaches the surface and the depth at which upward propagation ceases (Rubin, 1992; Schultz *et al.*, 2004). The dyke width is controlled by the interaction between the excess pressure in the dyke magma and the regional horizontal extensional tectonic stress deviator prior to dyke injection, as described by Rubin and Pollard (1987). Water migrates through pathways provided partly by the propagation of the dyke and partly by the graben boundary faults.

In summary, multiple factors can account for the inferred catastrophic discharges of water from fissures. Given evidence for recent volcanism and tectonism on Mars, a combination of factors is reasonable, and different mechanisms may have operated at different locations or times.

10.5 Thermal and mechanical aspects of water release

If the origin of a fracture through the cryosphere is purely tectonic, the thermodynamics of water rise through the fracture is relatively straightforward. If the water being released is derived from just below the base of the cryosphere, thus having a temperature only just above its freezing point, and if it rises only slowly, adiabatic cooling may cause partial freezing (Gaidos and Marion, 2003). However, given the large discharge rate and water volume estimates (at least $\sim 10^6$ to 10^7 $m^3\ s^{-1}$ and up to at least 20 000 km^3, respectively; Wilson *et al.*, this volume Chapter 16) it seems likely that water is tapped from a wide range of aquifer depths. Then even in the absence of volcanic heat sources the mean temperature of the released water may be substantially above the freezing point. As shown in Wilson *et al.*, hindered convection in the pore spaces occupied by aquifer water (Ogawa *et al.*, 2003) probably reduces the vertical temperature difference across a few-kilometer-thick aquifer by a factor of ~ 2 below the ~ 100 °C implied by the geotherm. As a result, the mean water temperature in an aquifer with a few kilometres' vertical extent may be up to at least ~ 30 °C and possibly as much as ~ 50 °C (Wilson *et al.*, this volume Chapter 16).

If a dyke intrusion is involved, there are more factors to consider. On the basis of a detailed morphological analysis, Leask *et al.* (2006) proposed that the Mangala Fossa graben acting as the source of the Mangala Valles is the consequence of two dyke intrusions, each dyke having an implied width of ~ 250 m with its top located at a depth varying between ~ 150 m and ~ 550 m below the local surface. These width and depth values lie within the range predicted theoretically by Wilson and Head (2002) for dykes sourced from the crust–mantle boundary on Mars. Leask *et al.* (2006) ascribed part of the present depth of Mangala Fossa to two episodes of subsidence of the graben floor as heat released from the dykes melted nearby cryosphere ice, with water escape to the surface allowing compaction to occur. Through a heat-sharing calculation, Leask *et al.* (2006) showed that if the cryosphere contained 10% by volume ice, a plausible amount based on the crust model of Hanna and Phillips (2005), and if all of the heat released by the dyke were contained in a hydrothermal system confined by the graben faults, the maximum temperature reached by the hydrothermal water would have been ~ 40 °C. The volume of that water, summed over the ~ 200 km length along strike of the graben, would have been ~ 155 km^3 in each event, a total of ~ 310 km^3. If the ice melting had extended to a greater lateral distance than the graben faults, then the absolute maximum amount of water that could have been generated by cooling each dyke to a temperature infinitesimally above the melting point would have been ~1650 km^3, providing a total of ~3300 km^3 of water just above its freezing point, though this efficiency of heat transfer from magma to ice could never be approached in practice. From measurements of the volume of rock eroded to form the Mangala Valles channels and an estimate (Komar, 1980) of the maximum sediment-bearing capacity of a water flood (≤40% by volume), calculations of the minimum amount of water required to flow through the system give estimates ranging from 8600 km^3 (Hanna and Phillips, 2006) to between 13 000 and 30 000 km^3 (Ghatan *et al.*, 2005). Thus it is extremely unlikely, in this case at least, that melted cryosphere ice contributed more than a small fraction of the water released, and tapping of a pre-existing sub-cryosphere aquifer seems required.

Along most of its length the Mangala Fossa graben is ~2 km wide, and this implies that the graben boundary faults, dipping at ~60° near the surface, extended downward for ~1.5 km into the ~4 km thick cryosphere. Thus it seems inevitable that water was transferred through the lower part of the cryosphere along one or both of the margins of the dyke. Thermal interaction between chilling magma and melting ice is likely to be inherently unstable, and it is not surprising that evidence of phreatomagmatic explosive activity is seen at the eastern end of Mangala Fossa (Wilson and Head, 2004). The fact that such explosive activity is not seen at other sites of water release hypothesised to have resulted from dyke intrusion suggests that the distribution of ice-bearing pore space in the cryosphere is heterogeneous. A stable contact between dyke margin and crustal rock may then exist in some places, with minimal or no cryosphere ice melting and upward aquifer water flow, whereas water may rise relatively freely in other places through cavities created by an initial violent but short-lived dyke–cryosphere interaction. Alternatively, an absence of evidence for explosive activity at other sites may suggest an amagmatic water-release mechanism.

The typical widths of the open pathways required for water discharge, together with the flow speeds of water through them, can be estimated from the relationship between pressure in the aquifer system and pathway wall friction. Using varying assumptions about the origin of excess pressure in the aquifer, Head *et al.* (2003) and Manga (2004) find average pathway widths in the range 1–2.5 m and water rise speeds up to ~60 m s^{-1}. Combining these rise speeds, pathway widths and the ~30–40 km length along strike of the pathways, Head *et al.* (2003) and Manga (2004) show that the water discharges of order 1–2 × 10^6 m^3 s^{-1} estimated for the Athabasca Valles channels by Burr *et al.* (2002a) can readily be provided. A similar analysis for the Mangala Fossa/Mangala Valles system yields a typical pathway width of 2.3 m and water rise speed of ~20 m s^{-1} along the ~200 km horizontal extent of the graben to provide the ~10^7 m^3 s^{-1} estimated discharge (Leask *et al.*, 2006). All of these water rise speed and pathway width combinations lead to minimal cooling of the water during its ascent.

Water rise speeds of ~20 to more than 50 m s^{-1} imply that water emerging at the surface would have formed a fountain with a height of at least 50 to as much as ~400 m. The mechanics of water spreading away on the surface from such a fountain has much in common with the formation of pyroclastic density currents from fountains of gas and particles in some explosive volcanic eruptions (Wilson and Heslop, 1990). Thus, as it falls onto the surface, the water from the fountain causes a dynamic pressure equal to the stagnation pressure of the water flow, one-half of the water density times the square of its speed. The speed on reaching the ground is essentially the same as the speed with which water is projected upward into the fountain, ~20–50 m s^{-1}, and so the pressure in the vicinity of the release point will be in the range ~0.2 to ~1.5 MPa. These pressures will suppress water vapour formation and will minimise release of any dissolved carbon dioxide in the core of the fountain and in the immediate vicinity of the release point. However, on the outer edges of the fountain, and on the top surface of the water flowing away from the fountain, both processes would occur, and would lead to instabilities in the water–atmosphere interface and formation of water droplets. In the case of fountains formed from water at a temperature very close to the triple point, extraction of latent heat of evaporation could lead to freezing of some of the droplets, as discussed by Gaidos and Marion (2003). Some of these issues, together with the dynamics of water flow away from its source, are considered by Wilson *et al.* (this volume Chapter 16).

10.6 Summary and implications

The four most recent flood channels on Mars all head at tectonic features. These channels can be grouped in space and time (Table 10.1). Mangala Valles are the oldest of the four, flowing during the late Hesperian and early Amazonian Epochs (Tanaka and Chapman, 1990; Zimbelman *et al.*, 1992). Modelling suggests that the channels may have formed through repeated short-duration, high-discharge events (Hanna and Phillips, 2006; Manga, 2004), whereas geological mapping suggests only one period of flow (Ghatan *et al.*, 2005). The channels originate from the Memnonia Fossae graben system located off the western flanks of the Tharsis rise at an elevation of about 0 m.

The other three channels are located to the east, north and west of the Cerberus plains and originate from the Cerberus Fossae. The elevations at their origination sites range from approximately −3100 m to −2100 m.

Water flowed in the channels at different times during the late to very late Amazonian Epoch (Burr *et al.*, 2002b; McEwen *et al.*, 2005). During the Amazonian, the loci of flow progressed from the east (Marte Vallis) to the west (Athabasca Valles) ends of the fossae. This situation mirrors the migration of volcanic activity on the fossae, which has been interpreted on the basis of superposition of lava flows to have moved westward with time (Lanagan and McEwen, 2003). This correlation is consistent with (but does not require) a volcanic triggering mechanism and admits the possibility that the regional stress field in this region is not static.

The likely sources of the floodwater are subsurface aquifers with thicknesses of at least a few kilometres and temperatures of a few tens of degrees Celsius. Mechanisms that may have triggered groundwater release from these aquifers to the surface could have been volcanic, tectonic or a combination of the two. Geomorphological evidence, or lack thereof, suggests that each of these mechanisms may have operated at different times and locations. Upon reaching the surface, the water would likely have fountained at least a few tens of metres above the surface, suppressing water vapour formation and carbon dioxide release within the fountain but promoting water and/or ice droplet formation at the margins. Continued discussion of the resultant surface processes is provided in Wilson *et al.* (this volume Chapter 16).

In large part, Mars exploration is the search for evidence of life. Because life as known requires liquid water, sites of astrobiological interest are those where liquid water existed, including catastrophic flood channels (National Academy of Sciences, 2007). These flood channels must have had sources within subsurface aquifers, the depths of which may be consistent with modelled water table depth (see Burr *et al.*, 2002b). At present, the surface of Mars is inhospitable to life (e.g. Carr, 1996; National Academy of Sciences, 2007) but these source aquifers, located in volcanic terrain, may provide or have provided both liquid water and energy to sustain life. To the extent that the flood channels are draped in lava (see Jaeger *et al.*, 2007), their utility as an astrobiology target (see Fairén *et al.*, 2005) may be reduced. However, sediments from the subsurface deposited in post-magmatic flooding may provide chemical or spectroscopic evidence of a putative subsurface biosphere, to match the geophysical modelling of the type provided here. Ongoing exploration of the surface of Mars will continue to deepen our understanding of the aqueous subsurface as well.

Acknowledgements

All MOLA topography figures (Plates 23 through 26) were made in Generic Mapping Tools (Wessel and Smith, 1998) with data from the Planetary Data System (Smith *et al.*, 2003). We thank Jeffery Andrews-Hanna and Jim Zimbelman for constructive and helpful reviews.

References

Bargery, A. S. and Wilson, L. (2006). Modelling water flow with bedload on the surface of Mars. *Lunar and Planetary Science Conference* XXXVII, Abstract 1218, Lunar and Planetary Institute, Houston, Texas (CD ROM).

Berman, D. C. and Hartmann, W. K. (2002). Recent fluvial, volcanic, and tectonic activity on the Cerberus Plains of Mars. *Icarus*, **159**, 1–17.

Berman, D. C., Hartmann, W. K. and Burr, D. M. (2001). Marte Vallis and the Cerberus Plains: evidence of young water flow on Mars. *Lunar and Planetary Science Conference XXXII*, Abstract 1732, Lunar and Planetary Institute, Houston, Texas (CD ROM).

Burr, D. M. (2005). Clustered streamlined forms in Athabasca Valles, Mars: evidence for sediment deposition during floodwater ponding. *Geomorphology*, **69**, 242–252.

Burr, D. M. and Parker, A. H. (2006). Grjotá Valles and implications for flood sediment deposition on Mars. *Geophysical Research Letters*, **33**, L22201, doi:10.1029/2006GL028011.

Burr, D. M., McEwen, A. S. and Sakimoto, S. E. H. (2002a). Recent aqueous floods from the Cerberus Fossae, Mars. *Geophysical Research Letters*, **29** (1), doi:10.1029/2001GL013345.

Burr, D. M., Grier, J. A., McEwen, A. S. and Keszthelyi, L. P. (2002b). Repeated aqueous flooding from the Cerberus Fossae: evidence for very recently extant, deep groundwater on Mars. *Icarus*, **159**, 53–73.

Burr, D. M., Carling, P. A., Beyer, R. A. and Lancaster, N. (2004). Flood-formed dunes in Athabasca Valles, Mars: morphology, modeling, and implications. *Icarus*, **171**, 68–83.

Burr, D. M., Soare, R. J., Wan Bun Tseung, J.-M. and Emery, J. P. (2005). Young (late Amazonian), near surface, ground ice features near the equator, Athabasca Valles, Mars. *Icarus*, **178**, 56–73.

Burr, D. M., Bruno, B. C., Lanagan, P. D. *et al.* (2008). Mesoscale raised rim depressions (MRRDs) on Earth: a review of the characteristics, processes, and spatial distributions of analogs for Mars. *Planetary and Space Science*, doi:10.1016/j.pss.2008.11.011.

Carr, M. H. (1979). Formation of Martian flood features by release of water from confined aquifers. *Journal of Geophysical Research*, **84**, 2995–3007.

Carr, M. H. (1996). *Water on Mars*. Oxford: Oxford University Press.

Chapman, M. G. and Tanaka, K. L. (1993). *Geologic Map of the MTM -05152 and -10152 Quadrangles, Mangala Valles Region of Mars, scale 1:500,000*. U.S. Geological Survey Investigations Series Map, I-2294.

Christensen, P. R. and 21 others (2003). Morphology and composition of the surface of Mars: Mars Odyssey THEMIS results. *Science*, **300** (5628), 2056–2061.

Clifford, S. M. (1993). A model for the hydrologic and climatic behavior of water on Mars. *Journal of Geophysical Research*, **98**, 10,973–11,016.

Clifford, S. M. and Parker, T. J. (2001). The evoluton of the martian hydrosphere: implications for the fate of a primordial ocean and the current state of the northern plains. *Icarus* 154, 40–79.

Craddock, R. A. and Greeley, R. (1994). *Geologic map of the MTM -20147 Quadrangle, Mangala Valles Region of Mars, scale 1:500,000*, U.S. Geological Survey Miscellaneous Investigation Series Map I-2310.

Di Achille, G., Ori, G. G., Reiss, D. *et al.* (2006). A steep fan at Coprates Catena, Valles Marineris, Mars, as seen by HRSC data. *Geophysical Research Letters*, **33** (7), CiteID L07204, doi:10.1029/2005GL025435.

Edgett, K. S. and Malin, M. C. (2003). The layered upper crust of Mars: an update on MGS MOC observations after two Mars years in the mapping orbit. In *Lunar and Planetary Science Conference XXXIV*, Abstract 1124, Lunar and Planetary Institute, Houston, Texas (CD ROM).

Fairén, A. G., Dohm, J. M., Uceda, E. R. *et al.* (2005). Prime candidate sites for astrobiological exploration through the hydrogeological history of Mars. *Planetary and Space Science*, **53**, 1355–1375, doi:10.1016/j.pss.2005.06.007.

Fuller, E. R. and Head, III, J. W. (2002). Amazonis Planitia: the role of geologically recent volcanism and sedimentation in the formation of the smoothest plains on Mars. *Journal of Geophysical Research*, **107** (E10), 5081, 2002JE001842.

Gaidos, E. and Marion, G. (2003). Geological and geochemical legacy of a cold, early Mars. *Journal of Geophysical Research*, **108** (E6), 5005, doi:10.1029/2002JE002000.

Ghatan, G. J., Head, J. W. and Wilson, L. (2005). Mangala Valles, Mars: assessment of early stages of flooding and downstream flood evolution. *Earth Moon Planets*, **96** (1–2), 1–57, doi:10.1007/s11038-005-9009-y.

Greeley, R. and Fagents, S. A. (2001). Icelandic pseudocraters as analogs to some volcanic cones on Mars. *Journal of Geophysical Research*, **106**, 20527–20546.

Hanna, J. C. and Phillips, R. J. (2005). Hydrological modeling of the Martian crust with application to the pressurization of aquifers. *Journal of Geophysical Research*, **110**, E01004, doi:10.1029/2004JE002330.

Hanna, J. C. and Phillips, R. J. (2006). Tectonic pressurization of aquifers in the formation of Mangala and Athabasca Valles, Mars. *Journal of Geophysical Research*, **111**, E03003, doi:10.1029/2005JE002546.

Hartmann, W. K. (1999). Martian cratering VI: Crater count isochrones and evidence for recent volcanism from Mars Global Surveyor. *Meteoritics and Planetary Science*, **34**, 168–177.

Hartmann, W. K. (2005). Martian cratering 8: Isochron refinement and the chronology of Mars. *Icarus*, **174** (2), 294–320.

Hartmann, W. K. and Neukum, G. (2001). Cratering chronology and evolution of Mars. *Space Science Reviews*, **96**, 165–194.

Head, J. W. III, Wilson, L. and Mitchell, K. L. (2003). Generation of recent massive water floods at Cerberus Fossae, Mars by dike emplacement, cryospheric cracking, and confined aquifer groundwater release. *Geophysical Research Letters*, **30** (11), 1577, doi:10.1029/2003GL0117135.

Head, J. W. III, Marchant, D. R. and Ghatan, G. J. (2004). Glacial deposits on the rim of a Hesperian-Amazonian outflow channel source trough: Mangala Valles, Mars. *Geophysical Research Letters*, **31**, L10701, doi:10.1029/2004GL020294.

House, P. K., Pearthree, P. A. and Klawon, J. E. (2001). Historical flood and paleoflood chronology of the Lower Verde River, Arizona: stratigraphic evidence and related uncertainties. In *Ancient Floods, Modern Hazards: Principles and Applications of Paleoflood Hydrology*, Water Science and Application Volume 5, American Geophysical Union, pp. 267–293.

Jaeger, W. L., Keszthelyi, L. P., Burr, D. M. *et al.* (2003). Ring dike structures in the Channeled Scabland as analogs for circular features in Athabasca Valles, Mars. In *Lunar and Planetary Science Conference XXXIV*, Abstract 2045, Lunar and Planetary Institute, Houston, Texas (CD ROM).

Jaeger, W. L., Keszthelyi, L. P., Burr, D. M. *et al.* (2005). Basaltic ring structures as an analog for ring features in Athabasca Valles, Mars. In *Lunar and Planetary Science Conference* XXXVI, Abstract 1886, Lunar and Planetary Institute, Houston, Texas (CD ROM).

Jaeger, W. L., Keszthelyi, L. P., McEwen, A. S., Dundas, C. M. and Russell, P. S. (2007). Athabasca Valles, Mars: a lava-draped channel system. *Science*, **317**, 1709–1711, doi:10.1126/science.1143315.

Keszthelyi, L. P., McEwen, A. S. and Thordarson, Th. (2000). Terrestrial analogs and thermal models for martian flood lavas. *Journal of Geophysical Research*, **105**, 15,027–15,049.

Keszthelyi, L., Thordarson, Th., McEwen, A. *et al.* (2004a). Icelandic analogs to martian flood lavas. *Geochemistry, Geophysics, Geosystems (G 3)*, **5**, Q11014, doi:10.1029/2004GC000758.

Keszthelyi, L., Burr, D. M. and McEwen, A. S. (2004b). Geomorphologic/thermophysical mapping of the Athabasca Region, Mars, using THEMIS infrared imaging. In *Lunar and Planetary Science Conference XXXV*, Abstract 1657, Lunar and Planetary Institute, Houston, Texas (CD ROM).

Komar, P. D. (1980). Modes of sediment transport in channelized water flows with ramifications to the erosion of the Martian outflow channels. *Icarus*, **42** (3), 317–329.

Kossacki, K. J., Markiewicz, W. J., Smith, M. D., Page, D. and Murray, J. (2006). Possible remnants of a frozen mud lake in southern Elysium, Mars. *Icarus*, **181**, 363–374, doi:10.1016/j.icarus.2005.11.018.

Lanagan, P. D. (2004). Geologic history of the Cerberus Plains, Mars. Ph.D. thesis, University of Arizona, Tucson.

Lanagan, P. D. and McEwen, A. S. (2003). Cerberus Plains volcanism: constraints on temporal emplacement of the youngest flood lavas on Mars. *Sixth International Conference on Mars*, July 20–25 2003, Pasadena, California, Abstract 3215.

Lanagan, P. D., McEwen, A. S., Keszthelyi, L. P. and Thordarson, Th. (2001). Rootless cones on Mars indicating the presence of shallow equatorial ground ice in recent times. *Geophysical Research Letters*, **28**, 2365–2367.

Leask, H. J., Wilson, L. and Mitchell, K. L. (2006). Formation of Mangala Fossa, the source of the Mangala Valles, Mars: morphological development as a result of volcano-cryosphere interactions. *Journal of Geophysical Research*, **112**, E02011, doi:10.1029/2005JE002644.

Leask, H. J., Wilson, L. and Mitchell, K. L. (2007). Formation of Mangala Valles outflow channel, Mars: morphological development, and water discharge and duration estimates. *Journal of Geophysical Research*, **112**, E08003, doi:10.1029/006JE002851.

Malin, M. C. and Edgett, K. S. (2000). Sedimentary rocks of early Mars. *Science*, **290** (5498), 1927–1937.

Malin, M. C. and Edgett, K. S. (2001). Mars Global Surveyor Mars Orbiter Camera: interplanetary cruise through primary mission. *Journal of Geophysical Research*, **106** (E10), 23,429–23,570.

Manga, M. (2004). Martian floods at Cerberus Fossae can be produced by groundwater discharge. *Geophysical Research Letters*, **31**, L02702, doi:10.1029/2003GL018958.

McEwen, A. S. and Bierhaus, E. B. (2006). The importance of secondary cratering to age constraints on planetary surfaces. *Annual Review of Earth and Planetary Science*, **34**, 535–567, doi:10.1146/annurev.earth.34.031405.125018.

McEwen, A. S., Preblich, B. S., Turtle, E. P. *et al.* (2005). The rayed crater Zunil and interpretations of small impact craters on Mars. *Icarus*, **176**, 351–381.

McKenzie, D. and Nimmo, F. (1999). The generation of Martian floods by the melting of ground ice above dykes. *Nature*, **397**, 231– 233.

Milton, D. J. (1973). Water and the processes of degradation in the Martian landscape. *Journal of Geophysical Research*, **78**, 4037–4047.

Moller, S. C., Poulter, K., Grosfills, E. *et al.* (2001). Morphology of the Marte Valles channel system. In *Lunar and Planetary Science Conference XXXII*, Abstract 1382, Lunar and Planetary Institute, Houston, Texas (CD ROM).

Mouginis-Mark, P. (1990). Recent melt water release in the Tharsis region of Mars. *Icarus*, **84**, 362–373.

Mouginis-Mark, P. J. and Christensen, P. R. (2005). New observations of volcanic features on Mars from the THEMIS instrument. *Journal of Geophysical Research*, **110**, E08007, doi:10.1029/2005JE002421.

Mouginis-Mark, P. J., Wilson, L., Head, J. W. *et al.* (1984). Elysium Planitia, Mars: regional geology, volcanology, and evidence for volcano-ground ice interactions. *Earth, Moon, and Planets*, **30**, 149–173.

Murray, J. B. and 12 co-authors (2005). Evidence from the Mars Express High Resolution Stereo Camera for a frozen sea close to Mars equator. *Nature*, **434**, 352–356, doi:10.1038/nature03379.

National Academy of Sciences (2007). *An Astrogeological Strategy for the Exploration of Mars*. Washington DC: National Academies Press.

Neukum, G. and Ivanov, B. A. (1994). Crater size distribution and impact probabilities on the Earth from lunar, terrestrial-planet, and asteroid cratering data. In *Hazards Due to Asteroids and Comets*, ed. T. Gehrels. Tucson: University of Arizona Press, pp. 359–416.

Ogawa Y., Yamagishi, Y. and Kurita, K. (2003). Evaluation of melting process of the permafrost on Mars: its implication for surface features. *Journal Geophysical Research*, **108** (E4), 8046, doi:10.1029/2002JE001886.

Page, D. P. and Murray, J. B. (2006). Stratigraphical and morphological evidence for pingo genesis in the Cerberus plains. *Icarus*, **183**, 46–54, doi:10.1016/j.icarus.2006.01.017.

Plescia, J. B. (1990). Recent flood lavas in the Elysium region of Mars. *Icarus*, **88**, 465–490.

Plescia, J. B. (2003). Cerberus Fossae, Elysium Mars: a source for lava and water. *Icarus*, **164**, 79–95.

Rice Jr., J. W., Parker, T. J., Russell, A. J. and Knudsen, Ó. (2002). Morphology of fresh outflow channel deposits on Mars. In *Lunar and Planetary Science Conference XXXIII*, Abstract 2026, Lunar and Planetary Institute, Houston, Texas (CD ROM).

Rubin, A. M. (1992). Dike-induced faulting and graben subsidence in volcanic rift zones. *Journal of Geophysical Research*, **97**, 1839–1858.

Rubin, A. M. and Pollard, D. D. (1987). Origin of blade-like dikes in volcanic rift zones. In *Volcanism in Hawaii* eds. R. W. Decker, T. L. Wright and P. H. Stauffer. U.S. Geological Survey Professional Paper 1350, pp. 1449–1470.

Sakimoto, S. E. H., Reidel, S. and Burr, D. M. (2001). Geologically recent Martian volcanism and flooding in Elysium Planitia and Cerberus Rupes: plains-style eruptions and related water release? *Geological Society of America Abstracts with Program* (abstract 178–0).

Schultz, R. A., Okubo, C. H., Goudy, C. L. and Wilkins, S. J. (2004). Igneous dikes on Mars revealed by Mars Orbiter Laser Altimeter topography. *Geology*, **32** (10), 889–892.

Scott, D. H. and Chapman, M. G. (1991). Mars Elysium Basin: geologic/volumetric analyses of a young lake and exobiologic implications. In *Lunar and Planetary Science Conference XXI*, 669–677 (abstract).

Sharp, R. P. and Malin, M. C. (1975). Channels on Mars. *Geological Society of America Bulletin*, **86**, 593–609.

Smith, D. E., Zuber, M. T., Frey, H. V. *et al.* (1998). Topography of the northern hemisphere of Mars from the Mars Orbiter Laser Altimeter. *Science*, **279**, 1686–1692.

Smith, D., Neumann, G., Arvidson, R. E., Guinness, E. A. and Slavney, S. (2003). *Mars Global Surveyor Laser Altimeter Mission Experiment Gridded Data Record.* NASA Planetary Data System, MGS-M-MOLA-5-MEGDR-L3-V1.0.

Tanaka, K. L. (1986). The stratigraphy of Mars. *Journal of Geophysical Research*, **91**, 139–158.

Tanaka, K. L. and Chapman, M. G. (1990). The relation of catastrophic flooding of Mangala Valles, Mars, to faulting of Memnonia Fossae and Tharsis volcanism. *Journal of Geophysical Research*, **95**, 14,315–14,323.

Tanaka, K. L. and Scott, D. H. (1986). The youngest channel system on Mars. In *Lunar and Planetary Science Conference XVII*, 865–866 (abstract).

Weitz, C. M., Irwin, R. P., III, Chuang, F. C. Bourke, M. C., and Crown, D. A. (2006). Formation of a terraced fan deposit in Coprates Catena, Mars. *Icarus*, **184**, 436–451, doi:10/1016/j.icarus.2006.05.024.

Werner, S. C., van Gasselt, S. and Neukem, G. (2003). Continual geological activity in Athabasca Valles, Mars. *Journal of Geophysical Research*, **108** (E12), 8081, doi:10.1029/2002JE002020.

Wessel, P. and Smith, W. H. F. (1998). New, improved version of Generic Mapping Tools released. *EOS Transactions, American Geophysical Union*, **79** (47), 579.

Williams, R. M. E. and Edgett, K. S. (2005). Valleys in the Martian rock record. In *Lunar and Planetary Science Conference XXXVI*, Abstract 1099, Lunar and Planetary Institute, Houston, Texas (CD ROM).

Williams, R. M. E., Malin, M. C. and Edgett, K. S. (2005). Remnants of the courses of fine-scale, precipitation-fed runoff streams preserved in the Martian rock record. In *Lunar and Planetary Science Conference XXXVI*, Abstract 1173, Lunar and Planetary Institute, Houston, Texas (CD ROM).

Wilson, L. and Head, J. W. (2002). Tharsis-radial graben systems as the surface manifestation of plume-related dike intrusion complexes: model and implications. *Journal of Geophysical Research*, **107** (E8), 5057, doi:10.1029/2001JE001593.

Wilson, L. and Head III, J. W. (2004). Evidence for a massive phreatomagmatic eruption in the initial stages of formation of the Mangala Valles outflow channel, Mars. *Geophysical Research Letters*, **31**, L15701, doi:10.1029/2004GL020322.

Wilson, L. and Heslop, S. E. (1990). Clast sizes in terrestrial and martian ignimbrite lag deposits. *Journal of Geophysical Research*, **95**, 17309–17314.

Zimbelman, J. R. (1989). Geological mapping of southern Mangala Valles, Mars. In *Lunar and Planetary Science Conference XX*, 1239–1240 (abstract).

Zimbelman, J. R., Craddock, R. A., Greeley, R. and Kuzmin, R. O. (1992). Volatile history of Mangala Valles, Mars. *Journal of Geophysical Research*, **97**, 18,309–18,317.

Zimbelman, J. R., Craddock, R. A. and Greeley, R. (1994). *Geologic Map of the MTM-15147 Quadrangle, Mangala Valles Region of Mars, scale 1:500,000.* U.S. Geological Survey Miscellaneous Investigations Series Map I-2402.

11

Large basin overflow floods on Mars

ROSSMAN P. IRWIN III and
JOHN A. GRANT

Summary

Breaches of large natural basins, usually initiated by high runoff or meltwater production in their contributing watersheds, have been responsible for the most intense recognised terrestrial floods. Some of the many impact craters and intercrater basins in the Martian highlands also apparently overflowed during the Noachian Period (>3.7 Ga), forming relatively wide and deep outlet valleys. Broad, mid-latitude basins overflowed to carve Ma'adim Vallis and the Uzboi–Ladon–Morava Valles system, which are similar in scale to the terrestrial Grand Canyon but record much larger formative discharges. Other valley network stems of comparable size are also associated with smaller breached basins or broad areas of topographic convergence, and even the smaller basin outlets are typically deeper than other valleys in their vicinity. Little evidence for catastrophic (by terrestrial standards) meteorological floods has been recognised to date in Martian alluvial deposits. For these reasons, basin overflows may have been disproportionately important mechanisms for valley incision on Mars. Many of the Martian outflow channels also head in topographic settings that favoured ponding, including large canyons, impact or intercrater basins, chaotic terrain basins and grabens. Draining of this topography may have supported peak discharges of $\sim 10^6$–10^8 m^3 s^{-1}, particularly in the largest channels, but the basin overflow mechanism does not eliminate fully the need for large subsurface outflows.

11.1 Introduction

Periods of intense precipitation or snowmelt are a common cause of destructive floods, but the largest events have involved uncontrolled water releases from basins. Failures of glaciers or natural earthen dams have been implicated in all recognised terrestrial floods of >500 000 m^3 s^{-1} (O'Connor and Costa, 2004). Such large discharges are rare and often unprecedented in their watersheds and previously stable surfaces may be deeply eroded. In a typical scenario, impounded runoff or meltwater overtops and erodes an unconsolidated dam (e.g. glacier, moraine, landslide, alluvial fan or artificial dam). Piping failures are also common, where water emerges under pressure through the dam, widening a pipe until the overburden collapses. Discharge from the basin increases until the rate of downcutting (along with widening of the outlet) slows to the rate of lake level fall. Ultimately, the flood wanes because the dam is compromised fully and all stored water is released, or exposure of more resistant rock prevents further incision of the barrier (e.g. Gilbert, 1890; Jarrett and Costa, 1986; Costa, 1988; O'Connor, 1993; Waythomas *et al.*, 1996; Walder and O'Connor, 1997). O'Connor and Beebee (this volume Chapter 8) discuss examples of large dambreak floods and several of the companion papers in this volume describe their characteristic erosional morphology and sedimentary deposits. Ice dams that impound deep lakes are particularly susceptible to catastrophic failure. Large enclosed basins on Earth have a variety of origins but most are related to recent glaciation or tectonism and occur in regions where these processes have been active (e.g. Cohen, 2003, p. 25).

Large basins also have overflowed in the cratered highland landscape of Mars (Parker, 1985, 1994; Goldspiel and Squyres, 1991; De Hon, 1992; Grant and Parker, 2002; Irwin *et al.*, 2002, 2004, 2005b). Heavy impact bombardment during the Noachian Period, prior to $\sim$3.7 Ga (absolute ages herein follow Hartmann and Neukum, 2001), divided the Martian highlands into many enclosed craters and intercrater basins. Fluvial and aeolian erosion, impact gardening and other processes modified these craters throughout the time of heavy bombardment, whereas younger large craters retain a fresh morphology (e.g. Arvidson, 1974; Grant and Schultz, 1993; Craddock *et al.*, 1997; see Figure 11.1a of Irwin *et al.*, 2005b). Cross-sectional profiles and mass-balance analyses of modified Noachian craters in the equatorial highlands show rim erosion, widening and infilling that are most consistent with dominant fluvial erosion (Craddock and Maxwell, 1993; Craddock *et al.*, 1997; Craddock and Howard, 2002; Forsberg-Taylor *et al.*, 2004). Widespread valley networks and a smaller number of alluvial fans and possible deltas formed closer to the Noachian/Hesperian transition at $\sim$3.7 Ga (e.g. Malin and Edgett, 2003; Moore and Howard, 2005; Irwin *et al.*, 2005b; Di Achille *et al.*, 2006). Orbital imaging and topography from Mars also show some breached impact craters and intercrater basins with both contributing and outlet valleys, but no dissection of the basin floor (e.g. Forsythe and Zimbelman, 1995; Cabrol and Grin, 1999; see later section on small stem valleys). In many of these candidate palaeolakes, the outlet valley has a sharply defined, V-shaped cross-section, suggesting active downcutting during the last major epoch of erosion (Irwin *et al.*, 2005b). The lower gravity on Mars would make fluvial abrasion of

Megaflooding on Earth and Mars, ed. Devon M. Burr, Paul A. Carling and Victor R. Baker.
Published by Cambridge University Press.

bedrock less effective per unit discharge, so valley incision would depend more on high discharges (such as where a basin outlet is incised), longer-lived favourable paleoclimates, or other weathering processes, relative to similarly immature terrestrial valleys (Irwin *et al.*, 2008).

Whereas abundant evidence has been found for running water on early Mars, an environment perhaps comparable to terrestrial semi-arid or arid regions appears to have been short-lived at best and there is little evidence for catastrophic meteorological floods or high annual runoff production (by terrestrial standards). Fluvial activity did not fully integrate most of the larger or younger highland basins into continuous drainage networks (Carr and Clow, 1981; Grant, 1987; Irwin and Howard, 2002). Where alluvial fans occur, most appear to be dominated by fluvial gravels and sand rather than debris flows, based on distributary channel morphology, development of inverted channels by selective deflation of fines and relationships between contributing area and gradient (Moore and Howard, 2005). Meander chute cutoffs and some boulder transport on the Eberswalde crater delta reflect variable flow conditions (Wood, 2006; Howard *et al.*, 2007) but no evidence for catastrophic runoff has been observed in highland alluvial deposits. The few channels and meanders observed in Martian valley networks suggest dominant discharges of $\sim 10^2$–10^3 m^3 s^{-1} and runoff production up to a few centimetres per day, comparable to terrestrial mean annual floods in semi-arid basins of the same size (Irwin *et al.*, 2005a). Drainage densities usually are measured on the order of 10^{-2} to 10^{-1} km^{-1}, one to two orders of magnitude less than typical terrestrial values, as headwater tributaries are underdeveloped in most areas (e.g. Baker and Partridge, 1986; Tanaka *et al.*, 1998; Grant, 2000; Cabrol and Grin, 2001; Gulick, 2001; Irwin and Howard, 2002; Hynek and Phillips, 2003; Stepinski and Collier, 2004).

In this chapter the role of basin overflow floods in carving large valleys on Mars is explored. Although ice dam failures on early Mars are plausible, obvious glacial landforms are not found next to the outflow channels or the largest valley networks (except perhaps some in high southern latitudes (Head and Pratt, 2001)). Some of the larger outflow channels, which had inferred discharges of 10^6-10^8 m^3 s^{-1} (e.g. Baker, 1982; Carr, 1996; Wilson *et al.*, 2004), either emerge from, or cross-cut, previously enclosed basins. Drawdown of water temporarily stored in those basins may have augmented the subsurface outflow to produce the peak channel discharge, but it is shown below that basin overflows do not eliminate fully the need for substantial groundwater flows. Among branching valley networks, many of the largest stem valleys also emerge from basins, which may have augmented the hydrology and power of these rivers. Morphological evidence for large ($>10^5$ m^3 s^{-1}) floods has been recognised in only a few of these valleys, including Ma'adim Vallis and the Uzboi–Ladon–Morava (ULM) Valles system, which originate full-born from large mid-latitude basins and cross-cut other sizeable basins downstream (Parker, 1985; Grant and Parker, 2002; Irwin *et al.*, 2002, 2004). Lower on a continuum of size, many valleys that emerge from breached (but still partially enclosed) basins were incised more deeply than other nearby valleys around the Noachian/Hesperian transition (Irwin *et al.*, 2005b). Megafloods from basin overflows were rare on early Mars, as they are on Earth, but both larger and smaller floods had significant roles in establishing the longer continuous valleys through the cratered landscape.

11.2 Peak discharge from damburst floods on Mars

The peak discharge (Q_p) of damburst floods can be estimated using the Walder and O'Connor (1997) method (see also O'Connor and Beebee, this volume Chapter 8), which accounts for the influence of released water volume (V_o), water level decline in the source lake (h), breach erosion rate (k), breach width (b) and gravity (g). For application to Mars, the breach geometry and released volume are measurable using the Mars Orbiter Laser Altimeter (MOLA) topographic data returned by the Mars Global Surveyor orbiter (Smith *et al.*, 2001), and $g = 3.81$ m s^{-2}. For large lakes with a high ratio of released volume to breach depth, i.e., $V_o/h^3 > 10^4$, their model and data predict a negligible breach formation time with respect to the time needed to draw down the lake. The breach depth is thus a close approximation of the maximum flow depth, and peak discharge for large lakes becomes primarily a function of breach depth, breach width, and gravity:

$$Q_p = c(b/h)g^{1/2}h^{5/2} \tag{11.1}$$

where the constant c is based on outlet geometry as defined in O'Connor and Beebee (this volume, Table 8.3). The Walder and O'Connor (1997) method has some important limitations for use in Martian channels. It was designed for unconsolidated earthen dams, whereas large basin outlets on Mars may be partially entrenched into more resistant bedrock. Many outlets formed in broad divides, in contrast to the narrow dams for which the model was developed; and some long, low-gradient outlet valleys may have had a backwater effect on discharge. Because the original depth to bedrock and the subsequent bedrock erosion rates are not known, this dam breach model provides only an approximate upper limit to peak discharge for lake overflows where $V_o/h^3 > 10^4$. The width/depth ratio of terrestrial dam breaches typically falls within the range of 1–20 but may be higher in some Martian outflow channels. Enclosed basins along a flow path may have also influenced the peak discharge (negatively or positively) along different reaches of a channel.

For some overflows of very large basins on Mars discussed below, this method yields a peak discharge that is similar (to first order) to discharges estimated from channel dimensions. To produce a discharge of 10^6, 10^7 and $10^8\,m^3\,s^{-1}$ from a compromised dam with a rectangular breach where $c(b/h) = 20$, the model predicts that a breach of at least 60, 150 and 370 m, respectively, must form rapidly with respect to the decline of lake level. Terrestrial damburst floods have not exceeded $\sim 2 \times 10^7\,m^3\,s^{-1}$ for ice dams and $\sim 10^6\,m^3\,s^{-1}$ for earthen dams (O'Connor and Costa, 2004). The geology of Mars may be favourable for this type of overflow, because much of the highland surface likely consists of erodable impact ejecta, impact-fractured bedrock and sedimentary deposits (e.g. Tanaka *et al.*, 1992). Moreover, several of the basins that are identified below were much larger than the enclosed continental basins on Earth, so long-lived overflows may have entrenched deeper channels and valleys.

11.3 Basin influence on Martian outflow channels

Basin topography influenced all sizes of Martian rivers (e.g. De Hon, 1992), and the first considered are at the upper end of this scale. The enormous Martian outflow channels head in a variety of geological settings that favoured rapid groundwater emergence and/or ponding. These include large tectonic canyons (at the heads of Kasei and Simud Valles), basins in cratered terrain (Mawrth), chaotic terrain that formed within a larger pre-existing basin (Ares, Shalbatana), broad volcanic plains that may have received drainage from upslope (Maumee/Vedra/Bahram, Reull, Marte), chaotic terrain that formed its own sizeable basin (Ravi, Maja), and smaller grabens or collapse pits (Athabasca, Mangala) (Table 11.1). Subsurface connections between large basins and distal channel heads also are possible in some cases (e.g. Carr, 2006, pp. 120–121). It may be significant that the largest outflow channels (Simud, Kasei and Ares Valles, discussed in more detail below) all emerge from broad enclosures, whereas smaller ones (Maja, Ravi, Shalbatana, Mangala and Athabasca Valles) originate from more limited collapse pits or grabens. The enhancement of peak discharge by downcutting the margin of chaotic terrain or grabens would be considerably less than from breaching a large tectonic basin.

A suite of characteristics supports a catastrophic flood origin for the outflow channels, including sinuous streamlined walls and islands, anastomosis, lateral and vertical confinement of the eroded surface, and a high ratio of width to depth. The most diagnostic features are the erosional channel bedforms, including scabland, recessional headcuts, deeply scoured areas at constrictions, and longitudinal grooves, which are not found in volcanic channels (e.g. Baker and Milton, 1974; Baker, 1982; Coleman and Baker, this volume Chapter 9). Exposure of most of the channel bed suggests that flows with high sediment concentrations or debris flows, if they occurred initially, gave way to less concentrated flows that could leave the channel bedforms exposed.

11.3.1 Ares and Mawrth Valles

The Ares and Mawrth outflow channels emerge from Margaritifer basin (~0° S, 16° W), a broad confluence plain located along the Chryse Trough between cratered highlands to the southeast and west (Plate 27). This basin received the ULM system from the south, the well-integrated Samara and Parana–Loire Valles that drained the southeastern flank of the Chryse Trough, and many smaller tributaries (Grant, 1987, 2000; Grant and Parker, 2002). An unnamed channel emerging from Hydaspis Chaos (which is also at the head of Tiu Valles) also breached the northwestern divide and drained into Ares Vallis before it was fully incised (point 3 in Plate 28; streamlined obstructions in the channel bed indicate this flow direction). The basin contains smooth plains punctuated by inliers of cratered uplands (Edgett and Parker, 1997), the Margaritifer and Iani chaotic terrains, and the head of Ares Vallis (Grant, 1987; Grant and Parker, 2002). The topography of the Margaritifer basin, prior to incision of its divide by the outflow channels, suggests that an overflow may have been the previously unrecognised source for Mawrth Vallis (Plate 28), and that some of the erosive power of Ares Vallis came from draining the basin.

If the unnamed inlet from Hydaspis Chaos and the two outlets at and to the west of Ares Vallis are filled to the top of the channel banks, the reconstructed basin divide would allow ponding to about –2000 m elevation (red outline in Plate 27, approximate measurements relative to the MOLA datum, Smith *et al.*, 2001) at the Late Hesperian time of the Ares outflow (Tanaka, 1997). The unnamed channel entered Margaritifer basin across a higher part of the divide at –1800 m (point 3 in Plate 28), and there is an undissected pass to the west of Ares Vallis at –1900 m (between points 2 and 3 in Plate 28). The outlets incised the two lowest points in the divide at about –2000 m (points 2 and 4 in Plate 28). Ares also dissects another divide at –2000 m (points 5 and 6 in Plate 28), headward within Margaritifer basin near the contributing outlet of Aram Chaos (a highly modified impact crater).

The floor of Ares Vallis is incised to –3200 m at the head breach and –3700 m at the downstream breach. Most of the decline of the channel bed occurs within the upper 120 km of its ~1500 km length, and the downstream gradient is flat to slightly climbing (by ~100 m) north of that descending reach. The water must have been hundreds of metres deep to allow erosive flow along this channel bed, unless the downstream incline were due to uplift or secondary infilling by a later outflow of Simud Valles. Much of the water supplied to Ares Vallis came from sub-basins

Table 11.1. Outflow channels and large valleys on Mars

Valley	Head location		Source type[a]	Notes
Al Qahira	22° S	160° E	T	Valley widens downslope, multiple major tributaries.
Ares	0° N	17° W	C/B	Heads in Iani Chaos within Margaritifer basin (the terminal basin for narrower ULM Valles). Tributaries come from Hydaspis and Aram Chaotes.
Athabasca	10° N	157° E	G	Source graben on SE flank of Elysium Mons.
Auqukuh	25° N	61° E	B	The larger head basin (400 km crater) overflowed at <740 m and the smaller intermediate basin (110 km crater) overflowed at ~470 m.
Bahram	20° N	59° W	C/P	Possibly related to ponding on Lunae Planum during the Maja/Maumee/Vedra Valles outflow.
Dao/Hamarkhis	33° S	95° E	C	These valleys have discontinuous collapse pits near their heads.
Huo Hsing	26° N	67° E	B	The head basin, an irregular collection of several sub-basins, overflowed at <500 m.
Kasei	15° N	75° W	Y	Echus Chasma.
Licus	4° S	127° E	T/B	A 30 km crater within a broader partial topographic enclosure is the source basin for Licus Vallis. Valley widens downslope.
Loire	21° S	14° W	B	Parana Valles flow into a basin that was the source for the much larger Loire Vallis.
Ma'adim	28° S	178° E	B	Ma'adim Valles originate at two spill points of a large mid-latitude basin.
Maja	2° S	61° W	C/Y	Juventae Chasma.
Mamers	29° N	26° E	B	Heads in a large intercrater basin in Arabia Terra, supplied by Naktong/Scamander and Indus Valles.
Mangala	18° S	149° W	G	Heads in a graben of Memnonia Fossae.
Marti	5° N	177° E	P	Related to shallow ponding on Elysium Planitia.
Maumee/Vedra	18° N	56° W	C/P	Related to ponding on Lunae Planum during the Maja/Maumee/Vedra Valles outflow.
Mawrth	17° N	12° W	B	Heads at the margin of Margaritifer basin.
Nanedi	4° N	51° W	B & P	Northern branch heads in a breached crater at >300 m, southern branch heads on volcanic plain near Shalbatana Vallis head. Likely outflow with smaller groundwater-fed tributaries.
Nirgal	27° S	46° W	P	Likely outflow on plains NW of valley head with smaller groundwater-fed tributaries.
Ravi	1° S	43° W	C	Aromatum chaos.
Reull	37° S	113° E	B/P	The valley has three major segments that may have diverse origins. The upper one originates in an enclosed basin south of Hesperia Planum.
Samara	32° S	8° W	T	Multiple major tributaries.
Shalbatana	0° N	45° W	C	Heads in a 115 km crater that has been modified into deep chaotic terrain.
Simud	5° S	36° W	Y	Originated in Valles Marineris.
Tiu	3° S	27° W	C/Y	Hydaspis chaos, possibly sourced from Valles Marineris.
ULM	35° S	35° W	B	Uzboi–Ladon–Morava system heads in Argyre basin, incising its rim.

[a]C, chaotic terrain; B, highland crater or intercrater basin; Y, canyon; G, graben; T, area of topographic convergence; P, volcanic plain. Where two source types are given, the ultimate upstream source is followed by the terrain type immediately at the valley head. Many highland basins have tributaries that are much smaller than the basin outlet valley. Some basins (Loire, Ma'adim, Simud) contain chaotic terrains that may or may not be a water source.

Source: Modified from Table 3.1 of Baker (1982)

(which also contain the chaotic terrains) of Margaritifer basin south of the divide at points 5 and 6 in Plate 28. Komatsu and Baker (1997) calculated a maximum possible discharge of 5.7×10^8 m^3 s^{-1} by assuming that the channel was full to the banks of a constricted reach. However, the actual depth at peak flow, the channel floor elevation at that time, and the slope of the flow surface were difficult to constrain and remain so with the advent of MOLA data. As in terrestrial bedrock channels, bankfull conditions cannot be assumed, which may have led to overestimation of discharge by an order of magnitude in some cases (Wilson *et al.*, 2004).

Tanaka (1997) determined a Late Noachian age for Mawrth Vallis using crater counts, making it the oldest outflow channel. Its head occurs at −1880 m on the northern divide of the Margaritifer basin (Plates 27 and 28). The location is not the low point of the reconstructed divide, however, given the three lower points at and just to the west of the main Ares Vallis outlet (points 2 and 4, and a pass southwest of point 2 in Plate 28) but no other surface or subsurface water source for Mawrth is evident. The channel head does not have chaotic terrain or glacial landforms that would favour the ice damburst or groundwater megaflood alternative hypotheses, and a water source immediately southwest of the Mawrth head would drain to Ares rather than to Mawrth Vallis in the current topography. One possible explanation is that some tilting of the Chryse Trough occurred with the growth of Tharsis during the Late Noachian Epoch, lowering the area of Ares Vallis by ~100–200 m with respect to Mawrth, which originates ~800 km to the northeast. Such a tilt would leave the Mawrth outlet abandoned and would favour development of Ares Vallis a few hundred million years later. This scenario is speculative and difficult to test with available data, and although the sense of tilting in this region is consistent with a geophysical model result, the timing and magnitude of the tilt are not known precisely (Phillips *et al.*, 2001). The ULM system and other valleys debouched into Margaritifer basin during the Noachian Period, providing a large contemporary water source that may be related. In the following section two other examples are discussed, where outlet valleys that apparently conveyed large floods head at the breached rims of large basins but the valleys did not emerge at the modern low point of their basin divides.

11.3.2 Kasei Valles

Kasei Valles are the longest and largest continuous channel system on Mars (Baker, 1982; Carr, 1996, p. 63). The anastomosing channels head in the tectonic Echus Chasma and drained it to the north and east through broad pre-existing valleys of Sacra Fossae, which had been incised by other means into the Lunae Planum high volcanic plains (Sharp and Malin, 1975; Williams *et al.*, 2000) (Figure 11.1). These irregular scarps are streamlined by the flow only where the channel developed adjacent to them, whereas broad, undissected plains separate the incised channel bed from the scarps in other areas (e.g. to the north of North Kasei Vallis, Figure 11.1).

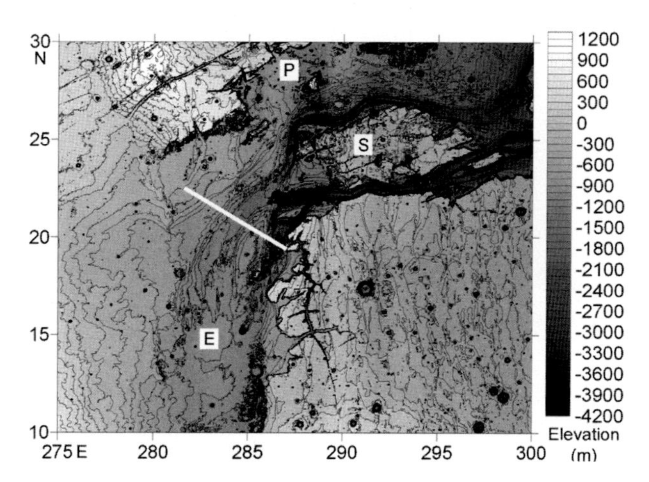

Figure 11.1. The head region of Kasei Valles in 128 pixel/degree gridded MOLA topography, 100 m contour interval. The white bar marks the approximate overflow area of Echus Chasma (E). For scale, this line is 330 km long. Sacra Fossae troughs occur in the region of (S), and this older topography apparently conveyed water to the northeast, so the basin need not have filled to the plateau surface. Undissected plains (P) separate the channel from scarps to the north.

Before the head of Kasei Valles was incised, the northern floor of Echus Chasma stood at ~−550 m elevation, forming a barrier to flow from the deeper southern part of the canyon (Robinson and Tanaka, 1990; Figure 11.1). This broad plain of northern Echus Chasma apparently could not confine the large overflow into a narrow channel, so Kasei Valles developed in two coaeval branches, of which the southern one incised more deeply and advanced farther to the south. A possible second stage of flow would have followed only the southern route (Williams *et al.*, 2000). As with Ares Vallis, peak discharge estimates depend on the water depth and width but vary from ~8×10^4 (Williams *et al.*, 2000) to ~10^9 m^3 s^{-1} (Robinson and Tanaka, 1990). The volume eroded to form the channel is only twice the water volume that could be impounded behind the barrier, so the lake must have been actively replenished during the outflow. Thus, although Mawrth Vallis may have formed entirely by a palaeolake overflow, and Ares Vallis combined a lake overflow with groundwater discharge, Kasei Valles appear to have been dominated by flow into and through Echus Chasma from the south. Alternatively, Coleman and Baker (this volume Chapter 9) suggest that flow may have entered the basin from the Tharsis plateau to the west. Incision of this barrier lowered the lake level by ~250 m, augmenting the discharge

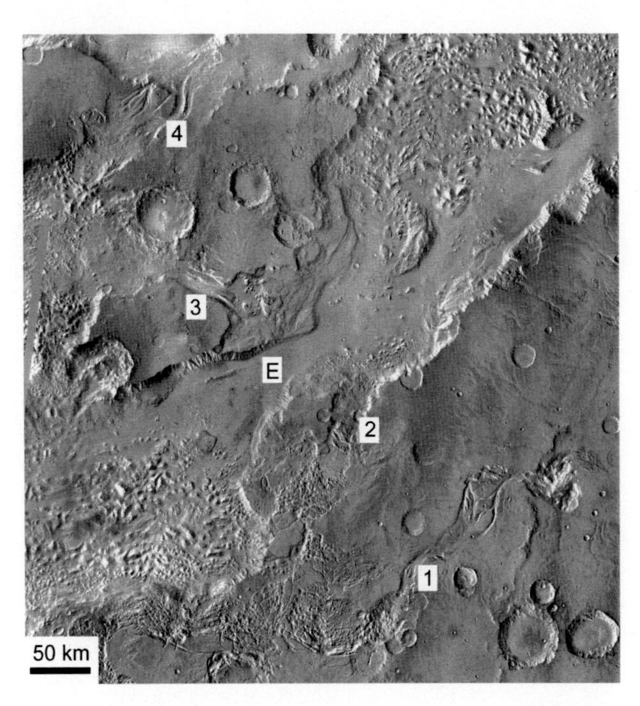

Figure 11.2. Possible outlet channels for Valles Marineris, numbered 1–4, and the main outlet through Eos Chasma. Flow direction was toward the east. THEMIS global mosaic, centred at 13.1° S, 40.7° W.

from underground sources but the contribution of the dam incision to the peak discharge is difficult to constrain without some knowledge of how quickly the barrier failed.

11.3.3 Simud Valles

Simud Valles are the outlet channel for Valles Marineris (Plate 27), the largest and deepest tectonic basin on Mars. Much of the eastern canyon wall and floor has collapsed to form chaotic terrain (Figure 11.2) but some older floor sections have longitudinal grooves and streamlined landforms that record past incision by floods. The history of basin flooding from Valles Marineris is difficult to interpret; because tectonic extension, mass wasting, collapse, and flooding have all been highly effective in the region; and their relative roles in creating and modifying the canyons have not been quantified.

Eos Chasma is the eastern outlet to Valles Marineris (E in Figure 11.2). Longitudinal grooves on the canyon floor at –3600 to –3900 m elevation occur at higher levels than the distal Simud Valles floor and suggest a through-flowing stream. Small outflow channels also occur at several locations high on the plateau surface to the north and south of Eos and Capri Chasmata (Lucchitta *et al.*, 1992; Coleman *et al.*, 2007) and four of these appear to radiate from the canyons (Figure 11.2). It is not clear, however, whether these are former surface outlets to a smaller ancestral Valles Marineris, or if they represent large outflows from the subsurface before the canyon grew to its full extent, incorporating the source areas. Coleman *et al.* (2007) favour the former explanation, based on large estimated discharges of 10^6–$>10^8$ $m^3\ s^{-1}$ followed by inferred stability of the channel heads. Herein the tops of the channel sidewalls are used for an estimate of the maximum possible water surface in Valles Marineris, although precise elevations of the past spillways usually are difficult to obtain. The unnamed Channel 1 (Figure 11.2) incised a divide that originally stood at ~1050 m. Channels 2 and 3 (the latter is named Daga Vallis) dissect a plateau that may have been as high as 1150 m. Channel 4 (Columbia Valles) emerges at ~1050–1100 m to the north of this zone of collapse. These four potential outlets of Valles Marineris may have overflowed at ~1100 m but all four also have chaotic terrains at their heads that have been incorporated within the canyon system. Moreover, if these outlets were reconstructed to ~1100 m, Valles Marineris would not overflow at these points but through a broad pass at 900 m to the Nirgal Vallis drainage basin to the southeast. The geometry of Valles Marineris must have been very different at the time of these floods, if they are overflows (Coleman *et al.*, 2007), and there is no unique evidence that Simud Valles drained a palaeolake in Valles Marineris of >5 km depth, however a smaller damburst flood through Eos Chasma is consistent with catastrophic flood bedforms there and to the north. As with other outflow channels, the peak discharge is difficult to constrain at Eos Chasma due to poor constraints on the incision rate into bedrock and (in this case) the depth of the breach.

11.4 Intermediate-scale basin overflows on Mars

Two large valleys are described here that emerge from enclosed basins and preserve compelling evidence of megafloods. These valleys are distinguished from the typical outflow channels by their smaller (but still substantial) size, relatively deep incision per unit width, and more localised exposure of erosional bedforms. Experimental work by Shepherd and Schumm (1974) showed that longitudinal grooves in bedrock channels tend to become less well-expressed as the central channel is deeply incised with time, so longer-lived flows, late-stage sedimentary infilling, and subsequent modification may account for the relatively feature-poor valley floors discussed below. However, more persistent features of these valleys indicate one or more high-magnitude water discharges.

11.4.1 Uzboi–Ladon–Morava system

The segmented Uzboi–Ladon–Morava (ULM) mesoscale outflow system emerges at full size from the northern margin of Argyre basin and traverses northward along the southwestern flank of the Chryse Trough in Margaritifer Terra (Saunders, 1979; Baker, 1982; Phillips

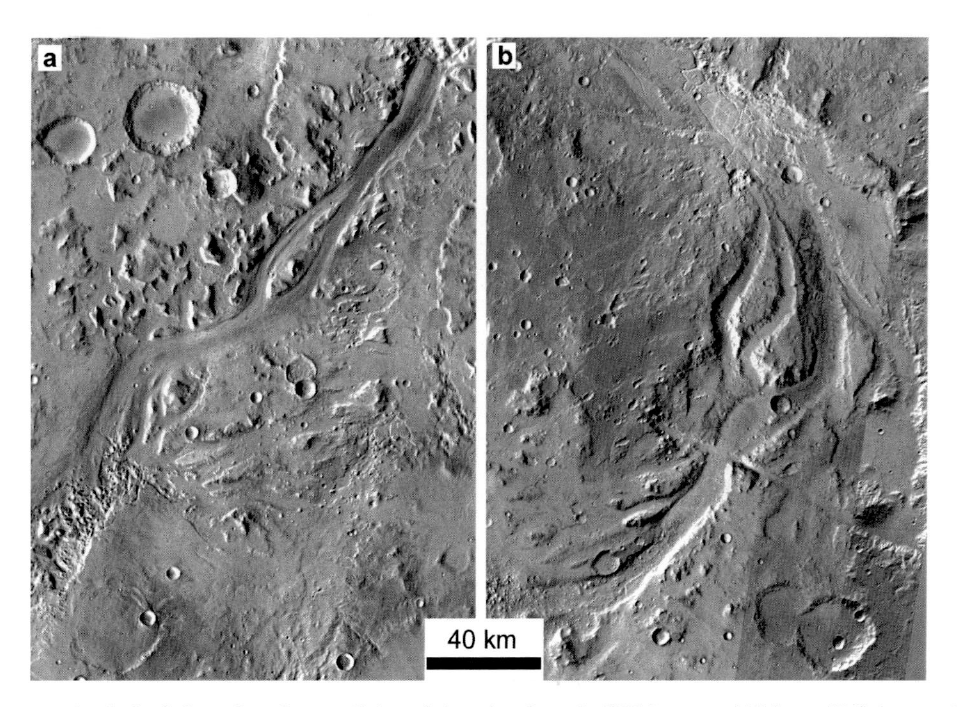

Figure 11.3. Anastomosing incised channels and streamlining of obstacles along the ULM system. (a) Morava Valles, centred at 12.8° S, 23.6° W. (b) Ladon Valles, centred at 22.6° S, 29.3° W. Morava Valles occur north and downstream of Ladon. (See also Grant and Parker (2002).) These areas are located immediately west of M and L, respectively, in Plate 27. (THEMIS global mosaic supplemented with recent images.)

et al., 2001) (Plate 27). Several large valley systems that head even farther to the south may have fed the drainage through Argyre (Parker, 1985, 1994), resulting in a total watershed of more than $\sim 11 \times 10^6 \, km^2$, or about 9% of Mars (Banerdt, 2000; Phillips *et al.*, 2001). The terrain adjacent to the Uzboi Vallis head stands a few hundred metres higher than the low point of the Argyre rim, which occurs at ~320 m elevation to the east of the Uzboi Vallis outlet (Plate 27). It is uncertain whether Uzboi exploited a local low point in the Argyre rim or if the basin was later tilted toward the east, perhaps with uplift of the nearby Thaumasia highlands to the northwest. The Noachian craters Bond and Hale overlie portions of proximal Uzboi Vallis and have made any spillway from Argyre indistinct, although the crater ejecta do not completely fill the older basin outlet.

The ULM system is characterised by deeply incised trunk segments of 15–20 km width, separated by depositional plains that partially fill the Early Noachian Holden and Ladon multi-ringed impact basins (Saunders, 1979; Schultz *et al.*, 1982; Frey, 2003). From the incised margin of Argyre, ULM descends >1800 m in a series of longitudinal steps from south to north. Nirgal Vallis joins Uzboi Vallis more than halfway between Argyre and Holden basins at an elevation of around –760 m. Uzboi Vallis then drops to approximately –1275 m to the south of Holden crater (154 km in diameter) where the crater ejecta partly fill the valley. Before it was interrupted by the Holden crater impact, ULM dropped to the floor of the multi-ringed Holden basin at an elevation of –1700 m to –1800 m, before declining to –2000 m through Ladon Valles and out onto the floor of Ladon basin (Figure 11.3b). The system finally drops an additional 480 m through Morava Valles to the floor of Margaritifer basin (Figure 11.3a), described earlier, at an elevation close to –2480 m.

Geological history of the ULM system Geological mapping in Margaritifer Terra constrains the general timing of activity along ULM with respect to regional geomorphic events (Saunders, 1979; Hodges, 1980; Parker, 1985, 1994; Scott and Tanaka, 1986; Grant, 1987, 2000; Grant and Parker, 2002). As summarised by Grant (2000) and Grant and Parker (2002), the Early Noachian Argyre basin and the degraded Ladon and Holden multi-ringed impact basins (Schultz *et al.*, 1982) are the oldest features crossed by the ULM. These basins imparted considerable structural and topographic influence on the course and character of the ULM drainage. Generally, incised segments of the ULM system are radial to the basin rims/rings and are separated by intervening deposition within basin centres (Schultz *et al.*, 1982; Boothroyd, 1983; Parker, 1985; Figure 11.4). These ancient impacts were followed by evolution of the diverse cratered upland surface that included three general resurfacing events between the Early and Middle-to-Late Noachian Epochs. A fourth, more localised resurfacing

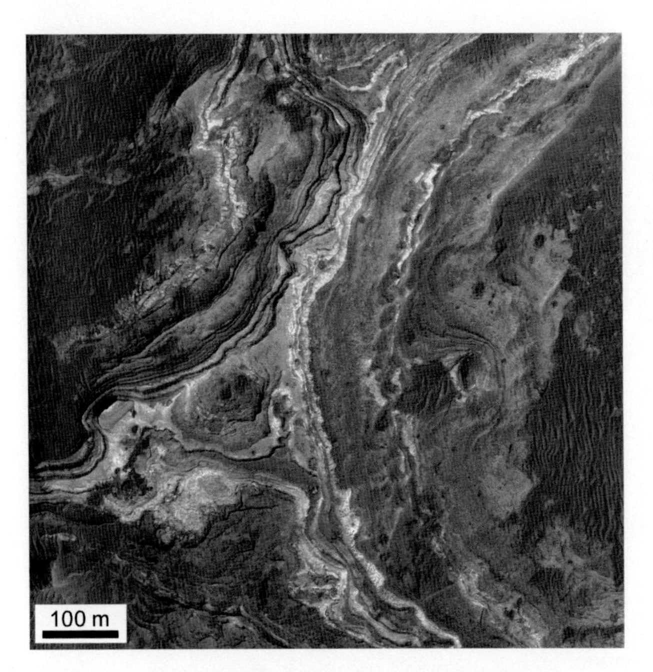

Figure 11.4. Extensive layering exposed in Ladon Valles where it debouches to Ladon basin, near point L in Plate 27. (Subset of Mars Reconnaissance Orbiter High Resolution Imaging Science Experiment (HiRISE) PSP_006637_1590, centred at 20.83° S, 29.68° W.)

event of Early-to-Middle Hesperian age emplaced materials that embay many valleys and channels (Grant, 1987). The ULM system dissects the surfaces associated with the third regional resurfacing event but it is usually embayed by materials emplaced during the fourth and final resurfacing event.

Holden crater likely formed during the Late Noachian Epoch (Scott and Tanaka, 1986; Pondrelli *et al.*, 2005) and interrupted the older ULM system (Figure 11.5a). Following an overflow of the Holden rim by water ponded in Uzboi Vallis (which likely came from the contemporary Nirgal Vallis tributary), the crater floor at –2200 to –2300 m became the terminal basin for Uzboi and Nirgal Valles (Figure 11.5) (Grant and Parker, 2002; Irwin *et al.*, 2005b; Pondrelli *et al.*, 2005; Grant *et al.*, 2008). There is no outlet breach of the crater rim that would allow continued flow through the remainder of the ULM system. The bulk of the through-flowing drainage along the ULM occurred during the Late Noachian but flow along some segments of the system may have reoccurred or persisted into at least the Early and perhaps the Late Hesperian Epochs (Rotto and Tanaka, 1995; Williams *et al.*, 2005). Collapse of Margaritifer and Iani Chaotes disrupted ULM deposits in Margaritifer basin and their age supports cessation of channel and valley formation by the Late Hesperian. Multiple studies indicate that these chaos regions served as the source for Ares Vallis (Carr, 1979; Carr and Clow, 1981; Rotto and Tanaka, 1995; Grant and Parker, 2002) but it appears that the bulk of the ULM discharge into the area occurred prior to onset of widespread collapse.

Indications of large floods in the ULM system Multiple large-discharge events shaped the ULM system, as indicated by several erosional and depositional forms. The size of Uzboi Vallis relative to other valleys in the area is difficult to explain without a basin overflow, as its tributary network is very poorly developed. The valley emerges from Argyre basin with its full width of 15–20 km. The Holden and Ladon basins have multiple, anabranching outlet valleys, so the influx of water greatly exceeded the evaporation potential from these large basins, and the initial basin-rim topography could not confine the overflowing water into a single channel (Figure 11.3). Some small obstructions to flow are streamlined. Hanging relationships between side channels and the main stem suggest that multiple overflow points remained active until the central one was incised deeply enough to confine the entire flow at that later stage (Figure 11.3). There are a number of possible terraces along Uzboi Vallis, at least five distinct terraces along Ladon Valles (Boothroyd, 1983; Parker, 1985; Grant, 1987; Grant and Parker, 2002) and a complex set of deep distributaries and terraces at varying elevations where Morava Valles enter Margaritifer basin (Figure 11.3a). In addition, the floor of Ladon basin is relatively flat (maximum relief 0.27 km over 350 km) and coarsely layered, suggesting that multiple depositional events contributed to infilling (Figure 11.4). Evidence for comparable low depositional relief across Holden basin was destroyed by formation of Holden crater and adjacent chaotic terrain to the northeast. Despite this evidence, however, it is not known whether multiple flows were separated widely in time or occurred over a relatively short interval.

Profiles along and across Ladon and Morava Valles provide estimates of the channel floor gradient and cross-section below possible fluvial terraces. A combination of the unit-balanced continuity and semi-empirical Manning equations can be used to calculate the formative discharge Q (m^3 s^{-1}) of a channel:

$$Q = HWV = H^{5/3}W(g_{\text{me}}S)^{1/2}n^{-1}, \tag{11.2}$$

where H is mean flow depth (m) for channels with high width/depth ratio, W is channel width (m), V is velocity (m s^{-1}), g_{me} is the ratio of Martian to terrestrial gravity (0.38), S is gradient, and n is the Manning roughness (s m$^{-1/3}$; the square root of the gravity ratio is sometimes included in n, increasing the Manning roughness by a factor of 1.62 for Mars relative to an equivalent terrestrial stream, but is separate in Equation (11.2)). A high degree of uncertainty attends such calculations, because the terraces may not reflect past flow depths, the valley floor gradient may

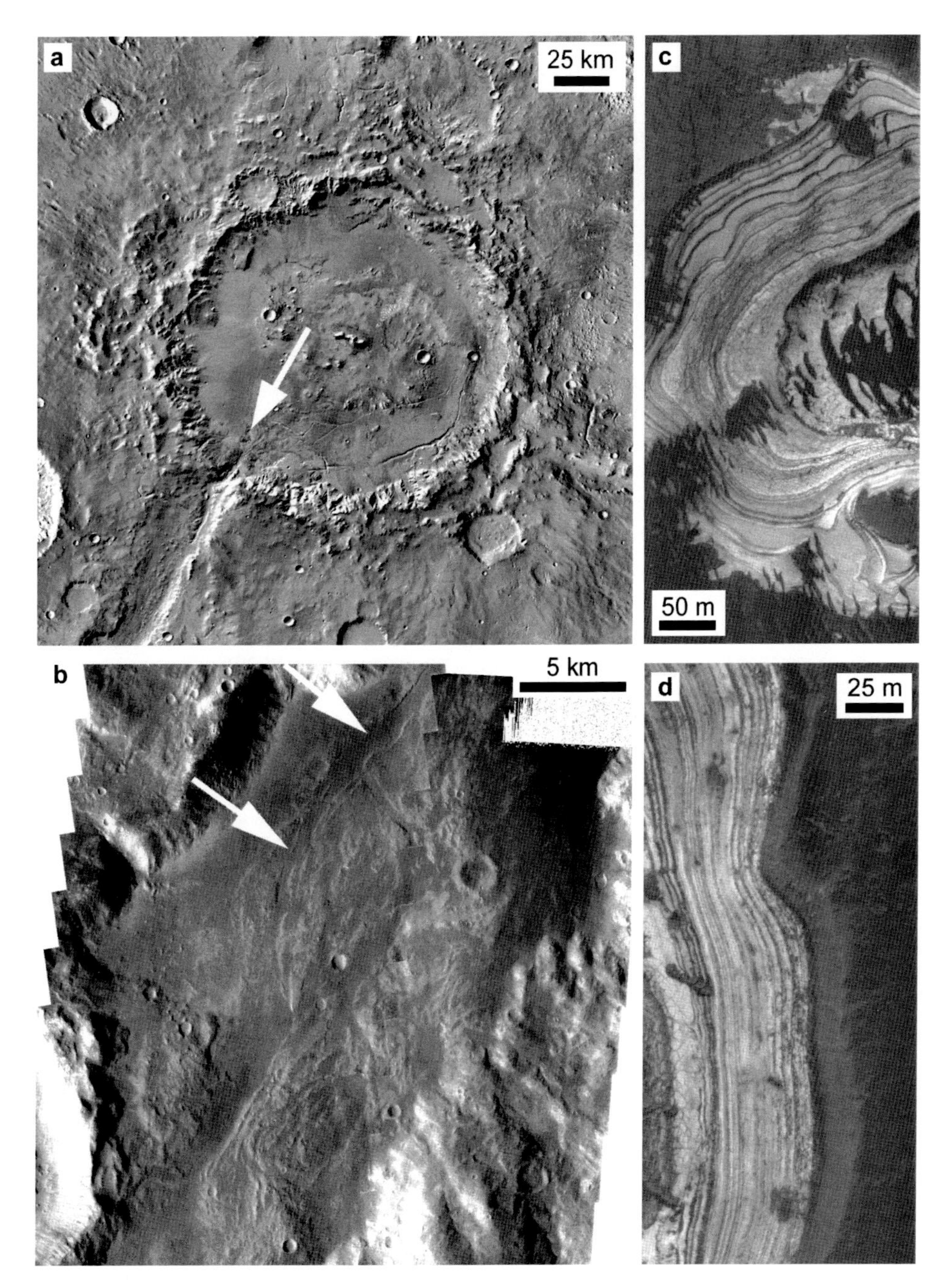

Figure 11.5. (a) Holden crater, centred at 26.0° S, 34.2° W, with Uzboi Vallis entering the crater from the SW. (THEMIS global mosaic 2.0.) Arrow shows the approximate location of (c) and (d). (b) A channel is partially exposed on the floor of Uzboi Vallis (arrows) and may be related to later flow from the Nirgal Vallis tributary. (THEMIS V15891005 and V06643002, centred at 29.4° S, 37.0° W.) (c, d) Light-toned layered deposits on the southwestern floor of Holden crater may reflect an earlier low-energy depositional setting. (Subsets of HiRISE PSP_002088_1530.)

not parallel the water surface during the flow and roughness estimates are imprecise. Hence, derived values for flow velocity and discharge are only first-order estimates and they may not reflect an earlier peak flood that was not confined within the deepest valley.

As summarised by Grant and Parker (2002), measured slopes along the channel thalwegs average 0.0008 (0.046°) and 0.0006 (0.034°) for Ladon and Morava Valles, respectively. The flow depth of 28–88 m was estimated using the difference in elevation from the channel bottom to the lowest detectable terrace. When combined with a Manning roughness value of 0.05, these channel slopes and flow depths yield mean flow velocities of 3.2–6.7 m s^{-1} and 2.5–5.3 m s^{-1} for Ladon and Morava

Valles, respectively (Grant and Parker (2002) used $n = 0.04$ with a slightly different form of the discharge equation (Carr, 1979; Komar, 1979; Baker, 1982); the inclusion of an explicit gravity term in Equation (11.2) adjusts this to 0.05). Channel width of 1.6 to 6.4 km was measured at the level of the inferred terraces used for estimating flow depth and cross-sections were assumed to be rectangular. Resulting discharge estimates vary but most are between 150 000 $m^3\,s^{-1}$ and 450 000 $m^3\,s^{-1}$ (Grant and Parker, 2002). Such discharge rates are 5–10 times higher than discharge from the Mississippi River (Komar, 1979) and on the lower end of discharge from the Channeled Scabland (Baker and Nummedal, 1978; Baker, 1982). For comparison, estimated Ares Vallis discharge ranges between 1.5×10^6 and $7 \times 10^7\,m^3\,s^{-1}$ (Carr, 1979; Baker, 1982), with maximum possible discharge of $5.7 \times 10^8\,m^3\,s^{-1}$ (Komatsu and Baker, 1997).

Whereas the ultimate source and the cause for water release through the ULM are ambiguous, it must have been voluminous, and an overflow of Argyre basin is the only adequate source with observational support. Moreover, broad similarity in the estimated discharge from Ladon and Morava Valles, coarse layering in Ladon and Holden basin fill and graded alluvium from the mouth of Ladon to the head of Morava Valles argue that all segments were simultaneously active. Some discharge may have occurred along sections of the ULM system after Holden crater formed (e.g. Grant and Parker, 2002; Pondrelli *et al.*, 2005) but there is no evidence for significant through-flowing discharge after that impact.

Holden crater and late-stage flooding along ULM The formation of Holden crater (26.0° S, 34.2° W) interrupted the ULM system and blocked/destroyed the lower reaches of Uzboi Vallis, where it formerly debouched into Holden basin (Figure 11.5a). The crater initially drained internally with no outside contributions of surface water. Late Noachian fluvial erosion formed deep alcoves and gullies primarily in high-standing sections of the crater rim, which were the source of the prominent alluvial fans and a bajada along the western wall (Moore and Howard, 2005). This alluvium overlies at least a 150-m section of light-toned layers (Malin and Edgett, 2000; Grant and Parker, 2002; Grant *et al.*, 2008), which exhibit bedding of <1 m thickness, lateral continuity over distances of kilometres, a widespread occurrence within the crater, spectral signatures indicating phyllosilicates, and poor erosional resistance to wind (Grant *et al.*, 2008). These characteristics suggest an origin as lacustrine or distal alluvial deposits (Grant *et al.*, 2008) and contrast with the coarser layering in Ladon basin (Figure 11.4). A smaller channel emerging from Nirgal Vallis into the middle section of Uzboi Vallis (30° S, 38° W) indicates that the Nirgal system was also active after Holden formed (Figure 11.5b) but this small channel is not continuous to the breach into Holden crater, possibly because of the positive northward gradient on the Holden ejecta deposited in Uzboi. Discharge from Nirgal Vallis was relatively small by comparison to earlier discharge along the ULM (approximately 4800 $m^3\,s^{-1}$, see Irwin *et al.*, 2005a) but it ponded in Uzboi Vallis against the southwest rim of Holden and eventually fully breached the crater rim (Grant and Parker, 2002; Pondrelli *et al.*, 2005). Boulder-rich, poorly weathered alluvial deposits cap the light-toned, layered deposits in the southwestern portion of the crater (Parker, 1985; Grant and Parker, 2002; Pondrelli *et al.*, 2005; Moore and Howard, 2005; Irwin *et al.*, 2005b; Grant *et al.*, 2008), feeding a late-stage lake with a depth of ~50–140 m (Grant and Parker, 2002; Irwin *et al.*, 2005b, Grant *et al.*, 2008). Multiple alluvial wedges at varying distance from the crater entrance breach and inward fining of deposits imply that multiple or perhaps variable discharges occurred. Layered deposits in Holden fall exclusively below the −1800 m floor of Holden basin, so late discharge into Holden did not continue as surface or subsurface flow along the ULM system.

11.4.2 Ma'adim Vallis

Ma'adim Vallis is similar in many respects to the ULM system, with a wide and deep stem valley, multiple breached basins along its length, local anastomosis and longitudinal grooves, a Late Noachian age and a source at the incised divide of a large enclosed basin to the south. The valley is 920 km long, 8–25 km wide and up to 1.5 km deep, making it one of the largest valley network stems on Mars (Plate 29). The head basin for Ma'adim Vallis, which Irwin *et al.* (2004) informally called the Eridania basin, is ~$10^6\,km^2$ in area and contains multiple sub-basins with atypically concave floors (Plate 29b). The main valley head is incised ~250 m into and ~25 km south of the Eridania basin divide. The steep upper reach of the valley descends northward from the valley head at 950 m elevation to the floor of the intermediate basin, an ancient impact basin of 500 km diameter (Plate 29a). The intermediate reach of the valley is incised only ~400 m into the floor of this basin, where the floor widens from 3 km to 5 km as the gradient declines. Along this reach a similarly wide, anastomosing tributary joins the main valley, having originated at a second overflow point to the west of the main valley head. A steep, narrow knickpoint separates the intermediate reach from the lower reach, which is the longest, widest and deepest section. It breaches fully the intermediate basin, allowing contiguous drainage to the north. The lower reach has terraces that bound an interior valley or channel, which migrates to the outside of bends and has the same width as the intermediate reach. Detailed topography shows a step-pool sequence in the lower reach, with some shallow enclosed depressions and an increase in the cross-sectional concavity of the floor at the steps. Ma'adim exits the intermediate basin and continues down the cratered slope of the crustal dichotomy boundary into Gusev crater (160 km in diameter), which it enters at a level ~400 m above the crater floor (Irwin *et al.*, 2004). A collection of mesas are on grade with the lower reach thalweg and may be the remains of a delta (Schneeberger, 1989; Grin and Cabrol, 1997) from late-stage flow. The northern rim of Gusev is also breached at this level, so the flow may have continued through the crater, but this is difficult to evaluate because younger chaotic terrain has developed north of the gap (Irwin *et al.*, 2004). Gusev crater is the landing site for the Spirit Mars Exploration Rover, which found that basaltic lava had resurfaced much of the crater floor and buried any possible lake beds but that some aqueous alternation of older volcanic rocks had occurred in the Columbia Hills (e.g. Arvidson *et al.*, 2006). Milam *et al.* (2003) had predicted that the landing site would be a volcanic plain but this interpretation was not held widely to be certain (e.g. Grin and Cabrol, 1997; Cabrol *et al.*, 2003).

Geological history of the Ma'adim Vallis region Ma'adim Vallis dissects cratered highland terrain of Early to Middle Noachian age (Greeley and Guest, 1987). Erosion of these craters was slow by terrestrial standards (e.g. Carr, 1996; Craddock *et al.*, 1997), and thc older, larger impact basins still maintained an enclosed drainage pattern by the Late Noachian Epoch (~3.8–3.7 Ga). The Eridania basin consists of six major and several smaller sub-basins of impact origin, all of which were heavily modified during the Noachian. These craters have an unusual morphology, however, as their interior plains are highly concave-up and have up to a kilometre of internal relief (Plate 29b), in contrast to the typical flat floors of other Noachian craters. Two other nearby basins, Newton and Copernicus, have similar concave floors but no apparent surface connection to the Eridania basin. These concave depressions contain the lowest areas on the plateau of Terra Sirenum and Terra Cimmeria, and they may have tapped a deep groundwater table. The basin floor has a much lower gradient above the ~700 m contour, and valley networks are better developed above this level than on the steeper gradients at lower levels in the concave sub-basins. It is significant that the concave break in slope at the margins of these concave depressions is not dissected. Irwin *et al.* (2002, 2004) interpreted these features as evidence of long-term flooding below ~700 m.

During the Late Noachian epoch, the lake rose and overflowed to the north, filling the more heavily degraded intermediate basin with ~60 000 km^3 of water, a minimum necessary to overflow it and incise the lower reach of Ma'adim Vallis. At this point, the Eridania basin may have had an area of ~1.1×10^6 km^2 and a volume of 5.6×10^5 km^3, although it is uncertain to what degree the Noachian topography has changed, particularly with the development of Tharsis. An outlet at 1180 m was not exploited, whereas the overflow to the north occurred at 1200 to 1250 m, so two possible sub-basins to the southwest may have been topographically isolated. Incision of Ma'adim Vallis lowered the lake level by up to ~250 m, particularly in the main sub-basin next to the valley heads, releasing 110 000–250 000 km^3 of water (Irwin *et al.*, 2004). After the overflow, Hesperian airfall deposits blanketed the southern part of the region to ~400 m thickness, and a late stage of valley activity dissected these deposits (Grant and Schultz, 1990). The tributary network to Ma'adim Vallis developed in the intermediate basin and on the northward cratered slope near Gusev, but these tributaries have concave-up longitudinal profiles and are poorly graded to the Ma'adim Vallis floor (Irwin *et al.*, 2004).

Incision history of Ma'adim Vallis Irwin *et al.* (2004) reconstructed the Noachian topography before Ma'adim Vallis was incised by filling the MOLA topographic grid to the top of the valley walls. Overland flow from the Eridania basin would follow the Ma'adim Vallis course exactly. Impact crater topography guided overland flow into the intermediate basin, which filled and overflowed at its spill point. Older drainage basins in cratered terrain along the crustal dichotomy boundary guided the flow toward Gusev crater, whose rim was compromised by two overlapping smaller craters at the site of the entrance breach. Incision of the lower reach to the terrace level occurred first to allow drainage of the intermediate basin. This stage of flow likely included the peak discharge, as water drained from the intermediate basin at the same time that it was being rapidly replenished from the Eridania basin. Finally, the intermediate and upper reaches of the valley were incised to their present depth and the lower reach was entrenched below the terrace level. The base level of the valley at −1550 m was likely maintained by a lake in Gusev crater, whose floor occurs at −1900 m. The grade of the interior channel of the lower reach is nearly constant at 0.0012, except for small steps in the longitudinal profile.

Indications of large floods in Ma'adim Vallis Several lines of evidence suggest that Ma'adim Vallis conveyed one or more large floods, although unique evidence for multiple overflows has not been found (Irwin *et al.*, 2004). (1) The valley originates full-born at two gaps in the drainage divide of the Eridania basin, with a width that is similar to downstream channel dimensions. (2) Although the western tributary appears at a single point, it follows an anastomosing course that merges with the main valley at three locations (Plate 29a). The northern two are hanging

Figure 11.6. The floor of Ma'adim Vallis has medial longitudinal ridges in places, suggesting that the flow occupied the whole valley floor rather than a narrow interior channel. (THEMIS daytime IR image I17194002, centred at 24.0° S, 176.1° E.)

with respect to the main valley floor, whereas the southern branch has a well-graded confluence. These features are common where topography is unable to confine a large flood. (3) The inner channel maintains a nearly constant width of ~5 km through crater rims and other obstructions to flow. (4) A medial sedimentary bar of 200 m height occurs on the intermediate reach where the western tributary joins the main valley. (5) Longitudinal ridges occur on parts of the intermediate reach, suggesting that the flow occupied the entire width of the valley floor (Figure 11.6). (6) Small tributaries to Ma'adim Vallis have convex-up longitudinal profiles, which are common where a river entrenches more rapidly than its tributaries. (7) Interior channel discharge confined by the lower reach terraces is $\sim 10^6\ m^3\ s^{-1}$ (calculated from Equation (11.2)), which is consistent with the step-pool topography and the height of the medial bar. This discharge is consistent with a dam breach of ~100 m depth (less than half of the total breach depth) at the time the interior channel was incised. Peak floods may have been higher, as suggested by the wider lower reach above the terrace level. (8) Adequate water could be stored in the Eridania basin between the 1100 and the 950 m contours to carve the 14 000 km^3 of Ma'adim Vallis in a single flood, with a water to sediment ratio on the order of 10:1.

11.4.3 Smaller stem valleys with head basins and few tributaries

In Table 11.1 are listed many of the largest valleys on Mars, including the outflow channels and intermediate-sized stem valleys that are much wider and deeper than their tributaries. All of the listed valleys in the latter class emerge from enclosed basins (Auqukuh, Huo Hsing, Loire, Ma'adim, Mamers, Nanedi and the ULM system) or areas of broad topographic convergence (Al-Qahira, Licus and Samara) that focused surface runoff in the cratered highlands (e.g. Goldspiel and Squyres, 1991; Irwin and Howard, 2002). Other large valleys, including Naktong, Scamander, Indus, several of the smaller tributaries to Margaritifer basin and other unnamed valleys either emerge from or cross-cut impact craters or broad intercrater basins (Figure 11.7a). These depressions would have to fill with water to provide runoff to the valley heads, which are incised into low-relief topographic divides (Irwin *et al.*, 2005b). The more common, smaller basin outlet valleys typically have V-shaped cross-sections and record downcutting near the end of fluvial activity (e.g. Cabrol and Grin, 1999; Howard *et al.*, 2005; Irwin *et al.*, 2005b) (Figure 11.7b). The valleys are mentioned here only to establish that basin overflow was an important process at a variety of spatial scales. The association between breached basins and the widest, deepest and/or most V-shaped valley network stems on Mars is reasonable, because an overflow would provide higher, sustained discharge that may not have been possible by collection of runoff from a landscape that did not drain efficiently.

11.5 Conclusions

Despite evidence for past fluvial activity in the form of modified impact craters, valley networks, interior channels and alluvial deposits (e.g. Pieri, 1980; Craddock *et al.*, 1997; Grant, 2000; Moore and Howard, 2005; Irwin *et al.*, 2005a, 2005b), the surface of Mars records little evidence for meteorological floods that were more intense than terrestrial mean annual floods. Valley network planform, drainage density and watershed topography reflect immature development and impact cratering still dominates the highland landscape (e.g. Pieri, 1980; Carr and Clow, 1981; Grant, 1987; Carr and Malin, 2000; Irwin and

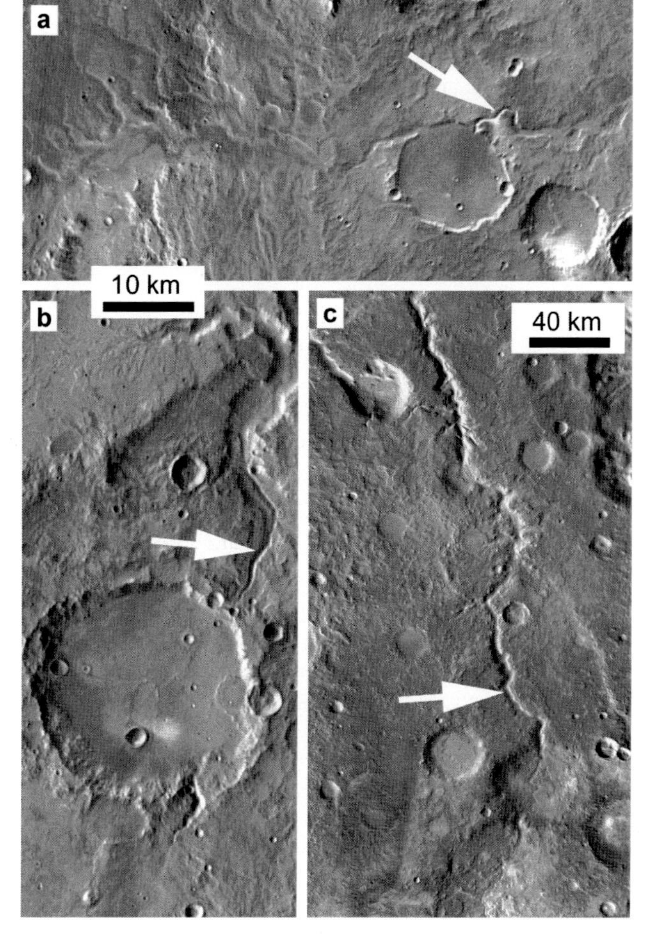

Figure 11.7. Pristine-appearing outlet valleys of enclosed basins (arrows). Flow direction is toward the east or north. (a) An unnamed valley network in Terra Cimmeria, centred at 12.4° S, 156.6° E. (b) A basin outlet tributary to Durius Vallis, centred at 17.1° S, 171.5° E. (c) Huo Hsing Vallis, centred at 29.5° N, 66.7° E. (THEMIS global mosaic.)

Howard, 2002). Overflows of enclosed basins augmented lower-magnitude runoff in some areas, providing the stream power needed to incise some of the largest and deepest valley network stems on Mars. The largest recognised examples of basin overflows outside of the Valles Marineris system (see Coleman and Baker, this volume Chapter 9) occurred in the Uzboi–Ladon–Morava Valles system (Parker, 1985; Grant and Parker, 2002) and Ma'adim Vallis (Irwin *et al.*, 2002, 2004), where suites of landforms implicate large floods on the order of $\sim 10^5 - 10^6\ m^3\ s^{-1}$. The largest outflow channels (Ares, Kasei, Simud-Tiu and Maja) also head in enclosed basins where groundwater inflows likely produced lakes that eventually breached their divides. The catastrophic release of water in these basins was probably an important to dominant mechanism for the erosion of the largest outflow channels, which were ultimately sourced by slower (but substantial) groundwater discharge.

References

Arvidson, R. E. (1974). Morphologic classification of Martian craters and some implications. *Icarus*, **22**, 264–271.

Arvidson, R. E., Squyres, S. W., Anderson, R. C. *et al.* (2006). Overview of the Spirit Mars Exploration Rover mission to Gusev crater: landing site to Backstay Rock in the Columbia Hills, *Journal of Geophysical Research*, **111**, E02S01, doi:10.1029/2005JE002499.

Baker, V. R. (1982). *The Channels of Mars*. Austin, Texas: University of Texas Press.

Baker, V. R. and Milton, D. J. (1974). Erosion by catastrophic floods on Mars and Earth. *Icarus*, **23**, 27–41.

Baker, V. R. and Nummedal, D. (1978). *The Channeled Scabland: A Guide to the Geomorphology of the Columbia Basin, Washington*. Prepared for the Comparative Planetary Geology Field Conference held in the Columbia basin June 5–8, 1978. Washington DC: NASA.

Baker, V. R. and Partridge, J. (1986). Small Martian valleys: pristine and degraded morphology. *Journal of Geophysical Research*, **91**, 3561–3572.

Banerdt, W. B. (2000). Surface drainage patterns on Mars from MOLA topography, *Eos Transactions AGU*, **81** (48), Fall Meeting Supplement, Abstract P52C-04.

Boothroyd, J. C. (1983). Fluvial drainage systems in the Ladon basin area: Margaritifer Sinus area, Mars. *Geological Society of America Abstracts with Programs*, **15**, 530.

Cabrol, N. A. and Grin, E. A. (1999). Distribution, classification, and ages of Martian impact crater lakes. *Icarus*, **142**, 160–172.

Cabrol, N. A. and Grin, E. A. (2001). Composition of the drainage network on early Mars. *Geomorphology*, **37**, 269–287.

Cabrol, N. A., Grin, E. A., Carr, M. H. *et al.* (2003). Exploring Gusev Crater with Spirit: review of science objectives and testable hypotheses. *Journal of Geophysical Research*, **108** (E12), 8076, doi:10.1029/2002JE002026.

Carr, M. H. (1979). Formation of Martian flood features by release of water from confined aquifers. *Journal of Geophysical Research*, **84**, 2995–3007.

Carr, M. H. (1996). *Water on Mars*. New York: Oxford University Press.

Carr, M. H. (2006). *The Surface of Mars*. Cambridge: Cambridge University Press.

Carr, M. H. and Clow, G. D. (1981). Martian channels and valleys: their characteristics, distribution, and age. *Icarus*, **48**, 91–117.

Carr, M. H. and Malin, M. C. (2000). Meter-scale characteristics of Martian channels and valleys. *Icarus*, **146**, 366–386.

Cohen, A. S. (2003). *Paleolimnology: The History and Evolution of Lake Systems*. New York: Oxford University Press.

Coleman N. M., Dinwiddie, C. L. and Baker, V. R. (2007). Evidence that floodwaters filled and overflowed Capri Chasma, Mars. *Geophysical Research Letters*, **34**, L07201, doi:10.1029/2006GL028872.

Costa, J. E. (1988). Floods from dam failures. In *Flood Geomorphology*, eds. V. R. Baker, R. C. Kochel and P. C. Patton. New York: John Wiley, pp. 439–463.

Craddock, R. A. and Howard, A. D. (2002). The case for rainfall on a warm, wet early Mars. *Journal of Geophysical Research*, **107** (E11), 5111, doi:10.1029/2001JE001505.

Craddock, R. A. and Maxwell, T. A. (1993). Geomorphic evolution of the Martian highlands through ancient fluvial processes. *Journal of Geophysical Research*, **98**, 3453–3468.

Craddock, R. A., Maxwell, T. A. and Howard, A. D. (1997). Crater morphometry and modification in the Sinus Sabaeus and Margaritifer Sinus regions of Mars. *Journal of Geophysical Research*, **102**, 13,321–13,340.

De Hon, R. A. (1992). Martian lake basins and lacustrine plains. *Earth, Moon, and Planets*, **56**, 95–122.

Di Achille, G., Marinangeli, L., Ori, G. G. *et al.* (2006). Geological evolution of the Tyras Vallis paleolacustrine system, Mars. *Journal of Geophysical Research*, **111**, E04003, doi:10.1029/2005JE002561.

Edgett, K. S. and Parker, T. J. (1997). Water on early Mars: possible subaqueous sedimentary deposits covering ancient cratered terrain in western Arabia and Sinus Meridiani. *Geophysical Research Letters*, **24**, 2897–2900.

Forsberg-Taylor, N. K., Howard, A. D. and Craddock, R. A. (2004). Crater degradation in the Martian highlands: morphometric analysis of the Sinus Sabaeus region and simulation modeling suggest fluvial processes. *Journal of Geophysical Research*, **109**, E05002, doi:10.1029/2004JE002242.

Forsythe, R. D. and Zimbelman, J. R. (1995). A case for ancient evaporite basins on Mars. *Journal of Geophysical Research*, **100**, 5553–5563.

Frey, H. V. (2003). Buried impact basins and the earliest history of Mars. *6th International Conference on Mars*, Abstract 3104, Lunar and Planetary Institute, Houston, Texas.

Gilbert, G. K. (1890). *Lake Bonneville*. U.S. Geological Survey Monograph 1.

Goldspiel, J. M. and Squyres, S. W. (1991). Ancient aqueous sedimentation on Mars. *Icarus*, **89**, 392–410.

Grant, J. A. (1987). The geomorphic evolution of eastern Margaritifer Sinus, Mars. In *Advances in Planetary Geology*, NASA Technical Memoir 89871, pp. 1–268.

Grant, J. A. (2000). Valley formation in Margaritifer Sinus, Mars, by precipitation-recharged ground-water sapping. *Geology*, **28**, 223–226.

Grant J. A. and Parker, T. J. (2002). Drainage evolution in the Margaritifer Sinus region, Mars. *Journal of Geophysical Research*, **107** (E9), 5066, doi:10.1029/2001JE001678.

Grant, J. A. and Schultz, P. H. (1990). Gradational epochs on Mars: evidence from west-northwest of Isidis Basin and Electris. *Icarus*, **84**, 166–195.

Grant, J. A. and Schultz, P. H. (1993). Degradation of selected terrestrial and Martian impact craters. *Journal of Geophysical Research*, **98**, 11,025–11,042.

Grant, J. A., Irwin III, R. P., Grotzinger, J. P. *et al.* (2008). HiRISE imaging of impact megabreccia and sub-meter aqueous strata in Holden crater, Mars. *Geology*, **36**, 195–198.

Greeley, R. and Guest, J. E. (1987). *Geologic Map of the Eastern Equatorial Region of Mars*. U.S. Geological Survey Geological Investigations Series Map I-1802-B, scale 1:15M.

Grin, E. A. and Cabrol, N. A. (1997). Limnologic analysis of Gusev crater paleolake, Mars. *Icarus*, **130**, 461–474.

Gulick, V. C. (2001). Origin of the valley networks on Mars: a hydrological perspective. *Geomorphology*, **37**, 241–268.

Hartmann, W. K. and Neukum, G. (2001). Cratering chronology and the evolution of Mars. *Space Science Reviews*, **96**, 165–194.

Head III, J. W. and Pratt, S. (2001). Extensive Hesperian-aged south polar ice sheet on Mars: evidence for massive melting and retreat, and lateral flow and ponding of meltwater. *Journal of Geophysical Research*, **106**, 12,275–12,299.

Hodges, C. A. (1980). *Geologic Map of the Argyre Quandrangle of Mars*. U.S. Geological Survey Miscellaneous Investigations Series Map I-1181, scale 1:5M.

Howard, A. D., Moore, J. M. and Irwin III, R. P. (2005). An intense terminal epoch of widespread fluvial activity on early Mars: 1. Valley network incision and associated deposits. *Journal of Geophysical Research*, **110**, E12S14, doi:10.1029/2005JE002459.

Howard, A. D., Moore, J. M., Irwin III, R. P. and Dietrich, W. E. (2007). Boulder transport across the Eberswalde delta. In *Lunar and Planetary Science Conference 38*, Abstract 1168, Lunar and Planetary Institute, Houston, Texas.

Hynek, B. M. and Phillips, R. J. (2003). New data reveal mature, integrated drainage systems on Mars indicative of past precipitation. *Geology*, **31**, 757–760.

Irwin III, R. P. and Howard, A. D. (2002). Drainage basin evolution in Noachian Terra Cimmeria, Mars. *Journal of Geophysical Research*, **107** (E7), doi:10.1029/2001JE00 1818.

Irwin III, R. P., Maxwell, T. A., Howard, A. D., Craddock, R. A. and Leverington, D. W. (2002). A large paleolake basin at the head of Ma'adim Vallis, Mars. *Science*, **296**, 2209–2212.

Irwin III, R. P., Howard, A. D. and Maxwell, T. A. (2004). Geomorphology of Ma'adim Vallis, Mars, and associated paleolake basins. *Journal of Geophysical Research*, **109**, E12009, doi:10.1029/2004JE002287.

Irwin III, R. P., Craddock, R. A. and Howard, A. D. (2005a). Interior channels in Martian valley networks: discharge and runoff production. *Geology*, **33**, 489–492.

Irwin III, R. P., Howard, A. D., Craddock, R. A. and Moore, J. M. (2005b). An intense terminal epoch of widespread fluvial activity on early Mars: 2. Increased runoff and paleolake development. *Journal of Geophysical Research*, **110**, E12S15, doi:10.1029/2005JE002460.

Irwin III, R. P., Howard, A. D. and Craddock, R. A. (2008). Fluvial valley networks on Mars. In *River Confluences, Tributaries and the Fluvial Network*, eds. S. Rice, A. Roy and B. Rhoads. Chichester, UK: John Wiley and Sons.

Jarrett, R. D. and Costa, J. E. (1986). *Hydrology, Geomorphology, and Dam-break Modeling of the July 15, 1982 Lawn Lake Dam and Cascade Lake Dam Failures, Larimer County, Colorado*. U.S. Geological Survey Professional Paper 1369.

Komar, P. D. (1979). Comparisons of the hydraulics of water flows in Martian outflow channels with flows of similar scale on Earth. *Icarus*, **37**, 156–181.

Komatsu, G. and Baker, V. R. (1997). Paleohydrology and flood geomorphology of Ares Vallis. *Journal of Geophysical Research*, **102**, 4151–4160.

Lucchitta, B. K., McEwen, A. S., Clow, G. D. *et al.* (1992). The canyon system on Mars. In *Mars*, eds. H. H. Kieffer, B. M. Jakosky, C. W. Snyder and M. S. Matthews. Tucson: University of Arizona Press, pp. 453–492.

Malin, M. C. and Edgett, K. S. (2000). Sedimentary rocks of early Mars. *Science*, **290**, 1927–1937.

Malin, M. C. and Edgett, K. S. (2003). Evidence for persistent flow and aqueous sedimentation on early Mars. *Science*, **302**, 1931–1934, doi:10.1126/science.1090544.

Milam, K. A., Stockstill, K. R., Moersch, J. E. *et al.* (2003). THEMIS characterization of the MER Gusev crater landing site. *Journal of Geophysical Research*, **108** (E12), 8078, doi:10.1029/2002JE002023.

Moore, J. M. and Howard, A. D. (2005). Large alluvial fans on Mars. *Journal of Geophysical Research*, **110**, E04005, doi:10.1029/2005JE002352.

O'Connor, J. E. (1993). *Hydrology, Hydraulics, and Geomorphology of the Bonneville Flood*. Geological Society of America Special Paper 274.

O'Connor, J. E. and Costa, J. E. (2004). *The World's Largest Floods, Past and Present: Their Causes and Magnitudes*. U.S. Geological Survey Circular 1254.

Parker, T. J. (1985). Geomorphology and geology of the southwestern Margaritifer Sinus-Northern Argyre region of Mars. Master's thesis, California State University, Los Angeles.

Parker, T. J. (1994). Martian paleolakes and oceans. Ph.D. thesis, University of Southern California, Los Angeles.

Phillips, R. J., Zuber, M. T., Solomon, S. C. *et al.* (2001). Ancient geodynamics and global-scale hydrology on Mars. *Science*, **291**, 2587–2591.

Pieri, D. C. (1980). Geomorphology of Martian valleys. In *Advances in Planetary Geology*, NASA TM-81979, pp. 1–160.

Pondrelli, M., Baliva, A., Di Lorenzo, S., Marinangeli, L. and Rossi, A. P. (2005). Complex evolution of paleolacustrine systems on Mars: an example from the Holden crater. *Journal of Geophysical Research*, **110**, doi:10.1029/2004JE002335.

Robinson, M. S. and Tanaka, K. L. (1990). Magnitude of a catastrophic flood event at Kasei Valles, Mars. *Geology*, **18**, 902–905.

Rotto, S. and Tanaka, K. L. (1995). *Geologic/geomorphic Map of the Chryse Planitia Region of Mars*. U.S. Geological Survey Miscellaneous Investigations Series Map I-2441, scale 1:5M.

Saunders, S. R. (1979). *Geologic Map of the Margaritifer Sinus Quadrangle of Mars*. U.S. Geological Survey Miscellaneous Investigations Series Map I-1144, scale 1:5M.

Schneeberger, D. M. (1989). Episodic channel activity at Ma'adim Vallis, Mars. *Lunar and Planetary Science Conference 20*, Abstract, 964–965.

Schultz, P. H., Schultz, R. A. and Rogers, J. (1982). The structure and evolution of ancient impact basins on Mars. *Journal of Geophysical Research*, **87**, 9803–9820.

Scott, D. H. and Tanaka, K. L. (1986). *Geologic Map of the Western Equatorial Region of Mars*. U.S. Geological Survey Miscellaneous Investigations Series Map I-1802-A, scale 1:15M.

Sharp, R. P. and Malin, M. C. (1975). Channels on Mars. *Geological Society of America Bulletin*, **86**, 593–609.

Shepherd, R. G. and Schumm, S. A. (1974). Experimental study of river incision. *Geological Society of America Bulletin*, **85**, 257–268.

Smith, D. E., Zuber, M. T., Frey, H. V. *et al.* (2001). Mars Orbiter Laser Altimeter: experiment summary after the first year of global mapping of Mars. *Journal of Geophysical Research*, **106**, 23,689–23,722.

Stepinski, T. F. and Collier, M. L. (2004). Extraction of Martian valley networks from digital topography. *Journal of Geophysical Research*, **109**, E11005, doi:10.1029/2004JE002269.

Tanaka, K. L. (1997). Sedimentary history and mass flow structures of Chryse and Acidalia Planitiae, Mars. *Journal of Geophysical Research*, **102**, 4131–4149.

Tanaka, K. L., Scott, D. H. and Greeley, R. (1992). Global stratigraphy. In *Mars*, eds. H. H. Kieffer, B. M. Jakosky, C. W. Snyder and M. S. Matthews. Tucson: University of Arizona Press, pp. 345–382.

Tanaka, K. L., Dohm, J. M., Lias, J. H. and Hare, T. M. (1998). Erosional valleys in the Thaumasia region of Mars: hydrothermal and seismic origins. *Journal of Geophysical Research*, **103**, 31,407–31,420, doi:10.1029/98JE01599.

Walder, J. S. and O'Connor, J. E. (1997). Methods for predicting peak discharge of floods caused by failure of natural and constructed earthen dams. *Water Resources Research*, **33**, 2337–2348.

Waythomas, C. F., Walder, J. S., McGimsey, R. G. and Neal, C. A. (1996). A catastrophic flood caused by drainage of a caldera lake at Aniakchak Volcano, Alaska, and implications for volcanic-hazards assessment. *Geological Society of America Bulletin*, **108**, 861–871.

Williams, R. M., Phillips, R. J. and Malin, M. C. (2000). Flow rates and duration within Kasei Valles, Mars: implications for the formation of a Martian ocean. *Geophysical Research Letters*, **27** (7), 1073–1076, doi:10.1029/1999GL010957.

Williams, K. K., Grant, J. A. and Fortezzo, C. M. (2005). New insights into the geologic history of Margaritifer Sinus and discovery of a phreatomagmatic event during late-stage fluvial activity. In *Lunar and Planetary Science Conference 36*, Abstract 1439, Lunar and Planetary Institute, Houston, Texas.

Wilson, L., Ghatan, G. J., Head III, J. W. and Mitchell, K. L. (2004). Mars outflow channels: a reappraisal of water flow velocities from water depths, regional slopes, and channel floor properties. *Journal of Geophysical Research*, **109**, E09003, doi:10.1029/2004JE002281.

Wood, L. J. (2006). Quantitative geomorphology of the Mars Eberswalde delta. *Geological Society of America Bulletin*, **118**, 557–566.

12

Criteria for identifying jökulhlaup deposits in the sedimentary record

PHILIP M. MARREN and
MATTHIAS SCHUH

Summary

A wide variety of sedimentary structures occur in modern jökulhlaup deposits and an important question arises when trying to identify jökulhlaup deposits in the sedimentary record: which sedimentary structures are distinctive of jökulhlaup deposition? A given sedimentary structure can be formed by more than one process, and in isolation cannot be used to distinguish a jökulhlaup deposit from those formed by other fluvial and flood processes. This chapter identifies those sedimentary structures that are thought to be unique or highly distinctive in jökulhlaups. Some structures are only formed by jökulhlaups in the proglacial environment, but can be found in other fluvial environments. Distinctive sedimentary features of jökulhlaup flows may include hyperconcentrated flow deposits, thick (greater than 5 m) upwards coarsening units formed by accretion during the rising stage of a flood and large gravel cross-beds (indicating formation by large gravel dunes) and flood bar deposits. Additional, non-distinctive indicators of jökulhlaup deposition include reactivation surfaces in gravel bedforms, widespread erosion surfaces and consistent palaeoflow indicators. Ice-block and rip-up clast related sedimentary structures are also non-unique, given that they can be formed under non-flood conditions, but their size and numbers are many times greater when formed by a jökulhlaup. To illustrate the points made in the review, this chapter presents a case study using ground-penetrating radar data that describe the sediments of a flood bar deposited during the November 1996 jökulhlaup in Iceland. Identifying jökulhlaup deposits will allow an assessment of the overall significance of high-magnitude flood events, relative to ablation controlled low-magnitude, high-frequency sedimentation.

12.1 Introduction

Jökulhlaups (glacial outburst floods) are significant agents of erosion, sediment transport and deposition in the proglacial zone, and can dominate the proglacial fluvial environment many kilometres downstream of the ice margin (Maizels, 1995; Marren, 2005). The jökulhlaups that occurred during the deglaciation of the Quaternary ice sheets constitute the largest floods known to have occurred on the surface of the Earth (Baker, 1973; Rudoy, 2002; Carrivick and Rushmer, 2006). Quaternary jökulhlaups can be reconstructed using geomorphological evidence (e.g. Baker, 1973; Carrivick *et al.*, 2004a), or with sedimentological evidence (e.g. Russell and Marren, 1998) or using a combination of the two. Hydraulic modelling of Quaternary jökulhlaups is used increasingly (Alho *et al.*, 2005; Alho and Aaltonen, 2008; Carrivick, 2007a), and is leading to better understanding of jökulhlaup hydraulics (Carrivick, 2007b). In bedrock rivers, erosional geomorphological evidence may dominate and the principle sedimentary evidence may take the form of slack-water deposits (Kochel and Baker, 1988; Baker, 1989). Long-term landscape evolution may destroy pre-Quaternary evidence for flooding in bedrock rivers. However, proglacial outwash plains (sandar) are largely aggradational depositional environments, in which the geomorphological evidence for an individual flood may be buried beneath subsequent deposition (Thompson and Jones, 1986; Maizels, 1979; Marren, 2002b, 2005). Many Quaternary and pre-Quaternary sedimentary successions can be preserved long after the morphological evidence has been removed. As such, sedimentary evidence may provide the best means of identifying jökulhlaups in many proglacial outwash plain deposits in the Quaternary and pre-Quaternary sedimentary record.

The relative importance of high magnitude jökulhlaup flows relative to 'normal' flood flows in shaping the landscape and sedimentary record can only be assessed if the effects of each discharge can be clearly identified and distinguished from the others (Marren *et al.*, 2002; Marren, 2005). More specifically, identifying extreme flood deposits in proglacial fluvial sediments can allow the identification of ice-dammed lakes to be inferred when their geomorphological evidence has been removed. In a similar vein, identifying evidence of jökulhlaups in glaciated volcanic settings can indicate periods of volcanic activity, and may even allow the identification of activity of different volcanoes at different times, as has recently been demonstrated for northern Iceland (Carrivick *et al.*, 2004b; Alho *et al.*, 2005).

Megaflooding on Earth and Mars, ed. Devon M. Burr, Paul A. Carling and Victor R. Baker.
Published by Cambridge University Press.

The aim of this chapter is to define a range of criteria that can be used to identify jökulhlaup deposits in the sedimentary record. This work expands upon and refines earlier studies that have made similar attempts to present criteria for identifying jökulhlaup deposits (Maizels and Russell, 1992; Maizels, 1997; Marren, 2002a, 2005). The first part considers exactly what makes a sedimentary structure formed by a jökulhlaup suitable for use as a *criterion* for identifying the origin of the deposit. The main body of the work presents a review of the literature identifying those features that qualify as criteria for identifying jökulhlaup deposits, whereas the final section is a case study of a field site in Iceland that illustrates many of the points made in the literature review, using ground-penetrating radar data which represent jökulhlaup bar architecture and the presence of large coarse-grained dunes.

12.2 Defining critical criteria

Recent floods in Iceland in 1996 (Russell and Knudsen, 1999), 1999 (Russell *et al.*, 2002) and 2002 (Rushmer, 2006), and from other parts of the world (e.g. Russell, 1993; Branney and Gilbert, 1995) have provided a wealth of examples of jökulhlaup deposits. Other examples come from Quaternary deposits interpreted as having a jökulhlaup origin (e.g. Maizels, 1989a, b, 1991; Carling, 1996a; Russell and Marren, 1998, 1999; Carling *et al.*, 2002; Carrivick *et al.*, 2004b). Many structures formed by jökulhlaups are similar to those formed by normal flood flows, although they may sometimes be larger in scale, reflecting the greater discharges and stream powers present in a jökulhlaup. For example, massive, coarse gravel units are frequently interpreted as flood units but could easily be the product of a very large river, or, in a glacial environment, an intense rainstorm (e.g. Browne, 2002; Davis *et al.*, 2003). Thus, there is a range of criteria for identifying jökulhlaups that are *scale dependent*. A jökulhlaup deposit may be essentially similar to that of a normal braided river, except much larger, reflecting the higher discharges (Russell and Marren, 1999; Marren *et al.*, 2002). The jökulhlaup origin of the deposit can be identified if the scale of the feature can be related to the water discharge. Clague (1975) has provided empirical relationships between glacier size and expected annual maximum and average discharges. These discharges can be compared with calculated palaeodischarges based on sizes of bars and channels (Wohl and Enzel, 1995). Similarly, if an ice-dammed lake is present or can be reconstructed, then the magnitude of jökulhlaups can be predicted using the empirical formulae of Clague and Mathews (1973), Costa (1988b) or Desloges *et al.* (1989) or the physically based models of Clarke (1982) and Walder and Costa (1996).

A small number of structures are unique to jökulhlaups and are therefore *jökulhlaup-specific* criteria. However, not all of the features ascribed to jökulhlaup deposits can be used to identify jökulhlaups in the sedimentary record, because many of the sedimentary structures formed in jökulhlaups can also be formed by normal floods. For example, hyperconcentrated flow deposits occur frequently in jökulhlaup deposits (e.g. Maizels, 1989a) but can also occur in a wide range of other fluvial environments (Batalla *et al.*, 1999; Benvenuti and Martini, 2002). This fact means that there is a range of criteria for identifying jökulhlaup deposits that are *context specific*. If a glacial context can be established independently, then many alternative explanations for the origin of a deposit can be discarded, and a jökulhlaup interpretation is likely, given the glacial context.

In practice, a *suite* of criteria can be used. Unique criteria should be used to positively identify the jökulhlaup origin of the deposit. Using a wider range of non-unique features and taking into account the context of the deposit can allow additional information on the characteristics of the flood to be elucidated, but this must be done once the jökulhlaup origin for the deposit has been established.

12.3 The nature of jökulhlaup flooding

Jökulhlaup floods from ice-dammed lakes represent some of the largest floods to have occurred on the surface of the Earth, and are becoming increasingly well documented (Church and Gilbert, 1975; Church, 1988; Tweed and Russell, 1999). In many cases, the discharges from ice-dammed lake outburst floods are many times larger than the normal peak summer discharges (Church and Gilbert, 1975; Cenderelli and Wohl, 2003). The effects of repeated glacial lake outburst floods can dominate landscapes (e.g. Baker, 1973). Other large jökulhlaups have been caused by the rapid melting of ice by subglacial volcanic eruptions (Tweed and Russell, 1999). Volcanically generated jökulhlaups are less common than floods from ice-dammed lakes. However, where they do occur, volcanically generated jökulhlaups potentially can have greater impact on the proglacial environment than jökulhlaups from ice-dammed lakes. Sediment availability in volcanically generated floods can be greater due to the generation of new volcanic material during the eruptive process. The very sudden 'linearly rising' jökulhlaups (Roberts, 2005) that have been identified recently may be more usually associated with volcanic generation.

Small floods, due to seasonal melting and rainstorms, are common in all glacial catchments, and relatively low discharge, frequent (possibly annual) jökulhlaups from ice-dammed lakes are common. These floods can be similar to normal seasonal flows and the results of an individual flood are rarely significant enough to have a large impact on proglacial deposits (Ashworth and Ferguson, 1986; Church, 1988; Warburton, 1994; Davis *et al.*, 2003). However, low-magnitude, high-frequency flows can have a geomorphological impact over time, and can rework the deposits of larger floods (Smith *et al.*, 2006).

Figure 12.1. Oblique aerial view of the Skeiðará river, Skeiðarársandur, Iceland, taken in summer, 2001. The bridge in the foreground is 1 km long. The bar was completely inundated by the November 1996 jökulhlaup (see Figure 12.8 for comparison). The margins of the bar were reworked, and the bar divided into two fragments by a minor flood in August 2000. The two outlined areas are the western (W) and eastern (E) bar fragments referred to in the case study at the end of the chapter. The geomorphology of the bar is discussed in Marren *et al.* (2002), and the sedimentology of the bar is discussed in the case study in this chapter.

12.4 Jökulhlaup sedimentation

12.4.1 Sediment texture

Poorly sorted cobble and boulder gravels dominate many jökulhlaup deposits. For example, Cenderelli and Wohl (2003) described the deposits of jökulhlaups in the Mount Everest region of Nepal as 'clast supported, moderately imbricated, moderately to very poorly sorted, and composed primarily of cobbles and boulders' (Cenderelli and Wohl, 2003, p. 391). Similarly, Russell (2009) and Harrison *et al.* (2006) described very poorly sorted, almost structureless cobble and boulder gravel units from jökulhlaup deposits in Greenland and Chile. Although common in jökulhlaups, poorly sorted cobble and boulder gravels are not in themselves distinctive, and could not be used as a diagnostic criterion. However, where some other means or additional information can be used to interpret a jökulhlaup origin for the deposit, poorly sorted gravel units provide much useful information on flood transport capacity and sediment sources and availability. In addition, the degree of sorting may provide some indication of the nature of the flood hydrograph, given that Ballantyne (1978) demonstrated that downstream fining trends are better developed in permanently occupied channels compared with intermittently flooded surfaces, and flume experiments by Rushmer (2007) have suggested that greater sorting occurs during the rising limb of a jökulhlaup with an exponentially rising hydrograph, in contrast to that which occurs in jökulhlaups with linearly rising hydrographs. Similarly, the duration of the waning phase has been shown to control the amount of any reworking of rising stage deposits that may occur (deposition may continue on the waning stage in many circumstances). Linearly rising jökuhlaups have longer waning stages than exponentially rising jökulhlaups, with a greater potential for sediment reworking during the flood (Russell and Knudsen, 1999, 2002).

Although the coarse-grained deposits typical of many jökulhlaups are not in themselves unique, the large scale of the flood deposit and the resulting alluvial architecture frequently is highly distinctive. Typically, jökulhlaup deposits occur within distinctive, large-scale bar forms, with equally large-scale alluvial architecture. Jökulhlaup bars are discussed below.

12.4.2 Channel bars

Bars formed by jökulhlaups are the largest individual structure used for the identification of catastrophic outburst floods in the sedimentary record. Entire sandar formed by jökulhlaups can exist. However, the identification of a sandur is reliant on identifying smaller-scale structures, including barforms, and it is the barforms and smaller-scale features that form critical criteria for identifying jökulhlaups in the sedimentary record.

Jökulhlaup flood bars commonly are scaled-up versions of bars found under normal flood equilibrium conditions (Marren *et al.*, 2002). In general, channel and bar geometry are adjusted to normal bankfull flood conditions. Bar dimensions increase with channel width and depth (height $\approx$ max depth, length $\approx$ 5 widths) although bars may not be in equilibrium with rapidly changing flows. However, where large jökulhlaups occur, bars can become stable features that persist between floods, scaled to flood-channel dimensions (Marren *et al.*, 2002; Marren, 2005; Figure 12.1), with smaller-scale bars scaled to normal flood dimensions in the channels between the jökulhlaup bars (Marren *et al.*, 2002). Fahnestock and Bradley (1973) describe bars in the Knik and Matanuska Rivers in Alaska. The bars in the Knik River, which had been formed by annual jökulhlaups for 40 years, were large (in some cases several kilometres long) and stable enough that channel positions were largely unchanged for 40 years. In contrast, the Matanuska River, which was not impacted by jökulhlaups, had small bars that rapidly changed position. Similar, semi-permanent barforms are described by Cenderelli and Wohl (2003) in Nepal. In Iceland, Nicholas and Sambrook Smith (1998) studied sediment transport and bar migration on the Virkisá river, and concluded that the demonstrated stability (sediment transport and bar migration were minimal, even under peak summer discharge flow conditions) indicated that the bar was most likely to have been formed by jökulhlaup flows. Jökulhlaup bar stability is related to two factors. Firstly, jökulhlaup deposits are frequently coarser grained than normal flood deposits, and consequently cannot be entrained by normal flows. Secondly, jökulhlaup flows are often far deeper than normal flows, forming thicker bars. Normal flows do not reach the elevation of the flood flows, and consequently the upper part of the bar is left undisturbed.

Where a proglacial river is dominated by jökulhlaups, there will be at least two scales of bar development. Large-scale jökulhlaup bars, up to several kilometres long and scaled to flood channel width and depth will be dominant. Smaller bars, scaled to normal flood flows will be found in the channels that flow around the jökulhlaup bars. There will be some reworking of the jökulhlaup bar margins between jökulhlaups, but typically the overall barform will persist. The reworking of the margins of jökulhlaup bars by normal floods means that the post-depositional form of jökulhlaup bars is controlled by erosional rather than depositional processes (Carson, 1984). In the sedimentary record of jökulhlaup-dominated sandur in Iceland, the deposits of smaller bars form a minor component of the total sedimentary succession (Russell and Marren, 1999; Marren, 2002c).

Where the proglacial zone is subjected to topographic constraints, flow separation may occur. The exact form of bars in semi-confined settings depends on the nature of the confining topography but examples in the literature have been referred to as expansion bars, pendant bars and eddy bars (Baker, 1973; O'Connor, 1993; Maizels, 1997). More commonly, in proglacial environments where a jökulhlaup exits the glacier via a single conduit and then expands onto an unconfined sandur, flow expansion and separation occurs (Russell and Knudsen, 1999, 2002). Flow expansion deposits are commonly associated with thick cross-bedded gravel and boulder units (Baker, 1973; O'Connor, 1993).

Proximal jökulhlaup sedimentary successions may consist of poorly structured, chaotic deposits (Maizels, 1993; Rushmer, 2006, 2007) due to rapid deposition on

Figure 12.2. Sedimentary section through jökulhlaup deposits exposed on the western side of Skeiðarársandur, which is thought to typify the large-scale stratification of jökulhlaup deposits. Note the thick sheets of sub-horizontal gravel, and the boulder unit at the base of the section. The upper dashed line marks the boundary between underlying flood deposits and overlying ice-marginal fan deposits, emplaced in association with a glacier advance which overrode the area. The dashed line below this marks a minor river channel which has cut into the flood deposits. The lowermost dashed line marks the boundary between a boulder unit, and the overlying gravel-boulder flood deposit. The surveying rod in the centre of the section is 4 m high. This section is discussed in detail in Russell and Marren (1999).

the rising limb or thick (5–>10 m) upward-coarsening gravel units, usually interpreted as representing progressively increasing flow competence during the rising limb of the flood hydrograph (Maizels, 1993, 1997; Russell and Knudsen, 2002). Sandur-wide flow and sedimentation may lead to strongly uniform flow directions of barforms, sedimentary structures and individual imbricated clasts within rising stage deposits (Dawson, 1989; Marren, 2002c).

Where deposition occurs on the waning stage, deposits may either fine upwards, representing decreasing flow competence (Maizels, 1993), or coarsen upwards, indicating reworking and winnowing of sediment (Russell and Knudsen, 2002). Alternatively, incision and terrace formation may occur (Maizels, 1997; Russell and Knudsen, 2002; Rushmer, 2006). The distinction is largely related to sediment availability, but the duration of the falling stage is also significant. For instance, in the January 2002 jökulhlaup at Kverkfjöll, Rushmer (2006) suggested that a prolonged falling stage allowed the development of well-defined terrace surfaces, and the exhumation of many of the structures formed on the rising stage.

There are no studies that systematically describe the sedimentology of a jökulhlaup bar that has been subjected to multiple floods. Russell (2009) and Russell and Marren (1999) described a relatively small jökulhlaup bar from Greenland, in which multiple floods are recorded as erosion or reactivation surfaces with gravel foresets. Russell and Marren (1999) also present details of a large (1 km long, 15 m high) sedimentary exposure in Iceland that probably is a cross-section through a large jökulhlaup bar (Figure 12.2). Here, the architecture reflects the movement of large units of sediment across the width of the hypothesised barform, with the sedimentology indicating hyperconcentrated flow conditions (see below).

12.4.3 Sediment concentration and flow rheology

The concentration of sediment within the flood flow controls the rheology of the flood flow, and the ultimate character of the deposit. Discussion of the role of sediment concentration in flood flows, and methods for the differentiation of the deposits of flows of varying rheology can be found in Costa (1984, 1988a) and Benvenuti and Martini (2002). The range of possible sediment concentrations form a continuum from 0 to over 70%, all of which can be found within jökulhlaup flows. However, three distinct types of flow have been identified, based on sediment concentration and the behaviour of the fluid–sediment mixture (Costa, 1988a).

Water floods (or stream flows) have 0–20% sediment concentration (by volume) and turbulent, Newtonian flow characteristics. Hyperconcentrated flows have sediment concentrations of 20–60% by volume. Hyperconcentrated flows are turbulent, and the water and sediment move as separate phases. Debris flows have high (>60%) sediment concentrations, and Bingham plastic, laminar flow characteristics.

Debris flows are common in ice-marginal environments but can also occur in some jökulhlaups, particularly during the initial phase, as sediment-rich flows first leave the glacier (Russell and Knudsen, 1999). In jökulhlaups, debris flows are usually quickly diluted as discharge increases, and debris flow deposits are not common. Hyperconcentrated flows are common in jökulhlaups, especially (but not exclusively) volcanically generated jökulhlaups, where sediment availability is particularly high (Maizels, 1989a, b, 1993, 1997; Russell and Knudsen, 1999, 2002; Figure 12.3). In the proglacial environment, hyperconcentrated flows have been described only in association with jökulhlaups. Hyperconcentrated flow deposits therefore provide a crucial means of identifying jökulhlaup deposits (although it must be borne in mind that hyperconcentrated flows are not exclusive to jökulhlaups).

Hyperconcentrated flow deposits are typically massive or crudely bedded and usually consist of poorly sorted granule–cobble sized material, with occasional boulders (Costa, 1988a; Carrivick *et al.*, 2004b). Matrix-supported units and 'openwork' gravels are associated commonly with hyperconcentrated flow deposits. A-axis parallel clast orientations may be more common than in turbulent flow deposits. Maizels (1989a, b, 1993) presented vertical profiles of hyperconcentrated flow deposits from a jökulhlaup, but did not provide details of their three-dimensional sedimentary architecture. Evidence from Icelandic

Figure 12.3. Detail of hyperconcentrated flow deposits, showing massive granule gravel with indistinct bedding marked by minor changes in grain size. This photograph is a detail shot of the thick tabular units shown in Figure 12.2. The lens cap for scale is 5 cm in diameter.

Quaternary sandur deposits suggests that hyperconcentrated flow deposits can occur either within jökulhlaup barforms, or as laterally extensive tabular sheets of poorly bedded gravel ranging in thickness between 2 and 5 m, and thinning down-sandur over distances of a few kilometres.

12.4.4 Flood bedforms

Probably the most common bedform found in jökulhlaup flows is the gravel bedload sheet (Russell and Knudsen, 1999; Russell *et al.*, 2006; Rushmer, 2006). However, such bedload sheets would be difficult to ascribe confidently to a jökulhlaup origin if encountered in the sedimentary record. Much more diagnostic are the gravel dunes that appear to be a common feature of many high-magnitude floods (Fahnestock and Bradley, 1973; Carling, 1996a; Maizels, 1997). Fine gravel dunes ($D_{50} = 2$–6 mm) have been reported from a number of fluvial environments (Carling, 1999; Carling and Shvidchenko, 2002) but coarse gravel dunes have only been encountered in some small and many large jökulhlaups. Jökulhlaups where the flows are very deep are where gravel dunes are found most often. Relatively small dunes from jökulhlaups include those described by Maizels (1989b) with wavelengths of 10 m and heights of 1 m, and dunes described by Clague and Rampton (1982), which had wavelengths of 200 m and heights of 10 m. Larger gravel dunes were formed by the Quaternary cataclysmic flooding from Lake Missoula and from similar floods in Siberia (Baker, 1973; Rudoy and Baker, 1993; Carling, 1996a; Carling *et al.*, 2002). Typically, these dunes have spacing of up to 200 m, and heights of up to 16 m. Dune dimensions are related to flow depth, (dune length $\approx$ 5 to 7 times flow depth; Yalin, 1992) so reconstructed

Figure 12.4. Large-scale cross bedding in flood gravels. The dashed lines indicate the angle of the cross-bedding. The vertical face is approximately 5 m high.

dune dimensions can provide palaeoflow information (Carling, 1996b). The internal sedimentology usually consists of coarse gravel in large foresets, dipping at angles of up to 30–35°. The dunes are thought to have formed both by the avalanching of the lee slope and by the migration of gravel sheets over the dune surface, because many dunes have cross-beds below the angle of repose (Carling, 1996a).

Gravel bedforms with large-scale foresets can be produced in several jökulhlaup sub-environments in addition to large-scale dunes, including expansion bars (Carling, 1996a). Only jökulhlaups appear to be capable of generating the flow depths and stream powers necessary to produce these structures. Similar sedimentary structures to those associated with gravel dunes can form as foresets on the front of bars (Lunt *et al.*, 2004). Their appearance in the sedimentary record would be similar to that of a small gravel dune although the relatively low angle of the cross-beds would indicate their migrating bar-form origin (e.g. Sieganthaler and Huggenberger, 1993). An example of large gravel foresets of this type formed during a Holocene jökulhlaup in Iceland is shown in Figure 12.4. Similar large-scale cross-stratification, produced by the 2002 jökulhlaup at Kverkfjöll, but typical of many Icelandic jökulhlaups, is described in Rushmer (2006) and Carrivick and Rushmer (2006) for those. Another example of gravel flood dunes found in Iceland is described in the case study at the end of the chapter.

The presence of large-scale (5–10 m) coarse-grained foresets is therefore a criterion for recognising high-magnitude flood deposits, which in a glacial environment would have to indicate a jökulhlaup. Reactivation surfaces have been described in gravel foresets by several authors including Baker and Bunker (1985) and Russell and Marren (1999). Reactivation surfaces can indicate that there was more than one flood, and suggest a proglacial environment dominated by repeated high-magnitude flooding, although flow pulsing and channel switching within a single flood can also produce reactivation surfaces.

12.4.5 Ice blocks and kettle holes

As fracturing of the glacier margin by jökulhlaup flows occurs, newly released ice blocks become entrained into the flood flows. These blocks can range in size from <1 m to over 30 m. Generally, ice blocks should float in the flood waters (90% submerged). However, the high sediment concentrations of some basal and marginal ice mean that its density is increased. This increased density can lead to greater submergence of the ice blocks and their incorporation into the deeper parts of the flow. Large ice blocks may be grounded and dragged along the bed by the flood. Such ice blocks are rarely transported more than a few hundred metres from the ice margin. In a rapidly aggrading flood, ice blocks can be buried within the sedimentary succession, rather than just being deposited on the surface of the flood deposit. Where an ice block is exposed above the floodwater, water flows around a stranded ice block. Under these conditions, water flow will accelerate around the ice block, generating scour at the upstream end of the ice block and along its sides. Flow separation in the lee of the ice block leads to deposition of sediment scoured from the upstream and sides of the block. These structures are termed ice block obstacle marks (Russell, 1993; Fay, 2002a). The size of an obstacle mark is related to ice-block size (Richardson, 1968; Paola, 1986).

Ice blocks are typically deposited on the flood surface in clusters, or lines of blocks, with the largest block at the upstream end, and smaller blocks in the lee of the larger block (Fay, 2002b). There is also a downstream fining trend, with the largest blocks and clusters of blocks deposited nearest the glacier. In comparison, ice emplaced in a sandur passively, by the stagnation of a retreating glacier, produces structures that do not show the distinctive downstream trends evident in jökulhlaup-deposited ice blocks. The structures formed by river ice and ice jams are also insignificant in comparison with jökulhlaup ice blocks (Collinson, 1971; Martini *et al.*, 1993), largely due to the size of jökulhlaup ice blocks compared to river-ice fragments, and the fact that jökulhlaup ice blocks are often formed on flood surfaces that stand above normal flood flows, whilst river-ice features are typically frequently reworked.

Kettle holes form from the melt-out of ice blocks that are either partially or totally buried within a flood deposit. Melting of the ice leaves a void, which may be an inverse cone (the 'classic' kettle hole) if the ice block was only partially buried. Steeper-sided kettle holes form if the ice block was buried entirely within the flood deposit. The sides of the kettle-hole wall collapse following the partial or complete melting of the ice block (Fay, 2002b). Commonly a buried ice block will melt slowly, with little disturbance of the overlying sediment, until eventually there is a catastrophic sediment collapse, instantly creating a steep-sided kettle hole, via one of the collapse mechanisms described by Branney and Gilbert (1995). Branney and Gilbert (1995) distinguish three types of collapse structure, depending on the depth of burial of the ice block and the cohesiveness of the surrounding sediment: (1) coherent collapse along ring fractures, typical of holes less than 2 m in diameter; (2) trapdoor collapse on horseshoe fractures; (3) down sag, with peripheral concentric fractures and crevasses, common in large holes up to 20 m in diameter. When an ice block is only partially buried, the exact form of the resulting kettle hole depends on the sediment content of the ice (Maizels, 1992). If the ice block was sediment rich, the resulting feature may actually be a mound, rather than a hole. 'Normal' kettle holes are formed by debris-poor ice (Maizels, 1977), whereas 'rimmed', 'crater' and 'till-fill' structures occur with increasing volumes of sediment in the glacier ice (Maizels, 1992). Commonly a stranded ice block will be partially buried with sediment. In these circumstances a hybrid structure is formed, part kettle hole, part obstacle mark, termed a 'kettle scour' by Fay (2002a).

Sedimentary structures related to obstacle marks include pseudo-anticlinal obstacle structures, and upstream-dipping bedding formed putatively by antidunes that occur alongside the ice block during the flood (Fay, 2002a; Alexander *et al.*, 2001). The presence of putative antidune structures indicates locally supercritical flow conditions around the ice block. Kettle holes are typically infilled by windblown or lacustrine sediment following their formation (Figure 12.5). For a kettle hole to be preserved in the sedimentary rather than geomorphological record, a subsequent aggradation event must take place.

Ice blocks are a common feature of modern jökulhlaups, although they have not been widely reported in the sedimentary record. The recognition of kettle holes and ice-block related sedimentary structures in the sedimentary record has the potential to provide good criteria for distinguishing jökulhlaup deposits (Russell *et al.*, 2006).

12.4.6 Rip-up clasts

During a jökulhlaup, large quantities of cohesive, possibly frozen, glacier bed or channel bank material can be eroded and incorporated into the flood flow (Russell and Marren, 1999; Russell and Knudsen, 1999, 2002; Russell *et al.*, 2006; Rushmer, 2006). These blocks of sediment then commonly are transported and deposited without being disaggregated, so that they form intraclasts within the jökulhlaup deposit (Figure 12.6B). These intraclasts of sediment are commonly referred to as 'rip-up clasts'. Typically, the high energy of a jökulhlaup means that most unconsolidated blocks of sediment transported in the flood flow are quickly broken down, and rarely survive more than a few hundred metres beyond the glacier margin. However, where the sediment block is frozen (Diffendal, 1984; Krainer and Poscher, 1990), or subglacial sediment erosion rates are high, and deposition is rapid, they may remain as cohesive blocks of sediment within the sedimentary succession, or scattered across the surface of a flood deposit (Russell and Marren, 1999; Russell and Knudsen, 1999, 2002; Russell *et al.*, 2006). Russell and Knudsen (2002) suggest that subglacial diamict intraclasts from the November 1996 jökulhlaup in Iceland may have travelled up to 2 km from the glacier margin in an unfrozen state. Figure 12.6A shows examples of diamict rip-up clasts on the surface of a jökulhlaup deposit and Figure 12.6B shows a block of channel back material incorporated into a jökulhlaup deposit and exposed in a sedimentary section.

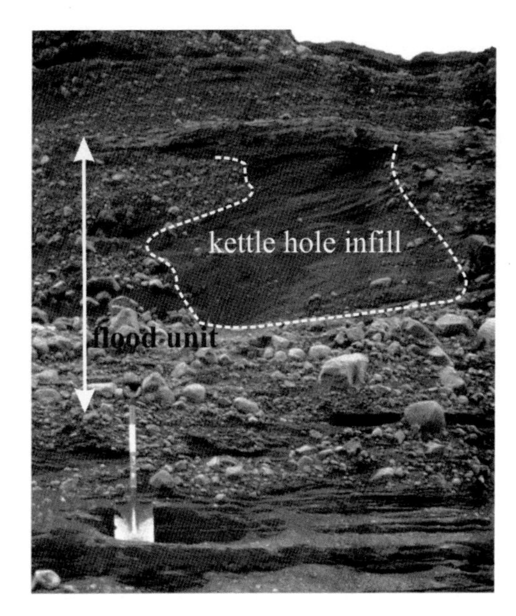

Figure 12.5. A steep-walled kettle hole filled with windblown sand. The sediment in the kettle fill dips from north to south, reflecting the prevailing wind direction. The surrounding flood deposits flowed from west to east (out of the picture; Marren, 2002c). Spade for scale.

Clay intraclasts or 'armoured mud balls' have been described in many fluvial environments (Picard and High, 1973). As such, clay-rich intraclasts are not unique to

Figure 12.6. (A) Bed-material rip-up clasts on the surface of a jökulhlaup deposit. The surveying staff in the centre of the picture is 2 m long. (B) Diamict rip-up clast within jökulhlaup deposits. Note hand holding trowel for scale.

jökulhlaups. Jökulhlaup rip-up clasts are unusually abundant, composed of unusually coarse-grained material and of unusually large size compared to the clay intraclasts found in other environments. Rip-up clasts have been found in almost all recent Icelandic jökulhlaups. These rip-up clasts have been composed of blocks of diamict or coarse-grained bank sediment, up to 3 m in diameter (Russell and Knudsen, 1999, 2002). The preservation potential for rip-up clasts is high, and they are easy to identify (Russell and Marren, 1999; Figure 12.6). Although rip-up clasts have been described from many other environments, those found in jökulhlaups are sufficiently distinctive to form useful criteria for identifying high-magnitude glacial flooding.

12.5 Skeiðarársandur jökulhlaup bar case study

The above section concerns those features of jökulhlaups that are, in some way, distinctive of high-magnitude glacial flooding. This next section will illustrate how these features might combine in the sedimentary record. This example uses previously unpublished data from the November 1996 jökulhlaup on Skeiðarársandur in southeast Iceland, and was selected because a large number of the features described above are present.

12.5.1 Study area

Three main rivers drain Skeiðarársandur outwash plain, the Súla in the west, the Gígjukvísl in the centre, and the Skeiðará on the eastern flank of the sandur (Figure 12.7). The Skeiðará is the largest of the three rivers, with a normal peak summer discharge of 200 to 400 $m^3 s^{-1}$ (Boothroyd and Nummedal, 1978; Snorrason *et al.*, 1997). Skeiðarársandur has a long history of jökulhlaups. During the twentieth century, jökulhlaups from the subglacial lake Grímsvötn occurred approximately every ten years until 1940, with discharges of 25 000–30 000 $m^3 s^{-1}$, and approximately every five years since 1940, with discharges of 1000 to 10 000 $m^3 s^{-1}$ (Guðmundsson *et al.*, 1995; Björnsson, 1997). In addition, large jökulhlaups related to subglacial volcanic eruptions, with discharges of 40 000 to 50 000 $m^3 s^{-1}$ occurred in 1934, 1938 and 1996 (Þórarinsson, 1974; Björnsson, 1997).

During the 1996 jökulhlaup, peak discharge in the Skeiðará was 15 000 to 20 000 $m^3 s^{-1}$ (Snorrason *et al.*, 1997). The Skeiðará river was the first part of the sandur to be inundated by flood waters during the 1996 jökulhlaup. Flood discharge in the Skeiðará began at 07:10 hrs on 5 November (Snorrason *et al.*, 1997). The flood reached the upstream end of the study reach at approximately 08:00 hrs and the Skeiðará bridge at 08:40–08:45 hrs. The channel width over the study reach was in excess of 2 km at flood peak.

The study reach is a 2 km long bar in the Skeiðará river, approximately 9 km downstream of the snout of Skeiðarárjökull (Figures 12.1, 12.7 and 12.8). The bar has been discussed previously by Marren *et al.* (2002) who suggested that the dimensions of the bar are scaled to the width of the channel during the 1996 jökulhlaup channel. Aerial photographs of the bar, taken before and after the November 1996 jökulhlaup, indicate that extensive surface reworking occurred during the 1996 jökulhlaup, replacing a network of small channels and bars (Figure 12.8A) with an extensive flood surface with numerous ice-block obstacle marks and kettle scours (Figures 12.7 and 12.8C). The bar margins and the downstream part of the bar have been reworked by post 1996 flows, including a small flood in August 2000. The bar was bisected by the August 2000 flood (Figure 12.7). These two fragments are referred to in this paper as the western and eastern bars. The upper

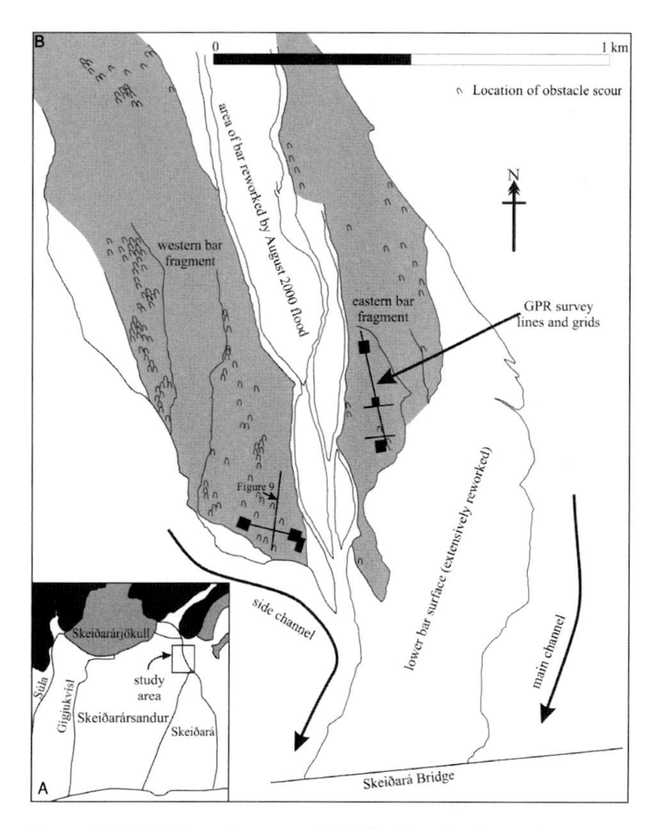

Figure 12.7. (A) Location map of Skeiðarársandur in southeast Iceland, showing the location of the Skeiðará river, and the study area in B. Black areas indicate basalt uplands, shaded areas indicate glaciers. (B) Map of the Skeiðará jökulhlaup bar, showing the location the western and eastern bar fragments (shaded areas, indicating areas of bar that have not been reworked since the 1996 jökulhlaup), the location of ice-block obstacle marks on the bar surface, and the location of the GPR survey lines.

surface of the bar is up to 4 m above average peak summer flows.

Sandur-scale topographic surveys by Smith *et al.* (2000, 2006) show that topographic levels in the proximal Skeiðará underwent only minimal change during the 1996 jökulhlaup compared to the proximal Gígjuvísl, which was dramatically effected due the disconnection of the glacier and the main sandur surface by glacier retreat in the years prior to the 1996 jökulhlaup (Russell and Knudsen, 1999; Gomez *et al.*, 2000; Smith *et al.*, 2000; Magilligan *et al.*, 2002). Relatively limited post 1996 jökulhlaup reworking of the proximal Skeiðará is also indicated by the surveys of Smith *et al.* (2006), with a net surface lowering over the study reach of 2.3 cm, mainly due to reworking of bar margins, but with little change to the bar surfaces. The findings of Smith *et al.* (2006) suggest that the long-term impact of the 1996 jökulhlaup will be confined largely to the proximal sandur, because surface reworking and lowering are more marked on the distal parts of Skeiðarársandur. This proximal to distal difference in the nature of jökulhlaup bar deposition and reworking is related to both grain size and the fact that only on the proximal sandur does the jökulhlaup flood surface occur a significant height above the normal peak summer flow elevations. Grain size on Skeiðarársandur decreases from coarse gravel near the glacier margin to fine sand and silt in the far distal reaches, a distance of approximately 40 km (Boothroyd and Nummedal, 1978). The coarse grain sizes present on the proximal sandur will inhibit bar reworking, whereas the potential for reworking will be greater downstream where the grain size is less. The occurrence of large barforms on proximal Skeiðarársandur was termed 'coarse-braiding' by Boothroyd and Nummedal (1978), although they attributed coarse-braiding solely to the coarse grain sizes dominant on the proximal sandur rather than formation by jökulhlaup flows.

12.5.2 Methods

The internal structure of the Skeiðará jökulhlaup bar was investigated using ground-penetrating radar (GPR) in the summer of 2001. Ground-penetrating radar has proved to be an effective tool for investigating coarse-grained rivers (Huggenberger 1993; Lunt and Bridge, 2004; Lunt *et al.*, 2004; Neal, 2004). The effective use of GPR on Skeiðarársandur jökulhlaup deposits has been demonstrated by Russell *et al.* (2001, 2006) and Cassidy *et al.* (2003), although all these studies focused on the central part of the sandur.

The GPR used in this section of the study was a GSSI SIR-10A system with a bistatic antenna configuration providing a centre frequency of approximately 150 Mhz. This set-up offers a vertical resolution of approximately 0.2 m in ideal conditions for this geological setting. Surveys were undertaken as a series of grids connected by long lines. On each of the two bar fragments a proximal to distal longitudinal baseline was measured down the centre of the bar. The baselines were orientated parallel to the flood flow direction, as indicated by the orientation of ice-block obstacle scours. Transverse lines and detailed grids provided three-dimensional detail (Figure 12.7). A topographic survey of the bar and all of the GPR lines using an electronic distance measurer was undertaken in order to correct the GPR data for topography.

12.5.3 Results

The data presented here are all taken from the survey of the western bar fragment. Figure 12.9A shows the long north to south GPR line from the western bar. Figure 12.9B is an interpretive panel showing the major radar facies identified in the GPR survey. The GPR section is 210 m

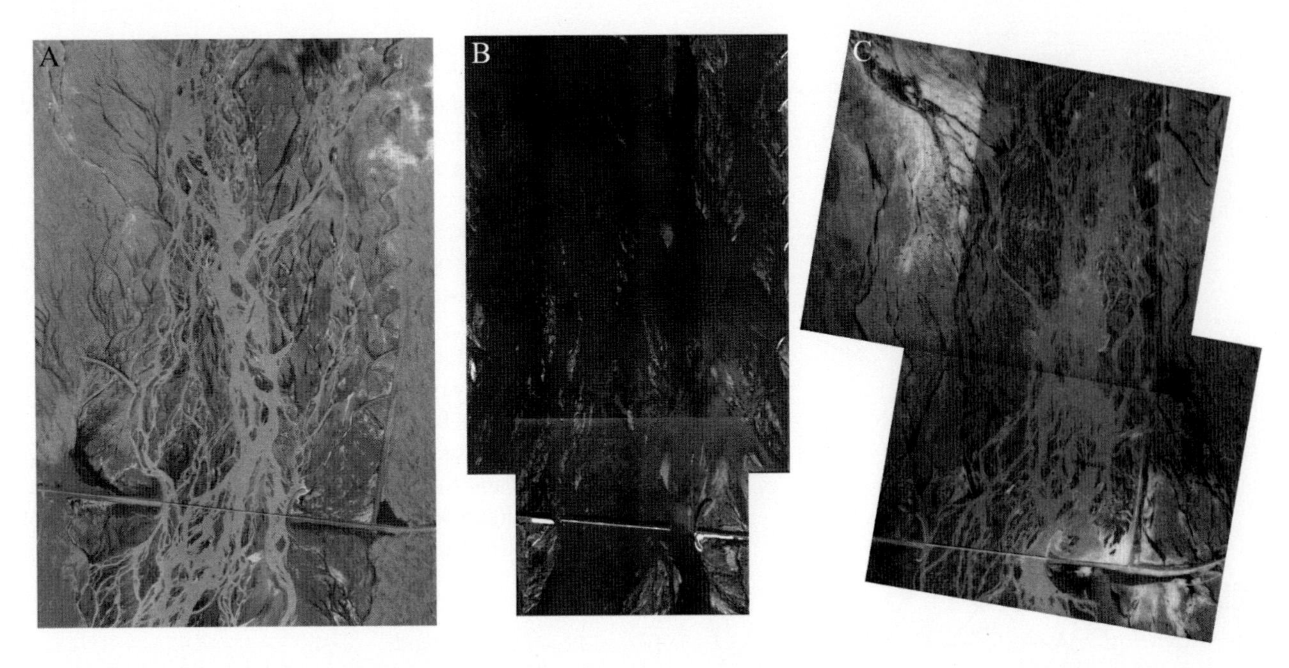

Figure 12.8. Repeat aerial photography of the Skeiðará river, upstream of the Skeiðará bridge. (A) Aerial photograph taken on 27 July 1992. Note the network of small abandoned channels and bars on the west side of the channel. (B) Aerial photograph taken on 6 November 1996, during the waning stage of the November 1996 jökulhlaup. (C) Aerial photograph taken during August 1997. Note that the active channel has shifted to the east due to growth of the jökulhlaup bar, and that to the west of the active channel the network of small channels and bars has been largely replaced by an extensive surface with numerous ice-block obstacle marks.

long, and 10 to 12 m deep. The depth of the water table is reflected in the radargrams by a significant attenuation of signal amplitudes at a depth of 2 to 3 m. The height of the water table is consistent with the height of the bar surface above the Skeiðará river at the time of the survey. All GPR data below this reflector are therefore from within the saturated zone. Reduction in the quality of the data below the water table, due to the high amplitude of the water table reflector, leads to a large number of multiple reflections from the water table that do not correspond to depositional structures. Nonetheless, a number of distinctive radar facies are apparent.

Four major radar facies can be identified in the GPR survey (Figure 12.10). Radar facies Sf (Figure 12.10) occurs from 0 to 80 m in the upper part of the section and consists of concave-upward structures with internal dipping reflectors (10–20° dip angle). These structures occur at depths of 2 to 5 m, although nearer the surface their structure is obscured by the water table reflector. The concave-upward structures overlap to the north, and the internal dipping reflectors dip to the north.

Radar facies sH (Figure 12.10) consist of low-angle (1–2° dip angle), north to south dipping reflectors, which occur along the length of the section at depths of between 5 and 7 m. These reflectors are laterally persistent and separate a succession of otherwise uniform radar signatures above and below the reflectors. The low-angle dipping reflectors divide thick (approximately 4 m) sub-horizontal sediment packages with relatively little internal structure apart from some undulating and locally concave-upward reflectors indicated by lower-amplitude reflectors.

Radar facies B (Figure 12.10) consists of three separate areas dominated by diffraction hyperbolae, each area approximately 20 to 30 m long and 2 to 3 m thick. Radar facies D consists of a high-amplitude reflector demarcating a series of asymmetrical hummocks at the base of the GPR profile (Figures 12.9 and 12.10). The high-amplitude reflector produces multiples which mask any internal structure within the hummocks. The hummocks occur along the full length of the GPR section and are 10 to 15 m long, and 1 to 2 m high. One of the hummocks is intersected by a transverse GPR line, allowing some insight into the three-dimensional geometry of the hummocks.

12.5.4 Interpretation

The four major radar facies are each interpreted as corresponding to distinctive sedimentary structures. The concave-upwards structures with internal dipping reflectors of radar facies Sf are interpreted as representing a migrating

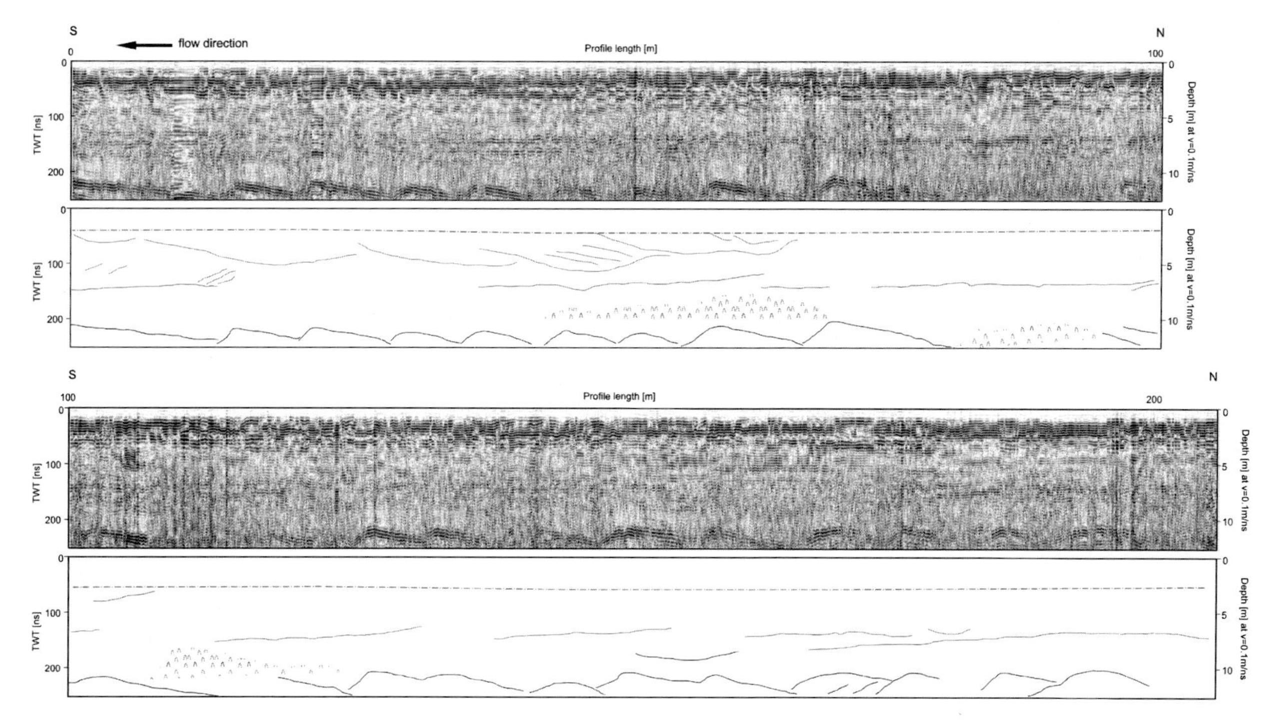

Figure 12.9. Long GPR profile from the western section of the Skeiðará jökulhlaup bar (the location of the line is indicated in Figure 12.7). Flow was from right to left (north to south). The upper panel is the processed GPR data, and the lower panel is an interpretive sketch showing the main sedimentary features. The dashed line indicates the water table; the upper lines indicate radar facies Sf; the middle lines indicate radar facies sH; the hyperbolae indicate radar facies B and the lower lines indicate radar facies D. Details of the facies interpretations are in Figure 12.10.

Examples from GPR transect	Stratal interpretation	Radar facies type	Description	Interpretation
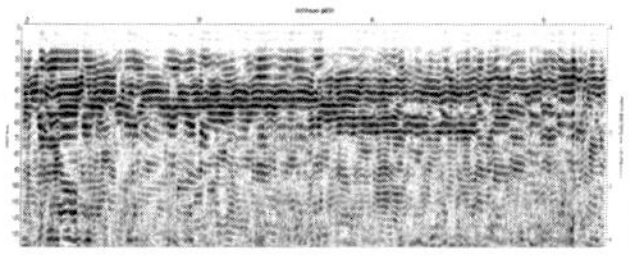	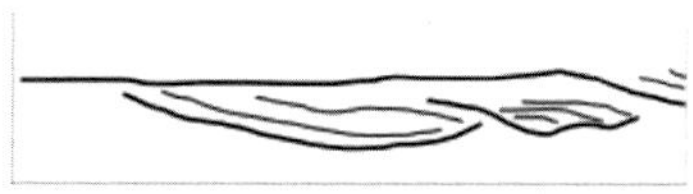	Sf	Concave upward basal reflectors confining internal dipping features (10-20° dip)	Channel / scour and fill structures / cross-strata
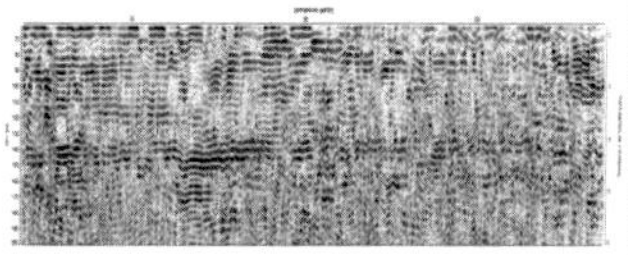	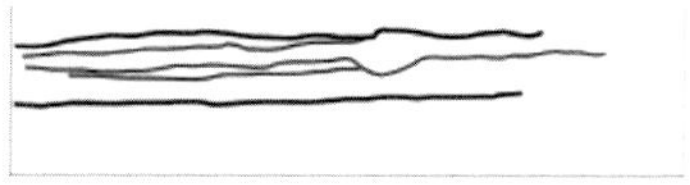	sH	Sub-horizontal laterally persistent reflectors, locally showing undulation	Sheet-like strata on a bar-wide scale
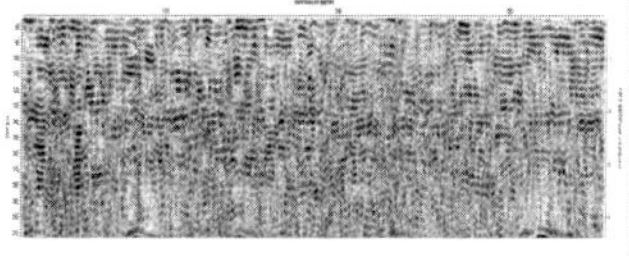	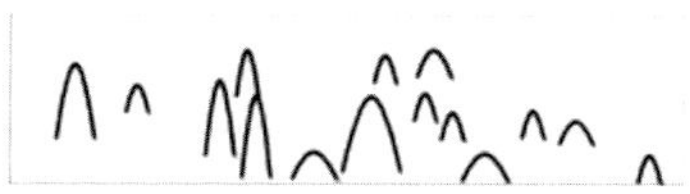	B	Laterally confined zones with abundant diffraction hyperbolae present	Large boulders and large-scale boulder lenses
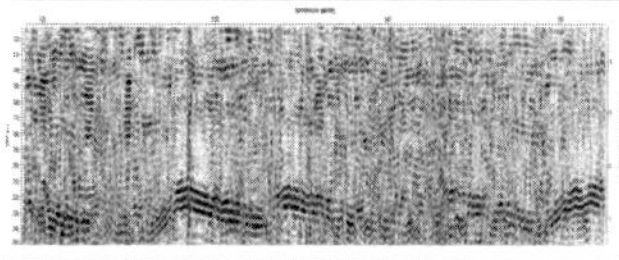	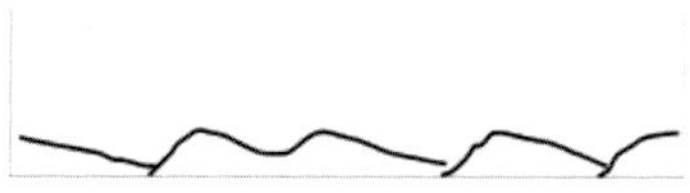	D	Asymmetrically shaped hummocky features near the base of the transect	'Dune trains' at the base of the depositional succession

Figure 12.10. Radar facies examples, their interpreted stratal features and their process interpretation. All examples are orientated such that flow was from north to south (right to left on the page).

channel and fill, and cross-channel bar deposits (Lunt *et al.*, 2004). Some of the dipping reflectors could represent cross-strata on bars or small dunes (Lunt *et al.*, 2004). No part of the bar surface covered by the GPR survey has been reworked since the 1996 jökulhlaup. Because these structures all occur at the top of the section and are relatively localised, they are interpreted as a waning-stage feature of the 1996 jökulhlaup, formed as the bar surface was revealed during the waning stage of the flood. Minor erosion, reworking and channel formation were documented throughout the waning stage of the 1996 jökulhlaup (Russell and Knudsen, 1999, 2002).

The laterally extensive nature of the low-angle dipping reflectors of radar facies sH indicates that they represent stratigraphic features at the bar-length scale. The reflectors are relatively low amplitude, and do not indicate a major change in radar signature above and below the reflector. As such, the reflectors are likely to indicate a widespread boundary between strata sets. This boundary could represent a change in the grain size of the deposit, due to factors such as unsteady flow, or changes in the flow direction during the course of the flood. The geometry of the strata sets bounded by the reflectors suggests that they are thick sub-horizontal strata sets of largely structureless sediment, except for some minor internal bedding. Laterally and longitudinally extensive structureless deposits are typical of many jökulhlaups, and thick strata sets of jökulhlaup sediment have been described from outcrops elsewhere on Skeiðarársandur (Russell and Marren, 1999; Russell and Knudsen, 1999; Marren, 2002c) and from numerous other studies of glaciofluvial deposits (e.g. Maizels, 1991; Russell and Marren, 1998; Marren, 2001; Rushmer, 2006; Kostic *et al.*, 2007). As such, the strata sets are likely to have been formed when the bar was fully inundated, at or close to the 1996 jökulhlaup flood peak.

The hyperbola structures of radar facies B are interpreted as boulder-rich deposits. Hyperbola structures are typically associated with subsurface features that are large with respect to radar wavelength (Neal, 2004), and in this environment these are likely to originate from either boulders or buried ice blocks (Cassidy *et al.*, 2003). The fact that the hyperbolas occur in distinct clusters suggests that they are more likely to be boulders, since this area of the sandur did not have such large numbers of ice blocks on it during the 1996 jökulhlaup, and does not have enough kettle holes on the present surface to indicate that there are large concentrations of buried ice blocks present. The three areas with a high concentration of boulders are all 2 to 3 m thick and 20 to 30 m long, forming large-scale boulder lenses. It is possible that these boulder lenses indicate the presence of boulder-filled channels. Similar boulder channel lenses are evident in the large section on western Skeiðarársandur described by Russell and Marren (1999).

Radar facies D, the asymmetrical hummocks at the base of the profile, indicates that these features are subaqueous dunes (Carling, 1999; Huggenberger *et al.*, 1998). The 'train' of dunes is over 200 m long. A pseudo-three-dimensional (3D) model based on data from a flow transverse GPR line at the south end of the baseline suggests that the dunes are predominantly transverse (two-dimensional dunes), albeit with some irregularities along the dune crest. The spacing of the dunes (10–15 m long) indicates flow depths of 2 to 3 m, although the dune height indicates depths of 3 to 6 m (Yalin, 1992). The high amplitude of the reflector between the dune structures and the overlying deposits suggests a marked change in radar properties between the two facies. This demarcation is indicative of a change in grain size, indicating an abrupt transition to coarser-grained sediment above the dune structures.

12.5.5 Implications of the Skeiðarársandur jökulhlaup bar GPR survey

The presence of the dune structures within the GPR section raises some questions regarding erosion and deposition during the 1996 jökulhlaup. For the dune morphology to be preserved, it is most likely the dunes were formed and buried during the same jökulhlaup. It is probable that if the dunes were formed during an earlier flood, they would be destroyed by the initial flood wave of a subsequent flood. Therefore it is likely that they were formed during the early stages of a flood and were rapidly buried by sediment during the same flood. Net elevation changes across the proximal Skeiðará in response to the 1996 jökulhlaup were relatively minor (Smith *et al.*, 2006). Given the 10 m of flood sediments which overlie the dunes within the Skeiðará flood bar, widespread scour and fill processes must have been operating over the bar during the 1996 jökulhlaup.

The data presented here illustrate a number of points made in the review section of the chapter regarding the distinctive features of jökulhlaup sedimentation. Firstly, the jökulhlaup bar surface featured numerous ice-block obstacle marks and kettles. Secondly, fast deep flow conditions can lead to the formation of large gravel dunes. Thirdly, the large-scale sedimentary structure of the bar was consistent with bar-wide flood sedimentation during the 1996 flood. Although dunes and cross-stratified barforms are not uncommon, large jökulhlaups appear to be characterised by laterally extensive sub-horizontal sheets of sediment over 4 m thick, which in GPR sections reveal relatively little internal structure except for minor horizontal stratification.

The jökulhlaup bar sedimentology present in the GPR is similar to the sedimentary section described in Russell and Marren (1999) and Marren (2005) located some 25 km to the west of the GPR survey site (Figure 12.2). This section has been shown to be the product of jökulhlaups from the early part of the twentieth century in the Súla River on western Skeiðarársandur, and appears to provide a close outcrop analogue for the GPR profile. The sediments in the sedimentary section form 3–4 m thick strata sets of crudely bedded gravels, and are interpreted by Russell and Marren (1999) as being largely the product of hyperconcentrated flows, with the exception of large boulder clusters, similar in extent to those interpreted from the GPR profile. The boulder clusters were found to contain numerous rip-up blocks of floodplain sediment and diamict (Russell and Marren, 1999). Taken together, the sedimentary section described in Russell and Marren (1999), the observations of jökulhlaup bar geomorphology in Marren *et al.* (2002) and Marren (2005), and the observations of jökulhlaup bar sedimentology described in this chapter form the basis for a preliminary model of jökulhlaup bar sedimentology.

12.6 Discussion: criteria for identifying jökulhlaups in the sedimentary record

The primary aim of sedimentary investigation into jökulhlaup deposits should be to reconstruct the discharge of the flows that deposited the flood, based on the size of reconstructed bars, and channels, dimensions of dunes, the clast sizes transported by the flood, and other palaeohydraulic indicators. The question of the degree to which jökulhlaups are scale-invariant has been a major focus of this chapter. Limited sedimentary exposure can make reconstruction of palaeodischarges difficult, and a lack of palaeogeographic context can make it difficult to assess whether reconstructed discharges are compatible with normal peak summer discharges. If jökulhlaups are to be distinguished in the sedimentary record, then it is advantageous that they are distinct from smaller braided rivers, and display a range of unique features.

This review and the case study have shown that there appears to be a large degree of scale invariance in jökulhlaup sedimentary depositional environments. Nonetheless, the fact that jökulhlaup discharges can typically be two or three orders of magnitude larger than ablation-controlled floods in the same river system (Marren, 2005) means that the contrast in scale between the deposits of the 'normal' flood flows and the 'extreme' flood flows will be so great that the jökulhlaup origin may be fairly easy to distinguish. In particular, the thick tabular sheets of relatively structureless gravel that comprise jökulhlaup bars appear to be a distinctive feature that should be recognisable in the sedimentary record. These sheets of flood gravel may be made up hyperconcentrated flow deposits, or they may form a thick coarsening-upwards succession, representing increasing flow competence on the flood rising limb.

The formation of large gravel dunes obeys the same hydrodynamic laws as for smaller dunes that occur in most rivers (Best, 1996; Carling, 1999). Large gravel dunes associated with jökulhlaup flows are a highly distinctive indicator of jökulhlaup deposition, because they are formed from coarse gravel, and because only jökulhlaup events appear to be capable of producing the flow depths and velocities necessary to create these features. The case study presented in this chapter has shown that large gravel dunes can be formed by jökulhlaups on unconfined alluvial sandur, and provides an indication of the conditions necessary for their preservation within the sedimentary record.

Ice-block obstacle marks and jökulhlaup kettle holes are distinctive to jökulhlaups, since in the fluvial environment only jökulhlaup flows are capable of breaking up the glacier margin in sufficient quantities to produce the large number of blocks encountered in floods such as the November 1996 jökulhlaup (in marine settings, ice margins can be broken up by the calving process). In addition, only jökulhlaup flows are capable of transporting large ice blocks and depositing them on floodplains in the distinctive downstream longitudinal and fining patterns witnessed in the 1996 jökulhlaup. Particular attention should therefore be paid to the sedimentary record left by ice-block obstacle marks and kettle holes.

Rip-up clasts can be formed in non-jökulhlaup flows; however, their occurrence in the size and number that are typically associated with jökulhlaup deposits appears to be unique. For this reason they are useful for identifying jökulhlaup deposits in the sedimentary record.

12.7 Conclusions

This chapter has highlighted a range of features that can be used to identify high-magnitude jökulhlaups in the sedimentary record. Rather than simply cataloguing all of the sedimentary structures that have been described in the literature in association with jökulhlaup deposits, the emphasis has been on identifying distinctive features that are confined to jökulhlaups in the proglacial environment. A case study using previously unpublished GPR data from the November 1996 jökulhlaup on Skeiðarársandur illustrates the points made in the review, and provides insights into how jökulhlaup deposits might be preserved in the sedimentary record.

The main criteria identified in the review are: (1) sedimentary architecture associated with 'jökulhlaup bar' scale or sandur-wide flow, particularly thick laterally extensive sub-horizontal tabular units of massive or poorly bedded gravel; (2) coarse-grained, upwards-coarsening sedimentary successions greater than 5 m in thickness; (3) evidence for hyperconcentrated flow processes; (4) large-scale coarse-grained bedforms, particularly gravel dunes; (5) sedimentary structures associated with ice-block obstacle marks and kettle holes; (6) widespread rip-up clasts of subglacial or floodplain material incorporated into the sedimentary succession.

Acknowledgements

The fieldwork in the case study was undertaken as part of a project funded by Earthwatch International. Dr Andy Russell is thanked for his role in coordinating the research in Iceland, which has shaped much of the thinking behind this chapter. John Bridge, Paul Carling and an anonymous reviewer provided constructive criticism of an earlier version of this chapter.

References

Alexander, J., Bridge, J. S., Cheel, R. J. and Leclair, S. F. (2001). Bedforms and associated sedimentary structures formed under supercritical water flows over aggrading sand beds. *Sedimentology*, **48**, 133–152.

Alho, P. and Aaltonen, J. (2008). Comparing a 1D hydraulic model with a 2D hydraulic model for the simulation of extreme glacial outburst floods. *Hydrological Processes*, **22**, 1537–1547.

Alho, P., Russell. A. J., Carrivick, J. L. and Käyhkö, J. (2005). Reconstruction of the largest Holocene jökulhlaup within Jökulsá á Fjöllum, NE Iceland. *Quaternary Science Reviews*, **24**, 2319–2334.

Ashworth, P. J. and Ferguson, R. I. (1986). Interrelationships of channel processes, changes and sediments in a proglacial braided river. *Geografiska Annaler*, **68A**, 361–371.

Baker, V. R. (1973). *Paleohydrology and Sedimentology of Lake Missoula Flooding in Eastern Washington*. Geological Society of America Special Paper 144.

Baker, V. R. (1989). Magnitude and frequency of paleofloods. In *Floods: Hydrological, Sedimentological and Geomorphological Implications*, eds. K. Beven and P. Carling. Chichester: John Wiley & Sons, pp. 171–183.

Baker, V. R. and Bunker, R. C. (1985). Cataclysmic late Pleistocene flooding from Glacial Lake Missoula: a review. *Quaternary Science Reviews*, **4**, 1–41.

Ballantyne, C. K. (1978). Variations in the size of coarse clastic particles over the surface of a small sandur, Ellesmere Island, N.W.T., Canada. *Sedimentology*, **25**, 141–147.

Batalla, R. J., De Jong, C., Ergenzinger, P. and Sala, M. (1999). Field observations on hyperconcentrated flows in mountain torrents. *Earth Surface Processes and Landforms*, **24**, 247–253.

Benvenuti, M. and Martini, I. P. (2002). Analysis of terrestrial hyperconcentrated flows and their deposits. In *Flood and Megaflood Processes and Deposits: Recent and Ancient Examples*, eds. I. P. Martini, V. R. Baker and G. Garón. Special Publication of the International Association of Sedimentologists, Vol. 32, pp. 167–193.

Best, J. L. (1996). The fluid dynamics of small-scale alluvial bedforms. In *Advances in Fluvial Dynamics and Stratigraphy*, eds. P. A. Carling and M. D. Dawson. Chichester, UK: Wiley, pp. 67–125.

Björnsson, H. (1997). Grímsvatnahlaup fyrr og nú. In *Vatnajókull: Gos og Hlaup*, ed. H. Haraldsson. Reykjavík: Vegargerðin, 61–77.

Boothroyd, J. C. and Nummedal, D. (1978). Proglacial braided outwash: a model for humid alluvial fan deposits. In *Fluvial Sedimentology*, ed. A. D. Miall. Canadian Society of Petroleum Geologists Memoir 5, pp. 641–668.

Branney, M. J. and Gilbert, J. S. (1995). Ice-melt collapse pits and associated features in the 1991 lahar deposits of Volcán Hudson, Chile: criteria to distinguish eruption induced glacier melt. *Bulletin of Volcanology*, **57**, 293–302.

Browne, G. H. (2002). A large-scale flood event in 1994 from the mid-Canterbury Plains, New Zealand, and implications for ancient fluvial deposits. In *Flood and Megaflood Processes and Deposits: Recent and Ancient Examples*, eds. I. P. Martine, V. R. Baker and G. Garzón. International Association of Sedimentologists Special Publication 32, pp. 99–109.

Carling, P. A. (1996a). Morphology, sedimentology and paleohydraulic significance of large gravel dunes: Altai Mountains, Siberia. *Sedimentology*, **43**, 647–664.

Carling, P. A. (1996b). A preliminary palaeohydraulic model applied to Late-Quaternary gravel dunes: Altai Mountains, Siberia. In *Global Continental Changes: The Context of Palaeohydrology*, eds. J. Branson, K. J. Gregory and A. Brown. Geological Society of London Special Publication 115, pp. 165–179.

Carling, P. A. (1999). Subaqueous gravel dunes. *Journal of Sedimentary Research*, **69**, 534–545.

Carling, P. A. and Shvidchenko, A. B. (2002). A consideration of the dune:antidune transition in fine gravel. *Sedimentology*, **49**, 1269–1282.

Carling, P. A., Kirkbride, A. D., Parnachov, S., Borodavko, P. S. and Berger, G. W. (2002). Late Quaternary catastrophic flooding in the Altai Mountains of south-central Siberia: a synoptic overview and an introduction to flood deposit sedimentology. In *Flood and Megaflood Processes and*

Deposits: Recent and Ancient Examples, eds. I. P. Martini, V. R. Baker and G. Garzón. International Association of Sedimentologists Special Publication 32, pp. 17–35.
Carson, M. A. (1984). The meandering braided river threshold: a reappraisal. *Journal of Hydrology*, **73**, 315–334.
Carrivick, J. L. (2007a). Hydrodynamics and geomorphic work of jökulhlaups (glacial outburst floods) from Kverkfjöll volcano, Iceland. *Hydrological Processes*, **21**, 725–740.
Carrivick, J. L. (2007b). Modelling coupled hydraulics and sediment transport of a high-magnitude flood and associated landscape change. *Annals of Glaciology*, **45**, 143–154.
Carrivick, J. L. and Rushmer, E. L. (2006). Understanding high-magnitude outburst floods. *Geology Today*, **22**, 60–65.
Carrivick, J. L., Russell, A. J. and Tweed, F. S. (2004a). Geomorphological evidence for jökulhlaups from Kverkfjöll volcano, Iceland. *Geomorphology*, **63**, 81–102.
Carrivick, J. L., Russell, A. J., Tweed, F. S. and Twigg, D. (2004b). Palaeohydrology and sedimentary impacts of jokulhlaups from Kverkfjoll, Iceland. *Sedimentary Geology*, **172**, 19–40.
Cassidy, N. J., Russell, A. J., Marren, P. M. *et al.* (2003). GPR-derived architecture of November 1996 jökulhlaup deposits, Skeiðarársandur, Iceland. In *Ground Penetrating Radar in Sediments*, eds. C. S. Bristow and H. M. Jol. Geological Society of London Special Publication **211**, pp. 153–166.
Cenderelli, D. A. and Wohl, E. E. (2003). Flow hydraulics and geomorphic effects of glacial-lake outburst floods in the Mount Everest Region, Nepal. *Earth Surface Processes and Landforms*, **28**, 385–407.
Church, M. (1988). Floods in cold climates. In *Flood Geomorphology*, eds. V. R. Baker, R. C. Kochel and P. C. Patton. New York: John Wiley & Sons, pp. 205–229.
Church, M. and Gilbert, R. (1975). Proglacial fluvial and lacustrine environments. In *Glaciofluvial and Glaciolacustrine Sedimentation*, eds. A. V. Jopling and B. C. McDonald. SEPM Special Publication 23, pp. 22–100.
Clague, J. J. (1975). Sedimentology and paleohydrology of Late Wisconsinan outwash, Rocky Mountain Trench, southeastern British Columbia. In *Glaciofluvial and Glaciolacustrine Sedimentation*, eds. A. V. Jopling and B. C. McDonald. SEPM Special Publication 23, pp. 223–237.
Clague, J. J. and Mathews, W. H. (1973). The magnitude of jökulhlaups. *Journal of Glaciology*, **12**, 501–504.
Clague, J. J. and Rampton, V. N. (1982). Neoglacial Lake Alsek. *Canadian Journal of Earth Sciences*, **19**, 94–117.
Clarke, G. K. C. (1982). Glacier outburst floods from "Hazard Lake", Yukon Territory, and the problem of flood magnitude prediction. *Journal of Glaciology*, **28**, 3–21.
Collinson, J. D. (1971). Some effects of ice on a river bed. *Journal of Sedimentary Petrology*, **41**, 557–564.
Costa, J. E. (1984). Physical geomorphology of debris flows. In *Developments and Applications of Geomorphology*, eds. J. E. Costa and P. J. Fleisher. New York: Springer-Verlag, pp. 268–317.
Costa, J. E. (1988a). Rheologic, geomorphic, and sedimentologic differentiation of water floods, hyperconcentrated flows and debris flows. In *Flood Geomorphology*, eds. V. R. Baker, R. C. Kochel and P. C. Patton. New York: John Wiley & Sons, pp. 113– 122.
Costa, J. E. (1988b). Floods from dam failures. In *Flood Geomorphology*, eds. V. R. Baker, R. C. Kocgel and P. C. Patton. New York: John Wiley & Sons, pp. 439–463.
Davis, T. R. H., Smart, C. C. and Turnbull, J. M. (2003). Water and sediment outbursts from advanced Franz Josef Glacier, New Zealand. *Earth Surface Processes and Landforms*, **28**, 1081–1096.
Dawson, M. R. (1989). Flood deposits present within the Severn main terrace. In *Floods: Hydrological, Sedimentological and Geomorphological Implications*, eds. K. Bevan and P. A. Carling. Chichester: John Wiley & Sons, pp. 253–264.
Desloges, J. R., Jones, D. P. and Ricker, K. E. (1989). Estimates of peak discharge from the drainage of ice-dammed Ape Lake, British Columbia. *Journal of Glaciology*, **35**, 349–354.
Diffendal, R. F. (1984). Armored mud balls and friable sand megaclasts from a complex early Pleistocene alluvial fill, southwestern Morrill County, Nebraska. *Journal of Geology*, **92**, 325–330.
Fahnestock, R. K. and Bradley, W. C. (1973). Knik and Matanuska Rivers, Alaska: a contrast in braiding. In *Fluvial Geomorphology*, ed. M. Morisawa. Binghamton Symposia in Geomorphology, Vol. 4, pp. 220–250.
Fay, H. (2002a). Formation of ice block obstacle marks during the November 1996 glacier outburst flood (jökulhlaup), Skeiðarársandur, southern Iceland. In *Flood and Megaflood Processes and Deposits: Recent and Ancient Examples*, eds. I. P. Martini, V. R. Baker and G. Garzón. International Association of Sedimentologists Special Publication 32, pp. 85–97.
Fay, H. (2002b). Formation of kettle holes following a glacial outburst flood (jökulhlaup), Skeiðarársandur, southern Iceland. In *The Extremes of the Extremes: Extraordinary Floods*, eds. Á. Snorrason and H. P. Finnsdóttir. IAHS Publication, Vol. 271, pp. 205–210.
Gomez, B., Smith, L. C., Magilligan, F. J., Mertes, L. A. K. and Smith, N. D. (2000). Glacier outburst floods and outwash plain development: Skeiðarársandur, Iceland. *Terra Nova*, **12**, 126–131.
Guðmundsson, M. T., Björnsson, H. and Pálsson, F. (1995). Changes in jökulhlaup sizes in Grímsvötn, Vatnajökull, Iceland, 1934–91, deduced from in-situ measurements of

subglacial lake volume. *Journal of Glaciology*, **41**, 263–272.

Harrison, S., Glasser, N., Winchester, V. *et al.* (2006). A glacial lake outburst flood associated with recent mountain glacier retreat, Patagonian Andes. *The Holocene*, **16**, 611–620.

Huggenberger, P. (1993). Radar facies: recognition of facies patterns and heterogeneities within Pleistocene Rhine gravels, NE Switzerland. In *Braided Rivers*, eds. J. L. Best and C. S. Bristow. Geological Society of London Special Publication 75, pp. 163–176.

Huggenberger, P., Carling, P. A., Scotney, T., Kirkbride, A. and Parnachov, S. V. (1998). GPR as a tool to elucidate the depositional processes of giant gravel dunes produced by late Pleistocene superflooding, Altai, Siberia. *Proceedings of the 7th International Conference on Ground Penetrating Radar, Vol. 1, Radar Systems and Remote Sensing*. University of Kansas, USA, pp. 279–28.

Kochel, R. C. and Baker, V. R. (1988). Paleoflood analysis using slackwater deposits. In *Flood Geomorphology*, eds. V. R. Baker, R. C. Kochel and P. C. Patton. New York: John Wiley & Sons, pp. 169–187.

Kostic, B., Süss, M. P. and Aigner, T. (2007). Three-dimensional sedimentary architecture of Quaternary sand and gravel resources: a case study of economic sedimentology (SW Germany). *International Journal of Earth Sciences*, **96**, 743–767.

Krainer, K. and Poscher, G. (1990). Ice-rich, redeposited diamicton blocks and associated structures in Quaternary outwash sediment of the Inn Valley near Innsbruck, Austria. *Geografiska Annaler*, **72A**, 249–254.

Lunt, I. A. and Bridge, J. S. (2004). Evolution and deposits of a gravelly braid bar, Sagavanirktok River, Alaska. *Sedimentology*, **51**, 415–432.

Lunt, I. A., Bridge, J. S. and Tye, R. S. (2004). A quantitative, three-dimensional depositional model of gravelly braided rivers. *Sedimentology*, **51**, 377–414.

Magilligan, F. J., Gomez, B., Mertes, L. A. K. *et al.* (2002). Geomorphic effectiveness, sandur development, and the pattern of landscape response jökulhlaups: Skeiðarársandur, southeastern Iceland. *Geomorphology*, **44**, 95–113.

Maizels, J. K. (1977). Experiments on the origin of kettle holes. *Journal of Glaciology*, **18**, 291–303.

Maizels, J. K. (1979). Proglacial aggradation and changes in braided channel patterns during a period of glacier advance: an Alpine example. *Geografiska Annaler*, **61A**, 87–101.

Maizels, J. K. (1989a). Sedimentology, paleoflow dynamics and flood history of jökulhlaup deposits: paleohydrology of Holocene sediment sequences in southern Iceland sandur deposits. *Journal of Sedimentary Petrology*, **59**, 204–223.

Maizels, J. K. (1989b). Sedimentology and palaeohydrology of Holocene flood deposits in front of a jökulhlaup glacier, south Iceland. In *Floods: Hydrological, Sedimentological and Geomorphological Implications*, eds. K. Bevan and P. A. Carling. Chichester: John Wiley & Sons, pp. 239–251.

Maizels, J. K. (1991). The origin and evolution of Holocene flood deposits in front of a jökulhlaup glacier, south Iceland. In *Environmental Change in Iceland: Past and Present*, eds. J. K. Maizels and C. Caseldine. Dordrecht: Kluwer Academic Publishers, pp. 267–302.

Maizels, J. K. (1992). Boulder ring structures produced during jökulhlaup flows: origin and hydraulic significance. *Geografiska Annaler*, **74A**, 21–33.

Maizels, J. K. (1993). Lithofacies variations within sandur deposits: the role of runoff regime, flow dynamics and sediment supply characteristics. *Sedimentary Geology*, **85**, 299–325.

Maizels, J. K. (1995). Sediments and landforms of modern proglacial terrestrial environments. In *Modern Glacial Environments*, ed. J. Menzies. Oxford: Butterworth-Heinemann, pp. 365–416.

Maizels, J. K. (1997). Jökulhlaup deposits in proglacial areas. *Quaternary Science Reviews*, **16**, 793–819.

Maizels, J. K. and Russell, A. J. (1992). Quaternary perspectives on jökulhlaup prediction. In *Applications of Quaternary Research. Quaternary Proceedings*, Vol. 2, ed. J. M. Gray. Cambridge: Quaternary Research Association, pp. 133–153.

Marren, P. M. (2001). Sedimentology of proglacial rivers in eastern Scotland during the Late Devensian. *Transactions of the Royal Society of Edinburgh: Earth Sciences*, **92**, 149–171.

Marren, P. M. (2002a). Criteria for distinguishing high magnitude flood events in the proglacial fluvial sedimentary record. In *The Extremes of the Extremes: Extraordinary Floods*, eds. Á. Snorrason, H. P. Finnsdóttir and M. Moss. IAHS Publication, Vol. 271, pp. 237–241.

Marren, P. M. (2002b). Glacier margin fluctuations, Skaftafellsjökull, Iceland: implications for sandur evolution. *Boreas*, **31**, 75–81.

Marren, P. M. (2002c). Fluvial-lacustrine interaction on Skeiðarársandur, Iceland: implications for sandur evolution. *Sedimentary Geology*, **149**, 43–58.

Marren, P. M. (2005). Magnitude and frequency in proglacial rivers: a geomorphological and sedimentological perspective. *Earth Science Reviews*, **70**, 203–251.

Marren, P. M., Russell, A. J. and Knudsen, Ó. (2002). Discharge magnitude and frequency as a control on proglacial fluvial sedimentary systems. In *The Structure, Function and Management Implications of Fluvial Sedimentary Systems*, eds. F. Dyer, M. C. Thoms and J. M. Olley. IAHS Publication, Vol. 276, pp. 297–303.

Martini, I. P., Kwong, J. K. and Sadura, S. (1993). Sediment ice rafting and cold climate fluvial deposits: Albany River, Ontario, Canada. In *Alluvial Sedimentation*, eds. M. Marzo and C. Puidefábregas. International Association of Sedimentologists Special Publication, Vol. 17, pp. 63–76.

Neal, A. (2004). Ground-penetrating radar and its use in sedimentology: principles, problems and progress. *Earth Science Reviews*, **66**, 261–330,

Nicholas, A. P. and Sambrook Smith, G. H. (1998). Relationships between flow hydraulics, sediment supply, bedload transport and channel stability in the proglacial Virkisa River, Iceland. *Geografiska Annaler*, **80A**, 111–122.

O'Connor, J. E. (1993). *Hydrology, Hydraulics and Geomorphology of the Bonneville Flood.* Geological Society of America Special Paper 274.

Paola, C. (1986). Skin friction behind isolated hemispheres and the formation of obstacle marks. *Sedimentology*, **33**, 279–293.

Picard, M. D. and High, L. R. (1973). *Sedimentary Structures of Ephemeral streams*. Developments in Sedimentology 17, Amsterdam: Elsevier.

Richardson, P. (1968). The generation of scour marks near obstacles. *Journal of Sedimentary Petrology*, **38**, 965–970.

Roberts, M. J. (2005). Jökulhlaups: a reassessment of floodwater flow through glaciers. *Reviews of Geophysics*, **43** (1), RG1002, doi:10.1029/2003RG000147.

Rudoy, A. N. (2002). Glacier-dammed lakes and geological work of glacial superfloods in the Late Pleistocene, Southern Siberia, Altai Mountains. *Quaternary International*, **87**, 119–140.

Rudoy, A. N. and Baker, V. R. (1993). Sedimentary effects of cataclysmic late Pleistocene glacial outburst flooding, Altay Mountains, Siberia. *Sedimentary Geology*, **85**, 53–62.

Rushmer, E. L. (2006). Sedimentological and geomorphological impacts of the jökulhlaup (glacial outburst flood) in January 2002 at Kverkfjöll, northern Iceland. *Geografiska Annaler*, **88A**, 43–53.

Rushmer, E. L. (2007). Physical-scale modelling of jökulhlaups (glacial outburst floods) with contrasting hydrograph shapes. *Earth Surface Processes and Landforms*, **32**, 954–963.

Russell, A. J. (1993). Obstacle marks produced by flow around stranded ice blocks during a jökulhlaup in west Greenland. *Sedimentology*, **40**, 1091–1111.

Russell, A. J. (2009). Jökulhlaup (ice-dammed lake outburst flood) impact within a valley-confined sandur subject to backwater conditions, Kangerlussuaq, West Greenland. *Sedimentary Geology*, **215**, 33–49.

Russell, A. J. and Knudsen, Ó. (1999). Controls on the sedimentology of the November 1996 jökulhlaup deposits, Skeiðarársandur, Iceland. In *Fluvial Sedimentology VI*, eds. N. D. Smith and J. Rogers International Association of Sedimentologists Special Publication, **28**, pp. 315–329.

Russell, A. J. and Knudsen, Ó. (2002). The effects of glacier outburst flood flow dynamics on ice-contact deposits: November 1996 jökulhlaup deposit, Skeiðarársandur, Iceland. In *Flood and Megaflood Processes and Deposits: Recent and Ancient Examples*, eds. I. P. Martini, V. R. Baker and G. Garzón. International Association of Sedimentologists Special Publication 32, pp. 67–83.

Russell, A. J. and Marren, P. M. (1998). A Younger Dryas (Loch Lomond Stadial) jökulhlaup deposit, Fort Augustus, Scotland. *Boreas*, **27**, 231–242.

Russell, A. J. and Marren, P. M. (1999). Proglacial fluvial sedimentary sequences in Greenland and Iceland: a case study from active proglacial environments subject to jökulhlaups. In *The Description and Analysis of Quaternary Stratigraphic Field Sections*, eds. A. P. Jones, M. E. Tucker and J. K. Hart. Quaternary Research Association Technical Guide, Vol. 7, pp. 171–208.

Russell, A. J., Knudsen, Ó., Fay, H. *et al.* (2001). Morphology and sedimentology of a giant supraglacial, ice-walled, jökulhlaup channel, Skeiðarársandur, Iceland. *Global and Planetary Change*, **28**, 203–226.

Russell, A. J., Tweed, F. S., Knudsen, Ó. *et al.* (2002). The geomorphic impact and sedimentary characteristics of the July 1999 jökulhlaup on the Jökulsá á Sólheimasandi, Mýrdalsjökull, southern Iceland. In *The Extremes of the Extremes: Extraordinary Floods*, eds. A. Snorasson, H. P. Finnsdóttir and M. Moss. IAHS Publication 271, pp. 249–254.

Russell, A. J., Roberts, M. J., Fay, H. *et al.* (2006). Icelandic jökulhlaup impacts: implications for ice-sheet hydrology, sediment transfer and geomorphology. *Geomorphology*, **75**, 33–64.

Siegenthaler, C. and Huggenberger, P. (1993). Pleistocene Rhine gravel: deposits of a braided river system with dominant pool preservation. In *Braided Rivers*, eds. J. L. Best and C. S. Bristow. Geological Society Special Publication, Vol. **75**, pp. 147–162.

Smith, L. C., Alsdorf, D. E., Magilligan, F. J. *et al.* (2000). Estimation of erosion, deposition, and net volumetric change caused by the 1996 Skeiðarársandur jökulhlaup, Iceland, from SAR inferometry. *Water Resources Research*, **36**, 1583–1594.

Smith, L. C., Sheng, Y., Magilligan, F. J. *et al.* (2006). Geomorphic impact and rapid subsequent recovery from the 1996 Skeiðarársandur jökulhlaup, Iceland, measured with multi-year airbourne lidar. *Geomorphology*, **75**, 65–75.

Snorrason, Á., Jónsson, P. and Pálsson, S. *et al.* (1997). Hlaupið á Skeiðarársandi haustið 1996: Útbreiðsla, rennsli og aurburður. In *Vatnajókull: Gos og Hlaup*, ed. H. Haraldsson. Reykjavík: Vegargerðin, pp. 79–137.

Thompson, A. P. and Jones, A. (1986). Rates and causes of proglacial river terrace formation in southeast Iceland: an application of lichenometric dating techniques. *Boreas*, **15**, 231–246.

Þórarinsson, S. (1974). Vötnin Stríð. *Saga Skeiðarárhlaupa og Grímsvatnagosa.* Reykjavík: Bókaútgáfa Menningarsjóðs.

Tweed, F. S. and Russell, A. J. (1999). Controls on the formation and sudden drainage of glacier-impounded lakes: implications for jökulhlaup characteristics. *Progress in Physical Geography*, **23**, 79–110.

Walder, J. S. and Costa, J. H. (1996). Outburst floods from glacier-dammed lakes: the effect of mode of lake drainage on flood magnitude. *Earth Surface Processes and Landforms*, **21**, 701–723.

Warburton, J. (1994). Channel change in relation to meltwater flooding, Bas Glacier d'Arolla, Switzerland. *Geomorphology*, **11**, 141–149.

Wohl, E. E. and Enzel, Y. (1995). Data for palaeohydrology. In *Global Continental Palaeohydrology*, eds. K. J. Gregory, L. Starkel and V. R. Baker. New York: John Wiley, pp. 23–59.

Yalin, M. S. (1992). *River Mechanics*. Oxford: Pergamon Press.

13 Megaflood sedimentary valley fill: Altai Mountains, Siberia

PAUL A. CARLING, I. PETER MARTINI,
JÜRGEN HERGET,
PAVEL BORODAVKO
and SERGEI PARNACHOV

Summary

During the Quaternary, the Altai Mountains of south-central Siberia sustained ice caps and valley glaciers. Glaciers or ice lobes emanating from plateaux blocked the outlet of the Chuja–Kuray intermontane basins and impounded meltwater to form large ice-dammed lakes up to 600 km^3 capacity. On occasion the ice dams failed and the lakes emptied catastrophically. The megafloods that resulted were deep, fast-flowing and heavily charged with sand and gravel, the sediment being sourced from the lake basins and also entrained along the course of the floodways. The floods were confined within mountain valleys of the present-day rivers Chuja and Katun but large quantities of sediment were deposited over a distance of more than 70 km from the dam site in tributary river-mouths, re-entrants in the confining valley walls and on the inside of major valley bends. The main depositional units that resulted are giant bars, which blocked the entrances to tributaries and temporarily impeded normal drainage from the tributaries into the main-stem valley such that minor lakes were impounded within the tributaries behind the bars. Fine sediment from the tributaries accumulated in these lakes as local lacustrine units. Later the bars were breached by the tributary flows and the local lakes were drained. Sections of the giant bar sediments and the local lacustrine units are used to describe the nature of the megaflood valley fill, which was deposited primarily during marine isotope stage 2. Although there is evidence of the Chuja–Kuray Lake being in existence within marine isotope stage 4 there are no flood sediments unequivocally ascribed to this period. Descriptions of the sedimentology and stratigraphy of the valley fill are interpreted within a context of proposed flow mechanisms associated with deposition of the various facies and thus provide some indication of the flood dynamics.

13.1 Introduction

Extensive cataclysmic Pleistocene (46 ka BP to 13 ka BP) flooding in south-central Siberia has been documented in recent decades (Rudoy, 1988, 1990, 1998; Rudoy *et al.*, 1989). These floods produced a suite of alluvial landforms, including gravel dunes (Carling, 1996), giant bars and flood-scoured channelways on a scale similar to the landforms associated with the draining of glacial Lake Missoula in North America (Baker and Bunker, 1985). The source of the Siberian floods was vast ice-dammed lakes, of a maximum volume totalling 594 km^3, impounded within intermontane basins of the Altai Mountains of south-central Siberia (Figures 13.1 and 13.2). An overview of the outburst floods from lakes in the Kuray and Chuja basins of the Altai has been provided by Baker *et al.* (1993), Rudoy and Baker (1993) and more comprehensively by Carling *et al.* (2002) and Herget (2005). Baker *et al.* (1993) concluded that there was one major Chuja–Kuray flood and a number of minor floods; the peak flow of the largest seemingly consistent with the sudden collapse of an impounding ice barrier at maximum lake level – hence the term 'cataclysmic flooding'. Peak flows were of the order of 10×10^6 m^3 s^{-1} with velocities up to 60 m s^{-1} and floodwater depths were up to 300 m (Baker *et al.*, 1993; Herget, 2005; Carling *et al.*, in press). Floods of this magnitude are termed 'megafloods' (Baker, 2002) and, as will be demonstrated below, these events deposited large quantities of sediment as valley fill along the course of the floodway. The detailed sedimentological studies reported herein provide good evidence of three large floods (Carling *et al.*, 2002) although a number of subordinate and unrecorded large floods may also have occurred. As will be shown below, the sedimentary records of these floods were generated at the end of the last glacial cycle, primarily during marine oxygen isotope stage 2 (MIS2) and these sediments dominate the valley fill. To date, no valley fill older than late Quaternary has been identified. The sedimentary expression is largely a series of giant flood bars that line the sides of the main valleys. As will be noted below, it is not clear if the present morphology of the bars is similar to when they were deposited, or if they are remnants of a sediment mass, deposited by megafloods, that largely filled the valleys but was eroded subsequently leaving the bars as residual fill. Later Holocene deposits are largely associated with terraces, apparently cut into the Pleistocene megaflood-deposited valley fill.

The terrace sediments have not been studied in any detail and their main characteristics are mentioned only to provide context. Rather, the primary objective is to record the nature of the Quaternary sedimentary valley fill as

Megaflooding on Earth and Mars, ed. Devon M. Burr, Paul A. Carling and Victor R. Baker.
Published by Cambridge University Press.

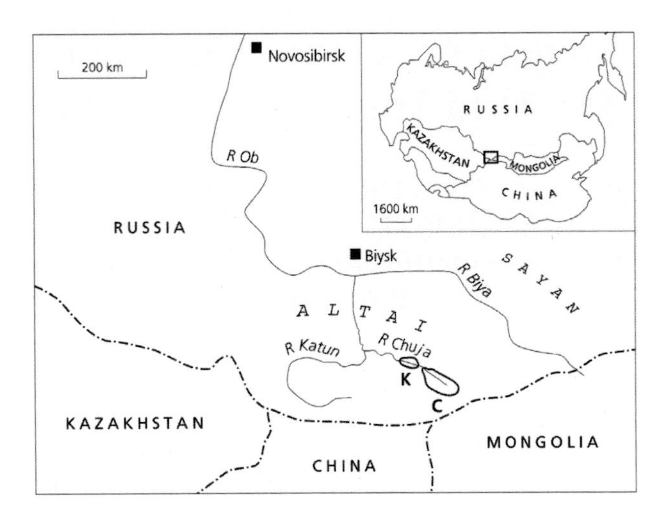

Figure 13.1. Location map of study area in southern Siberia (Russian Federation). The Kuray (K) and Chuja (C) basins are shown in the headwater of the River Chuja.

recorded in sections of the giant bars within the floodway. The nature of the fill is then used to interpret the detailed hydrodynamics of the depositing flows to shed light on the megaflooding process. Thus the first objective also allows the detail of the flood dynamics of the largest Chuja–Kuray flood(s) to be elucidated, and the sequence of flooding to be defined. These aspirations are achieved largely through a study of the sediment sequences found within prominent giant gravel bars in the Katun River valley with reference also to the Chuja River valley, set within the geological and geomorphological context that includes an appreciation of other elements of valley fill such as terraces. In a seminal paper, Smith (1993) refined earlier hydraulic interpretations of sediment sequences in Missoula bars with the primary purpose of defining the number of flood events responsible. Towards the end of the present chapter, the similarities and differences in the process of bar formation are noted for the Kuray and Missoula floods. However, the greater significance of this investigation is that the interpretation of the bar sediments is used to construct a simple conceptual model of the flood dynamics and morphological response of the Siberian bars. It is hoped that this model may spur comparative studies of other giant gravel bars worldwide to clarify the nature and the contribution of cataclysmic flooding to the landscapes of formerly glaciated regions.

13.2 Methods

The principal methodology consisted firstly of a review of the geology of the area and a geomorphological survey of the area conducted over several summers such that over 70 giant bars were visited and their locations and altitudes recorded using a hand-held GPS. Locations are accurate to ± 10 m, whereas altitudes are ± 5 m. The altitudes were calibrated twice a day by recording the altitude of spot heights marked on the topographic maps of the area using a barometric altimeter. If there was any drift during the day a linear interpolation of changes in altitude with time was used to correct the altitudes. The fieldwork was supplemented by consideration of (1) panchromatic 1:30 000 scale air photographs, (2) various sheets of the 1:200 000 scale topographic map produced by the Federal Service of Geodesy and Cartography, Russia, 1992, (3) the 1:1 000 000 scale pre-Quaternary geological map produced by the USSR Ministry of Geology, 1978, (4) 1 m resolution Ikonos satellite images, and (5) a variety of miscellaneous multispectral imagery. Notes were made of the sedimentology of numerous minor exposures and measurements and sedimentological descriptions supported by photographs were developed for several major exposed sections. The thickness of laminae in some deposits was measured in the field, and the grain size distributions of selected samples were determined through sieve analysis. Age control for the sediments is from radiocarbon, luminescence and cosmogenic methods. Where laboratory reference numbers are not cited for C^{14} dates it is because the information was not included in the primary publications. Further details are provided below.

13.3 Perspective

13.3.1 Geology

The area of study has a complex geological history but the key issue is the presence of two intermontane basins (the Chuja and Kuray basins) surrounded by mountain ranges up to 4000 m in height that sustained ice caps and valley glaciers during the Quaternary until the present day. During the Cenozoic the area underwent considerable SW–NE compression associated with the Indo-Eurasian collision (Buslov *et al.*, 1999; Ota *et al.*, 2007; Chikov *et al.*, 2008). The Chuja basin developed initially as a middle Miocene–lower Pleistocene graben with vertical relative movement of the order of 2500 m occurring primarily during the Pliocene–Pleistocene transition. The area is dissected by numerous reverse and thrust faults, and strike-slip faults with vertical components. Further motion occurred again at the beginning of the middle Pleistocene and during the late Pleistocene, and may have included isostatic movements owing to deglaciation and draining of large lakes (Carling *et al.*, 2002; Herget, 2005). Cenozoic deposits in-filling the Chuja and Kuray basins include Quaternary glacial moraine and glaciolacustrine deposits at the margins. Within the Chagun–Uzun River valley (50° 3′ 15.4″ N; 88° 25′ 58.78″ E), at the western end of the Chuja basin, are spectacular and extensive exposures of Quaternary lacustrine and glaciolacustrine sediment sequences associated with very large basin-filling

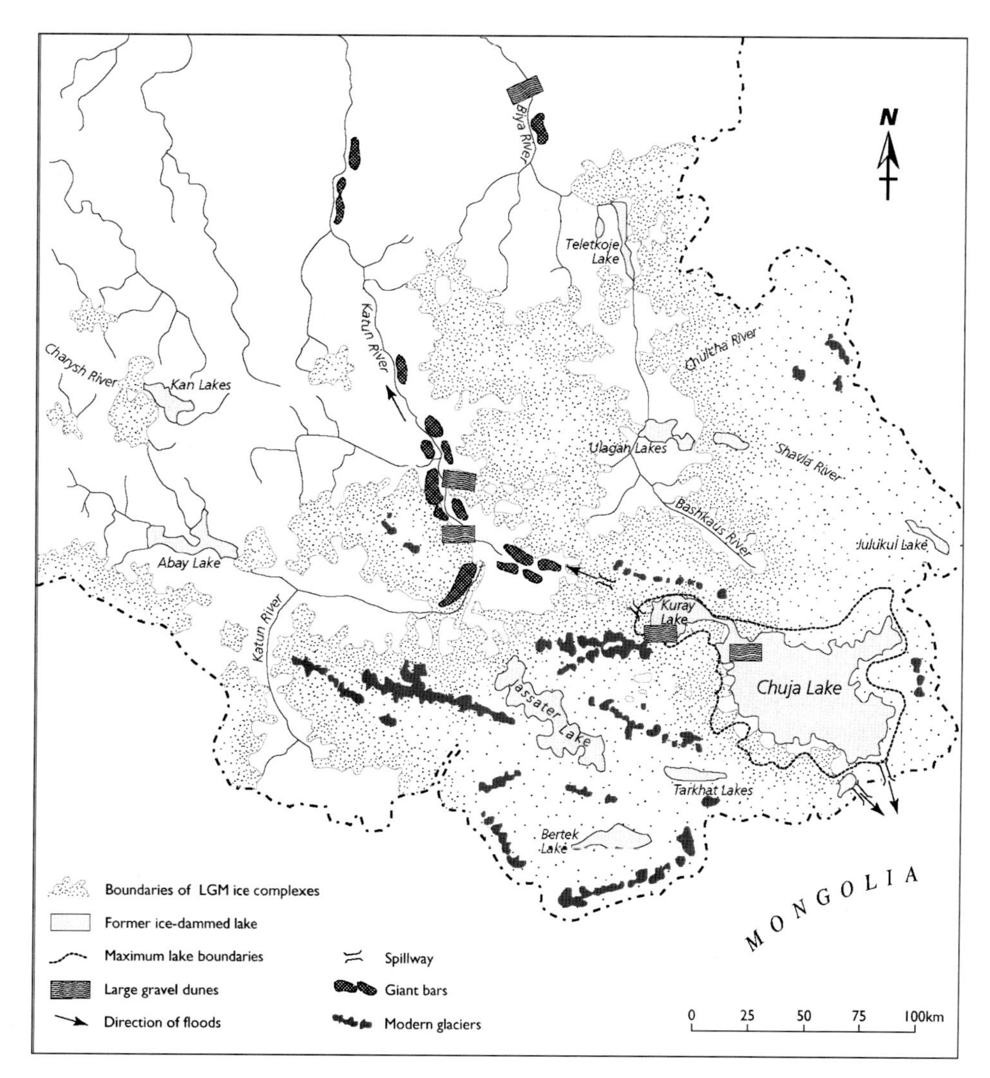

Figure 13.2. Schematic palaeogeography of the Altai region during the late Quaternary: MIS2. (Modified after Rudoy (1998).) Approximate extent of ice-impounded lakes is based on strandline evidence and locations of glaciolacustrine sedimentary complexes. Maximum potential highstands are constrained by the modern altitude of potential outflow channels (spillways) without any correction for tectonic adjustments. Features in the Katun, Chuja and upper Bashkaus valleys and in the Kuray and Chuja basins have been verified by the authors. Note: the boundaries of the ice complexes are approximate for the Sartan glaciation as different authors do not agree on the extent and timing of glaciation (see Herget, 2005, his Figure 13.11). Importantly all authors agree that the Kuray and Chuja basins were ice free.

lakes, which constitute parts of one of the most complete and important sections of the Cenozoic within central Asia (Zykin and Kazanskii, 1995; Sheinkman, 2002). The mountains framing the basins are extremely prominent with narrow piedmont zones across which Quaternary alluvial fans are well developed, in places amalgamating laterally to form a piedmont; the lower portions of these fan surfaces were sublacustrine at times of high lake stands. The flanks of the basins and the Chuja River and Katun River valleys are delimited by highly fractured bedrocks of a variety of lithologies. Ice-dammed lakes formed in both the Chuja and Katun tectonic depressions (Carling *et al.*, 2002) during the Pleistocene and, during high lake stands, the water bodies in both basins were conjoined as one large lake.

The history of glaciation and hence the history of ice-dammed lakes, within the Siberian Altai Mountains during the Pleistocene is not clear (Carling *et al.*, 2002). Regional context is provided by Gillespie *et al.* (2008) and Koppes *et al.* (2008) and references therein. The two Late Pleistocene glacial stades in the Siberian Altai are termed the Ermakovian or Ermakovo (MIS4: *c.* 100 to 50 ka) and the Sartan (MIS2). These stades were separated by the MIS3 Karginian (Karginskiy) interstadial. The exact timing and duration of the Sartan stade is uncertain, usually being ascribed to *c.* 22 to 13 ka in the Siberian Altai but with evidence in central Mongolia, for example, for a different duration of between 32 to 20 ka. The significance of this uncertainty is that the timing of the development of glaciers in the Altai is equivocable and thus the relationship of glacier development and the timing of the formation of large ice-dammed lakes and outburst floods also is uncertain.

13.3.2 System morphology

The flood-dominated drainage system consists of a previously ice-impounded basin (the source of water), the locus of the former glacial barrier (dam site) and mountain valleys acting as routes for flood flows. The valleys experienced both local erosion and deposition superimposed on an overall downvalley trend in styles of sedimentation from the megafloods.

Water was impounded in the glacial-dammed, intramontane Chuja and Kuray basins (Figures 13.1 and 13.2). The two basins are joined by a gorge (Figure 13.2) cut through faulted and highly fractured bedrock. Multiple remnant shorelines indicate that lakes filled the basins (Carling *et al.*, 2002) to 2100 m. The elevation of the present outlet of the Chuja basin is now at 1724 m altitude and that of the Kuray basin is at 1472 m. Consequently, the Chuja basin had a lake up to 300 m deep whilst the Kuray basin had a lake up to 600 m deep with a maximum water impoundment of *c.* 594 km^3 for both basins. At full capacity the Chuja and Kuray basins formed a single ancient lake but there is no rock incised spillway at 2100 m as water drained over or through the ice dam.

The portion of the Katun River valley considered in detail is immediately downstream of the confluence with the Chuja River (Figures 13.1 and 13.3), and shows no evidence of glaciation and much of the lower Chuja River valley downstream of the town of Aktash (Plate 30A) also may have been ice free during the Last Glacial Maximum (Figure 13.2; Herget, 2005: his Figures 11 and 12). Nevertheless the Chuja valley near Aktash contained glacial ice. Several converging valley glaciers in the vicinity of Aktash, notably emanating from the Mashej, Chibitka and Belgebash valleys (Figure 13.4B), had to turn sharply to the northwest to flow down the Chuja valley. The geometry of this constriction in ice flow, coupled with evidence of moraines, is believed to have caused a major ice barrier, at times up to 600 m in height in the vicinity of point 2264 m, which temporarily dammed the lake to the north and east of the present-day steep Chuja gorge (Figure 13.4B, Plate 30). On occasion ice lobes from glaciated plateaux of surrounding mountains also may have converged at this location from both the northeast and the southwest (Plate 30).

It is not known what triggered floods, but upon ice-dam failure, floods propagated probably along the line of the present Chuja gorge to the south (Figure 13.4B),

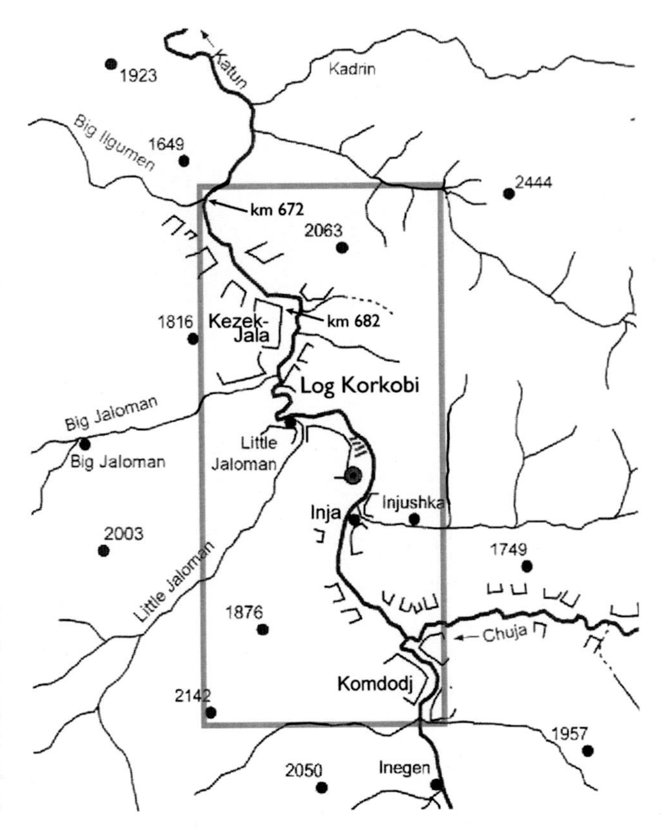

Figure 13.3. Schematic map of the main study area within the Katun River valley showing place names mentioned in text and location of the giant bars shown as heavy-line open boxes. (After Carling *et al.* (2002).)

rather than in the vicinity of Aktash to the north. Evidence for this supposition is that the largest regional ice mass lay to the northeast, such that the bulk of the ice dam was likely to have been on the northern flank of the valley. The present elevation of the Chuja River at the upstream end of the Chuja gorge is about 1424 m. The valley elevation drops to around 1134 m some 16 km downstream at the end of the gorge with an average slope of about 18 m km^{-1} (*c.* 1.8%) (Figure 13.4A). The gorge is fault-aligned but may have been deepened by megafloods (Rudoy and Baker, 1993). There is no evidence for flood deposits for a distance of 13 km downstream of this point owing to later, postflood, readvance of glaciers, which have erased flood evidence as far as the village of Ereapyk (Figure 13.4B) where late-glacial terminal moraines occur.

The floods travelled down the Chuja River valley that is generally narrow (*c.* 1 km wide; Herget, 2005), steep (average 6.6 m km^{-1}, *c.* 0.6%) with especially steep (*c.* 18.5 m km^{-1}) reaches just upstream from the junction with the Katun River (Figure 13.4A). The valley has bedrock flanks including prominent spurs, several Holocene alluvial terraces, and downstream of Ereapyk,

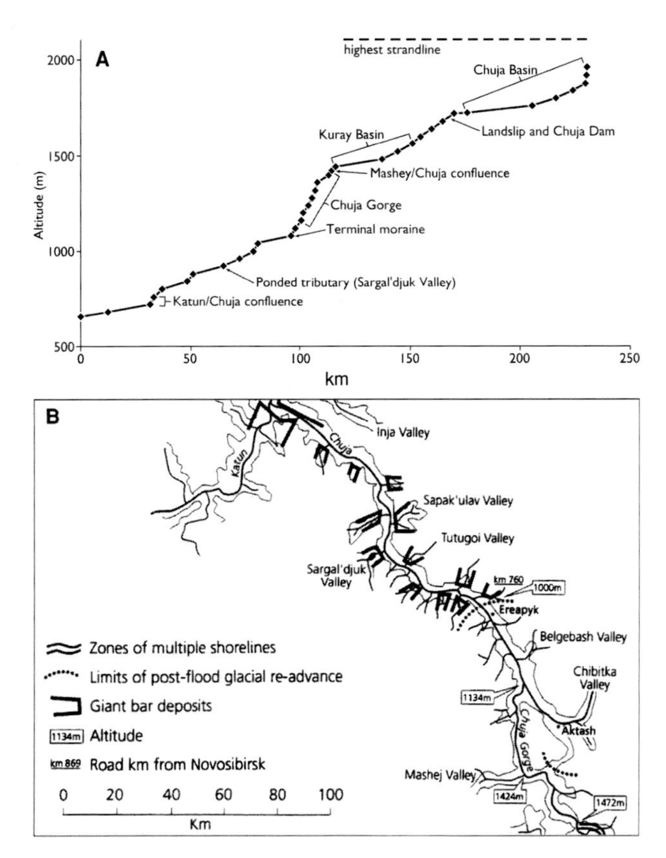

Figure 13.4. (A) Topographic profile from the Chuja basin to the Chuja–Katun river confluence. (B) Schematic map of the Chuja River showing major tributary valleys and location of the giant bars shown as heavy-line open boxes. (After Carling *et al.* (2002).)

large Pleistocene deposits (giant bars) fill and block the entrances to all the tributary valleys as well as filling other re-entrants (alcoves) in the valley side-wall alignments (Carling *et al.*, 2002) (Figure 13.4B).

The Katun River downstream from the Chuja River junction to the junction with the Big Ilgumen valley has, on average, a relatively low gradient of 2.2 m km^{-1}. The Katun valley is generally one and a half times to twice as wide as the Chuja River valley but, at the confluence with the Big Ilgumen valley (km 672, Figure 13.3), the Katun River turns sharply to the north and enters a narrow and steep gorge and the valley width decreases by a factor of two. It is likely that this constriction resulted in backwater effects on megaflood discharges, temporarily ponding water in the Katun valley (Carling *et al.*, 2002) such that sediment preferentially might be deposited on rising flood hydrographs rather than on the waning limbs of flood waves.

13.3.3 Morphology of giant bars

Downstream of the Chuja gorge a series of prominent gravel bars occur within the Chuja River valley from the village of Ereapyk at road km marker 760 and along the course of the Katun River until road km 672 (Figure 13.4B). These deposits have been regarded as high terraces of uncertain origin by Novikov and Parnachev (2000) and by earlier authors. Rudoy and Baker (1993), Carling *et al.* (2002) and Herget (2005) describe these surfaces as flood bars, deposited in embayments in the valley-wall alignment and at tributary junctions. An alternative proposal is that a complete sedimentary valley fill attributed to megafloods has been eroded to leave the 'bars' as remnant features. These various ideas are considered later. At the confluence with the Big Ilgumen River (road km 672 noted above) alluvial deposits extend to the northwest into the Big Ilgumen valley at an altitude of 800 to 840 m near road km 669. Within the 88 km between Ereapyk and road km 762, 70 giant flood bars have been visited and exposures of the sedimentary sequences examined. The course of the Katun River north of km 672 has not been investigated in any detail but there are giant bars on either side of the river close to the town of Chemal (for example at 51° 21′ 9.91″ N; 86° 2′47″ E) and large fluvial gravel dunes occur near the village of Platovo (Carling, 1996; Carling *et al.*, 2002). Thus presumed megaflood sedimentary deposits may be traced into the lower piedmont zone of the Katun River.

The most massive of the giant bars within the whole system, the Komdodj bar (Figures 13.3 and 13.4B), blocks the Katun valley immediately upstream of the Chuja confluence. Other downstream Katun-valley bars either infill small re-entrants (alcoves) which flank the main valley or form barriers across larger side valleys (e.g. Inja and Little Jaloman bars) or occur on the inside of main valley bends (e.g. Log Korkobi bar) (Figure 13.3). These bars have steep (outer) margins facing the main Katun valley (typically 20° to 35°), rise 80 m to 120 m above the highest river terrace, and are individually up to 5 km in length (Carling *et al.*, 2002). The majority of the bars formed across side valleys do not have near-planar tops sloping gently down valley (as might be expected for terrace surfaces), rather the surfaces often dip into the tributaries and may exhibit steep (inner) margins within the tributary valleys (Carling *et al.*, 2002, their Figures 10 and 11). The blocking of tributary valleys by large bars meant that in some cases tributary river flow was impounded and small lakes formed temporarily until the tributaries had time to cut through the bars and drain the local lakes. Thus local lacustrine units are found behind the bars in the entrances to some of the tributary streams joining the Chuja and Katun valleys. Reference in this text to these small-scale but important lake deposits should not be confused with the major lacustrine deposits found in the Kuray and Chuja basins. The former are evidence for megaflood-related bar deposition in the valleys rather than major ice-dammed impoundments. For example, a large bar close to the villages of Inja and Little

Jaloman (Figure 13.3 and Plate 31) forms a barrier across the Inja River valley, a right-bank tributary, in which there are three local lacustrine deposits, separated by units of flood gravels, that reflect repeated local ponding of water behind a giant bar in the tributary valley. The Little Jaloman bar forms a broad point-bar-like morphological body (Plate 31A), that extends across the entrance to the Little Jaloman River valley, a left-bank tributary, in which there are no local lacustrine deposits. Distinctive cone-shaped hollows (some tens of metres wide and a few metres deep), with associated drainage gullies, on the top of the Little Jaloman bar are illustrated and interpreted by Carling *et al.* (2002) as dewatering features, and these have been noted on the tops of one or two additional bars as well. They have been ascribed to the melting of ice blocks that were stranded contemporaneously within the flood sediment deposits and compared with kettle holes by Carling *et al.* (2002; their Figure 8).

The sides of the bars facing the Katun River have been modified by lateral cutting of the Katun River, by rilling and by other slope processes. Often the lateral cutting of the upstream portions of the giant bars has revealed a concentration of boulders within the flanks of the bars that is not evident in downstream sections of individual bars. These boulders may also form a lag on the neighbouring terrace surface as boulder trains or boulder berms (Plate 31A, B) and scour hollows around boulders attest to flood flows (Herget, 2005). However, it should be borne in mind that these latter features, along with fluvial gravel dunes, ornament terrace surfaces and might be related to floods that occurred later than those that deposited the giant bars.

The portions of the bars deposited within the side valleys have been extensively dissected by tributary flows. The longitudinal surface (relative to the Katun River) of each bar top in a gross sense is planar but most display local concavity and convexity and local bar-top surfaces may dip towards or away from the main river. In some locations there is a prominent topographic low between the bar surface and the adjacent valley wall. In many cases this topographic low is an enclosed trough that cannot be related to gully erosion by water draining from the flanking mountain slopes. This topographic low is a region of non-deposition of bar sediments and is here termed a *fosse* – an appellation first applied to similar features on bar tops in the Missoula flood tract (Bretz *et al.*, 1956) but earlier described and illustrated by Bretz (1928, his Figure 10). The longitudinal slope of many bar tops may vary from dipping down system to dipping up system. In the case of examples where the bar tops dip up system, extensive examples of so-called 'run-up' deposits (Carling *et al.*, 2002; Herget, 2005) are found smeared into valley-wall gullies and alcoves at the updip end of the bars, at heights of up to 150 m above the bar tops. These run-up deposits may have very steep valley-facing slopes.

13.4 Sediment characteristics of the giant bars

There are two styles of giant-bar deposition related to whether the bars are in the steep Chuja River valley, or in the more distal lower-gradient Katun River valley.

Within the Chuja River valley, the giant bars primarily consist of pebble gravels some 100 m thick, often with cobble beds or boulder beds. Although good exposures are few, the bars consist primarily of stacks of multiple, subparallel, gravel sheets; each some decimetres to 2 m thick. Individually, often the planar sheets are inclined, upwards and obliquely, into each tributary valley (Carling *et al.*, 2002, their Figures 10 and 11). Pebble imbrication indicates bedload transport was primarily updip and also obliquely into the tributary valleys. The original structure of the inner margins of these bars, within the tributary valleys, is not known but the present-day inner margins occasionally are extremely steep, forming impressive barriers across the tributaries. Flood flow velocities fluctuated in a downstream direction, but throughout the flood tract very high velocities occurred at some narrow and steep sections (40 to 60 m s^{-1}) with lower velocities in intervening wider reaches (Carling *et al.*, in press). Although the distribution of streampower down system has yet to be considered in detail, it seems likely that the relative coarseness of the deposits in the Chuja valley reflects the proximity of the source of the load and the lack of time and distance down the system for sorting and abrasion to fine the flood-transported load. Further downstream, within the Katun valley, bars consist of finer material, which is probably evidence of downstream fining.

Within the Katun River valley, the giant bars consist mainly of fine pebble gravels, granules and coarse sands, but average grain size reduces down valley to granules and sand near the confluence with the Big Ilgumen River (road km 674). In the vicinity of Little Jaloman and Inja the valley gradient is lower (Figure 13.4A) and the valley widens, and here the largest bars are found. In contrast to the Chuja valley bars, the bars in the Katun valley have gentle inner margins with deposits extending some hundreds to (exceptionally) thousands of metres up the tributaries (Plate 32A). Trenching by tributary streams has resulted in good exposures of the bar sediments at locations both closer to the main valley and at farther positions within the tributary valleys (Plate 32B, C). The bar-top surfaces and major bedding planes dip back into the tributary valleys at angles of a few degrees to 10°. Major bed sets are thickest (10 m to 20 m) proximally at the base of the bar and thin distally into the tributary valleys. Thus, although the giant bars usually achieve their greatest height and thickness close to the junction of a tributary with the main trunk valley, the wedge-like geometry of the deposits extend and pinch-out up-tributaries at a distance of up to 2 km (Carling *et al.*, 2002; their Figures 11 and 14) (Plate 32A).

13.4.1 Age of sediments

At present, there is insufficient dating control to be sure of the timing and sequence of sedimentation events within the Chuja and Katun valleys. The variety of dated units is summarised by Carling *et al.* (2002) and by Herget (2005). Here only the most pertinent and possibly the most reliable dates are reported. Cosmogenic dating of the age of three boulders exposed on the top of the Little Jaloman bar yielded a mean date of 15.7 ka ± 1.8 ka BP. Similar dates were obtained from three boulders within the Kuray and Chuja Lake basins (Reuther *et al.*, 2006), from which Reuther *et al.* (2006) concluded that the basins finally were drained at this time. The coincidence of dates on the bar top and in the lake basins was used by Reuther *et al.* (2006) to argue that the last flood was among the largest, inundating the Little Jaloman bar top and depositing the boulders on the bar top. This conclusion would indicate a high ice dam (*c.* 600 m high) at *c.* 15.7 ka, presumably associated with a large glacier advanced at least as far as the present town of Aktash and possibly several kilometres down the Chuja valley as far as the village of Ereapyk. A general narrative is emerging concerning the extent of glaciers in the Siberian and Mongolian Altai during MIS2 but the timing of the maximum advances of the glaciers of central Asia during MIS2 remains equivocal (Gillespie *et al.*, 2008). Nevertheless, a reconstruction of the late-glacial palaeo-environment in the vicinity of the former ice dam (Blyakharchuk *et al.*, 2004) indicates that a dam failure at that time could result from climatically induced downwasting of the glaciers. Low-level strandlines in both basins and a low-level Chuja strandline C^{14} date of 2580 a ± 105 a BP (Carling *et al.*, 2002), indicate that a minor pond may have persisted after *c.* 15.7 ka BP, during which time pingos developed in exposed lake sediments in both basins (Carling *et al.*, 2002). Thus, it is evident that during the Holocene the basins contained no large lakes that could generate megafloods.

Dating the sediment within the giant bars has proven more difficult owing to the general absence of organic matter for C^{14} dating and problems of incomplete bleaching of samples submitted for luminescence dating. The absolute dates obtained from optically and infra-red stimulated luminescence (OSL and IRSL) dating of the Little Jaloman sequence are considered unreliable (see Herget, 2005, his Figure 30). An accelerator mass spectrometry C^{14} date of 20 050 a ± 80 a BP obtained from organic material in a sand-silt layer at an elevation of 765 m within the Little Jaloman sequence is in accord with a suite of dates obtained from inter-layered flood gravels and tributary lake deposits within the neighbouring Inja bar. At the latter location, the middle of three lacustrine units has provided a C^{14} date of 23 350 a ± 400 a BP and an IRSL date of 22 400 a ± 2300 a BP with an upper lacustrine unit date by C^{14} at 22 274 a ± 370 a BP. The three lacustrine units were used by Carling *et al.* (2002) to argue for three flood events whereby three large floods sequentially deposited gravel to impound small temporary lakes in the Inja River tributary valley. Thus the evidence, in summary, points to multiple episodes of significant valley sedimentation during and towards the end of the Sartan glaciation.

13.4.2 Facies associations (sequences) and architecture of selected giant bars

Major exposures of giant-bar sediments occur at km 682 (roadside exposure of the Kezek-Jala bar – Figure 13.3), at Log Korkobi, Inja and Injuska and at the Komdodj bar immediately upstream of the Chuja–Katun confluence (Figure 13.4B).

1. The Komdodj bar is described first in the following text and contains important evidence for flood flow both up the Katun valley and down the Katun valley.
2. The Kezek-Jala bar is a major exposure with evidence of basal large-scale clinoforms overlain by flood bar sediments.
3. The Log Korkobi bar is immediately downstream of a prominent bedrock valley-wall spur on the inside of sharp bend in the Katun valley. Here isolated and local groups of angular boulders and cobbles are located just downstream of and thin away from the spur. The boulders tend to dip upvalley but pebble clusters are imbricated downvalley.
4. The Inja bar has sections near Injuska village that include intercalated lake deposits from temporary lakes impounded behind the giant bar at Inja (Carling *et al.*, 2002) and are very important because the sequences provide dated evidence for at least three major floods down the Katun valley.
5. The Little Jaloman bar has the most complete proximal to distal transect of giant-bar sediments that extends into a tributary with outcrops (Plate 32A, B) along the right flank of the Little Jaloman River (Plate 31A). Importantly, these outcrops allow inspection and interpretation of both the complete depositional sequences in the vertical succession and also the facies variation from the outer margins of the bar to the distal portions of the bar 2 km within the tributary valley. In the tributary valley, nine large and well-exposed sections of the sediment pile were examined. These, in composite, represent the full height of the bar at this location (*c.* 70 to 100 m) and, in the main, form the basis of the description of the flood-depositional succession and the hydraulic interpretation advanced below. The visible succession of bar deposits at Little Jaloman consists of a stack of at least six distinct sequences (Figure 13.5). The facies composition of each sequence are; similar to each other, except in terms of thickness. Close

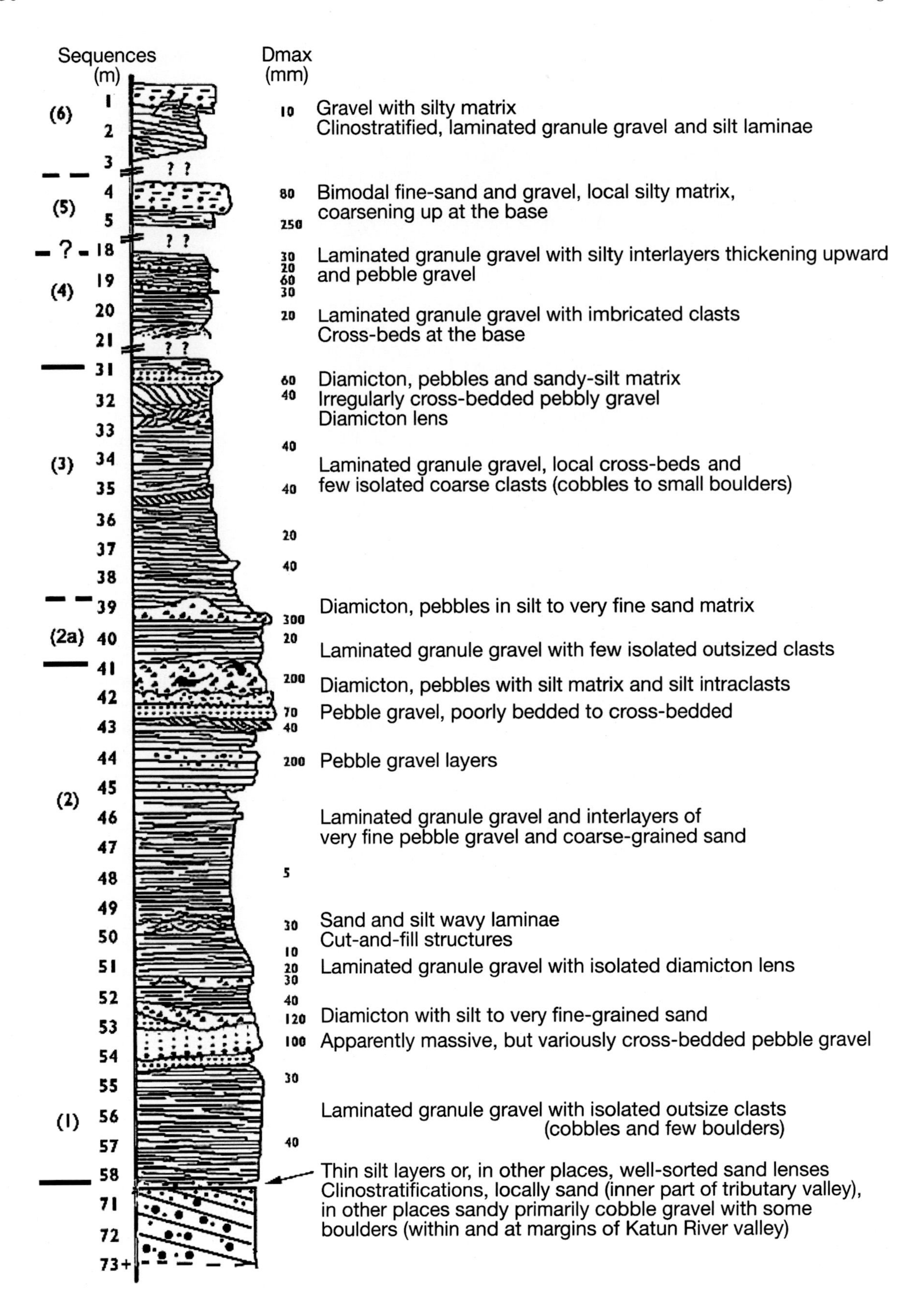

Figure 13.5. Composite stratigraphic column summarising the the stratigraphy of the Little Jaloman bar. A number of depositional sequences can be identified within the succession, each being capped usually by diamicton (Dm) or coarse gravels (Gt). The largest grain size present in each unit is shown as Dmax (mm) values.

to the outer margin of the bar the lowest sequence is 10 m to 20 m thick (Plate 32B) but all individual sequences thin both vertically and towards the inner margins within the side valleys.

Numerous other bars occur along the studied reach of the Katun River as well as the Chuja River and several have good exposures. Similarities occur in the type of facies, particularly the omnipresence of the granule (**Ggl**) and, to a lesser extent, pebble (**Gd/cl**) gravels (Table 13.1); however, the stacking of the facies (facies associations ~ sequences) vary from bar to bar. Other than the Inja and the Little Jaloman bars, no other bars exhibit sedimentary evidence for multiple floods. However, the lack of evidence may reflect the lack of exposure rather than an inherent absence of evidence.

Komdodj bar

Description This bar complex is the largest of the studied system. Some basal, limited outcrops occur just above the bedrock base of the modern river and show the local clinostratifications (**Gbp**, as described below), inclined down the Katun valley (Plate 33A). Above the clinoforms, the principal, approximately 50 m high exposure of the bar shows a 1 m thick basal layer of massive silt (**Fm**) to fine-grained ripple cross-laminated sand (**Fr**), partially deformed and overlain by granule gravel (**Ggl**) grading up into a very thick (approximately 20 m) fine-grained pebble gravel (**Gd/cl**) (Plate 34A) with few sparse cobbles and boulder stringers. From the evidence of weak imbrication, the inferred flow direction of the **Ggl** facies is upstream along the Katun valley above the confluence. This facies in turn is capped by large, sub-horizontal parallel to roughly clinostratified (dipping *c.* 20° down the Katun valley), thick cobble and boulder beds (1 to 2 m thick with some exceptional individual boulders to 6 m in diameter) consisting of angular clasts (**Gbp**) (Plate 33B). Thus when comparing the **Ggl:Gd/cl** facies (upstream flow direction) with the basal **Gbp** and capping **Gbp** facies (downstream flow direction), the inferred flow direction is reversed. To reinforce the interpretation that reversing flood flows occurred (in the upper Katun valley upstream of the confluence with the Chuja River), fluvial gravel dune cross-strata within terrace surfaces either side of the river upstream of the Komdodj bar are dipping upvalley towards the base of the dunes and downvalley near their crests.

Interpretation The exposure above a thin silt to cross-laminated sand layer shows a single, coarsening-upward, very thick, gravel sequence, likely formed during a single megaflood event. These deposits were formed near the 'T' junction between the Chuja and the Katun Rivers. It can therefore be inferred that cataclysmic floods flowing down the Chuja valley may have impinged against the bedrock bastion of the left-bank shoulder of the Katun River at the confluence and were diverted both upstream and downstream along the Katun valley. Flood-wave modelling has shown how this is possible. The main flood went down the Katun valley but a significant volume of water progressed up the Katun valley above the confluence before ponding and then flowing back down the Katun valley as the flood recession began (Carling *et al.*, in press). The fine gravel facies **Ggl** is a result of this upstream diverted flow. The coarser **Gbp** facies dips downvalley. The reason for the coarse clast sizes of the upper unit might be that the flood impact against the bedrock wall would have been enormous and, once the initial wave pressure reduced, the fractured bedrock may have slumped into the valley, where it could be partially reworked to produce the coarse-grained gravel clinostratification capping the deposit. This argument is inference, but the setting and the proposed flow-sediment behaviour are similar, albeit much larger in scale, to a recently documented flood in Vajont (Italy) where floodwaters emanating from a tributary valley owing to a catastrophic discharge from a reservoir coursed both upstream and downstream in the Piave River valley (Jaegar, 1980).

Kezek-Jala bar

Description The lowermost part of a giant bar (**Gc/bp**) is well exposed just south of the confluence of the Big Jaloman and Katun Rivers (within a large valley-side re-entrant at km 682 rather than within a tributary mouth) (Figure 13.3 and Plate 35A). Here a gravelly clinostratified unit (**Gg/bp**) is overlain by thick, sandy-granule gravel (**Ggl**) (Plate 35B, C). The transition is characterised by an erosional, channelised contact with remnants of massive to plane-bedded sand (**Sm/l**), overlain by medium-scale undulating, to cross-bedded, sandy-granule gravel with some pebbles and cobbles along the bedding planes (Plate 35B, C; Table 13.1). The latter in turn is gradationally overlain by a thick interval of plane-bedded granule gravel (**Ggl**) unit (Plate 35B).

Interpretation Considering the probable flood-flow paths, this location was relatively exposed to the initial flood flow, down the main valley, carrying coarse gravel such that the large-scale clinostrata are similar in style to those observed in the neighbouring terraced deposits of the Katun valley. The deposition of coarse gravel was followed by the deposition of laminated fine-grained gravels from a high-sediment concentrated to hyperconcentrated flood flow that expanded somewhat into the valley-side embayment. The limited exposure shows only one such highly concentrated megaflood event.

Table 13.1. Sedimentary facies of terrace top and giant bars

Facies	Texture	Structure	Typical occurrence
A. Terrace top			
Gbb Block-field	Medium to large boulders (Herget, 2005)	Boulders strewn on terrace as lag deposits usually at proximal end of present outer slope of giant bar	Terrace fronting Little Jaloman bar
Gdp-t Cross-bedded gravel	Pebble gravel with some cobbles. Sandy matrix varies from abundant to absent in openwork framework lenses. Clasts are subrounded to subangular, locally flat (Carling, 1996)	Trough cross-beds and massive beds, locally capped by planar cross-beds	Within gravel dunes ornamenting main terrace of the Katun River at Little Jaloman
St Cross-bedded sand	Gravelly (pebble), coarse- to very coarse-grained sand	Large-scale trough cross-beds	Examples occur as part of canalised (cut-and-fill structures in braided river?) on top of terraces at Little Jaloman
Gbp Clinostratified coarse-grained gravel	Varied, poorly sorted, gravelly deposits. Clasts vary from granules to very large boulders, some being large fragments of bedrock beds. Boulders generally are angular; cobbles and pebbles are subangular to subrounded	Thick clinostratified deposit with alternating lensing beds of poorly sorted boulder gravel, sandy cobble gravel and sandy pebble gravel	Upper portion of the Komdodj bar
Gdt Trough cross-bedded gravel (minor massive gravels)	Pebble gravel, few cobbles. Sandy and fine pebbles matrix. Poorly sorted. Clasts vary from subrounded to subangular	Cut-and-fill structures and medium-scale trough cross-beds are most outstanding characteristics. Strong variation in sandy matrix between various sets	Typical occurrence in upper part of sedimentary cycles of giant bar deposits (well exposed at Little Jaloman)
Gd/cl Sandy pebble to cobble gravel with dispersed boulders	Clasts are subrounded	Thin to medium layer of sandy pebble gravel, generally plane bedded, but separated by shallow scours and by concentration or alignment of coarse clasts (cobbles and some boulders)	Typical occurrence in mid part of sedimentary cycles of giant bars deposits (well exposed at Little Jaloman)
Ggl Plane-bedded, fine-grained, sandy gravel	Primarily granule gravel with some disseminated fine pebbles and coarse- to very coarse-grained sand. Fine pebbles concentrate in some layers and at layer boundaries	Variations in particle size and distribution of small pebbles indicate an overall laminated, irregularly cyclic structure. In parts of sections, thicker (~40 cm) layers are composed of granules and fine-gained pebbles occur Layers are laterally extensive over tens of metres; laminae though change in thickness from 5 mm to 15 mm within a few metres	Typical occurrence is in 'giant bars' where it constitutes the predominant facies (well exposed at Little Jaloman)
Ggl/b Plane-bedded gravel with disseminated boulders	Similar to **Ggl**, except for presence of large clasts and large conglomerate pods. Clasts are mainly subrounded, elongated; some discoidal	Coarser clasts (cobbles to boulders) occur along layer boundaries of plane-bedded, fine-pebble to granule conglomerate. Few flatter coarser clasts show imbrication. Local erosion (irregular cut-and-fill structures) occur around some large conglomeratic fragments	Typical occurrence in lower part of the Inja giant bar at entrance to Inja valley

(cont.)

Table 13.1. (cont.)

Facies	Texture	Structure	Typical occurrence
Sl Sandy plane beds	Very coarse-grained sand with few scattered pebbles. Local cut and fill structures and medium-scale (< 1 m thick) through to 2 m thick cross beds	Alternation of very coarse-grained sand with some granules and coarse-grained sand laminae. Pebbles occur isolated along bedding surfaces. Beds and laminae are rather continuous, but some thin and thicken by a few millimetres along the outcrop. Some evidence of small- to medium-scale cut-and-fill structures	Typical occurrence is in top, finer-grained sequences of Little Jaloman bar
Sm/l Massive to plane-bedded sand	Coarse-grained sand, fairly well sorted and in places with interspersed granules	Occurs rarely and seems to be concentration of grains as lag deposit from initial erosion of basal cobble clinoforms beneath giant bar fine gravels	Typically occurs at boundary of **Gc/bp** and **Ggl** in roadcut section along Katun River downvalley from Great Jaloman bar
Sp Planar clinostratified sand	Medium- to coarse-grained sand, moderately well sorted	Planar, large-scale cross-sets (clinostratifications)	Typical exposure at base of Little Jaloman giant bar near mouth of tributary valley
Fm/Fr Silt to very fine-grained sand	Light olive grey (5Y 6/1) to very pale orange (10YR 8/1 to 8/2) silt to very fine sand	Massive to laminated. Ripple cross-laminations (some climbing) are locally present in sandier beds	Typical occurrence at top of sequences in giant bars, such as at Little Jaloman. Occurs above **Gdt** or **Sp**. Often found in association with formation of **Dm**. Also present at base of Komdodj bar near junction of Chuja and Katun rivers
Dm Diamicton	Silt to fine sand with scattered clasts that are usually very angular and often red in colour	Deformed, bimodal mixture of cobble gravel with silty matrix (**Dm**) Locally, flame structures and other convolutions occur within this facies due to loading by coarser beds	Typical occurrence at top of each sequence of giant bars, such as at Little Jaloman. Overlies **Gdt** facies and underlies or is associated with **Fm** facies
Lac Lacustrine deposits	**Lac1** Silt (glacial flour)	Massive to finely laminated light grey silt, locally intensely folded (Carling *et al.*, 2002, their Figure 17)	Typical lacustrine deposits of 'back-bar' areas
	Lac2 Silt alternating with intraformational gravel (intraclasts of the silt itself). Locally red staining (10 YR 8/8) in some silt layers	Laminated silt layers with massive to horizontal beds of intraformational gravel	
	Lac3 Coarse to very coarse sand and silt with disseminated pebbles	Cycles characterised by thin beds to laminae of moderately well-sorted, normally graded, coarse-grained sand alternating with silt to very fine-grained sand layers. Some sandy layers contain few pebble-size armoured silt balls. The silty sandy layers are either well sorted or contain disseminated small pebbles acquiring textural inversion. Several silt interlayers show deformation, such as flame structures, due to loading by coarse sands	

Key for labels: G–gravel, S–sand, F–silt, D–diamicton, Lac–lacustrine succession, b–boulder, c–cobble, p–pebble, g–granule, l–laminated/plane bed, t–medium- to large-scale trough cross-bed, p–large-scale planar cross-bed/clinostratification, r–ripple cross-lamination, m–massive.

Log Korkobi

Description The lower parts of a large section at the margin of this bar expose sediment sequences that may compose the adjacent river terrace and not the bar itself. The sediments in this lower sequence are quite distinct from descriptions of bar sediments advanced in this chapter and consist of a 40 m thick sequence of trough cross-bedded cobble beds interspersed with very poorly sorted coarser gravel beds that contain numerous boulders up to 2 m in diameter and thick units of angular diamicton, which appear to be large-scale grainflow deposits (Carling *et al.*, 2002). The giant-bar sediments themselves are exposed in poor sections that occur along a track cut in the side of the giant bar (Plate 36). This facies consists of coarse-granule and pebble gravels (**Ggl**) with a significant **Gg/bl** component (Table 13.1) which is a variant of the **Ggl** facies. The facies differs from most other **Ggl** examples in other bars because it has cobbles and small boulders sparsely to densely distributed along some bedding surfaces, many of which are dipping upvalley and contains pebble clusters dipping downvalley (Plate 36B, C).

Interpretation The lower sequence is believed to be Holocene terrace deposits infilling the valley and alongside the giant-bar sediments. However, the contact between the two facies is obscured. More detailed study of this location might help elucidate if the valley was filled with flood deposits and later entrenched or if the bars were deposited marginally and the terrace fill added later.

Inja bar

Description This large bar is well exposed where the Inja River cuts through it exposing a complex stratigraphy of gravel beds that contain large rounded pods of cobbles (Plate 32C) and sharp-edged intraclasts of laminated silts. The origin of pods is discussed below but they have been rarely described in the literature and so a brief description is provided here. A typical cobble pod has the appearance of an isolated rounded boulder at a distance but close-up it is seen to be a rounded, isolated mass of cobbles surrounded on all sides by finer gravels that are well-bedded away from the pod. However, usually the layering within the finer gravel is disrupted close to the pod indicating the pod was a 'solid' obstacle in a flow that deposited the finer gravel around it.

Behind the main bar in the tributary valley, lacustrine (**Lac**) deposits occur ponded by the main bar close to the village of Injuskha within the tributary valley (Figure 13.3, Plate 37A, B, C, D). Here the depositional sequence is less thick and not as high as the main bar but shows a thick succession (*c.* 50 m) of modified granule gravel (**Ggl/b**) with discontinuous single-clast thick, cobble to small-boulder lags along certain bedding planes and large fragments of conglomerate beds at the exposed base (Plate 32C, Plate 37B, Table 13.1). The cobble horizons suggests some rough separation of the deposits into repetitive units composed of thick **Ggl** alternating with thinner **Ggl/b** (Plate 32C) and with beds of **Gd/cl**. Repetitive sequences however are well defined with in the tributary valley behind the main Inja bar (Plate 38A, B) by the presence of three distinctive white-coloured lacustrine units, which are clearly interbedded with the gravel beds described immediately above. The lake sediments consist of thinly laminated (varve-like) silt beds (**Lac** facies) from 1 m to 2 m thick interbedded with three distinct granule and pebble gravels layers (**Ggl** and **Gd/cl** respectively) (Plate 38C, D). Thus, the sequence within the Inja valley, behind the main bar, begins with a gravel unit at the base and this is followed by two further gravel units separated by two silty lacustrine units, the final gravel layer being topped by a third lacustrine unit that forms the present day land surface that is labelled as such in Plate 38A.

Interpretation The main bar exposure shows flood deposits at the base that may have been partially re-eroded after initial deposition, and large bed-fragments reworked and mixed into the megaflood granule to small pebble deposits. Within these deposits, some stringers of larger clasts suggest fluctuations in flow conditions such that large clasts group as a lag on a bed surface subject to short-term erosion before being buried by an influx of finer grains, but there is no strong sedimentological or chronological evidence to determine whether these thick gravel units have been the product of a single flood, surges within a single flood or multiple floods. The evidence instead is clearer in the back-valley lacustrine deposits. There the silty laminations indicate sedimentation from silt-rich glacial meltwater flowing down the tributary valley into temporary lakes that formed behind a gravel bar after each flood episode. The sequence of three gravel units records three megafloods that overrode or broke through the main Inja bar and discharged into the back-bar area, each depositing gravel above a limnic layer (Plate 38C, D; Table 13.1). These gravel deposits then blocked the exit from the Inja valley. The tributary stream deposited fine silts in shallow lakes behind the gravel barriers after each flood episode. The lakes however were transitory and were drained as soon as the tributary could cut down through the impounding bar.

Little Jaloman bar

Description This is the best exposed and therefore archetypical bar where nine sections were examined from near the junction with the Katun River to the furthest up-tributary-valley extension of the bar deposits. The sections varied in thickness, up to 70 m in the former locality, to a *c.* 10 m thick section up valley. The sections have been synthesised in the composite stratigraphic log shown in Figure 13.5. The sedimentary succession consists of a stack of at least six distinct sequences. The sequence motif remains approximately the same everywhere within the tributary valley, except for the lowermost unit that is underlain by clinostratified sand (**Sp**) near the Katun–Little Jaloman valleys junction. In the lower main part of the bar the facies sequences consist of primarily plane-bedded, fine-grained gravel (**Ggl**) gradually overlain in places by coarser-grained pebble gravel (**Gd/cl**) containing lenses (cut-and-fill structures) of cobbles and occasional boulders. Facies **Gd/cl,** or **Ggl** directly, are overlain by sharp, planar to locally shallowly scoured contacts by trough cross-bedded, locally apparently massive, sandy gravel (**Gdt**) that is in turn overlain by layers of silt (**Fm**) or very fine-grained sand, often deformed and mixed with neighbouring fine gravel layers (Plates 37C and 39A). The top of each sequence often is characterised by a distinctive diamicton (**Dm**) consisting of angular red clasts (Plates 37D and 39A) partly convoluted with neighbouring beds. The upper main part of the bar has similar motifs except for finer-grained deposits and thinner sequences (Plate 39B). Accordingly, sandy **Sl** facies tend to replace the granule gravel (**Ggl**), which in turn replaces pebble gravel (**Gd/cl**) from the base to the top of the exposures. The sandy and fine-granule facies are locally cross-bedded, with beds dipping into the tributary. In the topmost finer sequences, bimodal (fine-grained sand and pebble gravel) layers are present that take the position of the diamicton units of the lower stratigraphic levels. A few isolated outsize clasts may be present in the upper part of the succession.

Interpretation The Little Jaloman bar has developed on the inside of a major valley bend in the lee of a resistant bedrock valley-side wall where the flood flow could expand both along the Katun valley and up the left tributary, i.e. the Little Jaloman River valley (Figure 13.3 and Plate 31A). The flow most likely was highly concentrated to hyperconcentrated in the lower part of the flood, and rapid deposition of a 'side-bar' occurred. The bar grew rapidly, keeping pace with the water filling of the valley, and developed submerged fan-like surfaces that dipped back into the Little Jaloman valley, where granule and fine-grained pebble gravel (**Ggl** and **Gd/cl** respectively) units developed. Internal inhomogeneities in sediment concentration and turbulent flow fluctuations led to development of the various laminations and other structures. Laminations with inverse grading can develop under upper-flow-regime conditions as frequently as normal graded ones (Cheel, 1990). However, here they may attest to the occurrence of deposition from traction carpets, but the orientation and imbrication of clasts and local cut-and-fill structures indicate highly turbulent flows close to the depositional surface. The deposition of cross-bedded granule to pebble gravels (**Gdt**) higher in the sequences indicates the onset of a lower flow regime, perhaps during the waning stages of the megaflood(s). These sequences are similar to those recognised in relatively confined pathways of southern Iceland affected by meltwater floods (Maizels, 1993; Russell and Knudsen, 1999, their Figures 4 and 5; Russell *et al.*, 2005, their Figure 7.18). The lithologies of the two sites are somewhat different, nevertheless the strong similarity in motif of the successions suggests further that the Katun megafloods recorded in the Little Jaloman bar may have been relatively short-lived, single-event, high-peak hydrographs similar to those associated with Icelandic jökulhlaups (Maizels, 1993). Recent flow modelling has shown that the largest Katun flood could have lasted one day only (Carling *et al.*, in press).

The termination of every major flood sequence recorded at Little Jaloman is indicated by the thin silt to very fine-grained sand layers resting on top of the trough-cross-bedded gravels. These fine materials were most likely deposited during slack-water stages from floodwaters trapped behind the bar. The wet conditions and relatively unstable, slightly inclined depositional surfaces (subaqueous fan-like surface) may have led to local slumping and mixing of the underlying cobble–pebble gravels with the silt to form the convoluted layers that characteristically separate the various flood sequences. The mixing would readily occur as floodwater finally drained from the depositional surface during flood recession. Some diamicton contain the reddened exotic angular clasts as well as clay intraclasts, which attest to the admixture of weathered debris-flow material sloughing from flood-disturbed valley walls. It is also possible that some of the convolution and admixing in these beds at the top of sequences may have occurred during the early stages of the subsequent flood. However, the interface between these upper beds of each sequence and the overlying deposition units (**Ggl**) that form the base of each sequence are not erosional, at least within the inner part of the Little Jaloman bar.

Finally, the general trend to thinning and fining of the sequences both up the tributary valley and vertically in the Little Jaloman sections indicate that only progressively finer deposits could be carried onto these sites. This may have occurred simply because the bar barrier was progressively higher for each flood and only the uppermost part of larger megafloods containing finer suspended material could flood over the previous bar surface.

The repeating occurrence of similar sequences indicates that the Little Jaloman bar was formed by multiple floods, six of which seem to be clearly definable. The occurrence of multiple floods is further supported by the interdigitations of glacial lake deposits and three flood gravel deposits in the back of the Inja bar. The sections within deposits of other bars, though, do not show evidence of lacustrine deposits and the exposed sections often seem only to record one mega-event. These differences in the style of deposition have to do with the local environment of deposition: that is, the interactions between the floods, valley morphology and geology produce vertical sedimentary sequences at each locale that may not correlate well with neighbouring sites of deposition. As indicated above, each bar formed in different settings, and although some of the main facies are similar, their stacking (associations, sequences) differs. Some potential areas for deposition of giant bars may have been filled by the first flood so that the accommodation space was occupied completely by sediment such that there was no deposition by later floods and thus no record of later floods. Other locations may have undergone several cycles of deposition and erosion by subsequent floods such that the bars are composite records of multiple flood deposition. The last large flood(s) though the system must also be responsible for the appearance of the major terrace levels, although further Holocene modifications occurred mainly through incision. Later floods would have eroded previously deposited megaflood deposits giving rise to the steep sides to the giant bars and the local boulder concentrations (**Gbb**, Table 13.1) seen at the foot of bar margins. Crescentic erosional scours at the upstream ends of some boulders (Herget, 2005), the deposition of large trough cross-bedded sands (**St**, Table 13.1), and the formation of large gravel dunes (**Gdp-t**; Table 13.1) also may be associated with later floods. Post-megaflood deposits on the terraces consist of local reworking of deposits by the Katun River to form braidplain topography, slumps, sheetflow deposits, and some small alluvial fans associated with streams issuing from the bounding hills.

Summary of lithofacies and interpretation The facies of the alluvial deposits of the Katun River and the tributary valleys are summarised in Table 13.1. At this point the facies present and their summary interpretations are described that relate to the interpretation of the valley fill as a whole. The sediment facies consist essentially of sandy gravels varying from cobble and boulder beds (facies **Gbp**, **Gbb**, **Gc/bp**) to beds of fine pebbles and granules (**Gdp/Gdt**, **Gd/cl**, **Ggl**, **Ggl/b**); very coarse-grained sands often with a fine gravel component (**Sl**, **Sm/Sl**, **Sp**, **St**; minor silt beds (**Fm**), a few diamicton layers (**Dm**); and lacustrine successions (**Lac**) of silts, sands and gravels in the tributary valleys. All these facies, except the silt beds, contain outsized clasts. Although some deposits superficially appear massive on first sight, they contain a variety of sedimentary structures that may include graded beds, plane beds and small- to medium-scale cross-beds.

(A) The following major facies have been observed on the terraces.

(1) Bouldery **Gbb** that forms boulder fields on the surface of terraces (Table 13.1; Herget, 2005).
(2) Trough cross-bedded, finer pebble to cobble **Gdp-t** that typifies giant dunes (Table 13.1 and see Carling, 1996).
(3) Large-scale, trough cross-bedded sand (**St**) that represents channel fills oriented sub-parallel to the Katun River valley (Table 13.1).
(4) Large-scale clinostratifications of sandy gravels (**Gc/bp**, Table 13.1) and sand (**Sp,** Table 13.1) are exposed under some terrace surfaces and locally at the base of giant bars (Figure 13.9A). Gravelly clinostratifications **(Gc/bp)** have been observed along a road-cut exposure of the Kezek-Jala bar just south of Big Jaloman at the left margin of the Katun River valley (Plate 35B, C) and within the terrace on the right margin opposite Little Jaloman village, and at Iodro village in the Chuja valley (Carling *et al.*, 2002). Sandy clinoforms **(Sp)** have been observed at the base of the Little Jaloman bar near the mouth of the tributary valley and include examples that dip steeply into the tributary from the trunk valley. The large-scale clinostratified deposits record sedimentation in basal fluid-separation flood-flows zones. The examples developed in gravel are in areas of the main valley directly impacted by the flood strength and probably reflect deposition on the steep lee sides of fully submerged gravel bars developing on the bed of the Katun valley. These gravel bars would be of an order of magnitude smaller than giant bars that later would develop above these initial clinoforms. The sandy clinoforms tend to occur in flood-expansion areas that are more protected from the main flood strength, such as within the mouths of tributary valleys.

(B) The following are major facies observed within the giant bars of the Katun River.

(1) *Description* The tops of a few giant bars show isolated boulders and the large Komdodj giant bar has beds of clinostratified, thick, poorly sorted, angular, boulder to cobble gravels with angular clasts towards the upper part of the flood sequence (**Gbp**) (top part of Plates 33B and 34A, Table 13.1).
Interpretation **Gbp** at Komdodj mostly includes many large angular boulders of the same lithology as the local valley wall. These boulders are evidence for a partially flood-modified rock slump from the nearby bedrock bastion of the left shoulder of the Katun River.

Elsewhere, isolated angular boulders reflect small rock-fall events onto the bar tops.

(2) *Description* The inner parts of the giant bars are primarily characterised by sandy, granule to pebble gravels, with minor boulders and few fine sand to silt layers.

(2a) The **Ggl** facies, with local slight modifications, is the most typical one and occurs in all the studied exposures of the giant bars of the Katun River valley. It is primarily characterised by plane-bedded, fairly well sorted, sandy, granule to fine pebble gravels (Table 13.1; Plates 33A, 34A, B, C, D, E, 35B (top part of image) and 37A). Most clasts derive from schist and are angular to subangular, and only the larger polymictic ones have slightly smoothed edges and corners. Variation occurs in terms of grain size within laminae, which includes some inverse grading, variations from layer to layer (Plates 34 and 37A) and local occurrence of sedimentary structures and pebble fabrics. For instance, small cut-and-fill structures and pockets of fine pebbles occur, which introduce minor variation within an otherwise uniform, wavy-bed morphology along the bedding planes (Plate 34D, E).

(2b) **Gd/cl** is mostly characterised by plane-bedded pebble gravel with some cobbles (Plate 32B) and some cut-and-fill structures (Table 13.1). It is coarser than, but has similar structural appearance to and usually overlies gradationally the **Ggl** gravel (Plates 33A and 34A).

(2c) **Gg/bl** is a variant of the **Ggl** facies. It differs because it has cobbles and small boulders sparsely distributed along some bedding surfaces (Plates 32C, 35A and 37B). In places, such as at the base of the Inja bar exposure, it has incorporated very large boulders or rounded large (*c.* 1.5 m diameter) pods of pebble and cobble gravel (Table 13.1; Plates 32C and 37B) and small (*c.* 0.10 to 0.20 m) blocky intraclasts of laminated silts. Such intraclasts were probably ripped out by floods from frozen units of sediments (e.g. Fisher and Smith, 1993; Lønne, 1997) already in the valleys. The rounding of the pods indicates that large frozen masses of gravel were subject to rolling and rounding prior to deposition whilst the smaller blocky intraclasts were not rolled and rounded and must have been sourced locally. Additional examples dominated by large boulders are found at the upstream end of the Little Jaloman Bar and the Log Korkobi bar.

(2d) **Gdt** is a moderately well-sorted, sandy pebble gravel with some cobbles, primarily trough cross-bedded, locally apparently massive (Plate 39A; Table 13.1). It is best developed in the Little Jaloman bar where it generally overlies **Gd/cl** or directly **Ggl** gravels.

(2e) Sand is mainly exposed in the lower part of some bars and in the topmost units of Little Jaloman bar. In the lower parts it occurs as massive to plane-bedded, coarse-grained sand (**Sm/l**) in the Kezek-Jala bar roadcut downstream from the Great Jaloman bar (**Sm/l**), where it is found as an erosional remnant over the large gravely clinostratifications of **Gc/bp** (Table 13.1), as well-sorted, very fine-grained to silty, cross-laminated sand (**Fr**) in the Komdodj bar (Figure 13.9A), and, in a similar position, as a thin regular bed of finer-grained to massive silty facies (**Fm**) near the base of the Little Jaloman bar exposure. In the upper part of the Little Jaloman bar laminated coarse-grained sand (**Sl**, Table 13.1) in part replaces and, in part, is interbedded with the granule gravel (**Ggl**) facies. This provides a general sequence-thinning and an overall fining upward of the Little Jaloman bar (Figure 13.5).

(2f) Massive silt to very fine sand (**Fm**) is also found as thin layers, most often disrupted, at different levels in the Little Jaloman bar (Table 13.1; Plate 32B, Figure 13.5). There the sandy silt overlies trough cross-bedded sandy gravels (**Gdt**). Where disrupted it mixes with pebbles and cobbles similar to those of the underlying **Gdt** facies, to develop poorly sorted, texturally bimodal diamicton (**Dm**) that is always found in discontinuous layers and lenses (Table 13.1, Plate 37C). The diamicton characteristically may also contain angular clasts that are often red in colour and associated with thin wispy red-coloured clay and silt (Plate 37D).

Interpretation The **Ggl** and **Gd/cl** gravels slightly differ in particle size, but the similarity in structures suggests that they were deposited by similar highly concentrated to hyperconcentrated flows under overall critical flow regimes. The presence of various features such as: horizons with different grain size within each facies; the apparent lateral continuity of laminae that in effect pinch out in a few metres distance; the presence of internal grading within many laminae; the development of uniform wavy structures; the shallow scours with coarse material, some with flat pebbles with upflow imbrication in part enhanced by syndepositional scouring; all suggest variable, pulsating concentration of sediment within the flow. Sediment deposition from basal hyperconcentrated layers could then be modified by local turbulent eddies wherever and whenever a small drop in sediment concentration was occurring. This can be envisaged to be similar to the deposition of the so-called traction carpets reported in fluvial gravel by Todd (1989) and Benvenuti and Martini (2002) as well as within other highly concentrated flows (Lowe, 1982; Postma *et al.*, 1988; Shanmugan, 1996; Sohn, 1997). The particular facies case (see 2c) at Log Korkobi and at the upstream end of the Little Jaloman Bar are characterised by large boulders.

Thin layers of silt to very fine sand (**Fm**) indicate deposition in a low-energy setting. The reddening of many of the clasts and the silt suggests that surface weathering had occurred for a yet undetermined length of time. These reddened angular clasts are found in layers at the top of several cycles that occur throughout the flood deposits and so might mark the cessation of a flood when debris flow clasts were spread over the surface of the bar rather than being mixed into the flood deposits or periods of exposure to weather of the bar tops between flood events. Their mixing with rounded pebbles of the underlying higher energy **Gdt** gravel to form the diamicton (**Dm**) lenses indicates the action of syndepositional and post-depositional disturbing processes, including possible local slumping, loading and dewatering that eroded and mixed the various materials to generate the textural inversion in the **Dm** layers.

(C) Some tributary valleys to the Katun River expose typical glaciolacustrine and fluvial deposits.

(1) *Description* Characteristic glaciolacustrine facies consist of light grey, locally yellowish weathered silt that occurs massive to laminated in intensely folded layers (**Lac1**), or laminated in horizontal beds alternating with intraclast pebble gravel (**Lac 2**), or as massive thin beds with disseminated pebbles in thin cycles alternating with coarse-grained apparently massive sand thin beds (**Lac3**) (Table 13.1). The best exposure occurs in the Inja tributary valley behind the Inja giant bar (Plate 31A) where the **Lac** facies overlie and interbed (Plate 38C, D) with granule and pebble gravels (**Ggl** and **Gd/cl** respectively) of the giant bars. Worm trails are locally abundant in the silt layers; rare fish bones have also been reported (Ragosin, 1942) but examination of the fine sediments revealed no microscopic organisms so the lakes were relatively barren (Carling *et al.*, 2002).

Interpretation The silt and fine sand of the glaciolacustrine laminations derive from the outwash of a glacier that generated an end moraine a few kilometres up the tributary valley (Plate 38B). The interbedded granule and pebbly gravels represent the upvalley portions of flood gravels at the back of the giant bars.

(2) Other tributary valleys do not have outcrops of lacustrine deposits. The Little Jaloman valley contains the distal deposits of the giant bar, with upvalley thinning of the deposits characterised primarily by granule gravels (**Ggl**) overlain by trough cross-bedded, pebble gravel (**Gdt**). Furthest into the tributary valley the lower parts the **Ggl** gravel have been variously dissected and are overlain by pebble gravel though the work of the Holocene–Present stream (Plate 38E).

13.5 Discussion

The sedimentological analysis of the megaflood deposits of the Katun River valley leads to at least two topics that require further consideration and discussion. The first relates to the formation of thick, mostly plane-bedded facies (**Ggl, Gd/cl**) present in every bar; the second is the comparison and contrast between these deposits and others in the Altai Mountains area and elsewhere, such as the Missoula floods in North America and the historically recorded floods of Iceland.

13.5.1 Facies formation

The frequent textural homogeneity (at the scale of tens of metres); pronounced lateral continuity of beds of laminated granules; regular stratification within the coarsening-up and fining-up cycles; marked absence of major scours or frequent outsized clasts, all indicate rapid deposition in a short time, largely from high-concentration suspension fallout onto laterally extensive upper-stage

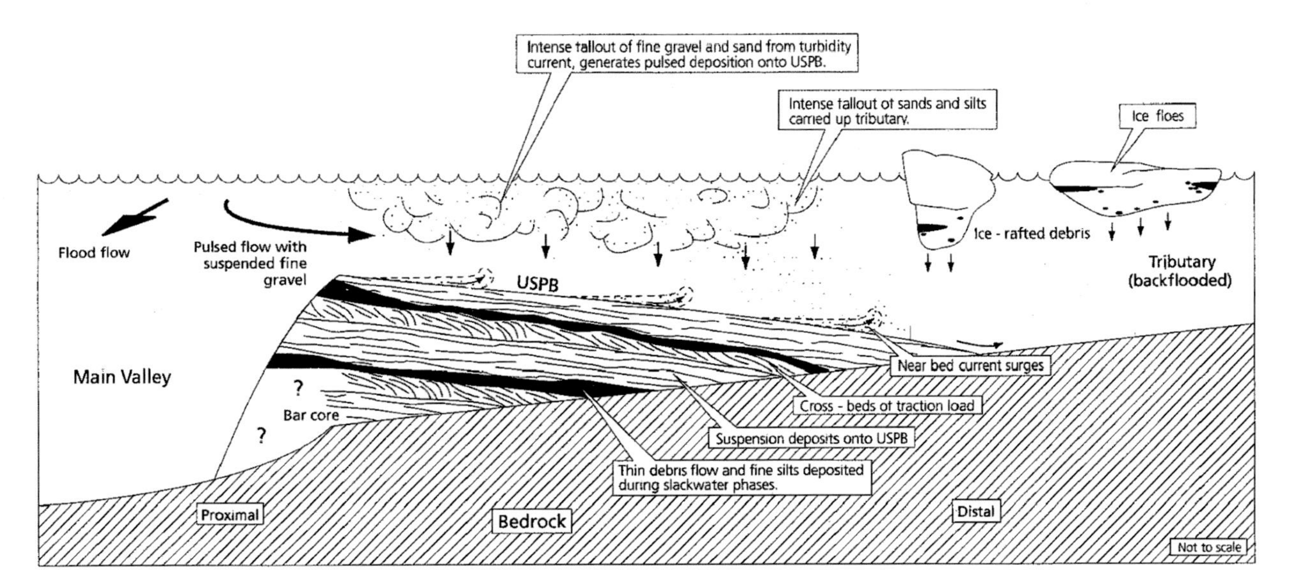

Figure 13.6. Model of formation of key sedimentary facies on inclined depositional surfaces of the Little Jaloman bar. The section represents a long profile from the main Katun valley into the backflooded tributary valley. In this simplified cartoon only three cycles of deposition are shown for simplicity although at least six have been recorded at this location. Fine gravel is deposited onto the flooded rapidly accreting bar surface largely from intense fallout from suspension and is then rearranged by unsteady (surges) traction transport within USPB flow. As flood flow wanes cross-beds form due to bedload sheet and minor ripple and dune migration across predominantly LSPB. Throughout the process coarse ice-rafted debris falls out to contribute the isolated outsized clasts observed at many sections (e.g. Plates 32C, 35A and 37B).

plane beds (USPBs), the competence of which declined distally up tributaries where lower-stage ripples are more commonly found. Frozen pods of cobbles and sharp-edged, tabular, layered slabs of silt must have been transported along the bed, but their preservation attests to a short transport distance and rapid burial. Thus the diffusively graded sediments, preservation of flame structures, soft-sediment intraclasts and preservation of 'originally-frozen' cobble pods attest to rapid sedimentation rates. Although rip-up clasts and 'floating boulders' can be 'buoyed-up' by turbidity currents in some environments (Postma *et al.*, 1988), in a proglacial environment they are not necessarily diagnostic of turbidity currents. Instead cobble pods, rip-up clasts and drop-stones can originate from melt-out of ice floes (Thomas and Connell, 1985; Lønne, 1997). Isolated pebbles and cobbles (Plates 32C, 35A and 37B) are believed to be drop-stones from ice floes in the flood waters.

At Little Jaloman, the vertical stacking of the various sequences in the large-scale cycles was generated by the recurrence of several megafloods, or pulses within a megaflood. The internal facies modulation within each sequence was constrained primarily by variability in the main flood flow competence controlling the supply of suspended sediment and the availability of various types of sediments. This primary control was mediated by additional traction load introduced to the flood flow by debris flow from the valley side-walls as was noted above and is explained below.

Outburst floods from ice-dammed lakes are frequently 'pulsed' (e.g. over periods of some hours) on the rising limb (Haeberli, 1983; Tweed and Russell, 1999) and this requires some consideration. Ice-dam failure may be due to overtopping such that an open channel progressively develops within the ice (Carling *et al.* in press). Alternatively, there may be no open channels, but rather conduits may develop within or at the base of the ice (Russell *et al.*, 2005). In either instance, a pulsed nature to the flood hydrograph may occur because tunnels or channels may enlarge through thermal erosion and corrasion from suspended load, but both are subject to sudden partial blocking by ice collapsing along the tunnel and channel margins. Equally, where the mode of failure is exceptionally fast (that is, catastrophic) such that the ice-dam disintegrates, the large quantity of ice debris may cause temporal and spatial series of ice jams to develop in the course of the floodwaters, which impound and then release floodwaters as a series of surges (Tweed and Russell, 1999).

Consequently, a series of flood pulses in the main river valley may have occurred during individual flood hydrographs introducing high concentrations of fine suspended gravels (**Ggl**) into the tributary valleys (Figure 13.6), the latter already having been backflooded by the rising limb of the flood hydrograph. Thus some temporal unsteadiness in the current within backflooded areas would be expected and this can account for some of the variation in the styles of sedimentation within individual sequences. During the periods of strong flood flow the coarsest boulders would be found primarily as bedload in the main valley but many would readily be broken down. In contrast, the finest gravels could move in traction or indeed in suspension and be deposited in 'backflooded' areas. At the same time, slope talus and other valley-side wedges of sediments would be undercut and slump into the floodwaters surging down the Katun valley. During the main flood pulses, this colluvial material would have been well mixed with the flood-borne suspended fine gravels, so that, of the debris-flow material, only large angular fragements and rip-up intraclasts can be recognised within the **Ggl** gravels. However, as flood flow weakened and water levels began to fall, hydrostatic support to the valley-wall slopes above flood levels would decrease such that the incidence of small regolith failures would increase (Morgenstern, 1963, 1967; Shaw, 1977). During this period the input of coarse-talus and debris-flow material would increase compared with the input during the passage of the main flood wave. However, in contrast to the main flood channel, within the relatively 'calm' waters of the embayments, mixing of the sloughed material with suspended fine sediments and with the previously deposited fine gravels would be poor, so that distinctive, coarse, angular bedload/debris-flow deposits were laid down at the top of each flood sequence effectively marking the end of a flood pulse or cycle and demarcating a boundary between each depositional event. During the recession limb of a flood wave or during the 'slackening flow' between flood pulses, thin drapes of clay-silt material (**Fm**) were deposited and mixed with debris-flow material (**Dm**) in the ponded water behind the bar. As water levels varied, strong fluctuations in hydrostatic pressure within the pore space of the unconsolidated gravels may have led to instability of the material, slumping and mixing of gravel and also may account for enrichment of silt and clay within some layers of the gravels.

Given the above arguments, it is evident that the distinct repetitive first-order cycles in the sediment pile are 'event-based', reflecting the pulsatory nature of deposition associated with major variations in the flood hydraulics. Such fluctuations in flow can be attributed to such mechanisms as unsteady overtopping and flow through a partially collapsed ice dam, followed by successive releases of water as the dam progressively failed. The resultant series of flood surges could also be affected by the temporary blocking by ice debris of constrictions in the Katun valley, as noted above. Such a model of progressive failure would result in a ramped increase in flood elevation with intervening short periods of static water level or draw-down followed by a sharp recession limb as the lake finally drained. This model is in accord with the rising limbs of monitored jökulhlaups, where ice-dam failure is progressive; the hydrograph rising as a series of pulses (events) before falling abruptly (e.g. Haeberli, 1983). Nevertheless, given the scale of flooding, it is probable that the dam eventually failed completely (Baker *et al.*, 1993).

An alternative model for the thin wispy **Dm/Fm** facies has been proposed by Gregory Brooks (personal communication, 2006). This model envisaged the **Ggl**, **Sl** and **Sp** facies being deposited by megafloods that fill the main valley with sediment. Hence, the Katun River and the tributaries (e.g. the Little Jaloman River) are locally graded to a higher base level after each flood, but then cut down through the unconsolidated fine gravel during periods of 'normal' river flow. For each period of higher base level, the Katun River and the tributary streams flood over the gravel flood plains during periods of seasonal ice melt depositing thin muds, thin cobble beds and thin debris-flow layers. These layers are readily disrupted by later megaflood gravels deposited on top of them by the next flood event, and become exposed as the **Dm/Fm** facies in the giant bar sections as the rivers incise. At present it is not possible to choose between this model and the one advanced earlier in this chapter.

Flow variability associated with lamination requires consideration. The small-scale (mm to a few cm) coarsening-upward or fining–coarsening–fining sets in mid-cycle are often amalgamated, indicative of rapid variations in sediment supply (Rust and Romanelli, 1975) and rapid deposition of the set sequences, and probably result from thin, high-density super-critical flows, traction carpets and bedload pulses (Postma, 1986; Postma and Cruickshank, 1988; Todd, 1989; Sohn, 1997). Cheel (1990) demonstrated that both fining-upward and coarsening-upward 1 mm thick laminae in fine sand beds can be related to burst–sweep cycles on USPBs, and the model of Sohn (1997) in particular can allow for stacks of thick laminations to develop. Sallenger (1979), Lowe (1982) and Hiscott and Middleton (1980) have demonstrated that a steady dispersive stress can readily generate coarsening-upward gravel units up to 100 mm thick on slopes of as little as 10° (Hiscott, 1994). Cycles of growth and collapse in the traction layer are mediated by the thickness of the layer and not fluctuations in the flow (Hiscott, 1994). On gentle slopes a temporary increase in flow competence can generate inverse grading (Allen, 1981), a process Kneller (1995) terms 'depletive waxing flow', which is particularly associated with regular pulsing of flood flow or the passage of turbulent coherent flow structures (Kneller, 1995). At the smallest scale, such near-bed velocity fluctuations might be related to the burst–sweep cycle inherent in fluid flow (Hiscott, 1994) and these coherent flow structures may account for the finer detail of individual laminae.

Generally these flow structures are temporally and spatially too small to account for the medium-scale (*c.* decimetre spaced) stratification observed. Consequently, unsteadiness in the main flood discharge propagating into the embayment is the preferred mechanism. The sweep–fallout model (Hiscott, 1994) requires a strongly pulsating short-period traction current and sustained rapid deposition from suspension via strong traction flows with 'freezing' at the bottom. The thin repeated bedding of medium-scale sets can then be related to flow pulsation associated with rapid deposition (Vrolijk and Southard, 1997), rather than processes peculiar to traction carpets per se (Hiscott, 1994). The presence of sand-infused gravel interstices above and below some open-work gravel sets might indicate that the sweep–fallout model is to be preferred. Fine sediment filled the interstices of deposited gravels during low-speed flow, but was held in suspension as flow strengthened so that only coarse gravel was deposited during the main flow pulse. As the flow waned again, fine sediment infiltrated into the upper part of the newly deposited bed (Sohn, 1997).

In considering the nature of megaflood dynamics it is appropriate to note similarities and differences between the well-studied Missoula bars and the Katun sediment sequences. Although a large number of localities have been examined by investigators of the Missoula floods (summarised by Waitt 1980; Smith 1993), for brevity these are referred to here generically as Missoula deposits.

13.5.2 Similarities

Although exact dates for the Katun and Chuja valley bars have still to be defined, they are most probably post-40 ka, which is consistent with Missoula bars that are generally younger than 20 ka and no older than *c.* 36 ka (Waitt, 1980; Smith, 1993). At both localities the bars appear to have been deposited concurrently with huge subfluvial transverse bars (steep-fronted deltas or subaqueous fans) in the main valleys (Waitt, 1980). Beds in both the Little Jaloman and Missoula bars dip up-tributary whilst grain size also grades in the same direction. Presumed drop-stones floating in finer matrix and reverse-flow indicators have been identified at both localities and similar-scale channelling of the top of individual sequences is equally attributed to subaerial exposure. More importantly, at both locations the cycles are monotonously repetitive, although they are thicker and fewer at Little Jaloman than at Missoula. Waitt (1980, p. 657), in particular, commented, 'the regularity and similarity of the ... [Missoula] ... cycles are far more impressive than their variations'. In the case of Missoula deposits, Moody (1988) describes unsorted massive gravel units 7 m to 37 m thick grading up into cyclic beds ranging in thickness from 0.04 m to 0.46 m. For example, near Wanapam Dam, Moody interprets three sequences as indicating surging within the flood rather than distinct flood episodes.

13.5.3 Differences

Despite similarities, important differences in styles of deposition occur. In the interpretation of the Little Jaloman sequences a coarse gravel/debris flow unit separates each sequence. Waitt (1980, p. 658), in contrast, noted that each Missoula cycle had 'a thick basal pebbly coarse sand to granule gravel ... conspicuously channelled into the underlying cycle' – thus Waitt clearly placed the coarsest unit at the base of each cycle. If the silt-infused coarse gravel/debris-flow layer separating each Jaloman cycle were also seen as a basal component then Missoula cycles and Jaloman cycles would be very similar stratigraphically. However, on the basis of process sequence, for the Katun deposits, it is preferred to envisage the debris-flow/gravel episode as the last phase of a cycle of deposition rather than marking the onset of the next cycle. At Little Jaloman in contrast to Missoula there is no evidence of biota/loess/tephra. The latter were used by Waitt (1980) to indicate decades of subaerial exposure for the bar top. Instead, at Little Jaloman no lengthy hiatus in deposition is indicated, from which it follows that each of the Little Jaloman cycles represented closely spaced floods into dry embayments. Pulsed flow in the main Katun valley sustained flooding of the the embayments and deposition of back-bar sedimentary units characteristic of unsteady flows and similar to turbidites. Only towards the end of each flooding cycle were 'ponded' environments of deposition created and fine silts and clays deposited. Whereas Baker (1973), amongst others, attributed Missoula deposits to the rise of ponded water, Waitt (1980, p. 672) argued that each of 40 or so Missoula cycles was not a turbidite per se but 'records an up-valley surge of a rapidly deepening flood over dry land ...'. To reconcile these conflicting models Smith (1993) has argued that at Missoula there is an important distinction between (a) deposits that indicate flooding by the rise of ponded water: termed *passive flooding*; and (b) deposits that indicate energetic flood surging: termed *dynamic flooding*. However, at Little Jaloman, from the sedimentological evidence, there seems no reason to disassociate the two mechanisms. Once the bar core was formed by initial flooding during the rising flood stage, the area behind the bar would be effectively flooded but this backwater would certainly not be still-water, being characterised by recirculating flows and also subject to series of surges as the flood level progressively rose. If filling of the embayments by sediment from each flood is regarded as a first-order event producing the main cycles in sedimentation, then surge effects would be secondary in comparison and may account for much of the local detail within sequences at Little Jaloman.

The morphological evidence of large primary bars filling every tributary and the lack of secondary bars 'inset' against the outsides of the primary bars suggest that one or more flood was large and largely filled the available accommodation space with sediment. Further detailed study of individual bars would be required to ascertain if they are contemporaneous. However, at Little Jaloman and at Inja the dated sequences young upwards and, moreover, the presence of inset flood units intercalated with three sets of lake deposits in the Inja valley indicate that subsequent floods both deposited sediments on the top of the primary bars and also breached some bars. Thus, in the Katun valley there is good evidence for three large floods (in the Inja tributary valley) and possibly six or more may have occurred in total when the evidence from the Little Jaloman tributary-valley sequences are considered.

13.6 Conclusions

Sedimentary valley fill within the Chuja and Katun valleys of the Altai Mountains of south-central Siberia consists primarily of valley-marginal giant-bar deposits and valley-floor terraced gravels. The sedimentology and stratigraphy may be best explained as owing to Quaternary megaflood deposition within the valley network modified by Holocene braidplain evolution and dissection.

Largely, or in part, the fill originally may have been complete across the valley floor, with the giant bars now representing marginal remnants of this fill trenched by later river action. Alternatively, gigantic gravel bars, some 120 m high (or about 300 m if related to the current and preflood valley bottom) and up to 5 km in length, may have developed across the entrances to valleys tributary to the Chuja and Katun rivers with sedimentation occuring in the main valley floors but at a lower elevation in comparison with the height of the bars. Presently, there is insufficient evidence to choose between these two styles of sedimentation. The origin of the floodwaters was the ice-impounded glacial Lake Chuja–Kuray, which emptied repeatedly. The evidence from the Inja valley deposits indicates at least three major floods occurred and the first deposited a series of *primary* bars down the Katun valley including the bar that closed off the Inja valley. Subsequent floods deposited sediments on top of the original bars and also breached the bars producing, at Inja, three major complexes of flood gravels intercalated with a series of three lake deposits. The sedimentary sections within the neighbouring Little Jaloman valley do not include any limnic units, which implies that the gravel bar that blocked the Little Jaloman valley was breached rapidly by the tributary stream and no lakes formed. It is difficult to imagine that the bars at Inja and Little Jaloman are not contemporaneous and so the sedimentary sequence at Little Jaloman results from the three floods recognised at Inja, in part masked by possibly other floods and the overprint of the effects of pulsed flow into the embayment during each main flood.

Thus, in summary, sedimentological evidence indicates that the stratigraphy of the bars may be attributed to a series of individual floods that either overtopped pre-existing flood bars or inundated the tributary valleys through the gaps in the bars that would have been cut by the tributary streams during the long periods between between main-valley flood events. Ramped increases in flood level in the main valley may have resulted in distinct flood surges within individual flooding episodes within embayments, which provide some of the local differences in styles of sedimentation. Each flood or flood pulse entering the tributary valleys caused the deposition of an individual sequence of sedimentation, which is here termed a 'cycle'. Sedimentological evidence indicates that there is no obvious hiatus in deposition between cycles of sedimentation, although the C^{14} dating control at Inja would suggest several thousand years separate some flood units. A simple conceptual model is proposed (Figure 13.6) to link environmental conditions and flood hydraulics with the observed sediment characteristics.

Although there are important differences, the timing, scale and basic style of bar deposition in the Altai Mountains of Asia are strikingly similar to those described for the Missoula-flood bars in North America, and the sedimentology is similar to that of some Icelandic jökulhlaup deposits. These observations raise the possibility of developing a general model for styles of megaflood deposition both modern and ancient; in particular, for those depositional landforms that were constructed at the end of the Quaternary period across the Northern Hemisphere. Deviations in the styles of sedimentation from such a model, especially comparing reaches with different morphological characteristics, then might be interpreted in terms of the morphological variability along the flood course and the hydraulic behaviour of the palaeoflood wave. Such detailed examination of facies variation could provide local information with respect to the detail of the flood progression and hydrodynamics.

Acknowledgements

Data were collected during expeditions to the area between 1994 and 2006 sponsored by the German Research Foundation (DFG), the Royal Society of London, the Royal Geographical Society of London, Tomsk State Pedogogical Institute, Tomsk University and the Tomterra Company. Professor A. Rudoy (TSPI) is thanked for introducing us to the diluvial landscapes of the Altai. Doctors Alistair Kirkbride, Christoph Siegenthaler, Suzanne Leclair and Greg Brooks provided stimulating discussion in the field. Doctors Greg Brooks, Timothy Fisher and Andrew Russell provided constructive reviews of earlier manuscripts, which assisted in the presentation of the final arguments.

References

Allen, J. R. L. (1981). Sediments and processes on a small alluvial stream-dominated Devonian alluvial fan, Shetland Islands. *Sedimentary Geology*, **29**, p. 31–66.

Baker, V. R. (1973). *Paleohydrology of Catastrophic Pleistocene Flooding in Eastern Washington*. Geological Society of America Special Paper, 144.

Baker, V. R. (2002). The study of superfloods. *Science*, **295**, 2379–2380.

Baker, V. R. and Bunker, R. C. (1985). Cataclysmic late Pleistocene flooding from glacial Lake Missoula: a review. *Quaternary Science Reviews*, **4**, 1–41.

Baker, V. R., Benito, G. B. and Rudoy, A. N. (1993). Paleohydrology of Late Pleistocene superflooding, Altay Mountains, Siberia. *Science*, **259**, 348–350.

Benvenuti, M. and Martini, I. P. (2002). Analysis of terrestrial hyperconcentrated flows and their continental deposits. In *Flood and Megaflood Processes and Deposits*, eds. I. P. Martini, V. R. Baker and G. Garzon. Sedimentology Special Publication 32, pp. 167–194.

Blyakharchuk, T., Wright, H. E., Borodavko, P. S., Van Der Knaap, W. O. and Ammann, B. (2004). Late Glacial and Holocene vegetational changes on the Ulagan high-mountain plateau, Altai Mountains, southern Siberia. *Palaeogeography Palaeoclimatology Paleoecology*, **209**, 259–279.

Bretz, J H. (1928). The Channeled Scabland of Eastern Washington. *Geographical Review*, **18**, 446–477.

Bretz, J H., Smith, H. T. U. and Neff, G. E. (1956). Channeled Scabland of Washington; new data and interpretations. *Geological Society of America Bulletin*, **67**, 957–1049.

Buslov, M. M., Zykin, V. S., Novikov, I. S. and Delvaux, D. (1999). The Cenozoic history of the Chuja Depression (Gonry Altai): Structures and geodynamics. *Geologiya i Geofizika*, **40** (12), 1720–1736 (in Russian).

Carling, P. A. (1996). Morphology, sedimentology and palaeohydraulic significance of large gravel dunes, Altai Mountains, Siberia. *Sedimentology*, **43**, 647–564.

Carling, P. A., Kirkbride, A. D., Parnachov, S., Borodavko, P. S. and Berger, G. B. (2002). Late-glacial catastrophic flooding in the Altai Mountains of south-central Siberia: a synoptic overview and introduction to flood deposit sedimentology. In *Flood and Megaflood Processes and Deposits: Recent and Ancient Examples*, eds. I. P. Martini, V. R. Baker and G. Garzon. Special Publication 32 of the IAS, Oxford: Blackwell Science.

Carling, P., Villanueva, I., Herget, J. *et al.* (in press). Unsteady 1-D and 2-D hydraulic models with ice-dam break for Quaternary megaflood, Altai Mountains, southern Siberia. *Global and Planetary Change*, Special Issue.

Cheel, R. J. (1990). Horizontal lamination and the sequence of bed phases and stratification under upper-flow-regime conditions. *Sedimentology*, **37**, 517–529.

Chikov, B. M., Zinoviev, S. V. and Deyev, E. V. (2008). Mesozoic and Cenozoic collisional structures of the southern Great Altai. *Russian Geology and Geophysics*, **49**, 323–331.

Fisher, T. G. and Smith, D. G. (1993). Exploration for Pleistocene aggragate resources using process-depositional models in the Fort McMurray region, NE Allberta, Canada. *Quaternary International*, **20**, 71–80.

Gilliespie, A. R., Burke, R. M., Komatsu, G. and Bayasgalan, A. (2008). Late Pleistocene glaciers in Darhad Basin, northern Mongolia. *Quaternary Research*, **69**, 169–187.

Haeberli, W. (1983). Frequency and characteristics of glacier floods in the Swiss Alps. *Annals of Glaciology*, **4**, 85–90.

Herget, J. (2005). *Reconstruction of Pleistocene Ice-Dammed Lake Outburst Floods in the Altai Mountains, Siberia*. Geological Soceity of America, Special Paper 386.

Hiscott, R. N. (1994). Traction-carpet stratification in turbidites: fact or fiction? *Journal of Sedimentary Research*, **A64** (2), 204–208.

Hiscott, R. N. and Middleton, G. V. (1980). Fabric of coarse deep-water sandstones, Tourelle Formation, Quebec, Canada. *Journal of Sedimentary Petrology*, **50**, 703–722.

Jaegar, C. (1980). *Rock Mechanics and Engineering*, Cambridge: Cambridge University Press.

Kneller, B. (1995). Beyond the turbidite paradigm: physical models for deposition of turbidites and their implications for reservoir prediction. In *Characteristics of Deep Marine Clastic Systems*, eds. A. J. Hartley and D. J. Prosser. Geological Society Special Publication 94, pp. 31–49.

Koppes, M., Gillespie, A. R., Burke, R. M., Thompson, S. C. and Stone, J. (2008). Late Quaternary glaciation in the Kyrgyz Tien Shan. *Quaternary Science Reviews*, **27**, 846–866.

Lønne, I. (1997). Facies characteristics of proglacial turbiditic sand-lobe at Svalbard. *Sedimentary Geology*, **109**, 13–35.

Lowe, D. R. (1982). Sediment gravity flows II. Depositional models with special reference to the deposits of high-density turbidity currents. *Journal of Sedimentary Petrology*, **52**, 279–297.

Maizels, J. (1993). Lithofacies variations within sandur deposits: the role of runoff regime, flow dynamics and sediment supply characteristics. *Sedimentary Geology*, **85**, 299–325.

Moody, U. L. (1988). Late Quaternary stratigraphy of the Channeled Scabland and adjacent areas. Unpublished Ph.D. thesis, University of Idaho, USA.

Morgenstern, N. R. (1963). Stability charts for earth slopes during rapid drawdown. *Geotechnique*, **13**, 121–131.

Morgenstern, N. R. (1967). Submarine slumping and the initiation of turbidity currents. In *Marine Geotechnique*, ed. A. F. Richards. University of Illinois Press, pp. 189–220.

Novikov, I. S. and Parnachev, S. V. (2000). Morphotectonics of late Quaternary lakes in river valleys and intermountain troughs of southwestern Altai. *Geologiya I Geofiska*, **2**, 227–238 (in Russian).

Ota, T., Utsunomiya, A., Uchio, Y. *et al.* (2007). Geology of the Gorny Altai subduction-accretion complex, southern

Siberia: tectonic evolution of an Ediacaran-Cambrian intra-oceanic arc-trench system. *Journal of Asian Earth Sciences*, doi:10.1016/j.jseaes.2007.03.001.

Postma, G. (1986). Classification of sediment gravity flow deposits based on flow conditions during sedimentation. *Geology*, **14**, 291–294.

Postma, G. and Cruickshank, C. (1988). Sedimentology of a late Weichselian to Holocene terraced fan delta, Varangerfjord, northern Norway. In *Fan Deltas: Sedimentology and Tectonic Settings*, eds. W. Nemec and R. J. Steel. Glassgow, London: Blackie and Son, pp. 144–157.

Postma, G., Nemec, W. and Kleinspehn, K. L. (1988). Large floating clasts in turbidites: a mechanism for their emplacement. *Sedimentary Geology*, **58**, 47–61.

Ragosin, L. A. (1942). Middle Katun valley terrace. In *Investigation and Exploration of the Productive Force of Siberia*, pp. 36–107. Proceedings of a Scientific Conference, Tomsk University, Tomsk (in Russian).

Reuther, A. U., Herget, J., Ivy-Ochs, S. *et al.* (2006). Constraining the timing of the most recent cataclysmic flood event from ice-dammed lakes in the Russian Altai Mountains, Siberia, using cosmogenic in situ 10Be. *Geology*, **34**, 913–916.

Rudoy, A. N. (1988). Regime of glacial-dammed lakes in the intermontane basins of South Siberia. *USSR Academy of Sciences Materials of Glaciological Research*, **61**, 36–42 (in Russian).

Rudoy, A. N. (1990). Ice flow and ice-dammed lakes of the Altai in the Pleistocene. *USSR Academy of Sciences Izvestiya, Seriya geograficheskaya*, **120**, 344–348 (in Russian).

Rudoy, A. N. (1998). Mountain ice-dammed lakes of southern Siberia and their influence on the development and regime of the intracontinental run-off systems of north Asia in Late Pleistocene. In *Palaeohydrology and Environmental Change*, eds. G. Beninto, V. R. Baker and K. J. Gregory. Chichester: John Wiley & Sons, pp. 215–234.

Rudoy, A. N. and Baker, V. R. (1993). Sedimentary effects of cataclysmic late Pleistocene glacial outburst flooding, Altay Mountains, Siberia. *Sedimentary Geology*, **85**, 53–62.

Rudoy, A. N., Galakov, V. P. and Danilin, A. L. (1989). Reconstruction of glacial drainage of the upper Chuja and the feeding of ice-dammed lakes in the Late Pleistocene. *All-Union Geographical Society Proceedings*, **121**, 236–244 (in Russian).

Russell, A. J. and Knudsen, Ó. (1999). An ice-contact rhythmite (turbidite) succession deposited during the November 1996 catastrophic outburst flood (jökulhlaup), Skeiðarárjökull, Iceland. *Sedimentary Geology*, **127**, 1–10.

Russell, A. J., Fay, H., Marren, P. M., Tweed, F. S. and Knudsen, Ó. (2005). Icelandic jökulhlaup impacts. In *Iceland: Modern Processes and Past Environments*, eds. C. Caseldine, A. Russell, J. Harðardóttir and Ó. Knudsen. Developments in Quaternary Science, 5, Amsterdam: Elsevier, pp. 153–203.

Rust, B. R. and Romanelli, R. (1975). Late Quaternary subaqueous outwash deposits near Ottawa, Canada. In *Glaciofluvial and Glaciolacustrine Sedimentation*, eds. A. V. Jopling and B. C. McDonald. SEPM Special Publication 23, pp. 177–192.

Sallenger, A. H. (1979). Inverse grading and hydraulic equivalence in grain-flow deposits. *Journal of Sedimentary Petrology*, **49**, 553–562.

Shanmugan, G. (1996). High-density turbidity currents: are they sandy debris flows? *Journal of Sedimentary Research*, **66**, 2–10.

Shaw, J. (1977). Sedimentation in an alpine lake during deglaciation; Okanagan Valley, British Columbia, Canada. *Geografiska Annaler*, **59**, 221–240.

Sheinkman, V. C. (2002). The diagnostic age of glacial sediments of the Altai, tested using sections of the Dead Sea. In *Data of Glaciological Studies*, Publication 93, Institute of Geography of the Russian Academy of Science, Glaciological Association, pp. 41–55. (In Russian with English summary.)

Smith, G. A. (1993). Missoula flood dynamics and magnitudes inferred from sedimentology of slack-water deposits on the Columbia Plateau, Washington. *Geological Society of America Bulletin*, **105**, 77–100.

Sohn, Y. K. (1997). On traction-carpet sedimentation. *Journal of Sedimentary Research*, **67A**, 502–509.

Thomas, G. S. P. and Connell, R. J. (1985). Iceberg drop, dump, and grounding structures for Pleistocene glacio-lacustrine sediments. *Journal of Sedimentary Research*, **55** (2), 243–249.

Todd, S. P. (1989). Stream-driven, high-density gravelly traction carpets: possible deposits in the Trabeg Conglomerate Formation, SW Ireland and some theoretical considerations of their origin. *Sedimentology*, **36**, 513–530.

Tweed. F. S. and Russell, A. J. (1999). Controls on the formation and sudden drainage of glacier-impounded lakes: implications for jökulhlaup characteristics. *Progress in Physical Geography*, **23**, 79–110.

Vrolijk, P. J. and Southard, J. B. (1997). Experiments on rapid deposition of sand from high-velocity flows. *Geoscience Canada*, **24**, 45–54.

Waitt, R. B. (1980). About forty last-glacial Lake Missoula jokulhlaups through southern Washington. *Journal of Geology*, **88**, 653–679.

Zykin, V. S. and Kazanskii, A. Y. (1995). Stratigraphy of Cainozoic (Pre-Quaternary) deposits of the Chuja depression in Gorny Altai. *Geologiya I Geofizika*, **36** (10), 75–84.

14

Modelling of subaerial jökulhlaups in Iceland

SNORRI PÁLL KJARAN,
SIGURÐUR LÁRUS HÓLM,
ERIC M. MYER,
TÓMAS JÓHANNESSON
and PETER SAMPL

Summary

The flow of subaerial jökulhlaups is in principle similar to other subaerial water floods, such as dam-break floods, although some jökulhlaups may carry so much suspended sediment and ice fragments that they would be more appropriately described as rapidly flowing debris flows or lahars. Many subaerial jökulhlaups start out as subglacial floods and propagate as subaerial floods below an outlet at the glacier terminus. Other jökulhlaups, in particular many outburst floods caused by volcanic eruptions, lead to a partial or almost complete breakup of the glacier along the flow path and become subaerial after flowing only a short distance subglacially. The dynamics of the subaerial part of jökulhlaups differs fundamentally from the dynamics of the part of the flood that flows along the bed of the glacier or ice cap. The estimated discharge of jökulhlaups observed at many locations in Iceland during the twentieth century ranges from 0.1 to $300 \times 10^3\ m^3\ s^{-1}$ and prehistoric jökulhlaups have been estimated to have reached on the order of $10^6\ m^3\ s^{-1}$. The subaerial propagation of jökulhlaups can cause widespread damage to buildings, roads, communication lines and farmland. Two-dimensional numerical modelling, based on a depth-integrated formulation of the dynamics of shallow water flow, has been used to study the flow of subaerial jökulhlaups at four locations in Iceland, two of which are described in this chapter. Model results include estimates of travel times, the most probable flood routes and the extent of lowland areas that might be flooded. This capacity has enabled the development of flood hazard management planning, which includes installation of early warning systems and determination of endangered areas and evacuation routes.

14.1 Introduction

Jökulhlaups and lahars represent one of the most significant dangers due to subglacial eruptions. The term 'jökulhlaup' in Icelandic is traditionally used to denote any type of outburst flood from a glacier or an ice cap irrespective of the amount of suspended sediment or transported ice blocks. The Javanese term 'lahar' is, on the other hand, most often used for debris flows composed of volcanic debris and water, where the concentration of unconsolidated sediments typically exceeds 50%, according to Iverson (1997). Lahars often are caused by volcanic eruptions but the term is also used for mass flows that arise when sediments on steep slopes become unstable and surge down the mountainside as water-saturated slurries, a well-known example being the lahars from Ruapehu mountain in New Zealand (Keys, 2007). Both terms have been adopted internationally in scientific discussion of outburst floods and landslides from glaciers and volcanoes. 'Jökulhlaup', in that connection, is used to denote rapid water floods arising from drainage of glacier-dammed lakes or impounded water within glaciers.

Most jökulhlaups from the Mýrdalsjökull and Eyjafjallajökull ice caps in southern Iceland are due to subglacial volcanic eruptions, but some of the smaller floods may be caused by accumulation of meltwater in cauldrons that are formed in the ice caps by subglacial geothermal areas. Many jökulhlaups have issued from the ice caps during the history of the settlement of Iceland (Guðmundsson and Högnadóttir, 2005) and prehistoric jökulhlaups have been inferred from geological evidence in the surrounding terrain (Larsen *et al.*, 2005). The most frequent jökulhlaups from Mýrdalsjökull and Eyjafjallajökull have originated in the Katla volcano under Mýrdalsjökull and have flowed eastward. The best known such flood occurred in 1918 and reached an estimated peak discharge on the order of $300\,000\ m^3\ s^{-1}$ (Tómasson, 1996). Large amounts of sediments are transported by some of these floods and the composition of the flood material has been a matter of discussion in studies of the 1918 jökulhlaup. The conclusion of this discussion is that the Katla flood in 1918 should be considered a water flood rather than a debris flow (Tómasson, 1996), and therefore the term jökulhlaup is used here. It is possible, however, that floods arising from eruptions at some locations will be so rich in sediments that they should be considered debris flows or lahars.

Jökulhlaups from Katla and Eyjafjallajökull endanger settled areas and are of particular concern to Icelandic civil defence authorities, who recently commissioned a study of potential hazards associated with eruptions

Megaflooding on Earth and Mars, ed. Devon M. Burr, Paul A. Carling and Victor R. Baker.
Published by Cambridge University Press.

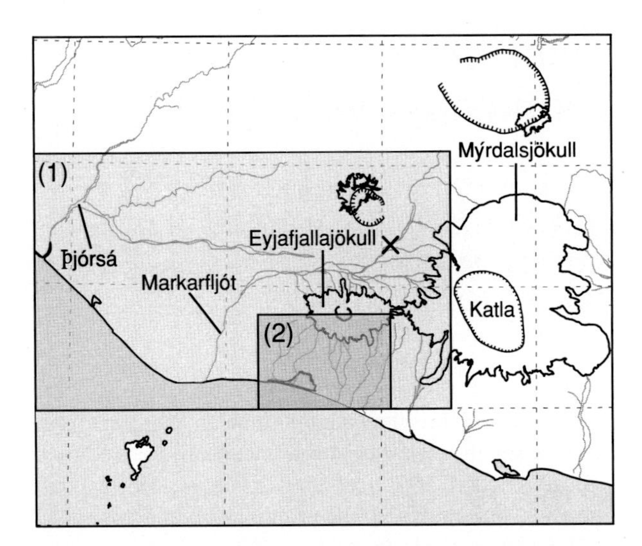

Figure 14.1. Location map of the Mýrdalsjökull and Eyjafjallajökull ice caps showing the model boundaries for the simulations of jökulhlaups to the west from Mýrdalsjökull (1) and to the south of Eyjafjallajökull (2). The location of the floodmarks at Fauskheiði used to calibrate the friction coefficient in the simulations of the floods from Mýrdalsjökull is indicated with a symbol (×).

from these volcanoes (Guðmundsson and Gylfason, 2005). This chapter describes numerical simulations of subaerial jökulhlaups from Katla to the west and to the south from Eyjafjallajökull (Figure 14.1). The purpose of the simulations was to obtain a general understanding of the nature of the floods and, in particular, to estimate the travel times of the floods down the mountainsides and to delineate the extent of the endangered areas.

The chapter starts with a discussion of the characteristics of jökulhlaups from Katla and Eyjafjallajökull followed by a summary of the dynamics of subaerial jökulhlaups and a description of the numerical formulation on which the simulations are based. The results of simulations of the floods from Katla and Eyjafjallajökull are then described and the chapter is concluded with a discussion of the results.

14.2 Jökulhlaups from Katla and Eyjafjallajökull

The amount of suspended sediment in volcanically induced floods from Katla and Eyjafjallajökull may be expected to vary considerably depending among other things on the ice thickness at the eruption site, the efficiency of the heat transfer from the erupted materials to melt the surrounding ice and the amount of loose material that is eroded by the flood on its way down the mountainside. According to Guðmundsson and Högnadóttir (2005) and Tómasson (1996), large floods from Katla, which are caused by eruptions under 400–700 m thick ice (Björnsson *et al.*, 2000), are expected to produce floodwaters with density in the range 1150–1200 kg m^{-3}. The density of floodwaters from Eyjafjallajökull, where typical ice thickness is on the order of 100 m (except in the top crater where it reaches 250 m) may be expected to be in the range 1250–1400 kg m^{-3}.

According to the above density estimates, the floodwaters of jökulhlaups from both ice caps consist mainly of water and may be expected to flow down the mountainside in a fluid-like manner. However, the composition of the floodwaters will vary with time, with the front having the highest concentration of suspended materials, as has been inferred to have been the case in the large jökulhlaup from Grímsvötn in Vatnajökull in 1996 (Snorrason *et al.*, 1997). In addition, numerous huge ice blocks and large amounts of smaller ice fragments are advected with the flood and are believed to play a role in determining the effective, frictional properties of the flood material, which seems to have higher frictional coefficients (Manning's n) than would be expected for flowing water (see for example discussion in Tómasson, 1996). Therefore, debris flow or lahar dynamics may be important for jökulhlaups from Katla and Eyjafjallajökull as mentioned above, although these aspects of the flood dynamics will not be pursued further here.

Large eruptions in Katla are expected to melt several hundred thousand cubic metres per second of water and sustain this rate of melting for many hours before heat transfer to melting becomes less efficient after the eruption penetrates the ice (Guðmundsson and Högnadóttir, 2005). Eruptions in Eyjafjallajökull are generally smaller than in Katla and penetrate the thinner ice there more quickly. The rate of melting of ice is, therefore, much smaller and the phase of effective heat transfer to melting of ice is shorter, perhaps only 1–2 hours, after which most of the heat flux of the eruption is lost to the atmosphere. The maximum discharge of the flood depends on the size of the eruption and on local conditions at the eruption site, such as the ice thickness and the basal and ice surface topography. Order of magnitude estimates for maximum discharge of jökulhlaups originating within different subglacial watersheds on the ice caps are given by Guðmundsson and Högnadóttir (2005, Table 14.2). The simulations described here are based on typical discharges for hypothetical, large floods from each ice cap, i.e. several hundred thousand cubic metres per second for jökulhlaups to the west from Katla and several thousand cubic metres per second for jökulhlaups to the south from Eyjafjallajökull.

14.3 Dynamics of subaerial jökulhlaups

The discharge of jökulhlaups from Katla and Eyjafjallajökull increases much more rapidly with time than in traditional jökulhlaups from Grímsvötn in Vatnajökull, which are well described by the classical jökulhlaup theory of Nye (1976). The formation of the subglacial water course in rapidly rising jökulhlaups is believed to be driven by high subglacial water pressure rather than subglacial melting of conduits (Björnsson, 1997; Jóhannesson, 2002; Björnsson, 2004; Flowers *et al.*, 2004). The jökulhlaups from Katla and Eyjafjallajökull flow comparatively long distances subaerially after they exit the ice caps. Settlements and other constructions in the flood paths are of course all located outside the glacier margins. Simulation of this subaerial phase of the flood is, therefore, of high practical relevance and it is this phase of the flood that is the main subject of this chapter. The initial discharge at the eruption site or at the glacier margin is predefined based on the history and nature of the volcano and ice cap as further described below and the subglacial phase of the flood is thus ignored.

The model simulations are based on a depth-integrated formulation of the conservation equations for incompressible, shallow water flow (see for example Gerhart *et al.*, 1992; Iverson, 1997) over complex terrain. The conservation of mass in the absence of entrainment may in this case be written as

$$\frac{\partial h}{\partial t} + \vec{\nabla} \cdot \vec{q} = 0, \tag{14.1}$$

where $h = z_s - z_b$ is flow thickness (perpendicular to the xy-plane, which need not be horizontal), z_b and z_s are the z-coordinates of the bedrock and of the surface of the flood, respectively, $\vec{q} = h\vec{u}$ is volume flux, $\vec{u}$ is depth-averaged velocity and $\vec{\nabla}$ denotes the gradient operation in the x- and y-directions only (the z-direction has been taken care of by the depth-averaging). Assuming free-surface flow and again ignoring entrainment, conservation of momentum is expressed with the equation

$$\frac{\partial (h\vec{u})}{\partial t} + \vec{\nabla} \cdot (h\vec{u}\vec{u}) = g \sin\psi h \vec{a}_\psi - g \cos\psi h \vec{\nabla} z_s - \rho^{-1} \vec{\tau}_b, \tag{14.2}$$

where g is the acceleration of gravity, ρ is the density of the fluid, ψ is the slope of the coordinate system, $\vec{a}_\psi = \cot\psi(\tan\psi_x \vec{a}_x + \tan\psi_y \vec{a}_y)$ is a unit vector in the direction of the slope of the coordinate system, ψ_x and ψ_y are the slopes along the x- and y-axes, $\vec{a}_x$ and $\vec{a}_y$ are unit vectors in the x-direction and y-direction, respectively, and $\vec{\tau}_b$ is shear stress at the bottom of the flood. The first term on the right-hand side of the momentum equation represents the component of gravity in the (possibly sloping) xy-plane, the second term is due to internal stresses within the moving body, which are assumed to be given by a hydrostatic pressure distribution, $p = g \cos\psi(z_s - z)$, due to the bed-perpendicular component of gravity, and the last term on the right-hand side represents retardation due to bottom friction, which is further discussed below.

Equations (14.1) and (14.2), for simplicity, are formulated here in a sloping, rectangular coordinate system $Oxyz$, which is assumed to be aligned approximately with the terrain over which the flood is flowing. The flow thickness, h, and undulations in the basal geometry, z_b, are assumed to be small in comparison with the longitudinal and transverse dimensions of the flood. In the case of a complicated large-scale geometry, the equations may be formulated in a curvilinear coordinate system, which approximates the large-scale terrain geometry, as described by Gray *et al.* (1999).

The shear stress $\vec{\tau}_b$ in the frictional term on the right-hand side of Equation (14.2) can be formulated on the basis of the Chézy and Manning formulae (Chow, 1959), as

$$\tau_b = \frac{\rho g u^2}{c^2}, \tag{14.3}$$

which may, in the case of shallow flow, be derived from Chézy's formula, $u = c\sqrt{rs}$, where u is the magnitude of the velocity, r is the hydraulic radius, s is channel slope and c is the Chézy coefficient, or

$$\tau_b = \frac{\rho g n^2 u^2}{h^{1/3}}, \tag{14.4}$$

which may be derived from Manning's formula, $u = r^{2/3}\sqrt{s/n}$, where u, r and s have the same meaning as before, and n is the Manning roughness coefficient. The friction in both cases is assumed to act in the direction opposite to the direction of the flow. It can be argued that the acceleration of gravity, g, should be replaced by the terrain perpendicular component, $g \cos\psi$, in Equations (14.3) and (14.4). Because this is a very small correction under the conditions considered here, the equations were used in the form given above.

Friction in subaerial jökulhlaups often has been formulated in terms of Manning's formula, both in discharge calculations for individual cross-sections based on flood marks (Pardee, 1942; Baker, 1973; Dorava and Meyer, 1994; Tómasson, 1996; Gröndal *et al.*, 2005), and in numerical simulations of whole flood paths (O'Connor and Webb, 1988; Eskilsson *et al.*, 2002; Alho *et al.*, 2005; Hólm and Kjaran, 2005). Manning's formula also has been used often in analyses of jökulhlaups in subglacial tunnels (Nye, 1976; Clarke, 1982). Chézy's formula has been used in a similar manner from the earliest studies of subaerial jökulhlaups, both in studies of discharge through cross-sections (Bretz, 1925, quoting D. F. Higgins; Pardee, 1942) and in numerical simulations (Eliasson *et al.*, 2005).

The flow of subaerial jökulhlaups in Iceland has been modelled at several locations based on Manning's formula as listed in Table 14.1. The table also lists ranges for the maximum discharge for the well-known jökulhlaups in Skeiðará from Grímsvötn and in Skaftá from two cauldrons in western Vatnajökull, showing the wide range of the discharge of these floods, which spans four orders of magnitude. The estimated Manning roughness coefficient ranges from 0.02 to 0.1 s m$^{-1/3}$ and

Table 14.1. Jökulhlaup discharges with frictional parameters used in modelling

Location	Max. discharge (10^3 m^3 s^{-1})	Manning's n (s m$^{-1/3}$)	Reference
Jökulsá á Fjöllum, prehistoric	900	0.025–0.06	Alho *et al.* (2005)
Katla, 1918	300	0.1	Tómasson (1996)
Skeiðará, 1996	50	0.02–0.07	Eskilsson *et al.* (2002)
Skeiðará	0.6–50	—	Björnsson (2002)
Skaftá	0.05–2	—	Zóphóníasson (2002)

it is, in the case of the Katla jökulhlaup in 1918, derived from flood marks based on an independent estimate of the maximum discharge (Tómasson, 1996) and, in the case of the Skeiðará jökulhlaup in 1996, based on direct observations of the flood including some discharge measurements (Snorrason *et al.*, 1997). Values of n near the higher end of the ranges were chosen near the ice caps for the jökulhlaups in Jökulsá á Fjöllum and Skeiðará (Eskilsson *et al.*, 2002; Alho *et al.*, 2005), where the flow may be expected to have been affected most by ice blocks and suspended sediment.

The Chézy coefficient can be written in terms of a non-dimensional (Darcy–Weisbach) friction factor as follows (Chow, 1959):

$$c = \sqrt{\frac{8g}{f}}, \tag{14.5}$$

and the corresponding expression for the Manning number is

$$n = \sqrt{\frac{f}{8g}}r^{1/6}. \tag{14.6}$$

For rough channels, the friction factor is just a function of the relative roughness, k/r, where k is a characteristic roughness height and r is the hydraulic radius (Chow, 1959). It can be argued that the relative roughness for many jökulhlaups may remain approximately constant throughout the flow domain. As the flood emerges from the glacier, the flood depth is relatively large and there is a high amount of suspended sediment and ice fragments. As the flood propagates downstream onto the wide flood plain, the water depth decreases as does the amount of suspended sediment and ice fragments. Therefore, the ratio of the characteristic roughness to water depth will vary less than the characteristic roughness itself. This observation indicates that a constant Chézy coefficient is a better approximation than a constant Manning number to describe the flow in sediment-laden streams of this type. Furthermore, considering floods of different magnitudes flowing over the same or similar terrain, it may be assumed that large floods will submerge wider flood paths with more varied roughness elements, and that the characteristic roughness height may, therefore, be assumed to increase with the flow depth. Thus, it is likely that it is possible to describe floods over a wide discharge range with similar values of the Chézy coefficient rather than with similar values of the Manning number. This procedure is, however, clearly a rough approximation, in particular at large distances from the ice margin where the flow may be expected to be less affected by ice fragments and suspended sediment.

14.4 Numerical formulation

Jökulhlaups from Katla towards the west flow over a large area with a comparatively flat geometry where complications due to steepness of the large-scale terrain are unimportant and the flow to great extent will remain subcritical. The simulations of these floods were carried out with the AQUARIVER model, where the dynamic equations (14.1) and (14.2) are formulated in a horizontal $Oxyz$ coordinate system. The equations are approximated using a Galerkin finite-element method on triangular elements. Continuous approximations are used for water elevation, linear within elements, but piecewise constant approximations for the velocities. Such a choice has been shown to lead to a spatially stable approximation. The numerical scheme for the time discretisation is either explicit or implicit using a wave equation formulation (Sigurðsson *et al.*, 1988; Sigurðsson, 1990, 1992 and 1994).

Jökulhlaups from Eyjafjallajökull flow over rugged, steep terrain with deep valleys between protruding ridges where the flow is likely to reach a supercritical state in the steeper parts of the path. The simulations of these floods were carried out with a specially adapted version of the SAMOS snow avalanche model (Zwinger *et al.*, 2003), which takes into account the large-scale geometry of terrain without assuming the large-scale terrain slope to be small and is able to handle both subcritical and supercritical flow. The depth-averaged formulation of snow avalanche dynamics, which was first proposed by Eglit (1983), is fundamentally similar to shallow water flow except for the formulation of the friction. The friction in the jökulhlaup simulations with the SAMOS model was implemented by specifying a zero dry-friction angle δ for the Coulomb friction at the bed. The formulation of the dynamic, velocity-dependent friction was modified from the formulation described by Zwinger *et al.* to

$$\tau_b = \rho C_D u^2, \tag{14.7}$$

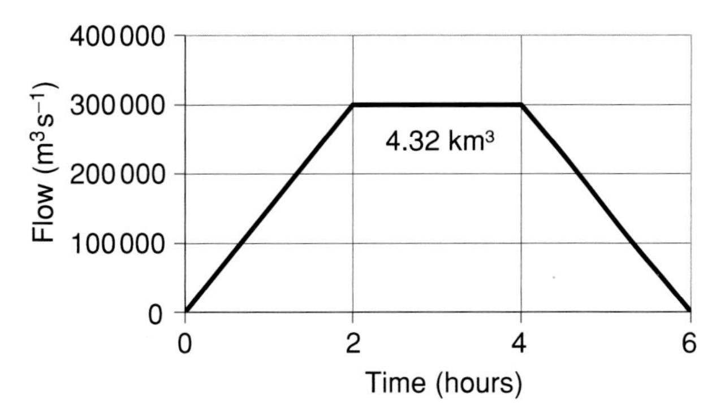

Figure 14.2. Assumed discharge as a function of time for a jökulhlaup to the west from Katla in Mýrdalsjökull.

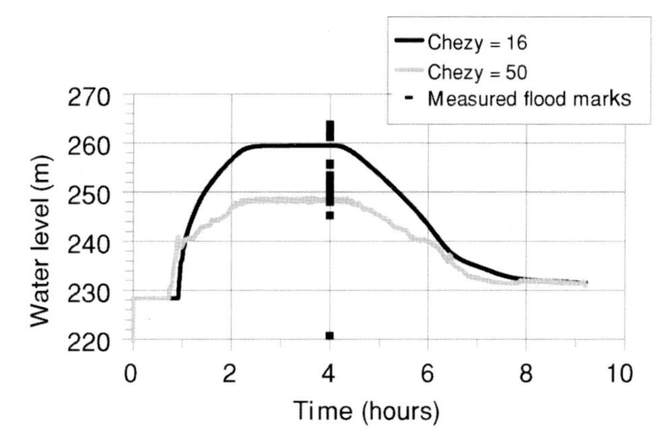

Figure 14.3. Water level at Fauskheiði according to two simulations with different values of the Chézy friction coefficient, c, compared with flood marks of prehistoric jökulhlaups from Katla (symbols).

where $C_D = gc^{-2}$ is a non-dimensional dynamic friction coefficient, which may be expressed in terms of the Chézy coefficient, c. The internal friction angle of the flood material was, furthermore, chosen to be $\phi = 0$, which leads to active/passive (or earth pressure) coefficients $K_{act/pass} = 1$, corresponding to a hydrostatic internal pressure distribution. The density of the flood material was set to $\rho = 1100$ kg m^{-3} but this does not affect the results of the simulations as ρ cancels out from the dynamic equations as they are formulated here. Definitions and discussion of the parameters δ, ϕ and $K_{act/pass}$ for snow avalanche simulations with the SAMOS model are given by Zwinger *et al.* (2003). The solution of the depth-averaged dynamic equations in the SAMOS model is based on a Lagrangian, finite-volume numerical formulation using an integral form of the conservation equations.

14.5 Model simulations: two examples

14.5.1 Katla

The model boundary is shown in Figure 14.1; it extends from the Entujökull outlet glacier at an elevation of approximately 400 metres above sea level down to the coast, and across to the Þjórsá river in the west.

It was decided that a jökulhlaup with a maximum discharge of 300 000 m^3 s^{-1} would be modelled based on Guðmundsson and Högnadóttir (2005) and Tómasson (1996). Figure 14.2 shows the assumed discharge at the glacier margin as a function of time for this flood. The flood takes two hours to reach its maximum flow rate, then continues at that rate for two hours, and finally declines for two hours until it is assumed to come to an end. This duration of the peak discharge is based roughly on descriptions of the 1918 jökulhlaup from Katla (Tómasson, 1996). The total water volume released during the flood is 4320 million cubic metres, which is on the same order of magnitude as the estimated volume of the 1918 jökulhlaup (Magnús Tumi Guðmundsson, personal communication, 2006).

The numerical model was calibrated against measured flood marks of prehistoric floods (Gröndal *et al.*, 2005; Larsen *et al.*, 2005) at Fauskheiði (see Figure 14.1 for location). Several model runs were made with varying values of the Chézy coefficient. Figure 14.3 shows the results of two of these runs. The figure shows that the run with the Chézy coefficient $c = 16\,\mathrm{m}^{1/2}\,\mathrm{s}^{-1}$ has a flow depth at Fauskheiði corresponding to the flood marks near the upper end of the range of the prehistoric floods. It was thus decided to use this value of the Chézy coefficient for the final model runs. This estimate of c is of course dependent on the assumed discharge but the discharge of the floods that produced the highest flood marks is unknown. Other combinations of discharge and Chézy coefficient can be found that are also consistent with the flood marks.

Plate 40 shows the extent and depth of the jökulhlaup after two and a half hours, four hours and six hours from the start of the flood. As seen on the maps, the jökulhlaup begins to spread out and the depth decreases as the flood progresses downstream. The upper part of the jökulhlaup path is a comparatively narrow valley with a deep canyon where the simulated flow depth reaches as high as 45 m and flow velocity is about 30 m s^{-1}. Farther down, the slope of the terrain is reduced and the course of the flood stretches over an almost flat sandur plain. The jökulhlaup reaches the main highway across the southern lowland after approximately four hours, and it reaches the sea after roughly six hours. The maximum depth of the flood at the location of the highway is about 8 m and the speed of propagation of the flood wave there is about 2.5 m s^{-1}.

The propagation speed of the flood across the sandur plain depends sensitively on the assumed constant value of the Chézy friction coefficient, c. As much of the ice fragments and largest-grained sediments will have stranded or sedimented out in the upper part of the path, it is possible that the adopted value of c leads to an underestimate of the propagation speed of the flood in the lower part of the path because the adopted value of c is rather low compared to recommended values for natural river courses (see Chow, 1959). This fact must be kept in mind in the interpretation of the results, but in view of the limited input data for calibration of the model, it was not considered possible to adopt different values of c for different parts of the flood path. The simulation described here with a constant value of c is equivalent to a simulation with Manning's formula with a friction coefficient n that becomes progressively lower as the water depth is reduced in the downstream direction. A five-fold reduction in flow depth, as found here when the flood propagates from the narrow channel close to the ice margin to the flood plain, with a constant value of Chézy's c, corresponds to a variation by a factor of approximately 1.3 in Manning's n (see Eqs. (14.5) and (14.6)). This is a smaller variation of n than adopted in some numerical simulations of jökulhlaups where a variable Manning's n has been used (Eskilsson *et al.*, 2002; Alho *et al.*, 2005), and smaller than the difference between the value of c adopted here and values that are typical for natural river courses (Chow, 1959).

14.5.2 Eyjafjallajökull

The model boundary is shown in Figure 14.1. It extends from the summit of the Eyjafjallajökull ice cap to the lowlands south of the mountain. Eruptions producing floods towards the south potentially can be located anywhere on the southern flank of the mountain. The flood will quickly become channellised into the main valleys in the rugged terrain below the ice margin so that the flow in the lower part of the flood path will be relatively independent of the location of the eruption within the watershed of each valley. In order to study all potential flood paths, the flood was released as a sheet of water of uniform thickness covering the whole elevation range 1100–1200 m a.s.l. and $\sim$1500 m a.s.l. near the summit of the ice cap. The starting area was approximately 2 km long in the downslope direction and approximately 8 km wide in the transverse direction along the mountainside. No effort was made to account for a possibly subglacial propagation of the flood in the upper part of the flood path. This setup of the simulations is of course not intended as a realistic description of the flow of any single jökulhlaup. Rather, it is intended to explore all potential eruption locations in a simple manner in one set of simulations, with a focus on the lower part of the mountainside below the ice margin. An actual eruption will probably be confined largely within the watershed of only one of the main valleys and, therefore, only produce a flood in one of them.

Several simulations with different thickness of the initial sheet of water were run for two choices of the Chézy friction coefficient, $c = 12$ and $24\,\mathrm{m^{1/2}\,s^{-1}}$ (Hákonardóttir *et al.*, 2005). The chosen values of c correspond to Manning numbers of approximately $n = 0.1$ and $0.05\,\mathrm{s\,m^{-1/3}}$ for flow depths of the order of 3 m, which corresponds to the estimated discharge of these jökulhlaups after they have become channellised in the valleys below the ice margin. This choice of the Chézy coefficient corresponds to the values $C_\mathrm{D} = 0.068$ and 0.017, respectively, of the non-dimensional friction coefficient defined in Equation (14.7) (see Hákonardóttir *et al.*, 2005). As expected, the floods quickly became channellised into the main valleys forming four or five main streams down the mountainside. The maximum depth of the flood and the maximum flow speed for an initial water depth of 0.7 m in the starting area and Chézy coefficient $c = 12\ \mathrm{m^{1/2}\ s^{-1}}$are shown in Plate 41. The discharge in the main valleys below the ice margin reached an approximate steady state during the passage of the main flood front with a maximum discharge of 2500 to 9000 $\mathrm{m^3\,s^{-1}}$ in the main gullies. This discharge corresponds to the order of magnitude estimate of the discharge due to eruptions through the thin ice cap on the southern flanks of Eyjafjallajökull as noted above.

Floods with initial water thickness of 0.3, 0.7 and 1.1 m and Chézy coefficient $c = 12\,\mathrm{m^{1/2}\,s^{-1}}$ reach maximum discharge in the range 1000 to 15500 $\mathrm{m^3\,s^{-1}}$, attain maximum flow depths of approximately 5 m and maximum flow speeds up to 15 $\mathrm{m\,s^{-1}}$. The floods take 20–30 minutes to flow from the source area on the ice cap down to the lowland along the main valley near the middle of the simulation domain (Hákonardóttir *et al.*, 2005). These floods reach a supercritical flow state in the steep gullies. Reducing the friction by increasing the Chézy coefficient to $c = 24\,\mathrm{m^{1/2}\,s^{-1}}$ shortens the propagation time down the mountainside from 25 to 15 minutes for runs with the central value of the initial water thickness.

14.6 Discussion

Jökulhlaups from Mýrdalsjökull and Eyjafjallajökull flow over terrain of different characteristics in terms of slope and channellisation. They span a wide range in discharge and may be expected to contain different amounts of suspended sediment and advected ice fragments. The simplified simulations presented here cover only a part of the potential range of these floods. It appears that available flood marks of prehistoric floods are simulated adequately by depth-integrated hydraulic models based on discharge estimates derived from heat-flow rates for the most likely sizes of eruptions of Katla through the ice cap. Flood marks for validating the simulations have only been considered for very large floods from Katla, but it is argued that similar values for the Chézy friction coefficient are likely to also represent smaller floods from the ice caps. In spite of the inherent uncertainty of the simulations, estimates of flow depth, flow speed and front propagation times are considered realistic enough to be of practical value for flood hazard management planning.

Acknowledgements

The work described in this chapter was funded by the Civil Protection Department of the National Commissioner of the Icelandic Police as a part of hazard zoning and preparations of an evacuation plan for settlements in southern Iceland endangered by eruptions and outburst floods from the volcanos Katla and Eyjafjallajökull. Data for the DTM for the Eyjafjallajökull simulations were made available by Eyjólfur Magnússon at the Institute of Earth Sciences of the University of Iceland.

References

Alho, P., Russell, A. J., Carrivick, J. L. and Käyhkö, J. (2005). Reconstruction of the largest Holocene jökulhlaup within Jökulsá á Fjöllum, NE Iceland. *Quaternary Science Reviews*, **24** (22), 2319–2334.

Baker, V. R. (1973). *Paleohydrology and Sedimentology of Lake Missoula Flooding in Eastern Washington*. Geological Society of Americal Special Paper 144.

Björnsson, H. (1997). Grímsvatnahlaup fyrr og nú [Past and present jökulhlaups from Grímsvötn]. In *Vatnajökull: Gos og hlaup 1996 [Vatnajökull. Eruption and Jökulhlaup 1996]*, ed. H. Haraldsson. Reykjavík: Icelandic Public Roads Administration, pp. 61–77.

Björnsson, H. (2002). Subglacial lakes and jökulhlaups in Iceland. *Global and Planetary Change*, **35**, 255–271.

Björnsson, H. (2004). Glacial lake outburst floods in mountain environments. In *Mountain Geomorphology*, eds. P. N. Owens and O. Slaymaker. London: Arnold Publishers, pp. 165–184.

Björnsson, H., Pálsson, F. and Guðmundsson, M. T. (2000). Surface and bedrock topography of the Mýrdalsjökull ice cap, Iceland: the Katla caldera, eruption sites and routes of jökulhlaups. *Jökull*, **49**, 29–46.

Bretz, J H. (1925). The Spokane flood beyond the Channeled Scabland. *Journal of Geology*, **33**, 97–115, 236–259.

Chow, V. T. (1959). *Open-channel Hydraulics*. New York: McGraw-Hill Book Company Inc.

Clarke, G. K. C. (1982). Glacier outburst floods from "Hazard Lake", Yukon territory, and the problem of flood magnitude prediction. *Journal of Glaciology*, **28** (98), 3–21.

Dorava, J. M. and Meyer, D. F. (1994). Hydrologic hazards in the lower Drift River basin associated with the 1989–1990 eruptions of Redoubt Volcano, Alaska. *Journal of Volcanology and Geothermal Research*, **62** (1–4), 387–407.

Eglit, M. E. (1983). Some mathematical models of snow avalanches. In *Advances in the Mechanics and the Flow of Granular Materials*, Vol. 2, ed. M. Shahinpoor. Houston, TX: Clausthal-Zellerfeld and Gulf Publ. Co., pp. 577–588.

Eliasson, J., Kjaran, S. P., Hólm, S. L., Guðmundsson, M. T. and Larsen, G. (2005). Large hazardous floods as translatory waves. In *Modelling, Computer-assisted Simulations and Mapping of Dangerous Phenomena for Hazard Assessment*, eds. G. Iovine, T. D. Gregorio, H. Miyamoto and M. Sheridan.

Eskilsson, C., Árnason, J. Y. and Rosbjerg, D. (2002). Simulation of the jökulhlaup on Skeiðarársandur, Iceland, in November 1996 using MIKE 21 HD. In *The Extremes of the Extremes*, eds. Á. Snorrason, H. P. Finnsdóttir and M. E. Moss. Proceedings of a symposium in Reykjavík in July 2000, IAHS Publication 271, pp. 37–43. Wallingford: IAHS Press.

Flowers, G. E., Björnsson, H., Pálsson, F. and Clarke, G. K. C. (2004). A coupled sheet-conduit model of jökulhlaup propagation. *Geophysical Research Letters*, **31**, L05401, doi:10.1029/2003GL019088.

Gerhart, P. M., Gross, R. J. and Hochstein, J. I. (1992). *Fundamentals of Fluid Mechanics*, 2nd edn. Reading, MA: Addison-Wesley.

Gray, J. M., Wieland, N. T. M. and Hutter, K. (1999). Gravity-driven free surface flow of granular avalanches over complex basal topography. *Proceedings of the Royal Society of London A*, **455**, 1841–1874.

Gröndal, G. O., Larsen, G. and Elefsen, S. (2005). Stærðir forsögulegra hamfaraflóða í Markarfljóti: mæling á farvegum neðan Einhyrningsflata [The size of prehistoric, catastrophic floods in Markarfljót: a measurement of flood paths below Einhyrningsflatir]. In *Hættumat vegna eldgosa og hlaupa frá vestanverðum Mýrdalsjökli og Eyjafjallajökli [Hazard Zoning for Eruptions and Jökulhlaups from the Western Part of the Mýrdalsjökull and Eyjafjallajökull Ice Caps]*, eds. M. T. Guðmundsson and Á. G. Gylfason. Reykjavík: Ríkislögreglustjórinn, Háskólaútgáfan, pp. 99–104.

Guðmundsson, M. T. and Gylfason, Á. G. (Eds.) (2005). *Hættumat vegna eldgosa og hlaupa frá vestanverðum Mýrdalsjökli og Eyjafjallajökli [Hazard Zoning for Eruptions and Jökulhlaups from the Western Part of the Mýrdalsjökull and Eyjafjallajökull Ice Caps]*. Reykjavík: Ríkislögreglustjórinn, Háskólaútgáfan.

Guðmundsson, M. T. and Högnadóttir, Þ. (2005). Ísbráðnun og upptakarennsli jökulhlaupa vegna eldgosa í Eyjafjallajökli og vestanverðum Mýrdalsjökli [Melting of ice and initial discharge of jökulhlaups due to volcanic eruptions in Eyjafjallajökull and the western part of Mýrdalsjökull]. In *Hættumat vegna eldgosa og hlaupa frá vestanverðum Mýrdalsjökli og Eyjafjallajökli [Hazard Zoning for Eruptions and Jökulhlaups from the Western Part of the Mýrdalsjökull and Eyjafjallajökull Ice Caps]*, eds. M. T. Guðmundsson and Á. G. Gylfason. Reykjavík: Ríkislögreglustjórinn, Háskólaútgáfan, 159–179.

Hákonardóttir, K. M., Jóhannesson, T. and Sampl, P. (2005). Líkanreikningar á jökulhlaupum niður suðurhlíðar Eyjafjallajökuls [Modeling of jökulhlaups down the southern side of the Eyjafjallajökull ice cap]. In *Hættumat vegna eldgosa og hlaupa frá vestanverðum Mýrdalsjökli og Eyjafjallajökli [Hazard Zoning for Eruptions and Jökulhlaups from the Western Part of the Mýrdalsjökull and Eyjafjallajökull Ice Caps]*, eds. M. T. Guðmundsson and Á. G. Gylfason. Reykjavík: Ríkislögreglustjórinn, Háskólaútgáfan, pp. 181–196.

Hólm, S. L. and Kjaran, S. P. (2005). Reiknilíkan fyrir útbreiðslu hlaupa úr Entujökli [A simulation model for the extent of outburst floods from Entujökull glacier]. In *Hættumat vegna eldgosa og hlaupa frá vestanverðum Mýrdalsjökli og Eyjafjallajökli [Hazard Zoning for Eruptions and Jökulhlaups from the Western Part of the Mýrdalsjökull and Eyjafjallajökull Ice Caps]*, eds. M. T. Guðmundsson and Á. G. Gylfason. Reykjavík: Ríkislögreglustjórinn, Háskólaútgáfan, pp. 197–210.

Iverson, R. M. (1997). The physics of debris flows. *Reviews of Geophysics*, **35** (3), 245–296.

Jóhannesson, T. (2002). Propagation of a subglacial flood wave during the initiation of a jökulhlaup. *Hydrological Sciences Journal*, **473**, 417–434.

Keys, H. J. R. (2007). Lahars from Ruapehu Volcano, New Zealand, and risk mitigation. *Annals of Glaciology*, **45**, 155–162.

Larsen, G., Smith, K., Newton, A. and Knudsen, Ó. (2005). Ummerki um forsöguleg hlaup niður Markarfljót [Evidence for prehistoric jökulhlaups along the Markarfljót river]. In *Hættumat vegna eldgosa og hlaupa frá vestanverðum Mýrdalsjökli og Eyjafjallajökli [Hazard Zoning for Eruptions and Jökulhlaups from the Western Part of the Mýrdalsjökull and Eyjafjallajökull Ice Caps]*, eds. M. T. Guðmundsson and Á. G. Gylfason. Reykjavík: Ríkislögreglustjórinn, Háskólaútgáfan, pp. 75–98.

Nye, J. F. (1976). Water flow in glaciers: Jökulhlaups, tunnels and veins. *Journal of Glaciology*, **17** (76), 181–207.

O'Connor, J. E. and Webb, R. H. (1988). Hydraulic modeling for paleoflood analysis. In *Flood Geomorphology*, eds. V. R. Baker, R. C. Kockel and P. C. Patton. New York: John Wiley and Sons, pp. 393–402.

Pardee, J. T. (1942). Unusual currents in Lake Missoula, Montana. *Geological Society of America Bulletin*, **53**, 1569–1600.

Sigurðsson, S. (1990). On the stability of staggered finite element schemes for simulation of shallow water free surface flow. In *Computational Methods in Water Resources VIII: Computational Methods in Surface Hydrology*. Springer Verlag, Computational Mechanics Publications, pp. 449–454.

Sigurðsson, S. (1992). Treatment of the convective term in staggered finite element schemes for shallow water flow. In *Computational Methods in Water Resources IX. Vol. 1: Numerical Methods in Water Resources*. Elsevier Applied Science, Computational Mechanics Publications, pp. 291–298.

Sigurðsson, S. (1994). Alternative approaches to the wave equation in shallow water flow. Modelling. In *Computational Methods in Water Resources X*, Vol. 2, Kluwer Academic Publishers, pp. 1113–1120.

Sigurðsson, S., Kjaran, S. P. and Tómasson, G. G. (1988). A simple staggered finite element scheme for linear shallow water. In *Computational Methods in Water Resources VII, Vol. 1: Modeling Surface and Subsurface Flows*. Elsevier, Computational Mechanics Publications, pp. 329–335.

Snorrason, Á., Jónsson, P., Pálsson, S. *et al.* (1997). Hlaupið á Skeiðarársandi haustið 1996. Útbreiðsla, rennsli og aurburður. [The jökulhlaup on Skeiðarársandur in the fall of 1996. Extent, discharge and suspended sediments]. In *Vatnajökull. Gos og Hlaup 1996 [Vatnajökull. Eruption and Jökulhlaup 1996]*, ed. H. Haraldsson. Reykjavík: Icelandic Public Roads Administration, pp. 79–137.

Tómasson, H. (1996). The jökulhlaup from Katla in 1918. *Annals of Glaciology*, **22**, 249–254.

Zóphóníasson, S. (2002). *Rennsli í Skaftárhlaupum 1955–2002*. Reykjavík, National Energy Authority, Report SZ-2002/01.

Zwinger, T., Kluwick, A. and Sampl, P. (2003). Numerical simulation of dry snow avalanche flow over natural terrain. In *Response of Granular and Porous Materials under Large and Catastrophic Deformations*. Lecture Notes in Applied and Computational Mechanics, Vol. 11, eds. K. Hutter and N. Kirchner. Berlin: Springer, pp. 160–194.

15

Jökulhlaups from Kverkfjöll volcano, Iceland: modelling transient hydraulic phenomena

JONATHAN L. CARRIVICK

Summary

Jökulhlaups, or glacier outburst floods, are complex flood phenomena, with hydraulics that vary considerably spatially and temporally. However, jökulhlaups occur too suddenly, are too powerful and often too infrequent and remote for direct measurements of hydraulics to be made. Thus various palaeohydraulic methods have been applied to reconstruct jökulhlaup hydraulics from geomorphological and sedimentological evidence. However, these techniques fail to sufficiently characterise transient jökulhlaup hydraulic phenomena, in both space and time. A detailed understanding of these transient hydraulics is important for understanding rapid landscape change, high-magnitude flood mechanisms of erosion, transport and deposition, and hence jökulhlaup hazard management.

Therefore this paper reconstructs transient jökulhlaup flow phenomena using boulder clusters, the slope-area method and a depth-averaged two-dimensional (2D) hydrodynamic model. Kverkfjöll volcano on the northern edge of Vatnajökull, Iceland, provides the study site. Jökulhlaups inundated anastomosing bedrock valleys and exhibited transient hydraulic phenomena including sheet or unconfined flow, channel flow, flow around islands, hydraulic jumps, multi-directional flow including backwater areas and hydraulic ponding. Reconstructions of these jökulhlaups indicate peak discharges of 50–100 000 $m^3\,s^{-1}$, which attenuated by ~65% within 20 km. Frontal flow velocities were ~1.6 $m\,s^{-1}$ but as stage increased velocities reached 5–15 $m\,s^{-1}$. Shear stress and stream power reached $1 \times 10^4\,N\,m^{-2}$ and $1 \times 10^5\,W\,m^{-2}$ respectively. Flows were largely supercritical due to steep channel gradients and shallow flows and highly turbulent due to high hydraulic roughness. Kverkfjöll jökulhlaups thus achieved geomorphological work comparable to that accomplished by 'megafloods'.

15.1 Introduction, background and rationale

High-magnitude outburst floods are sudden releases of water and sediment from temporary impoundments, with peak discharges that are at least several orders of magnitude greater than perennial flows. The very largest outburst floods known to have occurred on Earth, 'megafloods', were glacial outburst floods, 'jökulhlaups', from continental lakes impounded by Last Glacial Maximum (LGM) ice sheets (see references in Baker, 2002). In exceptional cases, jökulhlaup discharge rates could be comparable with those of ocean currents (Baker, 2002). Jökulhlaups can erode both unconsolidated sediments (e.g. Russell and Marren, 1999; Gomez *et al.*, 2002) and bedrock (e.g. Baker, 1988; Tómasson, 1996; Carrivick *et al.*, 2004a), and transport and deposit large amounts of sediment (up to 10^8 tonnes; Björnsson, 2002) over extensive areas. Jökulhlaups therefore constitute a serious threat to life, property and infrastructure (see Costa and Schuster (1988) for references) because of the volume, rapid onset and unpredictable nature of these flows. The overall poor understanding and lack of quantification of high-magnitude flow processes is a product, in part, of the difficulty of making direct measurements. For example, it is unlikely that standard hydrological monitoring and logging equipment could withstand such energetic flow processes and particle bombardment. Furthermore, high-magnitude floods often occur with little warning and usually reach peak flow within a matter of hours (e.g. Baker, 2000; Björnsson, 2002).

Recent contributions to address this problem utilise 1D dam-failure and unsteady-flow modelling, for example to estimate a flow hydrograph and peak discharge for a Pleistocene lava dam in the Colorado River (Fenton *et al.*, 2006), or physically based models, for example to examine Pleistocene flooding from Lake Agassiz (Clarke *et al.*, 2004). For reference, literature reviews on recent modelling approaches to high-magnitude outburst floods are given by Carling *et al.* (2003) and Herget (2005). To date, these models have focused on hydraulic properties, and have neglected to include sediment entrainment, transport and deposition. Consequently, high-magnitude fluvial erosion, transport and depositional mechanisms remain very poorly understood and very poorly quantified (Maizels, 1997; Whipple *et al.*, 2000).

Reconstructions of landscape changes due to jökulhlaups have been based on descriptions and classifications of landforms and sediments (e.g. Maizels, 1997; Tweed and Russell, 1999; Carrivick *et al.*, 2004a, b) and physical-scale modelling (Rushmer, 2007). These

Megaflooding on Earth and Mars, ed. Devon M. Burr, Paul A. Carling and Victor R. Baker.
Published by Cambridge University Press.

landforms and sediments are 'records' of flood hydraulics at a specific place and time during a flood. Fluvial landforms thus can permit flow reconstructions and can offer quantification of high-magnitude fluvial processes and mechanisms through the use of palaeohydraulic methods. Palaeohydraulic methods such as a palaeocompetence (Williams, 1984; Costa, 1983; Komar, 1989) and slope-area techniques (utilising the continuity equation) consider at-a-point and cross-section averages or one-dimensional (1D) hydraulic variations respectively. Alternatively, a source hydrograph can be used to initiate the (cellular) routing of floodwater over a Digital Elevation Model (DEM), incrementally computing depth-averaged hydraulics in two dimensions (2D). Two-dimensional modelling therefore offers new possibilities for studies of high-magnitude outburst floods, such as jökulhlaups (Carrivick, 2007b), by allowing hydraulic calculations that are independent of landforms preserved within flood routeways and that can consider both spatial and temporal variations of hydraulics. Carrivick and Rushmer (2006) provide a succinct review of both 'traditional' and 'new' palaeohydraulic methods.

15.2 Aim

The aim of this chapter is to demonstrate the high spatial and temporal variability of hydraulics within a high-magnitude flood, specifically a jökulhlaup, by using a number of flood reconstruction techniques. In particular, 2D hydrodynamic modelling will be applied to an example of a poorly understood and complex high-magnitude outburst flood; one occurring within a steep, confined and anastomosing bedrock channel system.

15.3 Study site

Hydraulic reconstructions are made of the largest Holocene jökulhlaup from Kverkfjöll; a glaciated stratovolcano on the northern margin of Vatnajökull (Jóhannesson and Saemundsson, 1989) (Figure 15.1). This study site is chosen because the proglacial area of the Kverkfjöll Volcanic System (KVS) (Kverkfjallarani, Figure 15.1) contains comprehensive field evidence of a high-magnitude flood (Carrivick *et al.*, 2004a, b; Carrivick, 2005; Rushmer, 2006) and because a high-resolution (10 m horizontal grid with sub-metre vertical accuracy) DEM of Kverkfjallarani exists (Carrivick and Twigg, 2005), which is necessary as input to a 2D hydrodynamic model.

Kverkfjallarani is dominated by a series of parallel volcanic pillow-hyaloclastite ridges, each typically 100 m high and several kilometres long (Jóhannesson and Saemundsson, 1989). Between these ridges, valley floors comprise subaerial lava flows and high-magnitude fluvial sedimentary deposits (Carrivick *et al.*, 2004b). High-magnitude outburst floods with cross-sectional discharges of up to 100 000 $m^3 s^{-1}$ flowed through Kverkfjallarani during the Holocene and inundated a series of anastomosing routeways (Figure 15.1) (Carrivick *et al.*, 2004a, b). These floods primarily flowed to the northeast along the 20 km long Hraundalur valley system and along a series of small steep valleys spilling westwards over low cols (Figure 15.1).

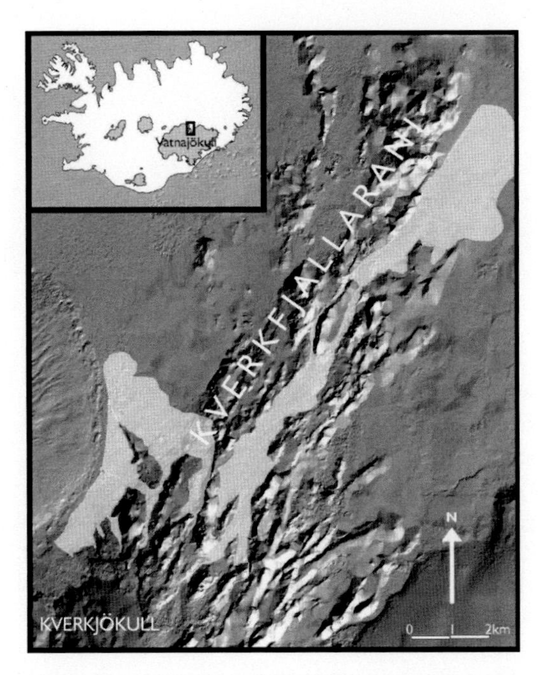

Figure 15.1. The location of Kverkfjöll within Iceland, and Holocene jökulhlaup routeways (shaded area).

15.4 Methodology

Three methods: boulder cluster fabric analysis, slope-area and 2D hydrodynamic modelling, are used to make palaeohydraulic reconstructions of jökulhlaups from Kverkfjöll (Table 15.1). Results from each of these methods emphasise strengths and weaknesses of each and, because they use different input data (Table 15.1), also allow independent support for results obtained from 2D modelling.

Clast fabric data were collected for a sample of 50 boulder clusters (Brayshaw, 1984), both within a confined valley reach and upon an unconfined outwash fan. Clast fabric data comprise cluster location (measured using a handheld GPS), cluster orientation (measured by a magnetic compass bearing), number of clasts within a cluster, axes lengths of each clast and orientation and dip of the *a–b* plane for each clast (Figure 15.2).

Slope-area methods frequently have been used to model jökulhlaups and other large flood flows but rely upon resistance factors quantified for lower-magnitude floods. For the largest floods known to have occurred on Earth, 'megafloods', slope-area methods have been used (usually HEC-2, or HEC-RAS in later studies), in order to

Table 15.1. Comparison of the requisite data, assumptions and output of the palaeohydraulic methods used in this chapter

Method	Requisite data	Major assumptions	Output
Palaeocompetence	Boulder diameter and/or fabric	Unlimited sediment availability (calibre) Validity of equations derived from lower-magnitude flows	At-a-point hydraulics most probably pertaining to near-peak stage
Slope-area	Cross-sectional geometries Roughness Palaeostages	Problem of roughness parameterisation, which is also stage dependent	1D (mean cross-sectional) velocity and discharge
2D hydrodynamic modelling	Initial hydrograph DEM	Fixed bed (no sediment transport) Most likely DEM pertains to postflood channel	2D (depth-averaged) hydraulics, with incremental time output

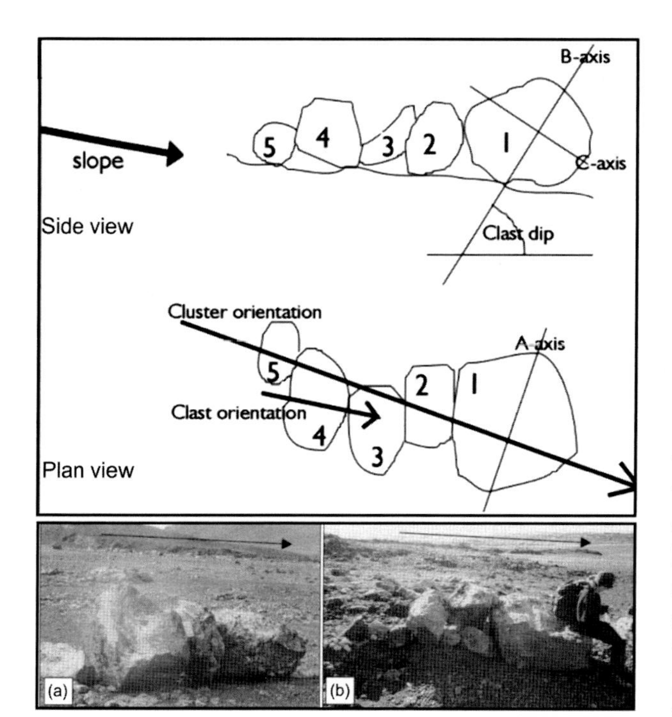

Figure 15.2. Schematic representation of clast fabric and boulder cluster measurements in profile (side) view and plan view. Examples of boulder clusters are given in (a) and (b). Notebook for scale in (a) and person in (b). Boulders clusters were observed with up to 12 clasts.

iteratively account for transmission or energy losses between successive cross-sections. These studies include Missoula, Bonneville and Altai palaeohydraulic reconstructions (Jarrett and Malde, 1987; O'Connor and Baker, 1992; Baker *et al.*, 1993; O'Connor, 1993; Benito, 1997; Herget, 2002, 2005; Rudoy, 2002).

All slope-area calculations are based on the continuity equation ($Q = AU$), where discharge, Q, is a product of the cross-sectional area, A, and flow velocity, U. A is determined by channel geometry and an identification of PalaeoStage Indicators (PSIs), and U is a function of the channel slope (as a proxy for the water energy slope), acceleration due to gravity (g), and resistance factors, which will be discussed below.

This study makes use of use cross-sectional geometry derived from a DEM of Kverkfjallarani. This DEM was geo-referenced from 29 ground control points that were surveyed using a differential Global Positioning System (dGPS) relative to the Icelandic Geodetic Network (Carrivick and Twigg, 2005). The DEM was processed using digitised aerial photographs and Erdas Imagine Orthobase software (Carrivick and Twigg, 2005). The resultant DEM has a 10 m horizontal resolution and a sub-metre vertical accuracy and therefore allows cross-sectional areas of jökulhlaup channels to be computed efficiently and at high resolution. The location and altitude of slackwater deposits and of other wash limits were identified and measured by field surveys using a combination of (Sokkia) Total Station (TS) surveying methods and a handheld GPS. It should be noted that these slope-area calculations pertain to the largest jökulhlaup along each valley and assume that bed elevation has not changed. Because some postflood infilling has occurred from perennial slope processes, reconstructed slope-area peak discharges may therefore be taken as minimum estimates. Bed elevation necessarily was assumed to remain constant during a jökulhlaup due to the absence of data pertaining to erosion rates or sediment depths to bedrock, for example.

Hydraulics are reconstructed in this study without iterative consideration of longitudinal transmission energy losses, which can be largely attributed to resistance factors. Therefore, with caution, herein four different methods are used to parameterise resistance factors. It should be noted that resistance can be parameterised at different scales. The original method of Strickler (1923) incorporates a grain resistance factor of transported sediment. It is a grain size parameter that is used to modify Manning's 'n' function within numerous equations that are based upon that of Strickler (1923):

$$n = kD^{1/6}, \tag{15.1}$$

where n is Manning's 'n', D is a measure of bed clast size (m) in the form of a percentile value of the cumulative frequency distribution of clast sizes and k is a coefficient. Limerinos (1970) adds a relative roughness element to the grain resistance factor. The method of Thompson and Campbell (1979) also includes a term for the relative protrusion of resistance elements into the flow depth. The method developed by Jarrett (1984) relies upon the hydraulic radius parameter rather than specific form and grain roughness values for its resistance term. Slope-area calculations of flow velocity (U), together with the known parameters of flow depth (d) and channel/energy slope (S) allow estimates of shear stress (τ), from the DuBoys relationship:

$$\tau = \rho g d S, \tag{15.2}$$

where ρ is the density of fluid (kg m^{-3}), g is the acceleration due to gravity (m s^{-2}), d is the flow depth (m) and S is the energy gradient. Shear stress (τ) therefore has units of N m^{-2}. Stream power (ω) is calculated per unit area as

$$\omega = \tau U, \tag{15.3}$$

where τ is shear stress (N m^{-2}) and U is velocity (m s^{-1}). Stream power per unit area therefore has units of W m^{-2}.

For further analysis, this chapter computes the dimensionless Froude number, Fr, which examines the ratio of inertial to gravitational forces, as given by

$$Fr = \frac{U}{\sqrt{gd}}, \tag{15.4}$$

where U is flow velocity and g and d are as defined above. Values of $Fr < 1$ represent subcritical flow conditions and values of $Fr > 1$ represent supercritical flows. The transition from subcritical to supercritical flow conditions is a hydraulic jump and this is a zone of intense turbulence, shear and vorticity.

The 2D model used is SOBEK, developed by WI Delft-Hydraulics (http://delftsoftware.wldelft.nl/). SOBEK is a graphically orientated depth-averaged hydrodynamic model. Two-dimensional models based upon a regular grid are ideal for integration with DEMs, can handle unsteady flows and are therefore suitable for modelling large distributed jökulhlaups. The full St Venant hydrodynamic equations, as used by SOBEK, are described by Sleigh and Goodwill (2000). A complete model therefore contains boundary data, constants, a topographic grid (a DEM), a roughness grid and specified outputs (Table 15.1). The roughness grid was of a uniform value over the entire study area because all of the valley floors in the area are bedrock (although some reaches have a veneer of alluvial material, which would be rapidly removed by a high-energy flow). The uniform roughness value was a Mannings 'n' of 0.05, which is the mean of that computed by slope-area methods. SOBEK outputs increment flow depth and velocity grids. These grids were imported to a GIS, along with the DEM, to compute grids of shear stress and stream power, using Equations (15.2) and (15.3). Sediment transport or bed elevation changes were not considered. Therefore, when fluid density was altered, to model high sediment loads, governing equations did not change to incorporate clast interactions or a rheological change. The hydrograph used was produced by analysis of the source and likely trigger mechanism and also by iteratively running SOBEK to best-fit modelled flow depths with PSI data (Carrivick, 2005, 2007a).

A few general problems were encountered with the model whilst simulating jökulhlaups through Kverkfjallarani. Excessively long model run times (up to 13 days) were experienced. These long model runs were due to a very large DEM (over 5 000 000 grid cells) and to a large volume of water (~50 000 000 m^3). Additionally, to retain model robustness, i.e. to conserve mass and energy throughout the simulation, SOBEK automatically reduces the computational time step. Conservation of mass and energy demanded that a small DEM grid size was necessary, given the large volumes of water being modelled. Nonetheless, computational time steps reached as low as 0.001 s due to the number of active grid cells within a single model. For note, an improved version of this model has been applied to the Kverkfjöll jökulhlaups by Carrivick (2007b).

15.5 Results and interpretation

The results presented below are for each method (previous section), with particular reference to demonstrating spatial and temporal variation in hydraulics, rather than being an exhaustive description of the Kverkfjöll jökulhlaups. Aspects of the gross generalised hydraulics of the Kverkfjöll jökulhlaups have been described from geomorphological and sedimentological evidence by Carrivick *et al.* (2004a, b), and by means of several palaeohydraulic reconstructions by Carrivick (2005, 2006, 2007a, b).

15.5.1 Boulder clusters

Boulder clusters (Brayshaw, 1984) within Kverkfjallarani (Figure 15.2) have been described and interpreted to be the product of jökulhlaup bedload transport by Carrivick *et al.* (2004a). Typically, the clusters are located immediately downstream of valley constrictions and upon tracts of scoured and smoothed subaerial lava. Boulder clusters in Kverkfjallarani can extend up to 10 m in length and comprise up to 12 clasts. Clasts usually display imbrication with each clast a-axis transverse to the valley axis

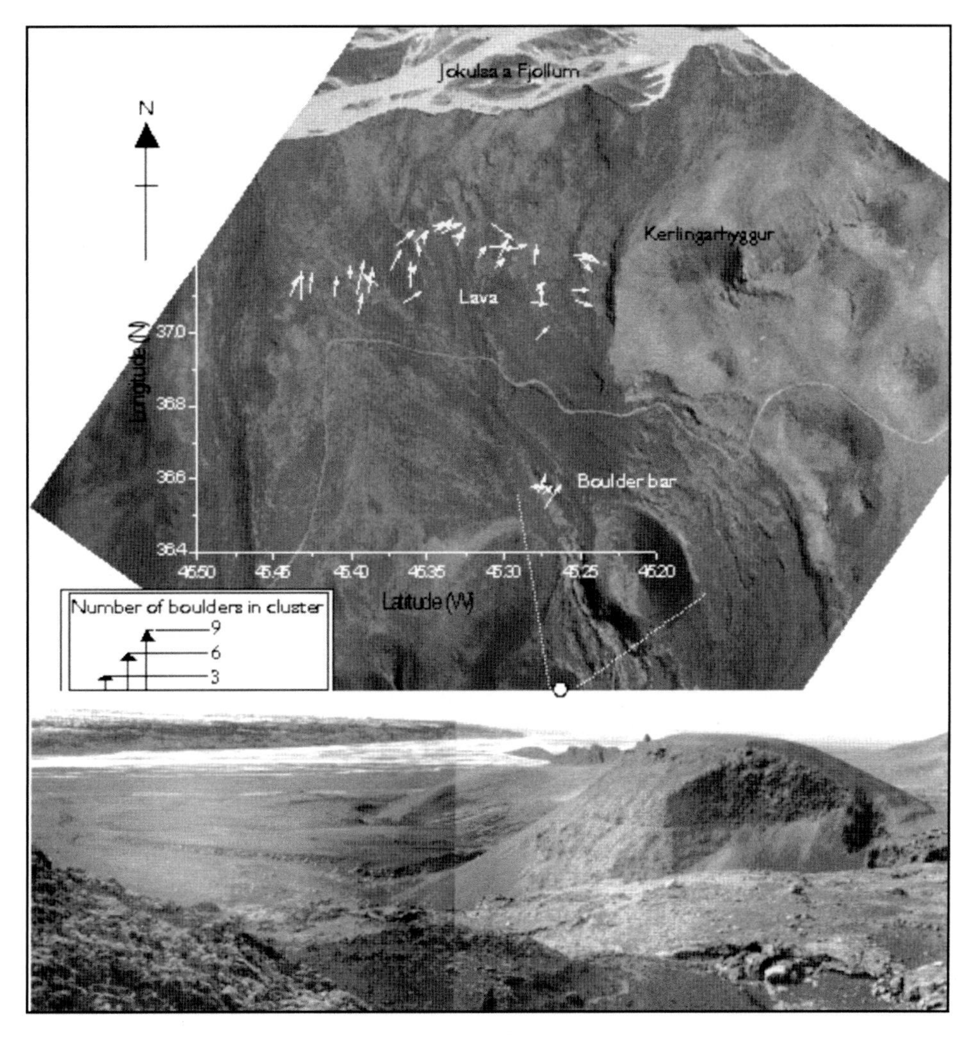

Figure 15.3. Location of imbricated boulder clusters upon an unconfined fan surface. Circle and lines denote location and field of view of photograph.

(Figure 15.2). Individual clast *a*-axes are typically ~110–160 cm diameter although this is highly variable, even within an individual cluster.

Principles and assumptions of palaeohydraulic reconstructions from boulders are given by Maziels (1983). Palaeocompetence calculations have been made of isolated boulders within Kverkfjallarani by Carrivick (2007a). Palaeocompetence equations and calculation results are not repeated here, for brevity, and because they only pertain to an instant in time: that of clast deposition (previous section). However, palaeocompetence calculations do show that hydraulics associated with boulder deposition were highly spatially variable (Carrivick, 2007a).

This part of the study solely considers boulder clusters. A sample of 50 imbricated boulder clusters was examined upon an unconfined outwash fan, underlain by subaerial lava and situated immediately east of Kerlingarhyggur (Figure 15.3). Cluster orientations trend along the local slope and therefore initially westwards, then in a northerly direction (Figure 15.3). These orientations suggest a spatial variability in fluid flow direction, specifically divergence over small cliffs and subaerial lava steps and convergence in shallow depressions. Individual clasts are well rounded, tending towards sphericity or an equant shape (Figure 15.4a). This shape is interpreted to be the product of considerable erosion of individual particles, probably due to abundant sediment load within the flow and high-impact bed collisions. This rounding is because boulder clasts originated from basaltic lava (Carrivick *et al.*, 2004b; Carrivick, 2005) and so had a columnar shape prior to erosion. The original columnar structure and vertical jointing aided erosion by plucking (Carrivick *et al.*, 2004b). Clast dip, of the *a*–*b* plane, also varies considerably (Figure 15.4e) and is independent of clast shape (Figure 15.4b). It should be noted that due to the equant shape of many clasts (Figure 15.4a), some doubt has to be cast on the ability to

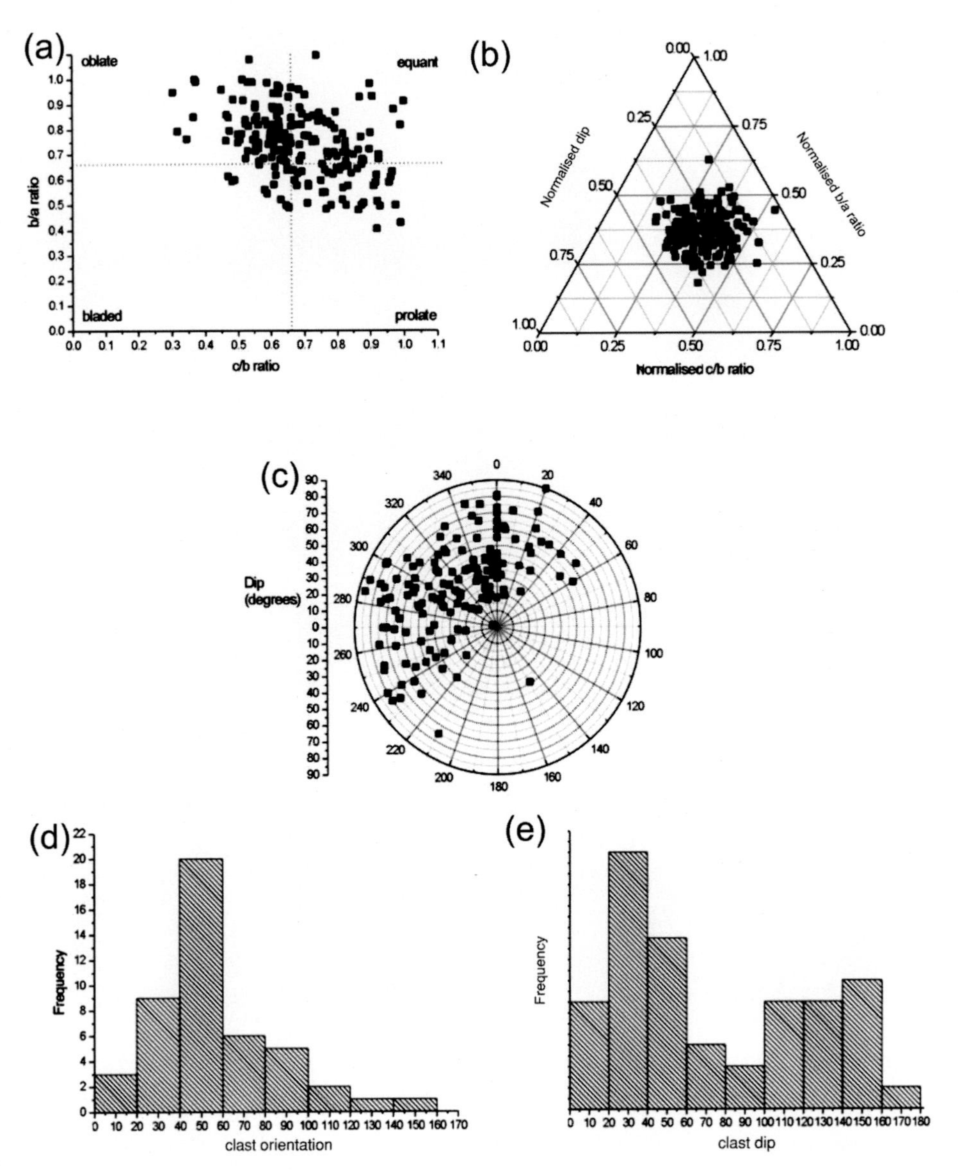

Figure 15.4. Properties of boulder clusters upon an unconfined fan surface. (a) Individual clast shapes. Adapted from Zingg (1935). (b) Variation of clast dip with clast shape. (c) Variation of individual clast orientations, where zero degrees is an *a–b* plane aligned perpendicular to due north, and clast dips, where 90° is a vertical *a–b* plane. (d) Variation of clast orientations within a cluster. (e) Variation of clast dips within a cluster.

accurately determine the *a–b* plane, as evidenced by the apparent bimodality of clast dip (Figure 15.4e). Nonetheless, these *a–b* plane dip observations imply that boulders were predominantly rolling along the bed, irrespective of clast size or shape, because clasts within a given cluster are imbricated, or stacked up against each other (Figure 15.2). Imbricated boulder clusters thus represent grain by grain deposition from traction load within a high-energy turbulent flow (Brayshaw, 1984). Furthermore, clast orientations trend between north and west and vary by less than 60 degrees within an individual cluster (Figure 15.4c, d). This lack of variation within a cluster suggests uniform flow directions, in space and time, and clearly refutes a hypothesis that these boulders are derived from slope processes without fluvial transport. Finer-grained clasts, such as cobbles and pebbles, are absent from these reaches but the flows must have had a bimodal sediment content originally, as a suspended load must have been present to induce the observed intense abrasion of imbricated clasts. The finer load has been entirely removed from the reach. Similarly, there is no excavation of sediment around boulder clusters and the few that are on cobble and pebble sediments stand proud of that surface. Some boulders are 'perched' upon pebbles and cobbles.

Figure 15.5. Location of imbricated boulder clusters within a confined bedrock valley. Note that cluster orientation is aligned to the valley slope. Circle and lines denote location and field of view of photograph.

A sample of 50 imbricated boulder clusters was also taken upon a subaerial lava within a confined valley, located immediately due east of Virkisfell (Figure 15.5). Some of these boulders are upon locally elevated areas of lava. Cluster orientations trend along the local slope and therefore in a northerly direction (Figure 15.5). These orientations preclude boulder clusters from being slope material and firmly establish them as products of gravity-driven flows (Carrivick *et al.*, 2004b; Carrivick, 2005). Boulder clusters within this confined valley predominantly are situated upon subaerial lava surfaces and increase in abundance downslope (Figures 15.5 and 15.6a, b, c). It is therefore significant to note that this valley decreases in gradient and widens downslope (Figure 15.5), implying that clusters formed as flows progressively lost energy. Unlike clusters on the unconfined surface, valley boulder clusters have the largest clasts located at the downslope end of most clusters, and fine upstream (Figure 15.6a, b, c). This pattern implies that clusters formed on the waning stage (Brayshaw, 1984). Larger boulders grounded first and formed obstacles to subsequent smaller boulders, thus developing a train, or cluster. Clast dips are unrelated to distance downslope (Figure 15.6d) and clast shapes are well rounded, tending towards sphericity or an equant shape (Figure 15.6e). Clast dips are irrespective of clast shape (Figure 15.6f), and are highly varied within an individual cluster (Figure 15.6i). These observations suggest that clasts predominantly were transported by rolling in a fluid flow imparting a bed shear stress, rather than by non-Newtonian mechanisms, in which case *a*-axes would most likely become longitudinal to valley slope (Carrivick, 2005). Clast orientations vary by 30–60 degrees within an individual cluster, and trend northwards (Figure 15.6g, h).

In both the unconfined outwash-fan setting and within the confined valley, boulders do not fine downstream, but the number of boulder clusters increases. Jökulhlaup sedimentation is therefore interpreted to have been capacity driven rather than competence driven

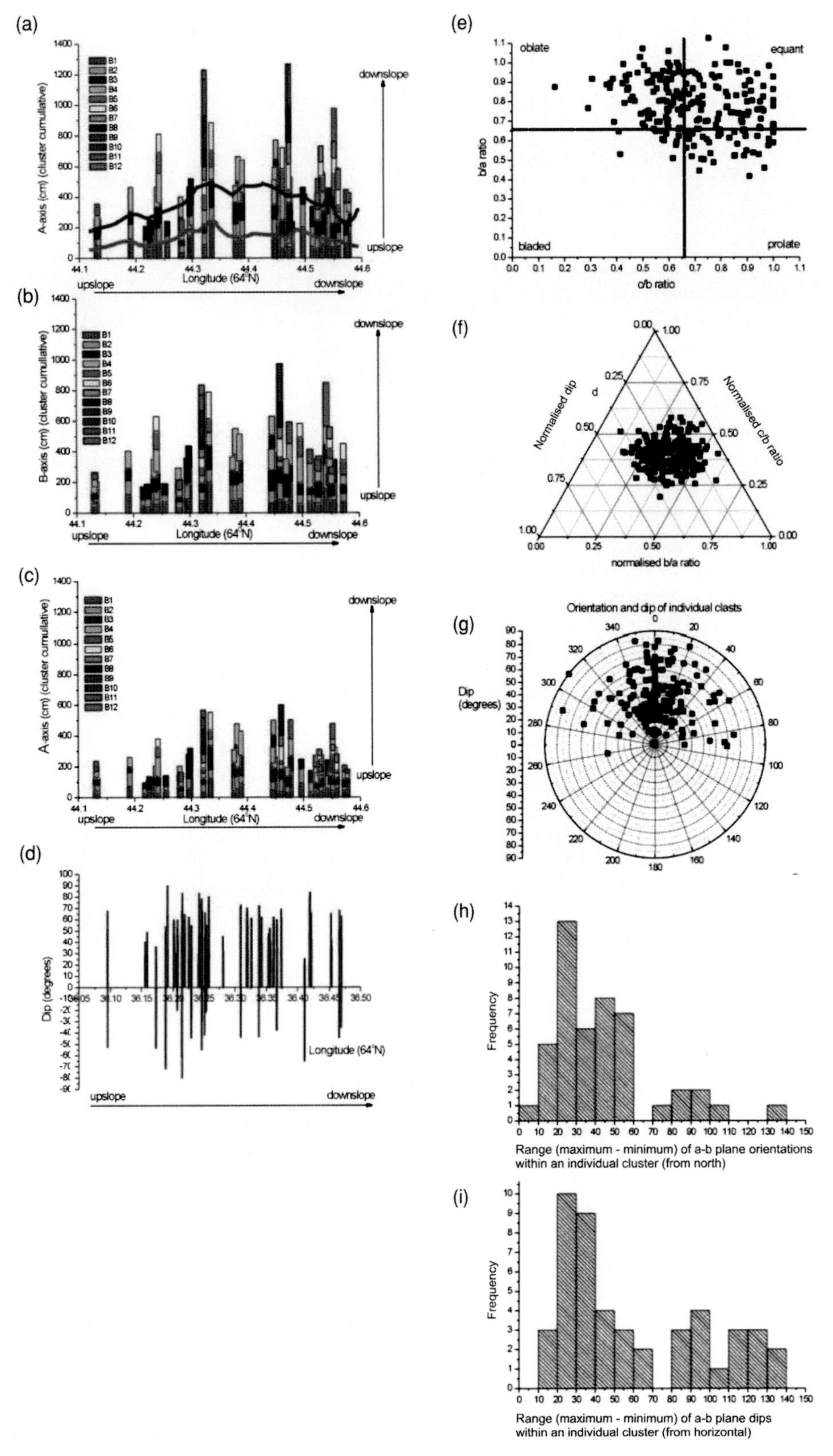

Figure 15.6. Properties of boulder clusters within a confined bedrock valley. Boulder clasts within a single cluster are numbered from one (B1) to twelve (B12), as given in Figure 15.2. (a) Variation of individual clast *a*-axis and cluster length, with distance downstream. (b) Variation of individual clast *b*-axis and cluster length. with distance downstream. (c) Variation of individual clast *c*-axis and cluster length, with distance downstream. (d) Variation of individual clast dips downstream. Note some clasts have inverse dips. (e) Individual clast shapes. Adapted from Zingg (1935). (f) Variation of clast dip with clast shape. (g) Variation of individual clast orientations, where zero degrees is an *a*–*b* plane aligned perpendicular to due north, and clast dips, where 90° is a vertical *a*–*b* plane. (h) Variation of clast orientations within a cluster. (i) Variation of clast dips within a cluster.

(Hiscott, 1994; Manville and White, 2003). The absence in these reaches of finer-grained sediment suggests that flows attenuated rapidly, without time for sediment sorting, waning stage incision, suspension settling and reworking of sediment.

15.5.2 Slope-area calculations

Slope-area calculations results span a range defined by maximum and minimum values calculated at each downstream cross-section (Figure 15.7). The range is due to the fact that four methods of parameterising resistance factors were used. Overall, slope-area results suggest that jökulhlaup velocities along Hraundalur, Kverkfjöll, ranged from ~6 to $20\,m\,s^{-1}$ and decreased with distance downstream (Figure 15.7a). Peak discharges, per valley cross-section, are estimated at $30\,000\,m^3\,s^{-1}$ to $115\,000\,m^3\,s^{-1}$ and attenuated with distance downstream, by ~50–75% in just 25 km (Figure 15.7b). The slope-area method suggests that shear stress and stream power were rather more uniform than flow velocity and peak discharge with distance downstream, with means of ~1500 $N\,m^{-2}$ and $25\,000\,W\,m^{-2}$ respectively (Figure 15.7c, d). Furthermore, slope-area reconstructions indicate that jökulhlaups along Hraundalur were characterised by predominantly supercritical flow conditions because the reconstructed Froude number is generally greater than 1 (Figure 15.7e). Some caution is drawn to this result because it is unusual for supercritical flow to be sustained for great longitudinal distances (Grant, 1997). However, it is clear that PSIs are poorly related to dramatic channel expansions (Figure 15.7f) and thus downstream attenuation and a supercritical regime are not simply a function of channel topography (both longitudinal slope and rapid cross-sectional shape changes). It is far more likely that flows developed hydraulic jumps and reverted to subcritical flow for some intervening distances. This behaviour would not be picked up by slope-area methods but is suggested by the frequent hydraulic jumps produced in SOBEK (next section).

15.5.3 Two-dimensional hydrodynamic modelling

The results of 2D hydrodynamic modelling with a best-fit hydrograph suggest that the main Hraundalur valley of Kverkfjallarani became inundated within 3.5 hours (Figure 15.8). Mean frontal velocity was therefore $1.6\,m\,s^{-1}$. Jökulhlaup 'main body' velocity was tremendously variable, with maxima of up to $15\,m\,s^{-1}$ occurring in constricted and steep zones of flow. Lowest velocities were associated with temporary ponding, e.g. immediately upstream of valley constrictions. Flow depths ranged from less than 1 m to ~10 m (Figure 15.8). Shear stresses ranged from $<100\,N\,m^{-2}$ in relatively wide reaches up to ~7500 $N\,m^{-2}$ within narrow and confined channels. Stream power per unit area ranged from ~2000 to $>50\,000\,W\,m^{-2}$ with an average value of $8500\,W\,m^{-2}$. Naturally, greater stream powers occurred within narrow channels of steeper gradient.

SOBEK results suggest that the Kverkfjallarani jökulhlaup along Hraundalur was characterised by frequent hydraulic jumps, between subcritical and supercritical flow regimes (Figure 15.9). The transition between the two regimes likely would have been a zone of turbulence, high pressure and potential for high-magnitude fluvial phenomena such as cavitation and plucking (Whipple *et al.*, 2000). Such zones are thought to be responsible for the production of cataracts, gorges and bedrock steps (e.g. Carrivick *et al.*, 2004a; Carrivick, 2005), but precise mechanisms and hydraulic parameters of bedrock erosion remain unquantified. It should be noted that the engineering literature suggests that the presence of high suspended sediment content, a dense traction load, and a rough bed can all interfere with cavitation development (e.g. Chincholle, 1994; Lee and Hoopes, 1996; Nie, 2001).

Supercritical flow, where $Fr > 1$, predominates in narrow, steep channels and subcritical flow, $Fr < 1$, predominates in wide, shallow reaches (Figure 15.9). However, supercritical flow also persists at some channel margins and around islands (Figure 15.9). Supercritical flow zones at channel margins in Kverkfjallarani correspond to a gap between valley floor lava flows and valley walls (Carrivick *et al.*, 2004a). Supercritical zones around islands relate to flow acceleration. That peak discharge through Kverkfjallarani apparently attenuated very rapidly, due to very high channel roughness (previous section), further is indicated by 2D modelling, because supercritical flows predominate in the most southerly or proximal reach and subcritical flows in the most northerly or distal reach (Figure 15.8) and because PSIs are poorly related to dramatic channel expansions (Figure 15.7f). The peak discharge of the Hraundalur jökulhlaup from Kverkfjöll therefore attenuated due to high longitudinal energy losses, predominantly caused by high hydraulic roughness: Manning's 'n' typically 0.05–0.1 (Carrivick *et al.*, 2004b) and to a lesser extent due to recirculation of water in backwater areas (Figure 15.8). The greatest contributing factor to hydraulic roughness would have been 'form roughness'. Form roughness quantifies the height and density of immobile protrusions into a flow, and is stage dependent. Form roughness in Kverkfjallarani comprises bedrock protrusions, which are typically 3–5 m and thus of the same order of height as flow depths. Therefore, jökulhlaups through Kverkfjallarani attenuated more rapidly in wide shallow distal valleys than in narrow deep proximal reaches (Figure 15.7b), not just due to changes in valley width and consequent energy dissipation but also due to greater relative roughness. However, the association of greatest discharge attenuation with valley expansions is also due to attempting to conserve flow velocity and

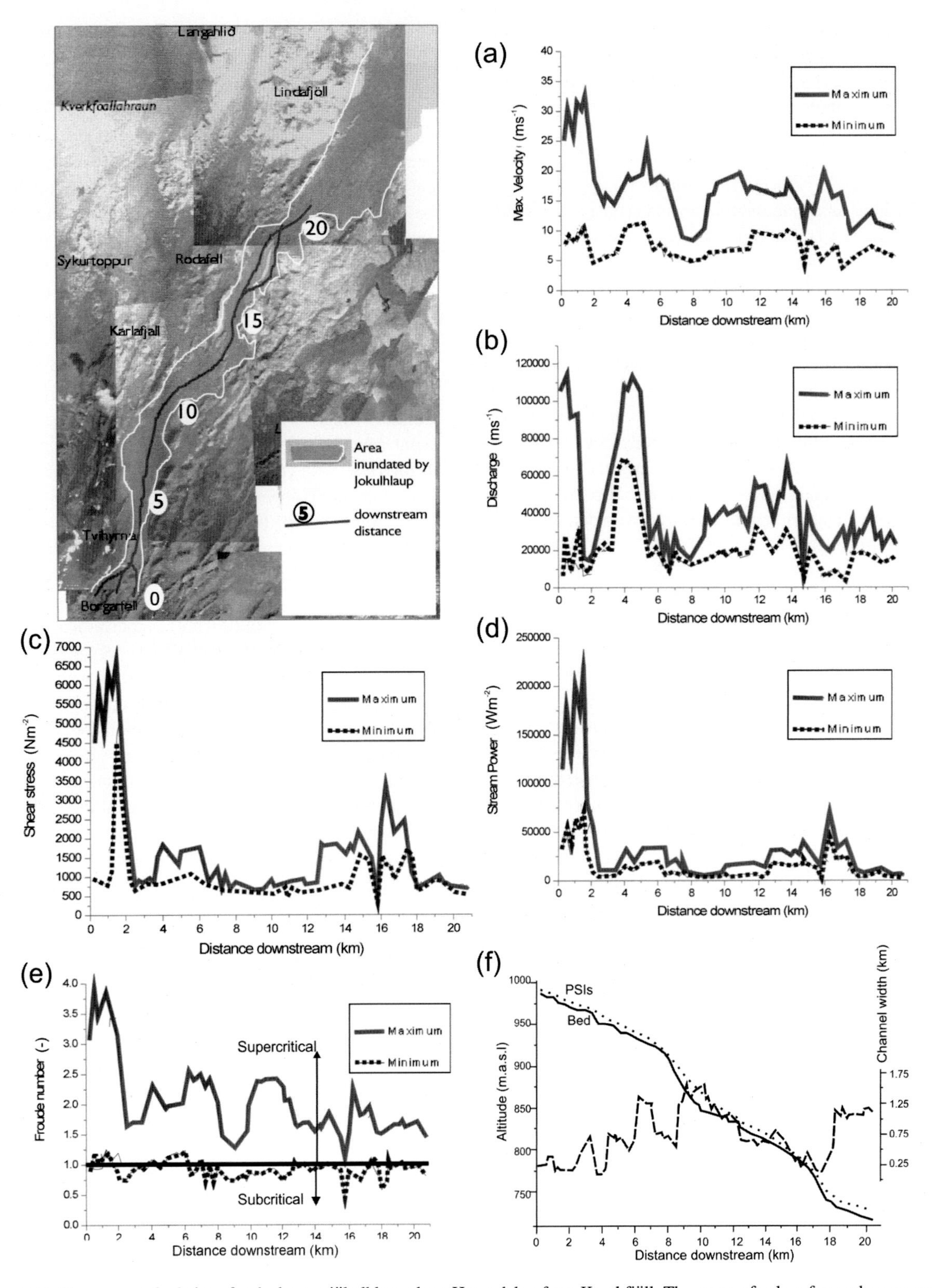

Figure 15.7. Slope-area calculations for the largest jökulhlaup along Hraundalur, from Kverkfjöll. The range of values for each cross-section is due to the fact that four methods of parameterising resistance factors were used (previous section). A 'segment' is one of multiple channels that are perceived to have been inundated simultaneously. (All adapted from Carrivick (2007a), except (f), which is adapted from Carrivick, 2005).)

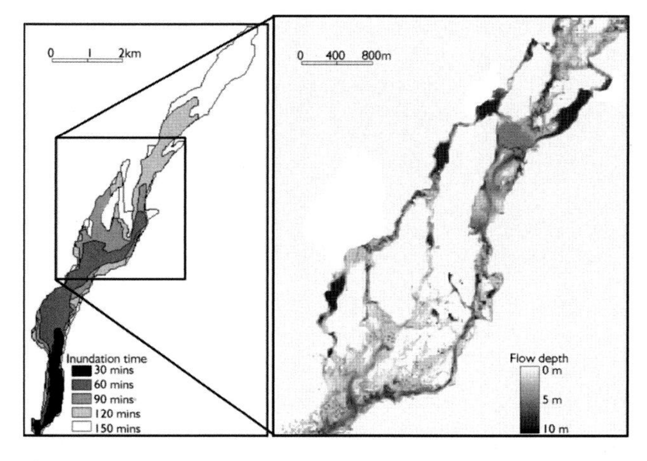

Figure 15.8. Two-dimensional hydrodynamic model of jökulhlaup flow front velocity and peak flow depth.

momentum in 2D, because secondary flow circulation is not accounted.

As discharge rises, channel or valley constrictions can cause upstream ponding, particularly where channel widths are reduced by the presence of mid-valley islands (Figure 15.7). Ponding depends not just on the magnitude of discharge but also on the rate of discharge rise. Ponding can be recognised in 2D-model output, either by zones of relatively deep flow (Figure 15.7) or by zones of relatively low flow velocity. Ponding in Kverkfjallarani could have been responsible for the deposition of valley-fill sediments (Carrivick *et al.*, 2004a, b; Carrivick, 2005). If ponding leaves PSIs, flood reconstructions using them could calculate erroneous flood hydraulics, by assuming that they relate to unponded channel flow. It should be noted that these 'backwater' effects are accommodated in computerised 1D hydraulic models, particularly the latest types with unsteady flow capabilities such as HEC-RAS.

An important result is that inundation of tributary valleys and other backwater sites depends on flow depth. Time series output from the 2D model shows progressive inundation and deepening flows within backwater areas (Figure 15.10), whilst main channel hydraulics remain relatively stable, despite rising stage, because backwaters accommodate water mass and thus help dissipate flow energy and attenuate main channel discharge. Backwaters can also develop a surging flow (Figure 15.10), as flow laps over-bank in successive waves. Waves may also be produced from upstream ponding, as continued discharge periodically causes overtopping of shallow topography. This unsteadiness is important for two reasons. Firstly, palaeocompetence, slope-area and other computerised 1D models can only consider peak stage flow conditions, which are assumed to be maximum values. This condition may not necessarily be the case and also it is the persistence of flow above a geomorphological threshold, the rate of rise to peak and the rate of discharge decrease that are the most important factors for landscape change (e.g. Magilligan, 1992). Secondly, it is clear that the 'megafloods' (next section) featured many instances of backwater flooding of tributaries and col overtopping, for example, and these are not accommodated except within 2D unsteady flow modelling.

15.6 Discussion

Palaeocompetence calculations in Kverkfjallarani consistently underestimate hydraulic parameters, in comparison with slope-area and 2D modelling methods (Table 15.2). This result suggests either that the palaeocompetence equations are not valid for this application (Maizels, 1983), or that Kverkfjallarani jökulhlaups were sediment supply-limited in terms of sediment calibre (Carrivick, 2007a). Slope-area calculations cannot account for energy transmission between cross-sections and only produce mean hydraulic values per cross-section. Hence, downstream values of hydraulics from the slope-area method are highly variable (Figure 15.7) and of limited use for geomorphological analyses. Two-dimensional modelling results of flow depth and velocity are of the same magnitude as those from both the palaeocompetence and the slope-area methods (Table 15.3), and the inundation extent and flow depths produced by the 2D model agree well with field evidence of inundation and PSIs (Carrivick *et al.*, 2004b).

The size of boulders within clusters suggests that Kverkfjöll jökulhlaups were sediment supply-limited in terms of clast size. This size limit is because the constituent materials of Kverkfjallarani are hyaloclastite breccia and subaerial lava (Karhunen, 1988; Jóhannesson and Saemundsson, 1989; Carrivick *et al.*, 2004a). The former readily breaks down to cobble fragments and the latter is characterised by vertical parallel cooling joints, where joint spacing (typically up to ~1 m) determines the maximum boulder diameter. Furthermore, finer-grained sediment, such as cobbles or pebbles, is not deposited alongside boulder clusters within the confined valley reach but forms a major component of the outwash fan farther downslope. This pattern implies that finer sediment was completely washed through the valley reach and that sedimentation in the upper bedrock reaches was competence-driven (first section). Boulders on the surface of outwash fans (e.g. Figure 15.4) are most likely to have been exposed due to waning flow stripping out intervening material (e.g. Manville and White, 2003), rather than by simply being products of late-stage deposition in a rapidly terminated flow (Maizels, 1983). Thus sedimentation on outwash fans was capacity-driven. This interpretation is based on the inference that jökulhlaups through Kverkfjallarani likely rapidly exhausted sediment supply and became capable of waning stage incision (Carrivick *et al.*, 2004b) and the observation that outwash-fan

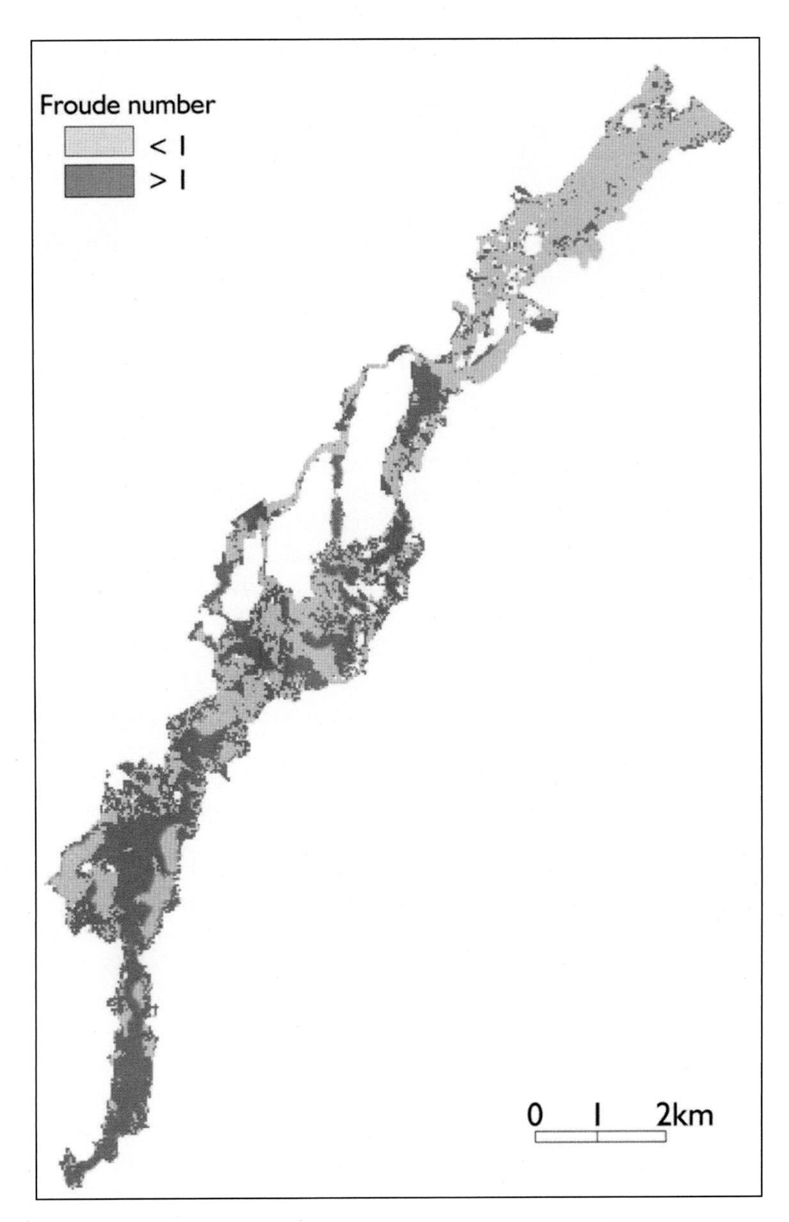

Figure 15.9. Two dimensional hydrodynamic model reconstruction of Froude number, at peak stage.

sediments predominantly comprise massive, poorly sorted sediments with 'floating' boulders throughout a given profile (Carrivick *et al.*, 2004b; Carrivick, 2005).

Slope-area estimates of peak discharge demonstrate a rapid downstream attenuation. This attenuation could be because a slope-area method cannot accommodate backwater areas or spill over topographic lows into neighbouring valleys. However, attenuation of peak discharge in Kverkfjallarani most likely is due to high channel roughness, imparted by immobile obstacles such as lava mounds (Carrivick *et al.*, 2004a) and high grain roughness, imparted by initially sediment-laden flows (Carrivick *et al.*, 2004b). In particular, form roughness was an important control on resistance to flow because bedrock protrusions into the flow were of similar scale to flow depths (1–10 m; previous section). Shallow depths were predominantly due to steep channel gradients, which also promoted supercritical flows (Grant, 1997).

Two-dimensional modelling also suggests relatively shallow flows and a predominantly supercritical flow regime. Whilst supercritical flows are more likely at shallow flow depths, a shallow flow exerts a lower bed shear stress than a deep flow of equivalent velocity and energy (Equation (15.2)). The 2D modelling therefore produces very high bed shear stresses (Figure 15.7c) in two situations: (i) where rapid changes in flow depth occur from a

Table 15.2. Summary of each of the hydraulic reconstructions of jökulhlaups from Kverkfjöll

	Depth (m)		Velocity ($\mathrm{m\,s^{-1}}$)		Shear stress ($\mathrm{N\,m^{-2}}$)		Stream power ($\mathrm{W\,m^{-2}}$)	
Method	range	mean	range	mean	range	mean	range	mean
Palaeocompetence (single point)	0.5–4.9	1.7	2.4–9.4	5.4	35–1000	312	170–8000	1200
Slope-area (peak discharge 1D cross-sectional mean)	*(input PSI data ~3–6)*	4	6–20	12	400–6500	1500	5000–20 0000	25 000
SOBEK (2D and temporal)	0.01–5.4		1–21	11.5	100–7500	1500	2000–800 000	28 000

Source: Adapted from Carrivick (2007a)

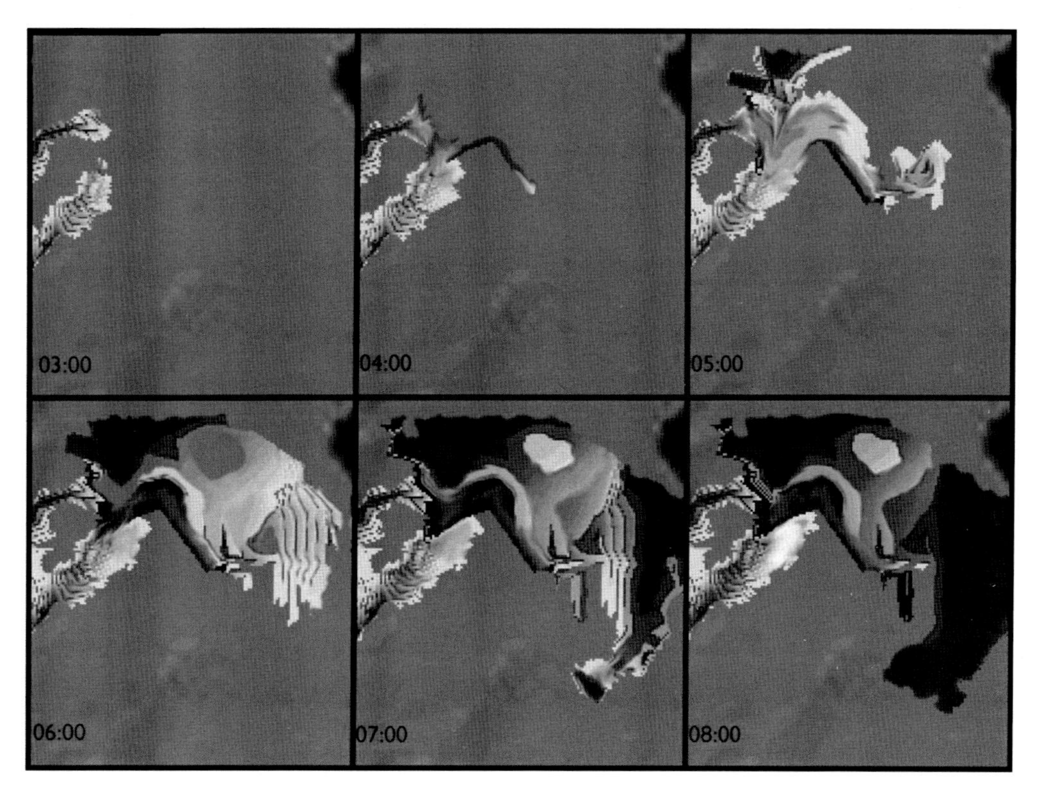

Figure 15.10. Example of backwater inundation by a jökulhlaup, as reconstructed by a 2D hydrodynamic model. Grey shades denote transitions from deeper flow (black) to shallower flow (white). Note apparent wave surges.

hydraulically ponded upstream reach; and (ii) with supercritical flow over a steep channel gradient, where the water surface slope parallels the bed gradient. The very similar estimates of downstream attenuation and flow regime obtained by 2D modelling to that of the slope-area method imply that channel roughness is a far more important control on attenuation than overspill or backwater recirculation (which were not accommodated by the slope-area method). The 2D model also reconstructs transient spatial and temporal flow patterns of sheet or unconfined flow, channel flow, flow around islands, hydraulic jumps, multidirectional flow including backwater areas and hydraulic ponding (Carrivick, 2005, 2006, 2007a, b). An example of these transient flow conditions is backwater inundation (Figure 15.10), where reconstructed flow surges over bank, and eddies and ponds develop within the mouth of a tributary valley.

Further analysis of 2D model output shows that main valley hydraulics predominantly were controlled by channel slope and channel width (e.g. Carrivick, 2005, 2006, 2007a, b; Figures 15.8 and 15.9). This conclusion is also indicated by the slope-area results, by comparison of Figure 15.7f to Figure 15.7b. Steep confined reaches were high energy and produced erosional landforms, such

Table 15.3. Comparison of the hydraulic parameters of the largest Kverkfjöll jökulhlaup with recent estimates of the largest known terrestrial outburst floods

Hydraulic parameter (maximum value)	Missoula flood O'Connor and Baker (1992) (step-backwater modelling); Benito (1997) (step-backwater modelling)	Bonneville flood Jarrett and Malde (1987) (step-backwater modelling); O'Connor (1993) (step-backwater modelling)	Altai flood Approaches by Baker *et al.* (1993) and Rudoy (2002) (step-backwater modelling); Agatz (2002) (step-backwater modelling and other methods). All summarised in Herget (2005)	Kverkfjöll jökulhlaup Summarised in this chapter
Width (km)	6	Not reported	2.5	1.6
Depth (m)	150	156	300–400	5.9
Velocity ($m\,s^{-1}$)	25	41	10–35	15
Discharge ($m^3\,s^{-1}$)	17×10^6	14×10^6	10×10^6	2×10^5
Bed shear stress ($N\,m^{-2}$)	1×10^4	2.9×10^3	0.4×10^5–2×10^5	1×10^4
Stream power ($W\,m^{-2}$)	2.5×10^5	1.2×10^5	$0.1 \times 10^6 - 1 \times 10^6$	1×10^5
Channel slope	0.01	0.0095	0.022–0.061	0.02–0.4
Roughness (Manning's n)	0.04–0.1	0.03–0.06 (0.1–0.9 for 'major flow obstacles')	0.03–0.05	0.03–0.1
Froude number	predominantly < 1	predominantly < 1	predominantly < 1	predominantly > 1

Source: Adapted from Carrivick (2007a)

as scoured bedrock and shallow cataracts, sometimes with superimposed boulder clusters. Shallow gradient-wide reaches were low energy, and produced depositional landforms such as outwash fans and bars (Carrivick *et al.*, 2004a, b).

The largest Kverkfjöll jökulhlaup, which routed along Hraundalur, had flow widths, depths and discharges that were orders of magnitude less than those parameters for the greatest terrestrial outburst floods (Table 15.3). Correspondingly, the geomorphological impacts of Kverkfjöll jökulhlaups scale by the same amount. However, the Kverkfjöll jökulhlaups generated bed shear stresses and stream powers comparable with those of the Missoula, Bonneville and Altai floods (Table 15.3) and thus can be termed 'megafloods'. The shear stress promoted intense and extensive bedrock erosion, with cavitation probably being an important mechanism of bedrock erosion (Figure 15.11). Cavitation in Kverkfjallarani may have occurred due to supercritical flow in relatively low flow depths and relatively low flow velocities (Figure 15.11). O'Connor presented evidence that cavitation occurred in the largely subcritical Bonneville flood, in relatively high flow depths and relatively high flow velocities (Figure 15.11). Therefore, it is suggested that high bed shear stresses were generated by the Kverkfjöll jökulhlaups as a product of high channel gradients that produced high-velocity but shallow flows. These shallow flows encountered high relative roughness (previous section) and consequently supercritical flows developed locally. Erosion by jökulhlaups flowing through Kverkfjallarani thus was concentrated into a series of discrete 'inner channels' and this contrasts with valley-wide lowering interpreted in deeper jökulhlaup flows, such as from Lake Missoula (e.g. Baker, 1988). The intensity of bedrock erosion by Kverkfjöll jökulhlaups was therefore disproportionate to peak discharge, or flow volume.

15.7 Conclusions

Jökulhlaups from Kverkfjöll exhibited highly transient hydraulic properties. These hydraulics characteristics were the product of high-magnitude flows within steep, confined and anastomosing bedrock channel systems. Through the application of several methods of reconstructing hydraulics, it can be summarised that:

1. Boulder clusters occur in reaches without waning stage deposits and illustrate considerable abrading of clasts, fluid bedload transport, and a spatially diverse flow pattern. Sedimentation within bedrock valleys was competence-driven.
2. Cobbles and pebbles do not occur alongside boulder clusters within a confined valley but form a major component of the downstream outwash fan. Sedimentation on outwash fans was capacity-driven.
3. Considerable downstream attenuation of peak discharge occurred primarily due to rapid channel

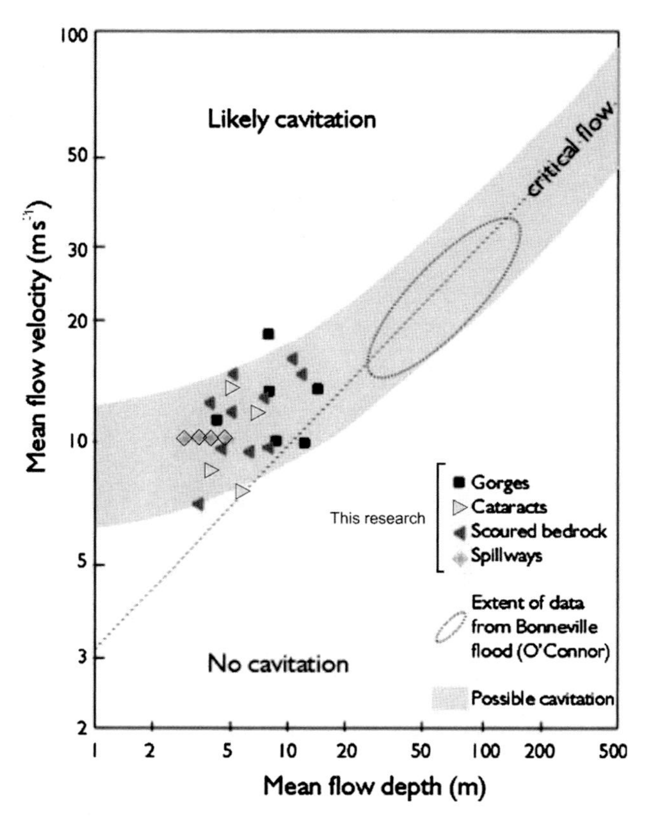

Figure 15.11. High-magnitude hydraulics associated with erosional landforms produced by jökulhlaups from Kverkfjöll. All erosional landforms are associated with supercritical flows (line denotes the critical Froude number) and were probably subjected to cavitation (grey shaded area). (Adapted from O'Connor (1993); Carrivick (2007a).)

expansions, contractions and high bed roughness, and to a lesser extent due to grain roughness, overspill to neighbouring valleys and backwater ponding.

4. Main valley hydraulics predominantly were controlled by channel slope and channel width. Steep confined reaches were high energy, and produced erosional landforms, sometimes with superimposed boulder clusters. Shallow gradient-wide reaches were low energy, and produced depositional landforms.
5. A supercritical flow regime predominated due to shallow flows, caused by steep gradients. Together with high bed roughness, widespread flow separation and shear layers developed, causing very high shear stress and stream power. These shear stress and stream power values were disproportionate to flow volume or peak discharge. Kverkfjöll jökulhlaups were therefore enabled to be as geomorphically effective as 'megafloods'.

Acknowledgements

The Icelandic Research Council and the Nature Conservation Agency gave permission to undertake research in Iceland. Grants from the BGRG, IAS and QRA are acknowledged. Adri Verwey of Delft granted two years use of the SOBEK model for academic research. David Twigg of Loughborough University produced the DEM. The Earthwatch Institute funded fieldwork during the summers of 2001–3. A very special thank you to the 2001–3 Earthwatch Team IV volunteers and to Oly Lowe, Hugh Deeming and Lucy Rushmer for invaluable and excellent field assistance. The Sigurdarskali wardens made us very welcome at Kverkfjöll. I am indebted to Andrew Russell and Fiona Tweed for their ideas, support and advice. Comments by V. Manville, J. Herget and P. Carling substantially improved this manuscript.

References

Agatz, H. (2002). EDV-gestützte paläohydrologische Modellierung der Ausbruchsflutwelle eines pleistozän gletschergestauten Sees im sibirischen Altai-Gebirge. Unpublished diploma thesis, Department of Geography, Universität Bochum.

Baker, V. R. (1988). Flood erosion. In *Flood Geomorphology*, eds. V. R. Baker, R. C. Kochel and P. C. Patton. New York: Wiley, pp. 81–95.

Baker, V. R. (2000). Palaeoflood hydrology and the estimation of extreme floods. In *Inland Flood Hazards: Human, Riparian and Aquatic Communities*, ed. E. E. Wohl. Cambridge: Cambridge University Press, p. 498.

Baker, V. R. (2002). High-energy megafloods: planetary settings and sedimentary dynamics. In *Flood and Megaflood Processes and Deposits: Recent and Ancient Examples*, eds. P. I. Martini, V. R. Baker and G. Garzon. Special Publications International Association of Sedimentologists, **32**, pp. 3–15.

Baker, V. R., Benito, G. and Rudoy, A. N. (1993). Palaeohydrology of Late Pleistocene superflooding, Altay Mountains, Siberia. *Science*, **259**, 348–350.

Benito, G. (1997). Energy expenditure and geomorphic work of the cataclysmic Missoula flooding in the Columbia River gorge, USA. *Earth Surface Processes and Landforms*, **22**, 457–472.

Björnsson, H. (2002). Subglacial lakes and jökulhlaups in Iceland. *Global and Planetary Change*, **35**, 255–271.

Brayshaw, A. C. (1984). Characteristics and origin of cluster bedforms in coarse-grained alluvial channels. In *Sedimentology of Gravels and Conglomerates*, eds. E. H. Koster and R. J. Steel. Canadian Society of Petroleum Geologists, pp. 77–85.

Carling, P. A., Hoffman, M. and Blatter, A. S. (2002). Initial motion of boulders in bedrock channels. In *Ancient Floods, Modern Hazards: Principles and Applications of Palaeoflood Hydrology*, eds. P. K. House, R. H. Webb, V. R. Baker and D. R. Levish. Washington, DC: American Geophysical Union, pp. 147–160.

Carling, P. A. *et al.* (2003). Palaeohydraulics of extreme flood events: reality and myth. In *Palaeohydrology:*

Understanding Global Change, eds. K. J. Gregory and G. Benito. Chichester: Wiley.

Carrivick, J. L. (2005). Characteristics and impacts of jökulhlaups (glacial outburst floods) from Kverkfjöll, Iceland. Unpublished Ph.D. thesis, Keele University, UK.

Carrivick, J. L. (2006). 2D modelling of high-magnitude outburst floods: an example from Kverkfjöll, Iceland. *Journal of Hydrology*, **321**, 187–199.

Carrivick, J. L. (2007a). Hydrodynamics and geomorphic work of jökulhlaups (glacial outburst floods) from Kverkfjöll volcano, Iceland. *Hydrological Processes*, **21**, 725–740.

Carrivick J. L. (2007b). Modelling coupled hydraulics and sediment transport of a high-magnitude flood and associated landscape change. *Annals of Glaciology*, **45**, 143–154.

Carrivick, J. L. and Rushmer, E. L. (2006). Understanding high-magnitude outburst floods. *Geology Today*, **22** (2), March–April 2006, 60–65.

Carrivick, J. L. and Twigg, D. (2005). Jökulhlaup-influenced topography and geomorphology at Kverkfjöll, Iceland. *Journal of Maps*, 2005, 17–27.

Carrivick, J. L., Russell, A. J. and Tweed, F. S. (2004a). Geomorphological evidence for jökulhlaups from Kverkfjöll volcano, Iceland. *Geomorphology*, **63**, 81–102. doi:10.1016/j.geomorph.2004.03.006.

Carrivick, J. L., Russell, A. J., Tweed, F. S. and Twigg, D. (2004b). Palaeohydrology and sedimentology of jökulhlaups from Kverkfjöll, Iceland. *Sedimentary Geology*, **172**, 19–40.

Chincholle, L. (1994). A new basic principle for a new series of hydraulic measurements: erosion by abrasion, corrosion, cavitation and sediment concentration. In *Fundamentals and Advancements in Hydraulic Measurements and Experimentation*, ed. C. A. Pugh. Proceedings of the Symposium held in Buffalo, New York, August 1–5, 1994. New York: ASCE, 0-7844-0036–9.

Clarke, G. K. C., Leverington, D. W., Teller, J. T. and Dyke, A. S. (2004). Paleohydraulics of the last outburst flood from glacial Lake Agassiz and the 8200 BP cold event. *Quaternary Science Reviews*, **23** (3–4). 389–407.

Costa, J. E. (1983). Palaeohydraulic reconstruction of flash-flood peaks from boulder deposits in the Colorado front range. *Geological Society of America Bulletin*, **94**, 986–1004.

Costa, J. E. and Schuster, R. L. (1988). The formation and failure of natural dams. *GSA Bulletin*, **100**, 1054–1068.

Fenton, C. R., Webb, R. H. and Cerling, T. E. (2006). Peak discharge of a Pleistocene lava-dam outburst flood in Grand Canyon, Arizona, USA. *Quaternary Research*, **65** (2), 324–335.

Gomez, B. S., Russell, A. J., Smith, L. C. and Knudsen, Ó. (2002). Erosion and deposition in the proglacial zone: the 1996 jökulhlaup on Skeiðarársandur, southeast Iceland. In *The Extremes of the Extremes: Extraordinary Floods*, eds. A. Snorrason, A. P. Finnsdóttir and M. Moss. Proceedings of a symposium at Reykjavík, July 2000. Publication 271, Wallingford, UK: IAHS Press, pp. 217–221.

Grant, G. E. (1997). Critical flow constrains flow hydraulics in mobile-bed streams: a new hypothesis. *Water Resources Research*, **33**, 349–358.

Herget, J. (2002). Reconstruction of ice-dammed lake outbursts, Altai Mountains (Siberia). Unpublished Habilitation thesis at Faculty of Earth Sciences, Ruhr-University Bochum.

Herget, J. (2005). *Reconstruction of Pleistocene Ice-dammed Lake Outburst Floods in the Altai Mountains, Siberia*. Geological Society of America Special Paper 386.

Hiscott, R. N. (1994). Loss of capacity, not competence, as the fundamental process governing deposition from turbidity currents. *Journal of Sedimentary Research*, **A64**, 209–214.

Jarrett, R. D. (1984). Hydraulics of high gradient streams. *Journal of Hydraulic Engineering*, **110**, 1519–1539.

Jarrett, R. D. and Malde, H. E. (1987). Palaeodischarge of the late Pleistocene Bonneville Flood, Snake River, Idaho, computed from new evidence. *GSA Bulletin*, **99**, 127–134.

Jóhannesson, H. and Saemundsson, K. (1989). *Geological Map of Iceland, 1:1500 000 Bedrock Geology*. Reykjavík: Icelandic Museum of Natural History and Icelandic Geodetic Survey.

Karhunen, R. (1988). *Eruption Mechanism and Rheomorphism During the Basaltic Fissure Eruption in Biskupsfell, Kverkfjoll, North-Central Iceland*. Reykjavik: Nordic Volcanological Institute.

Komar, P. D. (1989). Flow-competence evaluations of the hydraulic parameters of floods: an assessment of the technique. In *Floods: Hydrological, Sedimentological and Geomorphological Implications*, eds. K. Bevan and P. Carling. Chichester: John Wiley and Sons Ltd., pp. 107–134.

Lee, W. and Hoopes, J. A. (1996). Prediction of cavitation damage for spillways. *Journal of Hydraulic Engineering*, **122** (9), 481–488.

Limerinos, J. T. (1970). *Determination of the Manning Coefficient from Measured Bed Roughness in Natural Channels*. United States Geological Water Supply Paper, 1898-B.

Magilligan, F. J. (1992). Thresholds and extreme spatial variability of flood power during extreme floods. *Geomorphology*, **5**, 373–390.

Maizels, J. K. (1983). Palaeovelocity and palaeodischarge determination for coarse gravel deposits. In *Background to Palaeohydrology*, ed. K. J. Gregory. Chichester: John Wiley and Sons Ltd.

Maizels, J. K. (1997). Jökulhlaup deposits in proglacial areas. *Quaternary Science Reviews*, **16**, 793–819.

Manville, V. R. and White, J. D. L. (2003). Incipient granular mass flows at the base of sediment-laden floods, and the roles of flow competence and flow capacity in the deposition of stratified bouldery sands. *Sedimentary Geology*, **155** (1–2), 157–173.

Nie, Meng-Xi (2001). Cavitation prevention with roughened surface. *Journal of Hydraulic Engineering*, **127** (10), 878–880.

O'Connor, J. E. (1993). *Hydrology, Hydraulics and Geomorphology of the Bonneville Flood*. Geological Society of America Special Paper 274.

O'Connor, J. E. and Baker, V. R. (1992). Magnitudes and implications of peak discharges from glacial Lake Missoula. *Geological Society of America Bulletin*, **104**, 267–279.

Rudoy, A. N. (2002). Glacier-dammed lakes and geological work of glacial superfloods in the late Pleistocene, southern Siberia, Altai mountains. *Quaternary International*, **87**, 119–140.

Rushmer, E. L. (2006). Sedimentological and geomorphological impacts of the jökulhlaup (glacial outburst flood) in January 2002 at Kverkfjöll, northern Iceland. *Geografiska Annaler*, **88A**, 1–11.

Rushmer, E. L. (2007). Physical-scale modelling of jökulhlaups (glacial outburst floods) with contrasting hydrograph shapes. *Earth Surface Processes and Landforms*, **32**, 954–963.

Russell, A. J. and Marren, P. M. (1999). Proglacial fluvial sedimentary sequences in Greenland and Iceland: a case study from active proglacial environments subject to jökulhlaups. In *The Description and Analysis of Quaternary Stratigraphic Field Sections*, eds. A. P. Jones, J. K. Hart and M. E. Tucker. QRA Technical Guide 7, London: QRA, pp. 171–208.

Sleigh, P. A. and Goodwill, I. M. (2000). The St Venant equations. http://www.efm.leeds.ac.uk/CIVE/CIVE3400/stvenant.pdf, last visited 15 February 2007.

Strickler, A. (1923). Beiträge zur Frage der Geschwindigkeitsformel und der Rauhigkeitszahlen Für Ströme, Kanäle und geschlossene Leitungen. *Mitteilungen des Eidgenössischen Amtes für Wasserwirtschaft, Bern, Switzerland*, **16**.

Thompson, S. M. and Campbell, P. L. (1979). Hydraulics of a large channel paved with boulders. *Journal of Hydraulic Research*, **17**, 341–354.

Tómasson, H. (1996). The jökulhlaup from Katla in 1918. *Annals of Glaciology*, **22**, 249–254.

Tweed, F. S. and Russell, A. J. (1999). Controls on the formation and sudden drainage of glacier-impounded lakes: implications for jökulhlaup characteristics. *Progress in Physical Geography*, **23**, 79–110.

Whipple, K. X., Hancock, G. S. and Anderson, R. S. (2000). River incision into bedrock: mechanics and relative efficacy of plucking, abrasion, and cavitation. *GSA Bulletin*, **112** (3), 490–503.

Williams, G. P. (1984). Palaeohydraulic equations for rivers. In *Developments and Applications of Geomorphology*, eds. J. E. Costa and P. J. Fleisher. Berlin Heidelberg: Springer-Verlag, pp. 343–367.

Zingg, T. H. (1935). Beträge zur Schotteranalyse. *Schweizerische Mineralogische und Petrographische Mitteilungen*, **15**, 39–140.

16

Dynamics of fluid flow in Martian outflow channels

LIONEL WILSON,
ALISTAIR S. BARGERY
and DEVON M. BURR

Summary

The conditions under which large volumes of water may have flowed at high speeds across the surface of Mars are considered. To assess the likely ranges of initial water temperature and release rate, the possible conditions in subsurface aquifers confined beneath the cryosphere are explored. Then the transfer of water to the surface in fractures induced by volcanic activity or tectonic events is modelled and the physical processes involved in its release into the Martian environment are discussed. The motion of the water across the surface is analysed with standard treatments for fluvial systems on Earth, modified for Mars by taking account of the differing environmental conditions and removing what may be considered to be the unsafe assumption that most channels involved bankfull flows. The most commonly discussed environmental difference is the smaller acceleration due to gravity on Mars. However, an important additional factor may have been the initially vigorous evaporation of water into the low-pressure Martian atmosphere. This process, together with the thermal losses incurred by assimilation of very cold rock and ice eroded from the cryosphere over which the water travels, causes minor changes in the depth and speed of a water flood but, eventually, produces major changes in the flood rheology as the total ice and sediment loads increase. The roles of these processes in determining the maximum distance to which the water may travel, and the relative importance of erosion and deposition in its bed, are discussed.

16.1 Introduction

Since the recognition of the Martian outflow channels in the imaging data from the Viking missions (Sharp and Malin, 1975; Masursky *et al.*, 1977; Komar, 1980; Carr and Clow, 1981), and the realisation that, despite alternative suggestions (Milton, 1974; Hoffman, 2000, 2001; Leverington, 2004), flowing water was the most likely mechanism for their formation (Baker, 1979; Carr, 1979; Komar, 1979; Baker, 2001; Coleman, 2003), it has been appreciated that the great size of the channels probably implies extreme flow conditions. Depending on the assumptions made about the depths of water in the channels and the relevant flow regimes, various authors have obtained estimates of the peak discharges that mainly range between $\sim 10^6$ and 10^8 $m^3\,s^{-1}$ (Carr, 1979; Komar, 1979; De Hon and Pani, 1993; Komatsu and Baker, 1997; Ori and Mosangini, 1998; Williams *et al.*, 2000; Burr *et al.*, 2002a, b; Chapman *et al.*, 2003; Head *et al.*, 2003a; Coleman, 2004; Leask *et al.*, 2006a, 2007) but extend up to 10^{10} $m^3\,s^{-1}$ (Robinson and Tanaka, 1990; Dohm *et al.*, 2001a, 2001b). Table 16.1 summarises a representative range of examples. Here, the key factors involved in determining water flow conditions in outflow channels, including the limitations imposed by the likely nature of the subsurface water sources and the mechanisms of water release to the surface, are reviewed. The consequences for water flow on the Martian surface, under environmental conditions of low temperature, low atmospheric pressure, and low acceleration due to gravity are explored. Suggestions are made as to how some of the assumptions made in previous work may be improved, significantly reducing the largest implied discharges, and giving explicit estimates of discharge and flow conditions in several named channels.

16.2 Water sources

The morphologies of the source regions of most Martian outflow channels, including areas of chaos (Coleman and Baker, this volume Chapter 9) and graben (Burr *et al.*, this volume Chapter 10), suggest that the water forming the channels was released from a significant depth beneath the surface. In a few cases, release of water from ice-covered lakes has been suggested (Komatsu *et al.*, 2004; Woodworth-Lynas and Guigné, 2004). Elsewhere, a small number of deeply incised valley networks originate at large basins, implying flooding due to basin overflow as a result of runoff and/or groundwater inflow (Irwin and Grant, this volume Chapter 11). The largest flood channels on Mars, however, originate from chaotic terrain and graben.

Although graben systems can be formed by purely tectonic forces (Hanna and Phillips, 2006), an origin involving the additional stress change due to dyke intrusion has been proposed for many of them (Rubin, 1992; Chadwick and Embly, 1998; Wilson and Head, 2002, 2004; Leask *et al.*, 2006b), and an origin due to sill intrusion is a natural explanation for at least some areas of chaos (Scott and Wilson, 1999; Ogawa *et al.*, 2003; Nimmo and Tanaka, 2005; Leask *et al.*, 2006c), though geothermal melting of a

Megaflooding on Earth and Mars, ed. Devon M. Burr, Paul A. Carling and Victor R. Baker.
Published by Cambridge University Press.

Table 16.1. Outflow channel parameters

Valley	Source	Estimated flux ($m^3 s^{-1}$)	Fracture length (km)	Implied aquifer permeability (m^2)
(W. Tharsis)[a]	(Hidden)	10^9-10^{10}	1000?	8.0×10^{-6}?
Kasei[b]	Echus Chasma	$9–23 \times 10^8$	130–500?	$>1.4 \times 10^{-5}$?
Ares[c]	Ianni Chaos	10^8-10^9	360	2.2×10^{-6}
Hydraotes[d]	Hydraotes Chaos	$7–40 \times 10^7$	500?	1.1×10^{-6}
Tiu[d]	Hydraotes Chaos	$3–20 \times 10^7$	500?	4.8×10^{-7}?
Maja[e]	Juventae Chasma	9×10^7	240?	3.0×10^{-6}?
Mangala[f]	Memnonia Fossa	$1–8 \times 10^7$	223	4.0×10^{-7}
Simud[d]	Hydraotes Chaos	$1–5 \times 10^7$	500?	1.6×10^{-7}?
Mangala[g]	Memnonia Fossa	$8–40 \times 10^6$	210	3.2×10^{-7}
Ravi[h]	Aromatum Chaos	$3–30 \times 10^6$	50	4.8×10^{-7}
Athabasca[i]	Cerberus Fossae	$2–4 \times 10^6$	35	4.4×10^{-7}

[a] Dohm *et al.* (2001a, 2001b), [b] Robinson and Tanaka (1990), [c] Komatsu and Baker (1997), [d] Ori and Mosangini (1998), [e] De Hon and Pani (1993), [f] Ghatan *et al.* (2005), [g] Komar (1979), [h] Leask *et al.* (2004), [i] Burr *et al.* (2002a, 2002b); Head *et al.* (2003a).

buried crater-floor lake has been proposed for Aram Chaos (Oost-hoek *et al.*, 2007). At first sight it is tempting to suggest that the released water could consist entirely of ice melted by the heat from the intrusion (McKenzie and Nimmo, 1999). However, in the two cases that have been examined in detail, the Mangala Fossa graben source of the Mangala Valles channel system (Leask *et al.*, 2006b) and the Aromatum Chaos source of Ravi Vallis (Leask *et al.*, 2006c), it has been shown that the volume of water that could be melted by the relevant volcanic intrusion was too small by at least two orders of magnitude to explain the transport of the volume of crustal material eroded to form the channel. Thus it appears that, in most cases where a volcanic event was involved, its main physical role was to fracture the crust and release pre-existing liquid water trapped beneath the surface (though there may be other, chemical and thermal, consequences discussed in the next section).

An abundance of confined aquifers is to be expected on Mars as a consequence of the likely presence, for much of the history of the planet, of a cryosphere, in which any water is present as ice in pore spaces and fractures in the shallow crustal rocks (Rossbacher and Judson, 1981; Clifford, 1987, 1993; Clifford and Parker, 2001; Carr, 2002). Given plausible values for the geothermal heat flux at various times in Martian history of 20–30 mW m^{-2} (McGovern *et al.*, 2002) and for the thermal conductivity of the cryosphere ($\sim$2.5 W m^{-1} K^{-1} – Leask *et al.*, 2006c), the requirements that the temperature be close to the modern average value ($\sim$210 K) at the surface and $\sim$273 K at the top of the aquifer lead to estimates of the depths, D, to the tops of aquifers in the range 5 to 8 km, somewhat larger than the $\sim$2.3 to 6.5 km range in the preferred model of Clifford (1993). However, significantly shallower aquifers are possible locally in areas of abnormally high heat flow (Gulick, 1998, 2001; Coleman, 2005) or reduced thermal conductivity (Mellon and Phillips, 2001; Heldmann and Mellon, 2004; Edlund and Heldmann, 2006) and $D = 3.5$ km is used in calculations below. The levels of stress generated by tectonic events (Hanna and Phillips, 2006) and volcanic dyke and sill intrusions (Wilson and Head, 2004) on Mars appear to be quite adequate to fracture a several-kilometre thickness of cryosphere to provide pathways for water release.

16.3 Water ascent to the surface

To obtain realistic inputs to models of water flow on the surface of Mars the ascent of water to the surface from deep aquifer systems needs consideration. The speed at which water can flow up an open fracture is a function of the excess pressure at the base of the fracture and the fracture width. However, these variables are also linked to the conditions in the underlying aquifer and the stress field that induces the fracture. Modelling calculations using coupled models of fracture formation and aquifer properties have been made for the Cerberus Fossae source graben of the Athabasca Valles by Manga (2004) and Hanna and Phillips (2006), and for the Mangala Fossa source graben of the Mangala Valles channel system by Hanna and Phillips (2006). Initial water discharge rates are found to range between $\sim 10^5$ and somewhat in excess of 10^6 m^3 s^{-1}. Discharge is predicted to decay exponentially, on a time scale generally found to be of order hours, as a result of the flow limitations imposed by the permeabilities and compressibilities of the aquifers (Manga, 2004).

Although these discharges are consistent with the smallest values inferred from channel geometries (Table 16.1), they are orders of magnitude smaller than the largest estimates. This discrepancy can be resolved partly by noting that many of the discharge values in Table 16.1 probably are overestimated greatly, as will be shown below in the section 'Dynamics of water flow'. However, the very great extent of the discrepancy, and especially the inference (see below) that for the larger channels the high discharge rates were maintained for many weeks, not the hours suggested by the existing models, implies that the permeabilities generally inferred for the Martian subsurface in these models, at most $\sim 10^{-11}$ m^2 and decreasing to $\sim 10^{-15}$ m^2 at great depth (Manga, 2004; Hanna and Phillips, 2005), may be much too small. Martian aquifers may extend from the $\sim$4 km depth of the base of the cryosphere down to the $\sim$10 km depth where porosity becomes negligible, making their vertical extents $H \approx 6$ km. The horizontal extents of the fractures delineating some named outflow channel source areas are listed in Table 16.1, and equating the estimated fluxes listed to the product of the aquifer cross-sectional area and the Darcy velocity allows values to be calculated for the aquifer permeabilities that would be required to allow these discharges to take place (Wilson *et al.*, 2004). These values, given in the final column of Table 16.1, are at least $\sim 10^{-7}$ m^2 and are more similar to those commonly associated with loose gravel than to normal aquifers on Earth, though aquifers formed within the rubble tops and bottoms of layers of basaltic lava flows can have significantly greater permeabilities (Saar and Manga, 1999, 2004). If the typical permeability of the Martian crust is indeed much greater than that of the Earth, a possible explanation is that the crustal fractures defining the megaregolith, generated by the early impact bombardment of the planet, have been preserved over geological time. Unfortunately there is no information on the detailed structure of this or any other megaregolith, which on the Moon was detected only indirectly by the Apollo seismic network (Lognonne, 2005) via the scattering of seismic signals. However, it seems reasonable to assume that the Martian megaregolith may contain a network of relatively wide, interconnected fractures with both high porosity and high permeability. If so, an approach using the Darcy law may be inappropriate for modelling water flow through such a structure. Whatever the source of the high permeability, an additional implication would be that there has been no systematic infilling of the fracture systems by secondary minerals (Hanna and Phillips, 2005).

An important parameter determining the fate of water reaching the surface of Mars is its temperature on release, determined by its temperature within the aquifer and the amount of heat lost in transit. The water temperature depends strongly on whether convection is taking place in the aquifer. If it is assumed that no convection takes place, then with a geothermal heat flow of 20 mW m^{-2} and a bulk aquifer thermal conductivity of $\sim$1 W m^{-1} K^{-1} the temperature gradient will be 20 K km^{-1}. Over a 6 km vertical extent of aquifer the temperature increase will be 120 K, and because the top of the aquifer is at 0 °C the mean water temperature will be around 60 °C. The criterion for convection to occur (see Section 9–9 in Turcotte and Schubert, 2002) is that the aquifer permeability k must exceed a critical value k_c given by

$$k_c = (4\pi^2 \mu_w K)/(\alpha_w g \rho_w^2 c_w H^2 \, dT/dz), \tag{16.1}$$

where μ_w, α_w, ρ_w and c_w are the viscosity, volume thermal expansion coefficient, density and specific heat at constant pressure, respectively, of the water, K is the bulk thermal conductivity of the aquifer, g is the acceleration due to gravity (3.72 m s^{-2}), H is the aquifer thickness and dT/dz is the temperature gradient across the aquifer. Inserting appropriate values for water at 60 °C in an aquifer 6 km thick, ($\mu_w = 5 \times 10^{-4}$ Pa s, $\alpha_w = 5 \times 10^{-4}$, $\rho_w = 1000$ kg m^{-3}, $c_w = 4175$ J kg^{-1} K^{-1} and $K = 2.5$ W m^{-1} K^{-1}), together with the 20 K km^{-1} gradient in the absence of convection, the permeability must exceed $\sim 1 \times 10^{-14}$ m^2. This value lies towards the lower end of the 10^{-11} to 10^{-15} m^2 range expected by analogy with the Earth, and could be taken to imply that convection did not extend through the full vertical extent of typical aquifers. In contrast, if the much larger permeabilities suggested by the high discharges are correct then convection would easily occur throughout all aquifers. If convection does occur, it effectively increases the thermal conductivity of the aquifer, probably by a factor of two, based on simulations of volcanic intrusions producing hydrothermal systems (Ogawa *et al.*, 2003). This conclusion reduces the temperature gradient across the aquifer and acts to decrease the average water temperature, also by a factor of two, which would lead to a mean water temperature of about 30 °C in a 5 km thick aquifer. However, the temperature gradient is also one of the factors determining the critical permeability, and reducing the temperature gradient increases the minimum permeability that allows convection. Thus, to some extent the process is self-limiting, and without explicit knowledge of the actual permeabilities of aquifers on Mars it is difficult to do any better than to assert that aquifer water temperatures may lie in the range from just above the (weakly pressure- and hence depth-dependent) freezing point to at least 30 °C.

Reduction of the water temperature during its ascent will occur by two processes: adiabatic decompression and conduction through the fracture walls. The adiabatic temperature change ΔT_r of water rising a vertical distance $H = 6$ km through the crust, assuming a near-lithostatic pressure distribution in the water so that the excess pressure due to the water buoyancy drives the water motion, can be expressed as

$$\Delta T_r = (\alpha_w T_r g \rho_r H)/(c_w \rho_w), \tag{16.2}$$

Table 16.2. Conductive cooling of water rising in fractures

Fracture width W(m)	Rise speed U(m s^{-1})	Transit time τ (s)	Conductive cooling ΔT_c (K)
0.1	9.7	410	34.7
0.3	16.9	237	8.8
1.0	30.8	130	2.0
3.0	53.4	75	0.5

Note: Pressure gradients specified in text.

where T_r is the average water temperature and ρ_r is the density of the crust, ~3000 kg m^{-3}. Using values relevant to the water having a temperature somewhere in the range 0–30 °C ($\alpha_w = 2.5 \times 10^{-4}$ K^{-1}, $T_r = \sim 300 \pm 25$ K, $c_w = 4180$ J kg^{-1} K^{-1} and $\rho_w = \sim 1000$ kg m^{-3}) we find that ΔT_r will be only 1.2 K.

Cooling by conduction of heat into the walls of the fracture through which water rises may be more important. The speed U of the water in a fracture of width W, the motion being fully turbulent in all cases of interest, is given by

$$U = [(W\, dP/dz)/(f\rho_w)]^{1/2}, \qquad (16.3)$$

where f is a wall friction factor of order 10^{-2} and dP/dz is the pressure gradient driving the motion. At thc onset of water release, the pressure gradient will be the sum of contributions due to the buoyancy of the water relative to the crust, $g(\rho_r - \rho_w) = 7440$ Pa m^{-1}, and any excess pressure in the aquifer. Excess pressure may arise due to a topographic gradient in the aquifer (Head *et al.*, 2003a), due to tectonic pressurisation of the crust (Hanna and Phillips, 2006), or due to downward growth of the base of the cryosphere as the planet cools and the geothermal gradient decreases (Wang *et al.*, 2006). A topographic height difference of $Z = 2$ km, common in the highland areas of Mars (Neumann *et al.*, 2004) underlain by aquifers, would yield an excess pressure of $(\rho_w g Z) = 7.4$ MPa, similar to the 10 MPa tectonic excess pressures found by Hanna and Phillips (2006), implying an excess pressure gradient over the vertical extent of the fracture $D = \sim 3.5$ km of 2100 Pa m^{-1}. Combined with the pressure gradient due to buoyancy, this gives a total pressure gradient of 9500 Pa m^{-1}. This results in the water rise speeds as a function of fracture width shown in Table 16.2. Also given are the transit times, τ, of any given batch of water. The transit time controls the amount of heat lost by the water. At a time t after the opening of the fracture, a wave of warming will have penetrated a distance $\lambda = \sim(\kappa t)^{1/2}$ into the surrounding crust, where κ is the thermal diffusivity of cryosphere material, about 7×10^{-7} m^2 s^{-1}. The crustal temperature gradient controlling heat flow away from the fracture will then be $dT/dx = (T_r - T_c)/\lambda$, where T_r and T_c are the mean water and cryosphere temperatures. If T_c is 240 K, a conservative value of T_r would be 300 K. Consider a fracture of horizontal length L; the total instantaneous heat flux out of the two faces of the fracture is $(2KDL\, dT/dx)$, and so the heat lost during the transit time τ will be $(2\tau KDL\, dT/dx)$. The mass of water losing this heat is equal to $(\rho_w WLD)$, and so the temperature decrease in the water, $\delta T(t)$, will be equal to $[2\tau K(T_r - T_c)]/[\rho_w W c_w(\kappa t)^{1/2}]$. The average value of the temperature reduction, ΔT_c, due to conductive heat loss during the travel time, τ, of a given batch of water through the fracture is equal to $(1/\tau)\int \delta T(t)\, dt$. For the first batch of water rising through the fracture, for which heat loss is at a maximum because there has been no pre-heating of the fracture walls, the integral is to be evaluated between $t = 0$ and $t = \tau$, and is found to be

$$\Delta T_c = [4\tau^{1/2} K(T_r - T_c)]/[\rho_w W c_w \kappa^{1/2}]. \qquad (16.4)$$

Table 16.2 shows the values of ΔT_c; clearly, even the initial conductive losses are small for fracture widths in excess of one metre, and all conductive losses decrease with time as the fracture walls are progressively heated. For comparison, using various arguments to select the pressure gradient, Head *et al.* (2003a) found $W = 2.4$ m and $U = 64$ m s^{-1} for the fractures in the Cerberus Fossae graben system feeding the Athabasca Valles, and Manga (2004) found values of W up to 2 m implying U was up to ~40 m s^{-1} for the same system. These results would imply transit times of order 60–100 s and water temperature decreases of less than 0.5 K. These minimal heat losses during ascent imply that the temperature of water emerging at the surface of Mars may lie in the range from just above the freezing point to about 30 °C.

16.4 Water release at the surface

In lieu of other factors, a rise speed U of up to ~50 m s^{-1} implies that water reaching the surface of Mars through a fracture may be expected to form a fountain over the release site with a height of $U^2/(2g)$, in excess of 300 m. However, the issue may be more complicated if the water contains dissolved volatile compounds that come out of solution as the low ambient pressure at the surface is approached. Stewart and Nimmo (2002) use thermodynamical arguments to show that it is extremely unlikely that subsurface water on Mars will contain significant carbon dioxide (CO_2) derived from the atmosphere. However, in the case of a water-release event triggered by a volcanic intrusion, it is possible that CO_2 released from an intrusion might migrate into the aquifer as it was expelled from the cooling magma. The geometry would be particularly favourable for this if the intrusion were a sill. By analogy with Earth, a mafic mantle-derived magma could contain a CO_2 mass fraction as large as 0.0065 (Gerlach, 1986). If this magma was emplaced as a sill at the base of an aquifer at a depth of 10 km, where the ambient pressure is 120 MPa (the sum of a lithostatic pressure of 112 MPa and an excess water pressure of 7 to 10 MPa), the solubility of CO_2 in mafic magma (Harris, 1981; Dixon, 1997) is such that most of this CO_2 would be exsolved from the magma and would percolate up into the aquifer water where it would readily be dissolved. Taking plausible thicknesses of sills (about 200 m, Leask *et al.*, 2006c) and aquifers (6 km, see above) and of the volume fraction of an aquifer occupied by water (5%, based on the Mars crustal porosity model of Hanna and Phillips, 2005), the CO_2 would represent a mass fraction in the water of ~0.01 (Bargery and Wilson, submitted).

This amount of CO_2 would readily dissolve in the water at the base of the aquifer (Diamond and Akinfiev, 2003) and, furthermore, would still be retained if convection carried water to the top of the aquifer where the pressure is about 55 MPa. However, when fracturing of the overlying cryosphere allowed the water to ascend to the surface, essentially all of this CO_2 would be exsolved. A model (Bargery, 2007), analogous to the treatment of CO_2 release in overturning lake water (Zhang and Kling, 2006), of the progressive release of the gas to form bubbles as the water nears the surface shows that the volume fraction of CO_2 gas in the ensuing foam would reach 0.8, leading to foam collapse, at a depth of 10 m. Expansion of the CO_2 from the pressure at this depth to atmospheric pressure would provide 2600 J kg^{-1} of energy to add to the kinetic energy of the rising water before gas release, thus increasing the water emergence speed from a likely maximum (without dissolved CO_2) of 50 m s^{-1} to a maximum of 90 m s^{-1}. This increase in velocity would lead to formation of a fountain more than 1000 m high (Bargery, 2007). It should be stressed, however, that events like this may be rare, associated only with water release from areas of sill-related chaos, and that fountain heights up to about 300 m from fractures induced by dyke intrusions are likely to be more common.

The water at the outer edge of a fountain is exposed to an environment where the ambient temperature is on average 210 K (though it may, on rare occasions at the equator in the southern hemisphere summer, exceed 273 K) and the atmospheric pressure is 600 Pa at the mean surface elevation, 700 Pa in low-lying areas, 200 Pa on the flanks of large shield volcanoes and as little as 70 Pa at some of their summits. The triple point of pure water is at 273.15 K and 610.7 Pa and the vapour pressure increases rapidly with temperature (Figure 16.1). Given the estimates from the previous section of released water temperatures in the range from a few °C to 30 °C, i.e., 275 to 300 K, Figure 16.1 implies that, initially at least, vigorous evaporation of the water forming the outer envelope of the fountain will take place at all but the lowest elevations. The resulting vapour will expand and cool, ultimately reaching the surface in the solid form in a zone around the base of the liquid fountain. This process will be rendered less vigorous if aquifer water on Mars contains dissolved salts (e.g. Burt and Knauth, 2003), because these will reduce the vapour pressure at a given temperature and reduce the freezing point temperature. Nonetheless, geomorphological evidence indicates that this process may have occurred at the Mangala Valles fissure-headed outflow channel, where a restricted area of ridges and lobes on the rims of the source graben has been interpreted as moraines and tills deposited by locally derived ice and snow (Head *et al.*, 2004). The formation mechanism for this ice and snow is suggested to have been the rising and flooding by groundwater of the source graben, boiling and freezing of this ponded water, sublimation, and redeposition in solid form on the fissure rim. As discussed here, an alternative explanation for the formation of these features is through water vapour production at the outside of a water fountain, radiative cooling of the vapour to solid form, and deposition of this material on the surface as snow or ice. The first scenario is taken to imply that the Hesperian–Amazonian climate during formation of these moraine and till features was a hyper-arid cold desert similar to the climate of Mars today (Head *et al.*, 2004). The alternative scenario does not hold any implications about the palaeoclimate, but contains information about the depth and temperature of the

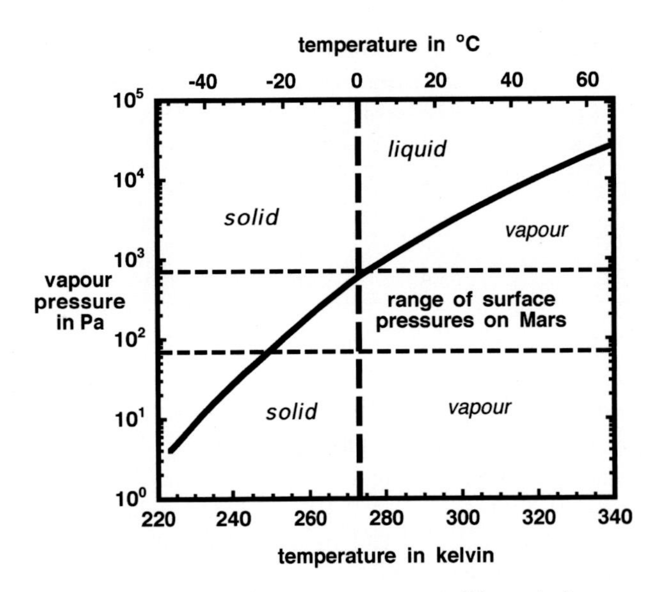

Figure 16.1. Variation of the vapour pressure (solid curve) of pure liquid and pure solid H_2O with temperature. The long-dashed line indicates the solid–fluid boundary and crosses the vapour pressure curve at the triple point, at 273.15 K and 610.7 Pa. The range of surface atmospheric pressures between low-lying plains and volcano summits on Mars is indicated by the short-dashed lines.

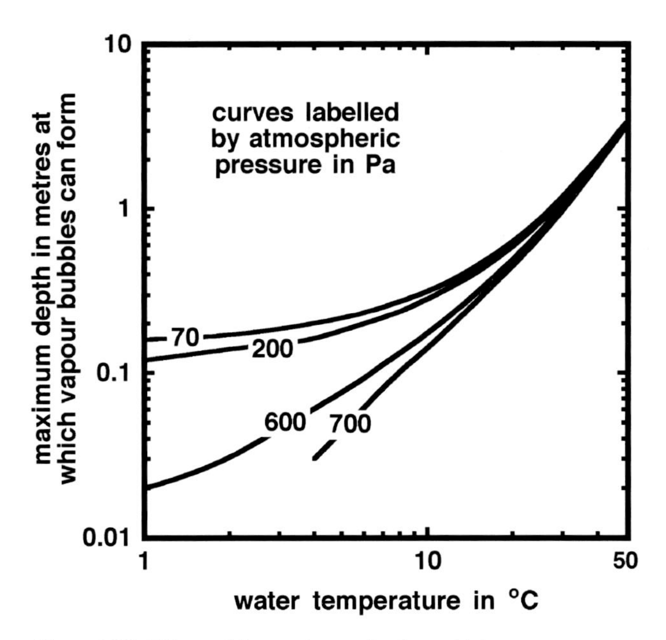

Figure 16.2. Values of the maximum depth at which water vapour bubbles form in a sheet of water as a function of the temperature at which it is released onto the surface of Mars and the ambient atmospheric pressure. Pressures of 700 and 600 Pa correspond to common elevations near Mars datum. Pressures of 200 and 70 Pa correspond to the flanks and summits, respectively, of large shield volcanoes. The absence of part of the 700 Pa curve indicates that no significant boiling occurs.

groundwater aquifer and the processes upon water release. The lack of observation of similar moraine and till features around other fissure-headed outflow channels that formed in the late Amazonian (Burr *et al.* this volume Chapter 10), during a cold climate conducive to glaciation (Neukum *et al.*, 2004) or of the formation of multi-layered surficial ice deposits (Head *et al.*, 2003b), suggests that climatic influence is not the controlling factor in the formation of these features.

Evaporation of water flowing away from a fountain will continue as long as the water remains warm enough for its vapour pressure to exceed the atmospheric pressure. Vapour bubbles will form in the water and rise up to burst at the surface. Using the temperature dependence of the vapour pressure (Kaye and Laby, 1995), the maximum depths below the surface at which vapour bubbles would form at various elevations are calculated and hence ambient pressures on Mars for a range of water release temperatures are derived, assuming that the overlying fluid is water with a density of 1000 kg m^{-3} (Figure 16.2). However, these are minimum estimates of these depths because in practice the fluid will be a mixture of liquid and vapour with a significantly smaller bulk density, so the actual depths of bubble nucleation may be greater than those given by a factor of up to 10.

The significance of water evaporation, whether by boiling or diffusion into the atmosphere, for subsequent water flow across the surface will be a function of the way the water cools. Cooling in turn will be influenced by the latent heat removed by the evaporation process itself and by the sensible heat lost to heating ice and silicates in cryosphere material eroded from the bed of the channel and entrained into the flow. These issues are addressed later, but first the basic dynamics of water flow in channels is considered.

16.5 Dynamics of water flow

16.5.1 Basic relationships governing water flow

The most important parameters determining the dynamics of water flow in channels on planetary surfaces are the bed slope α, the bed roughness, and the water depth d. If the width, w, of a channel is comparable to the depth, the influence of the banks cannot be neglected and, in general, the dimensions must be characterised by the hydraulic radius of the channel, R, defined as the cross-sectional area of the channel divided by its wetted perimeter, i.e.

$$R = (wd)/(w + 2d). \tag{16.5}$$

However, if w is much greater than d, as is commonly the case for large-scale natural water channels in general and for the Martian outflow channels in particular, then R is essentially equal to d.

Various equations are available to relate the above parameters to the mean water flow speed, u. The earliest is that due to Manning (1891); writing $S = \sin \alpha$, this relationship is

$$u = (R^{2/3} S^{1/2})/n, \tag{16.6}$$

where n is the Manning coefficient, a parameter with the dimensions of time divided by length$^{1/3}$, that characterises the combined effects of bed roughness, bed form and reach geometry. This equation, like a similar one due to Chézy (Herschel, 1897), has the disadvantage that it was developed for application only to the Earth, and does not explicitly account for the effects of the acceleration due to gravity, g. Various authors have attempted to use this equation for Martian channels by correcting for the differing values of g between Mars and Earth (see summary in Wilson *et al.*, 2004) but, even when done correctly, this leaves a residual problem. Gioia and Bombardelli (2002) showed that, in addition to the effects of gravity, the Manning coefficient involves a length scale that represents the typical physical scale of bed roughness. Thus, even when the differing gravities are accounted for, use of the resulting Manning coefficient may not be appropriate for Mars unless the bed roughness in Martian outflow channels is generally similar to that in the terrestrial rivers from which values of the

Manning coefficient are derived. Wilson *et al.* (2004) used the treatment of Gioia and Bombardelli (2002) to derive recommended values of the Manning coefficient for Martian channels, but also recommended against use of the Manning equation in favour of the treatment that follows.

A more fundamental relationship between the variables can be derived from dimensional analysis, the Darcy–Weisbach equation (ASCE, 1963):

$$u = [(8gRS)/f_c]^{1/2}, \tag{16.7}$$

in which f_c is a dimensionless friction factor that depends on the bed roughness relative to the flow depth for a given bedform. The friction factor plays the role of converting the shear velocity of the flow, $(gRS)^{1/2}$, to the average flow velocity. Equation (16.7) is applicable to fluid flows in channels and pipes of any cross-sectional shape provided that R is defined correctly. However, it should be noted that in some engineering applications a factor of two is adopted, rather than the value eight appearing in the above expression, leading to correspondingly smaller values for the friction factor. Bathurst (1993) gave relationships, based on a large body of experimental and field data, for the variation of f_c with bed roughness and water depth for a wide range of bed particle sizes and bedforms. In the case of channels with sand beds, Bathurst (1993) gave implicit expressions for f_c, and Wilson *et al.* (2004) manipulated these to yield the explicit formulae given as Equations (16.8) and (16.9) below. Equations (16.10), (16.11) and (16.12) are taken directly from Bathurst (1993). Equation (16.13) is adapted from engineering data on fluid flow in rough pipes (Knudsen and Katz, 1958) as a proxy for channels with fixed bed roughness. Finally, Equation (16.14) is a recent fit to a wide range of data from various sources given by Kleinhans (2005).

Channels with sand-dominated beds There are two main regimes for flow in channels with beds dominated by sand-size material (Hjülstrom, 1932): a lower regime (plane bed with sediment transport, or bed with ripples and dunes) and an upper regime (bed with sediment transport having antidunes and chutes and pools). A transition zone between these two regimes may have bedforms transitional between dunes and antidunes. Resistance formulae may take account separately of grain drag and bedform drag or combine the two together. The combined formulae are

$$(8/f_c)^{1/2} = 4.529(R/D_{50})^{0.02929} S^{-0.1113} \sigma_g^{-0.1606} \quad \text{lower regime,} \tag{16.8}$$

$$(8/f_c)^{1/2} = 7.515(R/D_{50})^{0.1005} S^{-0.03953} \sigma_g^{-0.1283} \quad \text{upper regime,} \tag{16.9}$$

where D_{50} is the channel-bed clast size such that 50% of clasts are smaller than D_{50} and σ_g is the geometric standard deviation of the bed clast size distribution, a dimensionless number equal to the ratio of the mean size to the size one standard deviation away from the mean.

Channels with gravel-dominated beds If the channel bed is dominated by gravel-size clasts, the grain size of the bed material is characterised by the parameter D_{84}, the channel-bed clast size such that 84% of clasts are smaller than D_{84}. Account is also taken of irregularities in channel depth by including the maximum channel depth d_m such that

$$(8/f_c)^{1/2} = 5.75 \log_{10}[(d_m^{0.314} R^{0.686})/D_{84}] + 2.8822. \tag{16.10}$$

Channels with boulder-dominated beds The relationship is

$$(8/f_c)^{1/2} = 5.62 \log_{10}[R/D_{84}] + 4. \tag{16.11}$$

Steep channels with fall and pool structures Steep channels on hillsides exist on Mars but these are generally very much smaller features than the outflow channels; however, the relevant formula is included here for completeness. The grain size of the bed material is incorporated via the parameter D_{90}, the channel-bed clast size such that 90% of clasts are smaller than D_{90}. Also, because all bed material is assumed to be constantly in motion, the depth parameter used is d_s, defined as the total depth of water plus sediment. The relationship is

$$(8/f_c)^{1/2} = 5.75\{1 - \exp[(-0.05d_s)/(D_{90}S^{1/2})]\}^{1/2} \times \log_{10}[(8.2d_s)/D_{90}]. \tag{16.12}$$

Channels with fixed bed roughness The following function, taken from engineering data on fluid flow in rough tubes (Knudsen and Katz, 1958), is included to characterise channels in which the bed roughness elements are fixed and cannot be moved by the fluid. Such conditions might arise if a channel previously created by an energetic outflow was re-used by a later, less energetic event. If r is the typical size of bed roughness elements (probably to be equated with D_{50}), then

$$(8/f_c)^{1/2} = 5.657 \log_{10}[R/r] + 6.6303. \tag{16.13}$$

Large terrestrial data set Kleinhans (2005) drew attention to the large uncertainties associated with relating friction factors of bed roughness and suggested the following function based on data relating to 190 water channels on Earth, including sand-bedded rivers with dune beds and 10 channels created by catastrophic glacial outburst events:

$$(8/f_c)^{1/2} = 2.2(R/D_{50})^{-0.055} S^{-0.275}. \tag{16.14}$$

Table 16.3. Morphological properties of outflow channels relevant to water discharge rates

Valley	Tangent of regional slope	Tangent of floor slope	Channel width (km)	Channel depth (m)
Athabasca	0.0004	0.0003	15–25	~100
Grjotá	0.0008	0.0006	several tens	~10
Mangala	0.0008	0.0005	~35 (proximal)	20–100
Marte	0.0003	0.0002	several tens	a few tens
Ravi	0.0050	0.0025	~20	~500

16.5.2 Morphological characteristics of outflow channels relevant to water flow speeds

Outflow channels on Mars have formed in regions where the large-scale topographic slopes have values of $\tan \alpha$ ranging from 3×10^{-4} to 5×10^{-3} (Wilson *et al.*, 2004; Burr *et al.*, this volume Chapter 10; see Table 16.3). As these channels developed by bed erosion, it appears that generally they were deepened most extensively in their proximal parts (though there are some exceptions dictated by very non-uniform preflood topography – see Burr *et al.*, this volume Chapter 10). Preferential proximal erosion is to be expected for two reasons. Firstly, particles removed from the bed by erosion must be transported by the water, and so the fraction of the moving fluid that consists of entrained solids increases with distance from the source up to some limiting value. Because the water must share its momentum with the entrained material, the entrained sediment acts to reduce the velocity and modify the turbulence regime in the water-sediment mixture and to reduce the erosion rate (Gay *et al.*, 1969). Indeed, if the bedload becomes large enough such that the flow is hyperconcentrated, the rheology of the fluid will cease to be Newtonian (Bargery *et al.*, 2005). The fluid then effectively becomes a mud flow or lahar, and will become depositional (Carling *et al.*, this volume Chapter 3). The issue of whether and how a flow may become hyperconcentrated is discussed later. Secondly, most of the surfaces into which outflow channels were incised on Mars probably consisted of cryosphere (Clifford, 1993), a mixture of very cold (e.g. 210 K) ice and rock in the volume proportions of 1 to 4, or 1 to 5 (Hanna and Phillips, 2005). Part of the erosion process included the melting of the ice component, and the sensible and latent heat required for this was extracted from the flowing water, causing progressive cooling with a consequent reduction in the down-flow ice-melting potential (Bargery *et al.*, 2005). Eventually ice crystal formation began, with the ice crystal volume fraction competing with silicate sediments for the carrying capacity of the remaining water. The net effect of the variation in bed erosion rate with distance from source has been to produce channels whose floors slope at values less than the local pre-erosion regional slopes: floor slope values are generally in the range 2×10^{-4} to 2×10^{-3} (Table 16.3).

Although no spacecraft has visited the floor of a Martian outflow channel, the Viking and Pathfinder spacecraft landed on geological units that may consist at least in part of materials washed through outflow channels (Golombek *et al.*, 1997), and so the rock size distributions on the surfaces at the Viking 1 and 2 and Pathfinder landing sites may be used as proxies for the size distributions of clasts in such channels. Using the data measured at these sites (Golombek and Rapp, 1997; Golombek *et al.*, 2003), Wilson *et al.* (2004) derived values for typical clast size distribution functions. Kleinhans (2005) has pointed out the need for a correction to the analysis of Wilson *et al.* (2004). Taking account of this, the values of the parameters of the bed clast size distribution needed for Equations (16.8) to (16.14) can be approximated as $D_{50} = 0.1$ m, $D_{84} = 0.48$ m, $D_{90} = 0.6$ m and $\sigma_g = 2.9$, respectively.

The depths of outflow channels commonly lie in the range 20 to 500 m (Table 3), though Smith *et al.* (1998) give depths up to 1300–1600 m for Ares Vallis. However, there have been significant differences of opinion in the literature about the depth of the water that typically flowed in these channels. Some authors have tacitly assumed that channels were always bankfull, though this is extremely unlikely for the following reason. It is clear that there was extensive, progressive erosion of the channel floors. Assume first that the erosion rate was constant. If a channel remained full, the water depth would have increased linearly with time. Equation (16.7) shows that the water flow speed would have increased as the square root of the depth and hence as the square root of the time. The volume flux, F, flowing through a channel of width w and depth d at speed u is

$$F = wdu, \tag{16.15}$$

so that F would also have had to increase with time. Even if the sides of the channel were vertical, so that the water did not get wider as it got deeper, the volume flux would have had to increase as time to the power 3/2. If, instead of being constant, the bed erosion rate increased, say linearly, with

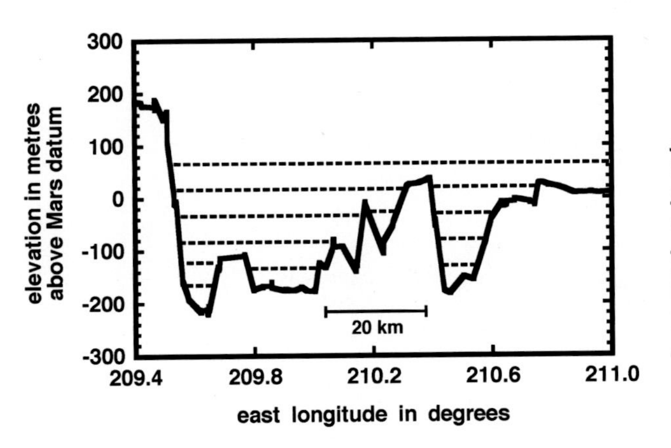

Figure 16.3. Cross-sectional profile at right-angles to water flow direction in Mangala Valles at 16.7° N using unsmoothed MOLA data. The presence of subvalleys on the floor of the main valley is clear, and the consequences of progressive water depth increases in ~50 m increments from the floors of each of these subvalleys is illustrated. It is inferred that water depths were ~50–100 m for most of the duration of the valley-forming event. (Based on Figure 9 in Leask *et al.* (2007); copyright (2007) American Geophysical Union.)

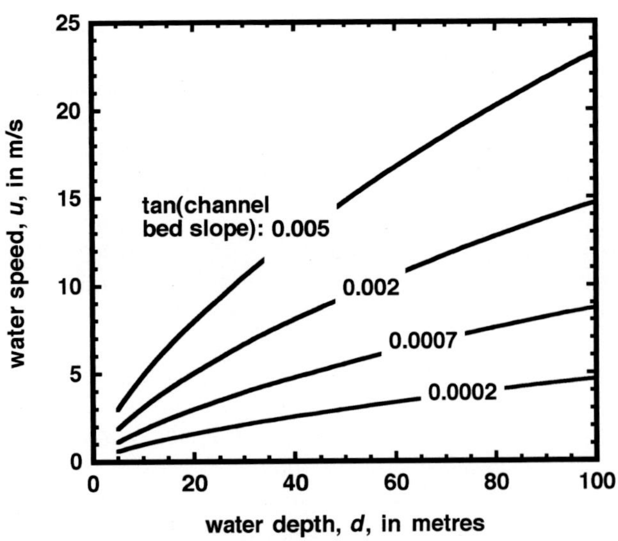

Figure 16.4. Variation of mean water flow speed, u, with water depth, d, in a channel on Mars for four values of the tangent of the channel-bed slope. At slopes this shallow, the tangent of the slope is essentially identical to the sine of the slope.

the flow speed, then F would have had to increase as time cubed. In practice, all reasonable scenarios for subsurface water release involve a rapid rise to a maximum discharge with a subsequent decline (Andrews-Hanna and Phillips, 2007), often roughly exponential (Manga, 2004). Thus we think it very likely that, for most of the time of their formation, the water levels in most of the channels were only a small fraction of the final channel depths, thus technically making these features valleys rather than channels, though we retain the latter term for consistency. This is underlined by the observation that, whereas the total widths of channels are commonly in the range 10 to 30 km (De Hon and Pani, 1993), channel floors are generally not flat but instead consist of a small number of subchannels with widths in the range 5–10 km. The depths of these subchannels are mainly in the range 10 to 50 m (e.g. Figure 16.3), and it may be inferred that values in this range are much better estimates of maximum water depths.

16.5.3 *Water flow speeds and predicted discharge estimates*

In a wide-ranging survey of possible modes of fluvial sediment transport and deposition on Mars, Kleinhans (2005) has suggested that the size distribution of the clasts moving through many Martian fluvial systems may have been distinctly bimodal, with roughly equal proportions of silt/sand and coarse gravel/cobbles. Given that the concern here is with large outflow channels, it is tacitly assumed to have involved short-lived catastrophic events involving energetic bed erosion, and so it is logical to use Equation (16.11) for channels beds dominated by boulders in generating values for friction factors, with the grain size parameters derived from the Viking and Pathfinder landing sites ($D_{50} = 0.1$ m, $D_{84} = 0.48$ m, $D_{90} = 0.6$ m and $\sigma_g = 2.9$). Combining the above maximum water depths (10–50 m) and channel-bed slope estimates ($\tan\alpha = 2 \times 10^{-4}$ to 2×10^{-3}), a range of likely maximum water flow speeds are derived as a function of water depth. These are shown in Figure 16.4; at each water depth the speed is given for each of four channel-bed slopes, 0.0002, 0.0007, 0.002 and 0.005. The three smallest of these span the range found for channel floors, and the largest represents a likely upper limit for pre-channel topography. Clearly, for mature channels that have reached their final floor slope, water flow speeds should lie mainly in the range 3 to 10 m s^{-1} if water depths do not exceed 50 m, and could be up to 15 m s^{-1} if water depths range up to 100 m. Unfortunately, there is no obvious way of estimating likely water depths on pre-channel topography, though Leask *et al.* (2006a) used the outermost terraces on the walls of Ravi Vallis to estimate an early-stage channel depth of at most 50 m on an unusually steep terrain slope of $\tan\alpha = \sim 0.01$; this implies a water speed of about 21 m s^{-1}. The range of water speeds is consistent with the sizes of clasts observed at the Viking and Pathfinder landing sites: flow speeds of order 10 m s^{-1} would be readily able to transport clasts approaching a metre in size (see Figure 16.7 in Kleinhans, 2005).

Using the observation that main channels are commonly 10–30 km wide and contain a small number of subchannels with widths in the range 5 to 10 km, an estimate of likely total channel discharges is obtained by combining the flow speeds in Figure 16.4 with a typical total

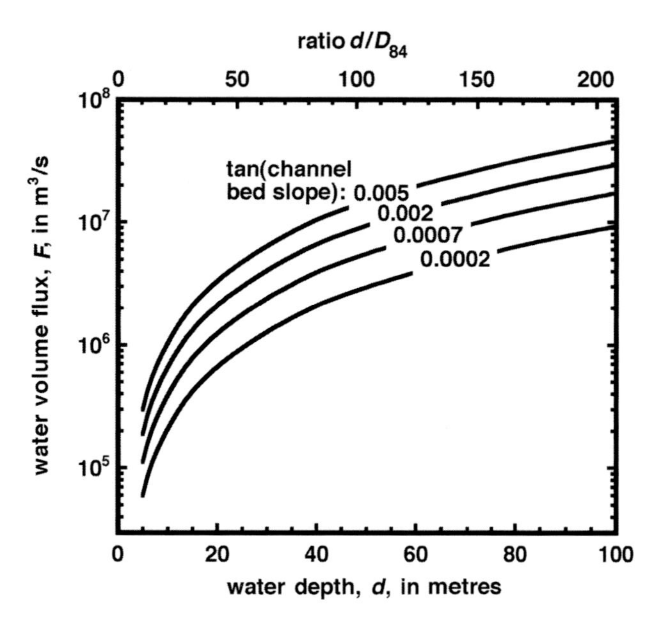

Figure 16.5. Water volume flux, F, as a function of water depth, d, on Mars in steep-sided channels 20 km wide, for four values of the tangent of the channel-bed slope. The volume flux is directly proportional to the channel width. The upper axis gives values of the water depth, d, divided by the bed roughness scale, D_{84}, to aid comparison with other channel systems.

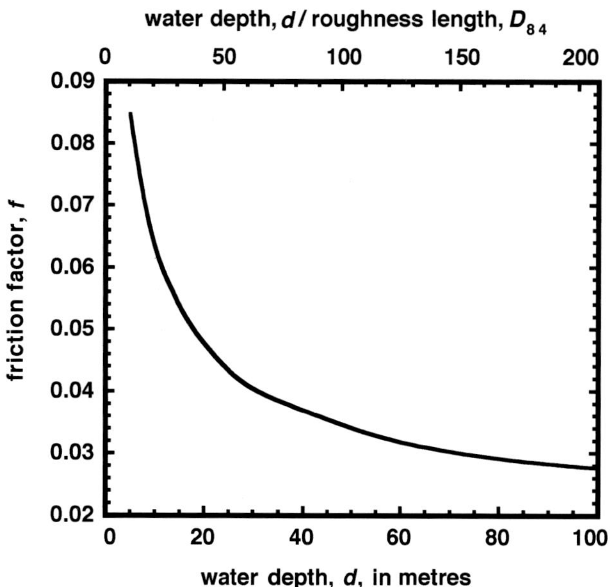

Figure 16.6. Bed friction factor, f, as a function of water depth, d, in channels on Mars in which the bed roughness is dominated by boulders, so that the friction factor is independent of bed slope. The upper axis shows the ratio of water depth to roughness scale, D_{84}, so that these results can be scaled to other roughness parameters.

active flowing water width of 20 km. This results in the variation of volume flux with water depth and bed slope given in Figure 16.5: fluxes in the range 10^5 to about 3×10^7 m^3 s^{-1} should have been common. These values are comparable to the smaller flux estimates summarised in Table 16.1.

16.5.4 Subcritical or supercritical flows?

An important parameter controlling the state of a fluid flow is its Froude number, *Fr*. Froude numbers in excess of unity correspond to highly unsteady, supercritical flow conditions, and fluids forced to flow under supercritical conditions commonly undergo hydraulic jumps after they have flowed for a short distance, with an increase in water depth and a decrease in velocity, in order to reduce the Froude number to less than unity. These flows also erode their beds over time in a way that reduces the bed slope so that the supercritical conditions are avoided (e.g. Grant, 1997; Hunt *et al.*, 2006). Froude number is defined by

$$Fr = u/(gd)^{1/2}. \quad (16.16)$$

Equation (16.16) can be combined with Equation (16.7) giving the flow speed in terms of the physical variables to yield

$$Fr = [(8 \sin\alpha)/f_c]^{1/2}. \quad (16.17)$$

If Equation (16.11) is used for the friction factor in channels with beds dominated by boulders, f_c is independent of channel-floor slope and is only a function of water depth; the relationship is shown in Figure 16.6. For a 50 m deep water flow, f_c is 0.034. Thus for floor slopes of $\tan\alpha =$ 0.0002, 0.0007 and 0.002, the range spanning mature channel floors, the implied Froude numbers are 0.22, 0.41 and 0.69, respectively, all values being subcritical. For a 100 m water depth the friction factor is 0.0276 and the corresponding Froude numbers for mature channels are 0.24, 0.45 and 0.76, again all subcritical values. However, for a pre-erosion topographic slope as steep as $\tan\alpha = 0.005$, the value measured in the case of Ravi Vallis (Table 16.3), the Froude numbers for 50 and 100 m water depths would have been 1.08 and 1.20, respectively, both supercritical values. All of these results are summarised in Figure 16.7; they lend support to the idea that the outflow channels developed in ways that tended to reduce their bed slopes and so eliminate potentially supercritical conditions that may have been present in the early stages of their formation, a process observed to operate in driving the development of the flow regimes in rivers on Earth (Grant, 1997; Kleinhans, 2005).

The most common bedforms observed in the Martian outflow channels are streamlined forms, which provide only imprecise information regarding flow conditions (see Carling *et al.*, this volume Chapter 3). However, observations of inferred subaqueous dunes support the idea from

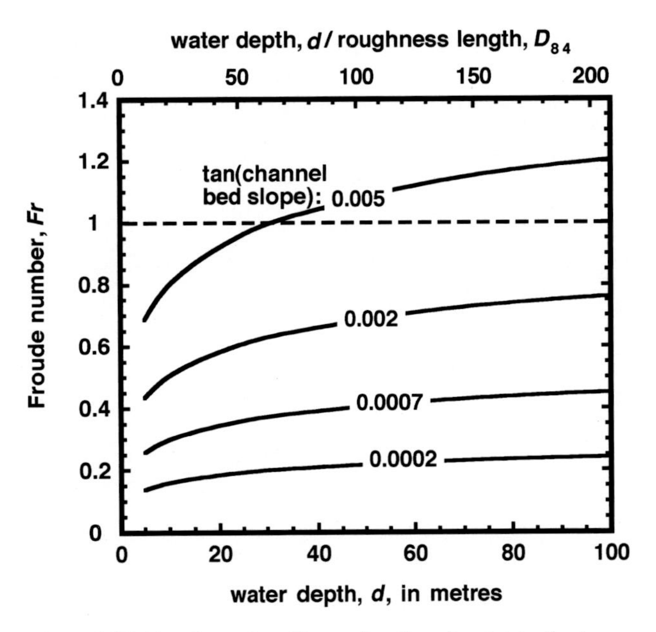

Figure 16.7. Froude number, Fr, as a function of water depth, d, in channels on Mars in which the bed roughness is dominated by boulders, for four values of the tangent of the channel-bed slope. As in Figure 16.5, the upper axis shows the ratio of water depth to roughness scale D_{84}. The horizontal broken line marks the transition between sub- and supercritical flow at $Fr = 1$.

the analysis above that flow in Martian outflow channels tends toward subcritical conditions. Dunes in plan-view morphology are elongate features oriented transverse to the direction of flow. Topographically, dunes are asymmetric, having a more shallowly sloping stoss (upstream) side and a more steeply sloping lee (downstream) side, which results from the difference in stoss and lee deposition processes (see Burr *et al.* (2004) or Carling *et al.*, this volume Chapter 3, for further description and explanation of dune formation). Subaqueous dunes have been identified in Athabasca (Burr *et al.*, 2004) and Maja Valles (Chapman *et al.*, 2003). The features observed in Maja Valles were classified as subaqueous dunes on the basis of their plan-view similarity to dunes in Jökulsá á Fjöllum, a Holocene-aged outflow channel in northern Iceland (Waitt, 2002). The transverse bedforms in Athabasca Valles were classified as dunes on the basis of their plan-view morphology and also their asymmetric vertical shape, which was similar to that observed for subaqueous dunes in Pleistocene-aged flood channels in the Altai Mountains of Siberia (Carling, 1996). In contrast to dunes, antidunes form within high subcritical (i.e., trans-critical) or supercritical flow (Fr > 1). Bedforms in one outflow channel have been proposed to be antidunes (Rice *et al.*, 2003), but the 3D morphology of the field of bedforms was found to be inconsistent with an antidune classification, and a duneform interpretation is preferred (Burr *et al.*, 2004).

16.5.5 Bed erosion

The great depths (tens to perhaps a hundred metres) of the Martian outflow channels imply an efficient bed erosion process. Baker (1979) discussed processes responsible for eroding the Missoula Flood channels on Earth and suggested the possible importance of cavitation – the production and collapse of vapour bubbles when transient velocity changes due to the presence of irregularities on the bed of a channel cause the local pressure to decrease below the saturation vapour pressure of the water. Baker (1979) used an analysis proposed by Barnes (1956). Re-deriving the equations of Barnes in their most general form, the following expression is derived for the mean flow speed u that, when increased locally by a factor Φ, will cause cavitation in a flow of depth d:

$$u = [2/(\Phi^2 - 1)]^{1/2}\{[(P_a - P_v)/\rho_w] + gd\}^{1/2}, \quad (16.18)$$

where P_a is the atmospheric pressure, P_v is the saturation vapour pressure of the water (a function of the water temperature) and as before ρ_w is the water density. The vapour pressure of water near its freezing point is comparable to the Martian atmospheric pressure, 600 Pa, whereas at 30 °C it is 4.2 kPa. Thus, depending on the temperature of the water as it is released and how much cooling it has undergone, $(P_a - P_v)$ could range from 3.5 kPa to a few tens of Pa. Then the term $[(P_a - P_v)/\rho]$ would range from 3.5 $m^2 s^{-2}$ to 0.1 $m^2 s^{-2}$. Above it has been argued that water depths in Martian channels are likely to be 10–50 m, so the value of (gd) will be 40–200 $m^2 s^{-2}$, almost always dominating the second term in the above equation. Inserting the smallest likely value for $[(P_a - P_v)/\rho]$, 0.1 $m^2 s^{-2}$, and corresponding pairs of values of u and d from Figure 16.4, it is found that for cavitation to occur would require Φ to be at least ~3.0 for $u = 3$ m s^{-1}, $d = 10$ m, and to be at least ~2.2 for $u = 10$ m s^{-1}, $d = 50$ m. Barnes (1956) asserted that obstructions on a stream bed on Earth could locally double the fluid velocity, i.e. cause $\Phi = 2$, and there is no reason why a similar argument should not hold for Mars. Thus, at least for the deeper, faster-flowing floods, cavitation should generally have been as effective a source of bed erosion in outflow channels on Mars as it has been suggested to have been on Earth. Equation (16.18) implies that the minimum value of Φ required to allow cavitation to occur increases as the flow speed decreases. Flow depth and hence flow speed are likely to decrease with distance from the point of release of a water flood on Mars due to both water vapour loss and momentum transfer to transported sediment. Thus if enhanced bed erosion is caused by cavitation it is likely to be greatest nearest the water release point, thus encouraging channels to decrease in depth with distance from source, as is observed generally. It is also possible that the erosion caused by cavitation would remove the bed irregularities increasing the flow speed, thus implying that the process would act preferentially in the early stages of channel formation.

16.5.6 Water volumes and event durations

Various authors (e.g. Komar, 1979; Ghatan *et al.*, 2005; Leask *et al.*, 2006a, 2007) have assumed that a minimum value for the volume of water passing through a Martian channel system can be found by estimating the volume of crustal material eroded and assuming that the floodwater carried some maximum sediment load, generally taken to be about 30–40% by volume. The volume V_s of cryosphere material eroded from a channel can be assessed from topographic profiles constructed using laser altimeter data from the Mars Orbiter Laser Altimeter on Mars Global Surveyor by making some simple plausible assumptions about the pre-erosion topography. If the mixture of water and eroded solids contains a volume fraction Q of solids, then the water volume V_w is equal to $[(1 - Q)/Q]V_s$, and with Q in the range 0.3 to 0.4, V_w will be 1.5 to 2.3 times the estimated missing cryosphere volume. Furthermore, dividing the water volume V_w by the water discharge-rate estimate for the same channel yields the minimum duration of the water release event. Using this method, Leask *et al.* (2006a) estimated that a total of a little more than 8000 km^3 of cryosphere was eroded from the Ravi Vallis–Aromatum Chaos system due to a water release rate averaging $\sim$10 to 15 $\times$ 10^6 m^3 s^{-1}, so that for sediment loads in the range 10–40%, the minimum duration was two to nine weeks. In a similar analysis for the Mangala Valles system, Ghatan *et al.* (2005) estimated that 13 000–20 000 km^3 of material was removed by a flow rate averaging 5 $\times$ 10^6 m^3 s^{-1}, implying a flow duration of at least one to three months. A somewhat larger discharge of 1 $\times$ 10^7 m^3 s^{-1} was estimated for Mangala Valles by Leask *et al.* (2007), implying that the 13 000–20 000 km^3 of material was removed in about one month at a 40% sediment load and up to three months at a 20% load.

The use of sediment loads, in percentages of multiples of ten, to estimate flood durations requires further consideration. As noted in a detailed analysis by Kleinhans (2005), a water flood eroding its bed cannot change smoothly from having a low concentration of washload (i.e. essentially fully suspended) solids to being hyperconcentrated; two stable washload solid concentrations are allowed for a given flow speed and depth, but the load cannot change smoothly from the one to the other by bed erosion alone unless the flow speed passes a very large value. These results, summarised by Kleinhans (2005, his Figure 16.12), imply that, for the typical flow conditions deduced above, Martian floods could not carry more than about 2% solids by volume as washload, very much less than the values up to 40% by volume assumed by many workers. A flow could only be hyperconcentrated if it were discharged from its source in a hyperconcentrated state or had a large sediment load injected into it, say by a massive collapse of the channel wall. Hyperconcentrated discharge from the source would require extraction of the water from an aquifer under conditions that damaged the structural fabric of the aquifer and elutriated large amounts of fine solids. It is hard to assess the plausibility of such a process but it might explain the apparently very large aquifer permeabilities implied by the larger estimates of discharges on Mars (Table 16.1). Injection of a sudden load by channel-wall collapse is plausible but would inevitably be a very transient event compared with the duration of channel formation. In the next section a critical issue for Mars is quantified; the low-temperature physical conditions eventually force part of the solids load (in this case the ice crystals) to increase monotonically with time and distance from the source. It is therefore possible that, although a smooth transition cannot occur in Martian channels by bed erosion alone, it may occur through ice-crystal growth. However, because ice crystals and silicate clasts would be competing in their contribution to the washload, this process would presumably reduce the amount of silicates capable of being fully suspended in the flow. Thus this process does not help the flow to carry a greater washload of eroded silicates from the bed, and has no bearing on the issue of the time taken to erode a given depth of channel. This conclusion is equally true of hyperconcentrated flows caused by an initial high solids content in the released water or by channel-wall collapse; if anything, these processes imply a reduced ability to carry material eroded from the bed in suspension. Thus it appears that the critical issue is the volume fraction of eroded clasts that could be carried as bedload or suspended solids. Kleinhans (2005) discusses the large uncertainties in the factors governing bed erosion rate and sediment transport as bed and suspended load, and his arguments suggest that it is unlikely that the combination of all three transport mechanisms would lead to a total volume fraction of sediment greater than about 5%. Adopting this value would increase the flood durations mentioned above by a factor of four.

Durations of at least weeks to months are very much greater than the time scales of at most hours on which aquifer flow rates are predicted to decrease significantly if aquifer permeabilities have values comparable to those on Earth (Manga, 2004). If Martian permeability values are indeed similar to those of Earth, so that initially high water release rates declined dramatically, or if the load-carrying capacities of floods were much less than the tens of percentage points commonly assumed, then discharges would have to have continued for even longer periods of time at smaller volume fluxes and hence lower flow speeds (Kleinhans, 2005). However, lower flow speeds would mean that a given batch of water in the system would have been exposed to Martian environmental conditions for a greater period of time while travelling a given distance, and this has a bearing on the maximum distance for which water can survive as a liquid on Mars, which is discussed below.

16.6 Water flow thermodynamics and rheology

The dynamics issues discussed in the previous section are clearly the major factors relevant to water flow in the proximal parts of outflow channels on Mars. However, two additional factors become progressively more important with distance from the source. These factors are both dictated by the environmental conditions of low ambient temperature and low atmospheric pressure. The low pressure can cause rapid rates of evaporation of water immediately after its release, enhancing initial cooling rates, but ultimately it is the low ambient temperature that dictates the final fate of the water, both via heat loss from the surface and through incorporation of the extremely cold surface materials at the base of the flow. The addition of the sediment load coupled with ice formation in the water eventually causes the rheology of the fluid to become non-Newtonian, with various implications for the water behaviour in distal parts of channels.

16.6.1 Water evaporation and consequent heat loss

Over much of the surface of Mars, the atmospheric pressure is low enough that water will boil spontaneously on being exposed to it, even if the water temperature is close to the freezing point (see Figure 16.2). The evaporation rate is proportional to the amount by which the vapour pressure of the water exceeds the local atmospheric pressure, and because the vapour pressure increases nearly exponentially with temperature, the evaporation rate is a very strong function of the water temperature. The latent heat of evaporation is extracted from the residual water and causes cooling. The cooling decreases the water vapour pressure and so the rate of cooling decreases as the temperature decreases. In low-lying areas on Mars the atmospheric pressure is high enough that water released at low to moderate temperatures will not boil, and even water that is hot enough to boil will cease to do so after it has cooled somewhat. Once boiling ceases, vapour loss by diffusion into the atmosphere still occurs, especially under windy conditions, but the mass loss rate and consequent heat loss rate are reduced greatly. Thus the detailed cooling history is strongly dependent on both the temperature of the water and the elevation relative to Mars datum of its release point.

As an example of the history of a flow strongly influenced by these issues, consider a source at an elevation where the surface atmospheric pressure is 610 Pa, just marginally less than the vapour pressure of water at the triple point, 610.5 Pa. Under these circumstances, water will continue to evaporate at a significant rate at all temperatures above the freezing point. Assume the water has a temperature of 30 °C and forms a flow with a depth of 50 m, then on a slope of $\tan^{-1}0.0007$ it flows away from its source at a speed of $5.5\ \mathrm{m\,s^{-1}}$ (see Figure 16.4). The initial vapour pressure of the water, P_v, is 4.23 kPa and so the difference between the water vapour pressure and the atmospheric pressure, P_a, is 3.62 kPa. The mass loss rate from the surface, dm/dt, is given by

$$dm/dt = z(P_v - P_a)[(M_w/2\pi G\,T_w)]^{1/2} \qquad (16.19)$$

(Kennard, 1938), where M_w is the molecular weight of water, G is the universal gas constant, T_w is the absolute water temperature and z is the coefficient of evaporation. The empirically determined value of z is 0.94 (Tschudin, 1946). Therefore dm/dt is equal to $3.63\ \mathrm{kg\,s^{-1}\,m^{-2}}$; the corresponding heat loss rate, the product of the mass loss rate and the latent heat of vaporisation, $1.78\ \mathrm{MJ\,kg^{-1}}$ at this temperature, is $6.46\ \mathrm{MJ\,s^{-1}\,m^{-2}}$. Thus the initial values of the rates of decrease of depth and temperature of the water are $3.63\ \mathrm{mm\,s^{-1}}$ and $0.011\ \mathrm{K\,s^{-1}}$, respectively. If these rates continued unchanged, the water would all have evaporated after 3.8 hours having travelled 76 km, and would have started to freeze after 45 minutes having travelled 15 km. In practice, the rates of change of depth and temperature both become smaller as the water vapour pressure decreases with the decreasing temperature, and the relevant equations describing these processes must be integrated numerically to follow the flow development. For this particular set of initial conditions, Figure 16.8a shows how depth and temperature evolve until the water freezing point is reached, and how the depth subsequently changes as evaporation continues at a constant temperature of 0 °C, so that progressive freezing of the water is taking place. The figure shows the changing water depth until the ice volume fraction in the water reaches 40%, at which point the presence of the ice crystals must have a significant effect on the rheology of the mixture, discussed in later sections. Figure 16.8b shows the equivalent information for water released at the same temperature, 30 °C, but in a flood initially 100 m deep, so that the initial flow speed is $8.7\ \mathrm{m\,s^{-1}}$ (see Figure 16.4). The illustrations in Figure 16.8 serve to underline how rapidly temperature and depth will change in the proximal part of a flood, and how long it will take for accumulation of ice crystals to significantly influence the water motion. These results, and analogous simulations for other water release temperatures and source elevations, would be directly applicable to Martian floods if the floods did not erode the surfaces over which they flowed. However, the observational evidence (see Burr *et al.*, this volume Chapter 10) suggests that erosion of cryosphere materials from channel beds, at least in the proximal parts of outflow channels, was a major factor in their development. This process is considered below.

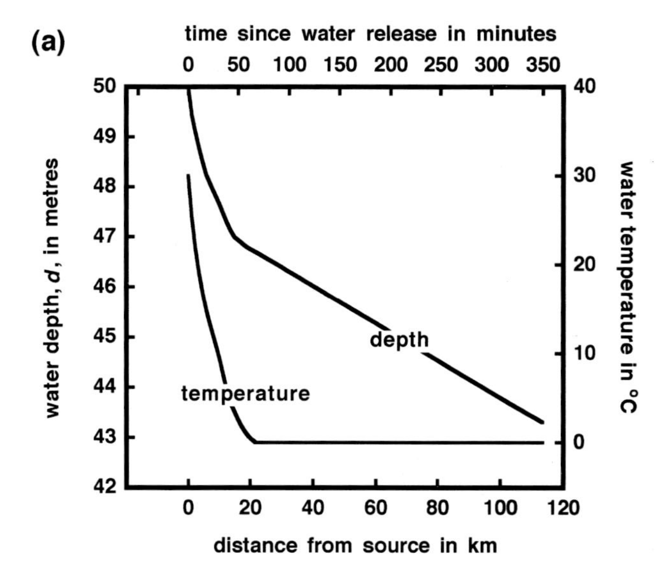

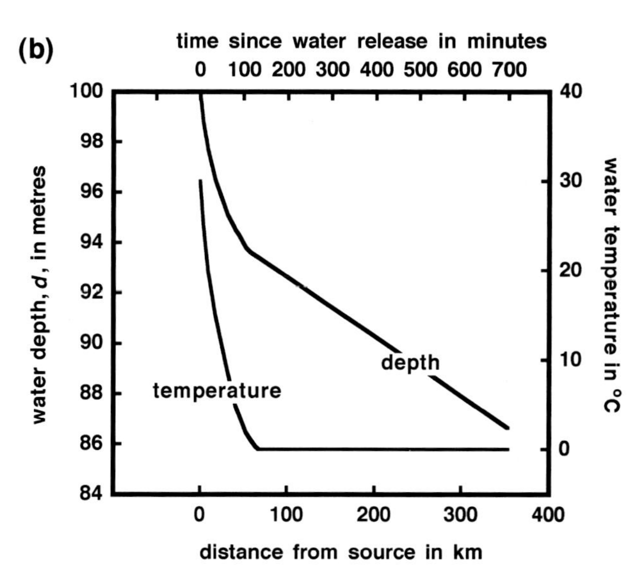

Figure 16.8. Variation of water depth and mean water temperature with distance from water release point in a channel on Mars. The time for which a given batch of water has been travelling is derived from the distance and the changing flow speed. The initial water temperature is taken as 30 °C. The elevation of the release point is such that the local atmospheric pressure is 610 Pa. In part (a) the initial water depth and speed are 50 m and 5.54 m s^{-1}, respectively; in part (b) they are 100 m and 8.69 m s^{-1}, respectively.

16.6.2 Thermal consequences of cryosphere erosion

The low Martian ambient temperature means that erosion of the surface involves not only the addition of a cold silicate sediment load to the water but also the addition of cryosphere ice. Transfer of sensible heat to both components, and of the latent heat needed to melt the ice, are both significant sources of water cooling, especially near channel sources where erosion rates are greatest. The calculations underlying Figure 16.8 imply that a source of proximal cooling in addition to evaporation will decrease the early-stage water evaporation rate. Thus, although the rate at which the temperature falls will increase, the rate at which the depth decreases will be reduced, relative to what is shown in the figure. When the water temperature reaches the freezing point, entrained cryosphere ice will no longer be melted, and the released floodwater will begin to freeze. The amount of freezing that takes place will be proportional to the volume fraction of cryosphere material entrained, and the total solid load carried by the water will be a mixture of eroded bedrock clasts, ice crystals derived from the cryosphere, and ice crystals produced by freezing of the original water. It is the total solid load volume fraction that will influence the bulk rheology of the fluid.

Consideration of the detailed consequences of eroding the cryosphere (Bargery, 2007) shows that the minimum water volumes found in the past by assuming that the water has a maximum sediment-carrying capacity are probably underestimates. To illustrate this, ignore the initial rapid cooling of the water and consider the addition of cryosphere materials after the water has cooled to its freezing temperature, T_f ($\sim$273.15 K). Consider a volume V_w of this water, a fraction ϕ of which freezes due to entrainment of a volume V_c of cryosphere, of which ice occupies a volume fraction q. The heat required to warm the cryosphere materials from the ambient temperature T_a ($\sim$210 K) to the water temperature T_f is $(T_f - T_a)V_c[q\rho_i c_i + (1-q)\rho_r c_r]$ where ρ_i and ρ_r are the densities of ice and rock, taken as 917 and 3000 kg m^{-3}, respectively, at cryosphere temperatures and c_i and c_r are the corresponding specific heats at constant pressure, 1900 and 700 J kg^{-1} K^{-1}. The heat released by partial freezing of the water is $\phi\, V_w \rho_w L_i$ where ρ_w is the water density, $\sim$1000 kg m^{-3}, and L_i is the latent heat of fusion of ice, $\sim 3.35 \times 10^5$ J kg^{-1}. Equating the two results in

$$V_w/V_c = \{(T_f - T_a)[q\rho_i c_i + (1-q)\rho_r c_r]\}/[\phi \rho_w L_i] = K'/\phi, \tag{16.20}$$

where K' is a constant. Inserting the relevant material properties specified earlier and using a value of $q = 0.15$, suggested by the cryosphere models of Hanna and Phillips (2005), produces $K' = 0.38\,575$. The water that freezes changes its volume because of the differing densities of water and ice and so the volume fraction of solids in the mixture after incorporation of cryosphere material and partial water freezing is s where

$$s = \frac{[1 + \phi(V_w/V_c)(\rho_w/\rho_i)]}{[1 + (1-\phi)(V_w/V_c) + \phi(V_w/V_c)(\rho_w/\rho_i)]}. \tag{16.21}$$

Substituting for (V_w/V_c) and solving this equation for ϕ:

$$\phi = sK'\rho_i/(sK'\rho_i + \rho_i - s\rho_i - sK'\rho_w + K'\rho_w). \tag{16.22}$$

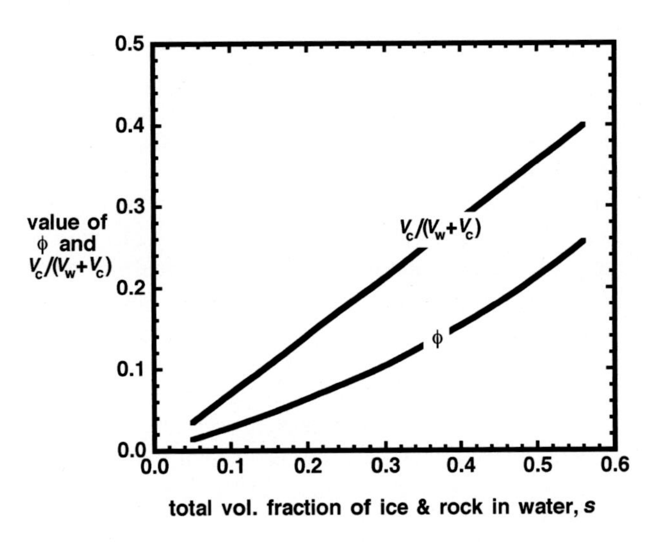

Figure 16.9. For water entraining eroded cryosphere material, the curves show the changes in the volume fraction of the original water that has frozen, ϕ, and the fraction of the total fluid volume that is cryosphere material, $[V_c/(V_w + V_c)]$, as a function of the total volume fraction of solids (ice and rock) in the water, s.

Figure 16.9 shows how s is related to $[V_c/(V_w + V_c)]$, the fraction of the total fluid volume that is cryosphere material. Consider the traditional assumption that the maximum sediment load that can be carried by a water flood is at least $[V_c/(V_w + V_c)] = 0.3$, i.e. 30% solids by volume. Ignoring the matter of cryosphere incorporation causing partial water freezing would have led to the conclusion that the volume of water passing through the channel was $(V_w/V_c) = 2.333$ times the missing volume eroded from the channel. However, when account is taken of the water freezing issue it is seen that if the true final volume fraction of rock *and ice* in the flowing fluid is $s = 0.3$, the true volume of water passing through the channel is $(V_w/V_c) = 3.701$ times the eroded volume, nearly 59% more water than the simple analysis would have predicted. If the upper limit for sediment transport is taken as 40% by volume, the traditional method would have implied $(V_w/V_c) = 1.500$ times the missing volume eroded from the channel, whereas using $s = 0.4$ implies $(V_w/V_c) = 2.517$ times the eroded volume, about 68% more water.

To illustrate the importance of the results presented immediately above, consider Figure 16.8a. The parameters used to generate this figure (an initial water depth of 50 m, a flow speed of 20 m s^{-1}, and a release location where the atmospheric pressure is close to the vapour pressure of water at the triple point) are similar to those relevant to the Mangala Valles channel system (Leask *et al.*, 2007). This channel shows strong indications of floor scouring for 150 km from its source (Burr *et al.*, this volume Chapter 10) and displays some erosional features for at least a further 100 km (Neather and Wilson, 2007). Thus, it seems likely that erosion continued in the case of this channel for, say, 250 km from the source. At this distance, Figure 16.8a implies that 25% of the water would already have been frozen due to evaporation alone. If 250 km from the source is identified as the distance at which the changing rheology causes a drastic change in erosive power and 40% total solids is adopted as the critical load that leads to that change in behaviour, a further 15% by volume of solids, a mixture of ice and rock, should have been added to the water. This corresponds to setting $s = 0.15$, for which $(V_w/V_c) = 8.4$; thus, in this case it may be concluded that the volume of water that must have passed through the channel to produce the observed features was not 2.5 times the eroded volume but instead more than 8 times that volume. These simple illustrations are imperfect in that they do not treat the two main sources of heat loss from Martian flood waters simultaneously; however, they serve to make it clear how important it is that a comprehensive model of conditions in outflow channels on Mars should be developed.

16.6.3 Influence of ice formation and sediment load on rheology

It was asserted above that the loading of water in an outflow channel by 30–40 volume per cent solids will change its rheology. The first-order effect is likely to be the generation of a yield strength causing an evolution from Newtonian to non-Newtonian behaviour. If the fluid behaves like the simplest non-Newtonian material, a Bingham plastic, the presence of a yield strength in turbulent flow (Malin, 1998) is likely to result in the formation of pods of fluid within which there is little deformation, shear being concentrated in the regions between these pods. The sizes of these pods will be a large fraction of the depth of the flow. The lack of deformation within the pods will encourage segregation of the solids from the liquid driven by buoyancy forces, silicates tending to sink and ice crystals tending to rise. However, the random tumbling of the pods in the turbulent flow will mean that no significant net segregation will occur.

The bulk density, β, yield strength, τ_y, and plastic viscosity, η, of the bulk fluid control two dimensionless groups, the Reynolds number (equal to $4Ru\beta/\eta$) and the Hedstrom number (a dimensionless resistance to shear, defined as $16R^2\tau_y\beta^2/\eta^2$), that characterise the internal fluid motion. When the sediment/ice load and hence the yield strength become large enough, the Hedstrom number will reach a critical value, itself a function of the Reynolds number, that causes a progressive transition towards laminar flow. If completely laminar conditions are reached, the fluid will develop plug flow, with an unsheared raft of fluid extending down from the surface and moving over a sheared zone beneath, extending down to the bed. The stresses exerted by the laminar sheared zone on the bed will be much smaller than those involved in turbulent flow, and so bed erosion should decrease dramatically or cease completely. If these conditions are approached, the unsheared material will no longer be tumbling, and gravitational sinking of the silicate clasts and flotation of ice crystals can begin. In this way a semi-continuous ice raft may begin to form on the flow surface. However, a rapid transition to complete elimination of turbulence is not expected, because if extensive segregation of the solids occurred, the resulting Newtonian water would immediately become turbulent again, and begin once more to erode and entrain bed material. Thus it should be expected that feedback loops cause conditions in the fluid to hover close to the critical Hedstrom number for as long as possible (Bargery and Wilson, 2006).

If ice raft formation does begin, there will not be a continuous raft, at least not initially; collisions with the banks and deformation wherever the bed slope changes will cause breaks in the coverage, and in these places water at 0 °C will be exposed to the atmosphere and will continue to evaporate and cool. However, on the undisturbed parts of the raft surface the temperature will decrease to the ambient, ~210 K, and at this temperature any mass and heat loss by evaporation will be essentially negligible. Thus, apart from a small amount of cooling through cracks in the ice (which may become smaller in number as the ice thickness increases downstream), significant heat loss from the upper surface of the flood will cease and only heat loss to the underlying cryosphere will be important. Furthermore, the cessation, or at least reduction, of turbulence and consequent settling of silicates is likely to cause a progressive change from bed erosion to deposition of a growing layer of solids at 0 °C on the bed. As a result, conduction from the body of the flow into the ground will also be minimised. This potential influence of the presence of ice crystals as well as suspended solids on the balance between bed erosion and sediment deposition may cause a systematic difference between the downstream variation of bedforms produced by major water floods on Earth and Mars. Although erosion of the bed may not occur where ice rafts are present, impacts of the ice rafts with the topography confining the flow will cause some erosion. Such erosion of the channel banks is observed in cold-climate (but non-glacial) rivers on Earth (e.g. Martini *et al.*, 1993).

Another geomorphic indicator of the presence of ice, which is commonly observed in terrestrial glacial settings, is kettle holes. Kettle holes are depressions that result from the stranding of large ice blocks (maximum size order tens of metres in diameter) during glacial meltwater flow (e.g. Maizels, 1977; Russell, 1993; Tómasson, 1996). In the largest contemporary floods, which occur from beneath glaciers (Björnsson, this volume Chapter 4), these blocks

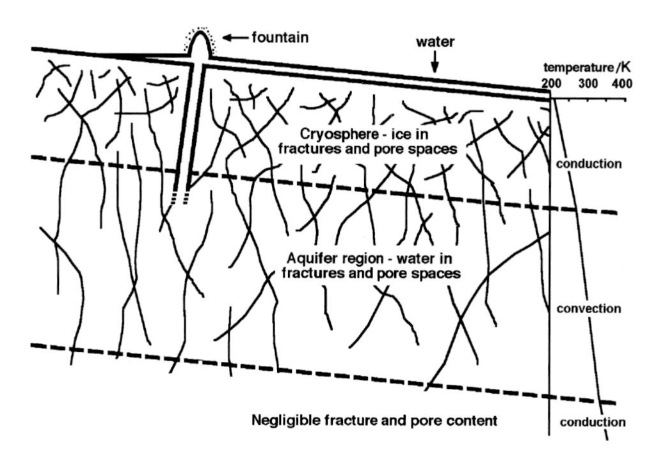

Figure 16.10. Diagram showing the key structures and processes involved in the release of water onto the surface of Mars from confined aquifers.

are commonly produced by breakage and collapse of the glacier margin during sub-, en-, or proglacial flooding (e.g. Sugden *et al.*, 1985). Subsequent melting of the ice blocks most commonly produces simple depressions, but these depressions may have rims or even contain mounds, depending on the amount of sediment carried by the block and its degree of burial (Maizels, 1992). Kettle holes have been inferred in two outflow channels on Mars, for a population of simple depressions in Maja Valles (Chapman *et al.*, 2003) and for a population of raised-rim depressions in Athabasca Valles (Gaidos and Marion, 2003), although the Athabasca Valles population has been hypothesised to have other origins (Burr *et al.*, this volume Chapter 10).

16.7 Closing summary

While the basic processes occurring in outflow channels on Mars and on Earth are likely to have been the same, there are numerous differences of detail imposed by the Martian environment (Figure 16.10). For much of the planet's history, major water reservoirs on Mars were probably located at depths of several kilometres below the surface, being released by tectonic and volcanic (and possibly impact) events that disrupted the cryosphere. Thus, despite the low surface temperature, quite a wide range of released water temperatures was possible. Exposure to the low-pressure atmosphere would have caused vigorous evaporation of warm water and rapid cooling; cooling would have been enhanced by the erosion and incorporation of rock and ice from the cryosphere. For a given water depth, water flow speeds on Mars would have been less than on Earth by the square root of the ratio of the respective accelerations due to gravity. This only represents a difference by a factor of 1.6, and so bed erosion processes, and the consequent evolution from supercritical to subcritical flow conditions, would probably have been similar. However, we have no direct observations of the formation of outflow channels by floods on the scale of the Martian outflows, the most recent on Earth being the prehistoric Missoula Flood. Thus, there are still uncertainties about the relative importance of various bed erosion processes, especially cavitation.

The importance of ice formation in Martian water floods cannot be overemphasized. On Earth, evolution of the rheology of flowing water is dictated by incorporation and transport of sediments, and changes in topographically driven flow speed control changes in erosion and deposition. On Mars, the rheological influence of increasing amounts of ice crystals must also be considered. A full understanding of water flow in Martian outflow channels awaits a comprehensive model incorporating all of the issues discussed in this review.

Even if all of the basic physics has been identified, there are still numerous uncertainties in the parameters required for such models. For example, the only *in situ* grain size determinations in what may be fluvial deposits on Mars are those from the Viking 1, Viking 2 and Pathfinder lander sites (Golombek and Rapp, 1997; Golombek *et al.*, 2003); however, camera resolution limitations mean that we have no reliable information on the sub-centimetre size fractions, which have a very great influence on flow conditions (Kleinhans, 2005). This problem could be remedied by suitably targeted next-generation landers with higher-resolution optics. The other most important uncertainties are in the physical properties of aquifers (permeability, porosity and mean water temperature) and of the cryosphere (permeability, porosity and ice content). Although some information on subsurface structure may be gleaned from orbiting RADAR systems (e.g. MARSIS on Mars Express – Picardi *et al.*, 2005), it is hard to see how major improvements can be made without ground-penetrating radar, active seismic experiments and, preferably, drilling from surface landers or penetrators.

Notation

Symbol	Definition	Units
D	depth of tops of aquifer, 5–8	km
D_{50}	clast size such that 50% of clasts are smaller than D_{50}, ~0.1	m
D_{84}	clast size such that 84% of clasts are smaller than D_{50}, ~0.48	m
D_{90}	clast size such that 90% of clasts are smaller than D_{50}, ~0.6	m
F	volume flux of water flowing through channel	$m^3\ s^{-1}$
Fr	Froude number for flow in channel	
G	universal gas constant, 8314	$J\ kmol^{-1}\ K^{-1}$
H	vertical extents of aquifer, ~6	km
K	bulk thermal conductivity of an aquifer, ~2.5	$W\ m^{-1}\ K^{-1}$
K'	constant appearing in Eq. (16.19), ~0.38575	
L_i	latent heat of fusion of ice, ~3.35×10^5	$J\ kg^{-1}$
M_w	molecular weight of water, 18.02	$kg\ kmol^{-1}$
P_a	atmospheric pressure, ~600	Pa
P_v	saturation vapour pressure of water	Pa
Q	volume fraction of solids in water–solids mixture	
R	hydraulic radius of the channel	m
S	sine of channel bed slope	
T_a	ambient temperature of shallow cryosphere, ~210	K
T_c	mean temperature of cryosphere, ~240	K
T_f	melting temperature of ice ~273.15	K
T_r	average temperature in water rising in fractures, ~300	K
T_w	mean water temperature in surface flow	K
U	rise speed of water in crustal fracture, ~40–50	$m\ s^{-1}$
V_c	volume of cryosphere material entrained by water	m^3
V_s	volume of cryosphere material eroded from channel	m^3
V_w	volume of water in water–solids mixture	m^3
W	width of crustal fracture, ~2	m
Z	topographic height difference	km
c_i	specific heat at constant pressure of cryosphere ice, ~1900	$J\ kg^{-1}\ K^{-1}$
c_r	specific heat at constant pressure of cryosphere silicate rock, ~700	$J\ kg^{-1}\ K^{-1}$
c_w	specific heat at constant pressure of water, 4175	$J\ kg^{-1}\ K^{-1}$
d	depth of water in channel	m
dm/dt	mass flux per unit area of evaporating water molecules	$kg\ s^{-1}\ m^{-2}$
dP/dz	pressure gradient driving water up fracture, ~950	$Pa\ m^{-1}$
dT/dx	temperature gradient controlling heat flow away from fracture	$K\ m^{-1}$

dT/dz	vertical temperature gradient across aquifer	$K\,m^{-1}$
f	friction factor for motion of water against wall of fracture, $\sim 10^{-2}$	
f_c	friction factor for motion of water in channel	
g	acceleration due to gravity on Mars, ~ 3.72	$m\,s^{-2}$
k	permeability of aquifer	
k_c	critical permeability of aquifer	
n	Manning coefficient for water flow in channel	$s\,m^{-1/3}$
q	volume fraction of ice in cryosphere, ~ 0.15	
r	typical size of bed roughness elements, $\sim D_{50}$	m
s	volume fraction of all solids in water–solid mixture in channel	
w	width of channel	m
z	empirical coefficient of evaporation efficiency, 0.94	
ΔT_a	adiabatic temperature change in water rising in fractures	K
ΔT_c	average decrease in temperature of water rising in fracture, ~ 0.5	K
Φ	factor by which flow speed must increase for cavitation, ~ 2	
α	slope of channel bed	radian
α_w	volume thermal expansion coefficient of water, $(2.5-5) \times 10^{-4}$	K^{-1}
β	bulk density of water–solid mixture in channel	$kg\,m^{-3}$
δT	decrease in temperature of water rising in fracture	K
η	plastic viscosity of water–solid mixture in channel	Pa s
κ	thermal diffusivity of cryosphere material, $\sim 7 \times 10^{-7}$	$m^2\,s^{-1}$
λ	depth to which heat penetrates into fracture wall from water	m
μ_w	viscosity of water	Pa s
ρ_i	density of ice, ~ 917	$kg\,m^{-3}$
ρ_r	density of silicate rock, ~ 3000	$kg\,m^{-3}$
ρ_w	density of water, ~ 1000	$kg\,m^{-3}$
σ_g	geometric standard deviation of the bed clast size distribution, ~ 2.9	
τ	transit time of water rising through fracture, $\sim 60-100$	s
τ_y	yield strength of water–solid mixture in channel	Pa
ϕ	fraction of water that freezes due to entrainment of cryosphere	

Acknowledgements

We are grateful to Maarten Kleinhans and Michael Manga for thoughtful and constructive comments on this chapter.

References

Andrews-Hanna, J. C. and Phillips, R. J. (2007). Hydrological modeling of outflow channels and chaos regions on Mars. *Journal of Geophysical Research*, **112** (E08001), doi:10.1029/2006JE002881.

ASCE (1963). Friction factors in open channels (Task force report). *Journal of the Hydraulics Division, Proceedings of the American Society of Civil Engineers*, **89** (HY2), 97–143.

Baker, V. R. (1979). Martian channel morphology: Maja and Kasei Valles. *Journal of Geophysical Research*, **84** (B14), 7,961–7,983.

Baker, V. R. (1982). *The Channels of Mars*. Austin, TX: University of Texas Press.

Baker, V. R. (2001). Water and the Martian landscape. *Nature*, **412**, 228–236.

Bargery, A. S. (2007). Aqueous eruption and channel flow on Mars during the Amazonian epoch. Unpublished Ph.D. thesis, Lancaster University, UK.

Bargery, A. S. and Wilson, L. (2006). Modelling water flow with bedload on the surface of Mars. In *Lunar and Planetary Science Conference XXXVI*, Abstract 1218. Houston, TX: Lunar and Planetary Institute.

Bargery, A. S., Wilson, L. and Mitchell, K. L. (2005). Modelling catastrophic floods on the surface of Mars. In *Lunar and Planetary Science Conference XXXV*, Abstract 1961. Houston, TX: Lunar and Planetary Institute.

Bargery, A. S. and Wilson, L. (Submitted). Dynamics of the ascend and eruption of water containing dissolved CO_2 on Mars. *Journal of Geophysical Research*, 2009 JE003403.

Barnes, H. L. (1956). Cavitation as a geological agent. *American Journal of Science*, **254**, 493–505.

Bathurst, J. C. (1993). Flow resistance through the channel network. In *Channel Network Hydrology*, eds. K. Beven and M. J. Kirkby. Chichester, UK: Wiley, pp. 69–98.

Burr, D. M., McEwen, A. S. and Sakimoto, S. E. H. (2002a). Recent aqueous floods from the Cerberus Fossae. *Geophysical Research Letters*, **29** (1), doi:10.1029/2001GL013345.

Burr, D. M., Grier, J. A., McEwen, A. S. and Keszthelyi, L. P. (2002b). Repeated aqueous flooding from the Cerberus Fossae: evidence for very recently extant, deep groundwater on Mars. *Icarus*, **155**, 53–73.

Burr, D. M., Carling, P. A., Beyer, R. A. and Lancaster, N. (2004). Flood-formed dunes in Athabasca Valles, Mars:

morphology, modeling, and implications. *Icarus*, **171**, 68–83.

Burt, D. M. and Knauth, L. P. (2003). Electrically conducting, Ca-rich brines, rather than water, expected in the Martian subsurface. *Journal of Geophysical Research*, **108** (E4), 8026, doi:10.1029/2002JE001862.

Carling, P. A. (1996). Morphology, sedimentology and palaeohydraulic significance of large gravel dunes, Altai Mountains, Siberia. *Sedimentology*, **43**, 647–664.

Carr, M. H. (1979). Formation of Martian flood features by release of water from confined aquifers. *Journal of Geophysical Research*, **84** (B6), 2,995–3,007.

Carr, M. H. (2002). Elevations of water-worn features on Mars: implications for circulation of groundwater. *Journal of Geophysical Research*, **107** (E12), 5131, doi:10.1029/2002JE001845.

Carr, M. H. and Clow, G. D. (1981). Martian channels and valleys: their characteristics, distribution, and age. *Icarus*, **48**, 250–253.

Chadwick Jr., W. W. and Embley, R. W. (1998). Graben formation associated with recent dike intrusions and volcanic eruptions on the mid-ocean ridge. *Journal of Geophysical Research*, **103** (B5), 9807–9825, doi:10.1029/1997JB02485.

Chapman, M. G., Hare, T. M., Russell, A. J. and Gudmundsson, M. T. (2003). Possible Juventae Chasma subice volcanic eruptions and Maja Valles ice outburst floods on Mars: implications of Mars Global Surveyor crater densities, geomorphology, and topography. *Journal of Geophysical Research*, **108** (E10), 5113, doi:10.1029/2002JE002009.

Clifford, S. M. (1987). Polar based melting on Mars. *Journal of Geophysical Research*, **92** (B9), 9,135–9,152.

Clifford, S. M. (1993). A model for the hydrologic and climatic behaviour of water on Mars. *Journal of Geophysical Research*, **98** (E6), 10,973–11,016.

Clifford, S. M. and Parker, T. J. (2001). The evolution of the Martian hydrosphere: implications for the fate of a primordial ocean and the current state of the northern plains. *Icarus*, **154**, 40–79, doi:10.1006/icar.2001.6671.

Coleman, N. M. (2003). Aqueous flows carved the outflow channels on Mars. *Journal of Geophysical Research*, **108** (E5), 5039, doi:10.1029/2002JE001940.

Coleman, N. M. (2004). Ravi Vallis, Mars-paleoflood origin and genesis of secondary chaos zones. In *Lunar and Planetary Science Conference XXXV*, Abstract 1299. Houston, TX: Lunar and Planetary Institute.

Coleman, N. M. (2005). Martian megaflood triggered chaos formation, revealing groundwater depth, cryosphere thickness, and crustal heat flux. *Journal of Geophysical Research*, **110** (E12S20), doi:10.1029/2005JE002419.

De Hon, R. A. and Pani, E. A. (1993). Duration and rates of discharge: Maja Valles, Mars. *Journal of Geophysical Research*, **98** (E5), 9,128–9,138.

Diamond, L. W. and Akinfiev, N. N. (2003). Solubility of CO_2 in water from −1.5 to 100 °C and from 0.1 to 100 MPa: evaluation of literature data and thermodynamic modelling. *Fluid Phase Equilibria*, **208**, 265–290, doi:10.1016/S0378-3812(03)00041-4.

Dixon, J. E. (1997). Degassing of alkalic basalts. *American Mineralogist*, **82**, 368–378.

Dohm, J. M. *et al.* (2001a). Ancient drainage basin of the Tharsis Region, Mars: potential source for outflow channel systems and putative oceans or paleolakes. *Journal of Geophysical Research*, **106** (E12), 32,943–32,958.

Dohm, J. M. *et al.* (2001b). Latent outflow activity for Western Tharsis, Mars: significant flood record exposed. *Journal of Geophysical Research*, **106** (E6), 12,301–12,314.

Edlund, S. J. and Heldmann, J. L. (2006). Correlation of subsurface ice content and gulley locations on Mars: testing the shallow aquifer theory of gulley formation. In *Lunar and Planetary Science Conference XXXVII*, Abstract 2049. Houston, TX: Lunar and Planetary Institute.

Gaidos, E. and Marion, G. (2003). Geological and geochemical legacy of a cold early Mars. *Journal of Geophysical Research*, **108** (E6), 5055, doi:10.1029/2002JE002000.

Gay, E. C., Nelson, P. A. and Armstrong, W. P. (1969). Flow properties of suspensions with high solids concentrations. *Journal of the American Institute of Chemical Engineers*, **15**, 815–822.

Gerlach, T. M. (1986). Exsolution of H_2O, CO_2, and S during eruptive episodes at Kilauea volcano, Hawaii. *Journal of Geophysical Research*, **91** (B12), 12,177–12,185.

Ghatan, G. J., Head, J. W. and Wilson, L. (2005). Mangala Valles, Mars: assessment of early stages of flooding and downstream flood evolution. *Earth, Moon and Planets*, **96** (1–2), 1–57, doi:10.1007/s11038-005-9009-y.

Gioia, G. and Bombardelli, F. A. (2002). Scaling and similarity in rough channel flows. *Physical Review Letters*, **88** (1), doi:10.1103/PhysRevLett.88.014501.

Golombek, M. P. and Rapp, D. (1997). Size-frequency distributions of rocks on Mars and Earth analog sites: implications for future landed missions. *Journal of Geophysical Research*, **102** (E2), 4117–4129.

Golombek, M. P., Cook, R. A., Moore, H. J. and Parker, T. J. (1997). Selection of the Mars Pathfinder landing site. *Journal of Geophysical Research*, **102** (E2), 3967–3988, doi:10.1029/96JE03318.

Golombek, M. P., Haldemann, A. F. C., Forsberg-Taylor, N. K. *et al.* (2003). Rock size-frequency distributions on Mars and implications for Mars Exploration Rover landing safety and operations. *Journal of Geophysical Research*, **108** (E12), 8086, doi:10.1029/2002JE002035.

Grant, G. E. (1997). Critical flow constrains flow hydraulics in mobile-bed streams: a new hypothesis. *Water Resources Research*, **33** (2), 349–358.

Gulick, V. C. (1998). Magmatic intrusions and a hydrothermal origin for fluvial valleys on Mars. *Journal of Geophysical Research*, **103** (E8), 19,365–19,387.

Gulick, V. C. (2001). Origin of the valley networks on Mars: a hydrological perspective. *Geomorphology*, **37**, 241–268.

Hanna, J. C. and Phillips, R. J. (2005). Hydrological modeling of the Martian crust with application to the pressurization of aquifers. *Journal of Geophysical Research*, **110**, E01004, doi:10.1029/2004JE002330.

Hanna, J. C. and Phillips, R. J. (2006). Tectonic pressurization of aquifers in the formation of Mangala and Athabasca Valles, Mars. *Journal of Geophysical Research*, **111**, E03003, doi:10.1029/2005JE002546.

Harris, D. M. (1981). The concentration of CO_2 in submarine tholeiitic basalts. *Journal of Geology*, **89**, 689–701.

Head, J. W., Wilson, L. and Mitchell, K. L. (2003a). Generation of recent massive water floods at Cerberus Fossae, Mars by dike emplacement, cryospheric cracking, and confined aquifer groundwater release. *Geophysical Research Letters*, **30** (11), 1577, doi:10.1029/2003GL017135.

Head, J. W., Mustard, J. F., Kreslavsky, M. A., Milliken, R. E. and Marchant, D. R. (2003b). Recent ice ages on Mars. *Nature*, **426**, 797–802, doi:10.1038/nature02114.

Head, J. W., Marchant, D. R. and Ghatan, G. J. (2004). Glacial deposits on the rim of a Hesperian-Amazonian outflow channel source trough: Mangala Valles, Mars. *Geophysical Research Letters*, **31**, L10701, doi:10.1029/2004GL020294.

Heldmann, J. L. and Mellon, M. T. (2004). Observations of martian gullies and constraints on potential formation mechanisms. *Icarus*, **168**, 285–304.

Herschel, C. (1897). On the origin of the Chézy formula. *Journal of the Association of Engineering Societies*, **18**, 2–51.

Hjülstrom, F. (1932). Das transportvermögen der flüsse und die bestimmung des erosionsbetrages. *Geografiska Annaler*, **14**, 244–258.

Hoffman, N. (2000). White Mars: a new model for Mars' surface and atmosphere based on CO_2. *Icarus*, **146**, 326–342.

Hoffman, N. (2001). Explosive CO_2-driven source mechanisms for an energetic outflow "jet" at Aromatum Chaos, Mars. In *Lunar and Planetary Science Conference XXXII*, Abstract 1257. Houston, TX: Lunar and Planetary Institute.

Hunt, A. G., Grant, G. E. and Gupta, V. K. (2006). Spatio-temporal scaling of channels in braided streams. *Journal of Hydrology*, **322**, 192–198.

Kaye, G. W. C. and Laby, T. H. (1995). *Tables of Physical and Chemical Constants*, 16th edn. Harlow, Essex, UK: Longman.

Kennard, E. H. (1938). *Kinetic Theory of Gases*, 1st edn. New York, NY: McGraw-Hill.

Kleinhans, M. G. (2005). Flow discharge and sediment transport models for estimating a minimum timescale of hydrological activity and channel and delta formation on Mars. *Journal of Geophysical Research*, **110**, E12003, doi:10.1029/2005JE002521.

Knudsen, J. G. and Katz, D. L. (1958). *Fluid Dynamics and Heat Transfer*, New York, NY: McGraw-Hill.

Komar, P. D. (1979). Comparisons of the hydraulics of water flows in Martian outflow channels with flows of similar scale on Earth. *Icarus*, **37**, 156–181.

Komar, P. D. (1980). Modes of sediment transport in channelized water flows with ramifications to the erosion of the Martian outflow channels. *Icarus*, **42**, 317–329.

Komatsu, G. and Baker, V. R. (1997). Paleohydrology and flood geomorphology of Ares Vallis. *Journal of Geophysical Research*, **102** (E2), 4,151–4,160.

Komatsu, G., Ori, G. G., Ciarcelluti, P. and Litasov, Y. D. (2004). Interior layered deposits of Valles Marineris, Mars: analogous subice volcanism related to Baikal Rifting, Southern Siberia. *Planetary and Space Science*, **52**, 167–187.

Leask, H. J., Wilson, L. and Mitchell, K. L. (2004). The formation of Aromatum Chaos and the water discharge rate at Ravi Vallis. In *Lunar and Planetary Science Conference XXXV*, Abstract 1544. Houston, TX: Lunar and Planetary Institute.

Leask, H. J., Wilson, L. and Mitchell, K. L. (2006a). Formation of Ravi Vallis outflow channel, Mars: morphological development, water discharge, and duration estimates. *Journal of Geophysical Research*, **111**, E08070, doi:10.1029/2005JE002550.

Leask, H. J., Wilson, L. and Mitchell, K. L. (2006b). Formation of Mangala Fossa, the source of the Mangala Valles, Mars: morphological development as a result of volcano-cryosphere interactions. *Journal of Geophysical Research*, **112**, E02011, doi:10.1029/2005JE002644.

Leask, H. J., Wilson, L. and Mitchell, K. L. (2006c). Formation of Aromatum Chaos, Mars: morphological development as a result of volcano-ice interactions. *Journal of Geophysical Research*, **111**, E08071, doi:10.1029/2005JE002549.

Leask, H. J., Wilson, L. and Mitchell, K. L. (2007). Formation of Mangala Valles outflow channel, Mars: morphological development, and water discharge and duration estimates. *Journal of Geophysical Research*, **112**, E08003, doi:10.1029/2006JE002851.

Leverington, D. W. (2004). Volcanic rilles, streamlined islands, and the origin of outflow channels on Mars. *Journal of Geophysical Research*, **109**, E10011, doi:10.1029/2004JE002311.

Lognonne, P. (2005). Planetary seismology. *Annual Review of Earth and Planetary Sciences*, **33**, 571–604.

Maizels, J. K. (1977). Experiments on the origin of kettle-holes. *Journal of Glaciology*, **18** (79), 291–303.

Maizels, J. (1992). Boulder ring structures produced during jökulhlaup flows: origin and hydraulic significance. *Geografiska Annaler*, **A74** (1), 21–33.

Malin, M. R. (1998). Turbulent pipe flow of Herschel-Bulkley fluids. *International Communications in Heat and Mass Transfer*, **25** (3), 321–330.

Manga, M. (2004). Martian floods at Cerberus Fossae can be produced by groundwater discharge. *Geophysical Research Letters*, **31**, L02702, doi:10.1029/ 2003GL018958.

Manning, R. (1891). On the flow of water in open channels and pipes. *Transactions of the Institute of Civil Engineers of Ireland*, **20**, 161–207.

Martini, I. P., Kwong, J. K. and Sadura, S. (1993). Sediment ice rafting and cold climate fluvial deposits: Albany River, Ontario, Canada. *Special Publications of the International Association for Sedimentology*, **17**, 63–76.

Masursky, H., Boyce, J. M., Dial, A. L., Schaber, G. G. and Strobel, M. E. (1977). Classification and time of formation of

Martian channels based on Viking data. *Journal of Geophysical Research*, **82** (28), 4,016–4,038.

McGovern, P. J., Solomon, S. C., Smith, D. E. *et al.* (2002). Localized gravity/topography admittance and correlation spectra on Mars: implications for regional and global evolution. *Journal of Geophysical Research*, **107** (E12), 5136, doi:10.1029/2002JE001854.

McKenzie, D. and Nimmo, F. (1999). The generation of Martian floods by the melting of ground ice above dykes. *Nature*, **397** (6716), 231–233, doi:10.1038/16649.

Mellon, M. T. and Phillips, R. J. (2001). Recent gullies on Mars and the source of liquid water. *Journal of Geophysical Research*, **106** (E10), 23,165–23,179.

Milton, D. J. (1974). Carbon dioxide hydrate and floods on Mars. *Science*, **183**, 654–656.

Neather, A. and Wilson, L. (2007). Morphological features and discharge estimates for the medial part of the Mangala Valles channel system, Mars. In *Lunar and Planetary Science Conference XXXIII*, Abstract 1265. Houston, TX: Lunar and Planetary Institute.

Neukum, G., Jaumann, R., Hoffmann, H. *et al.* and the HRSC Co-Investigator Team (2004). Recent and episodic volcanic and glacial activity on Mars revealed by the High Resolution Stereo Camera. *Nature*, **432**, 971–979, doi:10.1038/nature03231.

Neumann, G. A., Zuber, M. T., Wieczorek, M. A. *et al.* (2004). Crustal structure of Mars from gravity and topography. *Journal of Geophysical Research*, **109** (E8), 8002, doi:10.1029/2004JE002262.

Nimmo, F. and Tanaka, K. (2005). Early crustal evolution of Mars. *Annual Review of Earth and Planetary Sciences*, **33**, 133–161.

Ogawa, Y., Yamagishi, Y. and Kurita, K. (2003). Evaluation of melting process of the permafrost on Mars: its implication for surface features. *Journal of Geophysical Research*, **108** (E4), 8046, doi:10.1029/2002JE001886.

Oost-hoek, J. H. P., Zegers, T. E., Rossi, A., Foing, B., Neukum, G. and the HRSC Co-Investigator Team (2007). 3D mapping of Aram Chaos: a record of fracturing and fluid activity. In *Lunar and Planetary Science Conference XXXVIII*, Abstract 1577. Houston, TX: Lunar and Planetary Institute.

Ori, G. G. and Mosangini, C. (1998). Complex depositional systems in Hydraotes Chaos, Mars: an example of sedimentary process interactions in the Martian hydrological cycle. *Journal of Geophysical Research*, **103** (E10), 22,713–22,723.

Picardi, G. *et al.* (2005). Radar soundings of the subsurface of Mars. *Science*, **310** (5756), 1925–1928, doi:10.1126/science.1122165.

Rice, J. W., Christensen, P. R., Ruff, S. W. and Harris, J. C. (2003). Martian fluvial landforms: a THEMIS perspective after one year at Mars. In *Lunar and Planetary Science Conference XXXIV*, Abstract 2091. Houston, TX: Lunar and Planetary Institute.

Robinson, M. S. and Tanaka, K. L. (1990). Magnitude of a catastrophic flood event at Kasei Valles, Mars. *Geology*, **18**, 902–905.

Rossbacher, L. A. and Judson, S. (1981). Ground ice on Mars: inventory, distribution, and resulting landforms. *Icarus*, **45**, 39–59.

Rubin, A. M. (1992). Dike-induced faulting and graben subsidence in volcanic rift zones. *Journal of Geophysical Research*, **97** (B2), 1839–1858.

Russell, A. J. (1993). Obstacle marks produced by flow around stranded ice blocks during a glacier outburst flood (jökulhlaup) in west Greenland. *Sedimentology*, **40**, 1091–1111.

Saar, M. O. and Manga, M. (1999). Permeability-porosity relationship in vesicular basalts. *Geophysical Research Letters*, **26** (1), 111–114.

Saar, M. O. and Manga, M. (2004). Depth dependence of permeability in the Oregon Cascades inferred from hydrogeologic, thermal, seismic, and magmatic modeling constraints. *Journal of Geophysical Research*, **109**, B04204, doi:10.1029/2003JB002855.

Scott, E. D. and Wilson, L. (1999). Evidence for a sill emplacement on the upper flanks of the Ascraeus Mons shield volcano, Mars. *Journal of Geophysical Research*, **104** (E11), 27,079–27,089.

Sharp, R. P. and Malin, M. C. (1975). Channels on Mars. *Geological Society of America Bulletin*, **86**, 593–609.

Smith, D. E., Zuber, M. T., Frey, H. V. *et al.* (1998). Topography of the northern hemisphere of Mars from the Mars Orbiter Laser Altimeter. *Science*, **279** (5357), 1686–1692.

Stewart, S. T. and Nimmo, F. (2002). Surface runoff features on Mars: testing the carbon dioxide formation hypothesis. *Journal of Geophysical Research*, **107** (E9), 5069, doi:10.1029/2000JE001465.

Sugden, D. E., Clapperton, C. M. and Knight, P. G. (1985). A jökulhlaup near Søndre Strømfjord, west Greenland, and some effects on the ice-sheet margin. *Journal of Glaciology*, **31** (109), 366–368.

Tómasson, H. (1996). The jökulhlaup from Katla in 1918. *Annals of Glaciology*, **22**, 249–254.

Tschudin, K. (1946). Rate of evaporation of ice. *Helveticae Physica Acta*, **19**, 91–102.

Turcotte, D. L. and Schubert, G. (2002). *Geodynamics*, 2nd edn. Cambridge: Cambridge University Press.

Waitt, R. B. (2002). Great Holocene floods along Jökulsá á Fjöllum, north Iceland. *Special Publications of the International Association for Sedimentology*, **32**, 37–51.

Wang, C. Y., Manga, M. and Hanna, J. C. (2006). Can freezing cause floods on Mars? *Geophysical Research Letters*, **33** (20), L20202.

Williams, R. M., Phillips, R. J. and Malin, M. C. (2000). Flow rates and duration within Kasei Valles, Mars: implications for the formation of a Martian ocean. *Geophysical Research Letters*, **27**, 1073–1076.

Wilson, L. and Head, J. W. (2002). Tharsis-radial graben systems as the surface manifestation of plume-related dike intrusion complexes: models and implications, *Journal of Geophysical Research*, **107** (E8), 10.1029/2001JE001593.

Wilson, L. and Head, J. W. (2004). Evidence for a massive phreatomagmatic eruption in the initial stages of formation of the Mangala Valles outflow channel, Mars. *Geophysical Research Letters*, **31** (15), L15701, doi:10.1029/2004GL020322.

Wilson, L., Ghatan, G. J., Head, J. W. and Mitchell, K. L. (2004). Mars outflow channels: a reappraisal of the estimation of water flow velocities from water depths, regional slopes, and channel floor properties. *Journal of Geophysical Research*, **109**, E09003, doi:10.1029/2004JE002281.

Woodworth-Lynas, C. and Guigné, J. Y. (2004). Extent of floating ice in an ancient Echus Chasma/Kasei Valles valley system, Mars. In *Lunar and Planetary Science Conference XXXV*, Abstract 1571. Houston, TX: Lunar and Planetary Institute.

Zhang, Y. X. and Kling, G. W. (2006). Dynamics of lake eruptions and possible ocean eruptions. *Annual Review of Earth and Planetary Sciences*, **34**, 293–324.

Index

abrasion 16
accommodation space 41, 256, 262
adiabatic cooling 203
ages
  channel ages 195–196
  exposure age 195
  formation age 195
age-dating (*see also* marine isotope stages) 194–196
  carbon-14 249
  cosmogenic dating
  crater counting 195, 196, 198
  epoch boundaries
  isochrons 195
  luminescence dating 249
  secondaries 195
  sediments 249
  size-frequency distribution 194
Aktash 246, 249
Alaska 227
Allegheny Plateau
Allegheny Vallis 187–188
alluvial deposits, alluvial fans 209, 210, 218, 220, 245
alluvial architecture 227
Alps, Europe 133
Al-Qahira Vallis 220
Altai Mountains, Siberia 14, 23, 38, 43, 44, 243, 245, 262, 300
Amazonian Epoch 194, 195, 204, 205, 294
Amazonis Planitia 196, 198, 199
American Cordillera 130
American Falls Lake 150, 155
anastomosis (*see also* morphology, plan view) 14, 66–67, 68, 107, 110, 115, 120
Aniakchak River 136
Aniakchak Volcano, Alaska 136, 298
Antarctica 44
antidunes (*see* bedforms)
Apollo seismic network (*see* seismic networks)
AQUARIVER model (*see* modelling)
aquifers 201, 202, 204, 205, 267, 291, 306
  confined (Mars) 291
  convection in 292, 294
  flow rates 301
  perched aquifers 197
  permeability/-ies 292, 301
  porosity 292
  temperature 292, 295
  thermal conductivitiy 292
  thermal gradient 292
Aram Chaos 178, 179, 211, 291
Ares Vallis 173, 174, 178–179, 211–213, 216, 218, 221, 297
Argyre basin 214, 215, 216, 218
Argyre Planitia 174, 178, 179
Armero, Columbia 131
Aromatum Chaos 184, 291, 301
Arsia Mons 267
Ascraeus Mons 194
Asia 245
association 36, 38
Athabasca Valles 36, 39, 42, 43, 44, 45, 195, 198, 199, 201, 203, 205, 211, 293, 300, 305
  age 196
  discharge 204
  morphology 199–201
  tectonism 200
  volcanism 200
Auqukuh Vallis 220
avalanches
  debris 131
  rock 129, 277

backwater (*see also* ponding) 281, 283, 285
Baetis Chaos 183, 287
Basin and Range Province, North America 134, 136
basins (*see also* lakes)
  Lake Superior basin 118
  Mediterranean basin 136
  Millican basin, Oregon, USA 134
  moraine basins 132–134, 144
  natural basins 128
  Pasco basin 145
  tectonic basins 134–136, 140, 141, 142
  volcanic basins 140, 141, 142
bars 33, 36, 38–40, 70–71, 80, 108, 114, 116, 117, 120, 229, 230, 232, 244, 251, 286
  architecture 249–258
  Bonneville Flood bars 146
  Chuja valley bars 261
  eddy bars 39
  expansion bars 38, 39, 230
  facies 249–258, 259, 260
  giant bars 38–40, 243, 244
  Katun valley bars 261
  Kezek-Jala bar 249, 251
  Komdodj bar 247, 249, 251
  Kuray Flood bars 147
  Inja bar 249, 254
  Little Jaloman bar 248, 249–251, 254–256, 261
  Log Korkobi bar 249, 254
  medial bars
  Missoula bars 147, 261, 262
  morphology 247–248
  pendant bars 39
  primary bars 262
  push bars 40
  secondary bars 267
  sedimentary characteristics 248–258
  streamlined bars (*see* bedforms, streamlined)
basaltic ring structures 200
bedding
  cross-bedding 251, 254, 255, 256, 257
  plane-bedding 251, 255, 257, 258
bedforms 305
  antidunes 13, 19–20, 21–22, 40–44, 230, 231, 300
  butte-and-basin scabland 71
  depositional forms 70–71, 73–74, 145–147, 287
  dunes 6, 8, 21, 22, 33, 36, 38, 40, 43, 44–45, 65, 73–75, 200, 229, 236, 243, 247, 248, 256, 299, 300
  erosional 15–16, 68–70, 71–73, 78, 80, 86, 89, 97, 114, 145–147, 285, 286, 287, 304
  flutes 17, 84, 88
  furrows 16, 17–19, 20
  irregularities 300
  lemniscate forms 70
  loess hills (*see also* streamlined forms) 71
  longitudinal s-forms 84, 85
  longitudinal forms 17, 19, 26, 44, 45, 71, 107, 110, 111, 196, 214
  macroforms 33, 68–71, 95–96, 229–230
  megaforms 96–97
  mesoforms 33, 71–74, 85–95, 114
  microforms 33, 80–85

non-directional forms 84
obstacle marks 22–23, 230, 231, 236
potholes 14, 16–17, 19, 26, 72, 84–85, 107, 119
scallops 16
scour 73, 107, 108, 110, 116, 117, 121, 195, 196, 248, 304
streamlined forms 14, 20, 22, 26, 39, 69–70, 78, 85–86, 95, 107, 108, 110, 114, 115, 116, 117, 196, 198, 199, 200, 299
transverse forms 22, 40–44, 85, 107, 110, 111, 300
transverse s-forms 84
trenched-spur buttes 68
undulating features 178–179
bedload transport 248
Bhutan 133
Big Ilgumen River valley 247, 248
Big Jaloman River 251
Big Stone moraine 114
Black Sea 5, 7–8, 154
Bond crater 215
Bonneville flood (*see* floods, Bonneville)
Bosphorus Strait 8, 136
boulders
armour 106
bars (*see* bedforms, bars)
berm 248
clusters 276–281, 283, 286
lag 108, 115, 116
braided morphology 196
Bretz, J Harlen 1, 2, 65, 176
British Columbia, Canada 133, 134
Brule spillway
burial 195
burst-and-sweep cycles 260
butte-and-basin scabland 71

C^{14} dating
caldera 52
ice-filled 60
calorimeter, natural 53
Calumet level 116, 118
canyons 211, 213, 214
ice 60
Capri Chasma 175–177, 214
carbon dioxide 204, 205, 293–294
Cascade Range, Oregon, USA 134
Caspian Sea 134
cataclysmic flooding 243, 244
cataracts (*see* dry cataracts, waterfalls) 14, 20, 21
catastrophism 1
cavitation 174, 300–302, 306
Cedar Buttes lava flow 132
Cenozoic 244, 245
Ceraunius Fossae 194
Cerberus Fossae 198, 199, 200, 201, 204, 291, 293
Cerberus plains 195, 198, 199, 200, 201, 204
Chagun–Uzun River 244
Chakachatna River, Alaska 132
Champlain/Hudson sub-lobe 120
Channeled Scabland 1, 2, 3, 13–14, 17, 19, 22, 23, 38, 39, 65–74, 107, 108, 176, 177, 218
channels
boulder dominated beds 296
circum-Chryse channels 194
depths 196, 198, 199, 200, 290, 295, 297, 298
flood 34
geometry 291
gravel-dominated beds 296
inner channels 17, 19, 107, 110, 111, 117, 119, 121, 220, 286
large-scale 17, 19, 295
length 200
Martian megaflood, outflow 23, 24, 27, 36, 297–298
Missoula Flood 300
outflow channels 290, 306
piping 137
roughness 295
sand-dominated beds 296
sapping
sides 297
slope 196, 198, 199, 200, 217, 276, 295, 297, 298, 299
small-scale 194
sources 290–291
steep channels 296
subchannels 298
tunnel channels 86–88, 90, 91, 92–94, 99
water level 298
width 198, 199, 200, 218, 298
channelised flow
chaos, chaotic terrain 23, 34, 197, 201, 211, 213, 214, 216, 219, 290, 294
Chemal 247
Cheney–Palouse Tract 66, 69
Chézy (friction) formula/coefficient 267, 268, 269, 270
Chia Crater 183
Chicago outlet 116, 117, 118
Chile 227
China 130
Chryse Colles 186
Chryse Planitia 172
Chryse trough 213, 214
Chuja basin 244, 246
Chuja palaeolake
Chuja River valley 23, 34, 246, 247, 248, 249, 251, 262
Chuja–Kuray floods
Chuja–Kuray palaeolake 5
chute–pool 20, 21
circum-Chryse channels (*see* channels)
Claritas Fossae 203
clasts (*see also* interclasts, rip-up clasts)
shape 277
Clearwater Lower Athabasca spillway
climate change 134, 136, 138
clinoforms 249, 251
clinostratification 251
CO_2 (*see* carbon dioxide)
cobble pods 254, 257, 259
collapse 201, 214, 301
Collier Glacier, Oregon, USA 144
Colorado River 14, 132, 134
Columbia Plateau 66, 67
Columbia River Gorge 65, 74, 145
Columbia Valles 175–177
composite forms 80
concave features 80
Connecticut River 2, 9
continuity equation 216
convection 203
convex features 80, 178
cooling 204
Copernicus crater 219
Coprates Chasm 198
Cordilleran Ice Sheet 3–4, 65
corrosion
Cottonwood spillway
coulees 14, 67
Coulomb friction 268
Coveville level 122
Crab Creek, Washington, USA 67
crater counting
Crater Lake, Oregon, USA
crater paleolakes (*see* lakes)
creationism 2
creation science 2
critical flow 21–22, 140
Crooked Creek, central Oregon, USA 134
cross-lamination (*see* lamination)
cross-stratification (*see* stratification)
cryosphere 197, 201, 203, 204, 291, 293, 294, 297, 301, 303, 305
erosion of 303–304, 305
cut-and-fill structures 255, 257

Dadu River, China 131
Daga Vallis 175–177
dams
beach berms (*see also* ice, dams)
compromised dams 211
constructed dams 129, 138, 143
earthen dams 137, 211
experimental breaching 137
failure 129, 137–138, 209, 210, 213, 220
glacial dams 209, 211
Gros Ventre landslide dam failure 137
hyaloclastite dam 132
incision 214

dams (*cont.*)
  Kuray ice dam 141
  landslide dams 129–131, 132, 137, 138, 140, 141, 142, 143, 144, 145
  lava-flow dams 129, 132
  Missoula ice dam 140
  moraine dams 50, 134, 137, 138, 144, 145, 146
  natural dams 128
  Nurek rockfill dam, Tajikistan 129
  rock-material dams 128
  stability 131, 137, 138, 143
  unconsolidated dams 209
  Usoi landslide dam, Tajikistan 129, 145
  volcanogenic 131–132, 142
Dana, James D.
Darcy law 292–295
Darcy velocity 292
Darcy–Weisbach equation 296
Darcy–Weisbach friction factors 296, 298, 299
dating (*see* age-dating)
debris flows 129, 131, 133, 134, 138, 144, 145, 229, 259, 265, 266
deglaciation (*see* glacial processes)
depletive waxing flow 260
deposition (*see also* sedimentation)
  debris flow deposits 255, 258, 260, 261
  flood sediment 131, 133, 146, 200
  glaciolacustrine deposits 244
  grainflow deposits 254
  hyperconcentrated flow deposits 200, 226–232
  lacustrine deposits 247, 249, 254, 256, 258
  lahar deposits 131
  landforms 38, 108–109
  limnic units 262
  Missoula deposits (*see* bars, Missoula)
  pyroclastic debris 131
  run-up deposits 39, 248
  stoss/lee deposition 300
  slackwater deposits 275
depth of flow (*see* flow, depth; channel, depth)
diamicton 254, 255, 256, 257
Dig Tsho, Nepal 144
digital elevation model (DEM) 274, 275, 276
diluvial 34
dimensionless blockage index 138
discharges 210, 211, 216, 301
  attenuation 281, 284, 285, 286
  duration 301
  estimates 109, 209, 210, 213, 214, 216, 217, 218, 219, 290, 298
  groundwater 203, 204
  of jökulhlaups 52, 54–55, 265, 266–269, 273
  peak 138–144, 145, 210–211, 213, 214, 275, 290, 298
  rate(s) 201, 218, 291–292, 301
dissolution 15, 16
downcutting, episodic outlet 106
downstream fining 248
drainage density 220
drainage integration 134, 136
drop-stones 261
drumlins 78, 80, 85–88, 89, 92, 94, 95, 96, 97–98
dry cataracts
  Dry Falls 72
  Martian 176, 181, 182–183
  Potholes 65
dunes (*see* bedforms, giant current ripples, megaripples)
dykes 174, 186, 203, 204, 267
  emplacement 195, 197, 200, 201, 203
dynamics (*see* flood, dynamics; flow, dynamics)

earthflows 129
earthquakes 131, 137
East Fork Hood River, Oregon 129
Eberswalde crater delta 210
echo sounding 52
Echus Chasma 213
El Chichón, Mexico 131
Elaver Vallis 188–190
Elysium Fossae 194
Elysium Mons 194
energy 205
English Channel 5
Entujökull (glacier) 269
environmental conditions, Mars 290, 294, 302, 305
Eos Chasma 214
Ephrata Fan 67, 73–74
Erbalyk 247, 249
Eridania basin 218, 219, 220
erosion (*see also* bedforms, spillways) 14–15, 196, 273
  abrasion 15
  bed 297, 298, 300–301, 305
  bedrock 281, 286
  cavitation 14, 15, 16, 286
  constrictions 19
  corrosion (*see also* dissolution) 15
  of cryosphere
  flaking 15
  fluid stressing 15
  melting 297 (Mars)
  particle erosion 277
  plucking 15, 16
  surfaces 96
error 195
esker(s) 78, 88, 92, 94, 95, 96, 99
Eurasian Ice Sheets 5
Europe 130
evaporation, Mars 294, 302, 303, 305
exhumation 195
explosive boiling 54
Eyjafjallajökull (ice cap) 265, 266, 267, 268, 270

facies
fall and pool structure 296
fans 67–68
  Ephrata Fan 67, 73–74
  terraced fan 198
faults, faulting (*see also* graben) 195, 202, 203, 246
fault blocks 199
Fauskheiði 269
Fennoscandian Ice Sheet 5
Finger Lakes 120
Fish Creek spillway
Fisher Caldera, Alaska 136
fissures 196, 200
flaking
flame structures 259
Flims rockslide 129
floods (*see also* jökulhlaups)
  Altai floods (*see also* Altai Mountains) 286
  Bonneville flood 132, 134, 145, 146, 286
  breached basin floods 132
  breached valley blockage floods 129
  Chuja–Kuray floods 129, 147, 244
  damburst, dam-failure floods 128, 137, 210, 211
  duration 146, 301
  dynamics 244
  glacier outburst (*see also* jökulhlaups) 50, 225, 265
  hazard management 131, 132, 265, 271
  hydrograph (*see* hydrographs, flood)
  marine floods 134, 136
  marks 134, 267, 268, 269, 270
  meteorological floods 146
  Missoula 19, 38, 39, 129, 145, 146, 147, 177, 244, 258, 286, 306
  Noah's Flood 9, 14
  sediment-laden floods 130
  sediment load 301
  self-censoring 194
  temperature 300
  water volume 301
flood geology 2
flows
  capacity 303
  channelised 196
  depth 210, 217, 276, 300
  dynamics 204, 295–296
  laminar 304
  momentum 300

obstacles 198
overland 196
plug 304
rheology 229, 302–305, 306
speed, (velocity) 109, 204, 216, 217, 295, 296, 297, 298, 299, 300, 301, 302, 305, 306
sub/supercritical 281, 299–300, 305
thermodynamics 302–305
turbulence 304
volume flux 297, 299
fluid stressing 15
flutes 17, 84, 88
Fort Ann levels 122
Fort Wayne moraine 115
fossea 39, 248
fountains 201, 204, 205, 293, 294
frazil ice 56
freezing point 203
friction (*see* Chézy (friction) formula/coefficient, Manning's friction coefficient, Darcy–Weisbach friction factors)
Froude number (*see also* flow, sub/supercritical) 73, 276, 281, 299–300
furrows 17, 84

Galerkin finite element method
Ganges Cavus 189–190
Ganges Chasma 175, 176, 177, 184, 185, 186, 187, 188, 190, 191
Genesee River Valley 120
geometry 34
geothermal areas
subglacial 265
geothermal heat, melting 50, 290, 291, 292
giant current ripples (*see also* dunes, megaripples) 6, 65, 73–75
Gilbert, G. K. 80
glacial lakes (*see also* lakes)
Glacial Lake Agassiz 33, 111–115, 118
eastern outlets 114–115
northwestern outlets 115
southern outlet 114
Glacial Lake Albany 120, 121
Glacial Lake Algonquin 118
Glacial Lake Avon 120
Glacial Lake Chicago 1, 68, 71, 74, 106, 116, 118
Glacial Lake Duluth 118
Glacial Lake Hind 7, 111
Glacial Lake Iroquois 119, 120, 121, 122
Glacial Lake Lesley 119, 120, 179
Glacial Lake Maumee 3, 67–70, 116
Glacial Lake McConnell 67, 115
Glacial Lake Milwaukee 67, 115
Glacial Lake Missoula 1, 3, 14, 22, 23, 33, 38, 44, 45, 65, 73, 78, 145, 229, 243, 244
Glacial Lake Ojibway 65
Glacial Lake Regina 7, 110, 111
outburst 5
Glacial Lake Saginaw 70, 74–75, 116, 118
Glacial Lake Souris 6, 111
Glacial Lake Vermont 120, 121, 123
Glacial Lake Warren 120
Glacial Lake Whittlesey 116
Grímsvötn 52–57, 232, 266
ice-dammed/glacier-dammed lakes 4–5, 52, 58, 226, 265
marginal lakes (*see* ice-dammed lakes)
proglacial 105
subglacial 52
under ice cauldrons 57–58
glacial processes 197, 200, 249, 295
advance 136
alpine/valley glaciers 129, 246
continental ice sheets 129
moraines 132
retreat 130, 133, 134
glaciolacustrine deposits
Glenwood level 117
Global Positioning System (GPS) 275
GPR (*see* radar, ground penetrating)
graben 196, 197, 199, 201, 203, 204, 211, 290
Grand Canyon 132
Grand Coulee 17, 19, 20, 66
Grand Valley 116
Granicus Valles 194
gravel dunes (*see* bedforms, dunes; giant current ripples, megaripples)
Great Salt Lake
Greenland 227, 228
Grímsvötn (*see* glacial lakes)
Grjotá Valles 39, 44, 198
age 195–196
morphology 199
Gros Ventre landslide dam failure
Grosswald, Mikhail G. 5
ground ice 203
ground-penetrating radar
groundwater 202
discharge 203, 213
piping 131
release 203, 205, 213, 294
table 219
Gulf of St Lawrence 119, 122
Gulf Stream 6
gullies 34
Gusev crater 219

Hale crater 215
hanging valleys 67
heat
frictional 55, 56
sources 203
heat flux 60
HEC-2 modelling 109
Hedstrom number 304–305
Heinrich events 6
Hells Canyon 134
Hesperian Epoch 194, 204, 209, 210, 211, 216, 219, 294
highland terrain 199
Himalayas 130, 134
Hitchcock, Edward 2
Holden crater 215, 216, 218
Holocene 133, 249, 262, 274
horseshoe vortices
Hou Hsing Vallis 220
Hraundalur 281
Hudson Bay 105
Hudson Valley 119, 120, 121, 123
hummocky terrain 88–89, 200
Huron–Erie lobe 105
hyaloclastite 132
hydrofracturing 56
Hydaspis Chaos 174, 178, 179, 211
hydrated sulfates 183
hydraulic formulae 167
hydraulic damming (*see* ponding)
hydraulic jumps 19, 40–44, 281
hydraulic radius 268
hydraulic seals 52, 54–55
hydrographs 55
exponentially rising 227
flood 138, 145, 228, 281
linearly rising 227
hydrothermal convection 53
hydrothermal fluid 50, 203
hydrothermal vents 52
hyperconcentrated flows 50, 108, 109, 111, 226–232, 237, 251, 255, 258, 297, 301
hyperpycnal flow 7
hypotheses
megaflood hypothesis 79–80
Noah's Flood Hypothesis 8, 9
working 2, 9

Iamuna Chaos 185–186
Icaria Fossae 203
ice (*see also* glacial processes)
barriers 52, 246
blocks 197, 230–231, 236, 266, 305
conduits 53
flow in 58
melting of 266
crystals 297, 303, 305, 306
dams 14, 129, 137, 141, 143, 146, 197, 210, 249
flotation 55
droplets

ice (*cont.*)
formation 197, 304, 306
raft 305
shelves 53
tunnels 54
expansion 55
Iceland 50, 131, 225, 228, 232, 258, 262, 265, 271, 273
Illinois Valley 116, 117
imbrication 248, 251, 258, 277–278
incision 134
infiltration 199, 201
Indus Vallis 220
Inja 248
Inja bar
inner channels (*see* channels)
InSAR 58
interclasts 254, 255, 257, 258, 260
interior layer deposits 174
inverse grading 260
isochrons
Issyk, Kazakhstan 131

Japan 130, 132
Johnstown Crater 190
jökulhlaups 13, 34, 50–61, 225–227, 232, 255, 260, 265, 273, 274
deposits 108, 226, 262
depth 270
dynamics 266–269
fissure swarm eruptions 60–61
hazard management planning (*see* flood, hazard management)
Hraundalur 281
Katla 60, 129, 265, 268, 269
Kverkfjöll, Kverkfjallarani 273, 274, 276, 281, 283, 284, 286
Lake Missoula 286
linearly rising 55–57, 226, 297
modelling
November, 1996, Iceland 226
prehistoric 265, 269, 270
release 54
sedimentation 227
speed 270
subaerial 266–271
Öraefajökull 60
Jökulsá á Fjöllum (river) 44, 268, 300
Juventae Chasma 181, 183

Kaministikwia River 115
Kankakee River 117
Kasei Valles 23–27, 174, 179–182, 211, 213–214, 221
North Kasei Vallis 213
Katla
flood (*see* jökulhlaups, Katla)
volcano 265, 266, 267, 268, 270
Katun River valley 246, 247, 248, 249, 251, 255, 256, 258, 262
Kerlingarhyggur 274, 277
kettle holes 200, 230–231, 236, 248, 305
Kezek-Jala bar (*see* bars)
Khvalyn palaeolake 5
knickpoints 17, 19, 25
Komdodj bar
Kuray basin 246, 249, 265
Kuray–Chuja floods (*see* Chuja–Kuray Floods)
Kuray palaeolake
Kverkfjöll (volcano), Kverkfjallarani (*see also* jökulhlaups, Kverkfjöll) 274, 276, 277, 281, 283

lacustrine deposits 247, 258
Ladon basin 215, 216, 218
Ladon Valles 215
lahars 131, 136, 265, 266, 297
Lake Michigan lobe 105, 118
Lake Superior basin
lakes (*see also* basins, glacial lakes)
American Falls Lake 132, 137
Chuja palaeolake 246
Chuja–Kuray palaeolake 5, 243, 246
caldera and crater lakes 128, 136–146
Crater Lake, Oregon, USA 136, 137
crater palaeolakes, Mars 209, 219
East Lake, Oregon, USA 137
Great Salt Lake 132, 134
Kuray palaeolake 246
Lake Agassiz (*see also* lakes, Glacial Lake Agassiz) 5, 8, 105, 108, 109, 111, 115
Lake Atitlán
Lake Alvord 134
Lake Bonneville 134
Lake Issyk 137
Lake Maumee 108, 115
Lake Nipigon 115
Lake Ojibway 111
Lake Tarawera, North Island, New Zealand 132
Lake Taupo, New Zealand 136
Lake Toba, Indonesia 136
Largo Argentina 44
level 55
Mansi Lake
Martian palaeolakes 172, 173, 174, 176, 177, 178, 179, 182, 183, 184, 185, 186, 187, 189, 190, 191, 197, 199
megaflood deposits in 109
meteor impact basins 128
moraine-dammed, moraine-rimmed lakes 128, 132, 133, 134, 141, 142, 144
Paulina Lake, Oregon, USA 136, 137
tectonic basins 128, 132
volcanic calderas, lakes 128, 132, 145
lamination (*see also* layering) 255, 257, 258, 260
cross-lamination 251, 257
inverse graded laminations 255
normal graded laminations 255
landscape change, evolution 131, 273, 283
landscape unconformity 78, 88, 92, 96–97
Largo Argentina 44
latent heat of evaporation 204
Laurentian Fan 80
Laurentide Ice Sheet 4–5, 8, 104–105, 109, 111, 115
lava flows 131, 132, 292
emplacement 202
platy-ridged material 200, 201
spatter-fed flows 201
layering (*see also* lamination) 200, 244
convoluted layering 255
light-toned layered deposits, Mars 218
lemniscate form 70, 108
Lenore Canyon 67
Lethe Vallis 200
Licus Vallis 220
life 205
Lillooet River, British Columbia, Canada 132
Little Ice Age 133
Little Jaloman 248
Little Jaloman bar
Little Jaloman River valley 248
Lobo Vallis 181
location 34
Log Korkobi bar
Loire Vallis 220
longitudinal grooves (*see* bedforms)
longitudinal vortices
Lower Fort Ann level
luminescence dating
Lunae Planum 213
Lyell, Charles 1, 2

Ma'adim Valles 39, 210, 218–220, 221
macroturbulence 14
Mae Chaem River, Thailand 19
magma
chambers 53
intrusions 50
magmatic melting of ground ice 197
Maja Valles 181, 182–183, 211, 221, 300, 305
Mamers Valles 220
Mangala Fossa 196, 197, 203, 204, 291
Mangala Valles 199, 201, 202, 203, 204, 211, 301, 304
age 195
discharge 301
ice processes 197
morphology 196–198
release mechanism 203

sinks 197–198
source 201, 291, 294
Mansi Lake
Manning's equation/formula/roughness/coefficient/number 109, 179, 216, 267, 268, 270, 276, 281, 295–296
Manych spillway
Margaritifer basin 211, 213, 216
Margaritifer Terra 214
marine floods
marine isotope stages 243, 245, 249
Mars 18, 19, 23, 34, 209, 210–211, 214, 215, 216, 217, 218, 220, 221, 302, 305–306
exploration 205
Mars Exploration Rovers 219
Mars Orbiter Camera (MOC) 21, 195, 196, 198, 199, 200
Mars Orbiter Laser Altimeter (MOLA) 24, 179, 184, 195, 196, 198, 199, 200, 205, 210, 211, 301
Mars Pathfinder 179, 297, 298, 306
Marte Vallis 198, 199, 205, 211
age 195
Martian outflow channels 1, 6, 7, 108, 287, 290–306
mass movements
Mawrth Vallis 173, 211–213
Maumee Torrent 115, 116
Maumee Vallis 211
Medicine Lake Volcano, California, USA 136
Mediterranean basin
Mediterranean Sea 136
Medusae Fossae Formation 196, 198
megafloods 1, 6, 9, 33, 34, 37, 38, 39, 136, 137, 243, 246, 247, 249, 251, 254, 255, 256, 259, 260, 262, 286
Martian 6
terrestrial 3–6, 7
triggering mechanisms 201–203
megaripples (*see also* giant current ripples) 78, 89–92, 182, 183
Mekong River, Laos 22
Memnonia Fossae 195, 196, 203, 204
meteor impacts 128
Millican basin
Minnesota River spillway
Minnesota River valley 105, 108
Mississippi River 218
Missoula floods (*see* floods, Missoula)
MOC (*see* Mars Orbiter Camera)
MOLA (*see* Mars Orbiter Laser Altimeter)
modelling 199, 205, 267, 291–293
AQUARIVER model 268
BEED model 138, 143
breach-erosion models 143
BREACH model 138, 143
DAMBRK model 138
ERODE model 138, 143
FLDWAV model 145
forward modelling 37
Galerkin finite element method 268
HEC-2, HEC-RAS 145, 274, 283
inverse modelling 37
of jökulhlaups 55, 270
Mars outflow channels 302, 304, 306
SAMOS snow avalanche model 268, 269
slope area method 274–275, 276, 281, 283, 284
SOBEK 276, 281
two-dimensional (2-D) hydrodynamic modelling 274, 276, 281–283, 284–286
Mohawk Outlet 120
Mohawk Valley 119, 121
Mongolia 6, 246
Monterey East Channel 80
Moon 292
moraines 294
Rogen moraines 78, 89, 99
moraine dams (*see* dams, lakes)
Morava (formerly Margaritifer) Valles
Morella Crater 188–190
morphology
Athabasca Valles 199–201
Mangala Valles 196–198
Mars channels 196–201
Marte Vallis 198–199
outflow channels 290
plan-view morphology 199, 220
anastomosing 198, 286
braided 196, 256
Moses Coulee 66, 67
Mount Everest 227
Mount Mazama, Oregon 132, 137
Mount Ruapehu, New Zealand 136
Mount Saint Helens 131
Mount Spurr volcanic complex 132
Mount Toc
Mýrdalsjökull (ice cap) 52, 58, 60, 265, 270

Nanedi Vallis 220
Naranjo River 131
natural basins 128
Navy Submarine Fan 80
negative water balance 136
Nepal 133, 227, 228
Nevado de Colima (volcano) 131
Nevado del Ruiz, Columbia 131
New Zealand 130, 131, 136
Newberry Volcano, Oregon 136
Newton crater 219
Nirgal Vallis 218
Noachian debacle 2
Noachian Epoch 194, 209, 210, 213, 215, 216, 218, 219
Noctis Labyrinthus 174
North Fork Toutle River 131

obstacle marks (*see* bedforms)
Okmok Caldera, Umnak Island, Alaska, USA 136
olivine 178
Olympica Fossae 194
Olympus Mons 194
Ophir Catenae 187, 188, 191
Ophir Cavus 187
Ophir Creek, Nevada 130
Öraefajökull volcano 60
Öraefajökull floods
Orcus Patera 198, 199
Oregon, USA 134
outburst floods (*see also* jökulhlaups) 128, 131, 145, 259, 296
Glacial Lake Regina 110
outflow channels (*see* channels)
outlet geometry 137
outlet valleys 209, 210, 211, 213, 214, 219, 220
outwash plains (*see also* sandur) 60, 67
overfitness 14, 67–68
overflow of basins 210, 211, 213, 214, 216, 218, 219, 220, 221
overland flow
Oxia Chaos 184–186

palaeocompetence 274, 277, 283
palaeohydraulics 74–75, 129, 146, 274, 277
palaeolakes (*see* lakes)
palaeostage indicators (PSI) 275, 276, 281, 283
Pardee, Joseph Thomas 3
partial freezing 203
passive flooding 261
Pathfinder (*see* Mars Pathfinder)
Patagonia
Paulina Lake
Peru 133, 137
phreatomagmatism 197, 201, 204
Pinatubo, Philippines 136
pingos 200, 249
pitted mounds 196, 200, 201
platy-ridged material
Pleistocene 65–74, 129, 133, 134, 243, 245
Pliocene–Pleistocene transition 244
plucking
pluvial periods 136
Pollalie Creek debris flow 129
polygons (*see also* hummocky terrain) 200
ponding (*see also* backwater) 198, 200, 201, 247, 261, 283
pool–riffle sequence 20, 21

pore pressure 202
pore space 204
Port Huron advance 116
Portage spillway
Portland Delta 67
potholes 119, 174, 177
Potholes Cataract 65
Preboreal Oscillation 115
pressure, subglacial water 202, 266
proglacial lakes (*see* lakes)
PSI (*see* palaeostage indicators)
pyroclastic deposits
pyroclastic density currents 204
pyroclastic flows 132

Qu'Appelle-Assiniboine spillway
Quaternary 229, 244, 262, 268
Quincy Basin 67

radar
  ground penetrating radar 226, 233
  orbiting 306
Ravi Vallis 184, 205, 211, 298, 299, 301
reactivation surfaces 228, 230
regression analysis, equations 138
release mechanisms 197, 201–203, 204, 205
  combined 203
  dyke-induced 201
  stress release 197
  tectonic 197, 202–203
  volcanic 201–202, 203–204
Reynolds number 73–74, 304
Rhine river, Switzerland 129
rheology (*see also* flow)
  Newtonian material 297
  non-Newtonian material (Bingham plastic) 304
rhythms, rhythmites 254, 258, 259, 260, 261, 262
rip-up blocks, rip-up clasts 231–232, 237, 259
river diversion 132
rock size distribution 297
rockslides 59
Rocky Butte 68
Rogen moraines
Rome outlet 121
rootless cones 196, 199, 200, 201
roughness 287
  characteristic roughness height 268
  form roughness 284
Ruapehu, New Zealand 265
Rush-Victor channels 120

Sacra Fossae 213
Samara Valles 220
SAMOS snow avalanche model
sandur (*see also* outwash plains) 50, 67, 227, 228
Sandyland moraine 111
scablands (*see also* bedforms) 78
  Cheney–Palouse Tract 66, 69
  definition 65
  Grand Coulee 66
  Moses Coulee 66, 67
  patterns 67–68
  Telford–Crab Creek Tract 66
  topography 120
scale 34, 38
scallops
Scamander Vallis 220
scour (*see* bedforms)
sea level rise 136
secondary production
sediments
  ice-volume fraction 302
  load 33, 37, 303, 304
  size-distribution 298
  supply 283
  transport 33, 34, 36–38, 301, 306
sedimentology 36
sedimentation
  capacity-driven 286
  competence-driven 286
  from jökulhlaups 226–232
seismic experiments, networks 292, 306
seismicity 54, 130
sequences 255, 256, 260, 261
  coarsening-upward 260
  fining-upward 260
Shalbatana Vallis 173, 186–187
Shaw, John 6, 8
shear stress 146, 267, 276, 281, 284, 286, 287
shorelines 246
Siberia (*see also* Altai Mountains, Siberia) 229, 243, 262
Sierra Nevada, California, USA 133
Simud Valles 173, 175–178, 211, 214, 221
sinks 196
  Athabasca Valles 201
  Grjotá Valles 199
  Mangala Valles 197–198
  Marte Vallis 199
Sirenum Fossae 203
size-frequency distribution
Skaftá (river) 267
Skeiðará (river) 267, 268
Skeiðarársandur 56, 232–237
slope (*see also* channel slope, water slope) 196
slope area method
slot canyons 19
Snake River, Idaho, USA 132, 134, 145
Snake River Plains, USA 134, 136
Soap Lake 19
SOBEK
Somerset Crater 190
sources 196, 201
  Athabasca Valles 199
  Mangala Valles 201
  Marte Vallis 198–199
  Mawrth Vallis 213
  subsurface 290
  tectonic 196
  ULM 218
spillways 105, 107
  Brule 118
  Clearwater Lower Athabasca 115
  Cottonwood 114
  erosional 106–107
  Fish Creek 114
  Manych 5
  Minnesota River 106
  Portage 118
  Qu'Appelle-Assiniboine 111
  St Croix 118
  Sheyenne 111
  Souris 110, 111
  Souris–Pembina 111
Spirit Lake, Idaho 73, 131
Spokane Flood 3
Staircase Rapids, Washington state 19
standing waves 40, 41
step–pool 20, 21
strandlines 249
stratification
  clinostratification 251, 256, 257
  cross-stratification 251
  stratigraphic log 255
streamlined hills (*see* bedforms)
streamlining 78, 86
stream power 146, 221, 276, 281, 287, 290–306
stress field 203, 205, 290
stress release
Straits of Gibraltar 136
Straits of Mackinaw 118
stratigraphy 36, 130, 195
subglacial megafloods 8–9
subglacial volcanic eruptions
supercooled water 57
superfloods 1
superposition 195
Susquehanna River 119, 120
sverdrups 1
Syracuse channels 120

Taylor channel 120
tectonic activity 136, 197, 200
  extensional 203
  pressurisation 293
tectonic sources
Tegermach River, Kirgizstan 129
Telford–Crab Creek Tract 66

temperature
  aquifer 203
  flood
  water
Terra Cimmeria 219
Terra Sirenum 219
terraces 195, 198, 218, 243, 244, 247, 248, 251, 256
Tharsis 96, 174, 191, 194, 196, 197, 202, 203, 213
Thaumasia Fossae 203
Thermal Emission Imaging Spectrometer (THEMIS) 196, 199
Þjórsá (river) 269
Tibet 130, 133
Tiu Valles 173, 175–178
traction carpets, traction load 255, 260, 261, 281
triple point 204
Tsangpo River gorge, Tibet 146
tunnel channels
tunnel valleys 95
turbidity currents 259
Turgai divide 5
Two Rivers advance 118

Uinkaret volcanic field 132
undulating features
uniformitarianism 1, 2, 3, 75
Upper Fort Ann levels
Usoi landslide dam
Uzboi–Ladon–Morava (ULM) Valles 210, 211, 213, 214–218, 220, 221

Vajont, Italy 130, 251
Valles Marineris 172–191, 214, 221
valley blockage 128
valley fill 243, 244, 247, 256
valley glaciers
Valley Heads moraine 120
valley networks (Mars) 209, 210, 218, 220, 221, 290, 298
vapour bubbles 295, 300–302
vapour pressure 302
varves 254
Vastitas Borealis formation 7, 177
Vatnajökull (ice cap) 5, 6, 8, 52, 60, 266, 267, 274
Viking 3–6, 198, 290, 298, 306
volcanic arcs 130
volcanic provinces 194
volcanism 130
  calderas and craters 136
  eruptions 50, 204, 226, 265
  fissures 60
  intrusions 290, 291, 293
  stratovolcanoes 60
  subglacial 52, 59–60, 129, 265
vortices
  horseshoe 80, 85, 89
  longitudinal 85, 86, 95, 96, 97

Wabash Valley 115, 116
Walla Walla Vallis 187–188
Wallula Crater, Mars 188
Wallula Gap, Washington 19
washload 36, 37, 301
waterfalls (*see also* dry cataracts) 14, 20, 40
water ascent, Mars 291–293
water–atmosphere interactions 204
water droplets 204
water release, Mars 292, 293–295
water-sheets 57, 58
weathering 15, 16
welded block-and-ash breccia 131
Whangahu River 136
Whewell, William 2
Williamson River 132
Wilmette bed 118
Wilson Creek 67, 68, 73
wrinkle ridge 199

Xanthe Chaos 186
Xanthe Terra 173, 184

Younger Dryas 6, 115, 118